THE RNA WORLD
Second Edition

The Nature of Modern RNA Suggests a Prebiotic RNA World

Edited by

Raymond F. Gesteland
University of Utah

Thomas R. Cech
University of Colorado

John F. Atkins
University of Utah
Trinity College Dublin

COLD SPRING HARBOR LABORATORY PRESS
Cold Spring Harbor, New York

THE RNA WORLD, SECOND EDITION

Monograph 37
© 1999 by Cold Spring Harbor Laboratory Press
Cold Spring Harbor, New York
All rights reserved
Printed in the United States of America

Project Coordinator Inez Sialiano
Production Editor Patricia Barker
Book Designer Emily Harste

Library of Congress Cataloging-in-Publication Data

The RNA world / edited by Raymond F. Gesteland, Thomas R. Cech, John
   F. Atkins. — 2nd ed.
       p.   cm. — (Monograph, 0270-1847 ; 37)
    Includes bibliographical references and index.
    ISBN 0-87969-561-7 (cloth); 087969-589-7 (paperback)
    1. RNA—Evolution.   2. Evolutionary genetics.   I. Gesteland,
Raymond F.   II. Cech, Thomas.   III. Atkins, John F. (John Fuller)
IV. Series: Cold Spring Harbor monograph series ; 37.
QP623.R6   1998
572.8'8—dc21   98-41849
                                                                        CIP

Authorization to photocopy items for internal or personal use, or the internal or personal use of specific clients, is granted by Cold Spring Harbor Laboratory Press, provided that the appropriate fee is paid directly to the Copyright Clearance Center (CCC). Write or call CCC at 222 Rosewood Drive, Danvers, MA 01923 (508-750-8400) for information about fees and regulations. Prior to photocopying items for educational classroom use, contact CCC at the above address. Additional information on CCC can be obtained at CCC Online at http://www.copyright.com/

All Cold Spring Harbor Laboratory Press publications may be ordered directly from Cold Spring Harbor Laboratory Press, 10 Skyline Drive, Plainview, New York 11803-2500. Phone: 1-800-843-4388 in Continental U.S. and Canada. All other locations: (516) 349-1930. FAX: (516) 349-1946. E-mail: cshpress@cshl.org. For a complete catalog of Cold Spring Harbor Laboratory Press publications, visit our World Wide Web Site http://www.cshl.org

Contents

Preface, ix

Foreword to the First Edition, xiii
Francis Crick

Prologue to the First Edition, xvii
James D. Watson

THE ORIGINS OF RNA AND RNA AT THE ORIGIN

1. **Before RNA and After: Geophysical and Geochemical Constraints on Molecular Evolution, 1**
   Stephen J. Mojzsis, Ramanarayanan Krishnamurthy, and Gustaf Arrhenius

2. **Prospects for Understanding the Origin of the RNA World, 49**
   Gerald F. Joyce and Leslie E. Orgel

3. **The Genomic Tag Hypothesis: What Molecular Fossils Tell Us about the Evolution of tRNA, 79**
   Nancy Maizels and Alan M. Weiner

4. **Probing RNA Structure, Function, and History by Comparative Analysis, 113**
   Norman R. Pace, Brian C. Thomas, and Carl R. Woese

5. **Re-creating an RNA Replicase, 143**
   David P. Bartel

6   **Did the RNA World Exploit an Expanded Genetic Alphabet?, 163**
Steven A. Benner, Petra Burgstaller, Thomas R. Battersby, and Simona Jurczyk

7   **Aminoacyl-tRNA Synthetases and Self-acylating Ribozymes, 183**
Michael Yarus and Mali Illangasekare

8   **On the Origin of the Ribosome: Coevolution of Subdomains of tRNA and rRNA, 197**
Harry F. Noller

9   **Introns and the RNA World, 221**
Walter Gilbert and Sandro J. de Souza

HOW TO BUILD A FUNCTIONAL RNA

10   **The Interactions That Shape RNA Structure, 233**
Mark E. Burkard, Douglas H. Turner, and Ignacio Tinoco, Jr.

11   **Small Ribozymes, 265**
David B. McKay and Joseph E. Wedekind

12   **The Role of Metal Ions in RNA Biochemistry, 287**
Andrew L. Feig and Olke C. Uhlenbeck

13   **Building a Catalytic Active Site Using Only RNA, 321**
Thomas R. Cech and Barbara L. Golden

14   **Ribonuclease P, 351**
Sidney Altman and Leif Kirsebom

15   **The RNA Folding Problem, 381**
Peter B. Moore

TRANSITION TO THE RNP WORLD

16   **RNA Interaction with Small Ligands and Peptides, 403**
Joseph D. Puglisi and James R. Williamson

17   **RNA Recognition by Proteins, 427**
Thomas A. Steitz

18   **Group I and Group II Ribozymes as RNPs: Clues to the Past and Guides to the Future, 451**
Alan M. Lambowitz, Mark G. Caprara, Steven Zimmerly, and Philip S. Perlman

19 The Growing World of Small Nuclear Ribonucleoproteins, 487
    Yi-Tao Yu, Elizabeth C. Scharl, Christine M. Smith, and Joan A. Steitz

20 Splicing of Precursors to mRNAs by the Spliceosomes, 525
    Christopher B. Burge, Thomas Tuschl, and Philip A. Sharp

21 tRNA Splicing: An RNA World Add-on or an Ancient Reaction?, 561
    Christopher R. Trotta and John Abelson

22 RNA Editing—An Evolutionary Perspective, 585
    Larry Simpson

23 Telomerase, 609
    Elizabeth H. Blackburn

24 Dynamics of the Genetic Code, 637
    John F. Atkins, August Böck, Senya Matsufuji, and Raymond F. Gesteland

APPENDICES

1 Structures of Base Pairs Involving at Least Two Hydrogen Bonds, 675
   Mark E. Burkard, Douglas H. Turner, and Ignacio Tinoco, Jr.

2 Schematic Diagrams of Secondary and Tertiary Structure Elements, 681
   Mark E. Burkard, Douglas H. Turner, and Ignacio Tinoco, Jr.

3 Reactions Catalyzed by RNA and DNA Enzymes, 687
   Gerald F. Joyce

4 Visualization of Elongation Factor Tu on the 70S *E. coli* Ribosome, 691
   Holger Stark, Marin van Heel, Marina Rodnina, Wolfgang Wintermeyer, and Richard Brimacombe

5 The Large Ribosomal Subunit from *H. marismortui* at 9 Å Resolution, 695
   Nenad Ban, Peter B. Moore, and Thomas A. Steitz

Index, 697

Preface

We are all fascinated with our origins. Among the diverse human societies on earth there is a wonderful richness of explanations—the Haida of Vancouver Island originating from small beings who crawled out of a clam shell onto the back of a raven to be carried to his wonderful land—the Aranda of Australia emanating as bandicoots from the armpits and navel of the "great father"—or the Yanomamo of Brazil coming from drops of blood dripping from a wound in the belly of the moon. But what was the origin of replicating molecules, those special molecules that ultimately gave rise to all extant life on earth?

Replicating the hereditary DNA molecules in all contemporary organisms requires protein catalysts. Early evolution of this mechanism poses the obvious dilemma of how to generate proteins without nucleic acid templates and how to replicate nucleic acids without protein catalysts. The notion that prebiotic evolution depended on RNA replication has intrigued many, since the early suggestions by Woese, Orgel, and Crick. However, even those of us especially interested in the current functions of RNA have been surprised by the richness of apparent "relics" from the RNA World that have been discovered. The stimulus for a new wave of interest in the RNA World came from the discovery of RNA molecules with catalytic activity. RNA can be both an informational molecule *and* a catalyst. Even natural catalytic RNAs, or ribozymes, can promote nucleotide joining reactions that could be a first step toward self-replication, and RNAs with further abilities have been identified by the very powerful technology of iterative selection. Further probing of these in vitro evolutionary mechanisms will illuminate how life may have originated. This is the only obvious route available to understand these early events, since there is little hope of finding direct evidence—molecules are not thought to leave any paleontological fossil tracks.

Iterative selection schemes, Darwinian evolution in a test tube, are increasing in sophistication, and the limits are not obvious. This approach has yielded molecules with specific catalytic functions and molecules that have specific binding abilities. For many catalytic RNA molecules, metal ions play a key role in catalysis, just as cofactors participate in reactions catalyzed by many protein enzymes. The notion of RNA catalysts having cofactors has been exploited by inclusion of small molecules such as histidine in iterative selection experiments. This new dimension is opening the door to a diversity of attainable functions. Another route to expanding the repertoire of iteratively selected molecules may be the introduction of bases with different chemistries, although this would have coding implications.

As several of the chapters in this book show, an RNA World is an attractive hypothesis to explain a key early stage in the evolution of life, even if it in turn was preceded by an even simpler system. Is it possible that DNA could be self-replicating? Iterative selections with DNA can also give catalytically active molecules, despite the advantage provided by the 2′ hydroxyl on RNA for achieving a diversity of folded shapes and expanding a limited catalytic repertoire. Although there are no known cases of naturally occurring DNA catalysts, the door is clearly open for comparison of the capabilities of the two nucleic acids.

Accurate replication of modern DNA genomes is maintained by sophisticated error correction mechanisms, whereas RNA replication (viral genomes) is not so endowed. Is it possible that if error correction had evolved early for RNA replication, RNA genomes would be the rule?

Since all extant life on earth came from a common ancestor, there is no opportunity for comparison of alternative origins. Even a few years ago, few took seriously the prospect of finding life elsewhere in the solar system. However, hope has been awakened by realization of the abundance of life deep in the Earth and discovery of water on Mars. Perhaps there will be a chance to see the results of an alternative experiment, or perhaps there is enough interplanetary travel via meteorites that life seeded on one planet has spread; it would be simultaneously thrilling and disappointing to find that Martian life also used DNA, RNA, and proteins.

The valuable Foreword and Prologue of the first edition, written by Crick and Watson (the Canons of the canonical pair), have been reprinted here.

The two editors of the first edition welcome Tom Cech as a coeditor. The chapters in this volume reflect our perspective on the modern world of RNA and how it relates to the RNA origin hypothesis. Obviously, space limitations prevented inclusion of some topics that belong here.

We thank the many people who made this book possible: the authors, the staff of Cold Spring Harbor Laboratory Press, especially Inez Sialiano, Patricia Barker, and John Inglis.

R. F. Gesteland
T. R. Cech
J. F. Atkins

Foreword to the First Edition

The term "RNA world" originally referred to a hypothetical time in the evolution of earthly life when there was no elaborate mechanism for protein synthesis such as we have today. As described by Joyce and Orgel (Chapter 1), there were speculations in the 1960s that RNA catalysts existed at that stage in evolution, that RNA was the sole genetic material, and that the standard Watson-Crick pairing was the basis of genetic replication. None of these early authors was smart enough to suggest that relics of these hypothetical catalytic RNAs might still be around today. Indeed, it was speculated that the original ribosomes might have been made solely of RNA but not that their main catalytic activity (the formation of the peptide bond) might still be performed today by RNA alone, as recent evidence seems to suggest.

This hypothesis of an RNA world without protein was largely forgotten but has now become fashionable again because of the remarkable discoveries by Altman and by Cech (described by Cech, Chapter 11) of RNA molecules that do indeed have catalytic activity of one sort or another. These discoveries have removed the chief objection to the RNA world—that RNA by itself cannot act catalytically.

This volume on the RNA world covers a somewhat wider field. It was realized in the 1960s that not all present-day RNA was messenger RNA or viral RNA. The ribosome was known to contain several distinct types of "structural" RNA molecules; tRNAs were familiar, and there were isolated examples of other small RNA molecules. What was not realized was the richness of the *present-day* RNA world, including snRNPs (as described by Baserga and Steitz, Chapter 14) and now many others. We have come to realize that RNA molecules can occur with many different shapes and can thus have many different functions. Several of the chapters in this book deal with what these structures are, what their functions may be, what makes them stable, and how to predict their structure. Just how many of these RNAs are, by themselves, truly catalytic remains to be seen. Some of them may use divalent cations to produce both their folding and their catalytic activities, as discussed by Pan et al. (Chapter 12).

Some of the properties of present-day RNA molecules were quite unexpected. One might perhaps have guessed that some limited form of messenger-RNA editing could occur very rarely as a freak, but who would have predicted that such editing would take so many distinct forms and occur as widely as Bass describes in Chapter 15?

That a retrovirus contains a sequence rather like part of a tRNA molecule was a surprise when it was first discovered. In 1987, Maizels and Weiner proposed that such structures tagged RNA genomes for replication in the early RNA world. In Chapter 23, they explore the many possible ramifications of this genome tag hypothesis. In particular, they propose that both tRNA and tRNA-aminoacylation activity were subject to selection *before* protein synthesis arrived on the scene. Whether one can safely extrapolate from a present-day property of rapidly evolving RNA viruses to as far back in time as the early RNA world remains to be seen.

A good case can be made (Gold et al., Chapter 19) that small lengths of RNA can often form relatively rigid structures more easily than can polypeptides of similar length. New techniques are now being used (Chapter 19 and Chapter 20 [Szostak and Ellington]) to explore this vast "space" of small RNA sequences for interesting properties such as the binding of one specific substrate or another and, in some cases, a particular type of catalytic activity.

The recombinant-DNA revolution of the 1970s was possible because molecular biologists could use the sophisticated products of billions of years of natural evolution—the replicases, restriction enzymes, and so forth—as precise and delicate chemical tools. These molecules were all proteins that acted on nucleic acids in one way or another. Now experimentalists (Chapters 19 and 20) are using a combination of biochemical processes, related to nucleic acid replication, which together embody the mechanism of natural selection. These powerful tools can provide them with new and desirable RNA molecules in the laboratory while they wait. Because these methods do not need whole cells, but only cellular components, they can handle many more individuals at one time and so explore the RNA "space" more quickly. The message is clear: Where possible, let Nature do the work, and if she proves a little slow, take your whip to her.

The exact nature of the early RNA world is now a matter of active debate. What can we learn from the present "relics" and, in particular, can we deduce the composition and nature of each particular type of early RNA from the detailed study of its many descendants today? Can we learn anything from small RNA-related molecules (such as coenzymes) about the general nature of the early RNA world? Early speculations had assumed that the RNA world was rather simple and rather inaccurate. Benner and his colleagues (Chapter 2) now believe it was much more complex and more accurate. As more sequence data accumulate, one would expect these suggestions to become less speculative.

It may be possible to deduce something about some of the activities in the RNA world by a detailed study of the present mechanisms of protein synthesis. For example, Weiss and Cherry (Chapter 3) have proposed an ingenious model for the origins of the smaller ribosomal subunit. They suggest that it was originally an RNA replicase that used oligonucleotides as a substrate, cannibalizing triplets from them for its own replication.

We may, in time, arrive at a rather plausible picture of this early stage in evolution, even if true molecular fossils (that is, actual specimens of the molecules that then existed) are forever unavailable to us because of the ceaseless battering of thermal motion over billions of years.

The details of the transition from the RNA world to the present protein-dominated world are also debatable. For example, was there some sort of protein synthesis before the existence of a primitive ribosome? Did these first ribosomes have some protein, or did the ribosomal proteins arrive a little later?

It may turn out that we will eventually be able to see how this RNA world got started. At present, the gap from the primal "soup" to the first RNA system capable of natural selection looks forbiddingly wide. Was there perhaps a pre-RNA world, with replication based on some even more primitive system?

Such a pre-RNA world might have been of several types. It would presumably have been based on some form of genetic polymer or sheet that was formed more easily than RNA under prebiotic conditions. It might eventually have transcribed its detailed genetic information directly onto RNA to form the RNA world. Alternatively, it might have acted solely as a midwife, setting up the basis of RNA synthesis for some use of its own, with RNA replication then taking over and replacing it. Cairns-Smith has already suggested that this pre-RNA system might have been based on clays, or on some organic polymer (Cairns-Smith, A.G., *Genetic Takeover and the Mineral Origins of Life*. Cambridge University Press [1982]).

This time we should be smart enough to ask if there are any relics of this *pre*-RNA world still with us today.

Finally, a word of heresy: As Moore has pointed out (Chapter 5), the real fossil record suggests that our present form of protein-based life was already in existence 3.6 billion years ago and evolved rather slowly for a billion or so years after that. This leaves an astonishingly short time to get life started. Moreover, the three main lines of descent (see Fig. 3 in Woese and Pace, Chapter 4) seem a very long distance from their hypothetical common ancestor.

The rather far-fetched hypothesis of Directed Panspermia would predict that life was sent here in the form of "bacteria" suitable for growth in anaerobic conditions, *and that several somewhat different forms would probably have been sent at the same time*, in the hope that at least one would survive. All are likely to have evolved originally from a common ancestor (on another planet) that existed some billions of years before the formation of our solar system. Therefore, the final question about the RNA world and the pre-RNA world (if it existed) is: *Where* did it occur? Are we totally confident that our form of life started here, or did it perhaps originate elsewhere in the universe? It might have been easier to start elsewhere because, for example, the atmosphere there was more reducing than the Earth's early atmosphere appears to have been.

The lively and authoritative chapters of this book deal comprehensively both with the hypothetical RNA world and with the complexities of RNA structure and function we find around us today. I recommend it to all molecular biologists and especially to anyone fascinated by the baroque complexity of the nucleic acids and of RNA in particular.

Francis Crick

Prologue to the First Edition

Early Speculations and Facts about RNA Templates

RNA first came alive to me during the fall of 1947 at Indiana University when I took Salvador Luria's course on viruses. There I first learned that whereas the then-known phage, pox and papilloma viruses, contained DNA, this molecule was totally absent in several purified plant viruses as well as in the viruses that caused encephalitis and polio, which instead contained RNA. Apparently a given virus had either RNA or DNA, in contrast to cells which contained both. But whether it was the nucleic acid component that carried their genetic specificity was still unclear. At that time, most scientists wanted Avery, MacLeod, and McCarty's experiment on pneumococcal transformation by purified DNA to be extended to other life forms before jumping on the nucleic acid bandwagon. Then in the spring of 1952 came the report from Al Hershey that the DNA component of phage T2 carried genetic specificity. This immediately thrilled me, but I remember well the audience's indifference when in mid-April I read Hershey's letter at an Oxford meeting of The Society for General Microbiology.

When I had arrived at Cambridge in the fall of 1951, I started taking seriously the work of Brachet and his collaborators in Brussels, who emphasized the correlation between the RNA content and the protein-synthesizing capacity of cells. Those cells making large amounts of protein possessed large numbers of virus-sized ribonucleoprotein particles, known initially as microsomal particles but since 1958 as ribosomes. Most importantly, these particles had been pinpointed as the actual sites of protein synthesis by means of the then just-developed cell-free systems for protein synthesis. Here the key lab was that of Paul Zamecnik at Massachusetts General Hospital. Equally important was Brachet and Chantrenne's demonstration that the nucleus, and hence DNA, had no direct participation in protein synthesis. To show this, they cut the giant alga *Acetabularia* in half and observed that the half without a nucleus could maintain almost normal protein synthesis for more than a month. Yet from the one gene–one enzyme (protein) results of Beadle and Tatum, the ultimate source of the genetic information that specifies the amino acid sequences of proteins had to be the genes found in the nucleus. I thus postulated a two-stage scheme for protein synthesis in which DNA first serves as a template for nucleus-located synthesis of RNA, and this RNA then in turn moves to the cytoplasm where it functions as the template for protein synthesis.

No one then had any compelling reason to take my hypothesis seriously, but by November 1952 I liked it well enough to print DNA→RNA→protein on a small piece of paper which I taped on the wall above my writing table in my rooms at Clare College. From the day of our first meeting, Francis Crick and I thought it highly likely that the genetic information of DNA is conveyed by the sequence of its four bases, but we knew it was premature to promote this idea before the structure of DNA was known. However, from the moment we first saw how to build a double helix out of the four base pairs, it was clear that the essential uniqueness of a gene must reside in its respective sequence of base pairs. Moreover, not only could base pairing provide the way for genes to be copied exactly during gene duplication, but it was also very likely to underlie the process by which the genetic information of a DNA molecule is transferred to its RNA product.

Still totally unclear, however, was how RNA might serve as the template for ordering the amino acids in their respective polypeptide products. Emboldened by our fantastic good luck in so simply finding the structural essence of the gene duplication process, I saw no reason not to take on the challenge of finding out what RNA molecules looked like in three dimensions. Such knowledge, I felt, would be indispensable to understanding how they functioned in protein synthesis. I took on this task when I moved to Pasadena in the fall of 1953. There I joined forces with Alex Rich, who had started working on DNA just before Francis and I found the double helix. There was no difficulty in getting him to move on to a better pasture, and soon I was collecting RNA samples and drawing them into fibers that Alex exposed to X-ray beams. But despite much travail, even those fibers displaying high birefringence never gave rise to ordered diffraction patterns like those of DNA. Although we thought we saw reflections that might have come from short sections of double helices, we could never be sure and saw no way to decide whether RNA was a one- or two-chain molecule. Here the base composition was a bad tease. Viral RNAs clearly did not show equivalence of A with U or G with C, but the RNA from cells had A/U and G/C ratios sometimes closely approaching 1/1. But we could see no difference in the general features of the X-ray diagrams from viral or cellular RNAs. To our annoyance, RNA, no matter from what source, showed identical X-ray diagrams characterized by strong reflections at 3.36 Å and 4.00 Å. Clearly, there was some ordered structure in RNA, but we saw no way to get to it. After some six months of such frustration, we gave up.

A very different approach to understanding protein synthesis came from the very clever Russian-born theoretical physicist George Gamow, who was struck by the fact that the 3.5-Å distance between adjacent amino acids in extended polypeptide chains was very similar to the 3.4-Å separation between adjacent base pairs in the B form of the double helix. Not then cognizant of RNA's primary role in protein synthesis, Gamow proposed that amino acids are directly ordered by contacts with the DNA base pairs, with the polypeptide products containing the same number of amino acids as there are base pairs in their DNA templates. To deal with the fact that DNA has only 4 letters in its alphabet, whereas 20 different letters (amino acids) are used to specify proteins, Gamow assumed that each amino acid must be coded by several adjacent base pairs. Because there are only 16 (4×4) combinations of the four bases, taken two at a time, Gamow made the assumption that each amino acid must be specified by groups of three adjacent base pairs (a codon) along DNA chains. To deal with the fact that there are 64 ($4 \times 4 \times 4$) such triplets, he assumed that many amino acids must be specified by more than one triplet (redundant triplets). In such an "overlapping" code, adjacent amino acid codons have two out of the three base pairs in common, thereby restricting which amino acids can lie adjacent to each other. Gamow's overlapping code was only the first of several that would later be devised, each leading to different combinations of forbidden amino acid neighbors.

When George first told me his scheme, I quickly dismissed it since DNA was not the template that ordered the amino acids. But he was having combinatorial fun, and besides I could not rule out the possibility that some RNA molecules were double helices. With time, moreover, I realized there was a real virtue to Gamow-like codes. In disproving them, the possibility of overlapping codes could be ruled out, pointing the way to codes in which adjacent groups of most likely three bases specified successive amino acids along a polypeptide chain. In fact, the first known amino acid sequence, that determined by Sanger for insulin, disproved the first code, although Gamow did not at first realize this because his initial list of the 20 amino acids had several embarrassing mistakes (e.g., he included both cystine and cysteine). Other such overlapping codes devised over the next year were also ruled out as more proteins became sequenced. It was during the first coding rush that Leslie Orgel and I, on a trip to Berkeley where Gamow was spending the spring of 1954, suggested that we form a club of 20 members whose purpose was to crack the RNA structure and in so doing reveal how the genetic code operated. Soon to be known as the "RNA Tie Club," with its members reflecting Gamow's eclectic taste, it never had a formal meeting, nor did all its members ever cough up the money to purchase their RNA ties and tiepins bearing their respective amino acid code letters. Gamow's tiepin sported ALA (alanine), and mine, PRO (proline). Much more important in the long run was the opportunity it provided to exchange ideas about the code through "Notes to the RNA Tie Club."

Several of these communications, the most important of which came from Francis Crick, Leslie Orgel, and Sydney Brenner, later became incorporated into published manuscripts. Among these was the definitive disproof by Brenner of any form of overlapping code, which he wrote in Johannesburg in the fall of 1956 just before he returned to England to join Francis Crick. Equally important was the communication by Crick, John Griffiths, and Orgel suggesting a comma-less code as a device to let nonoverlapping triplets be read in the appropriate reading frame. But the most influential paper intellectually was the RNA Tie Club's first Note, written by Francis Crick and sent out early in 1955 under the title *On Degenerate Templates and the Adaptor Hypothesis*. After spending the previous August in Woods Hole batting about potential codes with Gamow and Brenner, Crick began questioning the basic assumption that a nucleic acid template provided specific cavities complementary in shape and charge to the amino acid side groups. Here he argued that the nucleic acid bases want to hydrogen-bond and are not at all suitable for forming cavities that could attract the hydrophobic side groups of amino acids such as valine, leucine, or isoleucine. Equally tricky to imagine was any structural basis for degenerative codes in which many amino acid side groups are specified by more than one set of triplets. Faced with what he considered insuperable obstacles, Crick made the radical proposal that prior to peptide bond formation, amino acids are enzymatically attached to small "adaptor" molecules that have surface specifically tailored to bond to nucleic acid triplets. Here Crick suggested that the adaptor molecules might be tiny polynucleotides that base pair to RNA templates.

My reaction to the adaptor hypothesis was initially very negative, even though I had spent the fall of 1954 in futile efforts trying to fold RNA chains into shapes bearing cavities appropriate for the amino acid side groups. The adaptor idea seemed much too complicated to me to have ever gotten started at the origin of life several billion years ago. Francis, too, had his moments of doubt; he even concluded his now famous Tie Club Note with the phrase, "In the comparative isolation of Cambridge I must confess there are times when I have no stomach for decoding." In fact, by the time I returned to Cambridge for the year beginning in June 1955, Francis and Alex Rich were immersed in building new three-dimensional models for collagen to compete with one earlier proposed by Linus Pauling, this despite the fact that Francis and I had often referred to collagen as the most boring of macromolecules.

Equally frustrated, I reverted to thinking about plant viruses, in particular tobacco mosaic virus (TMV), whose helical construction I had worked out in the spring of 1952 after Francis and I had been told by Sir Lawrence Bragg to stop trying to work out the structure of DNA through model building. Always troublesome to me was the apparent necessity to postulate both genetic and protein synthesis roles for RNA. Knowing that only a tiny fraction of TMV particles are actually infectious, I speculated whether in fact these rare particles contained DNA, not RNA, chains. But this was ruled out when it became possible to reconstitute infectious TMV particles from their purified RNA and protein components. This experiment, first successfully accomplished in the spring of 1955 in Berkeley by Heinz Fraenkel-Conrat and Robley Williams, generated much newspaper publicity which gave uninformed readers the idea that life itself had been created. Francis, however, put the matter in its proper light, being quoted in the English press as saying that this was a finding he had anticipated.

Reconstitution by itself, however, did not answer the question of whether the protein component played any more than a protective-coat role for the genetic-information-bearing RNA component. Less than a year later, however, Alfred Gierer, working in Gerhard Schramm's lab in Tübingen, clearly showed that the RNA alone was infectious. This primacy of nucleic acids as bearers of genetic information then lay at the heart of the way Francis and I thought about cells and viruses. But this was far from an acceptable paradigm for many of the attendees at the key, late March 1956, CIBA Foundation meeting on the Structure of Viruses. They were not at home with the concept that information flows unidirectionally from nucleic acids to proteins and never backward. This was awkwardly shown when André Lwoff and I passed on to Robley Williams a telegram message supposedly from Wendell Stanley reading "TMV protein infectious—be cautious!" To our amazement, Robley didn't question the result until we revealed the hoax.

At this gathering of some 30 scientists, Francis presented our ideas on why the protective coats of viruses are made up of protein subunits. In our view, it was a consequence of the fact that no viral nucleic acid had sufficient coding capacity to specify a single polypeptide chain large enough to surround a more centrally located core of nucleic acid. The 2-million-molecular-weight RNA of TMV, for example, contains only 6000 bases and, assuming a coding ratio of three bases per amino acid, is only capable of specifying a 2000-amino-acid polypeptide or about 230,000 molecular weight. To make up the 38-million-molecular-weight protein coat, at least 165 subunits would be needed. In fact, the TMV subunit contains only some 150 amino acids, suggesting that the TMV RNA codes for several proteins or that the coding ratio is very much larger than three. Initially, we thought that tomato bushy stunt virus (TBSV), with a much higher RNA content, might be more useful in giving a realistic value for the coding ratio. It contains only four bases for every amino acid in the crystallographic subunit. Later, direct analysis of its protein subunit size suggested that each crystallographic repeat contains some five protein subunits, very likely implying that TBSV RNA, like TMV RNA, also codes for several proteins.

After that 1956 meeting, I again had a go at the RNA structure, taking advantage of the newly discovered enzyme polynucleotide phosphorylase that Marianne Grunberg-Manago and Severo Ochoa found could be used to make synthetic RNA molecules. By then back at the National Institutes of Health, Alex Rich had shown the month before that random AU and AGCU co-polymers gave X-ray diffraction patterns similar to those we had obtained using purified cellular and viral RNAs. I focused instead on poly(A) (adenine) fibers drawn from material prepared in the Molteno Instiute by Roy Markham and David Lipkin. To my delight, they generated clean helical X-ray diagrams that were best interpreted as base-paired double helices built up from two parallel poly(A) chains. Initially, I was disturbed by the fact that many of the key reflections overlapped with those generated by purified RNA, which by then we had every reason to believe was single-chained. Later, this apparent paradox was possibly resolved by the finding that sections of hydrogen-bonded hairpins form along most RNA chains. Conceivably, it is these short sections of imperfect double helices that generate the DNA-like feature of the RNA X-ray patterns.

By then discouraged that the study of purified RNA would lead toward understanding how protein synthesis occurs, I decided to concentrate on the structure of ribosomal particles, at that time believing that they must carry the genetic information for ordering amino acids in proteins. In the spring of 1956 I convinced Alfred Tissières, then a Fellow at Kings College working in the Molteno Institute on oxidative phosphorylation, to join me at Harvard, where I would be moving in the fall of 1956. Alfred had in fact already done some preliminary experiments on the then-called microsomal particles and was keen to follow them up.

I arrived at Harvard some six months before Alfred but was preoccupied most of this time trying to be an effective teacher for the seniors and beginning graduate students who were taking my course on viruses. As soon as my lectures were under control, however, I went across the Charles River to see Paul Zamecnik whom I had first met the year before while briefly stopping at Harvard on my way back to England. There, in the building where Fritz Lipmann also had his lab, I first appreciated the importance of the discovery made there a year earlier by Mahlon Hoagland of the activated high-energy acyl-amino acid intermediates for protein synthesis. Even more important, I first learned of the more recent observation by Mahlon, Mary Stephenson, and Paul of a soluble RNA (sRNA) fraction to which the activated amino acids are transferred prior to protein synthesis. Quickly, I realized that these sRNA molecules might be the polynucleotide adaptors postulated two years before by Francis in his first Note to the RNA Tie Club. Prior to my visit, Crick's ideas were unknown to the Massachusetts General group and they eagerly sought out Francis when they came together at the 1957 Gordon Conference on Nucleic Acids and Proteins.

Tissières and I commenced our molecular characterization of *Escherichia coli* ribosomes in the spring of 1957, suspecting that they might have structural plans like those of the small RNA viruses, whose isocahedral-shaped protein shells are formed by the regular aggregation of a single protein building block. We could not have been more wrong about how they are organized. To start with, we found that the *E. coli* ribosomes, like those from all other organisms, are formed by the aggregation of two RNA-containing subunits, the larger 50S subunit approximately twice the size of the smaller 30S subunit. Each subunit contains a single major RNA chain, with the 50S ribosomal subunits possessing 23S RNA chains and the 30S subunits possessing 16S RNA chains. Moreover, both subunits contain a large number of different small proteins that for the most part are subunit-specific. At low Mg^{++} levels, the 30S and 50S subunits do not associate with each other, but when the Mg^{++} concentration is raised, they come together to form the 70S complex that subsequent work has shown to be the ribosomal form that carries out protein synthesis.

Naively, at first, we assumed that either the 16S RNA, or the 23S RNA, or both, were the actual templates for protein synthesis. Puzzling to us, however, was why the templates existed in only two size classes while there was great variation in the sizes of their putative polypeptide products. Equally disturbing was why the base composition of ribosomal RNAs barely varied between bacterial species with highly different AT/GC ratios. A priori we had expected to find that the base compositions of the RNA templates would reflect those of their DNA templates. Luckily, there was one powerful exception. In 1956 the phage T2-specific RNA made after T2 infection of *E. coli* was shown by Volkan and Astrachan to have a T2 DNA-like base composition. Moreover, in contrast to the metabolically very stable ribosomal RNA chains, T2 RNA had been found to have a half-life of only several minutes.

In retrospect, Tissières and I should have gravitated early to T2 RNA, but in fact not until the fall of 1959 were its molecular properties investigated anywhere. Then, Masayasu Nomura and Ben Hall, working with Sol Spiegelman at the University of Illinois, provided tentative evidence for the incorporation of T2 RNA into abnormally small ribosomes. In trying to follow up this observation early in 1960, my graduate student Bob Risebrough came to a radically different conclusion. After T2 RNA is synthesized it does not become part of a ribosomal subunit by aggregating with newly made ribosomal proteins. What in fact happens is that in the presence of Mg^{++}, the T2 RNA becomes attached to the smaller 30S ribosomal subunit, which in turn binds the larger 50S subunit to form the >70S complex that actually carries out protein synthesis. This result instantly changed the way we visualized protein synthesis. Instead of serving template roles, ribosomes function as stable assemblage sites for protein synthesis. The true template had to be a new RNA class unknown until that moment both because it comprises such a small percentage of the total RNA and because it is heterogeneous in length. Later, these metabolically unstable templates, whose amounts respond to cellular needs, would be named messenger RNA (mRNA) by Jacques Monod and François Jacob.

The first lab outsider to whom I revealed this conceptual breakthrough was Leo Szilard, whom I had gone down to see in New York where he was successfully plotting the radiation therapy that would cure his bladder cancer. Leo's reaction was entirely negative, not being convinced that we had the right interpretation for Risebrough's experimental results. Predictably, he wanted us to show that mRNA existed in uninfected *E. coli* cells before he would change his mind-set. These were experiments that we had in fact already planned to start several weeks later, as soon as François Gros arrived from Paris to spend several months working with Alfred and me. Also soon to join this effort was Wally Gilbert, then still teaching theoretical physics to Harvard students, but who was increasingly tempted to move into molecular biology by our excitement about mRNA. By the time of the June Gordon Conference on Nucleic Acids we were virtually convinced that *E. coli* mRNA also existed, and by the summer's end we had data for a convincing publication. Already at the Gordon Conference we had heard rumors that Sydney Brenner and François Jacob, using very different arguments, had also postulated the existence of mRNA and that their idea was being tested by Sydney that week in Matt Meselson's lab at Caltech. Eventually, we were to publish our independent proofs for mRNA's existence early in 1961 in back-to-back articles in *Nature*.

With the basic scheme for how RNA participates in protein synthesis known, the path became open for definitive experiments on the exact nature of the genetic code. Using enzymatically synthesized RNA as messengers for in vitro protein synthesis, the correct assignments for all of the triplet codons were determined by early 1966. With this major goal achieved, the time had clearly come to ask how the DNA→RNA→protein flow of information had ever gotten started. Here, Francis was again far ahead of his time. In 1968 he argued that RNA must have been the first genetic molecule, further suggesting that RNA, besides acting as a template, might also act as an enzyme and, in so doing, catalyze its own self-replication. As the chapters in this book will show, how right he was!

J.D. Watson

1
Before RNA and After: Geophysical and Geochemical Constraints on Molecular Evolution

Stephen J. Mojzsis
Department of Earth and Space Sciences
University of California Los Angeles
Los Angeles, California 90024-1567

Ramanarayanan Krishnamurthy
Skaggs Institute for Chemical Biology
The Scripps Research Institute
La Jolla, California 92037

Gustaf Arrhenius
Scripps Institution of Oceanography
University of California San Diego
La Jolla, California 92093-0220

1. INTRODUCTION

This chapter offers a description of some of the physical and chemical settings for the origin of life on Earth. Considering the topic of this book, particular attention is given to the conditions for a precursor RNA World; an ab initio system based on phosphate-sugar backbone structures in linear polymers and currently with nitrogen bases as recognition molecules. Both the appeal and the uncertainties in the assumption of an RNA World are obvious. The biochemical advantage of this model has geochemical and cosmochemical complements such as the abundance in the universe of simple aldehydes, "the sugar of space." With their relatively high oxidation state, the aldehydes are compatible with plausible models of early terrestrial atmospheres dominated by CO_2, H_2O, N_2, and small mixing fractions of CO, CH_4 and other reductants. Further suggestions of an RNA World come from observed and inferred sources of active oligophosphates and from concentration mechanisms based on the molecular charge conferred by phosphate esters. Finally, the facile formation of ribose phosphate has been successfully modeled under mild aqueous conditions in the laboratory. Obstacles to progress in prebiotic synthesis leading to the inception of an RNA World yet remain. The concentration of neutral molecules such as formaldehyde and glycolaldehyde required to permit the phosphorylation and sugar phosphate formation found in the laboratory remains an impediment to modeling molecular evolution. Furthermore, the oligomerization of nucleosides or nucleotides without the aid of artificial activating groups, and the synthesis and attachment of nitrogen bases in a formaldehyde environment hostile to the stability and uninhibited reactions of hydrogen cyanide, is another complicated problem remaining to be resolved by further laboratory investigations. At least for the time being, these limitations retain the concept of an RNA World as a metaphor in the company of all current theories of the origin of life. This metaphor has, however, several redeeming geophysical and geochemical qualities that may be regarded as strengthening the case for an RNA World as an early phase in the emergence of life on Earth in the earliest Archean (i.e., before 3800 Ma; Ma = 1×10^6 years).

2. EARLY ATMOSPHERE AND HYDROSPHERE

2.1 General Aspects

Because there is no known preserved record on Earth for the state of the atmosphere and hydrosphere during the first half billion years, there remains considerable room for speculation about the nature of this system wherein life began (Fig. 1).

It is currently popular to assign early molecular evolution to hot springs or adjacent cooler regions at the surface as in the case of Yellowstone, Wyoming, or at great depth in the ocean as in the case of the submarine hydrothermal or low-temperature vents. The attractiveness of such schemes comes from the conducive environment rich in energy resources that is provided for chemoautotrophs, and from the fact that a few thousand meters of ocean water affords some protection against thermal shocks by meteorites and comets early in Earth's history (Sect. 4). The appeal also includes the assumption of a hydrothermal supply of organic compounds, suggested by theoretical considerations and laboratory experiments (Shock 1990, 1996; Shock et al. 1997). However, as of today, the analysis of effluents from bioorganically uncontaminated hot or cold springs has failed to show any organic molecules besides methane, perhaps because of insufficient analytical attention paid to this aspect. In contrast, hot springs debouching through sediments transport a variety of organic decomposition and polymerization products, including petroleum, that are the direct result of thermal breakdown of residual bioorganic matter from past life, now trapped in rocks.

A particularly demanding but neglected link in such schemes for biopoesis in hydrothermal vents is the need for a crustal source of reduced nitrogen available for prebiotic chemical reactions, made acute by the low solubility of molecular nitrogen in the water transfer medium. A solution may be found in the mechanism demonstrated by Summers and Chang (1993), showing that photochemically produced nitrite ion is converted to

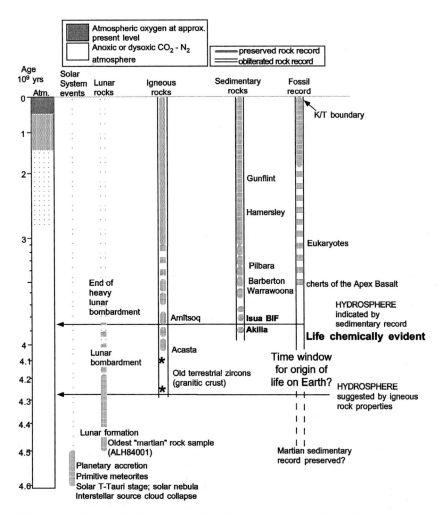

Figure 1 Time scale for events with relevance to the origin and early evolution of life.

ammonia by reduction with Fe^{++} in solution. The reaction is enhanced in the anion-concentrating reactive mineral green rust (ferroferric hydroxide). In contrast, inorganic carbon and hydrogen in the form of carbon dioxide and water are readily transported into oceanic crust and potentially (but for CO_2 not yet demonstrably) reduced there during hydrothermal alteration.

The fact that thermophilic Archea are found at the bottom of the phylogenetic tree (Woese 1987) has been considered an argument for the origin of life in a hydrothermal environment. However, present-day Archea are already highly evolved organisms with an elaborate polypeptide-based enzymatic machinery, protecting the otherwise unstable nucleic acid structures against decomposition at high temperature in the immediate vent environment. Though possessing deep phylogenetic roots, these organisms are still very far away from the origin of life. These circumstances tend to induce those advocating these deeply rooted hyperthermophiles as a remnant population from early molecular evolution to retreat from conduits of hot fluid at the vent proper. In the surroundings of hydrothermal vents, temperatures range all the way down to near freezing on the present sea floor, and as low or lower around some continental hot springs in cold climates. Hydrothermal advocates would also be drawn into an accretion camp that places a substantial reservoir of extraterrestrial organic solids, like cometary or carbonaceous meteorite material, in the early crust, available for hydrothermal digestion. If not depending on a source organic compounds from space, much will depend on the outcome of currently intensified analysis of virginal hot spring effluents. Expanded measurements of organic production at hot springs could testify to the efficiency of postulated inorganic–organic reactions. With these provisions and reservations, the hypothesis of hydrothermally assisted molecular evolution would seem to fall within the limits of geophysical and geochemical plausibility. The examples presented here point at the interdependence between different disciplines, and at avenues for verification.

2.2 The "Fertile Atmosphere" and the Concept of a Prebiotic Soup

Already early in the century, considerations of the possible nature of Earth's primordial atmosphere created compelling arguments for a neutral or weakly reducing composition, dominated by carbon dioxide, nitrogen, and water vapor. In his pioneering electrochemical experiments, Walther Löb (1906, 1914) actually used carbon dioxide and carbon monoxide as major components of the cold plasma discharge system (Fig. 2), which

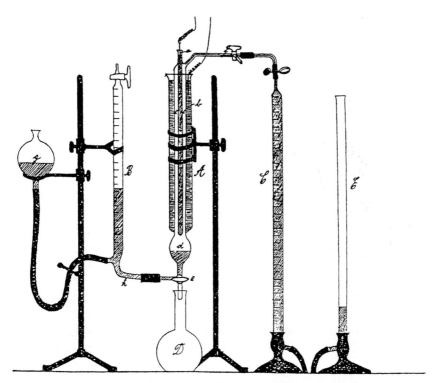

Figure 2 Apparatus used by Walther Löb for studying the biomimetic synthesis of organic compounds, including aldehydes (1906) and amino acids (1914) in cold plasmas, with CO_2, and NH_3 as reactants in various combinations. (*A*) Plasma generator ("elektrisator"); *a*, inner tube; *b*, outer tube; *c*, plasma discharge space; *d*, liquid bulb; *e* and *f*, three-way stopcocks; (*B*) burette with leveling funnel; (*D*), recipient flask for liquid with reaction products; (*C*) and (*E*), gas burettes. (Reprinted from Löb 1906.)

was found to produce aldehydes, including glycolaldehyde, and the amino acid glycine. Löb's experiments were intended as metaphors of the conversion of inorganic carbon and nitrogen to organic compounds by autotrophic life without apparent concern for their prebiotic significance.

In contrast, the later and more widely known experiments by Oparin (1924; Oparin and Clark 1959) deliberately aimed at modeling the origin of life. Since the formation of "proteinoids" (postulated then to have been important predecessors of life) demanded a reducing environment, Oparin proposed, without scrutiny of physical plausibilities, that a primordial atmosphere of the Earth must have consisted primarily of a mixture of hydrogen, methane, ammonia, and water vapor. A decisive event in the U.S.

efforts at scientifically addressing the problem of the origin of life was Harold Urey's entry into this field after World War II. Although with a better awareness of geophysical limitations (Urey 1952) and unaware of Löb's biomimetic experiments, Urey nonetheless championed Oparin's concept of a strongly reducing atmosphere despite objections from contemporary geochemists such as V.M. Goldschmidt (1952) and W.W. Rubey (1951, 1955) as well as organic chemists (Abelson 1966; Ferris and Nicodem 1972). The objections of these and other workers were put into the background by the impressive results of experiments by Miller (1953, 1987; Miller and Van Trump 1981) that expanded Löb's findings to a plethora of amino acids and other organic compounds forming under fertile Oparin-like conditions. The fact that a variety of amino acids, considered by science at the time to be the initial building blocks in biomolecular evolution, could be formed from a gas mixture with claims for modeling the primordial atmosphere made a lasting impression on organic chemists as well as the general public, and stimulated prebiotic chemistry as a worthy field of study. However, it is now held to be highly unlikely that the conditions used in these experiments could represent those in the Archean atmosphere. Even so, scientific articles still occasionally appear that report experiments modeled on these conditions and explicitly or tacitly claim the presence of resulting products in reactive concentrations "on the primordial Earth" or in a "prebiotic soup." The idea of such a "soup" containing all desired organic molecules in concentrated form in the ocean has been a misleading concept against which objections were raised early (see, e.g., Sillén 1965). Nonetheless, it still appears in popular presentations perhaps partly because of its gustatory associations.

The interest in planetary atmospheric physics and chemistry, rekindled in the era of space exploration, brought physical considerations back into focus in addressing the nature of the early atmosphere. Major contributions to the application of atmospheric dynamics to models for primordial planetary atmospheres were made by Walker (1980, 1982), Kasting and Walker (1981) and later by Kasting and collaborators (Kasting and Walker 1981; Kasting 1982, 1988, 1993; Kasting et al. 1983, 1984; Kasting and Ackerman 1986). These studies again indicated carbon dioxide, nitrogen, and water vapor as dominant components for Earth's primordial atmosphere and of materials such as H_2S, CO, CH_4 and H_2, ranging in concentration from ppm to a few hundred ppm, as minor components.

Melton and Ropp (1958) demonstrated the formation of cyanide ion by the reaction of atomic nitrogen on HCO_2^-, reactions that may have taken place in cold plasmas in the Archean atmosphere and ionosphere

(Sect. 2.3). Atomic nitrogen reacting with small amounts of methane in the upper atmosphere was also proposed by Zahnle (1986) as a potential source for hydrogen cyanide. Subsequent to Löb's (1906) work, several authors have continued to investigate the formation of aldehydes from CO_2, crucial for an RNA World and in what appears to be a realistic vision of the early atmosphere.

Modern views have thus returned the atmosphere to only a limited source of strongly reduced compounds for prebiotic synthesis, but still support the production of organic species at the oxidation level of aldehydes. Most importantly, the resulting low concentration in the hydrosphere of organic compounds made available for reaction in a weakly reducing atmosphere has focused the attention on the crucial need for selective concentration mechanisms in models for biopoesis.

A related casualty of the organic aridity of a near-neutral atmosphere is the concept of solution in the ocean, taken for granted almost automatically in much of the literature as the site of early chemical evolution toward complex biomolecules. The dilution in the ocean of soluble compounds from any weak source is forbidding; already sparse, unstable molecules introduced in a volume of 1.3×10^9 km^3 of seawater are mutually unreactive and hardly retrievable by evaporation or other means. Oceanic boundary reactions at the air–sea or crust–sea interface are entropically more advantageous than anything in the free water column and could have involved, for instance, large-scale processing in the hydrothermal plumbing system of the accumulated polymeric tar that constitutes the major fraction of organics from space. Such matter may have been continuously deposited on the ocean floor in the earliest Archean at a rate some 10^2–10^4 times higher than at present (see, e.g., Chyba et al. 1990; Chyba and Sagan 1992). It is conceivable that reactive concentrations of such pyrolysis and hydrolysis products could be maintained in low flow-rate regions of the oceanic crust. Semipermeable membranes of precipitating metal sulfides, observed in laboratory experiments, have been suggested by Russell et al. (1994) as containment for such solutions.

Another way to achieve both preconcentration and processing in schemes for biopoesis was quoted by Charles Darwin: a small lake or "pond" with soluble compounds accumulating from a surrounding drainage area. Solar evaporation in a Darwinian pond would be a questionable concentration process in view of the destructive effect of UV radiation on organic compounds. Selective sorption on surface-active sedimentary minerals may be more effective based on experimental results. In such an environment, wide local ranges of pH and temperature could be achieved (Arrhenius et al. 1997).

A conceptual representation of a number of processes proposed in the literature that could contribute to selective synthesis in the Darwinian environment is schematically illustrated in Figure 3. Extraterrestrial input by proposed higher rates of impact to the early Earth (Sect. 4) is represented by atmospheric ablation of meteors and a cometary impact pond (Clark 1988). Volcanic exhalation products issuing in a glacial environment (Schwartz and Henderson-Sellers 1983, quoting Spitzbergen as a modern counterpart) could be preconcentrated by eutectic freezout in the ice (Sanchez et al. 1966). Oligophosphates condensed in molten volcanic rock have been observed by Yamagata et al. (1991) to be carried into the hydrosphere by supercritical steam and to be effectively concentrated by bilateral surface-active minerals such as double-layer metal hydroxides and clays (Arrhenius et al. 1997). Condensed phosphates may also be produced at contact heating by lava flows over sedimentary deposits containing the protonated Mg-Ca phosphate that is expected as the abiotic product of phosphate precipitation (Sect. 6.1). Surface-active minerals

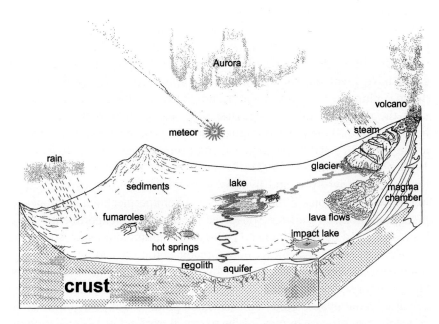

Figure 3 Physiographic diagram of a "Darwin pond" environment, visualizing proposals made to solve problems in biopoesis by authors referred to in the text. Condensed phosphates are emerging in hydrothermal steam and solution and also produced by lava flows in contact with sediment containing the condensed phosphate mineral, whitlockite, perhaps dominant in an abiotic environment.

forming from aqueous solution are capable of phosphorylation, selective concentration, and oligomerization of aldehydes and carboxylic acid anions (Pitsch et al. 1995; Kolb et al. 1997; De Graaf et al. 1998). Aside from the accumulation and reaction of prebiotically important organic molecules in lakes, porous groundwater aquifers with slowly diffusing and percolating solutions would offer voluminous and chemically diverse environments for selective concentration and reaction. Such a reacting groundwater environment would be important particularly for relevant chemistry taking place on bilateral surface-active minerals such as double-layer metal hydroxides and clays. In the "genetic takeover" concept of Cairns-Smith (1975, 1982), this would be the primary medium for emergence of the earliest life forms, with reactive minerals providing an integrated scaffolding, and a part of life itself.

The conceptual notions presented in Figure 3 should not be taken too literally from a geographic point of view, but they illustrate principles proposed by various authors in order to overcome some of the important geochemical obstacles perceived in molecular evolution in an Archean environment.

2.3 Synthesis of Source Molecules in the Interstellar Medium and the Atmospheric Electric Circuit

The nature of the space medium has increasingly been seen as controlling the production of compounds of prebiotic interest considered difficult to synthesize in an "infertile" terrestrial atmosphere. The favorable conditions for building large organic molecules in the space medium are due to the extremely low kinetic temperatures (10–50 K) and densities ($\leq 10^6$ cm^{-3}) with resulting long collision lifetimes (hours–days), coupled with high excitation, ion, and electron temperatures (10^3–10^6 K). The molecular products of such plasma reactions are gathered in solid condensates and brought to the planets by infalling solids. Such compounds, observed in meteorites, comets, and the interstellar medium, include (in meteorites) ethylene glycol, dihydroxyacetone, glycerol, and glyceric acid (Cooper 1996), hydroxyacids (Peltzer et al. 1984), phosphonates and amino acids (Cronin and Pizzarello 1997), in one case with chiral excess. These all are found in extremely low concentration (ppm range or even less), which emphasizes the problem of bringing these materials to Earth at sufficiently high concentrations for biopoesis.

Extended cold plasma regions in the atmosphere include the auroral electrojet (Fig. 4), the subauroral electron deposition region in the upper atmosphere (Fig. 5) and the transient sprites (Fig. 6) and jets in the tropo-

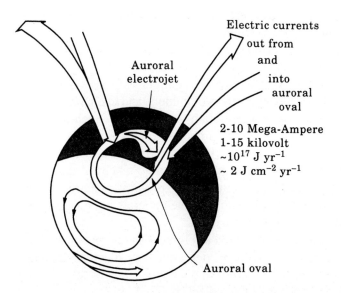

Figure 4 Electric current structure in the aurora. (Courtesy of C.G. Fälthammar.)

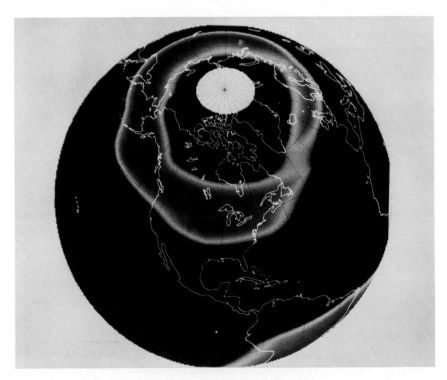

Figure 5 Boreal auroral oval seen from space. (NASA.)

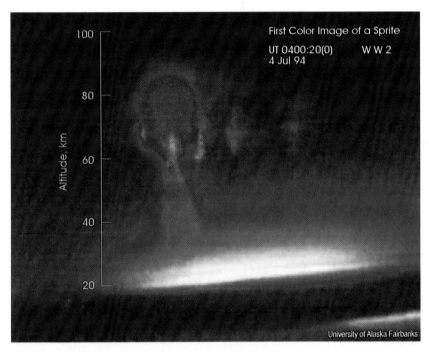

Figure 6 Red sprite extending above a thunderstorm through the stratosphere and into the ionospheric E-region, forming an electric connection between the ionosphere, the troposphere, and, in the addition to the aurora, a possible mechanism for generation of organic reaction products from cold plasma in a nonoxidizing atmosphere. (Courtesy of D.D. Sentman, University of Alaska.)

sphere-stratosphere. Sprites and jets are luminous transient plasma discharges that, because of the lack of visibility from the ground, have only recently been discovered and explored (Sentman and Wescott 1995; Sentman et al. 1995; Dowden et al. 1996; Hampton et al. 1996). Sprites extend to altitudes of 90–100 km above thunderclouds, forming a link in the atmospheric electric circuit; organic synthesis and transport of products in sprites may have played a role in the upper atmosphere of the Archean. The yield from organic reactions may have been limited by the short pulse duration, which may, however, contribute to efficient quenching of product molecules.

In the auroral oval, the currents feeding into and leaving the circuit are powered by the solar wind- magnetosphere dynamo; electrostatic double layers accelerate electrons into the upper stratosphere (<100 km altitude) with energies often exceeding 10 keV. In the high-energy range of the trajectories, the electron excitation of atmospheric molecules results in dissociation and positive ion formation; at thermalization, the electrons attach themselves to molecules, forming hydrated negative cluster ions. These, together with solid condensates, dominate the chemical reactions in the polar stratosphere, today including the formation and destruction of ozone. In an anoxic atmosphere, they presumably controlled the formation of cyanide, formaldehyde, glycolaldehyde, and other organic compounds.

In contrast to these cold plasma phenomena, lightning at the high gas density of the troposphere is associated with high kinetic temperatures and high energy density. Lightning is relatively effective in activating exothermic reactions such as the oxidation of nitrogen, but ineffective in synthesizing metastable polyatomic molecules. The high-voltage, low-current spark technique employed by Miller (1953) and numerous later workers as a simulation of lightning in fact bears little similarity to this natural process, and owes its relative efficiency to the narrowness of the discharge, thus enabling quenching of the numerous polyatomic species produced in a range of reducing gas mixtures (Schlesinger and Miller 1983). It is notable that the proportions of amino acids found in meteorites, with nitrile precursors most likely produced by ion–molecule reactions in the interstellar medium, are similar to the relative amounts found by Miller (1953, 1987) in the different spark discharge regime. Under present-day conditions, organic compounds deposited in the ionosphere and upper atmosphere by interstellar dust and vaporizing meteorites are destroyed by oxidation such that only the ashes of the meteors remain in this layer in the form of magnesium, calcium, iron, and other silicate mineral-derived ions and their condensates. However, in an Archean dysoxic atmosphere, meteoritic and cometary ablation products and atmospheric gases in the ionosphere and in the upper atmosphere may have provided significant sources for organic molecule formation.

The low density and kinetic temperature of the auroral plasma (<350 K), and the darkness during the polar night, would tend to protect reaction products against thermal and UV decomposition, which overcomes some problems faced by other production schemes. The polar night cone would then provide an efficient regime for molecular synthesis in an atmosphere dominated by N_2 and CO_2 with small mixing fractions of atmospheric reductants such as CO and CH_4, and with added organic pyrolysis products from vaporization of extraterrestrial material. Laboratory model experiments (Melton and Ropp 1958) indicate that excited atomic nitrogen under these conditions reacts with species such as HCO_2^- to form cyanide, an ion of crucial importance in biopoesis. The classic cold plasma model experiments by Löb (1906) cited above demonstrated the formation of formaldehyde and formic acid as main products, with glycolaldehyde as an additional component under similar conditions, and the amino acid glycine in the presence of ammonia. An important difference between the present Earth and the early Archean, is that the auroral current may have been increased by a factor of two if the inferred primordial 12-hour spin rate correspondingly increased the Earth's magnetic dipole field and beyond this if the magnetospheric current density was enhanced by coupling to an intensified solar wind.

2.4 The Carbon Dioxide Cycle and Greenhouse Effects in the Early Atmosphere

A question of importance for the radiation balance and thermal state of early Earth is how large a fraction of the total carbon reservoir of the planet would at any given time have been present as carbon dioxide in the atmosphere. As realized already in the last century (Högbom 1894; Arrhenius 1896; J.J. Ebelmen in Berner and Maasch 1996), carbon dioxide plays a primary role in the weathering of rocks. Carbonic acid in rain continually attacks exposed surfaces, which in the case of igneous rocks, consist of silicates with cations mainly from the alkali-alkaline earth and transition element groups. Weathering releases calcium, magnesium, divalent iron, and other cations into solution; in this process the silicates are converted to the corresponding carbonates together with free silica and low-energy hydrated silicates in the hydrosphere. As a result, the store of carbon dioxide in the atmosphere gradually becomes bound as sedimentary deposits of calcium, magnesium, and iron carbonates, which now represent about 80% of the total crustal reservoir of carbon. In the return branch of the geochemical cycle, carbonate rocks, brought to high temperature in the interior of the Earth, dissociate into carbon dioxide, metal oxides, and silicate. The cycle becomes complete when carbon dioxide is returned to the atmosphere by volcanic exhalations.

During the major part of the first 500 Ma, the time of emergence of life on Earth, pH in the hydrosphere has been estimated at 5–7 (Fig. 7), based on the speciation of minerals preserved in the oldest sedimentary precipitates such as banded-iron formations. A number of factors other than the partial pressure of CO_2 come into play in this process, in particular the total area of exposed continental rocks susceptible to weathering, and the increased rate of turnover of oceanic crust and its sediment load by enhanced sea floor spreading, releasing heat and volatiles and subduction of one crustal plate beneath another. Competition to CO_2 as a weathering agent is also offered by other stronger acids present in the primordial atmosphere related to increased volcanism, particularly by gaseous

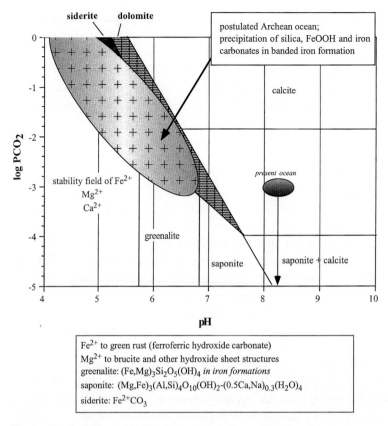

Figure 7 Stability of sedimentary minerals as function of pH and partial pressure of carbon dioxide, and estimates of the chemistry of Archean seawater. PCO_2, partial pressure of CO_2. (Modified from Mel'nik 1982.)

HCl and HF released by volcanism. These have been estimated by H.D. Holland (pers. comm.) to have kept pH of the initial hydrosphere as low as perhaps 2 or 3, with a titration time toward neutrality of perhaps a few million years.

Early greenhouse effects have to be considered in the context of the early history of the sun (Owen and Cess 1979). Studies of stellar evolution have made it possible to tie the formation of planets to a specific stage of evolution of young stars, represented by the T-Tauri objects, and to generalize stellar evolution in terms of luminosity. It is on this basis likely that the sun, during the early stages of the history of Earth before 4 billion years ago, would have been about 30% fainter than it is today (Newman and Rood 1977; Gough 1981). As an explanation of the postulated warmth of the early Earth regardless of the low incident solar radiation, Kasting and Ackerman (1986) have evaluated the effect of an enhanced CO_2 concentration. Without the compensating thermal blanketing by photoactive gases, the low solar power could have led to a completely frozen Earth with such a high albedo that the resulting cryosphere might have been prevented from ever returning to liquid state. This frozen world would have persisted to the present unless some later thermal event could have delivered sufficient energy to thaw out (Bada et al. 1994). This cycle is suggested to have been active in assisting in the emergence of life by permitting the production, accumulation, and reaction of hydrothermally produced organic molecules below the ice cover as proposed by Bada et al. (1994). Since the geological record appears to exclude sustained catastrophic impacts later than about 3900 Ma, such a freezing and thawing, which per se does not conflict with any geophysical principles, would need to be relegated to the earliest, unrecorded Archean (Hadean). Kasting, as illustrated in Figure 8, estimates the initial partial pressure of CO_2 to have been somewhere between 0.1 and 10 bar. The lower value for early Archean pCO_2 corresponds to the estimated minimum requirement for keeping the Earth from irreversibly freezing over, and thereby increasing the albedo of the planet and locking it into a permanent "ice house." With those less catastrophic accretion models that lead to moderate average temperatures of the Earth's surface, the reaction of the crust with rainout of dissolved carbon dioxide could have proceeded in pace with the late phases of accretion and resulted in atmospheric pressure of carbon dioxide in the lower range of Figure 8.

Oxygen isotope measurements (Knauth and Lowe 1978) and petrological observations (Costa et al. 1980) have been taken as indications of high water temperatures associated with the Archean banded-iron formations. The nature of the oldest known sedimentary rocks contributes testimony to a liquid hydrosphere persisting from the earliest recorded stages of the history of the Earth 3850 million years ago, or only 600 million years after the formation of the planet (Dymek and Klein 1988; Mojzsis 1997; Nutman et al. 1997), and giving rise to the conditions necessary for an RNA World that probably developed much earlier.

3. GEOCHEMISTRY OF MOLECULAR EVOLUTION

3.1 The Case for RNA

Since RNA can perform both catalytic and information-transfer functions, and is utilized in the most primitive organisms, it suggests molecules of similar construction as simpler predecessors. This observation provides the chemical paradigm for research into an RNA World and potential processes for the production and interaction of its components. The current structural design of RNA entails a backbone of pentose (furanosyl ribose) sugars linked by phosphodiester bonds, with nucleobases serving as letters in the genetic alphabet. A structure of this complexity is considered by many unlikely to appear by random molecular evolution. Simple (unidentified) nitriles may possibly have served as precursors of nucleobases in early forms of self-recognizing molecules. Organic phosphates

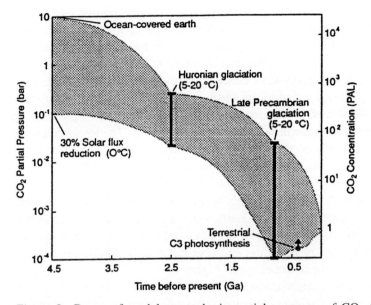

Figure 8 Range of model atmospheric partial pressures of CO_2 (*left vertical scale*) and concentration (*right-hand scale*; present atmospheric concentration as unit) as function of change in solar luminosity with time (units of 10^9 years [1000 Ma]; horizontal axis). With a 30% reduction in solar luminosity in the early Archean, CO_2 partial pressures of 10^{-1} bar 4500 million years ago would represent the minimum CO_2 concentration necessary to warm Earth to 0°C at that time. This ignores any thermal increment due to enhanced heat flow from Earth's interior. The initial CO_2 pressure of 10 bars refers to the model by Dymek and Klein (1988), assuming an ocean covered Earth, and corresponding to a calculated temperature of 85°C at 30% reduced solar luminosity. The figures 5°–20°C are estimates by Kasting for maximum and minimum temperatures during the Precambrian ice ages, 5°C being a probable limit below which the entire hydrosphere might freeze due to feedback effects; 20°C is the temperature estimate, derived from oxygen isotope measurements, for the time, about 30 million years ago, when the Cenozoic ice caps started to expand. (Courtesy of J.F. Kasting, Pennsylvania State University.)

are pervasive in a wide variety of present-day biochemical processes, implying, because of their unique chemical capabilities, similar functions already in ancient biochemistry (Westheimer 1987) and also in prebiotic chemistry (Arrhenius et al. 1997; Sect. 6.1). Processes for the formation of simple aldehydes as well as for their phosphorylation, concentration, and surface-induced oligomerization to form sugar phosphates have been accomplished experimentally. However the plausible formation of nucleobases and their attachment to the sugar phosphate backbone under abiotic conditions remains complicated, considering the interference in cyanide chemistry by formaldehyde. In contrast, the next step, the oligomerization (and potentially, replication) of nucleic acids, has been successfully approached (Orgel 1986; Ferris 1987, 1993; Bolli et al. 1997). Insight into the geochemical aspects of these precursor processes is being sought by laboratory model experiments and space observations, and may possibly be obtained from the oldest sedimentary systems on Mars and Earth. Mars is the only planet in our solar system that may have offered conditions for molecular evolution comparable to those on our planet, and it could contain a record of that critical time.

3.2 Search for Source Compounds

Simple source molecules, which could have served as building blocks for RNA or its possible precursor forms, include formaldehyde and glycolaldehyde, active phosphate species capable of phosphorylation, as well as reduced nitrogen species such as nitriles which serve as preferred components of hydrogen-bonding recognition molecules. For some compounds, like formaldehyde and phosphates, acquisition in concentrations that permit their use in natural synthesis reactions appears compatible with geochemical constraints. However, in other cases, like cyanide and ammonia, the accumulation to reactive concentration presents an enigma as far as terrestrial processes are concerned. Similar conflicts appear in the subsequent stages of molecular evolution between, on the one hand, the biopoetic wish for unlimited supplies of highly reduced organic raw materials, and on the other, the austere limits imposed by geochemical and geophysical preferences outlined here.

3.2.1 Oxiranecarbonitrile

Any compound expected to be reactive in a natural environment needs to overcome the threat of dilution in the hydrosphere. One of the most efficient ways to selectively achieve the necessary concentration is by excess

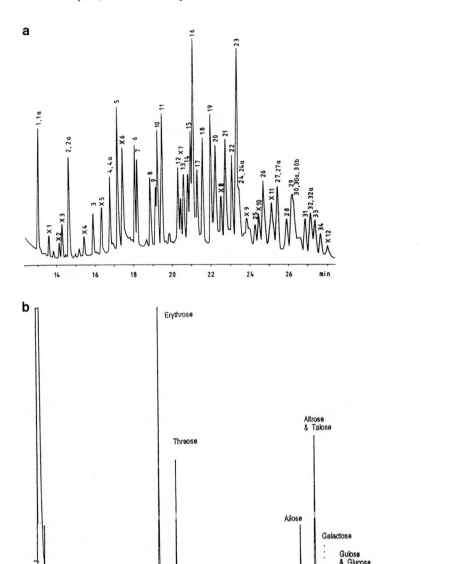

Figure 9 (*a*) Gas chromatogram of the (derivatized) large number of sugars and related compounds induced by a small amount of glycolaldehyde in the indiscriminate formose reaction, which would also require geochemically implausible reaction conditions. The inefficiency of this avenue to formation of specific sugars such as ribose, as well as the relative instability of unprotected sugars, has often been quoted in the literature as a major obstacle to the rise of an RNA World. (Reprinted, with permission, from Decker et al. 1982 [copyright Elsevier Science].)

charge, and this is most effectively accomplished by phosphorylation. Thus, the smallest and simplest (two-carbon unit) aldehyde phosphate (glycolaldehyde phosphate, GAP) has been investigated as a starting material in RNA-related biopoesis experiments. Studies have also been made of the analogous phosphonate acetaldehyde ester (De Graaf et al. 1997, 1998).

The epoxide oxiranecarbonitrile (*ocn*) has been considered as a source for GAP in nature (Pitsch et al. 1994). This compound has been shown experimentally to yield GAP by efficient reaction with orthophosphate in aqueous solution and could possibly react analogously on surface-active phosphate minerals or with the more soluble condensed phosphates. If effectively produced in the interstellar cloud medium, *ocn* could theoretically become incorporated in cometary condensates and subsequently be transported to Earth. It could also possibly have formed in the upper atmosphere of the Archean Earth, particularly in the electron deposition region below the auroral ring current (Fig. 4) in the Earth's nightcone, where constructive negative ion reactions are induced and the products are photoprotected. Atmospheric sprites and jets are other cold plasma regions where similarly favorable reaction conditions may have existed (Sect. 2.3).

In view of the chemical attractiveness of the *ocn* scheme, a radiotelescope search of the interstellar medium for its microwave signature has been initiated; the early results of this investigation show that the abundance is below the detection limit of the technique so far used (Dickens et al. 1996); condensation of low vapor pressure products like *ocn* may contribute to their paucity in the gas-plasma phase of the space medium, and their transfer to the interstellar dust component.

3.2.2 Cyanohydrin and Hexacyanoferroate

Because of the presumed prevalence of carbon dioxide and water vapor, formaldehyde is likely, next to carbon monoxide, to have been one of the

---

(*b*) In contrast to the formose reaction, the mineral (in this case hydrotalcite) catalyzed oligomerization of glycolaldehyde phosphate is highly selective, generating mainly erythrose-2,4-diphosphate as a precursor of altrose-2,4,6-triphosphate, the most abundant hexose phosphate product. This reaction also approaches more closely natural conditions; the mineral concentrates the reactant from ~10 μM solution and operates at near neutral pH (~7.5); the products, sorbed in the mineral, are stable under the reaction conditions. Together with glyceraldehyde phosphate, the mineral-induced reaction produces pentose phosphates in high yield, primarily ribose-2,4-diphosphate and arabinose diphosphate. The availability of this reaction path removes a frequently voiced reservation against the concept of an RNA World.

most abundant reduced carbon molecules in the Archean atmosphere–hydrosphere system. Activation mechanisms facilitating formaldehyde formation that have been evaluated experimentally include cold plasmas (Löb 1906), electric sparks (Miller 1953), UV irradiation (Hubbard et al. 1971; Pinto et al. 1980), and thermal shock (Chyba and Sagan 1992). A high abundance relative to hydrogen cyanide places formaldehyde potentially in control of Archean nitrile chemistry and thus of pathways to nucleobases (Arrhenius et al. 1994). Formaldehyde reacts spontaneously with hydrogen cyanide to form cyanohydrin, a well-known reaction that has vexed workers in the field of prebiotic chemistry relying on an unencumbered availability of HCN in high concentration to form a plethora of evolved molecules. Specifically, hydrogen cyanide in hypothetical isolation from formaldehyde has been implicated as a precursor of purines and pyrimidines, a scheme that appears realistic only if conditions can be found that provide protection against formaldehyde.

The initially forming monomeric cyanohydrin (glycolonitrile), when present in high concentration such as in eutectic freezeout and under basic conditions, undergoes rapid oligomerization to a dimer and a trimer (Arrhenius et al. 1994, 1997). Standard mechanistic analysis of the trimerization of cyanohydrin predicts three 6-membered heterocycles; an s-triazine, an oxazoline, and an amino-pyrimidine. The kinetically formed products comprise a dimer hydrate and a trimer based on the 5-membered oxazoline skeleton; the main hydrolysis product is glycolamide. Computational investigations predict that the pyrimidine isomer is thermodynamically favored and that the oxazoline actually found is a kinetic side product. These cyanohydric oligomerization products are of prebiotic interest because they could represent a geochemically plausible pathway to nitrogen bases circumventing the trap that formaldehyde sets for reactions of HCN with itself.

Under certain conditions, depending on pH and concentration, HCN would be expected to react with Fe^{++}, present in comparatively high concentration in the Archean hydrosphere due to the past low partial pressure of free oxygen (Keefe and Miller 1996). The high negative charge of the hexacyanoferroate ion [$Fe(CN)_6^{----}$] thus formed favors effective sorption in the interlayer of catalytically active double-layer metal hydroxide minerals (Arrhenius et al. 1997 and references therein). These minerals have been found to effectively concentrate charged molecules, such as phosphate, carbonate, and cyanide ions, believed to be significant in nucleoside and nucleotide synthesis, and also sulfur species and negatively charged amino acid monomers and oligomers such as polyaspartate, polyglutamate, and serine phosphate. Ferrocyanide (hexacyanoferroate) ion, when sorbed into the interlayer of the double-layer mineral green rust ($[Fe_2^{+++}Fe^{++} (OH)_6][A^{n-} \sim 4H_2O]$), undergoes further reaction with Fe (III) to form ferriferrocyanide (Prussian blue) (Arrhenius et al. 1993), or with Fe (II) in the green rust structure to form ferroferrocyanide (P.S. Braterman, pers. comm.). Such a concentrated mineral reservoir of cyanide could possibly have provided an escape from the dilemma of cyanide scavenging by formaldehyde.

3.2.3 Aldehydes and Sugars

As discussed above, formaldehyde is a natural precursor for selective synthesis of aldehydes because it was presumably, together with formic acid, the most abundant reduced carbon compound that could have formed from the Archean atmosphere. Based on known production mechanisms, formaldehyde was probably generally in excess over hydrogen cyanide, so HCN would have consumed only a minor fraction of the formaldehyde as cyanohydrin. Formaldehyde is a major monomeric component in the coma and tail of comets presumably originating from polyoxymethylene polymers in the cometary nucleus. H-C-O compounds, including formaldehyde, are also produced by photoreduction of carbonate ion with Fe^{++} in solution (Åkermark et al. 1980).

The classic formose reaction is often invoked in metaphoric schemes as a pathway to sugar formation from formaldehyde. However, it requires relatively high concentration, strongly alkaline conditions, or a suitable catalyst, and produces an indiscriminate mixture of sugars and sugar derivatives of which ribose is a minuscule fraction (Fig. 9a). In contrast to the formose reaction, the bilateral mineral-catalyzed aldolization reaction (Fig. 9b; Sec. 3.2.5) is highly selective and operates from dilute, neutral solution. The dimerization of formaldehyde to glycolaldehyde is an important step (Fig. 10), since the latter permits phosphorylation and glycolaldehyde phosphate has been shown to give rise to RNA-related sugar phosphates (Sect. 3.2.5). The demonstration of glycolaldehyde as a product of cold plasma reactions in addition to formaldehyde and formic acid was one of the early results of the electric discharge studies by Löb (1906). Photochemical pathways have also been shown for the formation of glycolaldehyde (Schwartz 1993; Schwartz and De Graaf 1993).

A remaining impediment to the further reactions of these simple aldehydes to form the sugar component of RNA is that the uncharged state of these molecules renders them unprotected against dilution in the hydrosphere. However, recent experiments have demonstrated efficient phosphorylation of glycolaldehyde in dilute aqueous solution (Krishnamurthy

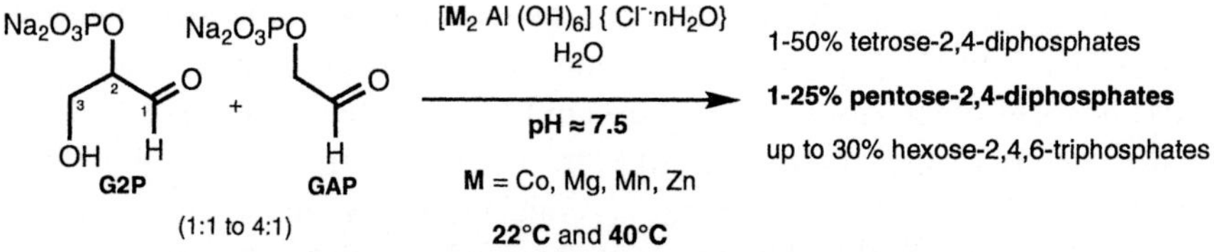

Na_2O_3PO — G2P (with OH, H) + Na_2O_3PO — GAP (with H)
(1:1 to 4:1)

$[M_2 Al(OH)_6]\{Cl^- \cdot nH_2O\}$
H_2O
$\xrightarrow{pH \approx 7.5}$
M = Co, Mg, Mn, Zn
22°C and 40°C

1-50% tetrose-2,4-diphosphates
1-25% pentose-2,4-diphosphates
up to 30% hexose-2,4,6-triphosphates

Figure 10 Reaction between glycolaldehyde phosphate (GAP) and glyceraldehyde-2-phosphate (G2P) in varying proportions in the presence of mixed valence double layer metal hydroxide minerals. Substitution of the metal cation in the main hydroxide sheets of the mineral can be made by divalent metal species (Fe^{++}, Mg^{++}, Mn^{++}...). The interlayer water content is generally in the range of 4–6 molecules per anion, giving an effective anion concentration of the order of 10 M in the reactive interlayer solution.

et al. 1998). The attachment of negative charge has proven to be an effective process for selective concentration of aldehydes by sorption on cationic minerals. As pointed out in a classic paper by Westheimer (1987), phosphorylation is nature's favored solution to the need for charge.

3.2.4 Phosphate Condensation, Activation, and Esterification

With the assumption of a fundamental role of soluble, condensed, naturally produced phosphates through early stages of molecular evolution, it becomes necessary to identify mechanisms by which phosphate ion can be coupled to the evolving molecular systems. Extensive studies pointing at such processes have been carried out, particularly by Orgel and collaborators (see, e.g., Österberg and Orgel 1972); an overview of these results is given by Kolb et al. (1997). A recent result is the achievement of efficient phosphate ester formation at near-neutral pH and low concentration of the reactants (Krishnamurthy et al. 1998).

Living organisms voraciously consume available phosphate and interfere with the inorganic phosphate reaction equilibria that must have controlled abiotic phosphate geochemistry. As outlined in Sect. 6.1, the thermodynamically stable phosphate mineral that forms from sterile solutions with the high ratio of magnesium to calcium characteristic of natural waters is the phosphate mineral, whitlockite ($HMgCa_9(PO_4)_7$). As a consequence of the protonation, whitlockite readily condenses upon heating and dehydration, forming a series of linear oligophosphates with enhanced solubility compared to the orthophosphate source. Because of the increasing charge with oligomer size, the condensed phosphate species are preferentially concentrated from dilute aqueous solution by the positively charged double-layer hydroxide minerals, found to be catalytically active in aldol phosphate condensation.

3.2.5 Sugar Phosphate Formation by Abiotic Processes

Considering the unique biochemical advantages of phosphate esters in synthesis, conservation, and function of information-carrying and energy-transducing biomolecules, Westheimer (1987) has pointed out the lack of any ionic group that could compete with phosphate species in fulfilling these functions. From the point of view of RNA synthesis, the initial phosphorylation of simple aldehydes would appear to be of central importance, and several arguments point to an early selection of aldose diphosphates in molecular evolution toward RNA-like molecules. One reason is evolutionary: Ribose diphosphate is now the backbone element in RNA, extending phylogenetically back in time to the most primitive organisms known. It must be said that although labeled "primitive," these organisms at the base of the tree are highly evolved biochemically and must have a long evolutionary pedigree, now seemingly lost from the geologic record on Earth.

Illustrating the efficiency of selective sugar phosphate formation, Müller et al. (1990) demonstrated the formation of rac.-pentose-2,4-diphosphates and hexose-2,4,6-triphosphates in strongly basic aqueous solution starting from glycolaldehyde phosphate and formaldehyde, resulting in rac.-ribose-2,4-diphosphate as the major product. It was shown that glycolaldehyde phosphate (GAP) is rapidly and efficiently sorbed in the interlayer of bilaterally active double-layer hydroxide minerals from dilute (≥ 10 μM) solution to form a reactive, concentrated (~10 M), quasi-two-dimensional solution constituting an isolated, diffusive, and exchangeable Stern layer in these minerals. Some of these minerals are widespread in nature and would have been particularly abundant in an anoxic Archean hydrosphere. GAP in this catalytic reaction space is rapidly (2 days, RT, pH 7.5) and in high yield (~60%) selectively aldolized to form tetrose-2,4-diphosphates. By further condensation (Fig. 9b), hexose- (predominantly altrose-) 2,4,6-triphosphates, in the external solution are produced (Pitsch et al. 1995). The sugar phosphates, intercalated in the mineral interlayer, are stable against hydrolysis (~5% in 10 months) and thus form a robust system, in contrast to unphosphorylated sugar in free solution.

The formation of pentose-2,4-diphosphates (up to 25% yield) was observed when a dilute aqueous solution of glycolaldehyde phosphate and glyceraldehyde-2-phosphate was exposed to the double-layer hydroxide minerals, again at near neutral pH of the external solution (Fig. 10). In this study it was also observed that different divalent metal ions in the double-layer hydroxide mineral had a different effect specifically on the yield of ribose-2,4-diphosphate. For example, with Mn^{++} as the divalent metal cation, the yield of ribose-2,4-diphosphate was as high as 48% of the pentose phosphates. The stability of the pentose phosphates parallels that of the hexose phosphates within the mineral interlayer.

3.2.6 The State of Metaphors for Natural Synthesis of RNA

As described above, there has been significant progress in the synthesis of the backbone units of RNA under conditions that bear some semblance to those expected in an abiotic world. Still missing are some critically important links in the chain leading to RNA. The natural synthesis of nitrogen bases or some other, undiscovered recognition molecules, is an open question; their attachment to the sugar phosphate backbone molecules is another unanswered problem. Further down the line another stretch of successful synthesis achievements follows: the oligomerization of (activated) nucleotides by J.P. Ferris and his group, the template-directed formation of complementary oligonucleotide strands by L.E. Orgel and collaborators, the demonstration by T. Cech and S. Altman of catalytic activity of RNA bound in ribozymes, and their evolutionary qualities shown by G.F. Joyce and J.W. Szostak. For most of the missing pieces there are obvious experimental search strategies, and the challenge of a natural synthesis of RNA now appears less daunting than other crucial problems in the emergence of life. One of these has been the widely accepted assumption that the terrestrial planets were heavily bombarded, sterilizing Earth as late as 3700 to 3800 Ma, a concept discussed below.

4. GIANT IMPACTS: A THREAT TO EARLY LIFE

4.1 Emergence of Life and the Late Heavy Bombardment

The extent and nature of impacts by asteroids, meteorites, and comets on Earth that occurred between the time of accretion and the first hints of the geological record in the early Archean has until recently been a subject relegated to speculation. This is based on events that happened late in solar system history (after 3900 Ma) recorded on the surface of the moon, which in contrast to Earth, has not been substantially altered and preserves some record of these earliest times. The overlap in time of proposed impacts on the early Earth, with the emergence of life, creates a problem for the sustainability of early ecosystems.

Following the planet-forming accretion period beginning at about 4600 Ma, the deposition of extraterrestrial matter has continued on all surfaces of the planets and their satellites, although today at an extremely attenuated rate. If uniformly extrapolated back in time on the basis of present rates, the amount of material of extraterrestrial origin that would have accumulated following the end of formation of the planets would be represented by only an approximately 10-meter-thick layer covering the Earth. Impacts by asteroid-size bodies have occasionally occurred with detrimental global effects on the biosphere, the latest notable event of this kind being the giant impact at the Cretaceous-Tertiary (K-T) boundary 65 million years ago. Relatively minor increases above the rate of small meteorite influx that we experience as normal are recorded at several occasions in the geological history. One such instance of enhanced concentration of cosmic dust and preserved meteorites in marine sediments is observed at 500 Ma during the Ordovician period and is correlated with a radiochemically dated breakup event in the asteroid belt that must have scattered debris into Earth-crossing orbits (Schmitz et al. 1997).

The chronological record on the moon, established from samples returned by the Apollo and Luna missions, shows that large impacts with high mass and at many times higher rates of impact than today occurred as late as 3800 Ma (Pappanastassiou and Wasserburg 1969; Tera et al. 1974); this phenomenon is often quoted as the "late heavy bombardment" (see, e.g., Chyba 1993). A currently prevalent interpretation of the lunar impact record assumes that the solar system during this time period was invaded by a swarm of marauding objects, bombarding all the planets and their satellites, including Earth and Mars, but with a retrievable record so far available only on the moon. A possible mechanism for this kind of assault has been presented by Kaula and Bigeleisen (1975) and Fernandez and Ip (1983), who suggested interactions with the outer planets that could have caused perturbation of asteroids into highly eccentric and ultimately planet-crossing orbits. Such a swarm of asteroids would, in a relatively short time period, hit all planets and satellites in the inner solar system. Analyses of meteorites for melting, shock, and exposure events provide evidence of collisions in the asteroid belt that extend from the time of formation of the moon at 4450 Ma up to as late as, in a few cases, 4000 Ma and throughout the history of the solar system.

On Earth, the presumed effects of energetic cometary and asteroidal bombardments up to and past 3800 Ma have been estimated to range from evaporation of the entire hydrosphere and melting the surface of the planet, to partial evaporation of the ocean (Sleep et al. 1989; Chyba et al. 1990). In the most catastrophic scenarios, the emergence of life would be prohibited and any existing life would be extinguished. In milder variants of this impact frustration theory, those life forms could have been spared that resided in the deepest reaches of the ocean and crust. The hyperthermophiles surviving thermal pulses from impacts by hiding in deep-sea vents could then, after the several extinction events implied, emerge as sole survivors in the last preserved ecosystem.

The extrapolation of the lunar impact record to the Earth with consequences for life stands in contrast to new observations on Earth and Mars. The early Archean banded iron formations consist of laminated sedimentary sequences testifying to undisturbed deposition (Sect. 5). Those studied in detail so far bear no evidence of interruptions by catastrophic events such as local impacts and their associated crushed-up rock formations (breccias), tsunamis from remote impacts, or complete or partial evaporation of the hydrosphere. Rather than being sterile, the sediments, extending in time beyond 3850 Ma, contain ubiquitous "chemofossils" (Mojzsis et al. 1996; Mojzsis and Arrhenius 1998) with isotopic composition suggestive of highly evolved enzyme systems—life seems to have been developing for an unknown, but probably considerable length of time before 3850 Ma (Fig. 1; Sect. 5).

On Mars, the record now available is limited to a few meteorites and crater size and density distributions correlated with the dated lunar record (Neukum and Hiller 1981). It is interesting to note that the igneous rock that forms the major part of the martian meteorite ALH84001 of potential biological renown (McKay et al. 1996), has a crystallization age of about 4500 Ma (Jagoutz et al. 1994). This suggests that the corresponding impact region of the martian crust from which this meteorite was derived has not been remelted in the same cataclysmic bombardment that affected the moon late in time. The late heavy bombardment ought to have been even more intense on Mars because of its larger gravitation and proximity to the asteroid belt. The same meteorite bears evidence of shock at 4000 Ma, somewhat earlier than the major phase of lunar bombardment peaking at 3850 Ma. This shock age of ALH84001 has been used to argue the case for impacting of Mars coincident with the terminal bombardment of the moon and thus the rest of the planets of the inner solar system (Ash et al. 1996).

Two interpretations of the late bombardment are possible. (1) The late lunar impactors came from external sources and affected the entire inner solar system, including Earth and Mars, but the impact events were sufficiently episodic so that the observed long undisturbed sedimentary sequences harboring life could be accumulated between catastrophic events and their aftermaths that have so far escaped attention. According to this model, life could, in spared niches, have survived and repeatedly repopulated Earth's hydrosphere. (2) Life may have been extinguished, restarted, and rapidly evolved to a high level of biochemical sophistication. Some impact rate scenarios illustrating these concepts are shown in Figure 11.

In a different interpretation, and on the basis of presently available evidence from the oldest known terrestrial record, the late bombardment of the moon could reflect the process of its own accretion or of its collision with other objects in orbit around the Earth. Such possible events, having dynamically limited effects on Earth, are discussed below.

4.2 Uncertainties about Lunar Origin, Impact Environments, and Hazards to Emerging Life

A currently popular scenario for the formation of the moon assumes that Earth in a relatively advanced state of accretion, including or followed by core formation, would have collided with a hypothetical planet in an Earth-crossing orbit and with a mass substantially larger than that of Mars (Wänke et al. 1984; Cameron and Benz 1991). Ejecta from the impact would have been placed in prograde orbit around Earth and eventually coalesced to form the moon.

Assuming this scenario to be within the realm of possibility, there are several reasons to believe that the last stages of accretion of the moon from its source ring of material could have extended the time of formation of the major impact features on the face of our satellite past 3800 Ma and perhaps thus explain the paradox of heavy bombardment and the presence of early life forms occurring simultaneously. One such reason for retardation of the terminal accretion would be the metastable retention of accretional material by 1:1 resonance in the Lagrangian points 60° before and after the moon in its orbit. Such resonance locking is a feature seen, for example, in Jupiter's orbit, where two clusters of asteroids, the Trojans, are bound in its two Lagrangian potential wells. In the case of the moon, these wells, because of its lower mass, would be much more shallow, and the rocks contained in them could be scattered out of resonance at an early time such as during a period ending past 3850 Ma. Their ghosts in the form of remaining matter, the Kordylevsky clouds, are claimed to have been confirmed by space observations (Roach 1975).

This explanatory legend is tied to the specific theory for the impact origin and evolution of the moon. In theories for the origin of the moon by capture (Gerstenkorn 1955, 1969; Singer 1972), alternative explanations are offered for the late lunar bombardment. One suggestion, based on the persistent generation of regular systems of proportionately small

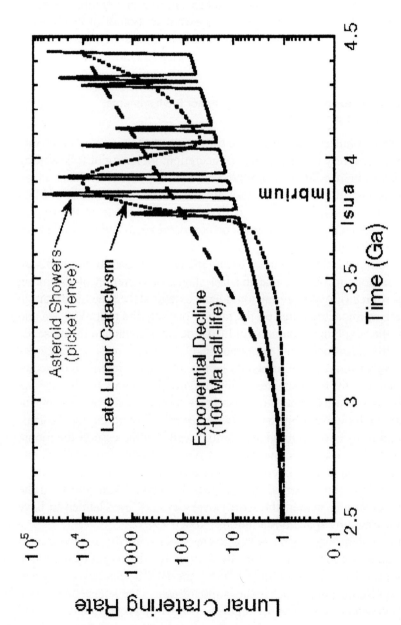

Figure 11 Three schematic views of the late lunar bombardment, and its relationship in time to the undisturbed 3800 Ma Isua sediments, southern West Greenland. (Diagram courtesy of K. Zahnle, NASA Ames Research Center, and S.J. Mojzsis, UCLA.)

satellites in the equatorial plane around all magnetized planets, assumes that Earth would also originally have been endowed with a satellite system of perhaps half a dozen moonlets swept up by the receding, abnormally large, and differently formed moon in its early recession from the Earth (Alfvén and Arrhenius 1969, 1972). Collision velocities of one or two km s^{-1} would have substantial cosmetic effects on the surface of our moon but would create ejecta generally slower than its escape velocity of 2.4 km s^{-1}. Such slow ejecta could have been retained in near-lunar orbit and recaptured by the moon with only a minor fraction overcoming the substantial energy barrier for diversion to Earth-impacting orbits.

All of these theories of lunar origin and collision history remain speculative. However, there are avenues toward verification of related circumstances. One consists in detailed examination of as long sequences as possible of the oldest laminated sediments such as those described from the Isua district in Greenland (Sect. 5), searching for and identifying depositional irregularities and embedded meteorites and interplanetary dust in them. Such quantitative elemental and mineralogical measurements are likely to give information about the existance, rate, and frequency of bursts of extraterrestrial accretion, and about their source.

The second approach toward verification involves analysis of lunar samples, chronologically identified with, and serving as the basis for, conclusions about the late bombardment on the moon. If the projectiles were interplanetary marauders, they would be expected to consist of undifferentiated cometary or meteorite type material, conferring their characteristic excess of platinum group elements (PGE; Ir, Os, Pd, Pt, Ru, Rh) to the ejecta, with allowance for dilution with lunar target material. If they were to be identified with partly differentiated asteroids, some impact material would be depleted in PGEs, but others, enriched in metal, would have a correspondingly enhanced PGE content. If, however, such a signature is found to be consistently absent in the samples representing these events, the most likely interpretation would be that the impacts were caused by metal-depleted low-density material of the kind that formed the moon and possibly original Earth satellites.

In conclusion, the necessarily speculative views of the hardships or even survival of life during the first quarter of the existence of Earth are now being replaced by observations in the oldest preserved records on our planet by the study of probably even older samples from ancient terranes on Mars, and by extended analysis of the lunar samples from the Apollo missions. Life on Earth instead of (as is often quoted) originating 3.5 to 3.8 billion years ago, is found to have developed to a high degree of autotrophic sophistication already before 3850 Ma. Carbon isotope fractionation mechanisms indicated by the geological record of even the oldest rocks are matched today only by the action of enzymes such as the ribose- and ribulose-phosphate carboxylases that mediate the first step in carbon fixation (Mojzsis et al. 1996).

5. THE RECORD OF LIFE'S APPEARANCE ON EARTH

5.1 Liquid Water and Life

Once the need for mutually supportive criteria for early Earth environments and organic reactions has been established, the emergence of life on Earth then becomes a geological issue as well as a biochemical one, and can be considered among the most important developments in the geochemical history of the surface of the planet after its formation. The initiation of biological processes could have occurred in the surficial layer of the planet comprising the hydrosphere, atmosphere, and lithosphere, so long as conditions persisted that satisfied the stability of liquid water on million- to billion-year time scales. Bioessential element cycling (H, C, N, O, P, S) is accomplished in the hydrosphere, by life. Therefore, the earliest evidence for liquid water at the surface of the Earth may be taken as the principal constraint on the timing of the emergence of life, and on the environmental conditions necessary for its sustainability, propagation, and evolution (Chang 1988).

After formation, the preservation of water oceans over geologic time scales is the main characteristic that separates the geophysical and geochemical evolution of Earth from the other terrestrial planets. The presence on a planet of a stable hydrosphere defines habitability. Liquid water on a planetary body is maintained under conditions such as atmospheric density, proximity to the central star, or the effects of another energy source such as tidal heating, and is used to define the habitable zone of the solar system within which a planet or satellite could harbor life (Kasting 1993; Sect. 2). A liquid water ocean hydrothermally alters oceanic crust during and following mid-oceanic ridge volcanism. This crust is then eventually drawn into subduction zones to form hydrated silica-saturated melts, the building material of granites and, ultimately, of continents. The effect of water, and the resulting generation of continental crust, on the bulk chemistry of the hydrosphere-atmosphere-lithosphere and biosphere system cannot be overstated. That a liquid water ocean has remained stable on Earth for the length of the known geological record implies that surface temperatures at the earliest recorded times were within the presently observed range. As has been emphasized previously, water has played a profound role in mediating tectonic styles on Earth, which makes possible the continued element cycling needed to maintain chemical equilibrium of the hydrosphere and atmosphere over billions of years. In one theory (Woese 1979; Woese and Wächtershäuser 1990), liquid water in the form of droplets is postulated as the locus for the origin of life in the cloud cover over a hot, uninhabitable crust, long before the rainout of a homogeneous hydrosphere. Liquid water, rather than steam, ice, or supercritical fluid, is essential for the origin and development of life. Early Archean sedimentary precipitates that must have formed in water and are older than 3800 Ma, such as cherts and banded-iron formations, signify that the terrestrial processes of accumulation and saturation of dissolved species like silica, carbonate, ferrous iron, and phosphate were operative at and before that time (Mojzsis et al. 1996; Mojzsis 1997; Nutman et al. 1997). The accumulation and concentration of such dissolved species, as reflected in the rocks, reveals the bioavailability of carbon and phosphorus in the early hydrosphere and has opened avenues to the search for evidence of life in the oldest known rocks.

The Earth's sedimentary rocks contain a relatively detailed record both of the planetary surface environment and of biological evolution only during the last 600 million years or so. The Earth lacks a record of its own accretion and of its earliest subsequent history, because of the actions of plate tectonics, weathering, and denudation. These processes, activated by water, recycle and modify crustal materials, and (until recently) the oldest known rocks of a clearly sedimentary origin extended back only into the range 3600–3700 Ma. Newly discovered deposits, as discussed below, extend into an age older than 3850 Ma. This is also the time at which the Earth was previously speculated to be uninhabitable due to heavy bombardment by objects raining in from space. New perspectives on this early period in the history of the Earth have been opened up by the finding instead of undisturbed sediment sequences containing evidence of biochemically advanced life forms in the oldest rocks. These recent observations have consequences for constraining environments for life also for the other planets and satellites of the outer planets in our solar system, and for the rapidity of life arising under the right conditions elsewhere. Due to the Earth's continuous self-digestion of crust, it is unlikely that a geologic record penetrating much deeper into the infancy of life can be found. Much hope is therefore attached to the exploration of the ancient surface of Mars, where remote sensing indicates minimal crustal recycling and the presence of an early, and in some regions, vigorous, hydrological cycle of weathering and deposition in liquid water. Such regimes on ancient Mars gave rise to extensive sedimentary deposits that are the principal targets for sample return missions, and are expected to provide crucial clues to the nature of the earliest planetary surfaces and prospects for past life, or perhaps even events leading to life on Mars.

5.2 The Incomplete Geologic Record

Age calculations based on radiogenic isotope distributions in primitive meteorites, the least evolved objects in the solar system, indicate that the events leading to the condensation, aggregation, and differentiation of the source material for the planets took place in an interval of approximately several tens of millions of years between 4600 and 4500 Ma. The impacting of the early Earth by meteorites and comets and the recycling mechanisms active from the outset add to the difficulty of arriving at an agreeable age of the first terrestrial crust. In related scenarios, assuming an early accretion of an iron-rich core (heterogeneous accretion) followed by growth of a mainly silicate mantle at low average surface temperature, the planet could have become habitable during or soon after planetary formation up to four and a half billion years ago, permitting several hundred million years for the emergence of life. Heterogeneous accretion theories for the formation of the Earth thus connect with biopoesis scenarios, which for intuitive reasons need the longest possible time available for the emergence of life on Earth. It is worth emphasizing that there is no factual evidence for the length of time actually required for biopoesis (Orgel 1998). Heterogeneous accretion of the planets is also favorable for models seeking to maximize the rate of generation of methane and hydrogen from the reducing power of available metallic iron while minimizing (by other means) their destruction and escape rate (Miller et al. 1998).

In contrast, homogeneous accretion theories assume accretion of mixed silicates and iron metal and subsequent core formation by gravitational settling of the metal to the center of the planet. The liberation of gravitational energy would be of such magnitude that the entire planet would melt, and all volatiles would form a dense atmosphere. A dense atmosphere would delay substantially, perhaps forever, the time when the surface would become inhabitable.

All known remnants of the oldest sedimentary record are locked in ancient (3600–4000 Ma) granitic gneiss complexes. In these terranes, the oldest sediments appear as strongly deformed rafts and enclaves between the much more voluminous igneous rocks, which have been severely altered by chemical changes induced by heat and pressure in the process of metamorphism. Sediments and volcanic rocks that are recognizable from the earliest Archean of Greenland (which are as old as 3900 Ma) include quartzites, originally precipitated as cryptocrystalline silica ("chert") directly from solution onto the early Archean seafloor; banded iron-formation (BIF) as alternating bands of silica (as quartz) and magnetite ($Fe^{++}Fe_2^{+++}O_4$), conglomerates probably formed in debris flows from river channels or submarine landslides; quartzitic sandstones that may have been the remains of stream and river run-off channel deposits; and pillow basalts and other volcanic rocks that are formed in a marine system of active volcanism, erosion, and sedimentation (McGregor 1973; Nutman et al. 1984).

Definite sediments older than 3600 Ma have not yet been discovered outside of southern West Greenland. The ages of these oldest Greenlandic sedimentary rocks range from about 3900 Ma (Nutman et al. 1997) to about 3600 Ma (McGregor 1973; Moorbath et al. 1973). A large section of the sedimentary and volcanic rocks of the Isua district of southern West Greenland consists of units that may have been formed in, or adjacent to, volcanic arcs (Dymek and Klein 1988), as exemplified in the present-day Western Pacific. Such an island arc sedimentary environment is a typical case preserved in the rocks of the extensive Archean Greenstone Belts (Eriksson et al. 1997) found throughout the world as massive accumulations and intercalations of sediments, volcanic rocks, and their metamorphosed remains. These preserved rocks probably represent a common style of marine sedimentation in the Archean and are the oldest preserved remnants of ecosystems yet recognized. Some types of sediments found in the oldest rocks of West Greenland may be the weathering products of emergent land masses such as volcanic islands and/or perhaps microcontinents, similar in some ways to present-day New Zealand. Such land masses, in an early CO_2-rich atmosphere, with water abundant and with no vegetative cover, would have experienced massive physical and chemical weathering contributing to the inventory of sediments (Sect. 2.4), including the carbonaceous remains of organisms accumulating on the ancient ocean floor. Many areas of the world, however, contain igneous rocks and their metamorphic equivalents with ages in excess of 3600 Ma. Such rocks are found, for example, in the northeastern Sino-Korean Craton (Song et al. 1996), the Napier Complex in Antarctica (Black et al. 1986), the Acasta Lake area, Northwest Territories, Canada (Bowring et al. 1989), and southern Africa (see, e.g., Compston and Kröner 1988). Some of these regions, like Antarctica, are little explored and are likely to contain exposures of as-yet undiscovered very old rocks, possibly including those hosting sediments and chemical traces of early life.

The oldest relatively unaltered sediments are found in the approximately 3500 Ma Warrawoona Group of the Pilbara Craton, Western Australia (Pidgeon 1978; Barley et al. 1979; DiMarco and Lowe 1989) and the Onverwacht Group of Swaziland, South Africa (see, e.g., Lowe and Bryerly 1986). These sediments were deposited in a shallow marine setting with abundant and well-preserved structures of diverse type, including those representative of shallow water deposition at continental and/or island margins. These rocks suggest ice-free oceanic conditions in the early Archean before 3500 Ma, at least in their local environment of deposition, even given that the sun was less luminous at that time. Well-preserved stromatolitic structures and yet rarer microfossils, recognizable by their shapes (morphofossils) in early Archean cherts of the Apex Basalt, Warrawoona Group (Schopf and Packer 1987) are strong evidence for advanced microbial (possibly even cyanobacterial) life then (Awramik et al. 1988; Schopf 1993).

Alternative views to the interpretation of the age relationships of these rocks and the genetically differentiated nature of fossilized organisms are given by Buick (1990) and Doolittle et al. (1996), respectively. In addition to the morphological evidence for life put forth from the interpretation of microfossils by micropaleontologists, stable carbon isotopic analyses reported from these rocks have, as in the older Greenland rocks, provided strong evidence for life in the early Archean that is independent of the presence of microstructures (for review, see Hayes et al. 1983; Strauss and Moore 1992; Sect. 6.2). It has long been recognized that the early Archean morphofossils represent life forms that must have required a considerable period of time to have evolved from inchoate life to the metabolically sophisticated and structurally complex organisms inferred by Awramik and coworkers (1988) and Schopf (1993) solely from their shapes. These observations must therefore relegate the emergence and first appearance of life on Earth, and certainly the hydrosphere, to the time period before 3800 Ma (Schidlowski 1988; Mojzsis et al. 1996; Nutman et al. 1997).

6. THE PHOSPHATE-GRAPHITE ASSEMBLAGE AS A BIOMARKER

6.1 The Importance of Being Phosphate

Phosphate is an essential constituent of all life, and its controlling importance in biochemical functions is likely to extend back to the origin of life itself. Aside from its role in the structure of RNA and DNA, the unique chemistry of phosphate is utilized in a vast variety of biochemical functions (Westheimer 1987). As a consequence, phosphate is a prominent component of bioorganic matter in marine sediments, and phosphatic biominerals appear in intimate association with isotopically light carbonaceous matter, consistent with its biological origin. This characteristic association of the remains of organisms with phosphate, primarily in the form of basic calcium phosphate (the mineral apatite), is found in presently forming sediments and throughout the geologic record to the oldest preserved sedimentary rock so far identified.

To understand the interaction between abiotic phosphate mineral formation and phosphatic biomineralization, knowledge of the solubilities and thermodynamic stabilities is crucial. The notable disequilibrium effects exerted by organisms on phosphate mineralization may be utilized for indicating the presence or absence of cellular life: Controlled synthesis experiments suggest that the deviation from equilibrium expressed by the formation of apatite ($Ca_5(PO_4)_3$ (OH, F, Cl)) instead of the equilibrium phase whitlockite ($HMgCa_9$ $(PO_4)_7$) in natural waters is due to the ability of cellular organisms to dynamically segregate Mg^{++} into solution while selectively retaining calcium, and in this environment to nucleate apatite by interaction with specific cellular polypeptides. In controlled sterile experiments, whitlockite is formed at Mg/Ca above 0.04 (LeRoux et al. 1963), at pH <8.5, and at 0–100°C. Whitlockite, rather than apatite, is the thermodynamically stable phase of the calcium phosphates under these conditions (Verbeeck et al. 1986; Driessens and Verbeeck 1990; Elliot 1994; Gedulin et al. 1993). Absence of life would therefore be expected to be reflected in the appearance of whitlockite (at pH >8.5 together with apatite) as the sedimentary phosphate mineral, and without association with bioorganic carbonaceous remains in an ancient sedimentary record (Arrhenius et al. 1993). Additional evidence is provided by the fact that biogenic apatite, in modern as well as ancient sediments, is found to be intergrown with residual organic matter from the organisms mediating the deposition of the mineral. In sedimentary rocks from the Archean, including the oldest known metasediments on Earth, the organic matter component of these microaggregates has been completely altered to crystalline graphite, enveloping or enclosed in the recrystallized apatite component (Fig. 12). Such characteristic apatite–graphite aggregates are found to be ubiquitous in sediments of increasing age such as the Gunflint chert (2100 Ma), Hamersley BIF (2500 Ma), Isua Supracrustal Belt (3600–3810 Ma), and what is the oldest currently known metasediment, the BIF of Akilia island in Greenland (older than 3850 Ma; Fig. 1). It would thus appear that evolved cellular life has existed on Earth well beyond 3850 Ma, leaving an interval of no more than 600 Ma for the emergence and early evolution of life on Earth following planetary formation and the time of the first isotopic evidence for biological activity.

6.2 Carbon Isotopic Record of Life

The most extensive alteration by life arises from the release of free oxygen by the splitting of water during oxygenic photosynthesis, which is catalyzed by phosphate-based enzymatic mechanisms. This creates the reducing power needed for the synthesis of biogenic carbonaceous matter, which becomes enriched in ^{12}C to a unique extent in this process. A record

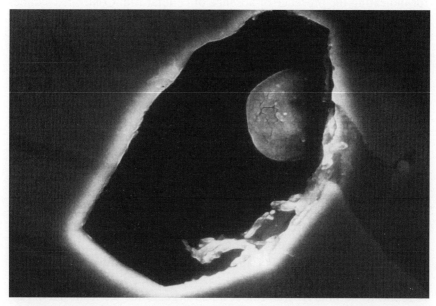

Figure 12 At elevated temperature and high pressure during metamorphism of rocks, the individual clusters of apatite microcrystals merge into a single crystal of the mineral, and the carbonaceous material remains of organisms convert to polycrystalline graphite (elemental carbon). Due to its surface tension on the phosphate, the originally disseminated carbonaceous matter coalesces into rounded inclusions in, and surface coatings on, the host apatite crystals. This scanning electron micrograph of a polished section from the >3850-million-year-old Akilia Island sedimentary rock shows a void, left after dissolving an apatite crystal by 30-min etching in 2% HNO_3, revealing an acid-resistant carbonaceous inclusion. This carbon is a chemofossil typical of those analyzed in Fig. 13 as "oldest terrestrial sediments," and with a carbon isotope composition indicating a bioorganic origin. The cross-section of the inclusion is ~10 μm: (Photograph by S.J. Mojzsis, UCLA.)

of this alteration and partial takeover of the global carbon cycle by life becomes captured in the carbonaceous remains of organisms in sedimentary rocks where they can be identified from their stable carbon isotopic signature, frequently in close association with biominerals. Earlier analyses (Fig. 13) (Schidlowski 1988) provided suggestive evidence on this basis for a biogenic origin of carbonaceous remains in some sediments of the Isua district in Greenland (3810 Ma), but were regarded as inconclusive because of the low resolution of the technique used and uncertainties about the sources of carbon described. More precise data have since been obtained by analyzing microscopic carbonaceous inclusions in apatite crystals in the sedimentary Akilia island rocks with an age >3850 Ma, objects believed to be of cellular origin (Fig. 12). The minute dimensions (5–15 μm) of the carbonaceous inclusions require that the mass spectrometric measurements to determine their isotopic compositions be carried out by high-resolution ion microprobe analysis.

The stable isotopes of carbon (^{12}C and ^{13}C) are partitioned as a result of both equilibrium exchange reactions and kinetic effects due to the metabolic mechanisms of organisms as well as inorganic processes. Inorganic processes such as evaporation, diffusion, and condensation are of far less magnitude in fractionating the isotopes than biogenically induced fractionations. In the standard delta notation, $\delta^{13}C$ is defined as the ratio of the $^{13}C/^{12}C$ of the sample relative to the ratio of the conventional PeeDee Belemnite (PDB) standard expressed in parts per thousand (per mil): $\delta^{13}C = \{[(^{13}C/^{12}C)_{sample} / (^{13}C/^{12}C)_{PDB}] - 1\} \times 1000$. High-resolution mass spectrometric measurements indicate a strong fractionation of $^{13}C/^{12}C$ ($\delta^{13}C = -23$ to -50%), a range uniquely limited to bioorganic carbon and well separated from the range (around 0–10‰), of inorganic sources on Earth (Fig. 13) and in consonance with the bioorganic origin suggested by the phosphate mineral phase association and crystal chemistry (Mojzsis et al. 1996).

7. CONCLUSION: THE WEB OF STRATEGIES

Problems related to biopoesis, the creation of life, are approached in the literature in three stages with growing constraints. At the most fundamental level, an inquiry is made about probabilistic aspects and the generation and transfer of information. Although inspiration at this stage is derived from the chemical structure of the modern replication system, no detailed inferences are necessarily made about structures or reactions participatory in the origin of life; in the words of one of the leaders in the field: "the chemists will take care of that."

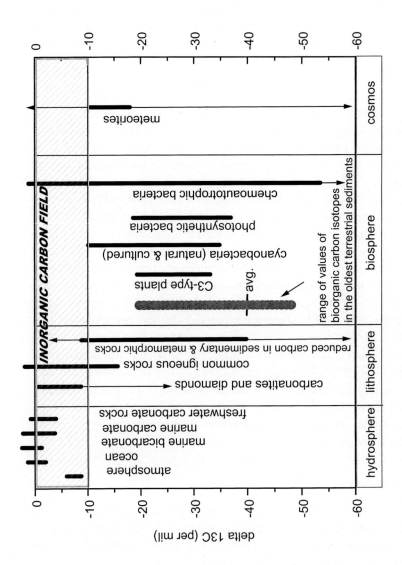

Figure 13 The distribution of carbon isotopic compositions in nature, showing the enrichment of the light isotope ^{12}C by autotrophic organisms (expressed as negative values of ^{13}C). This marked fractionation takes place at the enzymatically mediated transfer from the pool of inorganic carbon (mainly CO_2) and identifies bioorganic carbon in fossilized carbonaceous remains of organisms, such as in Fig. 12. Reduced carbon in sedimentary and metamorphic rocks is largely of such organic origin.

The organic chemists thus enter at the next stage in the interpretation of what it takes to make life by attempting to identify any stepwise organic chemical process that, coupled with various energy sources, could possibly have led from simple source molecules to complex structures. Such complex structures must have become functionally similar to those active in present-day biosystems. In order not to miss any opportunities at this level, most actors in the origins-of-life field purposely avoid side glances at complicating environmental constraints. Thus, if an appealing but clearly unnatural process is found, adaptation to natural conditions is left behind as a secondary problem to be dealt with downstream. Experimental proof is generally demanded for proposals at this level to be taken seriously.

At the third level, the question is raised whether chemically attractive reactions are compatible with environmental constraints imposed by geophysical considerations. Since natural processes are frequently highly complex, and since there is no cohesive geological record from the first 500 million years on Earth, a rather wide latitude for chemical speculation remains yet possible. However, with the combined criteria of the laws of physics and relative plausibilities of geochemical models, in addition to the important and increasing access to planetary materials and the oldest rocks, a more limited set of likelihoods is produced for the conditions present on Earth at the time of the origin of life. A typical example is presented by past chemical theories requiring a high atmospheric partial pressure of molecular hydrogen for desired reactions to proceed. This condition is fulfilled on massive gas-giant planets of the outer solar system like Jupiter. However, the light gases hydrogen and helium rapidly escape from the Earth's weaker gravitational field, and hydrogen generation sources on a planetary scale, capable of counterbalancing the escape in a sustained fashion, are unknown.

We have followed the record of the Archean Earth and prebiotic chemical schemes within the context of early life and the postulated RNA World. We see that the accretion history and early thermal state of the Earth remain subjects of widely varying hypotheses and conjectures, based only on fragmentary observational evidence that provides few concrete constraints. The body of such evidence is, however, increasing, deriving from measurement of magnetization and chemical separation in the interstellar medium, the observation of planet formation around protostars, radiochemical timing of the formation and development of the internal zones of Earth and Mars, and uncovering of increasingly older sedimentary records on Earth. All of these fields are in rapid development based on deliberate research efforts aimed at understanding the early

Earth and the origin of life. In the present situation it is, however, possible to claim initial thermal states of the Earth ranging from a low average temperature established already during the later stages of accretion, to a planetary ocean of molten magma blanketed by a hot and dense atmosphere, and requiring some kind of corrective acts to convert to a habitable crust and hydrosphere. These events have to take place in time to permit the evolution of biochemically advanced life to appear less than 650 million years after the accretion of the Earth; whether the earliest hints of life so far recognized were part of an RNA World remains unknown.

In this situation it is tempting for organic chemists, pursuing all chemically possible but otherwise unrestricted alternatives, to claim that as long as geologists, planetary physicists, and astrochemists cannot make up their minds, and continue to vacillate between such extremes as discussed above, there should be complete liberty not only for exploring all conceivable chemical pathways, but also to claim, without reservation, their validity on "the primitive Earth." However, as discussed above, additional and much more specific restrictions arise from consideration of the early state of the atmosphere and hydrosphere based on geophysical constraints. This, together with the question of life struggling against all odds in a fragile peace with impacts or other environmental catastrophes in the earliest times brings out the unsolved problem of how much time is needed for life to follow the "cosmic imperative" of de Duve (1995); to emerge from an inorganic state that, according to Descartes "does not act" but merely exists.

ACKNOWLEDGMENTS

The authors acknowledge support from NASA grants NAG5-4563 and NAGW-1031 (G.A.), and National Science Foundation grant EAR-9704651 (S.J.M.). Comments on the manuscript by C.E. Salomon are gratefully appreciated. We have benefited from discussions with P.S. Braterman, C.F. Chyba, M. Eigen, C. Engrand, A. Eschenmoser, M. Fayek, H.D. Holland, J.F. Kasting, G. Marklund, L.E. Orgel, and K. Zahnle. We also thank T. Cech for constructive editorial comments, and the invitation to participate in this volume.

REFERENCES

Abelson P. 1966. Chemical events on the primitive earth. *Proc. Natl. Acad. Sci.* **55:** 1365–1372.

Åkermark B., Eklund-Westlin U., Baeckström P., and Löf R. 1980. Photochemical, metal-promoted reduction of carbon dioxide and formaldehyde in aqueous solution. *Acta Chem. Scand.* **34B:** 27–30.

Alfvén H. and Arrhenius G. 1969. Two alternatives for the history of the Moon. *Science* **165:** 11–17.

———. 1972. Origin and evolution of the Earth-Moon system. *The Moon* **5:** 210–230.

Arrhenius T., Arrhenius G., and Paplawsky W.J. 1994. Archean geochemistry of formaldehyde and cyanide and the oligomerization of cyanohydrin. *Origins Life Evol. Biosph.* **24:** 1–19.

Arrhenius G., Gedulin B., and Mojzsis S. 1993. Phosphate in models for chemical evolution. In *Chemical evolution: Origin of life* (ed. C. Ponnamperuma and J. Chela–Flores), pp. 25–50. A. Deepak Publishers, Hampton, Virginia.

Arrhenius G., Sales B., Mojzsis S.J., and Lee T. 1997b. Entropy and charge in molecular evolution—The case of phosphate. *J. Theor. Biol.* **187:** 503–522.

Arrhenius S. 1896. On the influence of carbonic acid in the air upon the temperature of the ground. *Phil. Mag.* **41:** 237–276.

Ash R.D., Knott S.F., and Turner G. 1996. 4000 Gyr shock age for a martian meteorite and implications for the cratering history of Mars. *Nature* **380:** 57–59.

Awramik S.M., Schopf J.W., and Walter M.R. 1988. Carbonaceous filaments from North Pole, Western Australia: Are they fossil bacteria in ancient stomatolites? A discussion. *Precamb. Res.* **39:** 303–309.

Bada J.L., Bigham C., and Miller S.L. 1994. Impact melting of frozen oceans on the early Earth: Implications for the origin of life. *Proc. Natl. Acad. Sci.* **91:** 1248–1250.

Barley M.E., Dunlop J.S.R., Glover J.E., and Groves D.I. 1979. Sedimentary evidence for Archaean shallow-water volcanic-sedimentary facies, eastern Pilbara Block, Western Australia. *Earth Planet. Sci. Lett.* **43:** 74–84.

Berner R.A. and Maasch K.A. 1996. Chemical weathering and controls on atmospheric O_2 and CO_2: Fundamental principles were enunciated by J.J. Ebelmen in 1845. *Geochim. Cosmochim. Acta* **60:** 1633–1637.

Black L.P., Williams I.S., and Compston W. 1986. Four zircon ages from one rock: The history of a 3930 Ma-old granulite from Mount Sones, Enderby Land, Antarctica. *Contrib. Mineral. Petrol.* **94:** 427–437.

Bolli M., Micura R., and Eschenmoser A. 1997. Pyranosyl RNA: Chiroselective self-assembly of base sequences by ligative oligomerization of tetranucleotide-2′,3′-cyclophosphates. *Chem. Biol.* **4:** 309–320.

Bowring S.A., Williams I.S., and Compston W. 1989. 3.96 Ga gneisses from the Slave province, Northwest Territories, Canada. *Geology* **17:** 971–975.

Buick R. 1990. Microfossil recognition in Archean rocks; an appraisal of spheroids and filaments from a 3500 m.y. old chert-barite unit at North Pole, Western Australia. *Palaios* **5:** 441–459.

Cairns-Smith A.G. 1975. A case for an alien ancestry. *Proc. R. Soc. Ser. B* **189:** 249–274.

———. 1982. *Genetic takeover and the mineral origins of life.* Cambridge University Press, Cambridge, United Kingdom.

Cameron A.G.W. and Benz W. 1991. Origin of the Moon and the single impact hypothesis IV. *Icarus* **92:** 204–216.

Chang S. 1988. Planetary environments and the conditions of life. *Phil. Trans. R. Soc. Lond. Ser. A* **325:** 601–610.

Chyba C.F. 1993. The violent environment of the origin of life: Progress and uncertainties. *Geochim. Cosmochim. Acta* **57**: 3351–3358.
Chyba C. and Sagan C. 1992. Endogenous production, exogenous delivery and impact-shock synthesis of organic molecules: An inventory for the origins of life. *Nature* **355**: 125–132.
Chyba C., Thomas P., Brookshaw L., and Sagan C. 1990. Cometary delivery of organic molecules to the early Earth. *Science* **249**: 366–373.
Clark B.C. 1988. Primeval procreative comet pond. *Origins Life Evol. Biosph.* **18**: 209–238.
Compston W. and Kröner A. 1988. Multiple zircon growth within early Archaean tonalitic gneiss from the Ancient gneiss complex, Swaziland. *Earth Planet. Sci. Lett.* **87**: 13–28.
Cooper G. 1996. Polyhydroxylated compounds in the Murchison meteorite. *Origins Life Evol. Biosph.* **26**: 332–333.
Costa U.R., Fyfe W.S., Kerrich R., and Nesbitt H.W. 1980. Archean hydrothermal talc: Evidence for high ocean temperatures. *Chem. Geol.* **30**: 341–349.
Cronin J.R. and Pizzarello S. 1997. Enantiomeric excesses in meteoritic amino acids. *Science* **275**: 951–955.
Decker P., Schweer H., and Pohlmann R. 1982. Identification of formose sugars, presumable prebiotic metabolites, using capillary gas chromatography/gas chromatography-mass spectrometry of n-butoxime trifluoroacetates on OV-225. *J. Chromatogr.* **244**: 281–291.
de Duve C. 1995. *Vital dust: Life as a cosmic imperative*. Basic Books, New York.
De Graaf R.M., Visscher J., and Schwartz A.W. 1997. Reactive phosphonic acids as prebiotic carriers of phosphorus. *J. Mol. Evol.* **44**: 237–241.
De Graaf R.M., Visscher J., Xu Y., Arrhenius G., and Schwartz A.W. 1998. Mineral catalysis of a potentially prebiotic aldol condensation. *Origins Life Evol. Biosph.* **28**: (in press).
Dickens J., Irvine W., Ohishi M., Arrhenius G., Bauder, A., Eschenmoser A., and Pitsch S. 1996. A search for oxirane carbonitrile; a potential source for primitive aldehyde phosphates. Origins Life Evol. Biosph. **26**: 97–110.
DiMarco M.J. and Lowe D. 1989. Stratigraphy and sedimentology of an early Archean felsic volcanic sequence, eastern Pilbara Block, Western Australia, with special reference to the Duffer Formation and implications for crustal evolution. *Precambrian Res.* **44**: 147–169.
Doolittle R.F., Feng D.-F., Tsang S., Cho G., and Little E. 1996. Determining divergence times of the major kingdoms of living organisms with a protein clock. *Science* **271**: 470–477.
Dowden R.L., Brundell J.B., Lyons W.A., and Nelson T. 1996. Detection and location of red sprites by VLF scattering of subionosperic transmissions. *Geophys. Res. Lett.* **23**: 1737–1740.
Driessens F.C.M. and Verbeeck R.M.H. 1990. *Biominerals*. CRC Press, Boca Raton, Florida.
Dymek R.F. and Klein C. 1988. Chemistry, petrology, and origin of banded iron-formation lithologies from the 3800 Ma Isua supracrustal belt, West Greenland. *Precambrian Res.* **39**: 247–302.
Elliot J.C. 1994. *Structure and chemistry of the apatites and other calcium orthophosphates*. Elsevier, Amsterdam.
Eriksson K.A., Krapez B., and Frahlich P.W. 1997. Sedimentologic aspects of greenstone belts. In *Greenstone belts* (ed. M. de Wit and L. Ashwal), pp. 33–54. Clarendon Press, Oxford, United Kingdom.

Fernandez J.A. and Ip W.-H. 1983. On the time evolution of the cometary influx in the region of the terrestrial planets. *Icarus* **54:** 377–387.
Ferris, J.P. 1987. Prebiotic synthesis: Problems and challenges. *Cold Spring Harbor Symp. Quant. Biol.* **52:** 29–35.
———. 1993. Catalysis and prebiotic RNA synthesis. *Origins Life Evol. Biosph.* **23:** 307–315.
Ferris J.P. and Nicodem D.E. 1972. Ammonia photolysis and the role of ammonia in chemical evolution. *Nature* **238:** 268–269.
Gedulin B. and Arrhenius G. 1994. Sources and geochemical evolution of RNA precursor molecules: The role of phosphate. In *Early life on Earth—Nobel symposium 84* (ed. S. Bengston), pp. 91–110. Columbia University Press, New York.
Gerstenkorn H. 1955. Über Gezeitenreibung beim Zweikörper Problem. *Z. Astrophys.* **36:** 245–274.
———. 1969. The earliest past of the Earth-Moon system. *Icarus* **11:** 189–207.
Goldschmidt V.M. 1952. Geochemical aspects of the origin of complex organic molecules on the Earth as precursors to organic life, 1947. *New Biol.* **12:** 97–105.
Gough D.O. 1981. Solar interior structure and luminosity variations. *Solar Phys.* **74:** 21–34.
Hampton D.L., Heavner M.J., Wescott E.M., and Sentman D.D. 1996. Optical spectral characteristics of sprites. *Geophys. Res. Lett.* **23:** 82–92.
Hayes J.M., Kaplan I.R., and Wedeking K.W. 1983. Precambrian organic chemistry, preservation of the record. In *Earth's earliest biosphere* (ed. J.W. Schopf), pp. 53–134. Princeton University Press, Princeton, New Jersey.
Hoff S. 1952, 1980. Cartoon. *The New Yorker Magazine.*
Högbom A.G. 1894. Om sannolikheten för sekulära förändringar i atmosfärens kolsyrehalt. *Sven. Kem. Tidskr.* **6:** 169–177.
Hubbard J., Hardy J., and Horowitz N. 1971. Photocatalytic production of organic compounds from CO and H_2O in a simulated martian atmosphere. *Proc. Natl. Acad. Sci.* **68:** 574–578.
Jagoutz E., Sorowka A., Vogel J.D., and Wänke H. 1994. ALH84001: Alien or progenitor of the SNC family? *Meteoritics* **29:** 478–479.
Kasting J.F. 1982. Stability of ammonia in the primitive terrestrial atmosphere. *J. Geophys. Res.* **87:** 3091–3098.
———. 1988. How climate evolved on the terrestrial planets. *Sci. Am.* **258:** 90–97.
———. 1993. Earth's early atmosphere. *Science* **259:** 920–925.
Kasting J. and Ackerman T. 1986. Climatic consequences of very high carbon dioxide levels in the Earth's early atmosphere. *Science* **234:** 1383–1385.
Kasting J.F. and Walker J.C.G. 1981. Limits on oxygen concentration in the prebiological atmosphere and the rate of abiotic fixation of nitrogen. *J. Geophys. Res.* **86:** 1147–1158.
Kasting J., Pollack J., and Crisp D. 1984. Effects of high CO_2 levels on surface temperature and atmospheric oxidation state of the early Earth. *J. Atmos. Chem.* **1:** 403–428.
Kasting J., Zahnle K., and Walker J. 1983. Photochemistry of methane in the Earth's early atmosphere. *Precambrian Res.* **20:** 121–148.
Kaula W.M. and Bigeleisen P.E. 1975. Early scattering by Jupiter and its collision effects in the terrestrial zone. *Icarus* **25:** 18–33.
Keefe A.D. and Miller S.L. 1996. Was ferrocyanide a prebiotic reagent? *Origins Life Evol. Biosph.* **26:** 111–129.

Knauth L.P. and Lowe D.R. 1978. Oxygen isotope geochemistry of cherts from the Onverwacht group (3.4 billion years), Transvaal, South Africa, with implications for secular variations in the isotopic composition of cherts. *Earth Planet. Sci. Lett.* **41:** 209–222.

Kolb V., Zhang S., Xu Y., and Arrhenius G. 1997. Mineral-induced phosphorylation of glycolate ion—A metaphor in chemical evolution. *Origins Life Evol. Biosph.* **27:** 485–503.

Krishnamurthy R., Pitsch S., and Arrhenius G. 1998. Mineral induced formation of pentose 2,4-diphosphates. *Origins Life Evol. Biosph.* (in press).

LeRoux, J., Baratali, T., and Montel, G. 1963. Influence des certaines impuretés sur le nature et les propriètès du phosphate tricalcique precipité. *C.R. Acad. Sci.* **256:** 1312–1314.

Löb W. 1906. Studien über die chemische Wirkung der stillen elektrischen Entladung. *Z. Elektrochem.* **11:** 282–316.

———. 1914. Über das Verhalten des Formamids unter der Wirkung der stillen Entladung. Ein Beitrag zur Frage der Stickstoff-Assimilation. *Ber. Dtsch. Chem. Ges.* **46:** 684–697.

Lowe D.R. and Bryerly G.R. 1986. Sedimentological and stratigraphical evolution of the southern part of the Barberton greenstone belt: A case of changing provenance and stability. In *Workshop on the tectonic evolution of greenstone belts,* pp. 72–74. LPI Technical Report, Houston, Texas.

McGregor V.R. 1973. The early Precambrian gneisses of the Godthåb district, West Greenland. *Phil. Trans. R. Soc. Lond.* A **273:** 343–358.

McKay D.S., Gibson E.K., Thomas-Keprta K.L., Vali H., Romanek C.S., Clemett S.J., Chillier X.D.F., Maeschling C.R., and Zare R.N. 1996. Search for past life on Mars: Possible relict biogenic activity in martian meteorite ALH84001. *Science* **273:** 924–930.

Mel'nik Y.P. 1982. *Precambrian banded iron-formations.* Elsevier, Amsterdam.

Melton C. and Ropp G. 1958. Studies involving isotopically labeled formic acid and its derivatives. IV. Detection of cyanide in a mixture of formic acid and nitrogen in the mass spectrometer. *J. Am. Chem. Soc.* **80:** 5573.

Miller S.L. 1953. A production of amino acids under possible primitive Earth conditions. *Science* **117:** 528–529.

———. 1987. Which organic compounds could have occurred on the prebiotic Earth? *Cold Spring Harbor Symp. Quant. Biol.* **52:** 17–27.

Miller S.L., Lyons J.R., and Chyba C.F. 1998. Organic shielding of greenhouse gases on early Earth. *Science* **279:** 779.

Miller S.L. and Van Trump J.E. 1981. The Strecker synthesis in the primitive ocean. In *Origin of life* (ed. Y. Olman), pp. 135–141. Reidel, New York.

Mojzsis S.J. 1997. "Ancient sediments of Earth and Mars." Ph.D. thesis, Scripps Institution of Oceanography, University of California, San Diego.

Mojzsis S.J. and Arrhenius G. 1998. Phosphates and carbon on Mars: Exobiological implications and sample return considerations. *J. Geophys. Res. Planets* (in press).

Mojzsis S.J., Arrhenius G., McKeegan K.D., Harrison T.M., Nutman A.P., and Friend C.R.L. 1996. Evidence for life on Earth before 3800 million years ago. *Nature* **384:** 55–59.

Moorbath S., O'Nions R.K., and Pankhurst J.R. 1973. Early Archean age for the Isua iron-formation. *Nature* **245:** 138–139.

Müller D., Pitsch S., Kittaks A., Wagner E., Wintner C., and Eschenmoser A. 1990. Chemie von Alpha-Aminonitrilen. Aldomerisierung von Glycolaldehyd-phosphat zu racemischen Hexose-2,4,6–triphosphaten und (in Gegenwart von Formaldehyd) racemischen Pentose-2,4–diphosphaten: rac-Allose-2,4,6–triphosphat und rac-Ribose-2,4–diphosphat sind die Reaktionshauptprodukte *Helv. Chim. Acta* **73:** 1410–1468.

Neukum G. and Hiller K. 1981. Martian ages. *J. Geophys. Res. Planets* **86:** 3097–3121.

Newman M.J. and Rood R.T. 1977. Implications of solar evolution for the Earth's early atmosphere. *Science* **198:** 1035–1037.

Nutman A.P., Mojzsis S.J., and Friend C.R.L. 1997. Recognition of >3850 Ma water-lain sediments in West Greenland and their significance for the early Archaean Earth. *Geochim. Cosmochim. Acta* **61:** 2475–2484.

Nutman A.P., Allaart J., Bridgwater D., Dimroth E., and Rosing M. 1984. Stratigraphic and geochemical evidence for the depositional environment of the early Archean Isua supracrustal belt, southern West Greenland. *Precambrian Res.* **25:** 365–396.

Oparin A.I. 1924. *Proiskhozdenie Zhizny.* Moskovski'i Rabochii, Moskva.

Oparin A.I. and Clark F. 1959. *The origin of life on Earth.* Pergamon Press, New York.

Orgel L.E. 1986. RNA catalysis and the origins of life. *J. Theor. Biol.* **123:** 127–149.

———. 1992. Molecular replication and the origins of life. *Nature* **358:** 203–209.

———. 1998. The origin of life—How long did it take? *Origins Life Evol. Biosph.* **28:** 91–96.

Österberg R. and Orgel L.E. 1972. Polyphosphate and trimetaphosphate formation under potentially prebiotic conditions. *J. Mol. Evol.* **1:** 241–248.

Owen T. and Cess R.D. 1979. Enhanced CO_2 greenhouse to compensate for reduced solar luminosity on early Earth. *Macmillan J.* **277:** 640–642.

Papanastassiou D.A. and Wasserburg G. 1969. Initial strontium isotopic abundances and the resolution of small time differences in the formation of planetary objects. *Earth Planet. Sci. Lett.* **5:** 361–376.

Peltzer E.T., Bada J.L., Schlesinger G., and Miller S.L. 1984. The chemical conditions on the parent body of the Murchison meteorite: Some conclusions based on amino-, hydroxy- and dicarboxylic acids. *Adv. Space Res.* **4:** 69–74.

Pidgeon R.T. 1978. 3540 M.y. old volcanics in the Archaean layered greenstone succession of the Pilbara Block, Western Australia. *Earth Planet. Sci. Lett.* **37:** 421–428.

Pinto J.P., Gladstone G.R., and Yung Y.L. 1980. Photochemical production of formaldehyde in Earth's primitive atmosphere. *Science* **210:** 183–185.

Pitsch S., Pombo-Villar E., and Eschenmoser A. 1994. Chemistry of α-aminonitriles. Formation of 2–oxoethyl phosphates ("glycolaldehyde phosphates") from rac-oxiranecarbonitrile and (formal) constitutional relationships between 2–oxoethyl phosphates and oligo (hexo- and pentopyranosyl-) nucleotide backbones. *Helv. Chim. Acta* **77:** 2251–2285.

Pitsch S., Eschenmoser A., Gedulin B., Hui S., and Arrhenius G. 1995. Mineral induced formation of sugar phosphates. *Origins Life Evol. Biosph.* **25:** 294–334.

Roach J.R. 1975. Counterglow from the Earth-Moon libration points. *Planet. Space Sci.* **23:** 173–181.

Rubey W. 1951. Geologic history of sea water: An attempt to state the problem. *Bull. Geol. Soc. Am.* **62:** 1111–1147.

———. 1955. Development of the hydrosphere and atmosphere, with special reference to probable composition of the early atmosphere. *Geol. Soc. Amer.* (special paper) **62:** 631–650.

Russell M.J., Daniel R.M., Hall A.J., and Sherringham J. 1994. A hydrothermally precipitated catalytic iron sulphide membrane as first step toward life. *J. Mol. Evol.* **39:** 231–243.

Sanchez R.A., Ferris J.P., and Orgel L.E. 1966. Conditions for purine synthesis: Did prebiotic synthesis occur at low temperatures? *Science* **153:** 72–73.

Schidlowski M. 1988. A 3,800-million-year isotopic record of life from carbon in sedimentary rocks. *Nature* **333:** 313–318.

Schlesinger G. and Miller S.L. 1983. Prebiotic synthesis in atmospheres containing CH_4, CO, and CO_2. II. Hydrogen cyanide, formaldehyde and ammonia. *J. Mol. Evol.* **19:** 383–390.

Schmitz B., Peucker-Ehrenbrink B., Lindström M., and Tassinari M. 1997. Accretion rates of meteorites and cosmic dust in the early Ordovician. *Science* **278:** 88–89.

Schopf J.W. 1993. Microfossils of the early Archean Apex chert: New evidence of the antiquity of life. *Science* **260:** 640–646.

Schopf J.W. and Packer B.M. 1987. Early Archean (3.3-billion to 3.5-billion-year-old) microfossils from the Warrawoona Group, Australia. *Science* **237:** 70–73.

Schwartz A.W. 1993. Nucleotide analogs based on pentaerythritol—An hypothesis. *Origins Life Evol. Biosph.* **23:** 185–194.

Schwartz A.W. and De Graaf R.M. 1993. Photoreductive formation of acetaldehyde from aqueous formaldehyde. *Tetrahedron Lett.* **34:** 2201–2202.

Schwartz A.W. and Henderson-Sellers A. 1983. Glaciers, volcanic islands and the origin of life. *Precambrian Res.* **22:** 167–174.

Sentman D.D. and Wescott E.M. 1995. Red sprites and blue jets—Thunderstorm-excited optical emissions in the stratosphere, mesosphere, and ionosphere. *Phys. Plasmas* **2:** 2514–2522.

Sentman D.D., Wescott E.M., Osborne D.L., and Hampton D.L. 1995. Preliminary results from the *Sprites94* aircraft campaign. I. Red sprites. *Geophys. Res Lett.* **22:** 1205–1208.

Shock E.L. 1990. Geochemical constraints on the origin of organic compounds in hydrothermal systems. *Origins. Life Evol. Biosph.* **20:** 331–367.

———. 1996. Hydrothermal systems as environments for the emergence of life. In *Evolution of hydrothermal ecosystems on Earth (and Mars?). Ciba Found. Symp.* **202:** 40–60.

Shock E.L., Sassani D.C., Willis M., and Sverjensky D.A. 1997. Inorganic species in geologic fluids: Correlations among standard molal thermodynamic properties of aqueous ions and hydroxide complexes. *Geochim. Cosmochim. Acta* **61:** 907–950.

Sillén L. 1965. Oxidation state of Earth's ocean and atmosphere. I. A model calculation on earlier states. The myth of the "probiotic soup." *Ark. Kemi* **24:** 431–456.

Singer S.F. 1972. Origin of the Moon by tidal capture and some geophysical consequences. *The Moon* **5:** 206–209.

Sleep N.H., Zahnle K.J., Kasting J.F., and Morowitz H.J. 1989. Annihilation of ecosystems by large asteroid impacts on the early Earth. *Nature* **342:** 139–143.

Song B., Nutman A.P., Liu D., and Wu J. 1996. 3800 to 2500 Ma crustal evolution in the Anshan area of Liaoning province, northeastern China. *Precambian Res.* **78:** 79–94.

Strauss H. and Moore T.B. 1992. Abundances and isotopic compositions of carbon and sulfur species in whole rock and kerogen. In *The proterozoic biosphere* (ed. J.W. Schopf and C. Klein), pp. 709–798. Cambridge University Press, New York.

Summers D.P. and Chang S. 1993. Prebiotic ammonia from reduction of nitrite by iron (II) on the early Earth. *Nature* **365:** 630–633.

Tera F., Papanastassiou D.A., and Wasserburg G.J. 1974. Isotopic evidence for a terminal lunar cataclysm. *Earth Planet. Sci. Lett.* **14:** 281–304.

Urey H.C. 1952. *The planets, their origin and development.* Yale University Press, New Haven, Connecticut.

Verbeeck R.M.H., DeBruyne P.A.M., Driessens F.C.M., Terpstra R.A., and Verbeek F. 1986. Solubility behavior of Mg-containing β-$Ca_3(PO_4)_2$. *Bull. Soc. Chim. Belg.* **95:** 455–476.

Walker J. 1980. Atmospheric constraints on the evolution of metabolism. *Origins Life Evol. Biosph.* **10:** 93–104.

———. 1982. The earliest atmosphere of the Earth. *Precambrian Res.* **17:** 147–171.

Wänke H., Dreibus G., and Jagoutz E. 1984. Mantle chemistry and accretion history of the Earth. In *Archaean geochemistry* (ed. A. Kroner et al.), pp. 1–24. Springer-Verlag, New York.

Westheimer F.H. 1987. Why nature chose phosphates. *Science* **235:** 1173–1178.

Woese C.R. 1979. A proposal concerning the origin of life on the planet Earth. *J. Mol. Evol.* **13:** 95–101.

———. 1987. Bacterial evolution. *Microbiol Res.* **51:** 221–271.

Woese C.R. and Wächtershäuser G. 1990. Origin of life. In *Palaeobiology, a synthesis* (ed. D.P.C. Briggs), pp. 3–9. Blackwell Scientific, New York.

Yamagata Y., Watanabe H., Saitoh M., and Namba T. 1991. Volcanic production of polyphosphates and its relevance to chemical evolution. *Nature* **352:** 516–519.

Zahnle K.J. 1986. Photochemistry of methane and the formation of hydrocyanic acid (HCN) in the Earth's early atmosphere. *J. Geophys. Res.* **91:** 2819–2834.

2
Prospects for Understanding the Origin of the RNA World

Gerald F. Joyce
Departments of Chemistry and Molecular Biology
The Scripps Research Institute
La Jolla, California 92037

Leslie E. Orgel
The Salk Institute for Biological Studies
San Diego, California 92186

The general idea that, in the development of life on the earth, evolution based on RNA replication preceded the appearance of protein synthesis was first proposed about 30 years ago (Woese 1967; Crick 1968; Orgel 1968). It was suggested that catalysts made entirely of RNA are likely to have been important at this early stage in the evolution of life, but the possibility that RNA catalysts might still be present in contemporary organisms was overlooked. The unanticipated discovery of ribozymes (Kruger et al. 1982; Guerrier-Takada et al. 1983) initiated extensive discussion of the role of RNA in the origins of life (Pace and Marsh 1985; Sharp 1985; Lewin 1986) and led to the coining of the phrase "the RNA World" (Gilbert 1986).

The RNA World means different things to different authors, so it would be futile to attempt a restrictive definition. All RNA World hypotheses include three basic assumptions: (1) At some time in the evolution of life, genetic continuity was assured by the replication of RNA; (2) Watson-Crick base-pairing was the key to replication; (3) genetically encoded proteins were not involved as catalysts. RNA World hypotheses differ in what they assume about life that may have preceded the RNA World, about the metabolic complexity of the RNA World, and about the role of low-molecular-weight cofactors, possibly including peptides, in the chemistry of the RNA World.

It should be emphasized that the existence of an RNA World as a precursor of our DNA/protein world is a hypothesis. We find it an attractive hypothesis and believe that it derives some support from the results of experiments that it has inspired. The demonstration that the peptide-bond-forming step of protein synthesis is catalyzed by largely protein-free ribosomal RNA is particularly striking (Noller et al. 1992). We recognize, however, that not everyone will find the available evidence compelling.

In our initial discussion of the RNA World we will accept The Molecular Biologist's Dream: "Once upon a time there was a prebiotic pool full of β-D-nucleotides... ." We now consider what would have to have happened to make the dream come true. This discussion triggers The Prebiotic Chemist's Nightmare: how to make any kind of self-replicating system from the intractable mixtures that are formed in experiments designed to simulate the chemistry of the primitive earth.

THE MOLECULAR BIOLOGIST'S DREAM

Abiotic Synthesis of Polynucleotides

In this section we discuss the synthesis of oligonucleotides from β-D-nucleoside-5'-phosphates. Two fundamentally different chemical reactions are involved. First, the nucleotide must be converted to an activated derivative; for example, a nucleoside 5'-polyphosphate. Next the 3'-hydroxyl group of one nucleotide molecule must be made to react with the activated phosphate group of another. Synthesis of oligonucleotides from nucleoside 3'-phosphates is not discussed because activated nucleoside 2'- or 3'-phosphates in general react readily to form 2', 3'-cyclic phosphates. These cyclic phosphates are unlikely to oligomerize efficiently because the equilibrium constant for dimer formation is only of the order of 1.0 liter mole^{-1} (Erman and Hammes 1966; Mohr and Thach 1969).

In enzymatic RNA and DNA synthesis, the nucleoside 5'-triphosphates (NTPs) are the substrates of polymerization. Polynucleotide phosphorylase, although it is a degradative enzyme in nature, can be used to synthesize oligonucleotides from nucleoside 5'-diphosphates. Nucleoside 5'-polyphosphates are, therefore, obvious candidates for the activated forms of nucleotides. Although nucleoside 5'-triphosphates are not formed readily, the synthesis of nucleoside 5'-tetraphosphates from nucleotides and inorganic trimetaphosphate provides a reasonably plausible prebiotic route to activated nucleotides (Lohrmann 1975). A number of other more or less plausible prebiotic syntheses of nucleoside 5'-polyphosphates have been reported (Handschuh et al. 1973; Osterberg et al. 1973). Nucleoside 5'-polyphosphates, although they are high-energy phosphate esters, are relatively unreactive in aqueous solution. Consequently, little is known about their nonenzymatic oligomerization reactions.

In a different approach to the activation of nucleotides, the isolation of an activated intermediate is avoided by employing a condensing agent such as a carbodiimide (Khorana 1961). This is a popular method in organic synthesis, but its application to prebiotic chemistry is problematical. Potentially prebiotic molecules such as cyanamide and cyanoacetylene activate nucleotides in aqueous solution, but the subsequent condensation reactions are inefficient (Lohrmann and Orgel 1973).

Most attempts to study nonenzymatic polymerization of nucleotides in the context of prebiotic chemistry have used nucleoside 5'-phosphorimidazolides. Although phosphorimidazolides can be formed from imidazoles and nucleoside 5'-polyphosphates (Lohrmann 1977), they are only marginally plausible as prebiotic molecules. They were chosen because they are prepared easily and react at a convenient rate in aqueous solution.

Nucleotides contain three principal nucleophilic groups, the 5'-phosphate, the 2'-hydroxyl, and the 3'-hydroxyl, in order of decreasing reactivity. The reaction of a nucleotide or oligonucleotide with an activated nucleotide, therefore, normally yields 5', 5'-pyrophosphate-, 2', 5'-phosphodiester-, and 3', 5'-phosphodiester-linked adducts, in order of decreasing abundance (Fig. 1a) (Sulston et al. 1968). Thus, the condensation of several monomers would likely yield an oligomer containing one pyrophosphate and a preponderance of 2', 5' phosphodiester bonds (Fig. 1b). There is little chance of producing entirely 3', 5'-linked oligomers from activated nucleotides unless a catalyst can be found that increases the yield of 3', 5' phosphodiester bonds. Several metal ions, particularly Pb^{++} and UO_2^{++}, catalyze the formation of oligomers from nucleoside 5'-phosphorimidazolides (Sleeper and Orgel 1979; Sawai et al. 1988). However, the product oligomers always contain a large proportion of 2', 5' linkages.

What kinds of prebiotically plausible catalysts might lead to the production of 3', 5'-linked oligonucleotides directly from nucleoside 5'-phosphorimidazolides or other activated nucleotides? It is unlikely, but not impossible, that a metal ion or simple acid-base catalyst would provide sufficient stereospecificity. The most attractive of the other hypotheses is that adsorption to a specific surface of a mineral might orient activated nucleotides rigidly and thus catalyze a highly stereospecific reaction.

The work of Ferris and coworkers provides some support for this hypothesis. They studied the oligomerization of nucleoside 5'-phosphorimidazolides and related activated nucleotides on the clay mineral, montmorillonite (Ferris and Ertem 1993; Kawamura and Ferris 1994). They found that the mineral is an effective catalyst, promoting the formation of oligomers even from dilute solutions of activated nucleotide substrates.

Furthermore, the mineral profoundly affected the regiospecificity of the reactions for some but not all of the four activated nucleotides. The oligomerization of adenosine 5′-phosphorimidazolide, for example, gives predominantly 3′, 5′-phosphodiester linkages on montmorillonite, but predominantly 2′, 5′ linkages in aqueous solution. Successive "feedings" of activated monomers to oligoadenylates that had been adsorbed on hydroxylapatite led to the accumulation of mainly 3′, 5′-linked oligomers up to 40 subunits in length (Ferris et al. 1996).

The detailed analysis of this work on catalysis by montmorillonite strongly suggests that oligomerization is catalyzed at a limited number of structurally specific active sites. These sites are subject to inhibition by substrate analogs, and the regiospecificity of oligomerization is highly sensitive to the identity of the substrate nucleotide. It is important to dis-

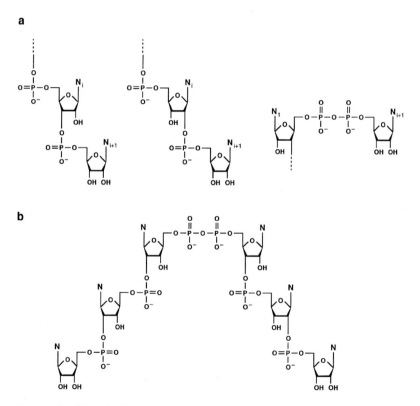

Figure 1 Phosphodiester linkages resulting from chemical condensation of nucleotides. (*a*) Reaction of an activated mononucleotide (N_{i+1}) with an oligonucleotide ($N_1 - N_i$) to produce a 3′, 5′-phosphodiester (*left*), 2′, 5′-phosphodiester (*middle*), or 5′, 5′-pyrophosphate (*right*). (*b*) Typical oligomeric product resulting from chemical condensation of activated mononucleotides.

cover whether the nucleotide–montmorillonite system is unusual or if other mineral catalysts of oligonucleotide synthesis are often comparably stereospecific.

Nonenzymatic Replication of RNA

If we suppose that a mechanism existed for the polymerization of activated nucleotides, it would have generated a complex mixture of product oligonucleotides, differing in both sequence and length. The next stage in the evolution of an RNA World would have been the replication of some of these molecules, so that a process equivalent to natural selection could begin. The reaction central to replication of nucleic acids is template-directed synthesis; that is, the synthesis of a complementary oligonucleotide under the direction of a preexisting oligonucleotide. A good deal of work has already been done on this aspect of nonenzymatic replication. Progress up to about 1987 has been reviewed previously (Joyce 1987), so only a summary of the results is given here.

The first major conclusion is that most activated nucleotides do not undergo efficient, regiospecific template-directed reactions. In general, only a small proportion of template molecules succeed in directing the synthesis of a complement, and the complement contains a mixture of 2′, 5′- and 3′, 5′-phosphodiester bonds. After a considerable search, a set of activated nucleotides was found that undergo efficient and highly regiospecific template-directed reactions. Working with guanosine 5′-phospho-2-methylimidazolide (2-MeImpG), it was shown that poly(C) can direct the synthesis of long oligo(G)s in a reaction that is highly efficient and highly regiospecific (Inoue and Orgel 1981). If poly(C) is incubated with an equimolar mixture of the four 2-MeImpNs (N = G, A, C, or U), less than 1% of the product consists of noncomplementary nucleotides (Inoue and Orgel 1982).

Random copolymers containing an excess of C residues can be used to direct the synthesis of products containing G and the complements of the other bases present in the template (Inoue and Orgel 1983). The reaction with a poly(C,G) template is especially interesting because the products, like the template, are composed entirely of C and G residues. If these products in turn could be used as templates, it might allow the emergence of a self-replicating sequence. Self-replication, however, is unlikely, mainly because poly(C,G) molecules that do not contain an excess of C residues tend to form stable self-structures that prevent them from acting as templates (Joyce and Orgel 1986). The self-structures are of two types: the standard Watson-Crick variety based on C·G pairs, and a quadrahelix

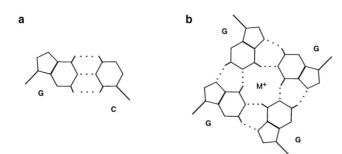

Figure 2 Two types of self-structure interactions expected for poly(C,G) templates. (*a*) Standard G·C Watson-Crick pair. (*b*) G quadrahelix pairing, stabilized by a monovalent cation (M^+). Dotted lines indicate hydrogen-bonding interactions.

structure that results from the association of four G-rich sequences (Fig. 2). As a consequence, any C-rich oligonucleotide that can serve as a good template will give rise to G-rich complementary products that tend to be locked in self-structure and so cannot act as templates. Overcoming the self-structure problem is very difficult because it requires the discovery of conditions that favor the binding of mononucleotides in order to allow template-directed synthesis to occur, but suppress the formation of long duplex regions that would exclude activated monomers from the template.

Some progress has been made in discovering defined-sequence templates that are copied faithfully to yield complementary products (Inoue et al. 1984; Acevedo and Orgel 1987; Wu and Orgel 1992a). Successful templates typically contain an excess of C residues, with A and U residues isolated from each other by at least three C residues. Runs of G residues are copied into runs of C residues, so long as the formation of self-structures by G residues can be avoided (Wu and Orgel 1992b). In light of the available evidence, it seems unlikely that a pair of complementary sequences can be found, each of which facilitates the synthesis of the other using nucleoside 5′-phospho-2-methylimidazolides as substrates. Some of the obstacles to self-replication may be attributable to the choice of reagents and reaction conditions, but others seem to be intrinsic to the template-directed condensation of activated mononucleotides.

Group I ribozymes catalyze transesterification reactions in which the 3′-terminal nucleotide of a preexisting oligonucleotide is added to the 3′ terminus of a growing polymer chain (Zaug and Cech 1986; Been and Cech 1988), e.g., the reaction

$$C_5 + C_i \rightarrow C_4 + C_{i+1}$$

Related chemistry could, in principle, be used to copy an oligonucleotide template. This suggestion, although it might be applicable to an established RNA World, is not very plausible for the nonenzymatic replication of RNA. The synthesis of the 3', 5'-linked oligonucleotide substrates, as we have seen, presents a very difficult problem. Furthermore, the use of an "energy-neutral" transesterification reaction would make it impossible to convert more than a small fraction of the substrate to high-molecular-weight products. The free energy of association of the template with its complement might alleviate this difficulty to some extent.

Another nonenzymatic replication scheme suggested by ribozyme-catalyzed reactions involves synthesis by the ligation of short 3', 5'-linked oligomers. This is certainly an attractive hypothesis, but it faces two major obstacles. The first is again the difficulty of obtaining the substrates in the first place. The second is concerned with fidelity. Pairs of oligonucleotides containing a single base mismatch, particularly if the mismatch forms a G·U wobble pair, still hybridize as efficiently as fully complementary oligomers, except in a temperature range very close to the melting point of the perfectly paired structure. Maintaining fidelity would therefore be difficult under any plausible temperature regime.

Despite these problems, template-directed ligation of short oligonucleotides may be a viable alternative to oligomerization of activated monomers. Ferris's work discussed above suggests that 3', 5'-linked oligonucleotides might form spontaneously from activated nucleotides on montmorillonite or possibly on some other mineral surface. Recent studies show that oligonucleotide 5'-triphosphates undergo slow but remarkably regiospecific nonenzymatic ligation on suitable templates (Rohatgi et al. 1996a,b). Combination of a similar pair of reactions might provide a replication scheme for polynucleotides starting with an input of activated monomers.

The First RNA Replicase

The molecular biologist's notion of the RNA World places emphasis on an RNA molecule that catalyzes its own replication. Such a molecule must function as an RNA-dependent RNA polymerase, acting on itself (or copies of itself) to produce complementary RNAs, and acting on the complementary RNAs to produce additional copies of itself. The efficiency of this process must be sufficient to produce "progeny" RNA molecules at a rate that exceeds the rate of decomposition of the "parents." The fidelity of polymerization must be sufficient to allow the functional copies to compete favorably against the distribution of inactive mutant copies.

Beyond these requirements, the details of the replication process are not highly constrained.

In the molecular biologist's dream it is assumed that a supply of activated β-D-nucleotides was available by some as yet unrecognized abiotic process. Furthermore, it is assumed that a catalyst existed to convert the activated nucleotides to a random ensemble of polynucleotide sequences, a subset of which had the ability to replicate. It seems to be implicit in the model that such sequences replicate themselves but, for whatever reason, do not replicate unrelated neighbors. It is not clear whether replication involves one molecule copying both itself and its complement or a family of molecules that together copy each other. We ignore these questions for the moment and discuss the question of whether an RNA molecule of reasonably short length can catalyze its own replication with reasonably high fidelity.

Accuracy and Survival

The concept of an error threshold, that is, an upper limit to the frequency of copying errors that can be tolerated by a replicating macromolecule, was first introduced by Eigen (1971). This important idea has been extended in a series of mathematically sophisticated papers by McCaskill, Schuster, and others (McCaskill 1984b; Eigen et al. 1988; Schuster and Swetina 1988). Here we attempt to give a pedestrian's guide to the subject.

We first consider an independently replicating RNA molecule of length n in an environment that contains a large excess of NTPs. We suppose that the probability of correct copying, q, is the same for all nucleotide positions in the molecule. Finally we suppose, unrealistically, that all mutations are lethal. Then it is clear, assuming that the mutations occur in both strands, that the average number of viable products of each replication is $2q^n$. Consequently, the maximum length of molecules in a stable clone is determined by the inequality $q^n > 1/2$. This simple model introduces several features of Eigen's treatment. It supposes that each molecule replicates independently of all others and that the mutation rate is the same at all positions in the sequence. The model ignores positions where mutations are without consequence, so n is to be interpreted as the "significant" length of the genome. In light of the result obtained, we should not be surprised if more realistic theories lead to inequalities of the form $q^n > f(n)$, governing the fidelity q and the maximum sustainable genome length n.

The model discussed above is unrealistic because it supposes that a mutation is either lethal or without consequence. In practice, we know that there would be a substantial proportion of mutations that reduced the rate and fidelity of replication but did not prevent replication completely. Furthermore, there might be some mutations that increase the rate and fidelity. This would not invalidate the result obtained above; in the absence of competition a clone would have a chance of persisting indefinitely if each parent molecule produced, on average, more than one identical descendant before disappearing from the ensemble. In fact, a clone might under some circumstances persist even if it produced less than one identical descendant, provided that a sufficient proportion of mutants were viable replicators. However, the existence of mutants with lower replication rate and fidelity becomes important once the supply of substrates becomes limiting. Then competition between the most efficient sequence, the master sequence, and its competitors that are less efficient replicators can lead to the disappearance of the master if the fidelity of replication is too low. The following discussion of the error threshold, therefore, is mainly relevant to the persistence of superior clones when competition is important and not to clonal expansion from a single replicator when the supply of substrates is in vast excess.

The Error Threshold

Eigen's model (1971) envisages a set of replicating polynucleotides that draw upon a limited supply of activated mononucleotides to produce additional copies of themselves. If a particular RNA sequence is to emerge and maintain a substantial representation in the population, then its rate of reproduction must exceed the mean rate of production of other RNA sequences in its locale. Satisfying these requirements places restrictions on the fidelity of replication, which in turn places restrictions on the length of the RNA that can be replicated.

The rate of synthesis of new copies of a self-replicating RNA molecule, R_i, is proportional to the concentration of R_i

$$d[R_i]/dt = W_i[R_i] \qquad (1)$$

The autocatalytic rate constant, W_i, is the difference between the rate of formation of error-free copies of R_i and the rate of decomposition of existing copies of R_i

$$W_i = A_i Q_i - D_i \qquad (2)$$

In this formulation, A_i refers to the rate of amplification of R_i, Q_i refers to the proportion of copies of R_i that are error-free, and D_i refers to the rate of decomposition of R_i. For simplicity, the nondiagonal error terms are neglected in this treatment, that is, the contributions to $[R_i]$ resulting from error copies of some other $R_{k \neq i}$ are ignored. In subsequent formulations (Eigen and Schuster 1977; McCaskill 1984a), the nondiagonal terms were included, leading to more precise results that are qualitatively similar to those described here.

In the simplified model, for an advantageous self-replicating RNA, R_m, to outgrow its competitors, $R_{k \neq m}$, it is necessary that

$$W_m > \overline{W}_{k \neq m}, \text{ where } \overline{W}_{k \neq m} = \Sigma W_k [R_k] / \Sigma [R_k] \tag{3a}$$

$$A_m Q_m - D_m > \overline{A}_{k \neq m} - \overline{D}_{k \neq m} \tag{3b}$$

Note that the term $\overline{Q}_{k \neq m}$ does not appear on the right side of Equation 3b because error copies of a particular $R_{k \neq m}$ give rise to some other $R_{k \neq m}$. Rearranging Equation 3B gives

$$Q_m > (\overline{A}_{k \neq m} - \overline{D}_{k \neq m} + D_m) / A_m \tag{4}$$

In other words, to maintain the genetic information contained within an advantageous individual, it is necessary that the fidelity of copying that information exceed the inverse of the relative advantage enjoyed by the advantageous individual compared to the rest of the population. This relative advantage, or "superiority," is expressed as

$$\sigma_m = A_m / (\overline{A}_{k \neq m} - \overline{D}_{k \neq m} + D_m) \tag{5}$$

Combining Equations 4 and 5 gives

$$Q_m > 1/\sigma_m \tag{6}$$

The proportion of copies of R_m that are error-free is determined by the fidelity of the component condensation reactions that are required to produce a complete copy. For simplicity, consider a self-replicating RNA that is formed by v condensation reactions, each having mean fidelity q. The probability of obtaining an error-free copy is the product of the fidelity of the component condensation reactions

$$Q_m = q^v \tag{7}$$

Combining Equations 6 and 7 and solving for v

$$\ln Q_m = v \ln q > -\ln \sigma_m \tag{8a}$$

$$v < -\ln \sigma_m / \ln q \tag{8b}$$

Equation 8b, the "error threshold," defines the trade-off between the fidelity of replication, q, and the maximum allowable number of component condensation reactions v. For a self-replicating RNA of length n that is formed by template-directed condensation of activated mononucleotides, a total of $2n - 2$ condensation reactions are required to produce a complete copy. This takes into account the synthesis of both a complementary strand and a complement of the complement.

Figure 3 depicts the relationship between q and n for two different values of σ_m. The maximum allowable genome length is highly sensitive to the fidelity of replication, but depends only weakly on the superiority of the advantageous self-replicating species. It should be recognized that a marked superiority of one sequence over all other sequences could not be maintained over evolutionary time because novel variants would soon arise to challenge the dominant species. However, a marked initial superiority may be important in allowing an efficient self-replicating RNA to

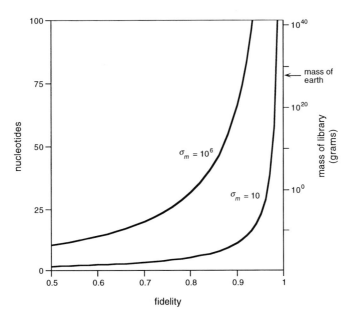

Figure 3 The Eigen error threshold, relating fidelity of replication to the maximum number of nucleotides, for two different values of superiority (σ_m). Fidelity (q) is that of the component condensation reactions in the replication cycle. The number of condensation reactions (v) is related to the number of nucleotides (n) by: $v = 2n - 2$. Scale at right indicates the mass of a combinatorial library containing one copy each of the 4^n possible sequences of length n.

emerge from a pool of less efficient replicators. Whether or not this can occur depends on the value of σ_m, the superiority of the emerging replicator. For example, if $\sigma_m = 10$, an RNA that replicates with 90% fidelity could be no longer than 12 nucleotides and one that replicates with 70% fidelity could be no longer than 4 nucleotides. It seems highly unlikely that any of the 17 million possible RNA dodecamers is able to catalyze its own replication with 90% fidelity and even less likely that a tetranucleotide could catalyze its own replication with 70% fidelity. However, if $\sigma_m = 10^6$, an RNA that replicates with 90% fidelity could be as long as 67 nucleotides and one that replicates with 70% fidelity could be as long as 20 nucleotides.

When self-replication is first established, fidelity is likely to be poor and there is strong selection pressure favoring improvement of the fidelity. As fidelity improves, a larger genome can be maintained. This allows exploration of a larger number of possible sequences, some of which may lead to further improvement in fidelity, which in turn allows a still larger genome size, and so on. Once the evolving population has entered the steep part of the curve shown in Figure 3, RNA-based life can become firmly established. Until that time, it is a race between evolutionary improvement in the context of a sloppy self-replicating system and the risk of delocalization of the genetic information due to overstepping the error threshold. If the time required to bootstrap to high fidelity and large genomes is too long, there is a risk that the population will succumb to an environmental catastrophe before it has had the chance to develop appropriate countermeasures.

It is difficult to state with certainty the minimum possible size of an RNA replicase ribozyme. An RNA consisting of a single secondary structural element, that is, a small stem-loop containing 12–17 nucleotides, would not be expected to have replicase activity, whereas a double stem-loop, perhaps forming a "dumbbell" structure or a pseudoknot, might just be capable of a low level of activity. A triple stem-loop structure, containing 40–60 nucleotides, offers a reasonable hope of functioning as a replicase ribozyme. One could, for example, imagine a molecule consisting of a pseudoknot and a pendant stem-loop that forms a cleft for template-dependent replication.

Suppose there is some 40-mer that enjoys a superiority σ_m of 10^3 and replicates with about 90% fidelity. This should be regarded as a highly optimistic but not outrageous view of what is possible for a minimum replicase ribozyme. Would such a molecule be expected to occur within a population of random RNAs? A complete library consisting of one copy each of all 10^{24} possible 40-mers would weigh about one kilogram.

Furthermore, there may be many such 40-mers, encompassing both distinct structural motifs and, more importantly, a large number of equivalent representations of each motif. As a result, even a small fraction of the total library, consisting of perhaps 10^{20} sequences and weighing about one gram, might be expected to contain at least one self-replicating RNA with the requisite properties. It is not sufficient, however, that there be just one copy of a self-replicating RNA. The above calculations assume that a self-replicating RNA can copy itself (or that a fully complementary sequence is automatically available; see below). If two or more copies of the same 40-mer RNA are needed, then a much larger library, consisting of 10^{48} RNAs and weighing 10^{28} grams, would be required. This amount is comparable to the mass of the earth.

At first sight, it might seem that one way to ease the error threshold would be for the replicase ribozyme to accept dinucleotide or trinucleotide substrates, so that a copy of an RNA of length n could be formed by $2(^n/_2 - 1)$ or $2(^n/_3 - 1)$ condensation reactions, respectively. Calculations show that, over a broad range of values for σ_m, RNAs that are required to replicate with 90% fidelity when using mononucleotide substrates would be required to replicate with roughly 80% fidelity when using dinucleotide substrates or roughly 70% fidelity when using trinucleotide substrates. Thus, the use of short oligomers is not likely to offer a significant advantage because lessening of the error threshold would be outweighed by the greater difficulty in achieving high fidelity when discriminating among the 16 possible dinucleotide or 64 possible trinucleotide substrates, rather than among the four mononucleotides.

If we accept the molecular biologist's dream that there was a prebiotic pool of random-sequence RNAs, and if we assume that that pool included a replicase ribozyme containing, say, 40 nucleotides and replicating with about 90% fidelity, then it is not difficult to imagine how RNA-based evolution might have started. During the initial period, a successful clone would have expanded in the absence of competition. As competition for substrates intensified, there would have followed a succession of increasingly efficient master sequences, each replicating within its error threshold. After a period of intensifying competition, the single master species would have been replaced by a "quasispecies," that is, a mixture of the master sequence and substantial amounts of closely related sequences that replicate almost as fast and almost as faithfully as the master (Eigen and Schuster 1977; Eigen et al. 1988). Under these conditions, the persistence of a dominant master sequence is no longer the problem, but one must understand the evolution of the composition of the quasispecies and the conditions for its persistence. This difficult problem has been partially solved by McCaskill (1984b). The general form of the solution is very similar to that obtained by Eigen (1971), but with different values for the constant in the inequality. Thus, concerns about the error threshold apply to the quasispecies as well as to the succession of individual master sequences. Practically speaking, however, once a quasispecies distribution of sophisticated replicators had emerged, the RNA World would have been on solid footing and unlikely to lose the ability to maintain genetic information over time.

Another Chicken-and-Egg Paradox

In the above discussion we have tried mightily to present the most optimistic view possible for the emergence of an RNA replicase ribozyme from a soup of random polynucleotides. It must be admitted, however, that we do not consider this model to be very plausible. Our discussion has focused on a straw man: the myth of a small RNA molecule that arises de novo and can replicate efficiently and with high fidelity under plausible prebiotic conditions. Not only is such a notion unrealistic in light of our current understanding of prebiotic chemistry, but it should strain the credulity of even an optimist's view of RNA's catalytic potential. If you doubt this, ask yourself whether you believe that a replicase ribozyme would arise in a solution containing nucleoside 5′-diphosphates and polynucleotide phosphorylase!

If one accepts the notion of an RNA World, one is faced with the dilemma of how such a genetic system came into existence. To say that the RNA World hypothesis "solves the paradox of the chicken and the egg" is correct if one means that RNA can function both as a genetic molecule and as a catalyst that promotes its own replication. RNA-catalyzed RNA replication provides a chemical basis for Darwinian evolution based on natural selection. Darwinian evolution is a powerful way to search among vast numbers of potential solutions for those that best address a particular problem. Selection based on inefficient RNA replication, for example, could be used to search among a population of RNA molecules for those individuals that promote improved RNA replication. But here we encounter another chicken-and-egg paradox: Without evolution it appears unlikely that a self-replicating ribozyme could arise, but without some form of self-replication there is no way to conduct an evolutionary search for the first, primitive self-replicating ribozyme.

One way that RNA evolution may have gotten started without the aid of an evolved catalyst might be by using nonenzymatic template-directed synthesis to permit some copying of RNA before the appearance of the first replicase ribozyme. Suppose that the initial ensemble of polymers was not produced by random copolymerization, but rather by a sequence of untemplated and templated reactions (Fig. 4), and further suppose that members of the initial ensemble of multiple stem-loop structures could be replicated, albeit inefficiently, by the template-directed process. This would have two important consequences. First, any molecule with replicase function that appeared in the mixture would likely find in its neighborhood similar (and complementary) molecules, related by descent, thus eliminating the requirement for two unrelated replicases to meet. Second, a majority of molecules in the mixture would contain stem-loop structures. If it is true that ribozyme function is favored by stable self-structure, and if the base sequences of the stems in stem-loop structures are relatively unimportant for function, this model might provide an economical way of generating a relatively small ensemble of sequences that is enriched with catalytic sequences.

How plausible is the assumption that replicases could act on sequences similar to themselves, while ignoring unrelated sequences?

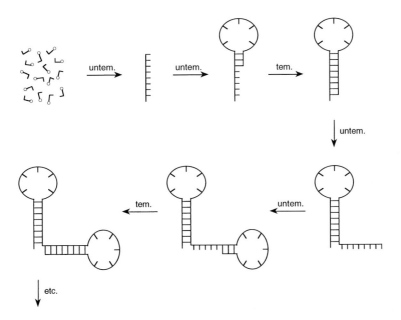

Figure 4 Nonenzymatic synthesis of multi-stem-loop structures as a result of untemplated (untem.) and templated (tem.) reactions. Template-directed synthesis is assumed to occur rapidly whenever a template and suitable primer are available. Once the complementary strand is completed, additional residues are added slowly, with random sequence.

This selectivity could be ensured by segregating individual molecules (or clonal lines) on the surface of mineral grains, on the surface of micelles, or within membranes. Closely related molecules might be segregated as a group through specific hydrogen-bonding interactions (the family that sticks together, replicates together). For any segregation mechanism, weak selection would result if the replicating molecules are sufficiently dispersed that diffusion over their intermolecular distance is slow compared to replication. Alternatively, the suggestions of Weiner and Maizels concerning "genomic tags" may be relevant (Weiner and Maizels 1987). Among self-replicating sequences, it is plausible that some are restricted to copying molecules with a particular 3'-terminal subsequence. A replicator that happened by chance to carry a terminal sequence that matched the preference of its active site would replicate itself while ignoring its neighbors.

Another resolution of the paradox of how RNA evolution was initiated without the aid of an evolved ribozyme is to abandon the molecular biologist's dream and suppose that RNA was *not* the first genetic molecule (Shapiro 1984; Joyce 1987, 1989; Orgel 1989). Perhaps RNA replication arose in the context of an evolving system based on something other than RNA (see section below entitled Alternative Genetic Systems). Even if this is true, all of the arguments concerning the relationship between the fidelity of replication and the maximum allowable genome length would still apply to this earlier genetic system. Of course, the challenge to those who advocate this approach is to demonstrate that there is an informational entity that is prebiotically plausible and is capable of initiating its own replication without the aid of a sophisticated catalyst.

Replicase Function in the Evolved RNA World

Although it is difficult to say how the first RNA replicase ribozyme arose, it is not difficult to imagine how such a molecule, once developed, would function. The chemistry of RNA replication would involve the template-directed polymerization of mononucleotides or short oligonucleotides, using chemistry in many ways similar to that employed by contemporary group I ribozymes (Cech 1986, 1987; Doudna and Szostak 1989). One important difference is that, unlike group I ribozymes, which rely on a nucleoside or oligonucleotide leaving group, an RNA replicase would more likely make use of a different leaving group that provides a substantial driving force for polymerization and that, after its release, does not become involved in some competing phosphoester transfer reaction.

The polymerization of activated nucleotides proceeds via nucleophilic attack by the 3′-hydroxyl of a template-bound oligonucleotide at the α-phosphorus of an adjacent template-bound nucleotide derivative (Fig. 5). The nucleotide is "activated" for attack by the presence of a phosphoryl substituent; for example, a phosphate, polyphosphate, alkoxide, or imidazole group. As discussed above, polyphosphates, such as inorganic pyrophosphate, are the most obvious candidates for the leaving group. The condensation reaction could be assisted by deprotonation of the nucleophilic 3′-hydroxyl, stabilization of the trigonal-bipyramidal intermediate, and protonation of the leaving group. All three of these tasks might be carried out by RNA, acting either alone or with the help of a suitably positioned metal cation or other cofactor (Cech and Herschlag 1996; Narlikar and Herschlag 1997).

The possibility that an RNA replicase ribozyme could have existed has been made abundantly clear by the recent work of Bartel and Szostak (Bartel and Szostak 1993; Ekland et al. 1995; Ekland and Bartel 1996). Beginning with a large population of random-sequence RNAs and utilizing in vitro evolution methods, they obtained an RNA ligase ribozyme, about 100 nucleotides in length, that catalyzes the joining of two template-bound oligonucleotides. Condensation occurs between the

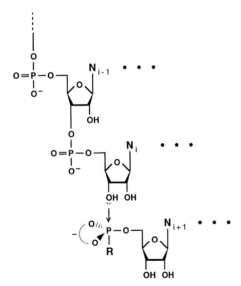

Figure 5 Nucleophilic attack by the 3′-hydroxyl of a template-bound oligonucleotide ($N_1 - N_i$) at the α-phosphorus of an adjacent template-bound mononucleotide (N_{i+1}). Dotted lines indicate base pairing to a complementary template. R is the leaving group.

3'-hydroxyl of one oligonucleotide and the 5'-triphosphate of another, forming a 3', 5'-phosphodiester linkage with release of inorganic pyrophosphate (Bartel and Szostak 1993). Subsequently it was shown that the ligase ribozyme catalyzes a polymerization reaction in which the 5'-triphosphate-bearing oligonucleotide is replaced by one or more NTPs (Ekland and Bartel 1996). This reaction proceeds with high fidelity, but the reaction rate drops sharply with successive nucleotide additions, becoming undetectable after the third addition.

The polymerase activity of this in vitro evolved ribozyme is fundamentally the same chemistry as would have been required of an RNA replicase ribozyme in the RNA World. There are, however, several important differences in the details. The error rate of the in vitro evolved ribozyme under optimal conditions is about 8% ($q = 0.92$) (Ekland and Bartel 1996). This is significantly better than would be expected for Watson-Crick pairing alone, but does not leave room under the error threshold for replication of a ribozyme that itself contains approximately 100 nucleotides. Furthermore, the polymerase ribozyme contains an internal template region of 1–3 nucleotides, whereas a replicase ribozyme would be expected to operate on an external template of much greater length.

An RNA replicase ribozyme must ensure that the polymerization reaction works efficiently for a broad range of template sequences. Differences in the templating properties of the four bases and in the local structure of the template must be "smoothed out" without sacrificing the overall efficiency and fidelity of the reaction. This might prove difficult because factors that promote efficient template-directed incorporation of a nucleotide irrespective of its base might be expected to lower the fidelity. On the other hand, factors that lower duplex stability to prevent template self-structure might enhance the replicase's ability to discriminate against base mismatch.

RNA replicase activity is probably not the only catalytic behavior that was essential for the existence of the RNA World. Maintaining an adequate supply of the four activated nucleotides would have been a top priority. Even if the prebiotic environment had contained a large reservoir of these compounds, the reservoir would have eventually become depleted, and some capacity for nucleotide biosynthesis would have been required. This metabolic complexity necessitates a more complicated genome. Rather than consisting of a single "gene" that provided replicase activity, the genome must have contained several genes, specifying both metabolic and replicative functions. This would have required the evolution of some form of control of gene expression.

THE PREBIOTIC CHEMIST'S NIGHTMARE

The molecular biologist's pool contains pure β-D-nucleotides. How close to such a pool could one hope to get without magic (or enzymes) on the primitive earth? Could one hope to achieve replication in a pool containing a more realistic mixture of organic molecules, including, of course, β-D-ribonucleotides?

The synthesis of a nucleotide could occur in a number of ways. The simplest, conceptually, would be to synthesize a nucleoside base, couple it to ribose, and finally phosphorylate the resulting nucleoside. However, a number of other routes are feasible; for example, the assembly of the base on a preformed ribose or ribose phosphate.

The classic prebiotic synthesis of sugars is by the polymerization of formaldehyde. It yields a very complex mixture of products including only a small proportion of ribose (Mizuno and Weiss 1974). This reaction does not provide a reasonable route to the ribonucleotides. An important discovery has been made by Eschenmouser and his colleagues. They have shown that the base-catalyzed aldomerization of glycoaldehyde phosphate in the presence of a half-equivalent of formaldehyde gives a relatively simple mixture of tetrose- and pentose-diphosphates and hexosetriphosphates, of which ribose 2, 4-diphosphate is the major component (Müller et al. 1990).

Subsequently it was found that reactions of this kind, in particular the reaction of glycoaldehyde phosphate with glyceraldehyde-2-phosphate, proceed efficiently when 2 mM solutions of substrates are incubated at room temperature and pH 9.5 in the presence of layered hydroxides such as hydrotalcite (magnesium aluminum hydroxide) (Pitsch 1992; Pitsch et al. 1995a). The phosphates are absorbed between the positively charged layers of the mineral. The reaction proceeds under mild conditions, presumably because of the high concentration of substrates in the interlayer and because the positive charge on the metal hydroxide layers favors enolization of glycoaldehyde phosphate.

Much remains to be demonstrated; for example, the plausibility of glycoaldehyde phosphate as a prebiotic precursor of sugar synthesis and the possibility of conversion of ribose 2, 4-diphosphate to a phosphate from which nucleotides could be synthesized, for example, ribose 5-phosphate or ribose 1, 5-diphosphate. However, ribose synthesis, although still problematical, is no longer the intractable problem it seemed 10 years ago.

The synthesis of the nucleoside bases is one of the success stories of prebiotic chemistry. Adenine is formed with remarkable ease from ammonia and hydrogen cyanide (Oró 1961). This synthesis has been described as "the rock of the faith" by Stanley Miller. Reasonably plausible syntheses of the other purine bases and of the pyrimidines has also been described (Sanchez et al. 1967). The coupling of the purine bases with ribose or its phosphates has been achieved under mild conditions, but in relatively low yield (Fuller et al. 1972). The formation of pyrimidine nucleosides in reasonable yield from the bases and ribose has not been achieved, and is likely to prove difficult.

We conclude that the direct synthesis of the nucleosides or nucleotides from prebiotic precursors in reasonable yield and unaccompanied by larger amounts of related molecules could not be achieved by presently known chemical reactions. The only remotely plausible route to the molecular biologist's pool would involve a series of mineral-catalyzed reactions, coupled with a series of subtle fractionations of nucleotide-like materials based on charge, stereochemistry, etc.

Even minerals could not achieve on a macroscopic scale one desirable separation, the resolution of D-ribonucleotides from the L-enantiomers. This is a serious problem because experiments on template-directed synthesis using poly(C) and the imidazolides of G suggest that the polymerization of the D-enantiomer is often strongly inhibited by the L-enantiomer (Joyce et al. 1984). This difficulty may not be insuperable; perhaps with a different activator group inhibition would be less severe. However, enantiomeric cross-inhibition is certainly a serious problem.

Scientists interested in the origins of life seem to divide neatly into two classes. The first, usually but not always molecular biologists, believe that RNA must have been the first replicating molecule and that chemists are exaggerating the difficulties of nucleotide synthesis. They believe that a few more striking chemical "surprises" will establish that a reasonable approximation to a racemic version of the molecular biologist's pool could have formed on the primitive earth, and that further experiments with different activating groups and minerals will solve the enantiomeric cross-inhibition problem. The second group of scientists are much more pessimistic. They believe that the de novo appearance of oligonucleotides on the primitive earth would have been a near miracle. (The authors subscribe to this latter view.) Time will tell which is correct.

ALTERNATIVE GENETIC SYSTEMS

The problems that arise when we try to understand how an RNA World could have arisen de novo on the primitive earth are sufficiently severe that we must explore other possibilities. What kind of alternative genetic systems might have preceded the RNA World? How could they have "invented" the RNA World? These topics have generated a good deal of speculative interest and some relevant experimental data.

Eschenmoser and his colleagues have undertaken a systematic study of the properties of analogs of nucleic acids in which ribose is replaced by some other sugar or in which the furanose form of ribose is replaced by the pyranose form (Fig. 6b) (Eschenmoser 1997). Strikingly, polynucleotides based on the pyranosyl analog of ribose (p-RNA) form Watson-Crick paired double helices that are more stable than RNA, and p-RNAs are less likely than the corresponding RNAs to form multiple-strand competing structures (Pitsch et al. 1993, 1995b). Furthermore, the helices twist much more gradually than those in the standard nucleic acids, which should make it easier to separate strands during replication. Pyranosyl RNA appears to be an excellent choice as a genetic system; in some ways it seems an improvement compared to the standard nucleic acids.

Peptide nucleic acid (PNA) is another nucleic acid analog that has been studied extensively (Fig. 6c). It was discovered by Nielsen and colleagues in the context of antisense oligonucleotides (Egholm et al. 1992, 1993; Wittung et al. 1994). PNA is an uncharged, achiral analog of RNA or DNA in which the ribose-phosphate backbone of the nucleic acid is replaced by a backbone held together by amide bonds. PNA forms very

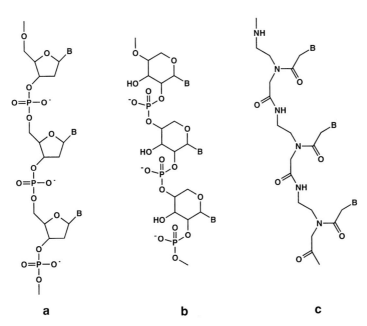

Figure 6 The structures of (*a*) DNA; (*b*) p-RNA; (*c*) PNA.

stable double helices with complementary RNA or DNA. Work in the Orgel laboratory has shown that information can be transferred from PNA to RNA or in the opposite direction in template-directed reactions, and that PNA/DNA chimeras are readily formed on either DNA or PNA templates (Schmidt et al. 1997a,b; Koppitz et al. 1998). Thus, it seems that a transition from a PNA world to an RNA world is possible.

The studies described above suggest that there are many ways of linking the nucleotide bases into chains that are capable of forming Watson-Crick double helices. It is not clear that it is easier to synthesize the monomers of PNA or p-RNA than to synthesize the standard nucleotides. However, it is possible that a Watson-Crick structure of this kind will be discovered that can be synthesized readily under prebiotic conditions. If so, it would be a strong candidate for the first genetic material.

The transition from an RNA-like world to the RNA World could take place in two ways. The transition might be continuous, if the pre-RNA template could direct the synthesis of an RNA product with a complementary sequence. Such a transition, for example, from PNA to RNA, would preserve information. RNA could then act as a genetic material in an initially PNA world. However, a transformation from p-RNA to RNA would not conserve function because single-stranded p-RNAs and RNAs would not adopt the same tertiary structure (Eschenmoser 1997). Functional RNA would have to develop ab initio in the new system.

The second type of transition can be described as a genetic takeover. A preexisting self-replicating system evolves, for its own selective advantage, the mechanism for synthesizing and polymerizing the components of a completely different genetic system, and is taken over by it. Cairns-Smith (1982) proposes that the first genetic system was inorganic, perhaps a clay, and that it "invented" a self-replicating system based on organic monomers. However, he clearly recognized the possibility of one organic genetic material replacing another (Cairns-Smith and Davies 1977). Genetic takeover does not require any structural relationship between the polymers of the two genetic systems. It suggests the possibility that the original genetic system may have been unrelated to nucleic acids.

The hypothesis of a genomic material completely different from a nucleic acid has one enormous advantage—it opens up the possibility of using very simple, easily synthesized prebiotic monomers in place of nucleotides. However, it also raises two new and difficult questions. Which prebiotic monomers are plausible candidates as the components of a replicating system? Why would an initial genetic system invent nucleic acids once it had evolved sufficient synthetic know-how to generate molecules as complex as nucleotides?

A number of prebiotic monomers that might have made up a simple genetic material have already been suggested. They include hydroxy acids (Weber 1987), amino acids (Orgel 1968; A. Rich and X. Zhang, pers. comm.), phosphomonoesters of polyhydric alcohols (Weber 1989), aminoaldehydes (Nelsestuen 1980), and molecules containing two sulfhydryl groups (Schwartz and Orgel 1985). The list could be expanded almost indefinitely. Here we discuss a small class of these monomers that we believe are particularly attractive in the light of recent work on enzyme mechanisms.

There is accumulating evidence that several enzymes that make or break phosphodiester bonds have two or three metal ions at their active sites (Cooperman et al. 1992). In the case of the editing site for phosphodiester hydrolysis in the Klenow fragment of *E. coli* DNA polymerase I, no other functional groups of the enzyme come close to the phosphodiester bond that is cleaved. This has led to the suggestion that the major role of the enzyme is to act as scaffolding on which to hang metal ions in precisely determined positions (Beese and Steitz 1991; Steitz 1998). A similar suggestion has been made for ribozymes on the basis of indirect evidence (Freemont et al. 1988; Yarus 1993; Steitz and Steitz 1993).

Perhaps these observations can be extended to suggest that, if informational polymers preceded RNA, they may also have been dependent on metal ions for their catalytic activity. If so, the range of prebiotic monomers that needs to be considered is greatly reduced. In addition to the functional groups that react to form the backbone, the monomers must have carried metal-binding functional groups. If the metal ions involved were divalent ions such as Mg^{++} and Ca^{++}, the side groups are likely to have been carboxylate or phosphate groups. If transition metal ions were involved, SH groups and possibly imidazole derivatives are likely to have been important.

Prebiotic monomers suitable for building polymers that bind Mg^{++} or Ca^{++} include aspartic acid, glutamic acid, and serine phosphate among biologically important amino acids. The β-amino acids, such as isoglutamic acid; hydroxydicarboxylic acids, such as α-hydroxysuccinic acid; and hydroxytricarboxylic acids, such as citric acid, are other possible candidates. A polymer containing D-aspartic acid, L-aspartic acid, and glycine as its subunits is typical of potentially informational copolymers that might, in the presence of divalent metal ions, both replicate and function as a catalyst. Transition-metal ions might play a corresponding role for polymers containing cysteine or homocysteine. The present challenge is to demonstrate replication, or at least information transfer in template-directed synthesis, in some such system.

What selective advantage could a simpler, metabolically competent system derive from the synthesis of oligonucleotides? This is a baffling question. Most arguments that come to mind do not stand up to detailed analysis. If, for example, one postulates that nucleotides were first synthesized as parts of cofactors such as DPN, one must explain why the particular heterocyclic bases and sugars were chosen. Even if we suppose that among the many "experiments" in secondary metabolism carried out by early organisms one happened by accident on a pair of complementary nucleotides that could form a replicating polymer, one must still explain how polymerization subsequently contributed to the success of the "inventor." The only glimmer of an explanation that we can see is that oligonucleotides, by hybridization, functioned as selective "glues" to tie together appropriate macromolecules.

The discussion so far, even though highly speculative, is still conservative in overall outlook. It supposes that the original information-accumulating system that led to the evolution of life on earth was either RNA or some linear copolymer that replicated in an aqueous environment in much the same way as RNA. There remains a lingering doubt that we are on the right track at all; maybe the original system was not an organic copolymer (Cairns-Smith 1982), or maybe it replicated in a nonaqueous environment and RNA is an adaptation that permitted invasion of the oceans. Perhaps systems of high complexity can develop without any need for a genome in the usual sense (Kauffman 1986; Wächtershäuser 1988; De Duve 1991).

Laboratory simulations of prebiotic chemistry are dependent on organic chemistry and can only explore the kinds of reactions understood by organic chemists. A good deal is known about reactions in aqueous solution, but less about reactions at the interface between water and inorganic solids. Very little is known about reactions in systems where inorganic solids are depositing from aqueous solutions containing organic material. It is hard to see how speculative schemes involving heterogeneous aqueous systems can be tested until much more is known about the underlying branches of chemistry.

CONCLUSIONS AND PERSPECTIVES

After dreaming of self-replicating ribozymes emerging from pools of random polynucleotides, and having had nightmares over the difficulties that must have been overcome for RNA replication to occur in a realistic prebiotic soup, we awaken to the cold light of day. The constraints that must have been met in order to originate a self-sustained evolving system are now reasonably well understood. One can sketch out a logical order of events, beginning with prebiotic chemistry and ending with DNA/protein-based life. However, it must be said that the details of this process remain obscure and are not likely to be known in the near future.

The presumed RNA World should be viewed as a milestone, a plateau in the early history of life on earth. So too, the concept of an RNA World has been a milestone in the scientific study of life's origins. Although this concept does not explain how life originated, it has helped to guide scientific thinking and has served to focus experimental efforts. Further progress will depend primarily on new experimental results, as chemists, biochemists, and molecular biologists work together to address problems concerning molecular replication, ribozyme enzymology, and RNA-based cellular processes.

ACKNOWLEDGMENTS

This work was supported by the NASA Specialized Center for Research and Training (NSCORT) in Exobiology.

REFERENCES

Acevedo O.L. and Orgel L.E. 1987. Non-enzymatic transcription of an oligodeoxynucleotide 14 residues long. *J. Mol. Biol.* **197:** 187–193.
Bartel D.P. and Szostak J.W. 1993. Isolation of new ribozymes from a large pool of random sequences. *Science* **261:** 1411–1418.
Been M.D. and Cech T.R. 1988. RNA as an RNA polymerase: Net elongation of an RNA primer catalyzed by the *Tetrahymena* ribozyme. *Science* **239:** 1412–1416.
Beese L.S. and Steitz T.A. 1991. Structural basis for the 3'-5' exonuclease activity of *Escherichia coli* DNA polymerase I: A two metal ion mechanism. *EMBO J.* **10:** 25–33.
Cairns-Smith A.G. 1982. *Genetic takeover and the mineral origins of life.* Cambridge University Press, Cambridge, United Kingdom.
Cairns-Smith A.G. and Davies C.J. 1977. The design of novel replicating polymers. In *Encyclopaedia of ignorance* (ed. R. Duncan and M. Weston-Smith), pp. 391–403. Pergamon Press, Oxford, United Kingdom.
Cech T.R. 1986. A model for the RNA-catalyzed replication of RNA. *Proc. Natl. Acad. Sci.* **83:** 4360–4363.
———. 1987. The chemistry of self-splicing RNA and RNA enzymes. *Science.* **236:** 1532–1539.
Cech T.R. and Herschlag D. 1996. Group I ribozymes: Substrate recognition, catalytic strategies, and comparative mechanistic analysis. *Nucleic Acids Mol. Biol.* **10:** 1–17.
Cooperman B.S., Baykov A.A., and Lahti R. 1992. Evolutionary conservation of the active site of soluble inorganic pyrophosphatase. *Trends Biochem. Sci.* **17:** 262–266.
Crick F.H.C. 1968. The origin of the genetic code. *J. Mol. Biol.* **38:** 367–379.
De Duve C. 1991. *Blueprint for a cell: The nature and origin of life.* Neil Patterson Publishers, Burlington, North Carolina.

Doudna J.A. and Szostak J.W. 1989. RNA-catalysed synthesis of complementary-strand RNA. *Nature* **339:** 519–522.

Egholm M., Buchardt O., Nielson P.E., and Berg R.H. 1992. Peptide nucleic acids (PNA). Oligonucleotide analogues with an achiral peptide backbone. *J. Am. Chem. Soc.* **114:** 1895–1897.

Egholm M., Buchardt O., Christensen L., Behrens C., Freier S.M., Driver D.A., Berg R.H., Kim S.K., Norden B., and Nielson P.E. 1993. PNA hybridizes to complementary oligonucleotides obeying the Watson-Crick hydrogen-bonding rules. *Nature* **365:** 566–568.

Eigen M. 1971. Selforganization of matter and the evolution of biological macromolecules. *Naturwissenschaften* **58:** 465–523.

Eigen M. and Schuster P. 1977. The hypercycle: A principle of natural self-organization. Part A: Emergence of the hypercycle. *Naturwissenschaften* **64:** 541–565.

Eigen M., McCaskill J., and Schuster P. 1988. Molecular quasi-species. *J. Phys. Chem.* **92:** 6881–6891.

Ekland E.H. and Bartel D.P. 1996. RNA-catalysed RNA polymerization using nucleoside triphosphates. *Nature* **382:** 373–376.

Ekland E.H., Szostak J.W., and Bartel D.P. 1995. Structurally complex and highly active RNA ligases derived from random RNA sequences. *Science* **269:** 364–370.

Erman J.E. and Hammes G.G. 1966. Relaxation spectra of ribonuclease. IV. The interaction of ribonuclease with cytidine 2':3'-cyclic phosphate. *J. Am. Chem. Soc.* **88:** 5607–5614.

Eschenmoser A. 1997. Towards a chemical aetiology of nucleic acid structure. *Origins Life* **27:** 535–553.

Ferris J.P. and Ertem G. 1993. Montmorillonite catalysis of RNA oligomer formation in aqueous solution. A model for the prebiotic formation of RNA. *J. Am. Chem. Soc.* **115:** 12270–12275.

Ferris J.P., Hill A.R., Liu R., and Orgel L.E. 1996. Synthesis of long prebiotic oligomers on mineral surfaces. *Nature* **381:** 59–61.

Freemont P.S., Friedman J.M., Beese L.S., Sanderson M.R., and Steitz T.A. 1988. Cocrystal structure of an editing complex of Klenow fragment with DNA. *Proc. Natl. Acad. Sci.* **85:** 8924–8928.

Fuller W.D., Sanchez R.A., and Orgel L.E. 1972. Studies in prebiotic synthesis. VI. Synthesis of purine nucleosides. *J. Mol. Biol.* **67:** 25–33.

Gilbert W. 1986. The RNA world. *Nature* **319:** 618.

Guerrier-Takada C., Gardiner K., Marsh T., Pace N., and Altman S. 1983. The RNA moiety of ribonuclease P is the catalytic subunit of the enzyme. *Cell* **35:** 849–857.

Handschuh G.J., Lohrmann R., and Orgel L.E. 1973. The effect of Mg^{++} and Ca^{++} on urea-catalyzed phosphorylation reactions. *J. Mol. Evol.* **2:** 251–262.

Inoue T. and Orgel L.E. 1981. Substituent control of the poly(C)-directed oligomerization of guanosine 5'-phosphoroimidazolide. *J. Am. Chem. Soc.* **103:** 7666–7667.

———. 1982. Oligomerization of (guanosine 5'-phosphor)-2-methylimidazolide on poly(C). *J. Mol. Biol.* **162:** 204–217.

———. 1983. A nonenzymatic RNA polymerase model. *Science* **219:** 859–862.

Inoue T., Joyce G.F., Grzeskowiak K., Orgel L.E., Brown J.M., and Reese C.B. 1984. Template-directed synthesis on the pentanucleotide CpCpGpCpC. *J. Mol. Biol.* **178:** 669–676.

Joyce G.F. 1987. Nonenzymatic template-directed synthesis of informational macromolecules. *Cold Spring Harbor Symp. Quant. Biol.* **52:** 41–51.

———. 1989. RNA evolution and the origins of life. *Nature* **338**: 217–224.
Joyce G.F. and Orgel L.E. 1986. Non-enzymic template-directed synthesis on RNA random copolymers: poly(C,G) templates. *J. Mol. Biol.* **188**: 433–441.
Joyce G.F., Visser G.M., van Boeckel C.A.A., van Boom J.H., Orgel L.E., and van Westrenen J. 1984. Chiral selection in poly(C)-directed synthesis of oligo(G). *Nature* **310**: 602–604.
Kauffman S.A. 1986. Autocatalytic sets of proteins. *J. Theor. Biol.* **119**: 1–24.
Kawamura K. and Ferris J.P. 1994. Kinetic and mechanistic analysis of dinucleotide and oligonucleotide formation from the 5'-phosphorimidazolide of adenosine on Na^+-montmorillonite. *J. Am. Chem. Soc.* **116**: 7564–7572.
Khorana H.G. 1961. *Some recent developments in the chemistry of phosphate esters of biological interest*, pp. 126–141. Wiley, New York.
Koppitz M., Nielsen P.E., and Orgel L.E. 1998. Formation of oligonucleotide-PNA-chimeras by template-directed ligation. *J. Am. Chem. Soc.* **120**: 4563–4569.
Kruger K., Grabowski P.J., Zaug A.J., Sands J., Gottschling D.E., and Cech T.R. 1982. Self-splicing RNA: Autoexcision and autocyclization of the ribosomal RNA intervening sequence of *Tetrahymena*. *Cell* **31**: 147–157.
Lewin R. 1986. RNA catalysis gives fresh perspective on the origin of life. *Science* **231**: 545–546.
Lohrmann R. 1975. Formation of nucleoside 5'-polyphosphates from nucleotides and trimetaphosphate. *J. Mol. Evol.* **6**: 237–252.
———. 1977. Formation of nucleoside 5'-phosphoramidates under potentially prebiological conditions. *J. Mol. Evol.* **10**: 137–154.
Lohrmann R. and Orgel L.E. 1973. Prebiotic activation processes. *Nature* **244**: 418–420.
McCaskill J.S. 1984a. A stochastic theory of molecular evolution. *Biol. Cybern.* **50**: 63–73.
———. 1984b. A localization threshold for macromolecular quasispecies from continuously distributed replication rates. *J. Chem. Phys.* **80**: 5194–5202.
Mizuno T. and Weiss A.H. 1974. Synthesis and utilization of formose sugars. *Adv. Carbohydr. Chem. Biochem.* **29**: 173–227.
Mohr S.C. and Thach R.E. 1969. Application of ribonuclease T_1 to the synthesis of oligoribonucleotides of defined base sequence. *J. Biol. Chem.* **244**: 6566–6576.
Müller D., Pitsch S., Kittaka A., Wagner E., Wintner C.E., and Eschenmoser A. 1990. Chemie von α-aminonitrilen. Aldomerisierung von Glykolaldehydphosphat zu *racemischen* hexose-2,4,6-triphosphaten und (in gegenwart von formaldehyd) *racemischen* pentose-2,4-diphosphaten: *rac.*-allose-2,4,6-triphosphat und *rac.*-ribose-2,4-diphosphat sind die reaktionshauptprodukte. *Helv. Chim. Acta* **73**: 1410–1468.
Narlikar G.J. and Herschlag D. 1997. Mechanistic aspects of enzymatic catalysis: Lessons from comparison of RNA and protein enzymes. *Annu. Rev. Biochem.* **66**: 19–59.
Nelsestuen G.L. 1980. Origin of life: Consideration of alternatives to proteins and nucleic acids. *J. Mol. Evol.* **15**: 59–72.
Noller H.F., Hoffarth V., and Zimniak L. 1992. Unusual resistance of peptidyl transferase to protein extraction procedures. *Science* **256**: 1416–1419.
Orgel L.E. 1968. Evolution of the genetic apparatus. *J. Mol. Biol.* **38**: 381–393.
———. 1989. Was RNA the first genetic polymer? In *Evolutionary tinkering in gene expression* (ed. M. Grunberg-Manago et al.), pp. 215–224. Plenum Press, London.
Oró J. 1961. Mechanism of synthesis of adenine from hydrogen cyanide under plausible primitive earth conditions. *Nature* **191**: 1193–1194.

Osterberg R., Orgel L.E., and Lohrmann R. 1973. Further studies of urea-catalyzed phosphorylation reactions. *J. Mol. Evol.* **2:** 231–234.

Pace N.R. and Marsh T.L. 1985. RNA catalysis and the origin of life. *Origins Life* **16:** 97–116.

Pitsch S. 1992. "Zur chemie von glykolaldehyd-phosphat: seine bildung aus oxirancarbonitril und seine aldomerisierumg zu den (racemischen) pentose-2,4-diphosphaten und hexose-2,4,6-triphosphaten." Ph.D. thesis, Eidgenössische Technische Hochschule, Zürich, Switzerland.

Pitsch S., Wendeborn S., Jaun B., and Eschenmoser A. 1993. Why pentose- and not hexose-nucleic acids? Pyranosyl-RNA ('p-RNA'). *Helv. Chim. Acta* **76:** 2161–2183.

Pitsch S., Eschenmoser A., Gedulin B., Hui S., and Arrhenius G. 1995a. Mineral induced formation of sugar phosphates. *Origins Life* **25:** 297–334.

Pitsch S., Krishnamurthy R., Bolli M., Wendeborn S., Holzner A., Minton M., Lesueur C., Schlönvogt I., Jaun B., and Eschenmoser A. 1995b. Pyranosyl-RNA ('p-RNA'): Basepairing selectivity and potential to replicate. *Helv. Chim. Acta* **78:** 1621–1635.

Rohatgi R., Bartel D.P., and Szostak J.W. 1996a. Kinetic and mechanistic analysis of nonenzymatic, template-directed oligoribonucleotide ligation. *J. Am. Chem. Soc.* **118:** 3332–3339.

———. 1996b. Nonenzymatic, template-directed ligation of oligoribonucleotides is highly regioselective for the formation of 3'-5' phosphodiester bonds. *J. Am. Chem. Soc.* **118:** 3340–3344.

Sanchez R.A., Ferris J.P., and Orgel L.E. 1967. Studies in prebiotic synthesis. II. Synthesis of purine precursors and amino acids from aqueous hydrogen cyanide. *J. Mol. Biol.* **30:** 223–253.

Sawai H., Kuroda K., and Hojo T. 1988. Efficient oligoadenylate synthesis catalyzed by uranyl ion complex in aqueous solution. In *Nucleic Acids Symp. Ser.* **19:** 5–7.

Schmidt J.G., Nielsen P.E., and Orgel L.E. 1997a. Information transfer from peptide nucleic acids RNA by template-directed syntheses. *Nucleic Acids Res.* **25:** 4797–4802.

Schmidt J.G., Christensen L., Nielsen P.E., and Orgel L.E. 1997b. Information transfer from DNA to peptide nucleic acids by template-directed syntheses. *Nucleic Acids Res.* **25:** 4792–4796.

Schuster P. and Swetina J. 1988. Stationary mutant distributions and evolutionary optimization. *Bull. Math. Biol.* **50:** 635–660.

Schwartz A.W. and Orgel L.E. 1985. Template-directed synthesis of novel, nucleic acid-like structures. *Science* **228:** 585–587.

Shapiro R. 1984. The improbability of prebiotic nucleic acid synthesis. *Origins Life* **14:** 565–570.

Sharp P.A. 1985. On the origin of RNA splicing and introns. *Cell* **42:** 397–400.

Sleeper H.L. and Orgel L.E. 1979. The catalysis of nucleotide polymerization by compounds of divalent lead. *J. Mol. Evol.* **12:** 357–364.

Steitz T.A. 1998. A mechanism for all polymerases. *Nature* **391:** 231–232.

Steitz T.A. and Steitz J.A. 1993. A general two-metal-ion mechanism for catalytic RNA. *Proc. Natl. Acad. Sci.* **90:** 6498–6502.

Sulston J., Lohrmann R., Orgel L.E., and Todd M.H. 1968. Nonenzymatic synthesis of oligoadenylates on a polyuridylic acid template. *Proc. Natl. Acad. Sci.* **59:** 726–733.

Wächtershäuser G. 1988. Before enzymes and templates: Theory of surface metabolism. *Microbiol. Rev.* **52:** 452–484.

Weber A.L. 1987. The triose model: Glyceraldehyde as a source of energy and monomers for prebiotic condensation reactions. *Origins Life* **17:** 107–119.

———. 1989. Model of early self-replication based on covalent complementarity for a copolymer of glycerate-3-phosphate and glycerol-3-phosphate. *Origins Life* **19:** 179–186.

Weiner A.M. and Maizels N. 1987. 3′ terminal tRNA-like structures tag genomic RNA molecules for replication: Implications for the origin of protein synthesis. *Proc. Natl. Acad. Sci.* **84:** 7383–7387.

Wittung P., Nielsen P.E., Buchardt O., Egholm M., and Norden B. 1994. DNA-like double helix formed by peptide nucleic acid. *Nature* **368:** 561–563.

Woese C. 1967. *The genetic code. The molecular basis for genetic expression.* pp. 179–195. Harper & Row, New York.

Wu T. and Orgel L.E. 1992a. Nonenzymatic template-directed synthesis on oligodeoxycytidylate sequences in hairpin oligonucleotides. *J. Am. Chem. Soc.* **114:** 317–322.

———. 1992b. Nonenzymatic template-directed synthesis on hairpin oligonucleotides. II. Templates containing cytidine and guanosine residues. *J. Am. Chem. Soc.* **114:** 5496–5501.

Yarus M. 1993. How many catalytic RNAs? Ions and the Cheshire cat conjecture. *FASEB J.* **7:** 31–39.

Zaug A.J. and Cech T.R. 1986. The intervening sequence RNA of *Tetrahymena* is an enzyme. *Science* **231:** 470–475.

3

The Genomic Tag Hypothesis: What Molecular Fossils Tell Us about the Evolution of tRNA

Nancy Maizels and Alan M. Weiner
Departments of Molecular Biophysics and Biochemistry, and of Genetics
Yale University School of Medicine
New Haven, Connecticut 06520-8024

INTRODUCTION

Holley's realization that tRNA could be folded into a two-dimensional cloverleaf posed more questions than it answered (Dudock et al. 1969). One of the most perplexing was whether the three-dimensional structure of tRNA would turn out to be an "integral fold" in which all parts were essential for the correct structure, or whether tRNA could be decomposed into smaller, structurally independent units. The crystal structure of tRNA immediately revealed that tRNA is composed of two perpendicular coaxial stacks (Quigley and Rich 1976): a stack of the acceptor stem on the dihydrouridine stem/loop (the "top half") and a stack of the TψC stem/loop on the anticodon stem/loop (the "bottom half") (see Fig. 1). Remarkably, the covalent connections between the middle of one helical stack and the middle of the other hardly distorted either helical stack: The top and bottom halves of tRNA appeared to be inserted into each other with surgical precision. A great deal of evidence has subsequently shown that the top and bottom halves of tRNA are indeed structurally and functionally independent units. This suggests that the two halves of tRNA could have evolved independently. Here we review the experimental evidence bearing on our hypothesis (Weiner and Maizels 1987) that the top half of tRNA evolved first as a 3' terminal "genomic tag" that marked single-stranded RNA genomes for replication in what Gilbert was first to call the "RNA World" (Gilbert 1986). The bottom half of tRNA would then have evolved separately as replication in the RNA World became more sophisticated, or as the advent of templated protein synthesis in the RNA World gave birth to the RNP World (Noller 1993; Schimmel et al. 1993).

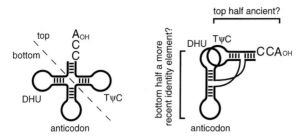

Figure 1 tRNA has two structural domains. The "top half" of tRNA (a coaxial stack of the acceptor stem on the TψC stem) is structurally and functionally independent of the "bottom half" (a coaxial stack of the dihydrouracil and anticodon stems). The top half may also be more ancient, and the bottom half may have evolved as an expansion loop which supplemented the function of the top half without disrupting it.

tRNA Plays a Surprising Number of Roles in Replication

First impressions are often lasting, and it is probably safe to say that most molecular biologists first encounter tRNA as a key component of the translation machinery. This is how tRNA is presented in elementary courses, and this is how the molecule is portrayed in textbooks. Yet because tRNA is commonly introduced as a component of the translation machinery, it is all too easy to think of translation as the *primary* or *proper* function of tRNA. In fact, as we discuss in detail, tRNA and tRNA-like molecules also play key roles in a wide variety of replicative processes including replication of single-stranded RNA viruses of bacteria, plants, and possibly mammals; replication of duplex DNA plasmids of fungal mitochondria; retroviral replication; and replication of modern chromosomal telomeres.

How did tRNA come to have so many different roles in replication? One possibility is that, for reasons which are not yet understood, tRNA or tRNA-like structures have been repeatedly and independently borrowed from translation to serve ad hoc roles in replication. Alternatively, tRNA or tRNA-like structures may be widespread in contemporary replication because tRNA played a central role in the replication of ancient RNA genomes, a role that has been conserved as well as subtly transformed as genomes evolved from RNA to duplex DNA.

We proposed that tRNA-like structural motifs first evolved as 3′-terminal structures that tagged RNA genomes for replication in the RNA World before the advent of protein synthesis (Weiner and Maizels 1987). This hypothesis provides a natural explanation for the ubiquity of tRNA-like motifs by suggesting that tRNA-like structures arose early and played an essential role in the earliest replicating systems. The central role of the early tRNA-like structures in replication makes it likely that the structural motif was conserved throughout subsequent evolution, and the antiquity of the motif would assure that it was used and reused in many different ways. This simple genomic tag hypothesis has surprising power to organize many apparently unrelated aspects of molecular biology into a coherent whole, and to suggest new relationships between areas of research that appear superficially to be unconnected. As genomic tags enter a second decade, no evidence has appeared that contradicts the hypothesis, and experimental support continues to mount.

Molecular Fossils, Coevolution, and Continuity

The genomic tag hypothesis explicitly argues that molecular evolution has been so conservative as to preserve a role for tRNA-like structures in replication through 3.5 billion years of genomic evolution. Given the sweep of evolution—from the first hopeful prebiotic smudge to the glories and follies of humankind—it may be tempting to believe that the transformations that occur during evolution obliterate the evidence of their own molecular origins, with the result that the most ancient aspects of cellular structure would be *least* likely to survive in recognizable form. However, the more we understand the full complexity of the biological machinery, the clearer it becomes that the components of a living cell interact in so many ways that a change in any key component requires compensatory changes in many others. Thus, most molecules cannot freely evolve to maximize the efficiency of a single function, but must instead *coevolve* with other physically and functionally interacting molecules. When a significant change in one molecule would entail an impossibly large number of simultaneous compensatory changes in others, then the necessity of coevolution can effectively freeze a molecule in time. It was clear even to Lucretius that "Natura non saltus fecit" (Nature never makes a leap). More recently, Orgel (1968) reformulated this old idea as the principle of continuity.

Skeletons, shells, feathers, leaves, and wood are preserved as fossils by death and mineralization. Molecular structures and functions are preserved because the complexity of the living process retards or prevents further significant evolution. White (1976) appears to have been the first to apply the term "fossil" to biochemical processes. We define a *molecular fossil* as any contemporary structure or function that is ancient in origin and provides us with clues about the history of life. We argue here that the ubiquity and conservation of tRNA-like structures strongly suggest that this motif defines a molecular fossil record of events dating back to the beginnings of life on earth.

The Explanatory Power of the Hypothesis

Molecular biologists are accustomed to experiments that produce instant yes or no answers. The genomic tag hypothesis does not suggest one particular "killer" experiment that could support or falsify it, but the hypothesis has nonetheless inspired a number of experiments and has proved relevant to many others in ways we could not have anticipated. Thus, in the years following the initial genomic tag hypothesis (Weiner and Maizels 1987), a remarkable number of unanticipated experimental results have emerged that directly support, were predicted by, or are consistent with, the hypothesis as originally stated. K.R. Popper (1963) coined the term "explanatory power" to describe the ability of a hypothesis to account for previously unrelated and apparently disparate observations. Not only does the genomic tag hypothesis appear to possess considerable explanatory power, but the hypothesis appears to be robust, comfortably making sense of new data rather than struggling to accommodate unwelcome experimental results. As discussed below, some of these new results include cleavage of modern genomic tags by RNase P (Green et al. 1988; Guerrier-Takada et al. 1988; Mans et al. 1990); identification of major tRNA identity elements within the acceptor stem (Musier-Forsyth and Schimmel 1992); division of tRNA synthetases into at least two unrelated classes (Eriani et al. 1990); a role for tRNA as *template* for reverse transcription of a retroplasmid genome (Akins et al. 1989; Saville and Collins 1990); and the existence of an internal tRNA-like template in telomerase (for review, see Blackburn 1991). A variety of results indicate that RNA can activate and polymerize amino acids: specific binding of an amino acid by RNA (for review, see Yarus et al. 1991); RNA catalysis of reactions at an aminoacylphosphoester center (Piccirilli et al. 1992; Illangesekare and Yarus 1995, 1997); and the ability of RNA to catalyze peptidyl transfer (Lohse and Szostak 1996) as rRNA itself is thought to do (Noller et al. 1992; Green and Noller 1997; Welch et al. 1997). The notion that the top half of tRNA evolved very early is supported by the fact that enzymes as different as RNase P (McClain et al. 1987), tRNA synthetases (Rould et al. 1989), ribosomal RNA (Noller et al. 1992), EF-Tu (Rasmussen et al. 1990), and the archaeal and eubacterial CCA-adding enzymes (Shi et al. 1998b) recognize primarily the top half of tRNA. We take the resilience and predictive power of the hypothesis as evidence that it has been fruitful and may be substantially correct.

THE GENOMIC TAG HYPOTHESIS

The Hypothesis in Outline

We proposed that ancient linear RNA genomes possessed 3′-terminal tRNA-like structures, which we called genomic tags (Weiner and Maizels 1987). Figure 2 shows the simplest form of a tRNA-like genomic tag, a stem and loop immediately followed by a 3′-terminal CCA. This resembles what is sometimes called the top half of tRNA or minihelix, i.e., a coaxial stack of the acceptor stem on the TψC arm (Fig. 1). Like the 3′-terminal tRNA-like motifs of contemporary bacterial and plant RNA viruses (Rao et al. 1989), and possibly animal picornaviruses (Pilipenko et al. 1992), the genomic tag would have served two main roles, providing an initiation site for replication and functioning as a simple telomere.

As an initiation site for replication, the tag would bind to the replicase, ensuring replication of genomic (as opposed to nongenomic and random) RNA molecules. In addition to conferring template specificity, the tag would also sequester subterminal RNA sequences in secondary structure, thereby forcing the replicase to initiate on the 3′-terminal CCA of the genomic RNA. Chemical considerations suggest that the 3′-terminal CC of the CCA sequence could in fact have been *selected* to facilitate efficient and faithful replication. Initiation with guanosine on the penultimate base

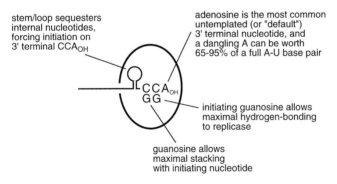

Figure 2 The simplest genomic tag. The tag functions as a simple telomere by sequestering subterminal RNA sequences in secondary structure, thus forcing the replicase to initiate on the 3′-terminal CCA of the genomic RNA. As shown in this figure and described in the text, the 3′-terminal CCA sequence may have been selected to facilitate efficient and faithful initiation of replication with the sequence 5′ GG. The duplex stem of the simplest tag corresponds to the top half of modern tRNA (Fig. 1). Stacking of the 3′-terminal A on the penultimate base pair can account for as much as 65–95% of the stability of a full base pair, and may account for the ability of many polymerases to add an untemplated A (see text).

of the template might be favored because G:C pairs are stronger than A:U pairs; because G has a greater potential than the other bases for hydrogen bonding with the polymerase; and because strong stacking of G on G in the 5' GG dinucleotide might help to compensate for the absence of a primer (Fig. 2).

As a telomere, the tag would also provide a site for *untemplated* nucleotide addition, ensuring that critical terminal regions of the genome were not lost during replication. For example, if replication initiated on the penultimate G (as Qβ replicase does today; Blumenthal and Carmichael 1979), the 3'-terminal A would be the minimal telomeric sequence, and untemplated addition of this A residue would be required to regenerate the genome before each new round of replication. Remarkably, addition of an untemplated 3'-terminal nucleotide (typically A but occasionally C) is a common activity of both RNA and DNA polymerases made of protein, and thus could also have been a property of an RNA replicase made of RNA. Qβ replicase possesses this terminal transferase activity (Blumenthal and Carmichael 1979); both the bacteriophage T7 (Milligan et al. 1987; Gardner et al. 1997) and SP6 RNA polymerases will add a single untemplated 3' nucleotide (Melton et al. 1984); and the ability of Taq I DNA polymerase to add an untemplated A to PCR products is notorious (see, e.g., Tse and Forget 1990). In fact, addition of an untemplated (default) 3'-terminal nucleotide may represent an attempt by polymerase to continue polymerization despite the absence of a template nucleotide. Although the incoming nucleotide cannot base-pair with the template strand, the ribose triphosphate moiety can still interact with the enzyme, and the base can still stack on the previous base pair. Stacking of such a "dangling" or untemplated A on the terminal base pair of an RNA duplex can account for as much as 65–95% of the stability contributed by a bona fide base pair (Freier et al. 1985; SantaLucia et al. 1990; Sugimoto et al. 1990; Limmer et al. 1993). Since binding of the ribose triphosphate and stacking of the base on the previous base pair must be universal properties of any replicative polymerase, addition of an untemplated 3'-terminal A (or C) might be expected to occur even if the polymerase were made of RNA.

Finally, although the CC of CCA might be explained by advantageous stacking and hydrogen-bonding interactions during the initiation of replication, and the 3'-terminal A of CCA by default addition of an untemplated nucleotide upon completion of replication, accumulating evidence indicates that the 3'-terminal NCCA sequence of tRNA (where N is the unpaired "discriminator" base) can assume a distinct structure when stacked upon the last base pair of the acceptor stem (Limmer et al. 1993; Puglisi et al. 1994; Shi et al. 1998a). Indeed, as we discuss below, the *Neurospora* retroplasmid reverse transcriptase mainly recognizes this CCA sequence (Chen and Lambowitz 1997; also see Fig. 4). Thus, the entire NCCA sequence may have been subject to an additional selection for the ability to form a structure immediately adjacent to a 3'-terminal stem/loop that could be easily recognized by the RNA replicase.

Some Implications of the Genomic Tag Hypothesis

The genomic tag hypothesis explains and relates many disparate roles of tRNA motifs in cellular metabolism. In this section, we review some aspects of the hypothesis considered in detail previously (Weiner and Maizels 1987) and discuss data from experiments inspired by or relevant to these facets of the hypothesis. In later sections, we develop new aspects of the hypothesis and discuss relevant new data.

RNase P

RNase P is a ribonucleoprotein enzyme that functions as an endoribonuclease to remove the 5' leader from tRNA precursors. This processing reaction occurs in all contemporary cells and organelles. In some sense this is a rather surprising reaction, since there is no obvious reason that the 5' end of tRNA could not be generated directly by transcription, as is known to be the case for *Xenopus laevis* selenocysteine tRNA (Lee et al. 1989).

We have suggested that the contemporary RNase P activity derives from an ancient activity which arose to convert genomic RNA molecules into functional subgenomic RNAs by removing the 3'-terminal tRNA-like tag, thereby enhancing the structural and enzymatic versatility of functional RNA molecules. Then, in a reversal of fortunes, this enzyme, which had once freed functional RNA from a tRNA-like 3' tag, survived as an enzyme for removing the nonfunctional 5' leader from tRNA precursors. Our suggestion that RNase P might have evolved to recognize 3'-terminal tRNA-like genomic tags led directly to experiments that asked, Is the 3' tRNA-like structure on contemporary plant viruses sufficiently conserved that it can be recognized by *Escherichia coli* RNase P? The answer, remarkably, was yes: *E. coli* RNase P can process the 3'-terminal tRNA-like pseudoknot of turnip yellow mosaic virus (TYMV) at a site corresponding to the 5' leader of a tRNA precursor (Green et al. 1988; Guerrier-Takada et al. 1988; Mans et al. 1990). The fact that a contemporary eubacterial processing enzyme recognizes the 3' structures of plant viruses suggests remarkably conservative coevolution of both the activity itself and the structure of its substrate.

Since the RNA component of RNase P is by itself enzymatically active under appropriate conditions (Guerrier-Takada et al. 1983), it is plausible that early forms of RNase P could have been composed entirely of RNA. Indeed, Alberts (1986) has argued that proteins are such efficient and versatile catalysts that the very presence of RNA in an enzyme may be taken as prima facie evidence that the enzyme is ancient, and might predate the advent of templated protein synthesis. Thus, both the function and composition of RNase P are consistent with preservation of this activity ever since a very early era in molecular evolution. The genomic tag hypothesis plausibly explains why an activity critical for tRNA processing would have arisen so early.

Was the CCA-adding Activity the First Telomerase?

While endonucleolytic cleavage by RNase P generates the mature 5′ end of tRNA, exonucleolytic processing resects the 3′ end of tRNA precursors (Li and Deutscher 1996), and the CCA-adding enzyme [ATP(CTP):tRNA nucleotidyltransferase] then reconstructs the CCA terminus (Sprinzl and Cramer 1979; Deutscher 1982). Once again, this is a rather surprising reaction because the CCA end is destroyed only to be rebuilt. The purpose of this activity becomes clear, however, when the CCA-adding activity is viewed as a molecular fossil. If the 3′ end of a tRNA-like structure served as the initiation site for an early replicase, a CCA-adding activity would have been necessary to restore nucleotides lost as a result of incorrect initiation. The CCA-adding activity therefore would have functioned as the first telomerase, and the modern CCA-adding activity may have evolved from an RNA enzyme into a protein enzyme by stepwise replacement of RNA by protein (White 1976, 1982; Visser and Kellogg 1978a,b). Indeed, if the CCA-adding activity is ancient and essential, it should be present in all three living kingdoms today. Consistent with this hypothesis, we found that archaea, like eubacteria and eukaryotes, also contain a CCA-adding enzyme (Yue et al. 1996). The archaeal, eukaryotic, and eubacterial CCA-adding enzymes all belong to the same nucleotidyltransferase superfamily (Holm and Sander 1995; Martin and Keller 1996; Yue et al. 1996) and are functionally very similar (Shi et al. 1998a), but the archaeal CCA-adding enzyme diverges completely from the closely related eukaryotic and eubacterial enzymes outside the nucleotidyltransferase active-site signature sequence (Yue et al. 1996). One explanation for sequence divergence at the protein level would be descent from a common ancestral ribozyme in the RNA World (Weiner and Maizels 1987; Benner et al. 1989). As discussed below, use of a telomeric C_mA_n motif by early RNA genomes may also explain why modern DNA telomeres employ very similar sequence motifs.

tRNA Synthetase Function and the Origin of Protein Synthesis

We have argued that the series of reactions required for aminoacylation of tRNA chemically resembles RNA polymerization (Fig. 3) (Weiner and Maizels 1987). This prompted us to suggest that 3′-terminal tRNA-like structures of ancient RNA genomes may have been aminoacylated in a reaction resembling the charging of contemporary tRNAs by aminoacyl tRNA synthetases. In making this suggestion, we explicitly postulated that the enzyme responsible for aminoacylation was an RNA, and that this RNA enzyme could bind both an amino acid and a mononucleotide and could catalyze reactions at an aminoacylphosphoester center. We also suggested that RNA, as a structured polyanion, would most likely bind basic amino acids, and that charging of a genomic tag with a basic amino acid would have the greatest effect on the structure or function of the RNA. The ability of group I introns to bind the mononucleotide guanosine was known at the time (for review, see Michel et al. 1989). It has since been shown that group I introns will bind L-arginine specifically (Hicke et al. 1989); that the group I guanosine-binding site can be rationally redesigned to bind other nucleotides (Michel et al. 1989); that a group I intron can, without significant redesign, function as a credible aminoacylphosphoesterase (Piccirilli et al. 1992); and that RNAs can be selected that resemble the suspected peptidyl transferase site on the large rRNA (Welch et al. 1997), aminoacylate the 3′ hydroxyl of RNA

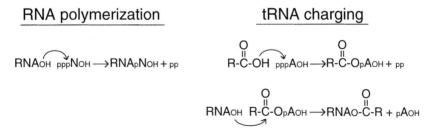

Figure 3 tRNA charging resembles RNA polymerization. Attack on the phosphoester bond proceeds by an in-line, S_N2 mechanism, resulting in a trigonal bipyramidal transition state, whereas attack on the aminoacylphosphoester bond occurs orthogonal to the π face, resulting in a tetrahedral intermediate. Despite these differences in stereochemistry, a ribozyme that catalyzes phosphoester bond transfer can also catalyze reactions at the carbon center, albeit at reduced efficiency (Piccirilli et al. 1992).

(Illangesekare et al. 1995, 1997), and possess peptidyl transferase activity (Lohse and Szostak 1996).

The genomic tag hypothesis suggests that aminoacylation initially conferred a *replication advantage* on molecules carrying a genomic tag. This could have occurred in any of several ways. Aminoacylation might have facilitated binding of the replicase to the 3' end of the genome, perhaps simply by countering the net negative charge of the RNA replicase with a positively charged (basic) amino acid. Aminoacylation could also have served as a regulatory mechanism for withdrawing a genomic RNA from the replicative pool by blocking binding of the tag to the replicase. A third possibility, suggested by Wong (1991), is that aminoacylation might be seen as a form of RNA modification (like methylation, thiolation, isopentenylylation) that would broaden the structural or catalytic range of the RNA bearing it.

Our model for the origin of protein synthesis is unique in postulating that key components of the translation apparatus—tRNA and tRNA aminoacylation activity—first evolved as essential components of the *replication* machinery before the advent of protein synthesis. With these two key components of the translation apparatus in place, the scene was set for the interdependent coevolution of replication and templated protein synthesis. Perhaps at first, random condensation of aminoacylated tRNAs generated short polycations. These simple polymers could have facilitated RNA-catalyzed reactions in a manner analogous to modern polyamines (Jay and Gilbert 1987; Maizels and Weiner 1987; Weiner and Maizels 1987) or they could have stabilized or promoted a particular RNA structure, much as the tract of basic amino acids within the HIV Tat protein shapes the structure of the TAR element (Puglisi et al. 1992). The ability of polyamines to regulate translational frameshifting on the ornithine decarboxylase antizyme mRNA also provides additional, albeit less direct, evidence that short polycations might have been useful in an RNA World (Matsufuji et al. 1995). Without several different species of charged tRNA, however, no mRNA would be necessary or (for that matter) useful. We therefore speculated that the original protoribosome evolved to facilitate synthesis of specific peptides by aligning charged tRNAs *before* the advent of mRNA, and that templating of peptide synthesis by mRNA was one of the *last* steps in the evolution of the modern translation apparatus (Maizels and Weiner 1987; also see Schimmel and Henderson 1994). This scenario is consistent with recent evidence that contemporary tRNAs comprise two separate structural domains (see below). We also emphasize that borrowing tRNA from replication for protein synthesis might constrain, but would not preclude, further evolution of the role of tRNA in replication.

The Unexpected Diversity of tRNA Synthetases

A further prediction of the genomic tag model for the origin of protein synthesis is that at least some tRNA synthetases were originally RNA enzymes (Weiner and Maizels 1987). We originally suggested that this might account for the remarkable diversity in structure of modern synthetases already apparent at the time (Schimmel 1987) because there is unlikely to be a unique pathway by which an RNA enzyme would evolve into a protein enzyme by gradual replacement of RNA structures with protein structures (White 1976, 1982; Visser and Kellogg 1978a,b). Thus, we explicitly proposed that the existence of apparently unrelated protein enzymes carrying out the same essential function could be taken as strong evidence for an ancestral RNA enzyme. This notion was further developed by Benner et al. (1989).

Despite the apparent diversity of tRNA synthetase subunit structure and sequences (Schimmel 1987), the suggestion that there might be more than one class of synthetase was not well received, perhaps because it was so difficult to imagine that 20 tRNA synthetases doing the same enzymatic job could have descended from more than one ancestral form. However, Eriani et al. (1990) subsequently found that tRNA synthetases could be partitioned into at least two classes based on mutually exclusive sets of sequence motifs. Class I has the Rossmann fold for nucleotide binding, class II an antiparallel β sheet (Cusack et al. 1990); class I acylates the 2' hydroxyl, class II the 3' hydroxyl; yet the two classes employ virtually identical reaction mechanisms involving an enzyme-bound aminoacyl-adenylate intermediate derived from ATP. From these structural and functional data one must conclude either that (1) tRNA synthetases made of protein evolved twice but independently adopted the same reaction mechanism; or, as we proposed originally, (2) the first tRNA synthetases were made of RNA and evolved stepwise by distinct pathways through RNP intermediates into two ancestral tRNA synthetases made of protein. In the second scenario, evolution of an RNA enzyme into a protein enzyme would have been constrained to preserve the two-step charging mechanism that first evolved in the RNA World (White 1976, 1982), but not charging of the 2' or 3' hydroxyl, as these two aminoacylated products equilibrate quickly.

tRNA as Genetic Punctuation in RNA Processing and Translation

If tRNA or tRNA-like structures functioned as 3'-terminal genomic tags in an RNA World, the principle of continuity suggests that tRNA would continue to play a role in genomic organization as RNA genomes evolved into transitional genomes made of both RNA and DNA, and eventually into DNA genomes. In fact, tRNA coding regions are well known to serve as intergenic "punctuation" in eubacterial rRNA genes and mitochondrial genomes: RNase P then cleaves the primary transcripts at the mature 5' end of each tRNA. This form of genomic organization could be either ancestral (reflecting an ancient origin) or derived (more recently acquired). tRNA punctuation in eubacterial rRNA genes would be ancestral if the 16S, 5S, and 23S components of the rRNA arose as separate RNA genomes each bearing a 3'-terminal genomic tag that was removed by RNase P. Genetic linkage of these Ur-rRNA subunits end-to-end either as one large RNA, or ultimately as DNA, might then preserve the original RNA processing strategy. On the other hand, tRNA punctuation in eubacterial organellar genomes may be more likely to represent "devolution" in which the complete genome of a free living endosymbiont was gradually reduced to rudimentary status, and RNA processing came to rely on a handful of essential enzymes.

tRNA can also serve as punctuation in translation. 10Sa RNA is a small, stable, highly conserved RNA found in diverse eubacteria and some eubacterial endosymbionts (Subbarao and Apirion 1989; Brown et al. 1990; Tyagi and Kinger 1992; Williams and Bartel 1996). The mature 10Sa molecule is generated by RNA processing from a larger transcript, and the 3' end of 10Sa RNA almost perfectly matches the conserved residues in the top half of tRNA including the CCA (Tyagi and Kinger 1992). This originally suggested the intriguing possibility that 10Sa RNA might be the genomic or subgenomic transcript of a new retroelement. However, 10Sa RNA was subsequently found to be a "tmRNA"—a single molecule combining both tRNA and mRNA functions (Komine et al. 1994; Tu et al. 1995; Keiler et al. 1996; Williams and Bartel 1996; Felden et al. 1997; Himeno et al. 1997). tmRNA is charged with alanine by alanine tRNA synthetase, enters the A site of a ribosome stalled at the 3' end of a broken mRNA, and adds a single untemplated alanine residue; the tmRNA then undergoes a conformational change that enables it to function as a bona fide mRNA encoding a 10-residue carboxy-terminal peptide tag that targets the potentially harmful amino-terminal protein fragment for rapid degradation. The 5' and 3' ends of 10Sa RNA come together to form the tRNA^{Ala}-like structure, with the decapeptide coding region in between (Williams and Bartel 1996; Felden et al. 1997). The remarkable ability of this tRNA-like structure to enter the ribosomal decoding site in the absence of the corresponding anticodon suggests that pairing between a 3' genomic tag and the 5' end of the same RNA could have served in the RNA World as a signal for the initiation of translation, and may have been the predecessor of modern initiator tRNA.

The Two Structural Domains of Contemporary tRNA May Have Evolved Independently

Contemporary tRNAs are composed of two structural domains, a top half consisting of a coaxial stack of the acceptor stem on the TψC arm, and a bottom half consisting of a coaxial stack of the dihydrouracil arm on the anticodon arm (Fig. 1). The genomic tag hypothesis suggested the intriguing possibility that these two structural domains evolved independently. When, how, or why the anticodon domain evolved is currently one—if not *the*—unsolved mystery of molecular evolution.

At present, all available experimental evidence is consistent not only with independent evolution of the top and bottom half domains, but also with the top half of the molecule arising early and the bottom half arising later. The fact that enzymes as different as RNase P (McClain et al. 1987), tRNA synthetases (Rould et al. 1989), ribosomal RNA (Noller et al. 1992), EF-Tu (Rasmussen et al. 1990), and the archaeal and eubacterial CCA-adding enzymes (Shi et al. 1998b) recognize primarily the top half of tRNA suggests that this is, indeed, the more ancient half of the molecule containing the most essential identity elements. Presuming that RNase P is ancient, the ability of RNase P to recognize the top half of tRNA alone (McClain et al. 1987) can be interpreted as evidence that the original genomic tags may have been as simple as the top half of the molecule (Fig. 2). Charging of such a simple genomic tag with different amino acids might have arisen initially to facilitate differential replication, modification, or processing of genomic RNAs; alternatively, differential charging could have arisen to facilitate synthesis of new (or more precise) peptides. In either case, there would have been selective pressure to distinguish different tags from each other, and this is consistent with functional (Francklyn et al. 1992) and structural studies (Rould et al. 1989) demonstrating that critical tRNA identity elements can lie within the top half of the molecule. In particular, minihelices corresponding to the top half of tRNA are sufficient in some cases for highly specific charging (Francklyn et al. 1992; Musier-Forsyth and Schimmel 1992), and a fragment of the top half is a substrate for the protein-free peptidyl transferase activity (Noller et al. 1992). Further evidence for the potential complexity of tRNA identity elements within the top half of the molecule comes from crystallographic work showing that critical identity elements are revealed by partial melting of the acceptor stem when *E. coli* glutamine tRNA binds to the cognate synthetase (Rould et al. 1989).

The bottom half of tRNA appears to be a more recent addition, used both by synthetases and by mRNAs to distinguish one species of tRNA from another. In this view, the bottom half of tRNA can be thought of as an expansion "loop," similar to those found in ribosomal RNA (see chapters 8 and 15). The notion that the bottom half of the tRNA molecule evolved after the top half is also consistent with our suggestion that templated protein synthesis, requiring tRNA–mRNA recognition mediated by the anticodon loop, could not have been selected until relatively late in the evolution of the translation apparatus when there were already several different species of charged tRNA to read the mRNA (Maizels and Weiner 1987).

TRACING GENOMIC TAGS FROM RNA PHAGE TO MODERN CHROMOSOMAL TELOMERES: THE NOTION OF TRANSITIONAL GENOMES

Although we suspected that tRNA-like genomic tags survived from an RNA World into a DNA World and were, in the process, transformed into the tRNA primers of retrovirus reverse transcription and the terminal C_mA_n motifs of modern chromosomal telomeres (Weiner and Maizels 1987), we were unable to make these connections explicit because there appeared to be one or more missing links in the molecular fossil record. Stunning and completely unanticipated results from the groups of Lambowitz (Kuiper and Lambowitz 1988; Akins et al. 1989; Chen and Lambowitz 1997), Saville and Collins (1990), Blackburn (for review, see Blackburn 1991), and Cech (Lingner et al. 1997) subsequently provided key molecular fossil evidence for genomes and genomic tags in transition from single-stranded RNA to double-stranded DNA. In this section we discuss the *Neurospora* retroplasmid and the *Tetrahymena* telomerase, two critical links that were missing from the molecular fossil record as it was known when we first formulated the genomic tag hypothesis (Weiner and Maizels 1987).

Neurospora Mitochondrial Retroplasmids: Evidence That Genomic Tags Survived the Transition from RNA to DNA Genomes

Mitochondria of some strains of *Neurospora crassa* contain double-stranded DNA plasmids that were originally detected because their replication causes a respiratory-deficient phenotype by competing with replication of mitochondrial DNA. Most surprisingly, replication of these plasmids requires production of a full-length RNA transcript of genomic DNA (Figs. 4 and 5) (Kuiper and Lambowitz 1988; Akins et al. 1989), and this RNA carries a 3´-terminal tRNA-like structure ending in CCACCA, implying that it functions as a tRNA-like genomic tag (Akins et al. 1989). An open reading frame within the genomic RNA encodes a

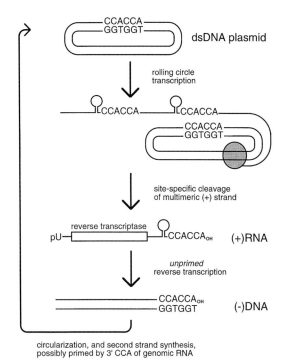

Figure 4 The *Neurospora* mitochondrial Varkud plasmid is a simple retroviral-like element. Note that first-strand cDNA synthesis uses the tRNA-like structure as template, whereas second-strand synthesis and/or circularization may reuse the tRNA-like structure as primer.

functional reverse transcriptase, and mitochondrial extracts of plasmid-containing strains can produce a full-length cDNA copy of the genomic RNA by de novo initiation of cDNA synthesis using the terminal A of the CCACCA as template (Chen and Lambowitz 1997). The 3'-terminal genomic tag therefore serves as the initiation site for replication as in Qβ, but Qβ replicase is replaced by the plasmid reverse transcriptase (Fig. 5). Whether the full-length RNA/DNA hybrid produced after the first step in plasmid replication circularizes directly, or is first converted into double-stranded DNA, is not yet known; the ability of the *Neurospora* reverse transcriptase to use both specific and nonspecific DNA primers (Chen and Lambowitz 1997) suggests that the enzyme itself might be responsible for circularizing the linear replicative intermediate. Once a circular, duplex DNA replicative intermediate has formed, rolling circle *transcription* yields multimeric plus strands, which are then cleaved into monomeric genomic RNAs. At least in the case of the Varkud-associated VSDNA

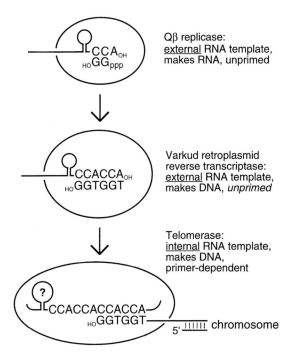

Figure 5 From Qβ replicase to telomerase. Qβ replicase, the *Neurospora* retroplasmid reverse transcriptase, and *Tetrahymena* telomerase all initiate on a tRNA-like template, suggesting that the ancestral forms of these enzymes represent a single line of evolutionary descent.

retroplasmid, a ribozyme contained within the RNA genome itself appears to perform this cleavage reaction (Saville and Collins 1990).

The fungal mitochondrial retroplasmids can be thought of either as linear RNA genomes that replicate through a circular DNA intermediate, or circular DNA genomes that replicate through a linear RNA intermediate. We therefore consider these retroplasmids to be examples of *transitional genomes*, contemporary representatives of an era in which genomic tags were being transformed from templates for RNA synthesis to templates for DNA synthesis.

Telomerase: A Genomic Tag Carried by the RNA Component of a Reverse Transcriptase

When we first suggested that the 3′-terminal CCA motif of tRNA-like genomic tags was related to the nearly universal $C_m A_n$ motif of eukaryotic nuclear telomeres (Weiner and Maizels 1987), we were unable to discern any hints in the molecular fossil record regarding the mechanism or chain of events by which a 3'-terminal CCA motif in RNA could be transformed into a 5'-terminal motif in DNA. Another missing link in the molecular fossil record emerged from detailed characterization of the *Tetrahymena* telomerase. This enzyme, first described by Greider and Blackburn (1985, 1989), possesses an unusual terminal nucleotidyltransferase activity that adds the species-specific T_nG_m repeats, one nucleotide at a time, to an appropriate T_nG_m primer (for review, see Blackburn 1991). Remarkably, telomerase is a ribonucleoprotein, and the RNA component is in fact an *internal* template for synthesis of the species-specific T_nG_m repeat. (For example, the telomeric repeat in the ciliate *Tetrahymena* is T_2G_4, and the internal template sequence in the *Tetrahymena* telomerase is 5'-CAACC-CCAA-3'.) One of the protein components of telomerase, known in *Saccharomyces cerevisiae* as *EST2* (*e*ver *s*horter *t*elomeres), is homologous to retroviral reverse transcriptases (Lingner et al. 1997). Telomerase can thus be seen as a specialized reverse transcriptase with an internal tRNA-like template (for review, see Blackburn 1991 and this volume). The secondary and tertiary structures of telomerase RNA have been studied in detail (Romero and Blackburn 1991; Bhattacharyya and Blackburn 1994; Lingner et al. 1994), and all known telomerase RNAs share a repetitive CCA-like internal template sequence; however, the rest of telomerase RNA is not well conserved across large evolutionary distances.

The Unusual Properties of G and G-rich DNAs

The unusual properties of G and of G-rich DNA suggest that chemical determinism may explain both the C_nA_m motif in RNA telomeres and use of the T_mG_n motif in modern DNA telomeres (Weiner and Maizels 1987). Telomeric T_nG_m repeats and other G-rich DNAs can spontaneously form unusual four-stranded structures in vitro that contain *intra*molecular non-Watson-Crick base pairs (Henderson et al. 1987). These structures, dubbed G quartets, may play a role in telomere aggregation or function in vivo (Sen and Gilbert 1988; Williamson et al. 1989; Sundquist and Klug 1989; Fang and Cech 1993a,b; for review, see Blackburn 1991; Zakian 1996). As detailed in Figure 2, the sequence CCA might have been selected in an RNA World as an efficient initiation site because the initiating guanosines stack more strongly and have more potential hydrogen-bonding interactions than any other bases. In effect, the CC of CCA would have been selected for the unusual stacking and hydrogen-bonding properties of its G-rich complement. Viewed in this way, the complementary

T_mG_n motif of modern telomeres was selected in the RNA World, but more fully exploited in the DNA World.

Functional Similarities between Telomerase and the Retroplasmid Reverse Transcriptase

The *Neurospora* retroplasmid reverse transcriptase copies genomic RNA into a cDNA by de novo initiation on the terminal nucleotide of the 3'-terminal CCACCA template sequence (Chen and Lambowitz 1997; see also Fig. 4). Telomerase uses telomeric DNA as primer to copy a CCACCA-like template within the enzyme itself into a cDNA. This provides a striking link between telomerase and the retroplasmid reverse transcriptase (Fig. 5). Both enzymes copy RNA into DNA and, like Qβ replicase, both initiate on the CCA motif of a tRNA-like template.

Might there be a single line of evolutionary descent for these enzymes and the genomes they replicate? Figure 5 shows how a tRNA recognition domain that is critical for initiation may be conserved among these enzymes, and how the polymerase domain would interact with this recognition complex. Figure 5 also suggests a simple pathway by which an enzyme like the retroplasmid reverse transcriptase could evolve into a specialized reverse transcriptase like telomerase. The retroplasmid enzyme would capture an internal template by stably binding a tRNA or tRNA-like structure; any primer terminus that could pair with this internal template would then allow the reverse transcriptase to enter directly into elongation mode. Indeed, as mentioned above, the *Neurospora* reverse transcriptase can bind the 3'-terminal CCA (inboard sequences help but are not essential) and can use specific as well as nonspecific primers (Chen and Lambowitz 1997). Thus, telomerase RNA appears to be derived from a captured 3'-terminal genomic tag, and telomerase action can therefore be regarded as abortive replication of a genome that has been reduced to a bare 3'-terminal C_nA_m motif. This may be mechanistically similar to abortive initiation by many RNA polymerases—the repeated synthesis of short initiating oligonucleotides before the polymerase enters elongation mode (McClure 1980). The ability of the *Neurospora* reverse transcriptase to initiate de novo, as well as to use specific and nonspecific primers (Chen and Lambowitz 1997), suggests that primer dependence would not be a major constraint on the evolutionary scenario (Fig. 5). Although retroviral reverse transcriptases require a primer and Qβ replicase does not, the initiating nucleotide can plausibly be regarded as a very short primer because it must occupy the same site on the enzyme as the primer terminus during elongation. Indeed, early work on *E. coli* RNA polymerase showed that a primer as short as a dinucleotide dramatically increases the efficiency and specificity of initiation (Downey et al. 1971; Maizels 1973), and more recent structural and mechanistic studies on DNA polymerase suggest that the initiating nucleotide occupies the same site on the enzyme as the primer terminus during elongation (Polesky et al. 1992).

Nonviral Retrotransposons: Parents or Children of Telomerase?

Eukaryotes from yeasts to humans have telomeric T_mG_n repeats templated by telomerase RNA, but diptera are a surprising and curious exception in which telomerase activity has apparently been lost (Levis et al. 1993; Biessmann and Mason 1997; Pardue et al. 1997). Instead, the chromosomes in *Drosophila melanogaster* are capped by either of two different kinds of nonviral retrotransposons, HeT-A (*het*erochromatin-*a*ssociated) and TART (*t*elomere-*a*ssociated *r*etro*t*ransposon). TART encodes a reverse transcriptase, but HeT-A does not; both retroelements are also found in pericentromeric heterochromatin. Complete Het-A and TART elements are transcribed to yield full-length polyadenylated genomic RNAs, and these in turn retropose to the 3' end of the chromosomes. The 3' end of chromosomal DNA may serve repeatedly as primer for reverse transcription of these polyadenylated genomic RNAs, thus generating tandem telomeric arrays of HeT-A and TART elements. Occasional retroposition of these 6-kb elements continually replenishes terminal chromosome sequences, balancing loss of about 50–100 bp of telomeric sequence per fly generation (Levis 1989). Thus, diptera have devised a mechanism for telomere maintenance that is strikingly similar to telomerase; in each case, terminal DNA sequences are regenerated by copying an RNA template—a complete genomic RNA in diptera, or an isolated genomic tag in telomerase.

Pardue et al. (1997) argue that cellular telomerase may have given rise to parasitic Het-A, TART, and similar non-LTR retrotransposons (Danilevskaya et al. 1997; Pardue et al. 1997). We favor the alternative scenario, equally consistent with the phylogenetic data (Eickbush 1997; Nakamura et al. 1997), that these retrotransposons are descendants of *autonomous* transitional genomes which had both RNA and DNA replicative forms (Fig. 6). Modern eukaryotic DNA chromosomes would then have been built by stepwise assembly of these smaller *independent* genomic elements. Indeed, the ability of Het-A and TART elements to generate tandem linear multimers (Danilevskaya et al. 1997) could be viewed as a continuation of this ancient chromosome assembly process. The his-

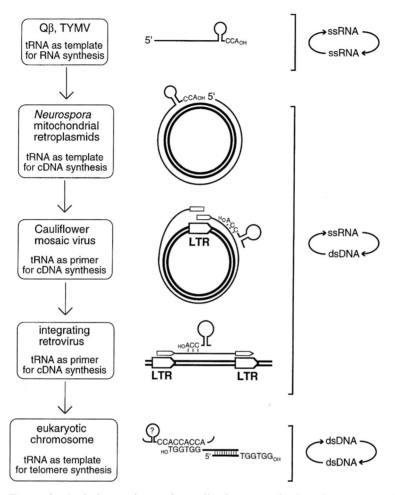

Figure 6 A phylogenetic tree for replication strategies based on conservation of tRNA-like structures in the initiation of genomic replication.

torical roots of modern DNA chromosomes in such transitional genomes would then be apparent in two conspicuous molecular fossils: RNA serves as *primer* for the initiation of DNA synthesis (at least in eubacteria), and RNA serves as *template* for the completion of chromosomal replication (eukaryotic telomeres). Indeed, as mentioned earlier, Alberts (1986) has argued that the presence of RNA in an enzyme suggests that the enzyme is ancient. A friendly amendment might be that involvement of RNA in any DNA transaction (replication, recombination, repair, or modification) suggests that the process is ancient, dating back to an RNA World or a world in transition from RNA to DNA genomes.

FROM TEMPLATE TO PRIMER

tRNAs Prime Retroviral Reverse Transcription

Initiation of retrovirus replication normally requires a tRNA primer (see Fig. 6); the hepadnaviruses are the only conspicuous exception (Ganem and Varmus 1987). tRNA priming of reverse transcription occurs not only in prototypical avian and mammalian retroviruses, but also in such lower eukaryotic retroviral elements as Ty1 in the yeast *S. cerevisiae* (Chapman et al. 1992), and *copia* in *D. melanogaster* (Kikuchi et al. 1986). To function as a primer, the 3' end of the tRNA partially unfolds to base-pair with the primer-binding site on the genomic RNA (Fig. 6). Additional interactions between the tRNA primer and nearby genomic sequences stabilize this initiation complex, providing multiple sequence determinants that restrict each retroviral element to a specific tRNA primer (Isel et al. 1995; Lanchy et al. 1996). As discussed below, tRNA also primes reverse transcription of circular extrachromosomal retroviruses such as cauliflower mosaic virus (CaMV) (Hohn et al. 1985; Covey and Turner 1986).

When we originally proposed that tRNA priming of retroviral replication was derived from the use of tRNA-like structures as templates for the initiation of RNA replication (Weiner and Maizels 1987), we were unable to cite any transitional forms in the molecular fossil record. The life cycle of the *Neurospora* retroplasmid has now revealed plausible missing links. Initially, a connection between tRNA and reverse transcription was established by the discovery that a 3'-terminal tRNA-like structure serves as template for first-strand cDNA synthesis on the retroplasmid genomic RNA (Figs. 4 and 5). More recently, Chen and Lambowitz (1997) established a connection between tRNA as template and tRNA as primer: The *Neurospora* retroplasmid reverse transcriptase can also use the 3'-terminal CCA of the RNA genome as a specific (base paired) or nonspecific (unpaired) primer (Fig. 7). Thus, first-strand synthesis by the *Neurospora* reverse transcriptase resembles Qβ replicase in using a tRNA-like structure as template; during second-strand synthesis and/or circularization, the same reverse transcriptase apparently reuses the tRNA-like structure as primer.

In Most Polymerases, Distinct Protein Domains Are Responsible for Template Specificity and Catalysis

How surprised should we be that one enzyme can use the same molecule as both template and primer? A wealth of molecular and structural data is consistent with the notion that the two key functions of any polymerase—template specificity and catalysis—are typically carried out by distinct

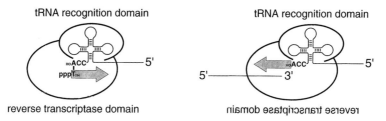

Figure 7 From template to primer. The transformation of tRNA from template for initiation of replication to primer requires that the tRNA-binding domain be functionally separate from, or flexibly tethered to, the catalytic domain. Reverse lettering in the right-hand panel indicates the opposite orientation of the tRNA recognition and reverse transcriptase domains. Note also that the *Neurospora* retroplasmid reverse transcriptase uses a tRNA-like structure both as template and as primer (Chen and Lambowitz 1997; also see Fig. 4).

protein domains. The classic examples are σ factors, which confer promoter specificity on eubacterial RNA polymerases (see, e.g., Decatur and Losick 1996). Separability of template specificity and catalysis is also a property of Qβ replicase and of the generic RNA replicase encoded by poliovirus. The Qβ replicase holoenzyme consists of four subunits (Blumenthal and Carmichael 1979), a phage-encoded replicase (subunit II) and three host-encoded subunits (EF-Tu, EF-Ts, and ribosomal protein S1). Subunit II exhibits a generic RNA replicase activity, whereas template specificity lies entirely within the other subunits, with EF-Tu presumably recognizing the 3′-terminal tRNA-like structure just as it recognizes tRNA during protein synthesis (Weiner and Maizels 1987, but see Brown and Gold 1996). Similarly, the poliovirus replicase polypeptide $3D^{pol}$ appears to require an additional specificity factor for initiation on the poliovirial minus-strand template (Andino et al. 1990; R. Andino, pers. comm.).

Additional evidence that the template specificity and catalytic functions of many polymerases are separable comes from analysis of sequence motifs shared by viral RNA-dependent RNA polymerases (replicases) and retroelement RNA-dependent DNA polymerases (reverse transcriptases). These enzymes display a wide range of template and primer specificity, but nonetheless share essentially invariant signature sequence motifs which are located at the active site rather than in the template recognition domains (Poch et al. 1989; Steitz 1998). Indeed, if template specificity and catalytic function were not at least partially separable, changes in template specificity would have obliterated these characteristic signatures. In contrast, a polymerase in which template specificity and catalysis are separate functions can easily adapt to new templates, thereby diversifying and ultimately producing more molecular descendants. Thus, as long as the tRNA recognition domain of reverse transcriptase is flexibly tethered to the active site, the enzyme might readily evolve from using tRNA as template to using it as primer. Although HIV reverse transcriptase is only distantly related to the *Neurospora* enzyme, the ability of the HIV enzyme to form a binary complex with primer tRNA (Barat et al. 1989) is consistent with the idea that separation of tRNA recognition and the polymerase module may be a general phenomenon.

Use of tRNA as primer also necessitates RNA helicase activity, because the enzyme would have to melt the top half of tRNA (a coaxial stack of the acceptor stem on the TψC arm) (Figs. 1 and 2) in order to allow the primer to base-pair with the genomic primer-binding site (see Fig. 6). However, this helicase activity would already be in place, because any enzyme that uses a genomic tag as template must possess an RNA helicase activity that can melt the top half of tRNA once initiation has occurred on the 3'-terminal CCA. Extensive complementarity between the tRNA primer and the template (Isel et al. 1995; Lanchy et al. 1996) may be a later refinement to assure a unique site of initiation and to stabilize the initiation complex.

Caulimoviruses Are Non-integrating One-LTR Retroviruses

We have discussed several reasons for thinking that mitochondrial retroplasmids may represent the ancestral form of modern retroviruses, but there is yet another missing link between duplex DNA plasmids and the prototypical integrating retrovirus with long terminal direct repeats (LTRs). A plant retrovirus, cauliflower mosaic virus (CaMV), may provide this connection. The CaMV genome is an extrachromosomal circular duplex DNA that lacks direct repeats and never integrates into chromosomal DNA. Transcription of viral DNA generates a full-length genomic RNA, but because the polyadenylation site for this transcript is located 180 bp downstream from the viral promoter, the transcript contains a 180-nucleotide terminal redundancy (Hohn et al. 1985; Covey and Turner 1986; also see Fig. 6).

Although CaMV (a pararetrovirus) differs from prototypical retroviruses in having one rather than two LTRs in the duplex DNA form, CaMV uses exactly the same strategy to generate a terminally redundant genomic RNA. In both cases, a polyadenylation site is ignored when it occurs too near the 5' end of the RNA, apparently because RNA processing of the nascent transcript is inhibited by proximity to the promoter (Sanfacon and Hohn 1990; but see DeZazzo et al. 1992). As in prototypical two-LTR retroviruses, the tRNA primer binding site in CaMV genomic RNA lies downstream from the 5'-terminal redundancy, so reverse transcriptase can effectively circularize the genome by switching templates from the 5' end of the genomic RNA to the corresponding position in the 3'-terminal redundancy.

These similarities in genomic structure and replication strategy imply that modern integrating retroviruses descended from an extrachromosomal retroviral element resembling CaMV. The obvious advantage of chromosomal integration is that it assures perpetuation of the element. The existence of two LTRs in a prototypical retroviral provirus could then be interpreted as an invention that preserved the established replication strategy by counterfeiting the circular topology of the ancestral genome.

The origin of LTRs remains mysterious. An intriguing scenario advanced by Danilevskaya et al. (1997) is that LTRs were first generated by repeated retroposition of a linear polyadenylated RNA such as HeT-A. As described above, contemporary HeT-A elements integrate head to tail, generating telomeric tandem repeats that are topologically equivalent to a circular HeT-A genome. Remarkably, the HeT-A promoter is located near the 3' end of the element, just upstream of the polyadenylation signal, in an arrangement reminiscent of modern LTRs (Fig. 6). As a result, each HeT-A element in a tandem array provides the promoter for the next element downstream (promoting in tandem) and generates a terminally redundant, polyadenylated primary transcript. Circularization of a single repeat unit from the tandem array could then generate an extrachromosomal LTR-containing CaMV-like retroelement. Circularization might occur at the DNA level by homologous recombination, or at the level of a hybrid RNA/DNA replicative intermediate as suspected for the *Neurospora* retroplasmid (Fig. 4). The key question, however, is the direction of time's arrow. Did linear HeT-A-like elements give rise to circular retroelements, or did autonomous circular elements generate multimers (perhaps by homologous recombination or runaround transcription) that then acquired the ability to retropose as linears? We favor the view that large DNA chromosomes were assembled stepwise from smaller autonomously replicating units carrying one or a few linked genes. In this view, small autonomous genomes—especially transitional genomes that replicate through both RNA and DNA intermediates—are more likely to be precursors than progeny of large modern chromosomes.

ARE MODERN VIRUSES MOLECULAR FOSSILS OF ANCIENT STRATEGIES FOR GENOMIC REPLICATION?

Most of us have come to have considerable confidence in the validity of sequence-based phylogenies relating organisms and organelles (Pace et al. 1986). Viruses, however, are quite another story. No single molecule, product, or function is common to all viruses, so there is no universal standard for evolutionary comparisons that is useful in quite the same way as small subunit ribosomal RNA sequences have been for organismal phylogenies. Furthermore, the interchangeability of functionally similar viral modules can undermine viral phylogenies based on a single molecule such as reverse transcriptase (but see Xiong and Eickbusch 1990; Eickbush 1997; Nakamura et al. 1997). Simple assumptions cannot be made about the regularity of viral molecular clocks, because generation times are short, burst sizes are large, and there is no single typical genome, but rather a population of quasispecies and defective interfering particles all propagating simultaneously (Eigen et al. 1981). Compounding these formidable problems, it can be difficult or impossible to trace the source of new viral genetic information, because viruses move horizontally between hosts, exchange genetic information with other viruses by recombination, and acquire new characteristics in each of a succession of hosts.

Although many aspects of viral life cycles violate the assumptions of sequence-based phylogenetic methods, viruses also present a unique opportunity for evolutionary analysis because certain aspects of viral life cycles appear to resist rapid change. Gene order and replication strategy, for example, are remarkably conserved between some plant and animal viruses. The stability of viral strategies for genomic replication and gene expression led Strauss and Strauss (1983) to propose that these properties of the viral life cycle are valuable and appropriate markers for virus evolution.

The attitude that viruses are cellular parasites, assembled in relatively recent evolutionary time from preexisting parts of the cell (Benner and Ellington 1988), is difficult to reconcile with data showing that many aspects of viral life cycles are stable over evolutionary time. In fact, the genomic tag hypothesis suggests an alternative view of viruses, as fossils that reveal the diversity of ancient replication pathways (see also Wintersberger and Wintersberger 1987). Considered in this way, viruses have not devised novel and subversive replication strategies; rather, they have conserved the useful features of more ancient forms of cellular chromosomal replication, even as cellular replication strategies continued to evolve. Contemporary viruses may thus preserve a record of the replication strategies used by more ancient chromosomes.

To illustrate how this view of contemporary viruses may be informative about molecular evolution, we show how RNA viruses, transitional genomes, and contemporary DNA genomes might be related using a phylogenetic tree based on the use of tRNA-like structures in replication (Fig. 6). The plus-strand RNA bacteriophages (Qβ) and plant viruses (brome mosaic virus) may represent molecular fossils of an RNA World in which cellular chromosomes were single-stranded RNA molecules with a 3′-terminal genomic tag. The *Neurospora* mitochondrial retroplasmid may represent a transitional stage from RNA to DNA genomes, in which duplex DNA functioned as the storage form for genetic information while genomic RNA with a 3′-terminal genomic tag served as a replicative intermediate. Cauliflower mosaic virus (CaMV) may be a molecular fossil of a more advanced transitional genome, which still replicated through a genomic RNA but in which the genomic tag had been transformed into a tRNA primer.

This view of viruses further suggests that modern duplex DNA genomes might be usefully regarded as retroviruses writ large. Eigen et al. (1981) originally observed that the role of RNA priming in the initiation of DNA synthesis might be a remnant of earlier RNA genomes. We are intrigued by the more specific possibility that the RNA primer for bacterial DNA replication is a degenerate form of the retroviral tRNA primer (see also Wintersberger and Wintersberger 1987), although it must be noted that eukaryotes generally use DNA primases that are capable of de novo initiation (for recent references, see Marini et al. 1997). RNA primases, for example, might turn out to be more closely related to reverse transcriptases than to RNA polymerases. Similarly, as discussed above, the telomerase responsible for completing replication of eukaryotic chromosomes is a specialized form of reverse transcriptase with an internal tRNA-like template (Fig. 5). Taken together, this evidence for the key role of RNA in both initiation and completion of chromosomal replication supports the notion that retroelements are the ancestors of modern eukaryotic chromosomes.

In arranging these selected viruses in this way, we in no sense imply that any one of these viruses is unchanged from ancient times. We do wish to point out that essential aspects of a viral replication strategy may remain stable over billions of years, and that the genomes of ancient organisms may survive in the form of viruses as the host replication strategy evolves (Maizels and Weiner 1994). This is of course consonant with the genomic tag hypothesis itself, which credits plus-strand RNA genomes similar to Qβ and modern plant viruses with the invention of tRNA and aminoacylation.

Viewed as predecessors rather than derivatives of modern chromosomes, viruses may have much more to tell us about molecular evolution than is commonly appreciated.

CONCLUSION

We originally conceived of the genomic tag hypothesis to explain the origin of protein synthesis (Weiner and Maizels 1987), but with time it became clear that this hypothesis had equally distinct implications for the evolution of replicative mechanisms, beginning in an RNA World and continuing to the present day. The new evidence that we have discussed for the central role of genomic tags in the evolution of RNA to DNA genomes strengthens the case that tRNA-like genomic tags arose early in an RNA World.

Translation today is a complex and sophisticated process, involving at least 2 ribosomal RNAs, more than 50 ribosomal proteins, 20 synthetases, tRNAs, initiation factors, elongation factors, etc. It is clear that the translation apparatus must have arisen stepwise, but it has been difficult to imagine how any single component could be useful by itself, or how additional components could each individually confer a further selective advantage. Central to the genomic tag hypothesis is the suggestion that the two key components of the translation apparatus—tRNA and tRNA aminoacylation activity—first evolved as essential components of the replication apparatus, and were subject to selection *before* the advent of protein synthesis. Once in place, these two key components of the translation apparatus could be coopted for other purposes, with the result that replication and templated protein synthesis were fated to coevolve forever after.

REFERENCES

Akins R.A., Kelley R.L., and Lambowitz A.M. 1989. Characterization of mutant mitochondrial plasmids of *Neurospora spp.* that have incorporated tRNAs by reverse transcription. *Mol. Cell. Biol.* **9:** 678–691.

Alberts B.M. 1986. The function of the hereditary materials: Biological catalysts reflect the cell's evolutionary history. *Am. Zool.* **26:** 781–796.

Andino R., Rieckhof G.E., and Baltimore D. 1990. A functional ribonucleoprotein complex forms around the 5′ end of poliovirus RNA. *Cell* **63:** 369–380.

Barat C., Lullien V., Schatz O., Keith G., Nugeyre M.T., Gruninger-Leitch F., Barré-Sinoussi F., LeGrice S.F., and Darlix J.L. 1989. HIV-1 reverse transcriptase specifically interacts with the anticodon domain of its cognate primer tRNA. *EMBO J.* **8:** 3279–3285.

Benner S.A. and Ellington A.D. 1988. Return of the "last ribo-organism." *Nature* **332:** 688–689.

Benner S.A., Ellington A.D., and Tauer A. 1989. Modern metabolism as a palimpsest of the RNA world. *Proc. Natl. Acad. Sci.* **86:** 7054–7058.

Bhattacharyya A. and Blackburn E.H. 1994. Architecture of telomerase RNA. *EMBO J.* **13:** 5721–5723.

Biessmann H. and Mason J.M. 1997. Telomere maintenance without telomerase. *Chromosoma* **106:** 63–69.

Blackburn E.H. 1991. Structure and function of telomeres. *Nature* **350:** 569–573.

Blumenthal T. and Carmichael G.C. 1979. RNA replication: Function and structure of Qβ-replicase. *Annu. Rev. Biochem.* **48:** 525–548.

Brown D. and Gold L. 1996. RNA replication by Qβ replicase: A working model. *Proc. Natl. Acad. Sci.* **93:** 11558–11562.

Brown J.W., Hunt D.A., and Pace N.R. 1990. Nucleotide sequence of the 10Sa RNA gene of the β-purple eubacterium *Alcaligenes eutrophus*. *Nucleic Acids Res.* **18:** 2820.

Chapman K.B., Bystrom A.S., and Boeke J.D. 1992. Initiator methionine tRNA is essential for Ty1 transposition in yeast. *Proc. Natl. Acad. Sci.* **89:** 3236–3240.

Chen B. and Lambowitz A.M. 1997. De novo and DNA primer-mediated initiation of cDNA synthesis by the Mauriceville retroplasmid reverse transcriptase involve recognition of a 3′ CCA sequence. *J. Mol. Biol.* **271:** 311–332.

Covey S.N. and Turner D.S. 1986. Hairpin DNAs of cauliflower mosaic virus generated by reverse transcription *in vivo*. *EMBO J.* **5:** 2763–2768.

Cusack S., Berthet-Colominas C., Hartlein M., Nassar N., and Leberman R. 1990. A second class of synthetase structure revealed by X-ray analysis of *Escherichia coli* seryl-tRNA synthetase at 2.5 Å. *Nature* **347:** 249–255.

Danilevskaya O.N., Arkhipova I.R., Traverse K.L., and Pardue M.L. 1997. Promoting in tandem: The promoter for telomere transposon HeT-A and implications for the evolution of retroviral LTRs. *Cell* **88:** 647–655.

Decatur A.L. and Losick R. 1996. Three sites of contact between the *Bacillus subtilis* transcription factor σF and its antisigma factor SpoIIAB. *Genes Dev.* **10:** 2348–2358.

Deutscher M.P. 1982. tRNA nucleotidyltransferase. *Enzymes* **15:** 183–215.

DeZazzo J.D., Scott J.M., and Imperiale M.J. 1992. Relative roles of signals upstream of AAUAAAA and promoter proximity in regulation of HIV-1 mRNA 3′ end formation. *Mol. Cell. Biol.* **12:** 5555–5562.

Downey K.M., Jurmark B.S., and So A.G. 1971. Determination of nucleotide sequences at promoter regions by the use of dinucleotides. *Biochemistry* **10:** 4970–4975.

Dudock B.S., Katz G., Taylor E.K., and Holley R.W. 1969. Primary structure of wheat germ phenylalanine transfer RNA. *Proc. Natl. Acad. Sci.* **62:** 941–945.

Eickbush T.H. 1997. Telomerase and retrotransposons: Which came first? *Science* **277:** 911–912.

Eigen M., Gardiner W., Schuster P., and Winkler-Oswatitsch H. 1981. The origin of genetic information. *Sci. Am.* **244:** 88–118.

Eriani G., Delarue M., Poch O., Gangloff J., and Moras D. 1990. Partition of tRNA synthetases into two classes based on mutually exclusive sets of sequence motifs. *Nature* **347:** 203–206.

Fang G. and Cech T.R. 1993a. The β-subunit of *Oxytricha* telomere-binding protein promotes G-quartet formation by telomeric DNA. *Cell* **74:** 875–885.
———. 1993b. Characterization of a G-quartet formation reaction promoted by the β-subunit of the *Oxytricha* telomere-binding protein. *Biochemistry* **32:** 11646–11657.
Felden B., Himeno H., Muto A., McCutcheon J.P., Atkins J.F., and Gesteland R.F. 1997. Probing the structure of the *Escherichia coli* 10Sa RNA (tmRNA). *RNA* **3:** 89–103.
Francklyn C., Shi J.P., and Schimmel P. 1992. Overlapping nucleotide determinants for specific aminoacylation of RNA microhelices. *Science* **255:** 1121–1125.
Freier S.M., Alkema D., Sinclair A., Neilson T., and Turner D.H. 1985. Contributions of dangling end stacking and terminal base-pair formation to the stabilities of XGGCCp, XCCGGp, XGGCCYp, and XCCGGYp helices. *Biochemistry* **24:** 4533–4539.
Ganem D. and Varmus H.E. 1987. The molecular biology of the hepatitis B viruses. *Annu. Rev. Biochem.* **56:** 651–693.
Gardner L.P., Mookhtiar K.A., and Coleman J.E. 1997. Initiation, elongation, and processivity of carboxyl-terminal mutants of T7 RNA polymerase. *Biochemistry* **36:** 2908–2918.
Gilbert W. 1986. The RNA world. *Nature* **319:** 618.
Green C.J., Vold B.S., Morch M.D., Joshi R.L., and Haenni A.L. 1988. Ionic conditions for the cleavage of the tRNA-like structure of turnip yellow mosaic virus by the catalytic RNA of RNase P. *J. Biol. Chem.* **263:** 11617–11620.
Green R. and Noller H.F. 1997. Ribosomes and translation. *Annu. Rev. Biochem.* **66:** 679–716.
Greider C.W. and Blackburn E.H. 1985. Identification of a specific telomere terminal transferase activity in *Tetrahymena* extracts. *Cell* **43:** 405–413.
———. 1989. A telomeric sequence in the RNA of *Tetrahymena* telomerase required for telomere repeat synthesis. *Nature* **337:** 331–337.
Guerrier-Takada C., van Belkum A., Pleij C.W.A., and Altman S. 1988. Novel reactions of RNAase P with a tRNA-like structure in turnip yellow mosaic virus RNA. *Cell* **53:** 267–272.
Guerrier-Takada C., Gardiner K., Marsh T., Pace N., and Altman S. 1983. The RNA moiety of ribonuclease P is the catalytic subunit of the enzyme. *Cell* **35:** 849–857
Henderson E., Hardin C.C., Walk S.K., Tinoco I. Jr., and Blackburn E.H. 1987. Telomeric DNA oligonucleotides form novel intramolecular structures containing guanine-guanine base pairs. *Cell* **51:** 899–908.
Hicke B.J., Christian E.L., and Yarus M. 1989. Stereoselective arginine binding is a phylogenetically conserved property of group I self-splicing RNAs. *EMBO J.* **8:** 3843–3851.
Himeno H., Sato M., Tadaki T., Fukushima M., Ushida C., and Muto A. 1997. In vitro *trans* translation mediated by alanine-charged 10Sa RNA. *J. Mol. Biol.* **268:** 803–808.
Hohn T., Hohn B., and Pfeiffer P. 1985. Reverse transcription in CaMV. *Trends Biochem. Sci.* **10:** 205–209.
Holm L. and Sander C. 1995. DNA polymerase β belongs to an ancient nucleotidyltransferase superfamily. *Trends Biochem. Sci.* **20:** 345–347.
Illangasekare M., Kovalchuke O., and Yarus M. 1997. Essential structures of a self-aminoacylating RNA. *J. Mol. Biol.* **274:** 519–529.
Illangasekare M., Sanchez G., Nickles T., and Yarus M. 1995. Aminoacyl-RNA synthesis catalyzed by an RNA. *Science* **267:** 643–647.

Isel C., Ehresmann C., Keith G., Ehresmann B., and Marquet R. 1995. Initiation of reverse transcription of HIV-1: Secondary structure of the HIV-1 RNA/tRNA(3Lys) (template/primer). *J. Mol. Biol.* **247:** 236–250.

Jay D.G. and Gilbert W. 1987. Basic protein enhances the incorporation of DNA into lipid vesicles: A model for the formation of primordial cells. *Proc. Natl. Acad. Sci.* **84:** 1978–1980.

Keiler K.C., Waller P.R.H., and Sauer R.T. 1996. Role of a peptide tagging system in degradation of proteins synthesized from damaged messenger RNA. *Science* **271:** 990–993.

Kikuchi Y., Ando Y., and Shiba T. 1986. Unusual priming mechanism of RNA-directed DNA synthesis in copia retrovirus-like particles of *Drosophila*. *Nature* **323:** 824–826.

Komine Y., Kitabatake M., Yokogawa T., Nishikawa K., and Inokuchi H. 1994. A tRNA-like structure is present in 10Sa RNA, a small stable RNA from *Escherichia coli*. *Proc. Natl. Acad. Sci.* **91:** 9223–9227.

Kuiper M.T.R. and Lambowitz A.M. 1988. A novel reverse transcriptase activity associated with mitochondrial plasmids of *Neurospora*. *Cell* **55:** 693–704.

Lanchy J.M., Isel C., Ehresmann C., Marquet R., and Ehresmann B. 1996. Structural and functional evidence that initiation and elongation of HIV-1 reverse transcription are distinct processes. *Biochimie* **78:** 1087–1096.

Lee B.J., Kang S.G., and Hatfield D. 1989. Transcription of *Xenopus* selenocysteine tRNA Ser (formerly designated opal suppressor phosphoserine tRNA) gene is directed by multiple 5′-extragenic regulatory elements. *J. Biol. Chem.* **264:** 9696–9702.

Levis R.W. 1989. Viable deletions of a telomere from a *Drosophila* chromosome. *Cell* **58:** 791–801.

Levis R.W., Ganesan R., Houtchens K., Tolar L.A., and Sheen F.M. 1993. Transposons in place of telomeric repeats at a *Drosophila* telomere. *Cell* **75:** 1083–1093.

Li Z. and Deutscher M.P. 1996. Maturation pathways for *E. coli* tRNA precursors: A random multienzyme process *in vivo*. *Cell* **86:** 503–512.

Limmer S., Hofmann H.P., Ott G., and Sprinzl M. 1993. The 3′-terminal end (NCCA) of tRNA determines the structure and stability of the aminoacyl acceptor stem. *Proc. Natl. Acad. Sci.* **90:** 6199–6202.

Lingner J., Hendrick L.L., and Cech T.R. 1994. Telomerase RNAs of different ciliates have a common secondary structure and a permuted template. *Genes Dev.* **8:** 1984–1998.

Lingner J., Hughes T.R., Shevchenko A., Mann M., Lundblad V., and Cech T.R. 1997. Reverse transcriptase motifs in the catalytic subunit of telomerase. *Science* **276:** 561–567.

Lohse P.A. and Szostak J.W. 1996. Ribozyme-catalysed amino-acid transfer reactions. *Nature* **381:** 442–444.

Maizels N. 1973. The nucleotide sequence of the lactose mRNA transcribed from the UV5 promoter mutant of *E. coli*. *Proc. Natl. Acad. Sci.* **70:** 3585–3589.

Maizels N. and Weiner A.M. 1987. Peptide-specific ribosomes, genomic tags and the origin of the genetic code. *Cold Spring Harbor Symp. Quant. Biol.* **52:** 743–749.

―――. 1994. Phylogeny from function: Evidence from the molecular fossil record that tRNA originated in replication, not translation. *Proc. Natl. Acad. Sci.* **91:** 6729–6734.

Mans R.M., Guerrier-Takada C., Altman S., and Pleij C.W. 1990. Interaction of RNase P from *Escherichia coli* with peudoknotted structures in viral RNAs. *Nucleic Acids Res.* **18:** 3479–3487.

Marini F., Pellicioli A., Paciotti V., Lucchini G., Plevani P., Stern D.F., and Foiani M. 1997. A role for DNA primase in coupling DNA replication to DNA damage response. *EMBO J.* **16:** 639–650.

Martin G. and Keller W. 1996. Mutational analysis of mammalian poly(A) polymerase identifies a region for primer binding and catalytic domain, homologous to the family X polymerases, and to other nucleotidyltransferases. *EMBO J.* **15:** 2593–2603.

Matsufuji S., Matsufuji T., Miyazaki Y., Murakami Y., Atkins J.F., Gesteland R.F., and Hayashi S. 1995. Autoregulatory frameshifting in decoding mammalian ornithine decarboxylase antizyme. *Cell* **80:** 51–60.

McClain W.H., Guerrier-Takada C., and Altman S. 1987. Model substrates for an RNA enzyme. *Science* **238:** 527–530.

McClure W.R. 1980. Rate-limiting steps in RNA chain initiation. *Proc. Natl. Acad. Sci.* **77:** 5634–5638.

Melton D.A., Krieg P.A., Rebagliati M.R., Maniatis T., Zinn K., and Green M.R. 1984. Efficient *in vitro* synthesis of biologically active RNA and RNA hybridization probes from plasmids containing a bacteriophage SP6 promoter. *Nucleic Acids Res.* **12:** 7035–7056.

Michel F., Hanna M., Green R., Bartel D.P., and Szostak J.W. 1989. The guanosine binding site of the *Tetrahymena* ribozyme. *Nature* **342:** 391–395.

Milligan J.F., Groebe D.R., Witherell G.W., and Uhlenbeck O.C. 1987. Oligoribonucleotide synthesis using T7 RNA polymerase and synthetic DNA templates. *Nucleic Acids Res.* **15:** 8783–8798.

Musier-Forsyth K. and Schimmel P. 1992. Functional contacts of a transfer RNA synthetase with 2′-hydroxyl groups in the RNA minor groove. *Nature* **357:** 513–514.

Nakamura T.M., Morin G.B., Chapman K.B., Weinrich S.L., Andrews W.H., Lingner J., Harley C.B., and Cech T.R. 1997. Telomerase catalytic subunit homologs from fission yeast and human. *Science* **277:** 955–959.

Noller H.F. 1993. On the origin of the ribosome: Coevolution of subdomains of tRNA and rRNA. In *The RNA world* (ed. R.F. Gesteland and J.F. Atkins), pp. 137–156. Cold Spring Harbor Laboratory Press, Cold Spring Harbor, New York.

Noller H.F., Hoffarth V., and Zimniak L. 1992. Unusual resistance of peptidyl transferase to protein extraction procedures. *Science* **256:** 1416–1419.

Orgel L.E. 1968. Evolution of the genetic apparatus. *J. Mol. Biol.* **38:** 381–393.

Pace N.R., Olsen G.J., and Woese C.R. 1986. Ribosomal RNA phylogeny and the primary lines of evolutionary descent. *Cell* **45:** 25–326.

Pardue M.L., Danilevskaya O.N., Traverse K.L., and Lowenhaupt K. 1997. Evolutionary links between telomeres and transposable elements. *Genetica* **100:** 73–84.

Piccirilli J.A., McConnell T.S., Zaug A.J., Noller H.F., and Cech T.R. 1992. Aminoacyl esterase activity of the *Tetrahymena* ribozyme. *Science* **256:** 1420–1424.

Pilipenko E.V., Maslova S.V., Sinyakov A.N., and Agol V.I. 1992. Towards identification of *cis*-acting elements involved in the replication of enterovirus and rhinovirus RNAs: A proposal for the existence of tRNA-like terminal structures. *Nucleic Acids Res.* **20:** 1739–1745.

Poch O., Sauvaget I., Delarue M., and Tordo N. 1989. Identification of four conserved motifs among the RNA-dependent polymerase encoding elements. *EMBO J.* **8:** 3867–3874.

Polesky A.H., Dahlberg M.E., Benkovic S.J., Grindley N.D.F., and Joyce C.M. 1992. Side chains involved in catalysis of the polymerase reaction of DNA polymerase I from *Escherichia coli. J. Biol. Chem.* **267:** 8417–8428.

Popper K.R. 1963. *Conjectures and refutations*, 4th edition. Routledge & Kegan Paul, London, United Kingdom.

Puglisi J.D., Tan R., Calnan B.J., Frankel A.D., and Williamson J.R. 1992. Conformation of the TAR RNA-arginine complex by NMR spectroscopy. *Science* **257:** 76–80.

Puglisi E.V., Puglisi J.D., Williamson J.R., and RajBhandary U.L. 1994. NMR analysis of tRNA acceptor stem microhelices: Discriminator base change affects tRNA conformation at the 3′ end. *Proc. Natl. Acad. Sci.* **91:** 11467–11471.

Quigley G.J. and Rich A. 1976. Structural domains of transfer RNA molecules. *Science* **194:** 796–806.

Rao A.L.N., Dreher T.W., Marsh L.E., and Hall T.C. 1989. Telomeric function of the tRNA-like structure of brome mosaic virus RNA. *Proc. Natl. Acad. Sci.* **86:** 5335–5339.

Rasmussen N.J., Wikman F.P., and Clark B.F. 1990. Crosslinking of tRNA containing a long extra arm to elongation factor Tu by *trans*-diamminedichloroplatinum(II). *Nucleic Acids Res.* **18:** 4883–4890.

Romero D.P. and Blackburn E.H. 1991. A conserved secondary structure for telomerase RNA. *Cell* **67:** 343–353.

Rould M.A., Perona J.J., Söll D., and Steitz T.A. 1989. Structure of *E. coli* glutaminyl-tRNA synthetase complexed with tRNA(Gln) and ATP at 2.8 Å resolution. *Science* **246:** 1135–1142.

Sanfacon H. and Hohn T. 1990. Proximity to the promoter inhibits recognition of cauliflower mosaic virus polyadenylation signal. *Nature* **346:** 81–84.

SantaLucia Jr. J., Kierzek R., and Turner D.H. 1990. Effects of GA mismatches on the structure and thermodynamics of RNA internal loops. *Biochemistry* **29:** 8813–8819.

Saville B.J. and Collins R.A. 1990. A site-specific self-cleavage reaction performed by a novel RNA in *Neurospora* mitochondria. *Cell* **61:** 685–696.

Schimmel P. 1987. Aminoacyl tRNA synthetases: General scheme of structure-function relationships in the polypeptides and recognition of transfer RNAs. *Annu. Rev. Biochem.* **56:** 125–158.

Schimmel P. and Henderson B. 1994. Possible role of aminoacyl-RNA complexes in non-coded peptide synthesis and origin of coded synthesis. *Proc. Natl. Acad. Sci.* **91:** 11283–11286.

Schimmel P., Giegé R., Moras D., and Yokoyama S. 1993. An operational RNA code for amino acids and possible relationship to genetic code. *Proc. Natl. Acad. Sci.* **90:** 8763–8768.

Sen D. and Gilbert W. 1988. Formation of parallel four-stranded complexes by guanine-rich motifs in DNA and its implications for meiosis. *Nature* **334:** 364–366.

Shi P.-Y., Maizels N., and Weiner A.M. 1998a. CCA addition by tRNA nucleotidyltransferase: Polymerization without translocation? *EMBO J.* **17:** 3188–3196.

Shi P.-Y., Weiner A.M., and Maizels N. 1998b. A top-half tDNA minihelix is a good substrate for the eubacterial CCA-adding enzyme. *RNA* **4:** 276–284.

Sprinzl M. and Cramer F. 1979. The -C-C-A end of tRNA and its role in protein biosynthesis. *Prog. Nucleic Acid Res. Mol. Biol.* **22:** 1–69.

Steitz T.A. 1998. A mechanism for all polymerases. *Nature* **391:** 231–232.

Strauss E.G. and Strauss J.H. 1983. Replication strategies of the single stranded RNA viruses of eukaryotes. In *Microbiology and immunology* (ed. M. Cooper et al.), vol. 105, pp. 2–98. Springer-Verlag, Berlin.

Subbarao M.N. and Apirion D. 1989. A precursor for a small stable RNA (10Sa RNA) of *Escherichia coli*. *Mol. Gen. Genet.* **217:** 499–504.

Sugimoto N., Hasegawa K., and Sasaki M. 1990. Stability of 3′ dangling ends on the core helix of AUGCAU at various Na⁺ concentrations. *Nucleic Acids Symp. Ser.* **22:** 107–108.

Sundquist W.I. and Klug A. 1989. Telomeric DNA dimerizes by formation of guanine tetrads between hairpin loops. *Nature* **342:** 825–829.

Tse W.T. and Forget B.G. 1990. Reverse transcription and direct amplification of cellular RNA transcripts by Taq polymerase. *Gene* **88:** 293–296.

Tu G.F., Reid G.E., Zhang J.G., Moritz R.L., and Simpson R.J. 1995. C-terminal extension of truncated recombinant proteins in *Escherichia coli* with 10Sa RNA decapeptide. *J. Biol. Chem.* **270:** 9322–9326.

Tyagi J.S. and Kinger A.K. 1992. Identification of the 10Sa RNA structural gene of *Mycobacterium tuberculosis*. *Nucleic Acids Res.* **20:** 138.

Visser C.M. and Kellogg R.M. 1978a. Biotin. Its place in evolution. *J. Mol. Evol.* **11:** 171–187.

———. 1978b. Bioorganic chemistry and the origin of life. *J. Mol. Evol.* **11:** 163–169.

Weiner A.M. and Maizels N. 1987. 3′ Terminal tRNA-like structures tag genomic RNA molecules for replication: Implications for the origin of protein synthesis. *Proc. Natl. Acad. Sci.* **84:** 7383–7387.

Welch M., Majerfeld I., and Yarus M. 1997. 23S rRNA similarity from selection for peptidyl transferase mimicry. *Biochemistry* **36:** 6614–6623.

White III, H.B. 1976. Coenzymes as fossils of an earlier metabolic state. *J. Mol. Evol.* **7:** 101–104

———. 1982. Evolution of coenzymes and the origin of pyridine nucleotides. In *The pyridine coenzymes* (ed. J. Everse et al.), pp. 1–17. Academic Press, New York.

Williams K.P. and Bartel D.P. 1996. Phylogenetic analysis of tmRNA secondary structure. *RNA* **2:** 1306–1310.

Williamson J.R., Raghuraman M.K., and Cech T.R. 1989. Monovalent cation-induced structure of telomeric DNA: the G-quartet model. *Cell* **59:** 871–880.

Wintersberger U. and Wintersberger E. 1987. RNA makes DNA: A speculative view of the evolution of DNA replication mechanisms. *Trends Genet.* **3:** 198–202.

Wong J.-T. 1991. Origin of genetically encoded protein synthesis: A model based on selection for RNA peptidation. *Origins Life Evol. Biosph.* **21:** 165–176.

Xiong Y. and Eickbush T.H. 1990. Origin and evolution of retroelements based upon their reverse transcriptase sequences. *EMBO J.* **9:** 3353–3362.

Yarus M., Illangesekare M., and Christian E. 1991. An axial binding site in the *Tetrahymena* precursor RNA. *J. Mol. Biol.* **222:** 995–1012.

Yue D., Maizels N., and Weiner A.M. 1996. CCA-adding enzymes and poly(A) polymerases are all members of the same nucleotidyltransferase superfamily: Characterization of the CCA-adding enzyme from the archaeal hyperthermophile *Sulfolobus shibatae*. *RNA* **2:** 895–908.

Zakian V.A. 1996. Structure, function, and replication of *Saccharomyces cerevisiae* telomeres. *Annu. Rev. Genet.* **30:** 141–172.

4

Probing RNA Structure, Function, and History by Comparative Analysis

Norman R. Pace and Brian C. Thomas
Departments of Plant and Microbial Biology, and Molecular and Cell Biology
University of California, Berkeley
Berkeley, California 94720

Carl R. Woese
Department of Microbiology
University of Illinois
Urbana, Illinois 61801

Life on this planet is a profusion of incredibly complex systems. From the biologist's perspective it is indeed fortunate that all these systems have sprung from a common ancestor. They have by their nature retained traces of their ancestries, and so are similar—homologous—to a greater or lesser extent. The science of biology has been built from its beginning upon cataloging similarities and differences among different living systems (and various states of the same system), the method that has become known as comparative analysis. It is only through this simple and sometimes tedious approach that biologists could begin to understand the complexity with which they are confronted, could begin to distinguish the important elements from the unimportant, and so reduce living systems to a set of understandable essentials. It is also through such comparisons, through measuring similarity and difference in degree and kind, that biologists have come to learn the genealogical relationships among all organisms.

Despite its compelling utility, comparative analysis has not been a commonly used tool in molecular biology. This is not because comparative analysis is without value at the molecular level: The secondary structures of the ribosomal RNAs, major accomplishments of modern molecular biology, are tribute to the comparative approach (Woese et al. 1980; Noller et al. 1981). The molecular biologist's aversion to comparative analysis would seem to lie in the molecular paradigm itself. Molecular biology arose from chemistry and physics. The entities with which these sciences deal, atoms and their relatively simple chemical combinations, do not have meaningful histories. Consequently, the conceptual and experimental outlook inherited from chemistry and physics effectively has no comparative, historical dimension. This may also explain why molecular biologists have tended to view evolution (apart from the origin of life issue) as a trivial part of biology, as merely a collection of relatively uninteresting historical accidents: No matter that evolution is the essence, the sine qua non of biology.

Our objective in this chapter is to review the various ways in which the comparative analysis of RNA at the molecular level has contributed and will contribute to biology. Comparative analysis contributes to molecular biology in four important ways: (1) in defining restraints and patterns from which molecular structure can be inferred; (2) as a necessary background for, and adjunct to, experimental analysis of molecular function and structure; (3) as the essence of genealogical analysis and classification; and (4) in providing a general conceptual and organizational framework for much of future biological research.

The changes that have occurred in microbiology over the past 15 years bear witness to the sharpening effect that a phylogenetic/comparative framework has on experimental design and interpretation of results. There can be little doubt that the deluge of genomic sequence information can only be handled effectively in a comparative framework.

COMPARATIVE ANALYSIS AND NUCLEIC ACID STRUCTURE

History: Molecular Biology's Flirtation with Comparative Analysis

The potential of a comparative approach to RNA higher-order structure became evident at the 1966 Cold Spring Harbor Symposium. By that time, the sequences of four tRNAs were known. The sequences reported had not been determined initially with a comparative approach in mind, but the fact that all four tRNA sequences could assume the same "clover leaf" configuration convinced most of those present at the meeting that all tRNAs had a common secondary structure. However, the general principle, that comparative analysis is an effective tool in the analysis of molecular structure, seems not to have been grasped, for the lesson had to be repeated when it came to 5S rRNA structure. Here, too, the initial attempts to determine secondary structure had not involved a systematic, comparative approach. Rather, structures were suggested on the basis of maximizing of base-pairing, for instance, or some other "first principle." The resulting structures tended to be awkwardly complex and unattractive (for review, see Erdmann 1976). A systemic comparative approach was eventually applied in this case, and the true structure of the 5S rRNA then began to emerge (Fox and Woese 1973).

When the sequence of the small subunit rRNA was in the process of being determined, the comparative lesson seemed once more to have been forgotten. Again, the search for secondary structure initially consisted of merely looking for stretches of nucleotides with the potential to form base pairs. The credibility of the resulting structures did not seem to be an issue. There is even an unsubstantiated rumor (worth repeating) that when Brosius, Noller, and their colleagues submitted the first complete 16S rRNA sequence (that of *Escherichia coli*) for publication, one of the reviewers questioned why they had not included the molecule's secondary structure as well! By this time, however, the comparative lesson had to some extent been assimilated, and Noller's group soon teamed up with Woese's group to produce a second sequence, for the specific purpose of using comparisons to identify among the 10,000 or so possible helical elements in any given small subunit rRNA a manageable number of probable ones (Woese et al. 1980). By the time the large subunit rRNA was sequenced (Brosius et al. 1980), it was generally accepted that comparative analysis was essential to determining its secondary structure, and the appropriate experimental steps, determination and comparison of sequences from different organisms, were taken to realize this (Noller et al. 1981).

Using the Comparative Approach to Analyze RNA Structure

Today, comparative analysis has become the method of choice for establishing higher-order structure for large RNAs. Regardless of their other virtues, none of the more direct, physical-chemical approaches can provide so detailed a picture of the relationship among bases in large RNAs.

Comparative analysis starts with the alignment of sets of homologous sequences. In its most stringent definition, homology refers strictly to common ancestry. The simply stated objective of alignment is to juxtapose related sequences so that the homologous residues in each occupy the same column in the alignment. Although there have been significant advances in computer-generated alignment in recent years, the most precise method of alignment is still essentially a manual one (with computer assistance). The reason for this is that a great deal more is involved in alignment than merely shifting nucleotides around until some maximum sequence similarity is achieved. Homology among individual sequence residues is best defined in the context of larger units of homologous structure and function. Therefore, the most useful and generally the most precise alignments are produced by iteratively invoking higher-order structure in the process. A practical condition imposed on sets of aligned sequences is that the sequences are arranged in an order that approximates their phylogenetic relationships to one another. This is the most useful context for comparative analysis and foreshadows a day when phylogenetic structuring will become an integral part of most if not all biological databases.

Given a phylogenetically ordered sequence alignment, one can begin to interpret Nature's evolutionary experimentation. Nature provides the results of those experiments that have worked, those in which molecular function remains optimized (within the limits defined by selective constraints); only rarely are we privy to natural experiments in which function has been adversely altered. Therefore, invariance in compositions of certain residues in a molecule identifies these as being important to the structure and function of the molecule. Conversely, areas of a molecule that frequently vary in composition from organism to organism, especially when length variation also occurs, mark themselves as probably of little or no direct functional significance. Rarely, a particular section of a molecule will become deleted, thereby identifying itself as some kind of functional or structural unit. Residues whose compositions covary (change in concert) must in some way be related, which in almost all cases means in direct physical contact.

Early comparative studies were limited by difficulties in accumulating large numbers of sequences for consideration. Consequently, comparisons were mainly limited to manual operations. With the expansion of rapid sequencing techniques, large data sets can be accumulated for computer comparisons seeking covariations of sequence or other structural features. Sufficiently large data sets (>100 sequences) can fruitfully be sifted for correlations using computer-based mutual-information algorithms (Gautheret et al. 1995). Homologous genes from different organisms are generally isolated from specific organisms, an arduous task for a thorough analysis. Another way to accumulate a large amount of sequence data for comparative analysis is by use of mixed populations of different organisms, for instance environmental samples, as sources of particular genes. The selected genes are obtained from community DNA by PCR using primers complementary to broadly conserved sequences in the genes (Brown et al. 1996; Hugenholtz et al. 1998). The source organism for a sequence obtained this way would not necessarily be known, but this is generally irrelevant to a comparative structure analysis.

Sequence comparisons reveal more than the simple existence of structural relationships. The pattern of a covariation may suggest the nature of an interaction. For example, if two positions in a sequence covary strictly according to the Watson-Crick rules (and cover all the allowable combinations), one can be fairly certain that they form a bona fide Watson-Crick base pair. On the other hand, a normal pairing geometry is unlikely in those cases exhibiting other than a canonical pattern of covariation. The pattern and frequency of variation at a given position is a far more subtle characteristic of that position than is generally appreciated. It is a refined measure of structural homology, for only if the full structural context, not merely the major contacts, is maintained from one group of organisms to the next, will the pattern (and frequency) of variation be similar in the groups.

Some Specific Examples Involving Ribosomal RNA

A principal use of comparative analysis in the study of rRNA has been to infer secondary structure. As used here, secondary structural elements are contiguous stretches of two or more nucleotides that form canonical (or G:U type) antiparallel pairs with one another. Comparative "proof" of such structures involves finding multiple phylogenetically independent instances in which the compositions of two potentially paired positions change in concert, according to the Watson-Crick pairing rules. Figure 1 gives some idea of what comparative analysis has accomplished in terms of rRNA structure, and the extent of the comparative evidence that supports the various helices (Gutell et al. 1993). About 60% of the nucleotides in the small subunit rRNA are involved in confirmed secondary structure, a proportion comparable to the 55% of nucleotides in tRNA that are known from its crystal structure to be involved in secondary structure (Kim 1979).

Comparative analysis shows that noncanonical pairs occur frequently in rRNAs, and that they occur in a variety of pairing geometries. The U:G type of pair is, of course, the most common of the noncanonical pairs, the next most common being the A:G type (Gutell et al. 1993). From their patterns of variation it is apparent that rRNA contains several different kinds of U:G pairs. One kind exhibits frequent compositional variation over the phylogenetic spectrum, but U:G remains its major composition. Perhaps the most striking example of this kind can be seen in the helix 829-40:846-857 shown in Figure 1. This structure, which typically comprises 10–12 pairs, almost always shows 5 or more U:G pairs in its central section. The helix is highly variable in composition, with U:G pairs often being replaced by other pairs (frequently G:U), and the exact location of the U:G pairs is somewhat variable (Gutell et al. 1993). It seems evident that at least some U:G pairs in rRNA have unique structural or functional significance.

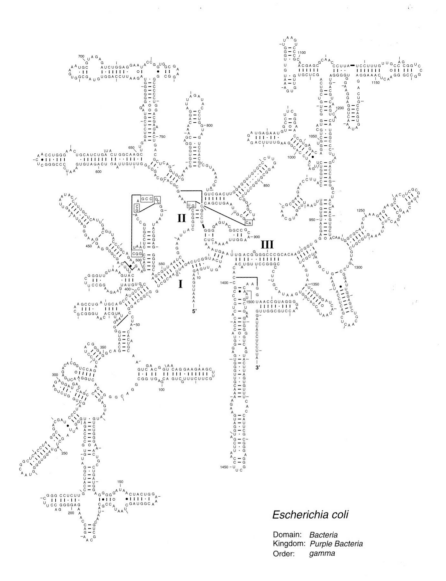

Figure 1 Higher-order structure of the 16S ribosomal RNA (Gutell 1994). As discussed in the text, this folded structure is based on comparative analysis. Watson-Crick pairs are connected with lines, G:U pairs are connected with filled circles, and A:G pairs are connected with open circles. Bases that are nonjuxtaposed as pairs, but connected, have yet to be proven phylogenetically. Boxed regions pair as indicated by the connecting lines. Every 10th position is indicated with a line and each 50th position is numbered. (Reprinted, with permission, from Gutell at http://pundit.colorado.edu.8080/)

A second kind of U:G pair is identified by the fact that its composition varies only slowly over evolutionary time, and when it does so, it almost always (or always) converts to a C:A pair (Gutell et al. 1993). This pattern of variation suggests a non-Watson-Crick pairing geometry.

Another interesting pattern of covariation is shown by the U:U pair that converts solely to C:C (and vice versa), again suggestive of an atypical pairing geometry (Gutell et al. 1993). Although the pattern of covariation does not allow us to infer the physical conformation of any pair with certainty, it does favor certain possibilities. For example, a U:U pairing that alternates (and is presumably isomorphic) with a C:C pairing could involve nucleotides in a syn-anti conformation. Direct physical measurement is needed to ascertain this.

One of the first unusual higher-order structures in the small subunit rRNA to be uncovered by comparative analysis was the so-called pseudoknot (Woese et al. 1983). Pseudoknot here describes a topology in which a stretch of nucleotides within a hairpin loop pairs with nucleotides external to that loop. Three such pseudoknots involve the pairing of positions 17-20:915-918, 505-507:524-526, and 570-571:865-866 (Fig. 1). In all three instances, the pseudoknot pairings tend to be nearly invariant in composition, suggesting that their three-dimensional context is stringently defined and that these structures are important to ribosome structure/function. Some of the proposed pseudoknot structures have been experimentally tested by mutagenesis: In all cases in which mispairings have been introduced, ribosome function was impaired, if not totally eliminated (Powers and Noller 1991).

Comparative analysis also suggested that the area of the small subunit rRNA between positions 500 and 545 (which contains the second of the three pseudoknot structures) is of major importance because of its highly conserved sequence and overall length (Woese et al. 1983). This particular structure is highly constrained: Of the 46 bases comprising it, only 14 are not known to be involved in secondary or tertiary interactions (see Fig. 1). Mutational analysis and rRNA probing studies show this region to be of functional significance (Noller et al. 1990; Powers and Noller 1991). The fact that the region's known structure is entirely self-contained (confined to these 46 nucleotides) makes it an excellent candidate for isolation and characterization by physical methods.

Another important RNA structural element identified by rRNA sequence comparisons is the so-called tetraloop (Woese et al. 1990b), a loop of four nucleotides that determines (caps) a double-helical stem of two or more base pairs. Such structures account for the majority of all hairpin loops in rRNA (Gutell et al. 1993). The most interesting characteristic of tetraloops is that the sequence in the loop is highly constrained. The naturally occurring examples are confined to an extremely limited subset of the 256 possible permutations of the four nucleotides. The compositions of the first and last bases in the loop are strongly coordinated, and to a lesser extent, the compositions of the inner two bases are as well. Three motifs cover the overwhelming majority of the naturally occurring tetraloops in rRNA; UNCG, CUYG, and GMRA (Woese et al. 1990b). In the small subunit of rRNA, the predominant compositions for each of the three major types are UUCG, CUUG, and GMAA (Woese et al. 1990b). The composition of the terminal base pair in the underlying stem is related to the sequence in the loop, especially for two of the three loop compositions: The UUCG loop strongly favors a C:G closure; CUUG loops are usually closed by a G:C pair, and the closing pair for the GCAA loop, although less constrained, tends to be R:Y (Woese et al. 1990b).

Some of the tetraloops in the small subunit rRNA are invariant or change only slowly in sequence over evolutionary time, whereas others change with remarkable frequency. In both cases, the variation is far from random, with the vast majority of variants tending to conform to one of the above three general types. The most frequently changing of the small subunit rRNA tetraloops is that located at positions 83-86 (Fig. 1). More than 50 phylogenetically independent alterations in loop sequence have been recorded among the Bacteria. The loop has a UUCG, CUUG, or GCAA sequence in 93% of cases (Woese et al. 1990b). Of the remaining 7%, most are themselves tetraloops whose compositions are related to one of the above three. In only 3% or so of cases does the size of the loop vary; and then, by addition or deletion of a single nucleotide (Woese et al. 1990b). The UUCG loop at position 83-86 shows a C:G closing pair 91% of the time, the CUUG loop here closes with a G:C pair in 95% of cases, and 86% of the GCAA loop closures have an R/Y composition (mainly A:U). Solution NMR spectroscopic analysis of two of the characteristic tetraloops, C(UUCG)G and C(GMAA)G (bases outside the parentheses form the closing pair), suggest structures in which the two terminal bases in the loop itself interact to form an unusual, noncanonical pair, which would seem to explain their rather strict covariation (Heus and Pardi 1991; Varani et al. 1991).

Tetraloops of the types described above are not confined to rRNAs; they clearly serve various other functions in the cell. They have been reported to occur in the context of controlling gene expression in T4 phage (Tuerk et al. 1988). Covariation analyses of the group 1 intron (Michel and Westhof 1990) and of ribonuclease P (Brown et al. 1996; Tanner and Cech 1995; Massire et al. 1997) also show that GNRA tetraloops are involved in long-range tertiary interactions. Comparative results are bolstered by a wealth of biochemical data (Jaeger et al. 1994; Murphy and Cech 1994; Pley et al. 1994; Abramovitz and Pyle 1997), and recently the crystal structure of a portion of the group 1 intron provided a direct visualization of such an interaction (Cate et al. 1996). The tetraloop/receptor helix docking interaction is now considered a common structural motif in large RNAs (Costa and Michel 1995; Abramovitz and Pyle 1997).

The Scope and Utility of Comparative Analysis

Comparative sequence analysis utilizes a highly abstract presentation of a molecule, a mere string of symbols representing the four nucleotides. In its own right, comparative analysis reveals only patterns involving these symbols; it says nothing about actual molecular structure (or function). However, when combined with structural knowledge derived from the more direct physical methods of measurement, comparative analysis takes on physical meaning, and so, becomes a more powerful and useful approach to analysis of structure. The potential of this link between comparative analysis and actual structure determination should be fully appreciated and exploited. When the physical chemists pay attention to the results of comparative analysis, it helps them to uncover biologically important and physically interesting molecular structures. What is not generally recognized is that as the physicist's repertoire of biologically important structural motifs increases, the potential for comparative analysis of primary structure to reveal the existence of physical structure increases proportionately. Thus, comparative analysis is increasingly useful as an adjunct to physical and chemical determinations of molecular structure.

One example of the interplay between physical characterizations and comparative analysis is given by work on G:A-type pairs. There are many instances in which otherwise canonical helices contain GpA doublets paired (antiparallel) with GpA, but never ApG paired with ApG. Such a bias among natural RNA indicates a profound difference between the two types of pairing. Based on these comparative data, Turner, Wilson, and their respective colleagues have examined thermodynamic and structural properties of RNA and DNA oligonucleotide duplexes containing the two types of paired doublets (SantaLucia et al. 1990; Li et al. 1991). They found that pairs formed from GpA are as energetically favorable as canonical pairs would be, but that the ApG pairing is unstable. Spectroscopic analysis revealed, moreover, that the more stable GpA pairing has a novel (non-Watson-Crick) geometry.

In a similar way, comparative analysis can serve as a guide in molecular genetic analyses. Much of the molecular biologist's approach to understanding and utilizing molecular structure and function turns upon manipulation of genetic sequences. Since sequence space is enormous, there is no way the molecular biologist can efficiently explore it without some kind of map. Taking random mutational "shots" at the ribosome, for example, is a most unproductive way to attempt to uncover the molecular basis for ribosome function. Comparative analysis can provide an initial guide and give essential clues that turn an intractable task into a manageable one. The following are some examples.

A common need in the study of macromolecules is the identification of features responsible for any activity; for instance, the structural elements involved in catalysis by ribozymes. Sequences responsible for the catalytic activity of the so-called hammerhead self-cleaving RNA (see Chapter 12) could be identified by their conservation throughout a collection of catalytic "satellite" RNAs (and their complements) with otherwise extremely variable sequences (Forster and Symons 1987a). Deletion analysis then demonstrated that the catalytic structure could be reduced to the only common structure in all the RNAs, about 50 nucleotides (Forster and Symons 1987b). Similar sequence comparisons coupled with deletion analyses have revealed functional aspects of self-splicing introns (see Chapter 13).

A more complex example of the identification of minimum catalytic RNA structure is that of ribonuclease P RNA. RNase P is the enzyme that cleaves 5'-leader sequences from precursor forms of tRNA in all cells. RNase P occurs as a complex holoenzyme in vivo, a protein–RNA complex. In vitro, however, at high ionic strength, the bacterial RNA is capable of catalyzing the reaction independently of the protein (Guerrier-Takada et al. 1983); thus, bacterial RNase P is a ribozyme. In contrast, the archaeal and eucaryal RNase P RNAs are inactive in the absence of the protein constituents of the holoenzyme. Substantial variation in sequence and length of RNase P RNAs from different organisms made alignment and comparative structure analysis difficult, but the variations are now reconciled in the context of a universally applicable RNase P RNA secondary structure (Fig. 2). The core bacterial secondary structural elements and several key base identities are also conserved in Archaea and Eucarya. This extent of conservation indicates that the archaeal and eucaryal versions of the RNA remain the catalytic center of RNase P, even though they no longer function independently of protein.

The approximately 130 diverse sequences of bacterial RNA that are now available constitute a comprehensive mutational survey of this cat-

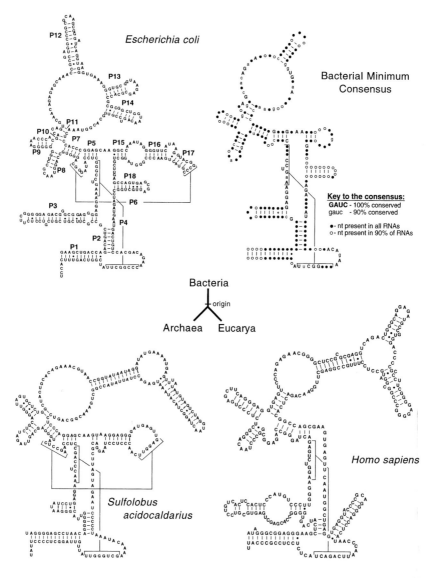

Figure 2 The universality of the core structure of RNase P RNA. Representatives of RNase P RNAs from the three domains of life are shown juxtaposed to a bacterial minimum consensus secondary structure, as described in the text. Helices are numbered for the *E. coli* RNA from 5' to 3' and designated with P ("pairing," e.g., P1, P2). Watson-Crick base pairs are represented with a line, and noncanonical base pairs are indicated with a full circle. The phylogenetic relationships of the organisms corresponding to the RNase P RNAs are shown in the central three-domain tree.

alytic RNA. The results are summarized in a "phylogenetic minimum consensus" structure of the bacterial RNase P RNA (Fig. 2), the minimum sequence and length present in each bacterial RNase P RNA. All RNase P RNAs contain sequence-length not present in some other instance of the RNA, so the length of the minimum consensus is only one-half to two-thirds that of the typical native RNA. Nonetheless, since all bacterial RNase P RNAs contain the minimum-consensus structure, it is expected that this structure should contain all the elements required for catalytic activity.

In constructing a phylogenetic-minimum RNase P RNA to test this notion, sequences in one type of the RNA were replaced with the shorter, corresponding sequences from the RNAs of other organisms. An early design was based on the minimum homologous sequence-lengths of *E. coli* and *Bacillus megaterium* RNase P RNAs, and resulted in Min 1 RNA, a highly active, 263-nucleotide RNA. When the RNase P RNA sequence from *Mycoplasma fermentans* became available, the size of the minimum RNase P RNA could be reduced even further. The RNase P RNA subunit from *M. fermentans*, at 276 nucleotides, is the smallest known RNase P RNA to date. This RNA, and those of *Chlorobium limicola* and *Mycoplasma hyopneumonia*, collectively, lack helices previously thought to be universally conserved in bacterial RNase P RNAs (Siegel et al. 1996). Assembly of minimum conserved sequence-lengths resulted in a new RNA, termed Micro P, shown in Figure 3. This 211-nucleotide RNA represents the phylogenetic-minimum RNase P RNA, a structure using only those sequences or secondary structural elements that are present in all RNase P RNAs (Fig. 3). The properties of Micro P RNA are similar to those of Min 1 RNA: It is highly active at a high ionic strength and in the presence of high concentrations of magnesium. A requirement for higher ionic strength than required by the native RNAs reflects a global destabilization of structure by removal of the peripheral helices. Nonetheless, the simplified, phylogenetic-minimum RNA retains all of the structural elements needed for catalysis. It is unlikely that a successful, piecemeal removal of nearly 50% of the molecular length of a native RNA could have been accomplished without approaching the problem from a comparative perspective.

Comparative Analysis as Interpretive Perspective

The interpretations of experimental results in molecular biology are often ambiguous because of the complexity of the systems and the consequent difficulty of differentiating meaningful from trivial information. Results

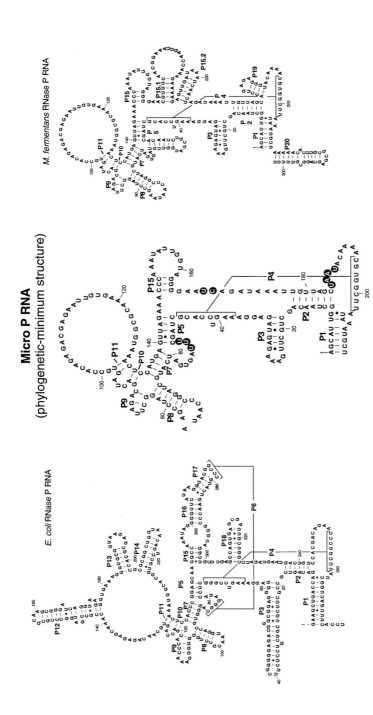

Figure 3 Design of Micro P—the smallest, functional RNase P RNA and a reflection of the phylogenetic-minimum bacterial consensus RNase P RNA. Secondary structures are shown for *E. coli* and *M. fermentans* RNase P RNAs, and the synthetic Micro P RNase P RNA (see text). Filled circles in the Micro P RNA identify nucleotides used to replace helices in the *M. fermentans* RNase P RNA in the design of the Micro P RNA. Base pairs in helix P4 and P6 are indicated with connecting lines. Helix numbering is as in Fig. 2.

derived from chemical cross-linking, footprinting, and nuclease or chemical structure-mapping experiments, for example, contain a wealth of information. However, it is often impossible to distinguish completely the significant data from the background of trivial data (idiosyncratic for the particular organism or an artifact of the particular experimental method) that invariably accompanies them. Such data are commonly presented with little attempt to evaluate their meaning and significance. Comparative studies with homologous molecules from different organisms add depth and precision to the interpretation of these kinds of data. Comparisons identify which elements in the data set correspond to general characteristics, universal properties, at the same time pointing out the possibly trivial results, the idiosyncratic ones.

One example of using the comparative approach in this manner involves a cross-linking analysis to locate the active site of RNase P RNA (Burgin and Pace 1990). In the study, an arylazide photoaffinity cross-linking agent was attached to the 5′-terminal phosphate in tRNA, the phosphate that is acted upon by RNase P. Ultraviolet irradiation cross-links the substrate to RNase P RNA, and sites of cross-linking can be identified by primer extension analysis. Three different RNase P RNAs, from *E. coli, Bacillus subtilis*, and *Chromatium vinosum*, were used in the study. These three RNAs differ extensively in sequence and in the presence or absence of some structural elements. Several nucleotides in each RNA formed cross-links with the photoagent-containing substrate. However, only a subset of these cross-linked nucleotides was found to be common to all three tRNAs. This subset is distributed in the core of all RNase P RNAs known and subsequently was shown to comprise the heart of the RNase P catalytic center (Harris and Pace 1995; Kazantsev and Pace 1998).

Similar site-specific photoaffinity cross-linking methodology with a comparative perspective has been used to identify the global architecture of the RNase P RNA. Strategic placement of photoagents at various sites in the *E. coli* and *B. subtilis* RNase P RNAs, with cross-linking and primer extension analysis, localizes regions of the RNAs relative to the rest of the molecule, and to the substrate. Photoagents at homologous positions in the two RNAs resulted in nearly identical cross-linking patterns, consistent with the notion that these two RNAs contain a common, core tertiary architecture (Fig. 4A, B) (Chen et al. 1998). Molecular modeling of the library of distance constraints has resulted in essentially coincident tertiary structure representations of the RNase P RNA–tRNA complexes (Fig. 4) (Harris et al. 1994; Harris et al. 1997; Chen et al. 1998). The models consist of two juxtaposed, multi-helix domains with the highly con-

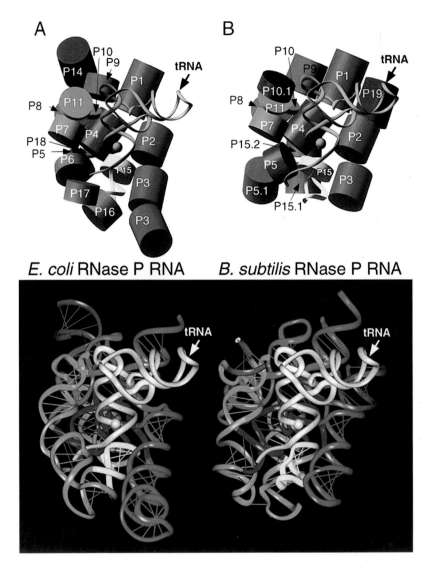

Figure 4 Tertiary structure models of *E. coli* (A) and *B. subtilis* (B) RNase P RNAs. The upper panel of the figure is a helix-barrel model where each cylindrical barrel presents the location of an A-form RNA helix of appropriate length. tRNA is displayed as a ribbon with a highlighted phosphate at the scissile bond. The lower panel is an extrapolation of the barrel model, showing the entire RNA backbone as a colored ribbon, with bars between ribbons idealizing base pairs. Helices P12, P13, and P14 in the *E. coli* model, and helices P10.1 and P12 in the *B. subtilis* model are not represented in this model due to the lack of experimental data available to position these conserved elements with the same degree of certainty as the rest of the helices shown in the model.

served regions of the molecules at the core of the models and the more variable helical elements near the periphery. Additionally, both models situate the scissile bond in a pre-tRNA immediately adjacent to the universally conserved P4 helix of the RNase P RNA. The agreement between the models of the two bacterial RNase P RNAs validates each of the proposals and testifies to the utility of combining phylogenetic-comparative analysis with biochemical experimentation.

RNA as Historical Record

The availability of rRNA sequences has allowed quantitative analysis of evolutionary relationships. Tapping the phylogenetic information in rRNA has had a revolutionary effect on microbiology, adding an evolutionary dimension where there had been none before. The prior lack of an evolutionary framework in microbiology was the unavoidable consequence of the fact that morphologies and physiologies of microorganisms are too simple and/or unpredictably variable to be of significant value in developing a natural classification. It is no wonder that, using those properties, microbiologists had been unable to produce a valid, natural (phylogenetic) classification from the bacteria. However, as Zuckerkandl and Pauling (1965) pointed out, at the molecular level there is no such problem; molecular sequences are historical records rich in readily interpretable geneaological information. Thus it was possible, by comparison of partial sequences of 16S rRNAs (oligonucleotide catalogs), to make initial sense of microbial genealogical relationships (Fox et al. 1980; Woese et al. 1985). The resulting phylogenetic framework, the natural classification, brings clarity and order to the day-to-day conduct of microbiology. Experimental progress is enhanced; deeper interpretations of results are now possible; new directions of research are indicated. Microbial ecology has come of age: The capacity to identify microorganisms phylogenetically using rRNA sequence-based techniques, even without cultivation (Pace et al. 1985; Pace 1997) and in situ (DeLong et al. 1989), gives microbial ecologists a power they have always lacked for comprehensive analysis of various niches. Not only can organisms be identified with greater precision and characterized in greater depth, but unculturable or previously unrecognized organisms in a particular niche can now be definitely related to those in other niches.

The early studies of microbial phylogeny based on rRNA sequences yielded the remarkable finding that the world of prokaryotes, which all biologists had taken to be phylogenetically unified (monophyletic), was indeed not so. There exist two distinct kinds of prokaryotes, now formally called the domains Archaea and Bacteria (Woese et al. 1990a), that are no more closely related to one another than either is to the eucaryotes. Although the early oligonucleotide cataloging approach could readily define and distinguish three primary groupings of organisms (Archaea, Bacteria, and Eucarya), it could not relate them to one another in any precise way. This joining had to await the technology that permitted determination of complete sequences of rRNAs (or their genes), which in turn allow construction of a universal phylogenetic tree. This tree is shown in Figure 5 (Pace 1997); its root has been inferred by the so-called Dayhoff strategy. In this method, an uprooted tree generated from a set of (homologous) sequences is rooted using a related (or paralogous) sequence. To root a tree that spans all extant life, it is necessary that the gene duplication that originally produced the paralogous genes occurred in the ancestral stem, prior to the initial phylogenetic radiations. The root of the universal tree is seen to lie between the Bacteria on the one hand and the common lineage that the Archaea and Eucarya share, on the other (Gogarten et al. 1989; Iwabe et al. 1989). The universal tree, in other words, predicts that the Archaea are specific relatives of the Eucarya (albeit at a deep level), to the exclusion of the Bacteria. This assessment is now borne out by many macromolecular sequences derived from the accumulating sequences of archaeal genomes (Olsen and Woese 1997). Since there exist characteristic archaeal, bacterial, and eucaryal versions for just about every universal macromolecular function so far characterized in the cell, there can be no doubt that all life on this planet is organized into three, very distinctive groupings. At the levels of the nucleic acid-based information transfer machinery, Archaea and Eucarya resemble one another more closely than either resembles Bacteria.

The Archaea

From a phenotypic perspective, based on relatively few instances of cultivated organisms, the Archaea would seem to be a rather strange and disparate collection. Unlike the Bacteria, they show only a few major phenotypes: the methanogenic; the extremely halophilic; the sulfate-reducing; and the sulfur-"dependent," extremely thermophilic phenotype (Woese 1987; Winker and Woese 1991). These four phenotypes are sufficiently dissimilar that, although they were known (except for the sulfate-reducing phenotype) before the Archaea were recognized as a phylogenetic unit, their relationship to one another was not recognized (this despite the existence of some molecular evidence suggestive of the relationship; e.g., the unusual archaeal ether-linked lipids [Winker and Woese 1991]). The four main archaeal phenotypes do not define four major taxa of equivalent taxonomic rank, however. As shown in Figure 6, cultivated Archaea comprise two major branches, two "kingdoms," the Euryarchaeota and the Crenarchaeota (Woese et al. 1990a). The Euryarchaeota encompass examples of all four archaeal phenotypes, with methanogens dominating the group. The methanogens comprise three major groups in addition to a very deeply branching lineage represented by the genus

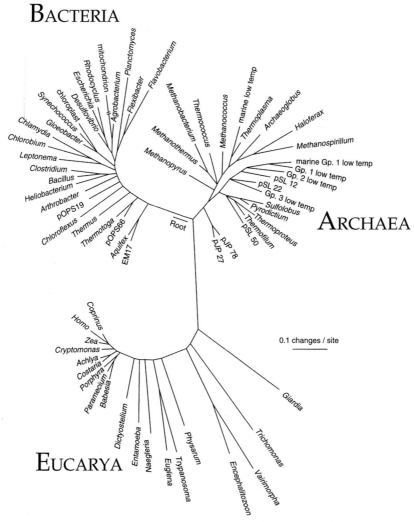

Figure 5 Rooted universal phylogenetic tree based on SSU rRNA sequences. The tree is based on maximum likelihood analysis of 64 rRNA sequences. (Reprinted, with permission, from Pace 1997 [copyright American Association for the Advancement of Science].)

Methanopyrus (Burggraf et al. 1991). Two of the three main methanogen lineages are phenotypically uniform, but the third, the *Methanomicrobiales* lineage, has spawned other phenotypes as well: the extreme halophiles, the sulfate reducers, and perhaps *Thermoplasma*. The remaining euryarchaeal lineage is the phylogenetically compact cluster of species that constitute the *Thermococcales*, a group phenotypically resembling the crenarchaeotes.

Our view of the potential phenotypic diversity of Crenarchaeota has recently expanded substantially with the development of methods for detecting uncultured organisms. These methods are PCR coupled with cloning to obtain naturally occurring rRNA genes and thereby to detect otherwise unknown organisms. Prior to the use of such methods to explore environmental Archaea, the taxon Crenarchaeota was considered phenotypically uniform; all of the sulfur-dependent, thermophilic type. With the application of molecular methods to the study of environmental organisms, however, numerous representatives of mesophilic Crenarchaeota recently have been discovered in marine (DeLong 1992; Fuhrman et al. 1992) and terrestrial (Ueda et al. 1995; Hershberger et al. 1996) environments. None of the low-temperature crenarchaeotes has yet been cultivated, so their metabolic basis remains unknown. The environmental surveys of Archaea additionally have revealed many thermophilic types of Crenarchaeota only distantly related to known organisms, including a third primary branch, Korarchaeota (Barns et al. 1996).

The distribution of phenotypes on the tree of Figure 6 strongly suggests the Archaea to be of thermophilic origin. All cultivated crenarchaeal species are thermophilic, some growing optimally at temperatures above 100°C; all the mesophilic Crenarchaeota originated from lineages that ancestrally were thermophilic. Additionally, the deepest branchings on the euryarchaeal side, the *Thermococcales* and the genus *Methanopyrus*, are thermophilic as well. This is true for the deepest branchings within all the major euryarchaeal sublineages. For the most part, Archaea are anaerobic; ability to grow aerobically, when it occurs, is generally facultative. Similarly, the Archaea are most often chemoautotrophic, particularly the deeply branching lineages. Thermophily, anaerobic growth, and chemoautotrophy, then, can be considered ancestral characteristics of the Archaea.

The surprising and scientifically inviting relationship of the Archaea to the Eucarya (Fig. 5) deserves comment. The sequences of many, but not all, archaeal genes resemble their eucaryotic homologs decidedly more than their bacterial homologs. Ribosomal protein sequences, for instance, are this way (Auer et al. 1989; Ramirez et al. 1989), as are the major subunits of the archaeal RNA polymerase (Pühler et al. 1989), the archaeal

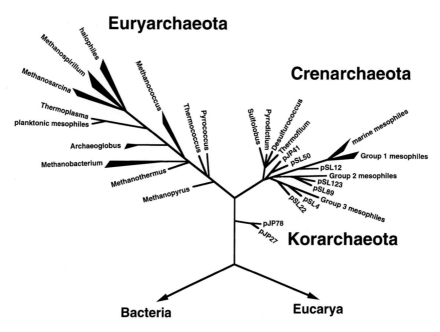

Figure 6 Phylogenetic tree of Archaea. The tree is based on maximum likelihood analysis of selected SSU rRNA sequences. (Figure provided by Scott C. Dawson.)

histone (Sandman et al. 1990), and a chaperone (Trent et al. 1991). Although the archaeal rRNA overall is an exception to this rule (in sequence and secondary structure it most resembles the bacterial type), the most highly conserved positions in the archaeal sequence do show a slight eucaryotic bias (Winker and Woese 1991).

Two examples of the above relationships between archaeal and eucaryal proteins are of particular interest, namely, RNA polymerase and histone. The archaeal RNA polymerase is more similar in sequence to both of the eucaryal polymerases II and III than these two are to each other (Pühler et al. 1989). Moreover, each of the ancestral sequence duplications that make up the archaeal TATA-binding protein are more like both of the corresponding human TATA-binding protein sequences than the human sequences are like one another. The case of the archaeal histone is even more striking. Its sequence is reported to be closer to the sequences of each of the four eucaryal histones, H2a, H2b, H3, and H4, than any of these four is to one another (Sandman et al. 1990). In other words, in these examples, the archaeal version of a molecular type appears to resemble the inferred common ancestor of a eucaryal family of proteins more closely than do any of the extant members of that eucaryal family.

Is this perhaps the tip of an iceberg? Do many families of eucaryotic genes have an archaeal homolog that resembles the (inferred) ancestor of that family more than its extant representatives do? Perhaps we can identify human gene families more readily by comparison with archaeal genes than by comparison of the members of the human gene families.

The Bacteria

Figure 7 shows a bacterial phylogenetic tree inferred from small subunit rRNA sequences. Anyone familiar with classic bacterial taxonomy will see immediately that the groupings defined by molecular sequence analysis bear little relationship to the groupings defined by classic methods (Woese 1987). For instance, microbiologists previously grouped all pho-

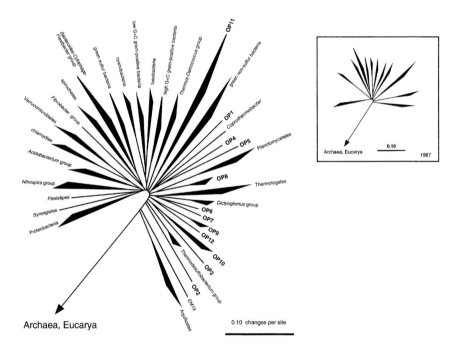

Figure 7 Phylogenetic tree of the Bacteria. The 36 division-level clades of Bacteria are depicted (Hugenholtz et al. 1998). The inset represents the known phylogenetic span of Bacteria in 1987 (Woese 1987), showing the 12 division-level clades known at the time. Filled sectors indicate that several representative sequences fall within the indicated depth of branching. Lines designated by OP represent one or more phylotypes that were identified in Obsidian Pool by means of molecular methods but have not been cultivated. (Reprinted, with permission, from Hugenholtz et al. 1998.)

totrophs into a single taxon, which contained few if any nonphotosynthetic species. In actuality, photosynthetic and nonphotosynthetic phenotypes are often intimately intermixed. Microbiologists formerly used morphology as a primary determinant of classification. Morphology turned out to be an extremely poor indicator of bacterial phylogenetic relationships, although there are a few notable exceptions, such as the spirochetes and the endospore-formers. Properties such as gliding motility, formerly used to group organisms into a few high-level taxa, also are phylogenetically widely dispersed. Gliding organisms, for instance, are intermixed with flagellated forms. The mycoplasmas, rickettsias, and a variety of other pathogens do not warrant the high-level taxonomic distinction previously accorded them on the basis of their parasitic character.

Our understanding of the breadth of bacterial diversity is rapidly changing due to molecular studies of rRNA sequences obtained from natural environments without cultivation. Of the approximately 40 bacterial division-level groups shown in Figure 7, representatives of only about 60% of them so far have been cultivated. Remarkably, since 1987 (Woese 1987), the number of recognized phylogenetic divisions has tripled (Pace 1997).

The distribution of phenotypes in the bacterial tree suggests that the Bacteria, like the Archaea, arose from a thermophilic ancestor: Thermophily is widespread in the bacterial tree, dominant in its deeper branches (Achenbach-Richter et al. 1987; Woese 1987). Moreover, no evidence exists to support the old notion that life began heterotrophically and that, consequently, the first organisms were heterotrophs; if anything, bacterial phylogeny is more consistent with an aboriginal autotrophy (Woese 1987). The fact that the most deeply divergent bacterial and archaeal lineages are thermophilic and chemoautotrophic indicates that the common ancestor of all life was of that nature, thermophilic and chemoautotrophic.

The Eucarya

The textbook picture of eucaryotes emphasizes four great "kingdoms": animals, plants, fungi, and protists. rRNA sequences tell a rather different story; see Figure 5. The animal, green plant, and fungal kingdoms are seen from that perspective to constitute some of the more superficial branchings on the tree, whereas the protist kingdom is a polyphyletic collection of (little-related) lineages, which together cover the full span of the eucaryal tree. Since mitochondria appear only among the higher branchings in the eucaryal tree, it seems likely that the aboriginal eucaryotic cell possessed none, a conclusion consistent with the fact that the earlier branchings on the tree presumably arose at a time when there was little or no free oxygen on this planet, prior to 2–2.5 billion years ago. Moreover, the eucaryal lineage originated before the ancestors of mitochondria, the α-purple Bacteria and relatives (Yang et al. 1985).

So far, the lower eucaryal branchings are defined only by mesophilic organisms, primarily parasitic species. These offer few clues as to the conditions under which their presumed free-living ancestors flourished, during the early, anaerobic phase in Earth history. Thus, the question of whether the eucaryotes, like the prokaryotes, arose from thermophilic ancestry cannot be addressed.

The Universal Ancestor

Biologists have long believed that all life ultimately arose from a common ancestor. The Universal Ancestor that biologists originally pictured was as simple as possible; it even lacked intermediary metabolism (Oparin 1964). It is only today, with a universal phylogenetic tree and a wealth of diverse sequence data, that we can begin to construct a realistic picture of this entity, one that can be experimentally tested and refined. The Universal Ancestor that we abstract from the genome sequence data is anything but simple, however. If properties that are present in all the modern domains are also in the Universal Ancestor, then the Ancestor would appear highly developed and metabolically rich, basically a modern cell. How else can we explain the spread of so many cellular functions, especially metabolic genes, across the full phylogenetic spectrum?

Yet, the idea that the Ancestor was basically a richly endowed, modern type of cell is not in keeping with the manner in which it appears to have evolved: that is, much more rapidly than have modern cells and in far more drastic ways. This is what would be expected for entities far simpler than modern cells, not ones equally or more complex (Woese 1987). Here, then, is a paradox, one that any consistent theory of the Universal Ancestor must resolve.

Life seems to have started in an RNA World—or at least in a world where polynucleotides played a far more prevalent role than they now do. The dynamic by which such simple living systems evolved must have been unique. It seems logical that the peculiar evolutionary dynamic that characterized the Universal Ancestor and its descent into the primary lineages reflects this earlier evolutionary world, when the cell itself, its basic mechanisms, were still in the throes of developing (Woese 1982, 1998). We suggest that the horizontal (lateral) gene transfer seen in the "modern" world is, as it were, a "background radiation" left over from an era when horizontal, not vertical, gene transfer dominated and defined the evolutionary dynamic (Woese 1998).

In the era of the Universal Ancestor, cells would have been simple and relatively ill-defined—in the number and kind of functions they possessed, in the size of their genomes, in the simplicity and imprecision of their information processing systems, in the sizes and types of proteins they possessed, and in their overall organization (Woese 1967, 1998; Woese and Fox 1977; Woese et al. 1983). Mutation rates were extraordinarily high. Lateral gene transfer was pervasive and pandemic; it applied to all genes and all cellular entities. In a sense there was universal genetic cross-talk among primitive cellular entities. They had yet to evolve to the stage where they were truly idiosyncratic in character. All shared a common evolutionary problem, the evolution of the basic cellular machinery; and its solution was a collective one. Any innovation that occurred in one cell line would readily be shared with other cell lines through lateral gene transfer (Woese 1998). It was in this way, not through vertical inheritance per se, that the basic cellular mechanisms evolved.

The primitive entities, called "progenotes" (Woese and Fox 1977), are taken to differ from one another metabolically. No one of them was sufficiently complex genetically to support a full metabolic capacity. Individual cells each with a specialized metabolic capacity could, however, form a loose-knit community that collectively was metabolically rich. It is such a diverse cellular community, in which individual cells (cell lines) communicated metabolically and genetically, that was the Universal Ancestor—not some particular organism or organismal lineage (Woese 1982, 1998). Modern analogs of this state can be seen in microbial consortia built on interdependent metabolisms.

When levels of lateral gene transfer are sufficiently high, organismal lineages cannot exist. Thus, at the stage of the Universal Ancestor there might have been relatively short-lived cell lines, but no true long-term organismal lineages (Woese 1998). The history of the Universal Ancestor was physical, not genealogical (Woese 1998). Only as cells and their componentry became more complex, idiosyncratic, and integrated did true lineages develop and become relatively refractory to the pervasive lateral gene transfer of the time (Woese 1998). From this point on, the cell (or subsystem) evolves vertically, i.e., has a genealogical history. This development resulted in the establishment of genealogical histories. Not all components in the cell became refractory to lateral gene transfer at the same stage; indeed the genes for some of them, e.g., the aminoacyl-tRNA synthetases and various metabolic enzymes, seem readily transferred laterally even today. However, complex componentries that are tightly integrated into the cell, such as the ribosome and the transcription apparatus, would have been among the first to become refractory to lateral gene transfer (Woese 1998).

One consequence of the early massive lateral gene transfer is that the universal rRNA tree is not a normal organismal tree. Its root and primary branchings are from a time when only a few of the cellular functions were refractory to lateral gene transfer, too few to constitute an organismal lineage. The rRNA tree grew only as increasing numbers of cellular functions became more or less refractory to lateral gene transfer. The rRNA tree became a true organismal one only in its more peripheral branches, as more and more components of the cell became refractory to lateral gene transfer. Thus, the universal tree does not start with an organism of the modern type, representing a specific modern cell. It starts at an earlier stage, near the end of the era of the Universal Ancestor (Woese 1998). This means, fortunately, that the tree provides an evolutionary framework that takes us back into the era when cells were still evolving.

ACKNOWLEDGMENTS

The authors' research activities are supported by grants from the National Science Foundation and the National Aeronautics and Space Agency (C.R.W.), and the National Institutes of Health (B.C.T. and N.R.P.). We thank Drs. Robin R. Gutell, Phil Hugenholtz, and Scott C. Dawson for figures.

REFERENCES

Abramovitz D.L. and Pyle A.M. 1997. Remarkable morphological variability of a common RNA folding motif: The GNRA tetraloop-receptor interaction. *J. Mol. Biol.* **266:** 493–506.

Achenbach-Richter L, Gupta R, Stetter K.O., and Woese CR. 1987. Were the original eubacteria thermophiles? *Syst. Appl. Microbiol.* **9:** 34–39.

Auer J, Lecher, K., and Böck A. 1989. Gene organization and structure of two transcriptional units from *Methanococcus* coding for ribosomal proteins and elongation factors. *Can. J. Microbiol.* **35:** 200–204.

Barns S.M., Delwiche C.F., Palmer J.D., and Pace N.R. 1996. Perspectives on archaeal diversity, thermophily and monophyly from environmental rRNA sequences. *Proc. Natl. Acad. Sci.* **93:** 9188–9193.

Brosius J., Dull, T.J., and Noller, H.P. 1980. Complete nucleotide sequence of a 23S ribosomal RNA gene from *Escherichia coli. Proc. Natl. Acad. Sci.* **77:** 201–204.

Brown J.W., Nolan J.M., Haas E.S., Rubio M.A.T., Major F., and Pace N.R. 1996. Comparative analysis of ribonuclease P RNA using gene sequences from natural microbial populations reveals tertiary structural elements. *Proc. Natl. Acad. Sci.* **93:** 3001–3006.

Burggraf S., Stetter K.O., Rouviere P., and Woese C.R. 1991. *Methanopyrus kandleri*: An archaeal methanogen unrelated to all other known methanogens. *Syst. Appl. Microbiol.* **14:** 346–351.

Burgin A. and Pace N.R. 1990. Mapping the active site of ribonuclease P RNA using a substrate containing a photoaffinity agent. *EMBO J.* **9:** 4111–4118.

Cate J.H., Gooding A.R., Podell E., Zhou K., Golden B.L., Kundrot C.E., Cech T.R., and Doudna J.A. 1996. Crystal structure of a group I ribozyme domain: Principles of RNA packing. *Science* **273:** 1678–1685.

Chen J.-L., Nolan J.M., Harris M.E., and Pace N.R. 1998. Comparative photocrosslinking analysis of the tertiary structures of *Escherichia coli* and *Bacillus subtilis* RNase P RNAs. *EMBO J.* **17:** 1515–1525.

Costa M. and Michel F. 1995. Frequent use of the same tertiary motif by self-folding RNAs. *EMBO J.* **14:** 1276–1285.

DeLong E.F. 1992. Archaea in coastal marine environments. *Proc. Natl. Acad. Sci.* **89:** 5685–5689.

DeLong E.F., Wickham G.S., and Pace. N.R. 1989. Phylogenetic stains: Ribosomal RNA-based probes for the identification of single cells. *Science* **243:** 1360–1363.

Erdmann V. 1976. Structure and function of 5S and 5.8S RNA. *Prog. Nucleic Acid Res. Mol. Biol.* **18:** 45–90.

Forster A.C. and Symons R.H. 1987a. Self-cleavage of plus and minus RNAs of a virusoid and a structural model for the active sites. *Cell* **49:** 211–220.

———. 1987b. Self-cleavage of virusoid RNA is performed by the proposed 5S nucleotide active side. *Cell* **50:** 9–16.

Fox G.E. and Woese C.R. 1973. 5S RNA secondary structure. *Nature* **256:** 505–507.

Fox G.E., Stackenbrandt E., Hespell R.B., Gibson J., Maniloff J., Dyer T.A., Wolfe R.S., Baich W.E., Tanner R., Magrum L., Zablen L.B., Blakemore R., Gupta R., Bonen L., Lewis B.J., Stahl D.A., Luehrsen K.R., Chen K.N., and Woese C.R. 1980. The phylogeny of prokaryotes. *Science* **209:** 457–463.

Fuhrman J.A., McAllum K., and Davis A.A. 1992. Novel major archaebacterial group from marine plankton. *Nature* **356:** 148–149.

Gautheret D., Damberger S.H., and Gutell R.R. 1995. Identification of base triples in RNA using comparative sequence analysis. *J. Mol. Biol.* **248:** 27–43.

Gogarten J.P., Kibak H., Dittrich P., Taiz L., Bowman E.J., Bowman B.J., Manolson M.F., Poole R.J., Date T.E., Oshima T., Konishi J., Denda K., and Yoshida M. 1989. Evolution of the vacuolar H^+-ATPase: Implication for the origin of eukaryotes. *Proc. Natl. Acad. Sci.* **86:** 9355–9359.

Guerrier-Takada C., Gardiner K., Marsh T., Pace N.R., and Altman S. 1983. The RNA moiety of ribonuclease P is the catalytic subunit of the enzyme. *Cell* **35:** 849–857.

Gutell R.R., Larsen N., and Woese C.R. 1993. Lessons from an evolving ribosomal RNA: 16S and 23S rRNA structure from a comparative perspective. In *Ribosomal RNA Structure, evolution, gene expression and function in protein synthesis* (ed. R.A. Zimmerman and A.E. Dahlberg). Tellford Press, Caldwell, New Jersey.

Gutell R.R. 1994. Collection of small subunit (16S- and 16S-like) ribosomal RNA structures: 1994. *Nucleic Acids Res.* **22:** 3502–3507.

Harris M.E. and Pace N.R. 1995. Identification of phosphates involved in catalysis by the ribozyme RNase P RNA. *RNA* **1:** 210–218.

Harris M.E., Kazantsev A., Chen J.-L., and Pace N.R. 1997. Analysis of the tertiary structure of the ribonuclease P ribozyme-substrate complex by site-specific photoaffinity crosslinking. *RNA* **3:** 561–576.

Harris M.E., Nolan J.M., Malhotra A., Brown J.W., Harvey S.C., and Pace, N.R. 1994. Use of photoaffinity cross-linking and molecular modeling to analyze the global architecture of ribonuclease P RNA. *EMBO J.* **13:** 3953–3963.

Heus H.A. and Pardi A. 1991. Structural features that give rise to the unusual stability of RNA hairpins containing GNRA loops. *Science* **253:** 191–194.

Hershberger K.L., Barns S.M., Reysenbach A.-L., and Pace N.R. 1996. Wide diversity of Crenarchaeota. *Nature* **384:** 420.

Hugenholtz P., Pitulle C., Hershberger K.L., and Pace N.R. 1998. Novel division level bacterial diversity in a Yellowstone hot spring. *J. Bacteriol.* **180:** 366–376.

Iwabe N., Kuma K., Hasegawa M., Osawa S., and Miyata T. 1989. Evolutionary relationship of archaebacteria, eubacteria and eukaryotes inferred from phylogenetic trees of duplicated genes. *Proc. Natl. Acad. Sci.* **86:** 9355–9359.

Jaeger L., Michel F., and Westhof E. 1994. Involvement of a GNRA tetraloop in long-range RNA tertiary interactions. *J. Mol. Biol.* **236:** 1271–1276.

Kazantsev A.V. and Pace N.R. 1998. Identification by modification—interference of purine N-7 and ribose $2'$-OH groups critical for catalysis by bacterial ribonuclease P. *RNA* **4:** (in press).

Kim S.-H. 1979. Crystal structure of yeast tRNA^{phe} and general structural features of other tRNAs. In *Transfer RNA: Structure, properties, and recognition* (ed. P.R. Schimmel et al.), pp. 83–100. Cold Spring Harbor Laboratory, Cold Spring Harbor, New York.

Li Y., Zon G., and Wilson W.D. 1991. Thermodynamics of DNA duplexes with adjacent G-A mismatches. *Biochemistry* **31:** 7566–7572.

Massire C., Jaeger L., and Westof E. 1997. Phylogenetic evidence for a new tertiary interaction in bacterial RNase P RNAs. *RNA* **3:** 553–556.

Michel F. and Westhof E. 1990. Modeling of the three-dimensional architecture of group I introns based on comparative sequence analysis. *J. Mol. Biol.* **216:** 585–610.

Murphy F.L. and Cech T.R. 1994. GAAA tetraloop and conserved bulge stabilize tertiary structure of a group I intron domain. *J. Mol. Biol.* **236:** 49–63.

Noller H.P., Moazed D., Stern S., Powers T., Allen P.N., Robertson J.M., Weiser B., and Triman K. 1990. Structure of rRNA and its functional interaction in translation. In *The ribosome: Structure, function and evolution* (ed. W.E. Hill et al.), pp. 73–92. American Society for Microbiology, Washington, D.C.

Noller H.P., Kop J., Wheaton V., Brosius J., Gutell R., Kopylov A.M., Dohme F., Herr W., Stahl D.A., Gupta R., and Woese C.R. 1981. Secondary structure model for 23S ribosomal RNA. *Nucleic Acids Res.* **9:** 6167–6189.

Olsen G.J. and Woese C.R. 1997. Archaeal genomics: An overview. *Cell* **89:** 991–994.

Oparin A.I. 1964. *The chemical origin of life* (transl. A. Synge). Charles C Thomas, Springfield, Illinois.

Pace N.R. 1997. A molecular view of microbial diversity and the biosphere. *Science* **276:** 734–740.

Pace N.R., Stahl D.A., Lane D.J., and Olsen G.J. 1985. Analyzing natural microbial populations by rRNA sequences. *Am. Soc. Microbiol. News* **51:** 4–12.

Pley H.M., Flaherty K.M., and McKay D.B. 1994. Model for an RNA tertiary interaction from the structure of an intermolecular complex between a GAAA tetraloop and an RNA helix. *Nature* **372:** 111–113.

Powers T. and Noller H.P. 1991. A functional pseudoknot in 16S ribosomal RNA. *EMBO J.* **10:** 2203–2214.

Pühler G., Leffers H., Gropp P., Palm P., Klenk H.-P., Lottspeich P., Garrett R.A., and Zillig W. 1989. Archaebacterial DNA-dependent RNA polymerases testify to the evolution of the eucaryotic nuclear genome. *Proc. Natl. Acad. Sci.* **86:** 4569–4573.

Ramirez C., Shimmin L.C., Newton C.H., Matheson A.T., and Dennis P.P. 1989. Structure and evolution of the L11, L1, L10, and L12 equivalent ribosomal proteins in eubacteria, archaebacteria and eukaryotes. *Can. J. Microbiol.* **35:** 234–244.

Sandman K., Krzycki J.A., Dobrinski B., Lurz R., and Reeve J.N. 1990. DNA binding protein HMf, isolated from the hyperthermophilic archaea *Methanothermus fervidus*, is most closely related to histones. *Proc. Natl. Acad. Sci.* **87:** 5788–5791.

SantaLucia J., Jr., Kierzek R., and Turner D.H. 1990. Effects of GA mismatches on the structure and thermodynamics of RNA internal loops. *Biochemistry* **29:** 8813–8819.

Siegel R.W., Banta A.B., Haas E.S., Brown J.W., and Pace N.R. 1996. Mycoplasma fermentans simplifies our view of the catalytic core of ribonuclease P RNA. *RNA* **2:** 452–462.

Tanner M.A. and Cech T.R. 1995. An important RNA tertiary interaction of group II introns is implicated in gram-positive RNase P RNAs. *RNA* **1:** 349–350.

Trent J.D., Nimmersgern E., Wall J.S., Harti F.-U., and Horwich A.L. 1991. A molecular chaperone from a thermophilic archaebacterium is related to the eukaryotic protein t-complex polypeptide-1. *Nature* **354:** 490–493.

Tuerk C., Gauss P., Thermes C., Groebe D.R., Gayle M., Guild N., Stormo G., D'Aubenton-Carafa Y., Uhlenbeck O.C., Tinoco I., Brody E.N., and Gold L. 1988. CUUCGG hairpins: Extraordinarily stable RNA secondary structures associated with various biochemical processes. *Proc. Natl. Acad. Sci.* **85:** 1364–1368.

Ueda T.Y., Suga Y., and Matsuguchi T. 1995. Molecular phylogenetic analysis of a soil microbial community in a soybean field. *Eur. J. Soil Sci.* **46:** 415–421.

Varani G., Cheong C., and Tinoco I., Jr. 1991. Structure of an unusually stable RNA hairpin. *Biochemistry* **30:** 3280–3289.

Wilson K.S. and Noller H.F. 1998. Molecular movement inside the translational engine. *Cell* **92:** 337–349.

Winker S. and Woese C.R. 1991. A definition of the domains archaea, bacteria and eucarya in terms of small subunit ribosomal RNA characteristics. *Syst. Appl. Microbiol.* **14:** 305–310.

Woese C.R. 1967. *The genetic code: The molecular basis of genetic expression*. Harper and Rowe, New York.

———. 1982. Archaebacteria and cellular origins: An overview. Zentbl. *Bakteriol. Mikrobiol. Hyg. Abt. 1 Orig. C* **3:** 1–17.

———. 1987. Bacterial evolution. *Microbiol. Rev.* **51:** 221–271.

———. 1998. The universal ancestor. *Proc. Natl. Acad. Sci.* **95:** 6854–6859.

Woese C.R. and Fox G.E. 1977. The concept of cellular evolution. *J. Mol. Evol.* **10:** 1–6.

Woese C.R., Kandler O., and Wheelis M.L. 1990a. Toward a natural system of organisms: Proposal for the domains Archaea, Bacteria, and Eucarya. *Proc. Natl. Acad. Sci.* **87:** 4576–4579.

Woese C.R., Winker S., and Gutell R.R. 1990b. Architecture of ribosomal RNA: Constraints on the sequence of tetra-loops. *Proc. Natl. Acad. Sci.* **87:** 8467–8471.

Woese C.R., Gutell R., Gupta R., and Noller H.P. 1983. Detailed analysis of the higher-order structure of 16S-like ribosomal ribonucleic acids. *Microbiol. Rev.* **47:** 621–669.

Woese C.R., Stackenbrandt E., Macke T.J., and Fox G.E. 1985. A phylogenetic definition of the major eubacterial taxa. *Syst. Appl. Microbiol.* **6:** 143–151.

Woese C.R., Magrum L.J., Gupta R., Siegel R.B., Stahl D.A., Kop J., Crawford N., Brosius J., Gutell R., Hogan J.J., and Noller H.P. 1980. Secondary structure model for bacterial 16S robosomal RNA: Phylogenetic, enzymatic and chemical evidence. *Nucleic Acids Res.* **8:** 2275–2293.

Yang D., Oyaizu Y., Oyaizu H., Olsen G.J., and Woese C.R. 1985. Mitochondrial origins. *Proc. Natl. Acad. Sci.* **82:** 4443–4447.

Zuckerkandl E. and Pauling L. 1965. Molecules as documents of evolutionary history. *J. Theor. Biol.* **8:** 357–366.

WWW RESOURCE

http://pundit.colorado.edu:8080/RNA/16S/eubacteria.html (eu)BACTERIA 16S rRNA Comparative Structure Database.

5
Re-creating an RNA Replicase

David P. Bartel
Whitehead Institute
and Department of Biology
Massachusetts Institute of Technology
Cambridge, Massachusetts 02142

The idea that certain RNA sequences can use the information from one RNA strand to accurately and efficiently produce another RNA is central to the RNA World hypothesis (Pace and Marsh 1985; Sharp 1985; Cech 1986; Orgel 1986). Such RNA-dependent RNA polymerase ribozymes would have been responsible for replicating the ribozymes of the RNA World, including themselves (via their complement sequences). The most extreme versions of the RNA World hypothesis suggest that life began with a self-replicating system seeded by a pair of replicase molecules, one serving as the template, the other as the enzyme. Less extreme versions of the RNA World hypothesis cite difficulties with prebiotic synthesis and stability of RNA on the early Earth and instead postulate that life originated in a "pre-RNA world"—an era dominated by an RNA-like polymer that could catalyze reactions and code for its own replication (Chapter 2). Because the pre-RNA polymer presumably facilitated production of RNA, issues regarding prebiotic availability of RNA are less problematic in this conservative RNA World view. However, the scenario calls for eventual transition to the RNA World, and thus still relies on some radical assumptions regarding the intrinsic catalytic capability of RNA—including the idea that there exist, somewhere among all RNA sequence possibilities, sequences that can promote the replicase reaction. The RNA replicase is one of the few assumptions common to all versions of the RNA World hypothesis.

Because nearly four billion years of evolution have obscured any vestiges of the presumed RNA replicases, our only recourse for testing the replicase assumption is to examine the intrinsic enzymatic properties of RNA and determine whether they are compatible with the replicase notion. To some extent, we are able to do this by studying extant ribozymes. However, the ultimate way to address the replicase assumption would be to (re)create an RNA polymerase ribozyme that can propagate itself and improved variants of itself. In addition to validating the replicase assumption, such a molecule would be the key ingredient for creating a self-sustained system capable of Darwinian evolution—providing a valuable working model for the RNA World, and breathing life back into RNA, the presumed parent of all contemporary biopolymers and life forms.

For a polymerase ribozyme to generate a new or improved copy of itself before it degrades, it must meet minimum standards for template fidelity, replication rate, and stability. These requirements are influenced by the length of the replicating molecule and are interdependent; for example, greater template fidelity compensates for slower replication or faster degradation. The work of Manfred Eigen and other theorists describing probable requirements of these parameters for sustained self-replication is outlined in Chapter 2.

This chapter reviews experimental efforts to address the replicase assumption. Catalytic RNAs with different subsets of the replicase attributes have been generated by three lines of experimentation—two starting from self-splicing intron ribozymes, and one starting from a large library of random sequences. I describe these three approaches and evaluate their potential for ultimately generating a replicase ribozyme. I also touch on the problem of complementary strand displacement, which will need to be solved before an RNA polymerase can become an RNA replicase.

Approach #1: Primer Extension by the Group I Ribozyme

Strategies to develop an RNA replicase in the lab must compensate for the fact that, unlike nature, the investigator does not have a 10^{24} milliliter test tube and 500 million years in which to explore RNA sequence space. For this reason, the initial studies of the capability of RNA to perform polymerase-like reactions focused on redesigning sequences already known to be catalytic. Despite the fact that investigators were limited to the seven ribozyme architectures that had been discovered in nature, this approach has been surprisingly fruitful.

The first polymerase engineering efforts built on the observation that a group I self-splicing intron from *Tetrahymena* can catalyze the following reactions (Zaug and Cech 1986; Kay and Inoue 1987):

1) $2 \text{ pCpCpCpCpC} \rightarrow \text{pCpCpCpC} + \text{pCpCpCpCpCpC}$
2) $\text{pCpU} + \text{GpN} \rightarrow \text{pCpUpN} + \text{G}$ (N = any of the four ribonucleosides)

The second reaction resembles the addition of one nucleotide during RNA polymerization; pCpU is elongated at the expense of the phosphoester bond of GpN rather than the α–β phosphoanhydride bond of an NTP. The ability of the ribozyme to catalyze this reaction can be explained by the reaction's similarities to the second step of splicing, a phosphodiester transfer reaction in which the 3' hydroxyl of the 5' exon attacks the phosphate at the 3' splice site, resulting in exon ligation with release of the intron (Fig. 1a). Here, pCpU corresponds to the 5' exon, and GpN corresponds to the 3' splice site where G represents the last nucleotide of the intron and pN represents the 3' exon (Fig. 1b).

Further studies demonstrated that the *Tetrahymena* ribozyme could extend a primer by successive addition of multiple nucleotides (Been and Cech 1988). When using GpC and GpU substrates to generate polypyrimidine products, a pCpCpCpCpC primer was efficiently extended by up to 6 nucleotides—clearly a milestone in demonstrating the possibility of RNA-catalyzed self-replication (Fig. 1c). Addition of at least two adenosine nucleotides was also observed using the GpA substrate. GpG substrate was not productive because cleavage rather than extension reactions dominated.

To study the degree to which a template could influence polymerization by the *Tetrahymena* ribozyme, the primer, template, ribozyme, and dinucleotide substrates were all modified (Fig. 1d) (Bartel et al. 1991). The primer and primer-binding site were redesigned so that they could stably pair in only one register. A defined template region was added just 5' of the primer-binding site so that the influence of a template could be examined. The guanosine-binding site of the ribozyme was altered to prefer 2-aminopurine (2AP) over G. With this ribozyme and 2AP-pN substrates, extension by pG could now be achieved because cleavage reactions involving binding of the 3'-terminal G residue of the substrate no longer dominated. Analysis of all 16 template–dinucleotide combinations revealed that the polymerization was influenced by the presence of a template, and in all cases, the primer was preferentially extended by the Watson-Crick match to the template. However, the overall error rate of extension was 0.35 per nucleotide, i.e., with a mixture of all four 2AP-pN substrates, 65% of the extension would be by the match to the template, and 35% would be by one of the three mismatches.

Fidelity and efficiency problems are not the only important limitations in these primer extension reactions. A true replicase must copy an external template, not a template attached to its 5' end. For extension reactions using GpN, detaching the template lowers the efficiency by 10,000-fold (Bartel et al. 1991), presumably due to inefficient formation of the enzyme·template·primer·GpN complex.

Approach #2: Oligonucleotide Assembly by the Group I Ribozyme

The external template problem was overcome by using a polymerization strategy in which the complementary strand is assembled from oligonucleotides rather than single nucleotides (Fig. 2) (Doudna and Szostak

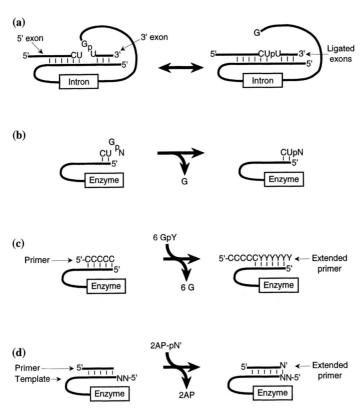

Figure 1 Engineering an RNA polymerase starting from a self-splicing ribozyme. Thick lines and boxes depict RNA; thin vertical lines represent pairing (Watson-Crick and G·U-wobble) involving substrate RNAs. (*a*) The exon ligation step of splicing by the *Tetrahymena* group I intron (Cech 1990). The 3′ hydroxyl of the 5′ exon attacks the phosphate of the 3′ splice site (GpU), leading to exon ligation with displacement of the 3′-terminal nucleotide of the intron (G). (*b*) The catalytic core of the intron promoting a reaction with dinucleotide cognates of the 5′ exon (CU) and 3′ splice site (GpN) (Kay and Inoue 1987). (*c*) Extension of a primer by successive addition of six pyrimidine (Y) nucleotides (Been and Cech 1988). Extension by two pA nucleotides was also observed when using GpA. (*d*) Template-dependent primer extension using each of the four 2AP-pN dinucleotides and the corresponding template-ribozyme constructs (Bartel et al. 1991). Prime (′) indicates the Watson-Crick match to the template RNA.

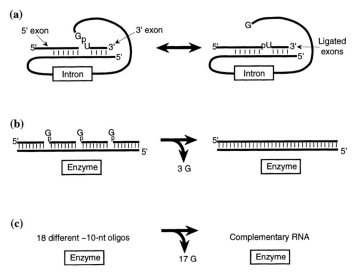

Figure 2 Ribozyme-catalyzed oligonucleotide assembly reactions. Thick lines and boxes depict RNA; thin vertical lines represent Watson-Crick pairing involving substrate RNAs. (*a*) The exon ligation step of splicing by the *Tetrahymena* group I intron (Cech 1990). (*b*) Assembly of a complementary RNA from four oligonucleotides aligned by pairing to the template RNA (Doudna and Szostak 1989). A fragmented *sunY* ribozyme was also shown to assemble five oligonucleotides complementary to one of its catalytic segments (Doudna et al. 1991). (*c*) Synthesis of the complete complementary strand of the ribozyme using 18 different RNA oligonucleotides (Green and Szostak 1992).

1989). Oligonucleotides bind to the template much more tightly than do mononucleotides, greatly facilitating formation of the enzyme·substrate quaternary complex. Another group I intron ribozyme, the ~180-nucleotide *sunY* ribozyme, is preferable to the *Tetrahymena* ribozyme for these oligonucleotide assembly experiments because it is only half the size of the *Tetrahymena* ribozyme and is more efficient at extending long primer·template substrates (Doudna et al. 1991). It appears that the *SunY* ribozyme can join oligonucleotides of any sequence, provided that they are aligned by pairing to the template strand. The RNA enzyme's ability to recognize an RNA helix without relying on sequence-specific contacts or covalent tethering is the most significant parallel between the oligonucleotide-assembly reactions and protein-catalyzed polymerization.

The template-directed oligonucleotide-assembly reactions provided a format for examining a ribozyme's activity not only as an enzyme but also as a template—another key property of an RNA replicase (Doudna et al.

1991). The *sunY* ribozyme can assemble its complete complementary strand by stringing together 18 oligoribonucleotides, each about 10 nucleotides in length (Fig. 2c) (Green and Szostak 1992). Shortening the *sunY* ribozyme by removing nonessential structured segments improved its fitness as a template by nearly 20 fold (Green and Szostak 1992).

Because fewer than 20 of the million possible ~10-nucleotide substrates are included in the *sunY*-copying reactions, these reactions are capable of copying only a very limited number of other template sequences. All possible substrate sequences would be difficult to provide, particularly at concentrations needed for efficient assembly. Therefore, more general polymerization requires the use of shorter oligonucleotide substrates. It would be feasible to supply all 64 substrates needed for triplet extension. With this in mind, oligonucleotide assembly has been examined using 2AP-$(pN)_3$ substrates, an external template, and an appropriate *sunY* derivative (Doudna et al. 1993). For some templates, the fidelity of triplet extension is impressive, with certain single mismatches between the triplet and template reducing extension by 100- to 1000-fold. However, for other templates, the matched trinucleotide adds only slightly more efficiently (1.25-fold) than do certain mismatched ones, with G·U wobble mismatches being particularly problematic. There is also a significant drop in efficiency for matched substrates when the substrate-template pairing is shortened from 10 to 3 base pairs. Therefore, in its current form, neither the fidelity nor the efficiency of the triplet extension reaction is thought to be sufficient for assembly reactions involving more than a few couplings (Doudna et al. 1993).

Prospects for Improved Polymerases Based on Group I Ribozymes

In terms of raw power to promote chemistry, the group I intron is impressive; with its preferred substrates in place it can catalyze phosphodiester transfer in less than a second (Herschlag and Cech 1990). Two divalent metal ions directly coordinate the axial oxygens of the transition state in a catalytic mechanism that appears strikingly similar to that of contemporary polymerases (Steitz and Steitz 1993; Pelletier et al. 1994; Doublié et al. 1998; Chapter 13). The most important challenges for improving the ribozyme's polymerase activity lie in substrate binding—more avid binding to the primer·template and much better discrimination between matched and mismatched monomers.

The group I ribozyme's innate ability to recognize substrate helices is a good start for primer·template binding but is still lacking in affinity. Much of this recognition is mediated by contacts to six 2′-hydroxyl groups of the primer·template (Bevilacqua and Turner 1991; Pyle and Cech 1991; Strobel and Cech 1993; Narlikar et al. 1997). Because these ribose contacts are not sequence-specific, the ribozyme can use them to bind a helix in different registers (Herschlag 1992; Strobel and Cech 1994). Likewise, contemporary DNA polymerases bind the primer·template through contacts to moieties that lie in identical positions in each of the four Watson-Crick pairs; i.e., phosphate oxygens of the helix backbone and purine N3 or pyrimidine O2 hydrogen-bond acceptors of the helix minor grove. T7 DNA polymerase uses a spectacular network of contacts involving seven minor groove moieties and each of the 12 phosphates of the first 6 base pairs in the primer·template helix (Doublié et al. 1998). RNA may not be able to match protein in forming such an extensive network of contacts. Nevertheless, the ribozyme's contacts to the six ribose hydroxyls of the template-primer, together with other possible contacts to the minor groove and to phosphate oxygens (perhaps mediated by water or hydrated metal ions), may form a rudimentary primer·template-binding motif. Among the six ribose hydroxyls of the template-primer, there is a spectrum of affinities (with energies ranging from 0.4 to 3.1 kcal/mole; Narlikar et al. 1997). If the lower-affinity contacts can be optimized to resemble the higher-affinity contacts, then sufficient primer·template binding would be achieved. Another strategy for improved primer·template binding could involve an RNA strand (or a more structured element) that loosely encircles the primer·template helix, forming a sliding clamp analogous to that seen with highly processive DNA polymerases (Kong et al. 1992; Krishna et al. 1994).

Improving template fidelity may be less tractable. The error rate of primer extension by the *Tetrahymena* ribozyme is 0.35, far worse than the ≤0.004 error rates of viral polymerases that synthesize RNA using RNA templates (Steinhauer and Holland 1986; Ward and Flanegan 1992). As with primer·template binding, this aspect of substrate binding need not necessarily match that of the protein enzymes. Consider hypothetical replicases the size of the *sunY* ribozyme (180 nucleotides), capable of 1000 couplings within their lifetime. For effective replication through complementary-strand then second-strand synthesis, 2 of the ~6 strands that each polymerase can produce must have the correct residues at all of the positions that contribute to function. If the identities of effectively 80% of the 180 nucleotides contribute to function, then these replicases would require an error rate ≤0.008 [$(1 - 0.008)^{144} = 2/6$]. This would represent a 44-fold improvement in the error rate of *Tetrahymena* ribozyme-catalyzed primer extension. It is questionable whether changes can be made so close to the heart of the active site that would sufficiently improve the error rate while retaining reasonable catalytic efficiency.

Further progress through rational engineering will likely require the crystal structure of a self-splicing ribozyme and a detailed understanding of its active site. In the meantime, strategies involving sequence randomization and selection in vitro have appeared more promising than pure rational design. Using in vitro selection (Fig. 3), sequence libraries of more than 10^{15} different variants can be sampled, and the rare variants with the desired catalytic properties can be identified. This approach was used to identify base changes that restored activity to a *sunY* variant designed to serve as a superior template in oligonucleotide assembly reactions (Green and Szostak 1992). However, attempts to isolate group I variants with more avid primer·template binding or more desirable catalytic properties have not succeeded (Hager et al. 1996), raising the question of whether such ribozymes exist among the group I variant possibilities.

Approach #3: Polymerase-like Ribozymes from Random Sequences

The ability to select in vitro very rare sequences with desired properties (Fig. 3) not only enables variants of a known ribozyme with new or improved activities to be identified, but also enables entirely new ribozymes to be isolated from random sequences (Bartel and Szostak 1993). With this method the catalytic repertoire of RNA can be explored far beyond the biochemistry found in contemporary biology, obviating the need to start with the extant ribozymes when examining whether the fundamental properties of RNA are compatible with the RNA World scenario. Ribozyme activities important in the transition from the RNA World to the protein-nucleic acid world have been isolated from random sequences. These activities include RNA aminoacylation (Illangasekare et al. 1995), amide bond formation (Lohse and Szostak 1996; Wiegand et al. 1997), and peptide bond formation (Zhang and Cech 1997). Ribozyme activities important for establishing an RNA World have also been isolated. These include nucleotide synthesis (Unrau and Bartel 1998), nucleotide phosphorylation (Lorsch and Szostak 1994), and the subject of this chapter—RNA polymerization (Fig. 4).

Because the effort to generate a polymerase started from random sequences rather than an existing ribozyme, the desired reaction could be modeled after protein-catalyzed polymerization. In particular, the leaving group during polymerization could be pyrophosphate—the same leaving group as in contemporary enzymatic polymerization. This is appealing because the choice of a pyrophosphate leaving group in modern-day poly-

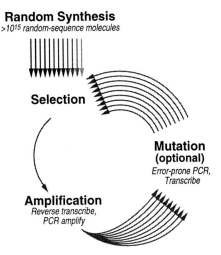

Figure 3 In vitro selection and evolution of RNA. Combinatorial synthesis generates a large pool of random-sequence DNA that is transcribed into RNA. RNA molecules are separated based on ability to perform a biochemical function, such as ligand binding or catalysis of a self-modification reaction. The selected material is amplified and selection-amplification is reiterated until active sequences dominate the pool (Ellington and Szostak 1990; Tuerk and Gold 1990). A mutagenesis step can be added to introduce additional sequence diversity, thus incorporating all of the salient features of evolution (Beaudry and Joyce 1992).

merization may be inherited from RNA-catalyzed polymerization of the RNA World in the same way that other nuances of contemporary metabolism may be relics of an RNA World (White 1976; Benner et al. 1989). Pyrophosphate also has a practical advantage as a leaving group for RNA synthesis in that phosphoanhydride activation of the α-phosphate is not prone to hydrolysis but still provides a chemical driving force for polymerization (Westheimer 1987). In contrast, self-splicing ribozymes promote disproportionation reactions, where one RNA becomes longer at the expense of a second RNA in a chemically neutral reaction with no net increase in RNA linkages. Replicase activity could, in principle, be based on RNA disproportionation if an adequate supply of dinucleotide (or larger oligonucleotide) substrates were provided and if the RNA leaving group were sequestered so that it did not participate in the reverse reaction. The chemical driving force would then be required "upstream" of polymerization; for instance, in the production of dinucleotide substrates using precursors activated with a leaving group such as pyrophosphate. However, direct polymerization of the activated

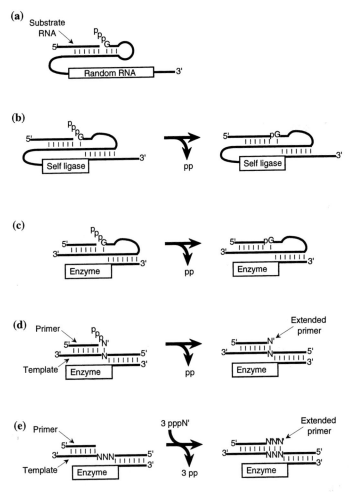

Figure 4 An RNA polymerase from random sequences. RNA is depicted as thick lines and boxes; also shown is Watson-Crick pairing involving substrate RNAs (thin vertical lines), the 5' triphosphate of the substrate RNA (ppp), and the inorganic pyrophosphate leaving group (pp). pppN' indicates the NTP that makes the appropriate Watson-Crick match to the template RNA. (*a*) Each of the >10^{15} different RNA sequences of the starting pool had a large random-sequence domain with 220 random positions. (*b*) New ribozymes that promote a self-ligation reaction were isolated from the pool of random sequences (Bartel and Szostak 1993). (*c*) The most promising lead for a polymerase ribozyme was subjected to further mutagenesis and selection (Ekland and Bartel 1995) and then engineered to catalyze a multiple-turnover reaction, where the two RNA substrates were no longer connected to the ribozyme (Ekland et al. 1995). (*d*) Fragmentation and partial deletion of the 5'-triphosphate-containing substrate RNA yielded a reaction in which an RNA primer was extended by a single nucleotide using an RNA template

precursors seems simpler than a scheme involving disproportionation reactions, because it would avoid these additional requirements.

The first step for developing a polymerase ribozyme was to find ribozymes that could catalyze a self-ligation reaction (Fig. 4b) designed to be analogous to chain elongation by one nucleotide during RNA polymerization in three respects: a hydroxyl of the growing strand attacks the α-phosphate of a 5' triphosphate, pyrophosphate is displaced with concomitant formation of a phosphodiester bond, and molecules that are joined are aligned by Watson-Crick pairing to a template (Bartel and Szostak 1993). At least 65 new ribozymes that promote self-ligation emerged from a pool of 1.4×10^{15} different RNA sequences, in which each sequence of the starting pool had a large random-sequence domain containing 220 random positions. The isolation of these new ribozymes demonstrated that RNA could catalyze a ligation reaction that has similarities to RNA polymerization, and it provided an indication of the abundance of ribozymes in arbitrary sequences—at least one in 20 trillion of the original pool molecules can promote this particular RNA ligation reaction.

Ribozymes that responded favorably to additional rounds of mutagenesis and selection represent at least three different structural classes (Ekland et al. 1995). One of these ribozymes, the class I ligase, promotes the attack of the 3' hydroxyl on the α-phosphate to form a normal RNA linkage and therefore was a viable starting point for a polymerase. The class I ligase was subjected to further mutagenesis and selection to generate functional sequence variants useful for deducing the important features of the ribozyme (Ekland and Bartel 1995). This procedure also generated much more active variants of the ligase.

The next steps toward a polymerase were to disconnect the "template" from the ribozyme (Fig. 4c) and to disconnect the NTP from the template (Fig. 4d). Then the central portion of the template was lengthened to permit extension by more than one nucleotide (Fig. 4e). Using NTPs and the information from an RNA template, this ribozyme was shown to extend an RNA primer by successive addition of three nucleotides (Ekland and Bartel 1996). Thus, an RNA catalyst can synthesize a short segment of RNA with the same reaction as that employed by contemporary RNA polymerases.

---

and the appropriate NTP (Ekland and Bartel 1996). (*e*) Lengthening the central portion of the template RNA converted the single-nucleotide addition reaction to an RNA polymerization reaction (Ekland and Bartel 1996).

In certain cases, extension by nucleotides complementary to the template was 1000 times more efficient than was extension by the mismatched nucleotides. The overall fidelity of this ribozyme (error rates of 0.08 and 0.12 for single-nucleotide and trinucleotide extension, respectively) represents a clear improvement over that of the self-splicing ribozymes and is much better than that expected from the intrinsic stabilities of Watson-Crick matches versus mismatches at the ends of RNA helices (expected error rate, 0.60) (Ekland and Bartel 1996). However, an overall error rate of 0.08 generally would not be sufficient for self-replication of a ribozyme the size of the class I ligase. For instance, 100-nucleotide replicases capable of 1000 couplings (i.e., synthesis of 10 RNA strands, each 100 nucleotides in length) within their lifetime require an error rate ≤0.02 for effective replication via their complementary strands ($[1 - 0.02]^{80} = 2/10$, assumes that the identities of effectively 80% of the 100 nucleotides contribute to function and an average of 2 [complement and second strand] of the ~10 strands synthesized per replicase molecule must have the proper residue at these particular 80 positions).

RNA polymerization by the class I ligase ribozyme demonstrates features essential for polymerization by an RNA replicase, such as template dependence, use of appropriately activated mononucleotides, regioselectivity for the formation of 3′, 5′-phosphodiester bonds, and absence of strict sequence requirements within the primer·template helix. However, in the case of this ribozyme, polymerization beyond three nucleotides is blocked by specific base-pairing of the ribozyme to the template (Fig. 4e). Polymerization beyond three nucleotides is only achieved in a special case where the ribozyme and template are designed to pair in multiple registers (Ekland and Bartel 1996).

Prospects for Improved Polymerase Variants

Like the group I ribozyme, the class I ligase is impressive in terms of power to promote chemistry. The class I ligase efficiently promotes multiple-turnover oligonucleotide ligation (Fig. 4c) with k_{cat} of 100 min^{-1}—a value far greater than that of most natural RNA catalysts and approaching that of comparable protein enzymes (Ekland et al. 1995). This k_{cat} represents an enhancement of about 10^9 over the uncatalyzed ligation rate of preformed substrate complex (Ekland et al. 1995). As with the group I intron, the most important challenges for improving the ribozyme's polymerase activity lie in substrate binding. Whereas improved template fidelity is the daunting substrate-binding challenge for group I polymerization, the most important challenge for the class I ligase is sequence-independent template binding.

How close is the polymerase ribozyme to an RNA replicase? At present, this ~100 nucleotide ribozyme can perform general polymerization of RNA corresponding to 3% of its length. Whether the cup is 3% full or 97% empty is debatable. An argument can be made that the cup is more than 97% empty. The polymerase ribozyme was selected on the basis of its ability to perform RNA ligation, not RNA polymerization. More extensive polymerization may be most readily achieved by starting from scratch with a ribozyme isolated from random sequences based on its ability to use NTP substrates and properly recognize the template. Furthermore, generating general, accurate, and efficient RNA polymerization may not be the most difficult challenge for demonstrating self-replication. For instance, replication requires not only polymerization but also dissociation between the polymerase strand and its complementary RNA; it may prove difficult to find ribozyme architectures and reaction conditions that facilitate strand dissociation. (The issue of strand dissociation is presented in more detail later in this chapter.)

An argument can also be made that the cup is more than 3% full. The rational engineering that converted the ligase into a polymerase was capable of sampling only a few obvious variants. Although the polymerase was born to perform RNA ligation, these crude engineering efforts quickly yielded a ribozyme with many attributes of a replicase, suggesting that the catalytic core of this ribozyme is a promising basis on which to build a replicase. In this view, finding an improved variant that performs general, template-dependent polymerization of 10 nucleotides will not be as difficult as finding the original ribozyme that could polymerize 3 nucleotides. The next step would be to use in vitro evolution, starting with an RNA library based on the core motif and selecting directly for polymerization within a context where the ribozyme does not recognize the template by base-pairing. Subsequent efforts can focus on evolving more avid NTP binding and increased template fidelity. An error rate of <0.02 could be achieved with only a single additional or enhanced interaction that further favors Watson-Crick matches by about 1 kcal/mole.

Strand Dissociation

Effective self-replication is not complete until the newly synthesized replicase strand dissociates from its complementary strand. Anyone who has tried to denature a 100-bp RNA duplex can appreciate the magnitude of this task. Helix denaturants such as urea and heat do not denature more than a small amount of the duplex at any one time. For instance, the only way to resolve long products of oligonucleotide assembly reactions on denaturing gels is to displace them from their template strands using denaturants in the presence of competitor RNA corresponding to the full-length product (Green and Szostak 1992).

A generic 100-bp RNA duplex with no mismatches has a predicted ΔG^0 of formation of about −190 kcal/mole at 37°C (Turner et al. 1988). Extrapolating from results of melting short helices, this ΔG^0 corresponds to a dissociation half-time exceeding 10^{100} years. Fortunately, the interconversion between two different RNA structures need not occur through a fully melted intermediate (LeCuyer and Crothers 1994). Long strands are expected to form intrastrand structure as the strands dissociate, lowering the activation energy of strand dissociation (Fig. 5a) (Cantor and Schimmel 1980). Sequences with very stable intrastrand structures would have significantly faster duplex dissociation rates, and certain exceptional sequences may have duplex dissociation rates faster than the RNA degradation rate. Some replicase architectures would be better than others in this regard. For instance, intrastrand structure would promote faster dissociation if it were predominantly local and thus could form concurrently with strand dissociation (Fig. 5a). In fact, an efficient self-replicator might avoid ever forming the long duplex shown at the top of Figure 5a, instead forming local structure as replication proceeds (Kramer and Mills 1981; Pace and Marsh 1985). A pulse of heat followed by "snap cooling" may also be useful to kinetically trap a fraction of the replicase·template duplexes in their dissociated state. Dilution of the RNA during this process would be of little value because the replicase and its complement must be in the same vicinity for self-replication.

Intrastrand structure is also important for preventing dissociated strands from re-annealing. Again, certain replicase architectures would be better than others. Preferred architectures would not present unpaired segments in a manner that favors re-nucleation of the replicase·template duplex. Otherwise, helix formation would likely be much faster than would polymerization, and the replicase would be inactivated by its complement before it accomplishes significant polymerization. Its disfavoring of duplex hybridization explains why self-structure of the template and product strands is critical for protein-catalyzed RNA replication by simple RNA-dependent RNA polymerases in the absence of helicases or single-stranded RNA-binding proteins (Dobkin et al. 1979; Priano et al. 1987; Konarska and Sharp 1990).

The intrastrand structure needed to generate and maintain strand dissociation does not come without a cost. A single strand with very stable

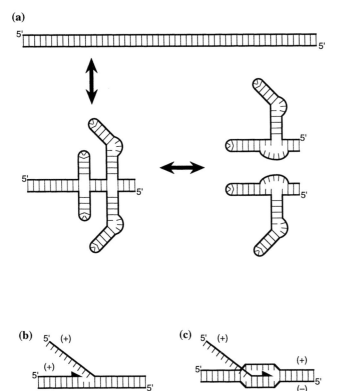

Figure 5 Three mechanisms for strand dissociation. (*a*) Schematic illustration of the role of local intrastrand structure during the interconversion of duplex RNA and dissociated strands. Folding intermediates with partial duplex and partial intrastrand structure lessen the activation energy of dissociation. For simplicity, symmetrical folding intermediates and products are depicted, although this need not be the case and would not be expected in regions with non-Watson-Crick interactions. (*b*) Permanent displacement of the resident (+) strand during semiconservative nucleic acid replication. (*c*) Temporary displacement of the resident (+) strand during conservative replication.

structure would not be an ideal template for complementary strand synthesis by a replicase ribozyme. Even evolved protein enzymes stall at highly structured template regions (Mills et al. 1978; Tuerk et al. 1988). For this reason, there have been efforts to reduce the intrastrand structure of templates for oligonucleotide assembly reactions (Doudna et al. 1991; Green and Szostak 1992). However, when considering synthesis of both strands, the advantages of a less structured template during synthesis of the ribozyme complement are likely to be more than offset by reduced template accessibility (Priano et al. 1987) and greater ribozyme inhibition during second-strand synthesis.

The restrictive requirement for very stable intrastrand structure would lead to selective pressure for more sophisticated strategies for coping with hybridization of complementary strands. One such strategy would involve a rudimentary double-stranded genome and displacement of the complementary strand during the act of polymerization (Fig. 5b). In this scenario, the replicase sense strand (+ strand) would be produced in excess over the (–) strand. Therefore, essentially all the (–) strands would be paired to (+) strands, trapping them in "genomic" helices, with the excess (+) strands fulfilling the enzymatic role. To produce more (+) strands the replicase would need to displace the resident (+) strand as it copies the (–) strand. This displacement could be modeled after that of leading-strand synthesis in DNA replication (Fig. 5b), or it could be in the form of a migrating "bubble," like that seen with cellular RNA transcription (Fig. 5c); both semiconservative and conservative replication strategies are seen among the double-stranded RNA viruses (Nemeroff and Bruenn 1986; Ewen and Revel 1988). Spontaneous fraying of the duplex at the replication fork may allow polymerization to proceed, which would consolidate the progress of template·complement dissociation. While binding the product· template, the polymerase can stabilize this helix, preventing displacement of the growing strand by the noncoding strand. This would favor polymerization in which the replicase remains associated with the product· template helix until polymerization reaches the end of the template; thus, with double-stranded templates, there is a premium for highly processive polymerization. It should also be mentioned that because double-stranded RNA is much less prone to spontaneous cleavage than is single-stranded RNA, it would have obvious advantages for genome maintenance (Usher and McHale 1976; Rohatgi et al. 1996).

CONCLUSION

From three lines of experimentation, we have learned much about the intrinsic ability of RNA to catalyze RNA polymerization: RNA can readily perform the type of chemistry needed to make new RNA linkages; RNA can recognize a primer·template helix without relying on sequence-specific contacts or covalent tethering; RNA can use an RNA template and appropriate monomers to synthesize a short segment of RNA. Nevertheless, a great deal more work needs to be done before the replicase assumption is proven. Currently, these three capabilities reside in different ribozymes and reactions. A true replicase ribozyme must embody each of these capabilities and carry out much more extensive and more accurate polymerization than has been witnessed so far.

If a self-replicating RNA is found, it will be interesting to examine how long replication is sustained in a serial dilution experiment where the experimenter provides continuous nourishment (primer and NTPs) and the ribozyme supplies progeny. With low template fidelity, many new replicase variants will arise. It has been suggested that some improved variants will develop a strategy for preferentially replicating their own kind, spontaneously generating Darwinian evolution (Joyce 1996). It seems more likely, however, that partitioning of the population (using, for example, membranes or surface adsorption) will be required both to enable the selfish replication needed for Darwinian evolution and to slow dissemination of parasitic species, such as sequences that are superior templates but deficient polymerases. If Darwinian evolution can be established, it will be fascinating to follow the development of these embryonic life forms.

ACKNOWLEDGMENTS

I thank Kelly Williams and Peter Unrau for helpful discussions, and Phil Zamore and others in the lab for useful comments on this manuscript. This work was supported by a grant from the National Institutes of Health.

REFERENCES

Bartel D.P. and Szostak J.W. 1993. Isolation of new ribozymes from a large pool of random sequences. *Science* **261:** 1411–1418.
Bartel D.P., Doudna J.A., Usman N., and Szostak J.W. 1991. Template-directed primer extension catalyzed by the *Tetrahymena* ribozyme. *Mol. Cell. Biol.* **11:** 3390–3394.
Beaudry A.A. and Joyce G.F. 1992. Directed evolution of an RNA enzyme. *Science* **257:** 635–641.
Been M.D. and Cech T.R. 1988. RNA as an RNA polymerase: Net elongation of an RNA primer catalyzed by the *Tetrahymena* ribozyme. *Science* **239:** 1412–1416.
Benner S.A., Ellington A.D., and Tauer A. 1989. Modern metabolism as a palimpsest of the RNA world. *Proc. Natl. Acad. Sci.* **86:** 7054–7058.
Bevilacqua P.C. and Turner D.H. 1991. Comparison of binding of mixed ribose-deoxyribose analogues of CUCU to a ribozyme and to GGAGAA by equilibrium dialysis: Evidence for ribozyme specific interactions with 2′ OH groups. *Biochemistry* **30:** 10632–10640.
Cantor C.R. and Schimmel P.R. 1980. *Biophysical chemistry*, part III: *The behavior of biological macromolecules*. W.H. Freeman, New York.
Cech T.R. 1986. A model for the RNA-catalyzed replication of RNA. *Proc. Natl. Acad. Sci.* **83:** 4360–4363.
———. 1990. Self-splicing of group I introns. *Annu. Rev. Biochem.* **59:** 543–568.

Dobkin C., Mills D.R., Kramer F.R., and Spiegelman S. 1979. RNA replication: Required intermediates and the dissociation of template, product, and Qβ replicase. *Biochemistry* **18**: 2038–2044.

Doublié S., Tabor S., Long A.M., Richardson C.C., and Ellenberger T. 1998. Crystal structure of a bacteriophage T7 DNA replication complex at 2.2 Å resolution. *Nature* **391**: 251–258.

Doudna J.A. and Szostak J.W. 1989. RNA-catalysed synthesis of complementary-strand RNA. *Nature* **339**: 519–522.

Doudna J.A., Couture S., and Szostak J.W. 1991. A multisubunit ribozyme that is a catalyst of and template for complementary strand RNA synthesis. *Science* **251**: 1605–1608.

Doudna J.A., Usman N., and Szostak J.W. 1993. Ribozyme-catalyzed primer extension by trinucleotides: A model for the RNA-catalyzed replication of RNA. *Biochemistry* **32**: 2111–2115.

Ekland E.H. and Bartel D.P. 1995. The secondary structure and sequence optimization of an RNA ligase ribozyme. *Nucleic Acids Res.* **23**: 3231–3238.

―――. 1996. RNA-catalysed RNA polymerization using nucleoside triphosphates. *Nature* **382**: 373–376.

Ekland E.H., Szostak J.W., and Bartel D.P. 1995. Structurally complex and highly active RNA ligases derived from random RNA sequences. *Science* **269**: 364–370.

Ellington A.D. and Szostak J.W. 1990. In vitro selection of RNA molecules that bind specific ligands. *Nature* **346**: 818–822.

Ewen M.E. and Revel H.R. 1988. In vitro replication and transcription of the segmented double-stranded RNA bacteriophage ø6. *Virology* **165**: 489–498.

Green R. and Szostak J.W. 1992. Selection of a ribozyme that functions as a superior template in a self-copying reaction. *Science* **258**: 1910–1915.

Hager A.J., Pollard J.D., and Szostak J.W. 1996. Ribozymes: aiming at RNA replication and protein synthesis. *Chem. Biol.* **3**: 717–725.

Herschlag D. 1992. Evidence for processivity and two-step binding of the RNA substrate from studies of J1/2 mutants of the *Tetrahymena* ribozyme. *Biochemistry* **31**: 1386–1399.

Herschlag D. and Cech T.R. 1990. Catalysis of RNA cleavage by the *Tetrahymena thermophila* ribozyme. 1. Kinetic description of the reaction of an RNA substrate complementary to the active site. *Biochemistry* **29**: 10159–10171.

Illangasekare M., Sanchez G., Nickles T., and Yarus M. 1995. Aminoacyl-RNA synthesis catalyzed by an RNA. *Science* **267**: 643–647.

Joyce G.F. 1996. Building the RNA world. Ribozymes. *Curr. Biol.* **6**: 965–967.

Kay P.S. and Inoue T. 1987. Catalysis of splicing-related reactions between dinucleotides by a ribozyme. *Nature* **327**: 343–346.

Konarska M.M. and Sharp P.A. 1990. Structure of RNAs replicated by the DNA-dependent T7 RNA polymerase. *Cell* **63**: 609–618.

Kong X.P., Onrust R., O'Donnell M., and Kuriyan J. 1992. Three-dimensional structure of the beta subunit of *E. coli* DNA polymerase III holoenzyme: A sliding DNA clamp. *Cell* **69**: 425–437.

Kramer F.R. and Mills D.R. 1981. Secondary structure formation during RNA synthesis. *Nucleic Acids Res.* **9**: 5109–5124.

Krishna T.S., Kong X.P., Gary S., Burgers P.M., and Kuriyan J. 1994. Crystal structure of the eukaryotic DNA polymerase processivity factor PCNA. *Cell* **79**: 1233–1243.

LeCuyer K.A. and Crothers D.M. 1994. Kinetics of an RNA conformational switch. *Proc. Natl. Acad. Sci.* **91:** 3373–3377.

Lohse P.A. and Szostak J.W. 1996. Ribozyme-catalysed amino-acid transfer reactions. *Nature* **381:** 442–444.

Lorsch J.R. and Szostak J.W. 1994. In vitro evolution of new ribozymes with polynucleotide kinase activity. *Nature* **371:** 31–36.

Mills D.R., Dobkin C., and Kramer F.R. 1978. Template-determined, variable rate of RNA chain elongation. *Cell* **15:** 541–550.

Narlikar G.J., Khosla M., Usman N., and Herschlag D. 1997. Quantitating tertiary binding energies of 2′ OH groups on the P1 duplex of the *Tetrahymena* ribozyme: Intrinsic binding energy in an RNA enzyme. *Biochemistry* **36:** 2465–2477.

Nemeroff M.E. and Bruenn J.A. 1986. Conservative replication and transcription of *Saccharomyces cerevisiae* viral double-stranded RNA in vitro. *J. Virol.* **57:** 754–758.

Orgel L.E. 1986. RNA catalysis and the origins of life. *J. Theor. Biol.* **123:** 127–149.

Pace N.R. and Marsh T.L. 1985. RNA catalysis and the origin of life. *Origins Life* **16:** 97–116.

Pelletier H., Sawaya M.R., Kumar A., Wilson S.H., and Kraut J. 1994. Structures of ternary complexes of rat DNA polymerase beta, a DNA template-primer, and ddCTP. *Science* **264:** 1891–1903.

Priano C., Kramer F.R., and Mills D.R. 1987. Evolution of the RNA coliphages: The role of secondary structures during RNA replication. *Cold Spring Harbor Symp. Quant. Biol.* **52:** 321–330.

Pyle A.M. and Cech T.R. 1991. Ribozyme recognition of RNA by tertiary interactions with specific ribose 2′-OH groups. *Nature* **350:** 628–631.

Rohatgi R., Bartel D.P., and Szostak J.W. 1996. Nonenzymatic, template-directed ligation of oligoribonucleotides is highly regioselective for the formation of 3′-5′ phosphodiester bonds. *J. Am. Chem. Soc.* **118:** 3340–3344.

Sharp P.A. 1985. On the origin of RNA splicing and introns. *Cell* **42:** 397–400.

Steinhauer D.A. and Holland J.J. 1986. Direct method for quantitation of extreme polymerase error frequencies at selected single base sites in viral RNA. *J. Virol.* **57:** 219–228.

Steitz T.A. and Steitz J.A. 1993. A general two-metal-ion mechanism for catalytic RNA. *Proc. Natl. Acad. Sci.* **90:** 6498–6502.

Strobel S.A. and Cech T.R. 1993. Tertiary interactions with the internal guide sequence mediate docking of the P1 helix into the catalytic core of the *Tetrahymena* ribozyme. *Biochemistry* **32:** 13593–13604.

———. 1994. Translocation of an RNA duplex on a ribozyme. *Nat. Struct. Biol.* **1:** 13–17.

Tuerk C. and Gold L. 1990. Systematic evolution of ligands by exponential enrichment: RNA ligands to bacteriophage T4 DNA polymerase. *Science* **249:** 505–510.

Tuerk C., Gauss P., Thermes C., Groebe D.R., Gayle M., Guild N., Stormo G., d'Aubenton-Carafa Y., Uhlenbeck O.C., Tinoco I., Jr., and et al. 1988. CUUCGG hairpins: extraordinarily stable RNA secondary structures associated with various biochemical processes. *Proc. Natl. Acad. Sci.* **85:** 1364–1368.

Turner D.H., Sugimoto N., and Freier S.M. 1988. RNA structure prediction. *Annu. Rev. Biophys. Biophys. Chem.* **17:** 167–192.

Unrau P.J. and Bartel D.P. 1998. RNA-catalyzed nucleotide synthesis. *Nature* (in press).

Usher D.A. and McHale A.H. 1976. Hydrolytic stability of helical RNA: A selective advantage for the natural 3′, 5′-bond. *Proc. Natl. Acad. Sci.* **73:** 1149–1153.

Ward C.D. and Flanegan J.B. 1992. Determination of the poliovirus RNA polymerase error frequency at eight sites in the viral genome. *J. Virol.* **66:** 3784–3793.

Westheimer F.H. 1987. Why nature chose phosphates. *Science* **235:** 1173–1178.

White H.3. 1976. Coenzymes as fossils of an earlier metabolic state. *J. Mol. Evol.* **7:** 101–104.

Wiegand T.W., Janssen R.C., and Eaton B.E. 1997. Selection of RNA amide synthases. *Chem. Biol.* **4:** 675–683.

Zaug A.J. and Cech T.R. 1986. The intervening sequence RNA of *Tetrahymena* is an enzyme. *Science* **231:** 470–475.

Zhang B. and Cech T.R. 1997. Peptide bond formation by in vitro selected ribozymes. *Nature* **390:** 96–100.

6

Did the RNA World Exploit an Expanded Genetic Alphabet?

Steven A. Benner, Petra Burgstaller, Thomas R. Battersby, and Simona Jurczyk
Departments of Chemistry and Anatomy and Cell Biology
University of Florida, Gainesville, Florida 32611 and
Sulfonics, Inc., Alachua, Florida 32615

SINGLE BIOPOLYMER LIFE FORMS BASED ON RNA

In terms of its macromolecular chemistry, life on Earth can be classified as a "two-biopolymer" system. Nucleic acid is the encoding biopolymer, storing information within an organism and passing it to its descendants. Nucleic acids also direct the biosynthesis of the second biopolymer, proteins. Proteins generate most of the selectable traits in contemporary organisms, from structure to motion to catalysis.

The two-biopolymer strategy evidently works rather well. It has lasted on Earth for several billion years, adapting in this time to a remarkable range of environments, surviving formidable geobiological (and perhaps cosmic) events that threatened its extinction, and generating intelligence capable of exploring beyond Earth.

The terrestrial version of two-biopolymer life contains a well recognized paradox, however, one relating to its origins. It is difficult enough to envision a nonbiological mechanism that would allow either proteins or nucleic acids to emerge spontaneously from nonliving precursors. But it seems astronomically improbable that both biopolymers arose simultaneously and spontaneously, and even more improbable (if that can be imagined) that both biopolymers so arose with an encoder-encoded relationship.

Accordingly, a variety of "single-biopolymer" models have been proposed as forms of life that antedated the two-biopolymer system. These (presumably) could have emerged more easily than a two-biopolymer system. Such models postulate that a single biopolymer can perform the catalytic and information repository roles and undergo the Darwinian evolution that defines life (Joyce 1994). For example, Rich (1962), Woese (1967), Orgel (1968), and Crick (1968) proposed that the first biopolymeric system that sustained Darwinian evolution on Earth was RNA. Usher and McHale (1976), White (1976), Visser and Kellogg (1978), and Benner et al. (1989) expanded on this proposal, recognizing that key elements of contemporary metabolism might be viewed as vestiges of an "RNA World" (Gilbert 1986), a time when the only encoded component of biological catalysis was RNA. The phenomenal discoveries by Cech, Altman, and their coworkers (Cech et al. 1981; Guerrier-Takada et al. 1983; Zaug and Cech 1986) showing that RNA performs catalytic functions in contemporary organisms has made the RNA World a part of the culture of contemporary molecular biology (Watson et al. 1987).

The notion that the RNA World was metabolically complex follows from the abundance of its vestiges in modern metabolism (Benner 1988; Benner et al. 1989). RNA fragments play roles in modern metabolism for which they are not intrinsically chemically suited, most notably in RNA cofactors such as ATP, coenzyme A, NADH, FAD, and S-adenosylmethionine. This suggests that these fragments originated during a time in natural history when RNA was the only available biopolymer, rather than by convergent evolution or recruitment in an environment where chemically better suited biomolecules could be encoded. If the RNA World developed ATP, coenzyme A, NADH, and S-adenosylmethionine, it follows that the RNA World needed these for some purpose, presumably for phosphorylations, Claisen condensations, oxidation-reduction reactions, and methyl transfers, respectively (White 1976; Visser and Kellogg 1978; Benner et al. 1989). This in turn implies complexity in the metabolism encoded by RNA-based life, implying in turn that RNA can catalyze a wide variety of chemical reactions. Conversely, the intellectual contribution of the RNA World model would be diminished were it not to embody a complex metabolism catalyzed by ribozymes, as there would then be no coherent explanation for the structures of contemporary RNA cofactors.

Accordingly, hopes were high when Szostak (1988), Joyce (1989a,b), Gold (Irvine et al. 1991), and their coworkers introduced in vitro selection as a combinatorial tool to identify RNA molecules within a pool that catalyze specific reactions. Elegantly conceived, the approach seemed likely to lead to the ultimate goal, the generation of an RNA (or DNA) molecule that would catalyze the template-directed polymerization of RNA (or DNA), a molecular system able to undergo Darwinian evolution. If selection procedures were appropriately designed, they should also produce RNA catalysts for almost any other reaction as well, at least if the RNA World model as elaborated above were a correct representation of natural history.

LIMITATIONS OF RNA AS A CATALYST

In contrast with these hopes (and only by this contrast), in vitro selection has been disappointing. RNA has proven to be an intrinsically poor matrix for obtaining catalysis, especially when compared with proteins. For example, to have a 50% chance of obtaining a single RNA molecule capable of catalyzing a template-directed ligation reaction by a modest (by protein standards) factor of 10,000, Bartel and Szostak estimated that one must sift through 2×10^{13} random RNA sequences 220 nucleotides in length (Bartel and Szostak 1993). Although many laboratories have tried, only a few have managed to extend the scope of RNA catalysis beyond the phosphate transesterification reactions where it was originally observed. For example, attempts to obtain an RNA catalyst for a Diels-Alder reaction using in vitro selection failed (Morris et al. 1994); the same reaction is readily catalyzed by protein antibodies (Gouverneur et al. 1993). Attempts to obtain RNA that catalyzes amide synthesis have succeeded, but with difficulty (Wiegand et al. 1997; Zhang and Cech 1997). The fact that such successes came only after many attempts is indicative of a relatively poor catalytic potential in oligonucleotides.

The comparison with peptides is instructive. For example, short (14 amino acids) peptides accelerate the rate-determining step for the amine-catalyzed decarboxylation of oxaloacetate by more than three orders of magnitude (Johnsson et al. 1990, 1993), not far below the acceleration observed for the first-generation ligases observed in the Bartel–Szostak selection beginning with 10^{13} random RNA sequences. Furthermore, the peptide is less than 10% the size of the RNA motif. Combinatorial experiments starting from this design (Perezpaya et al. 1996; L. Baltzer, pers. comm.) suggested that perhaps only 10^7 random sequences must be searched to get a similar catalytic effectiveness as is observed in a library of 10^{13} RNA molecules. This suggests that peptides are intrinsically a million-fold fitter as catalysts than RNA.

The comparison is imperfect, of course, because it involves different reactions and different design strategies. This imperfection characterizes most of the comparisons that can be made at present. Not surprisingly, ribozymes are most frequently sought for reactions where oligonucleotides are most likely to be effective catalysts (for example, where oligonucleotides themselves are substrates), whereas peptide catalysts are most frequently sought for reactions suited for peptide catalysts (for example, those that make use of functional groups found on amino acid side chains). This makes the comparison nonquantitative, but useful nevertheless as an estimate of how well oligonucleotides and oligopeptides, respectively, perform when challenged by their favorite target reactions.

THE CHEMISTRY OF FUNCTIONAL CATALYSIS

The apparent superiority of proteins as catalysts compared with RNA reflects (at the very least) the availability to proteins of a wider range of building blocks and catalytic functionality than available in RNA. RNA lacks the imidazole, thiol, amino, carboxylate, and hydrophobic aromatic and aliphatic groups that feature so prominently in protein-based enzymes and has only hydroxyl, polar aromatic, and phosphate groups. An uncounted number of studies with natural enzymes and their models has illustrated the use of this functionality by protein catalysts (Dugas 1989).

Proteins also have advantages as catalysts over nucleic acids in their greater propensity to "fold." As is well known from the statistical mechanics of polymers, the repeating negative charge of the polynucleotide backbone causes the polymer to favor an extended structure (Flory 1953; Richert et al. 1996). Accordingly, the most prominent physical characteristics of nucleic acids are their solubilities in water, their ability to bind other oligonucleotides following simple rules, and their constancy of physical behavior over a wide range of sequences. In contrast, the most prominent physical characteristic of peptides is their propensity to fold, best known as a propensity to precipitate (which is, of course, a type of folding, in that peptide interacts with peptide rather than with water). A catalyst must fold if it is to surround a transition state and be effective, providing another reason that peptides might be intrinsically better catalysts than RNA (Benner 1989).

If it is necessary to generate trillions of long, random RNA sequences in order to have a 50% likelihood of finding one that catalyzes even modestly a simple ligation (a reaction that itself assumes the preexistence of long RNA molecules that act as templates and substrates), how many more random sequences must be generated to obtain a template-directed RNA polymerase? We cannot say, because such a ribozyme has not been generated. An optimistic guess is 10^{20}. This, the difficulty of obtaining plausible prebiotic syntheses of RNA molecules (but see Müller et al. 1990), and the observation that racemic mixtures of RNA do not effectively undergo abiological polymerization (see, e.g., Schmidt et al. 1997) have prompted many to question the RNA World as a viable model for generating the first life on Earth (Joyce et al. 1987; Miller 1997). The critique acknowledges the premise that the single-biopolymer system is more plausible as a first life form than the two-biopolymer system. It continues, however, by holding that the chemical properties of RNA are such that it could not have been the first living biopolymer, as it is too difficult to generate under abiotic conditions and provides too little catalytic power even if it could be generated.

EXPANDING THE STRUCTURAL REPERTOIRE OF NUCLEIC ACIDS

A decade ago, the intrinsic limitations of standard nucleic acids as a biopolymer for obtaining functional behavior under conditions of Darwinian selection were discussed, and several solutions to these limitations were proposed (Benner et al. 1987; Benner 1988, 1989; Switzer et al. 1989; Piccirilli et al. 1990). Each of these involved an expedient by which additional functionality was provided to the RNA.

One expedient was obvious. RNA might gain functionality using cofactors, much as contemporary proteins gain the functionality that they lack through vitamins.

A second solution was to append functionality to the standard nucleotides. Prompting this suggestion was the observation that contemporary tRNA and rRNA contain much of the functionality found in proteins but lacking in contemporary encoded RNA, including amino, carboxylate, and aliphatic hydrophobic groups (Fig. 1) (Limbach et al. 1994). These functional groups are introduced by posttranscriptional modification of encoded RNA. Some of these might even be placed by parsimony in the protogenome, the reconstructable genome at the trifurcation in the evolutionary tree joining the archaebacterial, eubacterial, and eukaryotic kingdoms (Benner et al. 1989; Limbach et al. 1994).

The third approach to expand the functional diversity of nucleic acids pursued the possibility of expanding the number of base pairs from the four found in standard oligonucleotides to include some of the

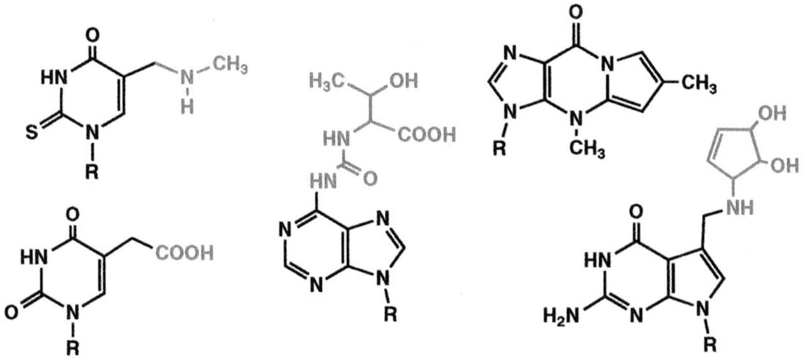

Figure 1 Transfer RNA contains a rich collection of functionalized standard nucleobases, created by posttranscriptional modification, that deliver functional groups (amino groups, carboxylic acid groups, aliphatic hydrophobic groups, in *green*) not found within unmodified RNA. Could these be vestiges of functionalized RNA originating in the RNA World?

nonstandard hydrogen-bonding patterns permitted by the geometry of the Watson-Crick base pair (Fig. 2) (Switzer et al. 1989; Piccirilli et al. 1990). Additional letters in the genetic alphabet could carry a richer diversity of functionality. Indeed, one might imagine a new type of biopolymer, one carrying functionalization like proteins but able to be copied like nucleic acids (Fig. 3) (Kodra and Benner 1997).

In a sense, the first approach had already been implemented in 1987. Most ribozymes require one or more metal ions to be effective catalysts. The metal ions are not encoded in the RNA sequence, provide a needed electrophilic center, and therefore compensate for the limited catalytic functionality of the biopolymer itself. Thus, metals can be considered to be "cofactors," and clearly improve the catalytic functionality of RNA. More recently, Breaker and his coworkers have expanded the approach to include organic molecules as second ligands in riboenzymes (Tang and Breaker 1997).

In contrast, the second and third approaches were far from implementation in 1987. Although standard bases carrying functionality were known to form stable base pairs and, in some cases, be accepted by polymerases (Prober et al. 1987), it was not clear that nonstandard bases would pair as expected, or whether polymerases would incorporate functionalized standard bases and nonstandard bases (Figs. 2 and 3) with sufficient speed and fidelity to be used in in vitro selection experiments. Furthermore, it was not known whether in vitro selection based on an expanded genetic alphabet might improve the binding and catalytic versatility of RNA.

Developing in vitro selection with an expanded genetic alphabet proved to be more difficult than developing in vitro selection with the standard nucleotides (A, T, G, and C), which was enabled by a rich collection of molecular biological tools. Nonstandard nucleobases needed to be synthesized (Switzer et al. 1989; Piccirilli et al. 1990; Vögel et al. 1993a; Vögel and Benner 1994). Their structures needed to be optimized for stability and pairing (Piccirilli et al. 1991a,b; Vögel et al. 1993b). New protecting group chemistry needed to be developed to permit automated synthesis of oligonucleotides containing them (Huang and Benner 1993; von Krosigk and Benner 1995). Polymerases were needed to catalyze their incorporation into oligonucleotides by the polymerase chain reaction (Horlacher et al. 1995; Lutz et al. 1996). These studies have been paralleled by work to append still more functionality onto standard nucleobases (Dewey et al. 1996; Kodra and Benner 1997). These experiments have established the chemistry of both functionalized standard and nonstandard nucleotides, and laid the ground for the first in vitro selection experiments using these.

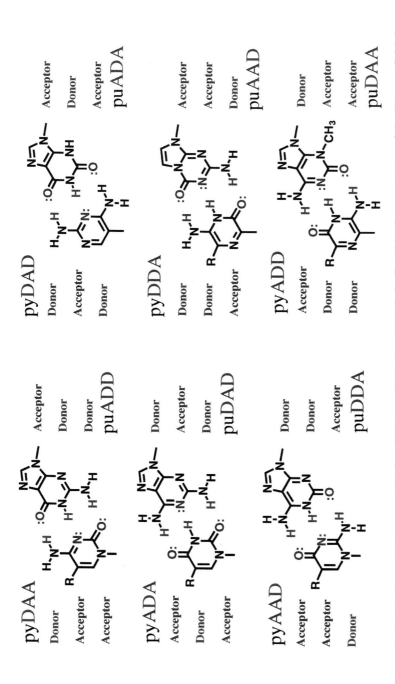

Figure 2 Twelve bases that are possible in a DNA- or RNA-based "alphabet" within the constraints of the Watson-Crick base-pair geometry. Pyrimidine base analogs are designated by "py," purine by "pu." The uppercase letters following the designation indicate the pattern of hydrogen bonding acceptor (A) (in *blue*) and donor (D) (in *red*) groups. Thus, cytosine is pyDAA, guanine is puADD, adenine is puDA-, (diaminopurine, puDAD, completes the Watson-Crick base pair), and thymine is pyADA. The remainder of the base pairs are joined by nonstandard hydrogen-bonding schemes.

THE RNA WORLD HAD THE MOTIVE TO EXPLOIT MODIFIED NUCLEOTIDES

With these chemical developments, it has been possible recently to make a convincing, if not compelling, argument that the RNA World had both the motive and the opportunity to exploit nonstandard and functionalized nucleobases. Three results are central to this argument.

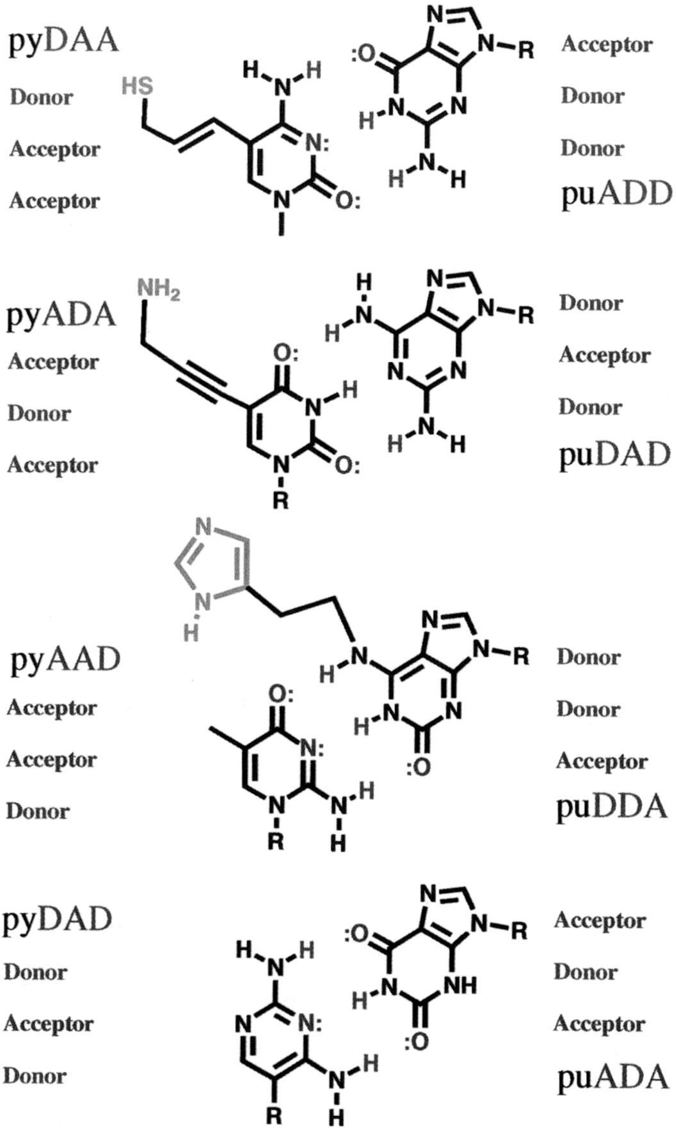

pyDAA
Donor
Acceptor
Acceptor

Acceptor
Donor
Donor
puADD

pyADA
Acceptor
Donor
Acceptor

Donor
Acceptor
Donor
puDAD

pyAAD
Acceptor
Acceptor
Donor

Donor
Donor
Acceptor
puDDA

pyDAD
Donor
Acceptor
Donor

Acceptor
Donor
Acceptor
puADA

Figure 3 Nonstandard and standard nucleobases with functionality (in *green*). Note that the pyDAD nucleobase can be protonated below pH 7 (pKa = 7.4).

First, functionality has been incorporated into an RNA molecule that catalyzes a Diels-Alder reaction (Tarasow et al. 1997), starting from a functionalized standard pyADA nucleobase (Fig. 4, right). A selection starting with a library that did not contain functionalized nucleotides failed to yield a catalyst (Morris et al. 1994). The successful experiment with the functionalized pyADA base selected directly for a Diels-Alderase, however, whereas the experiment on the unfunctionalized library sought a Diels-Alderase by selecting for RNA molecules that bound to a transition-state analog for the reaction. The different selection strategies prevent us from saying conclusively that this particular functionalized nucleoside improves the intrinsic power of RNA as a catalyst for Diels-Alder reactions. Experiments that bear on this question will undoubtedly emerge soon.

Another functionalized selection experiment does support this conclusion. Burgstaller, Jurczyk, Battersby, and Benner prepared a different functionalized implementation of the pyADA nucleobase (trivially designated "J," Fig. 4) and incorporated it into an in vitro selection experiment seeking receptors for an adenosine derivative (P. Burgstaller et al., unpubl.). This experiment was done in strict parallel with experiments done by Huizenga and Szostak (1995) using a standard, unfunctionalized DNA library.

The functionalized library containing J yielded new motifs as receptors for ATP, including the following (the randomized region is underlined):

GGTCGTCTAGAGTATGCGGTAG<u>GAACGJCAGJGGGGGGAGCA JAJGGJGJGAJACGCGA</u>CCGAAGAAGCJJGGCCCAJG

The motif prepared with unfunctionalized T replacing J does not bind ATP, suggesting that the ammonium functionality carried by J is essential for the binding properties of the new motif. A gel filtration experiment

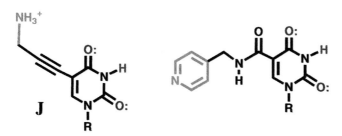

Figure 4 Functionalized standard bases that have been used in in vitro selections. Functional groups are shown in *green*, with the hydrogen-bonding acceptor and donor in *blue* and *red*, respectively.

was used to obtain an equilibrium binding constant (K_d) of 40 nM for affinity of this aptamer and ATP. This value is approximately 2 orders of magnitude greater than the reported K_d for the binding of RNA (Sassanfar and Szostak 1993) and DNA aptamers containing only standard bases to ATP (Huizenga and Szostak 1995). With the caveats that elution experiments permit only estimates of binding constants, and that further experiments with a wider range of ligands must be completed, these results suggest that introduction of a new functionality (an ammonium group bearing a positive charge) enhances the intrinsic value of a DNA library as a source of receptors by about two orders of magnitude.

These experiments make clear that functionalized oligonucleotides are superior to standard oligonucleotides as a matrix for generating receptors and catalysts. This would have given the RNA World a motivation to use functionalized oligonucleotides and an expanded genetic alphabet in its effort to generate diverse catalysts.

But did it? The third result comes from the field of "prebiotic chemistry," which seeks to discover ways by which the components of living systems might have emerged in the early Earth. Robertson and Miller (1995) showed how the intrinsic nucleophilicity of the 5-position of pyrimidines such as uracil might be exploited to generate functionalized uracil derivatives that carry positive charges at the 5-position under abiological conditions. Analogous chemistry can be used to generate other functionalized derivatives. The products resemble the amino group functionalized uracils found in some tRNA molecules (Fig. 1). This suggests that the RNA World may have had the opportunity to use some functionalized nucleosides when life first emerged on Earth.

Could nonstandard nucleobases (Fig. 2) also have been available during early episodes of life on Earth? The success of prebiotic chemists in generating organic species under prebiological conditions has expanded greatly the spectrum of molecules that might have been accessible to early life. Indeed, prebiotic chemistry might have been too successful, in that relatively simple prebiotic models can generate organic mixtures containing perhaps too many products (Khare et al. 1993). Contemporary prebiotic chemistry must become less an effort to show that a given moiety might be generated under prebiotic conditions, and more an effort to show how a useful moiety (such as a heterocycle or a ribose) arising under prebiotic conditions might be converted into one or more of its delicate derivatives (such as nucleosides) in the presence of organic gunk that emerges from a typical prebiotic experiment.

Notwithstanding these issues, several of the nonstandard nucleobases in Figure 2 do not appear to be less prebiotic than the standard nucleobases. The puADA nucleobase is, for example, a simple deamination product of the puADD base (also known as guanine). Thus, if guanine was generated on a prebiotic earth, puADA was a fortiori also generated on a prebiotic earth. Similar arguments can be made for the puDDA and pyAAD nucleobases. This suggests that if the RNA World had the opportunity to use the standard genetic alphabet, it may also have had the opportunity to use an expanded genetic alphabet.

CONTRADICTORY CHEMICAL REQUIREMENTS FOR CATALYSIS AND INFORMATION STORAGE

This evidence suggests that the RNA World had both access to a functionalized and/or expanded genetic alphabet and the motivation to use it. The case is made stronger by the functionalized nucleotides found in contemporary tRNA and rRNA (Fig. 1), presuming that these are vestiges of an RNA World.

Even assuming that further experimental work demonstrates the full catalytic potential of functionalized and expanded genetic alphabets, it is still not clear that they will support single-biopolymer systems of life, however. To support a self-sustaining chemical system capable of undergoing Darwinian evolution (Joyce 1994), a biopolymer must be able to search mutation-space independent of concern that it will lose properties essential for replication. We designate polymers that have this property as COSMIC-LOPER biopolymers (Capable Of Searching Mutation-space Independent of Concern over Loss Of Properties Essential for Replication), and comment briefly on the chemical constraints placed on biopolymers likely to have this property.

The need for the single biopolymer to be COSMIC-LOPER to support Darwinian evolution is nearly axiomatic. If a substantial fraction of the mutations possible within a genetic information system cause a biopolymer to precipitate, unfold, or otherwise no longer be recognizable by the catalyst responsible for replication, the biopolymer cannot evolve.

Curiously, catalysis on one hand and information storage on the other place competing and contradictory demands on molecular structure that make a single molecule that does both difficult to find. Specifically:

1. A biopolymer specialized to be a catalyst must have many building blocks, so that it can display a rich versatility of chemical reactivity. A biopolymer specialized to store information must have few building blocks, as a way of ensuring faithful replication (Szathmary 1992; Lutz et al. 1996).

2. A biopolymer specialized to be a catalyst must fold easily so that it can form an active site. A biopolymer specialized to store information should not fold easily, so that it can serve as a template.
3. A biopolymer specialized for catalysis must be able to change its physical properties rapidly with few changes in its sequence, enabling it to explore "function space" during divergent evolution. A biopolymer specialized to encode information must be COSMIC-LOPER, with its physical properties largely unchanged even after substantial change in its sequence, so that the polymer remains acceptable to the mechanisms by which it is replicated.

At the very least, a single biopolymer attempting to support Darwinian evolution must reflect some sort of structural compromise between these goals. No fundamental principle guarantees that a polymeric system will make this compromise in a satisfactory way, however. The demands for functional diversity, folding, and rapid search of function space might be so stringent, and the demands for few building blocks, templating ability, and COSMIC-LOPER ability so stringent, that no biopolymer structure achieves a suitable compromise.

Nor need a biopolymer exist that supports robust catalysis at the same time as it enables robust Darwinian evolution. If so, the single-biopolymer model for the origin of life would be unavailable as a solution to the "chicken-or-egg" paradox in the origin of two-biopolymer systems. Life would be scarce in the universe, and if a single-biopolymer system did arise, it would be poorly adaptable and easily extinguished by geobiological (and possibly cosmogenic) events. Conversely, if many polymeric systems exist that make an acceptable compromise between the demands of catalysis and the demands of information storage, life would have emerged rapidly via single-biopolymer forms and be abundant in the universe.

It is clear that proteins are not COSMIC polymers, even in cases where they can direct template-based replication (Lee et al. 1997). The physical properties of proteins (including their solubility) can change dramatically upon point mutation within the mutation space allowed by the 20 standard amino acids. Again, there are many examples of this phenomenon, but the peptides mentioned above that catalyze the decarboxylation of oxaloacetate are one. Altering their structure by a single acetyl group changes substantially their level of aggregation, and altering their internal sequence at a single residue changes substantially their helicity (Alleman 1989; Johnsson et al. 1990, 1993). If solubility and/or helicity are essential to the replicatability of a peptide template, a large range of plausible mutation would destroy it.

Natural oligonucleotides do not behave similarly. Indeed, molecular biologists rely on this fact. Every (or almost every) oligonucleotide will precipitate in ethanol. Every (or almost every, if we consider G-rich sequences [Wang and Patel 1994]) oligonucleotide will bind to its complement in a rule-based fashion. Every (or almost every) oligonucleotide will be a template for a polymerase. Every (or almost every) oligonucleotide will migrate as expected on an electrophoresis gel. This regularity is normal for oligonucleotides, but is exceptional for virtually every other class of organic molecule.

Even small steps taken from the natural backbone can destroy the COSMIC-LOPER properties of oligonucleotides. For example, work recently replaced the phosphate diester linkers in DNA and RNA by nonionic dimethylenesulfone linking units (Huang et al. 1991). The sulfone group is an "isosteric" and "isoelectronic" replacement for a phosphate. Nevertheless, these nonionic oligomers display some remarkable properties. First, they fold. For example, the octamer $ASO_2USO_2GSO_2GSO_2USO_2CSO_2ASO_2U$ folds in solution to give a folded form in water having a high melting temperature (~87°C) (Richert et al. 1996). Next, a synthetic intermediate leading to this oligosulfone was found to be a "catalyst" for a self-debenzoylation reaction (Richert et al. 1996). Still more remarkably, different oligosulfones evidently follow different strategies for folding and pairing. The dinucleotide analog GSO_2C in the crystal forms an antiparallel duplex approximately isomorphous with the analogous RNA (Roughton et al. 1995). In the crystal, the ASO_2T dinucleotide does not (Hyrup et al. 1995). The USO_2C dinucleotide forms a complex featuring backbone-to-backbone and backbone-to-nucleobase hydrogen bonds (C. Richert, pers. comm.). Even within a relatively small search of sequence space, these nonionic oligonucleotide analogs retain no conformational or physical property that could be a ready basis for a common mechanism for replication. In this respect, oligosulfone analogs of DNA and RNA behave much the same as peptides and conventional small organic molecules, not the nucleic acids upon which they are modeled.

These results suggest that the need for a COSMIC-LOPER behavior is a strong constraint on what biopolymers might serve as the basis for single-biopolymer life. They also suggest that a polyelectrolyte (polyanion or polycation) structure is important for the COSMIC-LOPER behavior that we see in standard nucleic acids (Richert et al. 1996):

1. Phosphate groups force the interaction surface between strands as far distant from the backbone as possible, to the Watson-Crick "edge" of the nucleobases. Without interstrand phosphate-phosphate repulsion,

sugar-sugar interstrand interactions, sugar-backbone interstrand interactions, interactions between the sugar and backbone groups of one strand and the Hoogsteen edge of the nucleobases on the other, Hoogsteen-Hoogsteen interstrand interactions, and Watson Crick-Hoogsteen interstrand interactions all become important, and the recognition phenomenon ceases to be rule-based.
2. Phosphates discourage folding in an oligonucleotide molecule. The statistical mechanical theory of polymers suggests that the polyanionic backbone will cause natural oligonucleotides to adopt an extended structure (Flory 1953; Brant and Flory 1965). Nonionic oligonucleotide analogs should (and do) fold like peptides. By discouraging folding, the repeating polyanionic backbone helps oligonucleotides act as templates.
3. Electronic distribution in a molecule is described as an infinite series (monopole + dipole + quadrapole + . . .). The first nonvanishing term dominates. The repeating monopole (charge) in DNA makes dipolar interactions (hydrogen bonding) secondary to its properties, allowing the DNA molecule to mutate without changing greatly its physical behavior.

Returning to functionalized and expanded genetic alphabets (see figures), this discussion suggests that one must be careful when "decorating" oligonucleotides with functionality. At some level of functionalization, the COSMIC-LOPER properties that enable DNA and RNA to serve as an evolvable Darwinian system will be lost. Preliminary data suggest, for example, that extensive functionalization with hydrophobic side chains destroys these properties. It remains to be seen whether the level of functionality that must be introduced into DNA and RNA to enable it to support a complicated metabolism is greater than that required to destroy its COSMIC-LOPER properties.

CAN A SINGLE-BIOPOLYMER LIFE BE FOUND TODAY IN THE SOLAR SYSTEM?

Single-biopolymer models for Darwinian chemistry have relevance to the search for extraterrestrial life. For example, biologists have noted that the microfossils in the Allan Hills meteorite, which are as small as 20–100 nanometers across, are too small to be living cells (Kerr 1997). After all, the argument is made, the ribosome is 25 nm across, and ribosomes are a basic requirement for life.

This argument is, of course, narrowly formulated. Ribosomes are a basic requirement for life based on two biopolymers. If a single biopolymer (such as RNA) can serve both genetic and catalytic functions, then ribosomes are not needed for life. Indeed, much of the metabolism of contemporary cells (aminoacyl tRNA synthetases and many amino acid biosynthesis enzymes, for example) comprising more than half of what is believed to be the core metabolism encoded by the protogenome (Benner et al. 1993) would also not be needed for life in an RNA World. A cell based on a single-biopolymer genetic system can be far smaller than one based on two biopolymers. This means that the fossils in the Martian meteorite structures are not too small to be remnants of a single-biopolymer form of life. Conversely, if the meteorite structures are indeed fossils, then they almost certainly are fossils of an organism that used only a single biopolymer as its molecular system capable of Darwinian evolution, and similar considerations should guide our search for nonterrean life.

The best place to search for single-biopolymer life may be here on Earth, however, assuming that terrestrial life originated here as a single-biopolymer Darwinian system. Whether such life remains on Earth depends on whether it was able to find a niche on the planet where it could compete with its descendants that developed two biopolymers. The superior power of proteins as catalysts provides presumptive arguments that a life form that did not exploit proteins as catalysts could not have competed with life that did. The biochemical innovation associated with translation almost certainly prompted an extinction more massive than the well-known extinctions at the end of the Cretaceous period.

A variety of ecological niches might provide single-biopolymer systems with an adaptive advantage over two-biopolymer systems, however, and may have provided ribo-organisms with the opportunity to survive on Earth even in the presence of two-biopolymer systems. For example, because cells containing single-biopolymer life can be much smaller than two-biopolymer cells, one-biopolymer life might have survived where small size offers a selective advantage. In subterranean matrices, for example, geological formations can have pore sizes that are too small to permit a two-biopolymer organism to live, but might permit a single-biopolymer cell to reside free from competition from its more adept protein-using cousins.

CONCLUSIONS

Experimental results suggest that the RNA World had both the opportunity and the motivation to use an expanded genetic alphabet. It remains to be seen how effectively functionalized oligonucleotides make a compromise between the structural demands for catalysis and the physical properties required for effective Darwinian evolution. Should experimental work show that they do so, we expect in vitro selections to provide effective new catalysts with the expanded genetic alphabet. In the most optimistic scenario, analogous single-biopolymer forms of life may be found elsewhere in the solar system, and perhaps in enclaves on planet Earth.

ACKNOWLEDGMENTS

We thank NASA, the Office of Naval Research, the National Institutes of Health, and the Swiss National Science Foundation for supporting some of the work described here. Petra Burgstaller acknowledges a fellowship from the Deutsche Akademische Austauschdienst.

REFERENCES

Allemann R.K. 1989. "Evolutionary guidance as a tool in organic chemistry." Ph.D. thesis no. 8804. Eidgenössische Technische Hochschule, Zürich, Switzerland.
Bartel, D.P. and Szostak J.W. 1993. Isolation of new ribozymes from a large pool or random sequences. *Science* **261:** 1411–1418.
Benner S.A. 1988. Reconstructing the evolution of proteins. In *Redesigning the molecules of life* (ed. S.A. Benner) pp. 115–175. Springer-Verlag, Heidelberg.
———. 1989. Enzyme kinetics and molecular evolution. *Chem. Rev* **89:** 789–806.
Benner S.A., Ellington A.D., and Tauer A. 1989. Modern metabolism as a palimpsest of the RNA World. *Proc. Natl. Acad. Sci.* **86:** 7054–7058.
Benner S.A., Cohen M.A., Gonnet G.H., Berkowitz D.B., and Johnsson K. 1993. Reading the palimpsest: Contemporary biochemical data and the RNA World. In *The RNA world* (ed. R. Gesteland and J. Atkins) pp. 27–70. Cold Spring Harbor Laboratory Press, Cold Spring Harbor, New York.
Benner S.A., Allemann R.K., Ellington A.D., Ge L., Glasfeld A., Leanz G.F., Krauch T., MacPherson L.F., Moroney S.E., Piccirilli J.A., and Weinhold E.G. 1987. Natural selection, protein engineering and the last riboorganism: Rational model building in biochemistry. *Cold Spring Harbor Symp. Quant. Biol.* **52:** 53–63.
Brant D.A. and Flory P.J. 1965. The configuration of random polypeptide chains. Experimental results. *J. Am. Chem. Soc.* **87:** 2788–2791.
Cech T.R., Zaug A.J., and Grabowski P.J. 1981. In vitro splicing of the ribosomal RNA precursor of *Tetrahymena*. Involvement of a guanosine nucleotide in the excision of the intervening sequence. *Cell* **27:** 487–496.
Crick F.H.C. 1968. The origin of the genetic code. *J. Mol. Biol.* **38:** 367–379.
Dewey T.M., Zyzniewski M.C., and Eaton B.E. 1996. The RNA world. Functional diversity in a nucleoside by carboxyamidation of uridine. *Nucleosides Nucleotides* **15:** 1611–1617.
Dugas H. 1989. *Bioorganic chemistry*, 2nd edition. Springer-Verlag, New York.
Flory P.J. 1953. *Principles of polymer chemistry*. Cornell University Press, Ithaca, New York.
Gilbert W. 1986. The RNA world. *Nature* **319:** 618.

Gouverneur V.E., Houk K.N., Depascualteresa B., Beno B., Janda K.D., and Lerner R.A. 1993. Control of the exo-pathway and the endo-pathway of the Diels-Alder reaction by antibody catalysis. *Science* **262:** 204–208.

Guerrier-Takada C., Gardiner K., Marsh T., Pace N., and Altman S. 1983. The RNA moiety of RNase P is the catalytic subunit of the enzyme. *Cell* **35:** 849–857.

Horlacher J., Hottiger M., Podust V.N., Hübscher U., and Benner S.A. 1995. Expanding the genetic alphabet: Recognition by viral and cellular DNA polymerases of nucleosides bearing bases with non-standard hydrogen bonding patterns. *Proc. Natl. Acad. Sci.* **92:** 6329–6333.

Huang Z. and Benner S.A. 1993. Selective protection and deprotection procedures for thiol and hydroxyl groups. *Synlett* 83–84.

Huang Z., Schneider K.C., and Benner S.A. 1991. Building blocks for analogs of ribo- and deoxyribonucleotides with dimethylene sulfide, -sulfoxide and -sulfone groups replacing phosphodiester linkages. *J. Org. Chem.* **56:** 3869–3882.

Huizenga D.E. and Szostak J.W. 1995. A DNA aptamer that binds adenosine and ATP. *Biochemistry* **34:** 656–665.

Hyrup B., Richert C., Schulte-Herbrüggen T., Benner S.A., and Egli M. 1995. X-ray crystal structure of a dimethylene-sulfone bridged ribonucleotide dimer crystallized at elevated temperature. *Nucleic Acids Res.* **23:** 2427–2433.

Irvine D., Tuerk C., and Gold L. 1991. Selexion. Systematic evolution of ligands by exponential enrichment with integrated optimization by non-linear analysis. *J. Mol. Biol.* **222:** 739–761.

Johnsson K., Allemann R.K., and Benner S.A. 1990. Designed enzymes. New peptides that fold in aqueous solution and catalyze reactions. In *Molecular mechanisms in bioorganic processes* (ed. C. Bleasdale and B.T. Golding), pp. 166–187. Royal Society of Chemistry, Cambridge, United Kingdom.

Johnsson K., Allemann R.K., Widmer H., and Benner S.A. 1993. Synthesis, structure and activity of artificial, rationally designed catalytic polypeptides. *Nature* **365:** 530–532.

Joyce G.F. 1989a. Amplification, mutation and selection of catalytic RNA. *Gene* **82:** 83–87.

———. 1989b. Building the RNA world. Evolution of catalytic RNA in the laboratory. *UCLA Symp. Mol. Cell. Biol. New Ser.* **94:** 361–371.

———. 1994. Foreword. In *Origins of life: The central concepts* (ed. D.W. Deamer and G.R. Fleischaker). Jones and Bartlett, Boston, Massachusetts.

Joyce G.F., Schwartz A.W., Miller S.L., and Orgel L.E. 1987. The case for an ancestreal genetic system involving simple analogs of the nucleotides. *Proc. Natl. Acad. Sci.* **84:** 4398–4402.

Kerr R.A. 1997. Ancient life on Mars? Putative Martian microbes called microscopy artifacts. *Science* **278:** 1706–1707.

Khare B.N., Thompson W.R., Cheng L., Chyba C., Sagan C., Arakawa E.T., Meisse C., and Tuminello P.S. 1993. Production and optimal-constants of ice tholin from charged-particle irradiation of (1/6) C_2H_6/H_2O at 77 K. *Icarus* **103:** 290–300.

Kodra J. and Benner S.A. 1997. Synthesis of an N-alkyl derivative of 2′-deoxyisoguanosine. *Synlett* 939–940.

Lee D.H., Severin K., Yokobayashi Y., and Ghadiri M.R. 1997. Emergence of symbiosis in peptide self-replication through a hypercyclic network. *Nature* **390:** 591–594.

Limbach P.A., Crain P.F., and McCloskey J.A. 1994. The modified nucleosides of RNA. Summary. *Nucleic Acids Res.* **22:** 2183–2196.

Lutz M.J., Held H.A., Hottiger M., Hübscher U., and Benner S.A. 1996. Differential discrimination of DNA polymerases for variants of the non-standard nucleobase pair between xanthosine and 2,4-diaminopyrimidine, two components of an expanded genetic alphabet. *Nucleic Acids Res.* **24:** 1308–1313.

Miller S.L. 1997. Peptide nucleic acids and prebiotic chemistry. *Nat. Struct. Biol.* **4:** 167–169.

Morris K.N., Tarasow T.M., Julin C.M., Simons S.L., Hilvert D., and Gold L. 1994. Enrichment for RNA molecules that bind a Diels-Alder transition-state analog. *Proc. Natl. Acad. Sci.* **91:** 13028–13032.

Müller D., Pitsch S., Kittaka A., Wagner E., Wintner C.E., and Eschenmoser A. 1990. Chemistry of alpha-aminonitriles: Aldomerisation of glycoaldehyde phosphate to rac-hexose 2,4,6-triphosphates and in presence of formaldehyde rac-pentose 2,4-diphosphates: Rac-allose 2,4,6-triphosphate and rac-ribose 2,4-diphosphate are the main reaction-products. *Helv. Chim. Acta* **73:** 1410–1468.

Orgel L.E. 1968. Evolution of the genetic apparatus. *J. Mol. Biol.* **38:** 381–393.

Perezpaya E., Houghten R.A., and Blondelle S.E. 1996. Functionalized protein-like structures from conformationally defined synthetic combinatorial libraries. *J. Biol. Chem.* **271:** 4120–4126.

Piccirilli J.A., Moroney S.E., and Benner S.A. 1991a. A C-nucleotide base pair. Methylpseudouridine-directed incorporation of formycin triphosphate into RNA catalyzed by T7 RNA polymerase. *Biochemistry* **30:** 10350–10356.

Piccirilli J.A., Krauch T., MacPherson L.J., and Benner S.A. 1991b. A direct route to 3-ribofuranosyl-pyridine nucleosides. *Helv. Chim. Acta* **74:** 397–406.

Piccirilli J.A., Krauch T., Moroney S.E., and Benner S.A. 1990. Extending the genetic alphabet: Enzymatic incorporation of a new base pair into DNA and RNA. *Nature* **343:** 33–37.

Prober J.M., Trainor G.L., Dam R.J., Hobbs F.W., Robertson C.W., Zagursky R.J., Cocuzza A.J., Jensen M.A., and Baumeister K. 1987. A system for rapid DNA sequencing with fluorescent chain-terminating dideoxynucleotides. *Science* **238:** 336–341.

Rich A. 1962. On the problems of evolution and biochemical information transfer. In *Horizons in biochemistry* (ed. M. Kasha and B. Pullman), pp. 103–126. Academic Press, New York.

Richert C., Roughton A.L., and Benner S.A. 1996. Nonionic analogs of RNA with dimethylene sulfone bridges. *J. Am. Chem. Soc.* **118:** 4518–4531.

Robertson M.P., and Miller S.L. 1995. Prebiotic synthesis of 5-substituted uracils: A bridge between the RNA world and the DNA-protein world. *Science* **268:** 702–705.

Roughton A.L., Portmann S., Benner S.A., and Egli M. 1995. Crystal structure of a dimethylene sulfone linked ribodinucleotide analog. *J. Am. Chem. Soc.* **117:** 7249–7250.

Sassanfar M. and Szostak J.W. 1993. An RNA motif that binds ATP. *Nature* **364:** 550–553.

Schmidt J.G., Nielsen P.E., and Orgel L.E. 1997. Enantiomeric cross-inhibition in the synthesis of oligonucleotides on a nonchiral template. *J. Am. Chem. Soc.* **119:** 1494–1495.

Switzer C.Y., Moroney S.E., and Benner S.A. 1989. Enzymatic incorporation of a new base pair into DNA and RNA. *J. Am. Chem. Soc.* **111:** 8322–8323.

Szathmary E. 1992. What is the optimum size for the genetic alphabet? *Proc. Natl. Acad. Sci.* **89:** 2614–2618.

Szostak J.W. 1988. Structure and activity of ribozymes. In *Redesigning the molecules of life* (ed. S.A. Benner) pp. 87–114. Springer-Verlag, Heidelberg.

Tang J. and Breaker R.R. 1997. Rational design of allosteric ribozymes. *Chem. Biol.* **4:** 453–459.

Tarasow T.M., Tarasow S.L., and Eaton B.E. 1997. RNA-catalyzed carbon-carbon bond formation. *Nature* **389:** 54–57.

Usher D.A. and McHale A.H. 1976. Hydrolytic stability of helical RNA: A selective advantage for the natural 3',5'-bond. *Proc. Natl. Acad. Sci.* **73:** 1149–1153.

Visser C.M. and Kellogg R.M. 1978. Biotin. Its place in evolution. *J. Mol. Evol.* **11:** 171–178.

Vögel J.J. and Benner S.A. 1994. Non-standard hydrogen bonding in duplex oligonucleotides. The base pair between the acceptor-donor-donor pyrimidine analog and the donor-acceptor-acceptor purine analog. *J. Am. Chem. Soc.* **116:** 6929–6930.

Vögel J.J., Altorfer M.M., and Benner S.A. 1993a. The donor-acceptor-acceptor purine analog. Transformation of 5-aza-7-deaza-*iso*guanine to 2'-deoxy-5-aza-7-deaza-*iso*guanosine using purine nucleoside phosphorylase. *Helv. Chim. Acta* **76:** 2061–2069.

Vögel J.J., von Krosigk U., and Benner S.A. 1993b. Synthesis and tautomeric equilibrium of 6-amino-5-benzyl-3-methylpyrazin-2-one. An acceptor-donor-donor nucleoside base analog. *J. Org. Chem.* **58:** 7542–7547.

von Krosigk U. and Benner S.A. 1995. pH-Independent triple helix formation by an oligonucleotide containing a pyrazine donor-donor-acceptor base. *J. Am. Chem. Soc.* **117:** 5361–5362.

Wang Y. and Patel D.J. 1994. Solution structure of the *Tetrahymena* telomeric repeat d(T_2G_4)$_4$ G-tetraplex. *Structure* **2:** 1141–1156.

Watson J.D., Hopkins N.H., Roberts J.W., Steitz J.A., and Weiner A.M. 1987. *Molecular biology of the gene,* 4th edition, p. 1115. Benjamin/Cummings, Menlo Park, California.

White H.B., III. 1976. Coenzymes as fossils of an earlier metabolic state. *J. Mol. Evol.* **7:** 101–104.

Wiegand T.W., Janssen R.C., and Eaton B.E. 1997. Selection of RNA amide synthases. *Chem. Biol.* **4:** 675–683.

Woese C.R. 1967. *The genetic code. The molecular basis for genetic expression.* Harper & Row, New York.

Zaug A.J. and Cech T.R. 1986. The intervening sequence RNA of *Tetrahymena* is an enzyme. *Science* **231:** 470–475.

Zhang B.L. and Cech T.R. 1997. Peptide bond formation by in vitro selected ribozymes. *Nature* **390:** 96–100.

7
Aminoacyl-tRNA Synthetases and Self-acylating Ribozymes

Michael Yarus and Mali Illangasekare
Department of Molecular, Cellular, and Developmental Biology
University of Colorado
Boulder, Colorado 80309-0347

The aminoacyl-tRNA synthetases (aaRS) are at the heart of modern translation, catalyzing the accurate biosynthesis of aminoacyl-tRNAs (aa-tRNAs), the immediate precursors for encoded peptides. However, the first catalysts that made aa-RNAs for coded protein synthesis probably appeared long before any protein aaRS, to serve a preexisting translation system (see below). It presently seems likely that this ancestral translation system relied on a molecule like RNA, and the proto-aaRS catalysts may themselves have been RNAs. As an exercise in the limits of RNA catalysis, and as data relevant to the existence of such an RNA World, we would like to know how closely RNA catalysis can approximate the essential capabilities of the modern aaRS. Below is, first, a brief introduction to these complex modern proteins, then similarly a description of the presently known self-acylating RNAs, and finally an assessment of the question of possible resemblance.

AMINOACYL-tRNA SYNTHETASE EVOLUTION

Current aaRS are uniformly large proteins (Schimmel and Söll 1979) and therefore could not have existed before translation itself. In fact, because aaRS share essential structures, like the nucleotide-binding Rossman fold, among themselves and with other proteins (Eriani et al. 1990), they presumably were derived from yet more ancient common ancestors. Thus, most aaRS cannot even have been among the first proteins.

Aminoacyl-tRNA synthetases from all species can be split into two equally sized protein families, termed type I and type II proteins, characterized by two different sets of conserved amino acid motifs in the active sites for amino acid activation and aa-tRNA synthesis (Eriani et al. 1990). Type I enzymes use the Rossman fold within their active sites, and type

II enzymes use an antiparallel sheet (Cusack et al. 1990) known elsewhere only in biotin synthetase (Artymiuk et al. 1994). As one consequential effect of these structural differences, type I aaRS first transfer amino acid to the 2' hydroxyl of the terminal ribose of tRNA, and type II aaRS *trans*-acylate initially to the 3' hydroxyl (PheRS is a type II exception [Eriani et al. 1990]). The reasons for the existence of the two families are mysterious, but trees relating the two types of aligned aaRS amino acid sequences remain disconnected back to deep and separate roots (Nagel and Doolittle 1995). Accordingly, one must deal with the implications of a period in which the aaRS themselves were evolving. Chemically similar amino acids like Tyr (type I) and Phe (type II), Glu (type I) and Asp (type II), Cys (type I) and Ser (type II) were assigned to different aaRS classes during a long period in which the new examples of two enzyme types must have been appearing side by side, evolving in part by peptide intercalations beside and between the conserved motifs (Eriani et al. 1990).

A prolonged period for evolutionary appearance of the protein aaRS again suggests the existence of a preexisting class of catalysts for aa-tRNA synthesis. Consider a time near the roots of the type I and II aaRS trees. There were, in this era, only one or two aaRS to activate amino acids for protein synthesis. Nevertheless, there necessarily were proteins with complex functions (the aaRS themselves) being synthesized. Because it is unlikely that one could make sophisticated catalysts containing only one or two amino acids, the other amino acids in these ancestral aaRS were activated by another, even more ancient, class of catalysts. Thus, it is difficult to avoid the idea that the protein aaRS and their predecessors coexisted and collaborated in the synthesis of proteins before there was a self-sufficient complement of protein aaRS. This notion in turn makes it very likely that the preexisting activities would have used the same or similar substrates: ATP and the standard amino acids are demonstrably ancient, and straightforward transition to the era of the protein aaRS would have been adaptive. The inference that the protein aaRS and their predecessors had very similar aminoacyl-RNA products seems even more secure, given that they had to collaborate, using the same translation apparatus, while the protein aaRS progressively appeared. Therefore, there seems to be good reason to seek ancestral aa-RNA-synthesizing enzymes.

AMINOACYL-tRNA SYNTHETASE REACTIONS

Modern protein aaRS achieve specific aa-tRNA synthesis in several steps, whose relative contribution to specificity varies from enzyme to enzyme.

Activation: In the first place, aaRS bind their amino acid selectively and appose its carboxyl for attack at the α phosphate of ATP.

$$aa + ATP + aaRS \longrightarrow (aa\text{-}AMP)\,aaRS + PP_i$$

ATP is bound for this amino acid activation reaction in a conformation that is characteristic for the protein type, 5′ oligophosphate extended for type I enzymes (Perona et al. 1993) and bent back to the base for type II (Cusack et al. 1990). However, the chemistry appears similar; in-line displacement of pyrophosphate from ATP by the amino acid carboxyl oxygen. The catalytic role of type I and type II proteins also appears similar, in that both act simply: entropically to tightly appose ATP and amino acids by making many specific bonds to these substrates using groups within the conserved type I and type II motifs. In addition, conserved elements in both type I and type II aaRS offer polar groups that specifically stabilize the pyrophosphate leaving group and the transition state for the formation of the adenylate (aa-AMP). Both types of aaRS prominently use both divalent metals and polar amino acids like arginine, lysine, and histidine in these roles. Within each type, the active sites superpose very well, and structural elements playing the same roles are easily identified (Arnez and Moras 1997).

Pretransfer editing: An amino acid activated by synthesis of its adenylate is not necessarily committed to later aa-tRNA formation. Some synthetases, possibly primarily type I synthetases, have an editing function that hydrolyzes the misactivated amino acid adenylate before transfer to tRNA (Fersht 1977).

$$(aa\text{-}AMP)\,aaRS \xrightarrow{tRNA} aa + AMP + aaRS$$

Whereas such specific hydrolytic discard of noncognate adenylate may require the concurrent presence of the cognate tRNA, as indicated by the tRNA in the scheme just above, the tRNA may not be a reactant (Fersht 1977; Lin et al. 1984); that is, it may be a conformational effector. Thus, aminoacylation of the tRNA need not be a required step in pretransfer editing, and the role of the tRNA may even be taken by a DNA oligomer selected for affinity to the synthetase (Hale and Schimmel 1996).

Aminoacyl transfer: Aminoacyl-tRNAs apparently arise from attack of a terminal 2′ (3′) ribose hydroxyl on the carbonyl carbon of the aaRS-bound aa-AMP (Perona et al. 1993; Cavarelli et al. 1994):

$$(\text{aa-AMP}) \text{ aaRS} + \text{tRNA} \longrightarrow \text{aa-tRNA} + \text{aaRS} + \text{AMP}$$

Posttransfer editing: However, even after aminoacyl transfer, aa-tRNA is not necessarily committed to a ribosomal fate. Hydrolytic editing of a mistakenly formed aa-tRNA (Eldred and Schimmel 1972; Yarus 1972) before its dissociation from the aaRS may clean up errors that have not been caught thus far. This last stage of editing may be either a large (Fersht and Kaethner 1976) or a small (Freist et al. 1996) factor in overall accuracy, depending on the aaRS, the particular noncognate amino acid, or even the conditions of reaction (Freist 1989).

$$(\text{aa-tRNA}) \text{ aaRS} \longrightarrow \text{aa} + \text{tRNA} + \text{aaRS}$$

To carry out these related activities, aaRS must clearly be complex catalysts. In particular, there are likely to be interactions among the amino acid, ATP, and RNA substrates of the aaRS, as might also be expected from first principles when three sites (one for a macromolecule) must converge to within a chemical bond length. For example, mobility of the ATP and amino acid sites has been implicated in activation (Schmitt et al. 1994), the amino acid site has long been known to be capable of interaction with the tRNA site (Yarus and Berg 1969), and the tRNA site can interact with amino acid selection (Yarus and Berg 1969; Ibba et al. 1996). The interaction between tRNA and amino acid sites can be sufficiently far-reaching that amino acid is not activated to form adenylate in the absence of tRNA, as for Gln-, Glu-, and Arg-RS (Schimmel and Söll 1979).

The above succession of specific steps gives overall discriminations against noncognate amino acids that range from about 200-fold to 500,000-fold with typical values of a few thousand-fold, as measured in vitro by specific aa-tRNA synthesis (Freist et al. 1996). This spectrum of specificities can be taken as the biochemical signature of a biologically fit aaRS activity, capable of serving a functional translation system.

SELF-AMINOACYLATING RNAs

These are RNA catalysts purified from pools of RNAs with internal randomized sequences by selection-amplification in vitro and cloning (Illangasekare et al. 1995). After incubation with adenylate, aminoacylated RNAs were made distinctly hydrophobic by derivatization of their unique α-amino group. The minority of self-acylated aa-RNA, now hydrophobic, were resolved by reverse-phase HPLC.

Phe-AMP + RNA $\longrightarrow$ Phe-RNA + AMP

Phe-RNA + naphthoxyacetyl-NHS $\longrightarrow$ naphthoxyacetyl-Phe-RNA + NHS

where NHS is N-hydroxy succinimide.

Self-aminoacylation is of interest outside the context of translation because it was among the first examples of RNA catalysis at carbon, and differs thereby from classic ribozymes that perform phosphate chemistry. Catalysis of self-aminoacylation shows that RNAs can stabilize the tetrahedral, polar transition state which is characteristic of nucleophilic attack at carbonyl carbon. Such a transition state is shared by an important class of reactions that includes attack of amino groups at acyl carbon; for example, peptide-bond synthesis is catalyzed by an rRNA-rich subribosomal particle (Noller et al. 1992), although this reaction has been found to be protein dependent (Noller 1993). However, the argument for ribozyme-catalyzed peptide synthesis is supported by other evidence. Independent selections establish that pure RNA can catalyze a similar aminoacyl transfer reaction (Lohse and Szostak 1996; Zhang and Cech 1997). Furthermore, large subunit rRNA is probably binding both A- and P-site reactants within a few bond lengths of the forming peptide bond—a peptidyl transferase transition state analog binds to a ribo-oligonucleotide containing conserved sequences from the rRNA peptidyl transferase loop (Welch et al. 1997). Hydrolysis, attack of OH^- or HOH at carbonyl carbon, leading to a similar transition state, is also weakly catalyzed (10- to 100-fold stimulation over spontaneous) by unselected RNAs acting on esters (Piccirilli et al. 1992) and amide bonds (Dai et al. 1995, 1996). Based on these examples in which RNAs host a similar tetrahedral transition intermediate, other ribozymes carrying out related reactions at carbonyl carbon will likely be found.

Six metalloribozymes: Six independently derived types of self-aminoacylating sequences have been recovered from the above selection (see Fig. 1) (Illangasekare et al. 1997). All are sensitive to periodate oxidation, a specific modification of the $2'(3')$ ribose hydroxyls. All are therefore probably analogous to natural aa-tRNA, forming terminal $2'(3')$ aminoacyl-ribose, as has been proven for the most prevalent family of sequences RNA 19, 29 (Illangasekare et al. 1995), and for RNA 77 (M. Illangasekare and M. Yarus, unpubl.). This product is chemically similar to the biological aa-tRNA product of the protein aaRS. Five of six types of RNA require calcium, and three of the calcium-requiring catalysts also require magnesium (Fig. 1). One RNA (#71) is functional with either calcium or magnesium alone. Thus, these are metalloribozymes,

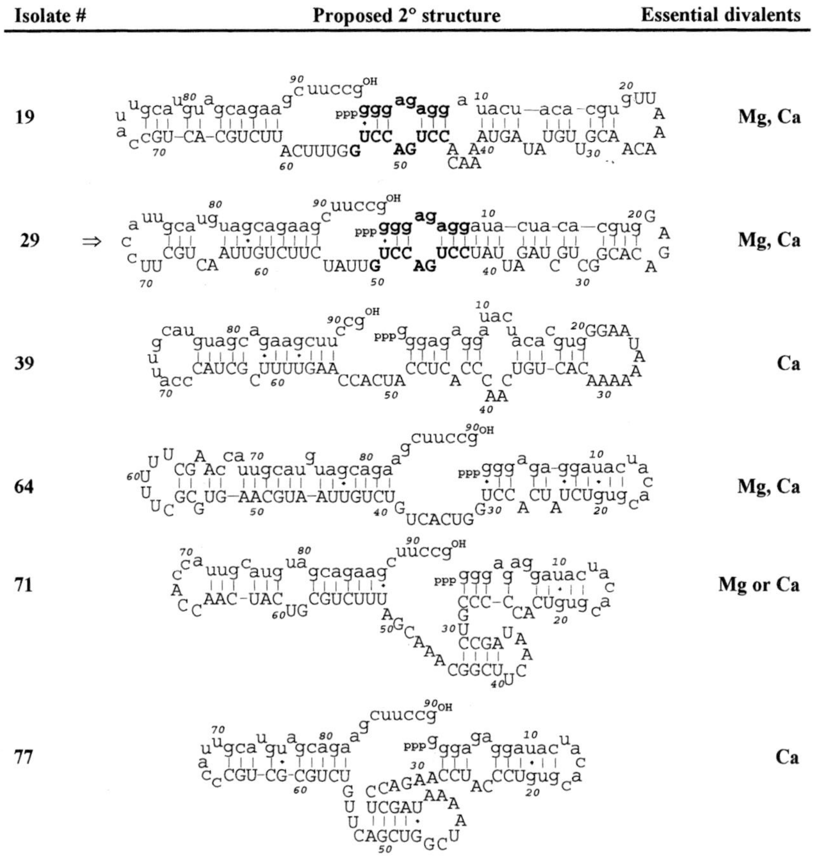

Isolate #	Proposed 2° structure	Essential divalents
19	(secondary structure diagram)	Mg, Ca
29	⇒ (secondary structure diagram)	Mg, Ca
39	(secondary structure diagram)	Ca
64	(secondary structure diagram)	Mg, Ca
71	(secondary structure diagram)	Mg or Ca
77	(secondary structure diagram)	Ca

Figure 1 The six independently derived families of self-aminoacylating RNAs. Lowercase letters indicate constant nucleotides, uppercase indicate initially randomized nucleotides. Bold nucleotides show a motif conserved in two independent isolates. The open arrow indicates the locus of a break in backbone continuity that has little effect on catalysis (Illangasekare et al. 1997).

with calcium having a uniquely prominent role in the aminoacylation reaction (Illangasekare et al. 1997).

The reaction: the acylation velocity of RNA #29 responds linearly at low concentrations of Phe-AMP, but can be observed to saturate at high concentrations of Phe-AMP, implying a Michaelis constant $K_M = 9$ mM and a $k_{cat} = 1.2$ min^{-1}. The rate of aa-RNA formation is proportional to hydroxyl ion concentration, consistent with the hypothesis that ionization of a group with a high pK, likely a ribose hydroxyl which attacks the amino acid carbonyl, is an essential step in the reaction (Illangasekare and Yarus 1997).

$$\text{RNA-G-2}'(3')\text{OH} \longrightarrow \underline{\text{RNA-G-O}^- + \text{NH}_3^+(\varphi)\text{CH-CO-AMP}} \longrightarrow$$

$$\text{RNA-G-O-Phe} + \text{AMP}$$

This in turn presumably implies that k_{cat} represents the chemical step of the reaction. If the amino acid is removed, the RNA will repetitively catalyze its own acylation with an undiminished rate, indicating that it is functionally unaltered by the acylation reaction.

Definition of the active site in the RNA 29 family: The independent isolates 19 and 29 (Fig. 1) form a family that shares sequences; sequence comparisons can therefore identify required and dispensable features. For example, the linking sequence (between the two helices) varies from RNA 19 to 29, and specific substitution experiments (Illangasekare et al. 1997) confirm that the linking sequence can be varied within limits. Similar substitutions suggest that the 3′ extension requires that the two terminal nucleotides –CG–OH must be specific bases, but that the extension may be substituted elsewhere.

Comparison of RNAs 19 and 29 also suggests that the proximal part of the 5′ bulged hairpin domain is conserved (bold nucleotides in Fig. 1) and therefore required for the reaction. This can be confirmed by truncation of this hairpin beyond the conserved region, and after replacement of these 36 nucleotides by a stabilizing tetraloop (29tr5; Fig. 2); the resulting molecule is as active as the 95-mer parent.

The 3′ hairpin is also poorly conserved in the two independent molecules, and it also can be abbreviated by removal of 27 distal nucleotides

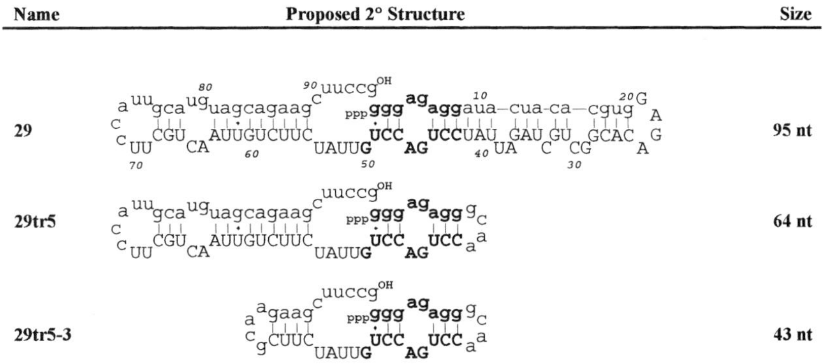

Name	Proposed 2° Structure	Size
29	(secondary structure diagram)	95 nt
29tr5	(secondary structure diagram)	64 nt
29tr5-3	(secondary structure diagram)	43 nt

Figure 2 Truncation of RNA 29. Notation as in Fig. 1, except that lowercase now includes nucleotides fixed arbitrarily during experiments on the initially selected sequence, as in the tetraloops.

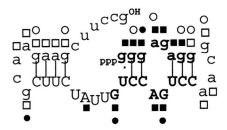

Figure 3 Modification-interference on RNA 29tr5-3. The results are indicated above or below the relevant purine, as follows: (□) DEPC modification that does not interfere with aminoacylation (≥70% of control); (■) DEPC modification that interferes; (○) DMS modification that does not interfere; (●) DMS modification that interferes.

which are replaced by a tetraloop (RNA 29tr5-3; Fig. 2). The resulting doubly deleted 43-nucleotide transcript retains full self-aminoacylation activity.

However, the 3' domain can be even more radically truncated. Not only is it not conserved in RNA 19 and 29, but modification-interference with DEPC and DMS suggests that although the 5' domain is crucial, purine modifications are often innocuous in the 3' hairpin of 29tr5-3 (Fig. 3) (O. Kovalchuke and M. Yarus, unpubl.). Therefore, we have entirely deleted the 3' hairpin. The three nucleotides that are boldfaced in RNA 29tr5-3-3 (Fig. 4) may be thought of as the bracketing nucleotides of the 3' hairpin, and a central nucleotide to replace the 3' hairpin. When these three positions were randomized, molecules with various sequences, having 3 of 4 nucleotides at each randomized position, were found to be active (M. Illangasekare and M. Yarus, unpubl.). In these 32-nucleotide molecules, the 3' hairpin is replaced, in effect, by a single nucleotide. The molecule shown, among the most active, is about 1/5 as active as parental sequences. This reflects primarily a decrease in k_{cat}, due to a slower chemical step and difficulty in fixing the 3' end.

The resulting 32-mer self-aminoacylating catalyst RNA 29tr5-3-3 (Fig. 4) retains the essential calcium and magnesium sites, the Michaelis complex with Phe-AMP, and the dependence on pH that were characteristic of the selected 95-mer. Thus, despite deletion of 63 nucleotides (2/3 of the initial sequence), the active site appears to be present virtually unaltered in this molecule.

Figure 4 superposes and combines all data, published and unpublished, on nucleotide requirements within the 32-mer catalyst. I assume in so doing that nucleotides in molecules at all stages of truncation play comparable roles.

RNA	Proposed 2° structure	Size

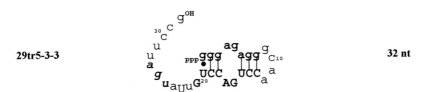

29tr5-3-3　　　　　　　　　　　　　　　　　　　　　　　　32 nt

Figure 4 RNA 29tr5-3-3 and the constraints on its nucleotides. Notation as in Fig. 1, except that lowercase indicates both arbitrarily fixed nucleotides and nucleotides that can be substituted with little effect on catalysis. Lowercase bold italics indicate nucleotides that were randomized in order to isolate 29tr5-3-3.

The resulting Figure 4 constrains the roles of 29 of the 32 nucleotides. In a hypothetical original molecule with constrained terminal sequences like those in RNA 29tr5-3-3, specification of as few as 9 randomized nucleotides (the conserved nonamer) or as many as 10 (including untested U_{23} as specific) would have created an effective catalyst. Put another way, there is evidence for only 11 nucleotides whose base must be specified. Its conservation in independent isolates implicates the nonamer, and substitution experiments implicate the 3′-terminal –CG–OH sequence as essential. If three untested nucleotides (U_{23}, U_{28}, U_{29}) are included to get a maximally conservative estimate, only 14 or fewer nucleotides may be essential to the creation of the bulged-helix-with-3′-tail which is the minimal self-aminoacylating RNA.

A transaminoacylation active site can therefore be assembled from few nucleotides. Because an active site must bring reactants together precisely, it is most likely assembled within the only conserved molecular environment in RNA 29tr5-3-3; that is, within the 5′ bulged helix, where the required 3′ end may be packed into the bulge loop, along with a molecule of Phe-AMP and perhaps the required Mg^{++} and Ca^{++} (c.f. Illangasekare et al. 1997).

Amino acid specificity: The above RNAs of the 29 family accelerate their reaction more than 10^5-fold, but are not amino acid specific. The apparent second-order rate constant for reaction of RNA and aa-AMP at subsaturating concentrations (a function of both k_{cat} and K_M) is usually measured to characterize reaction velocity (Table 1). RNA 29's minor substrate preferences are paradoxical; reaction is a few-fold faster with Ser-AMP and Ala-AMP (Table 1) than with Phe-AMP used to select them initially (Illangasekare et al. 1995). They bind the Phe-AMP, but they are

Table 1 Rates for RNA 29

Adenylate	Apparent 2nd order rate constant at 5 mM aa-AMP
Phe-AMP	80 M^{-1} min^{-1}
Ala-AMP	200
Ser-AMP	130

inhibited by AMP with $K_I^{AMP} \approx K_m^{Phe\text{-}AMP}$, but not inhibited detectably by free phenylalanine. These data taken together suggest that the adenylate substrate may be treated nonspecifically because it is bound almost exclusively via the AMP moiety.

Because amino acid specificity is one of the hallmarks of the protein aaRS, and a necessity for accurate translation, amino-acid-specific catalysis of aa-RNA synthesis would be of interest. RNA-binding sites that distinguish amino acid side chains exist naturally (Yarus 1988) and can also be selected (for review, see Yarus 1998), suggesting that a specifically self-acylating RNA could exist via the possession of a specific amino acid site.

In fact, amino acid side chain specificity has been observed (Table 2) for RNA 77 (Fig. 1), a Ca-specific ribozyme that strongly prefers the Phe-AMP used for its selection.

RNA 77 has the same 2'(3') terminal Phe-RNA product (M. Illangasekare and M. Yarus, unpubl.) but reacts about 900-fold more quickly with Phe-AMP than do the derivatives of RNA 29 (Table 2). Using an upper limit of 10^{-3} M^{-1} min^{-1} for the background acylation of randomized RNA by Phe-AMP (Illangasekare et al. 1995), RNA 77 accelerates its reaction more than 7.3×10^7-fold under its optimal conditions.

Despite its exceptional acceleration of Phe-RNA production, RNA 77 is even less reactive with Ala-AMP and Ser-AMP than is the RNA 29 family. In Table 2, the second-order rate constant $\approx k_{cat}/K_M$, and the rates of simultaneous reactions with competing substrates are in the ratio of their k_{cat}/K_M (Fersht 1985). Thus, RNA 77 selects about 4000-fold against alanine on the basis of rate, and 20,000-fold against serine.

There is yet another dimension to RNA 77 specificity, however. Only a minority of RNA 77 appears to react with Ala-AMP and Ser-AMP (Table 2). In contrast, the same RNA 77 in the presence of Phe-AMP is fast-reacting, kinetically homogeneous, and its reaction is almost complete, as seen before for the RNA 29 family (Illangasekare and Yarus 1997).

We suggest that both aspects of specificity reflect related underlying events. The fast reaction of RNA 77 to yield Phe-RNA is not a result of

Table 2 Rates for RNA 77

Adenylate	Apparent 2nd order rate constant	% aa-RNA formation	aa-AMP conc.
Phe-AMP	73,000 M^{-1} min^{-1}	85–90	5 μM
Ala-AMP	18	≤ 20	10 mM
Ser-AMP	3.6	≤ 20	10 mM

an exceptionally low K_M (e.g., compared to RNA 29), but instead a very fast k_{cat}, which we estimate as several hundred min^{-1}. Rate is also proportional to [OH$^-$], which probably means that rate reflects the chemical step (M. Illangasekare and M. Yarus, unpubl.). Thus, the superiority of the RNA 77 reaction rests on fast and accurate transfer of the aminoacyl group, rather than on selective binding of the adenylates.

Accordingly, we surmise that the active site requires a somewhat unusual conformation to perform aminoacyl transfer. On binding adenylates, the phenylalanine residue (e.g., the side-chain ring) facilitates rapid conversion (on a msec time scale) of the site to active form. Thus, Phe is transferred quickly, and alanine and serine slowly. Because free energy from binding of the phenylalanine side-chain is almost cancelled by the cost of the needed rearrangement, the phenylalanine advantage does not appear in K_M. A related, but more extensive, rearrangement explains why most molecules that have bound Ala-AMP and Ser-AMP do not make the transition to the active form at a measurable rate.

A COMPARISON

There are striking parallels between the aaRS and the self-acylating RNAs. The RNAs use the same ubiquitous biologically activated amino acid, the adenylate, and the RNAs produce the same terminally 2'(3') aminoacylated RNA as do the proteins. It is thought that the aminoacylation reaction, in which the carbonyl carbon of the adenylate is attacked by a ribose oxyanion, is fundamentally similar in the proteins (Arnez and Moras 1997) and the RNA (Illangasekare and Yarus 1997). Although the RNAs do not turn over rapidly, they are intrinsically capable of turnover; that is, multiple identical aminoacylations occur if the amino acid is serially removed from the RNA's 3' terminus.

Turnover in another sense, now meaning acylation of multiple RNAs, may also be possible. Molecules related to RNA 29 but formed from two base-paired pieces: a synthetic 3' acceptor fragment and a 5' enzyme, broken in the region of the open arrow (Fig. 1) are active (Illangasekare et al.

1997). In fact, with 6 or 7 base pairs between the two halves, the acceptor should bind substantially at experimental concentrations of the nucleic acids (as was observed) and the acceptor fragment should still turn over at several per minute, consistent with the overall rate of aminoacyl-RNA formation. Repetitive RNA-catalyzed formation of distinct aminoacyl-RNAs therefore seems achievable with known components.

Both protein and RNA catalysts can be specific. The aaRS proteins achieve specificity with a multistep process that sometimes includes two hydrolytic steps to destroy mistakes made earlier in the aminoacylation pathway. These hydrolytic editing steps were not selected in the case of the self-aminoacylating RNA; once formed, catalytic Ala-RNA and Ser-RNA have the expected stability in the absence of the adenylate (M. Illangasekare and M. Yarus, unpubl.). Therefore, the products are not actively hydrolyzed. However, the 4,000- to 20,000-fold specificity of RNA 77, presumably achieved by a simple binding-and-chemical-transfer mechanism, actually can exceed the overall selectivity of the well-investigated PheRS of baker's yeast, which selects 6300-fold against synthesis of Ala-tRNA^{Phe} and 3500-fold against Ser-tRNA^{Phe} (based on k_{cat}/K_M; Freist et al. 1996). Aminoacyl transfer by RNA 77 therefore already seems comparable to baker's yeast PheRS and therefore perhaps specific enough to serve an RNA translation system. This is more true if partial activity with noncognate amino acids is also taken into account.

The striking difference between the proteins and the present RNAs is amino acid activation. Several kinds of selections (G. Sanchez and M. Yarus; S. Zinnen and M. Yarus; F. Huang and M. Yarus, all unpubl.) for attack of the amino acid's carboxyl group on the 5′ triphosphate of randomized transcripts have failed to isolate the equivalent of an aaRS activation reaction:

$$NH_3^+ (R)CH\ COO^- + pppG\text{-}RNA \rightarrow NH_3^+ (R)CH\ COO\text{-}pG\text{-}RNA + PP_i$$

On one hand, such selections seek a thermodynamically unfavorable reaction with an unstable product, and in addition, the carboxyl may not be a particularly good nucleophile. Thus, there are particular impediments to selection that may be cited, even though protein aaRS catalyze a comparable reaction. However, triphosphorylated RNAs have been selected to allow attack at their own 5′ α-phosphate by a variety of phosphorylated nucleophiles (nucleotides, phosphorylated sugars, and phospho-amino acids), as well as allowing attack by water to release pyrophosphate (Huang and Yarus 1997a). It may now be possible to exploit this proven ability for RNA to make its triphosphate broadly reactive (Huang and

Yarus 1997b) to derive a ribozyme that activates the carbonyl of amino acids by biological means, formation of an amino acid adenylate or guanylate. Combination of such an activation reaction with the small RNA structures that specify aminoacyl transfer (reviewed above) would then presumably (re?)constitute an RNA aa-RNA synthetase.

ACKNOWLEDGMENTS

Thanks to members of my laboratory for help in the redaction of this manuscript; particularly to Oleh Kovalchuke for unpublished results, and to the National Institutes of Health for long-term support of the work described.

REFERENCES

Arnez J.G. and Moras D. 1997. Structural and functional considerations of the aminoacylation reaction. *Trends Biochem. Sci.* 22: 211–216.
Artymiuk P.J., Rice D.W., Poirette A.R., and Willett P. 1994. A tale of two synthetases. *Nat. Struct. Biol.* **1:** 758–760.
Cavarelli J., Eriani G., Rees B., Ruff M., Boeglin M., Mitschler A., Martin F., Gangloff J., Thierry J-C., and Moras D. 1994. The active site of yeast aspartyl-tRNA synthetase: Structural and functional aspects of the aminoacylation reaction. *EMBO J.* **13:** 327–337.
Cusack S., Berthet-Colominas C., Härtlein M., Nassar N., and Leberman R. 1990. A second class of synthetase structure revealed by X-ray analysis of *Escherichia coli* seryl-tRNA synthetase at 2.5 Å. *Nature* **347:** 249–255.
Dai X., Mesmaeker A., and Joyce G.F. 1995. Cleavage of an amide bond by a ribozyme. *Science* **267:** 237–240.
———. 1996. Amide cleavage by a ribozyme: Correction. *Science* **272:** 18–19.
Eldred E.W. and Schimmel P. 1972. Rapid deacylation by isoleucyl tRNA synthetase of isoleucine-specific tRNA aminoacylated with valine. *J. Biol. Chem.* **247:** 2961–2964.
Eriani G., Delarue M., Poch O., Gangloff J., and Moras D. 1990. Partition of tRNA synthetases into two classes based on mutually exclusive sets of sequence motifs. *Nature* **347:** 203–206.
Fersht A.R. 1977. Editing mechanisms in protein synthesis. Rejection of valine by the isoleucyl-tRNA synthetase. *Biochemistry* **16:** 1025–1030.
———. 1985. Specificity for competing substrates. In *Enzyme structure and mechanism*, 2nd edition, p. 112. W.H. Freeman, New York.
Fersht A.R. and Kaethner M.M. 1976. Enzyme hyperspecificity. Rejection of threonine by the valyl-tRNA synthetase by misacylation and hydrolytic editing. *Biochemistry* **15:** 3342–3346.
Freist W. 1989. Mechanisms of aminoacyl-tRNA synthetases: A critical consideration of recent results. *Biochemistry* **28:** 6787–6795.
Freist W., Sternbach H., and Cramer F. 1996. Phenylalanyl-tRNA synthetase from yeast and its discrimination of 19 amino acids in aminoacylation of tRNAPhe. *Eur. J. Biochem.* **240:** 526–531.

Hale S.P. and Schimmel P. 1996. Protein synthesis editing by a DNA aptamer. *Proc. Natl. Acad. Sci.* **93:** 2755–2758.

Huang F. and Yarus M. 1997a. 5′ RNA self-capping from GDP. *Biochemistry* **36:** 6557–6563.

———. 1997b. A calcium-metalloribozyme with autodecapping and pyrophosphatase activities. *Biochemistry* **36:** 14107–14119.

Ibba M., Hong K-W., Sherman J.M., Sever S., and Söll D. 1996. Interactions between tRNA identity nucleotides and their recognition sites in glutaminyl–tRNA synthetase determine the cognate amino acid affinity of the enzyme. *Proc. Natl. Acad. Sci.* **93:** 6953–6958.

Illangasekare M. and Yarus M. 1997. Small-molecule-substrate interactions with a self-aminoacylating ribozyme. *J. Mol. Biol.* **268:** 631–639.

Illangasekare M., Kovalchuke O., and Yarus M. 1997. Essential structures of a self-aminoacylating RNA. *J. Mol. Biol.* **274:** 519–529.

Illangasekare M., Sanchez G., Nickles T., and Yarus M. 1995. Aminoacyl-RNA synthesis catalyzed by an RNA. *Science* **267:** 643–647.

Lin S.X., Baltzinger M., and Remy P. 1984. Fast kinetic study of yeast phenylalanyl-tRNA synthetase: role of tRNA[Phe] in the discrimination between tyrosine and phenylalanine. *Biochemistry* **23:** 4109–4116.

Lohse P.A. and Szostak J.W. 1996. Ribozyme-catalyzed amino-acid transfer reactions. *Nature* **381:** 442–444.

Nagel G.M. and Doolittle R.F. 1995. Phylogenetic analysis of the aminoacyl-tRNA synthetases. *J. Mol. Evol.* **40:** 487–498.

Noller H.F. 1993. Peptidyl transferase: Protein, ribonucleoprotein, or RNA? *J. Bacteriol.* **175:** 5297–5300.

Noller H.F., Hoffarth V., and Zimniak L. 1992. Unusual resistance of peptidyl transferase to protein extraction procedures. *Science* **256:** 1416–1419.

Perona J.J., Rould M.A., and Steitz T.A. 1993. Structural basis for tRNA aminoacylation by *E. coli* glutaminyl-tRNA synthetase. *Biochemistry* **32:** 8758–8771.

Piccirilli J.A., McConnell T.S., Zaug A.J., Noller H.F., and Cech T.R. 1992. Aminoacyl esterase activity of the *Tetrahymena* ribozyme. *Science* **256:** 1420–1424.

Schimmel P.R. and Söll D. 1979. Aminoacyl-tRNA synthetases: General features and recognition of transfer RNAs. *Ann. Rev. Biochem.* **48:** 601–648.

Schmitt E., Meinnel T., Blanquet S., and Mechulam Y. 1994. Methionyl-tRNA synthetase needs an intact and mobile $_{332}$KMSKS$_{336}$ motif in catalysis of methionyl adenylate formation. *J. Mol. Biol.* **242:** 566–577.

Welch M., Majerfeld I., and Yarus M. 1997. 23S rRNA similarity from selection for peptidyl transferase mimicry. *Biochemistry* **36:** 6614–6623.

Yarus M. and Berg P. 1969. Recognition of tRNA by isoleucyl-tRNA synthetase: Effect of substrates on the dynamics of tRNA-enzyme interaction. *J. Mol. Biol.* **42:** 171–189.

Yarus M. 1972. Phenylalanyl-tRNA synthetase and isoleucyl-tRNA[Phe]: A possible verification mechanism for aminoacyl-tRNA. *Proc. Natl. Acad. Sci.* **69:** 1915–1919.

———. 1988. A specific amino acid binding site composed of RNA. *Science* **240:** 1751–1758.

———. 1998. Amino acids as RNA ligands; a Direct-RNA-template theory for the code's origin. *J. Mol. Evol.* (in press).

Zhang B. and Cech T.R. 1997. Peptide bond formation by in vitro selected ribozymes. *Nature* **390:** 96–100.

8

On the Origin of the Ribosome: Coevolution of Subdomains of tRNA and rRNA

Harry F. Noller
Sinsheimer Laboratories
University of California
Santa Cruz, California 95064

Translation is one of the most complicated of biological processes, involving literally hundreds of specific macromolecules. Not the least of this complexity is the structure of the ribosome itself, which even in the relatively simple *Escherichia coli* version, consists of over 50 different proteins and three RNA molecules, comprising more than 4500 nucleotides, giving an aggregate mass of around 2.5 million daltons (Hill et al. 1990; Matheson et al. 1995). The difficulty of imagining how such a structure evolved is eased somewhat by accepting the notion that the original ribosome was made solely of RNA, as tentatively suggested by Crick (1968) more than two decades ago, and asserted with increasing force (Woese 1980) and enthusiasm (Gesteland and Atkins 1993) since then. Adherents of RNA World scenarios imagine, to varying degrees, that something resembling protein synthesis was carried out by ribozyme-like proto-ribosomes prior to the advent of ribosomal proteins and translation factors (not to mention aminoacyl-tRNA synthetases). Such scenarios solve the chicken-or-the-egg problem of the molecular evolution of ribosomes but raise some difficult new questions. The most obvious is that rRNA itself has a vast and intricate structure containing thousands of nucleotides (see, e.g., Fig. 1), and is not likely to have evolved by chance in a few simple evolutionary steps. Compounding this problem is that, prior to the existence of protein synthesis, it is difficult to imagine what selective pressures could have driven its evolution; in other words, the RNA World could not have anticipated the invention of protein synthesis. More likely, it evolved from preexisting RNA functions. The questions are, what were these functions, and how could simpler, primitive RNAs have evolved into the vast, complex macromolecular machine that is our present-day ribosome?

198   H.F. Noller

This chapter is based on one that appeared in the previous incarnation of this volume (Noller 1993), incorporating new findings and their implications. I begin by summarizing the evidence that supports the view that rRNA plays a central role in translation. This is followed by a simplified view of translation that is centered on interactions between tRNA and rRNA. Our present understanding of the structural setting for these inter-

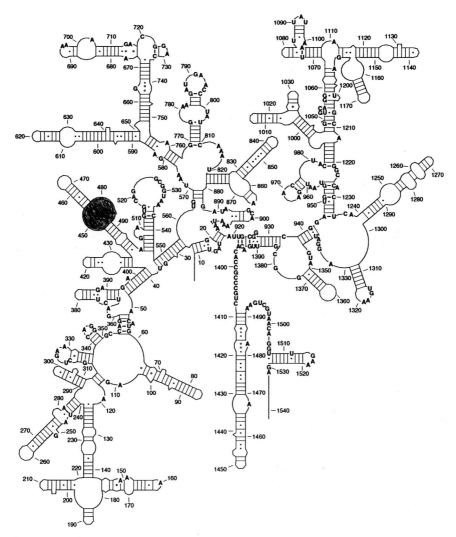

Figure 1 Secondary structure of 16S ribosomal RNA, showing (in uppercase letters) the bases that are universally conserved among the 16S-like rRNAs in all three primary kingdoms.

actions leads to the idea of three prototypical functional domains—two "RNA domains" that define essential aspects of the mechanism, and an "RNP domain" that is introduced later as a major refinement of a preexisting RNA mechanism. I then argue how each of the two RNA domains could have emerged from preexisting ribozymes and, specifically, that these ribozymes each contained both proto-tRNA and proto-rRNA elements. A central idea that emerges from this discussion is the independent evolution of the two ends of tRNA (the anticodon and acceptor ends) and their respective sites of interaction with rRNA (the decoding site and peptidyl transferase). Schimmel and Henderson (1994) also concluded that the two ends of tRNA evolved separately, on the basis of quite independent considerations of the origins of tRNA identity. Thus, different parts of tRNA and rRNA coevolved separately in the RNA World, eventually merging into larger, more complex, multifunctional, multi-domain molecules resembling the ones that we now find in ribosomes. Finally, some speculations are presented concerning details of structure, function, and mechanism.

EVIDENCE FOR THE FUNCTIONAL ROLE OF RIBOSOMAL RNA

There is extensive evidence, both biochemical and genetic, to support the involvement of rRNA in the fundamental mechanism of translation. This subject has been reviewed elsewhere (Noller 1991a; Green and Noller 1997), so only a few of the many results are summarized here. In retrospect, the results of early in vitro reconstitution studies should have raised suspicions about the correctness of the ribosomal protein paradigm. With few exceptions, omission of single proteins from 30S subunits resulted in particles that retained at least partial activity in in vitro translational assays (Nomura et al. 1970). Most of the exceptions were proteins whose omission caused major assembly defects. The finding that cleavage of a single covalent bond in 16S rRNA by colicin E3 abolished the activity of 30S subunits (Bowman et al. 1971; Senior and Holland 1971) also had surprisingly little impact on mainstream ribosomology at that time.

The colicin result, along with the inactivation of tRNA binding by kethoxal modification of 16S rRNA (Noller and Chaires 1972) and direct cross-linking of the tRNA anticodon to 16S rRNA (Prince et al. 1982), gradually led to a narrowly held perception that 16S rRNA was somehow important for tRNA–ribosome interaction. Affinity-labeling studies using reactive groups coupled to the aminoacyl end of tRNA showed that 23S rRNA was in the immediate environment of the peptidyl transferase catalytic center (Breitmeyer and Noller 1976; Barta et al. 1984, 1990).

Cleavage of a single bond in 23S rRNA by α-sarcin or ricin inactivated ribosomes, pointing again to the functional importance of the large subunit rRNA (Endo and Wool 1982; Endo et al. 1987). More often, however, the results of these and many other biochemical experiments were rationalized in terms of the protein paradigm (Fellner 1974).

Following the discovery of catalytic RNA (Kruger et al. 1982; Guerrier-Takada et al. 1983), the rRNA view gained wider acceptance, to the extent that at least one reviewer has cautioned that the pendulum may have swung too far (Moore 1990). Application of site-directed mutagenesis and other genetic approaches led to the study of the effects of a large number of mutations at potential functional sites in 16S and 23S rRNA (for summary, see Triman et al. 1998). Although analysis of the effects of such mutants is still in its infancy, the results obtained so far are consistent with the functional importance of rRNA. This includes a mutation that impairs EF-Tu-dependent binding of aminoacyl-tRNA (Powers and Noller 1993), and mutations that affect subunit association (Santer et al. 1990), initiation (Tapprich et al. 1989), antibiotic sensitivity (for review, see Cundliffe 1990), and translational accuracy (Allen and Noller 1991; Lodmell and Dahlberg 1997).

Among the most exciting findings is the recent discovery of a conformational "switch" in 16S rRNA (Lodmell and Dahlberg 1997). It was shown, by directed mutagenesis, that the sequence CUC912 can pair either with GGG885 or with GAG888, and that both pairings are required for proper ribosomal function (Fig. 2). Moreover, mutations that favor the 885 pairing confer increased translational error frequency (ram) pheno-

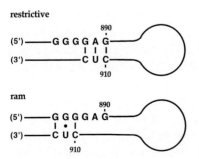

Figure 2 Two alternative pairings of the 890/910 region of 16S rRNA, as established by Lodmell and Dahlberg (1997). Mutations favoring the upper pairing confer increased translational accuracy (restrictive phenotype), and those favoring the lower pairing increase translational error frequencies (ram phenotype).

types, whereas those favoring the 888 pairing confer decreased translational error (restrictive) phenotypes. Among the many implications of these findings is the possibility that this switch may be of fundamental importance in the ribosome-mediated process of tRNA selection, and possibly in the mechanism of tRNA translocation.

Consistent with the notion that the mechanism of translation is fundamentally based on rRNA is the isolation of mutants lacking individual ribosomal proteins that are nevertheless viable (Dabbs 1986). It has been shown that peptidyl transferase activity survives vigorous extraction with phenol following treatment with detergents and proteinase K (Noller et al. 1992). The activity of the protein-depleted particles is inhibited by established peptidyl transferase inhibitors such as chloramphenicol and carbomycin, and is highly sensitive to ribonuclease. More recently, catalysis of peptide-bond formation by in vitro transcripts of *E. coli* 23S rRNA has been reported (Nitta et al. 1998). Activity required the presence of 0.5% SDS and was inhibited by chloramphenicol and carbomycin, and by digestion with ribonucleases. These results suggest that even peptide-bond formation, the single chemical reaction that is unambiguously catalyzed by the ribosome itself, may be a function carried out by rRNA.

FUNCTIONAL DOMAINS IN RIBOSOMAL RNA

tRNA-16S rRNA Interactions: The Decoding Site

Early studies showed that limited chemical modification of 30S ribosomal subunits with kethoxal caused loss of tRNA binding (most likely P-site binding), even though interactions between the ribosome and poly(U) message were largely unimpaired (Noller and Chaires 1972). Reconstitution experiments showed that the site of inactivation was 16S rRNA; proteins from the modified subunits were fully active when reconstituted with unmodified 16S rRNA. The subunits were protected from inactivation by bound tRNA, implying that modification of a small number of guanines in the tRNA-binding site was responsible for loss of function. We now know the identity of the protected guanines, as well as bases that are protected from other chemical probes by A- and P-site tRNA (Moazed and Noller 1986, 1990); Figure 3 shows their location in the secondary structure of 16S rRNA.

The bases protected by tRNA have the characteristics expected for functional sites: Most of them are completely invariant in the more than 4000 16S-like rRNA sequences now available, including examples spanning the extremes of all three primary kingdoms (Ribosomal Database Project, University of Illinois). Among these protected bases, the site of anticodon–codon interaction is almost certainly in the vicinity of the cluster of conserved nucleotides around positions 1400 and 1500 (Fig. 1). The site of direct cross-linking of the wobble base of the P-site anticodon, at C1400 (Prince et al. 1982), is among the protected P-site bases, and is near several others. Modification-interference experiments suggest that several

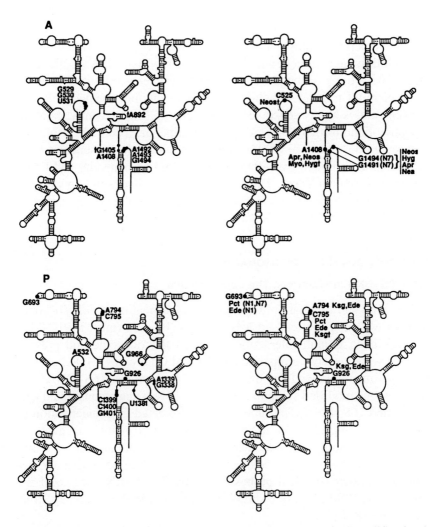

Figure 3 Bases in 16S rRNA that are protected from chemical modification by A-site (*upper left*) or P-site (*lower left*) tRNA (Moazed and Noller 1986, 1990). On the right are shown the bases that are protected by A-site- (*upper right*) and P-site-specific (*lower right*) antibiotics (Moazed and Noller 1987b; Woodcock et al. 1991).

of these bases are involved in tRNA–rRNA interactions (von Ahsen and Noller 1995). The site of colicin E3 cleavage, A1493 (Bowman et al. 1971; Senior and Holland 1971), is adjacent to a cluster of strong A-site protections. In addition, many drugs that cause miscoding protect bases in this same region (Fig. 3) (Moazed and Noller 1987a), and mutations that confer resistance to these drugs are also found here (Cundliffe 1990).

Studies using oligonucleotide models of the decoding site region of 16S rRNA provide further evidence for the direct participation of rRNA in ribosome function. Purohit and Stern (1994) showed that a 49-nucleotide construct analog of 1400/1500 region of 16S rRNA binds aminoglycoside antibiotics in much the same fashion as the intact ribosome. The molecular structure of a related oligonucleotide construct bound to the translational error-inducing antibiotic paromomycin has been solved by NMR spectroscopy, giving the first structural details of a functional interaction involving ribosomal RNA (Fourmy et al. 1996).

There is good evidence that interactions between tRNA and the small ribosomal subunit involve exclusively (or nearly so) the anticodon stem-loop region of tRNA. First, the binding constant of a 15-nucleotide anticodon stem-loop fragment for the 30S P site does not differ significantly from that observed for the intact tRNA (Rose et al. 1983). Second, this same fragment protects all of the same A- and P-site bases in 16S rRNA that are protected by intact tRNA (Moazed and Noller 1986). Finally, when bound to the 30S P site, the region of tRNA protected from hydroxyl radical attack corresponds almost precisely to the anticodon stem-loop (Fig. 4) (Hüttenhofer and Noller 1992). This indicates that the small subunit must contain a binding pocket that accommodates this stem-loop

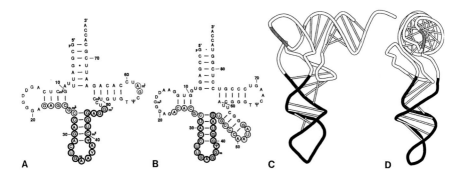

Figure 4 Protection by 30S ribosomal subunits of backbone ribose residues in tRNA from hydroxyl radical attack (Hüttenhofer and Noller 1992). Protected residues are circled (*A, B*) or shaded (*C, D*).

structure rather specifically and tightly, such that it becomes inaccessible even to solvent.

These results are consistent with what is known about the structure of the decoding site of the small subunit, from model-building studies (Stern et al. 1988; Malhotra and Harvey 1994; Fink et al. 1996; Mueller et al. 1997). According to these models, the P-site tRNA-protected nucleotides, although distributed widely in the 16S rRNA secondary structure, are localized in a compact neighborhood in three dimensions, as shown in the model for the 30S subunit shown in Figure 5. The anticodon stem-loop is enclosed in a cavity that is formed at the convergence of the head, body, and platform features of the 30S subunit. This is the same location (the "cleft") that has been shown experimentally to be the site of interaction of the anticodon end of P-site tRNA with the 30S subunit by difference electron microscopy (EM) (Wagenknecht et al. 1988: Agrawal et al. 1996; Stark et al. 1997). It has also been shown to be the location of the 1400 region of 16S rRNA (to which the anticodon has been cross-linked), by EM localization of a bound complementary oligonucleotide (Oakes et al. 1986).

tRNA-23S rRNA Interactions: The Peptidyl Transferase Site

Correspondingly, the other end (the CCA, or acceptor end) of tRNA interacts with the large ribosomal subunit; bases in 23S rRNA that are protected by tRNA in the A, P, and E sites are shown in Figure 6 (Moazed and Noller 1989). Almost all of the protected bases are located in domain V. Stepwise deletion of the aminoacyl moiety or the terminal A and C show that protection of specific bases depends on different structural components of the CCA terminus. Conversely, oligonucleotides such as CAACCA(f-Met), bound in the presence of sparsomycin and ethanol to the P site, protect almost exactly the same bases in 23S rRNA as intact tRNA (Moazed and Noller 1991). These results are persuasive evidence that interaction of tRNA with the 23S rRNA P site involves almost exclusively its CCA terminus.

Direct evidence for interaction between the acceptor end of tRNA and 23S rRNA in the peptidyl transferase P site has been obtained by in vitro genetics experiments (Samaha et al. 1995). These studies show that a Watson-Crick base pair is formed between C74 in the conserved CCA sequence of tRNA and the similarly conserved G2252 in domain V of 23S rRNA. A second conserved loop of 23S rRNA has been placed at the A site of the peptidyl transferase center (Green et al. 1998). A trinucleotide analog of aminoacyl-tRNA, 4-thio-dTpCpPuromycin, cross-links efficiently and specifically to G2553 in domain V of 23S rRNA. The

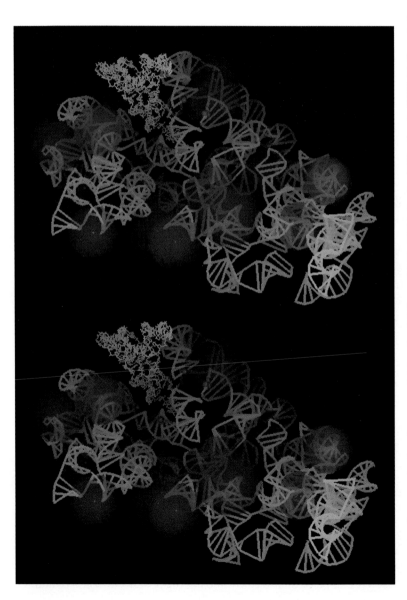

Figure 5 Stereo pair showing a model for the folding of 16S rRNA in the 30S subunit (H.F. Noller et al., unpubl.), based on biochemical and biophysical studies. The rRNA is shown as a *blue* backbone, and the ribosomal proteins as *magenta* spheres. P-site tRNA is shown as an all-atom structure in *salmon*, with its anticodon loop oriented toward the lower left. The positions of five 16S rRNA nucleotides that are protected from chemical probes by P-site tRNA (693, 794, 795, 926, and 1338) are shown as small *red* spheres.

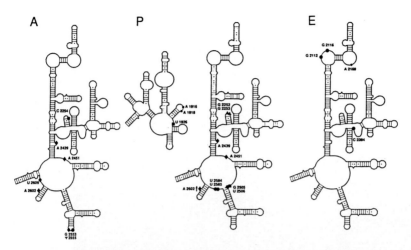

Figure 6 Protection of bases in 23S rRNA from chemical modification by binding tRNA to ribosomes in the A, P, and E sites (Moazed and Noller 1989).

covalently linked substrate is fully reactive in peptidyl transferase-catalyzed peptide-bond formation, unambiguously identifying this loop as a component of the peptidyl transferase A site. Intriguingly, cross-linking depends on occupancy of the P site by a tRNA carrying an intact CCA end, indicating that peptidyl-tRNA, directly or indirectly, helps to create the peptidyl transferase A site. Earlier studies by Barta and coworkers (Barta et al. 1984, 1990; Steiner et al. 1988) showed that a tRNA covalently linked to position 2451 in the central loop of domain V of 23S rRNA is active in peptide-bond formation. These studies, which rigorously localize the peptidyl transferase function to domain V of 23S rRNA, are supported by many other results, including localization of footprints and resistance mutations to antibiotic inhibitors of peptidyl transferase in this same region (Moazed and Noller 1987b; Cundliffe 1990). These three elements of domain V thus represent a minimal list of the RNA components that are part of the second primary functional domain, the peptidyl transferase site.

Elongation Factor–rRNA Interactions: An Intersubunit RNP Regulatory Domain

According to a strict RNA paradigm, the GTP-dependent elongation factors EF-Tu and EF-G would be late evolutionary additions to the translation apparatus. Indeed, under the appropriate in vitro conditions, aminoacyl-tRNA can be bound to ribosomes in the absence of EF-Tu, and translocation occurs in the absence of EF-G. It is likely that the basic mechanism is, therefore, embodied in the ribosome itself, and that the factors refine the process by enhancing its speed and accuracy.

Both EF-Tu and EF-G footprint the highly conserved loop around position 2660 of 23S rRNA (Moazed et al. 1988). This loop, commonly known as the α-sarcin loop, is the site of attack by the cytotoxins α-sarcin and ricin (Endo and Wool 1982; Endo et al. 1987; Wool et al. 1992). If GTP hydrolysis is blocked, the aminoacyl-tRNA.EF-Tu.GTP ternary complex is bound to the ribosome in a way that allows interaction of EF-Tu with the α-sarcin loop, but prevents access of tRNA to the 23S rRNA peptidyl transferase site (Moazed et al. 1988). This is probably the manifestation of a mechanism by which the accuracy of tRNA selection is regulated (Thompson 1988; Kurland et al. 1990). EF-G is known to interact with a second site, around position 1067 in domain II of 23S rRNA, which has been shown to be the site of action of the EF-G-dependent GTPase inhibitor thio-strepton (Thompson et al. 1982). A conserved feature of 16S rRNA, the 530 loop (Fig. 1), has been implicated in elongation factor-related function. Mutations of G530 carry a dominant lethal phenotype, due to specific impairment of EF-Tu-dependent binding of aminoacyl-tRNA (Powers and Noller 1993). Furthermore, mutations in the α-sarcin loop appear to interfere with release of EF-Tu, but only in S12 restrictive strains (Tapprich and Dahlberg 1990). Since S12 is known to modulate the conformation of the 530 loop (Stern et al. 1989), these results suggest that EF-Tu might mediate communication between the α-sarcin loop and the 530 loop.

Recently, the position of EF-Tu.tRNA.GTP complex in the ribosome has been found by EM reconstruction studies (Stark et al. 1997), and the EF-G.GDP complex has been localized by directed hydroxyl radical probing (Wilson and Noller 1998). In the latter study, specific positions on the surface of EF-G were localized with respect to many features of 16S and 23S rRNA. These include the aforementioned thiostrepton and α-sarcin loops of 23S rRNA, the decoding site of 16S rRNA, and others. The location and orientation of EF-G in the ribosome, inferred from these results, is shown in Figure 7 (Wilson and Noller 1998). Like the EF-Tu ternary complex (Stark et al. 1997), the factor is wedged between the two ribosomal subunits with its globular G domain between the L11 region of the 50S subunit and the S4 region of the 30S subunit, which has been placed near the base of the stem of the 530 loop. Its elongated domain IV, which appears to mimic the anticodon arm of tRNA in the EF-Tu ternary complex (Aevarsson et al. 1994; Czworkowski et al. 1994; Nissen et al. 1995), is oriented toward the decoding site of the 30S subunit.

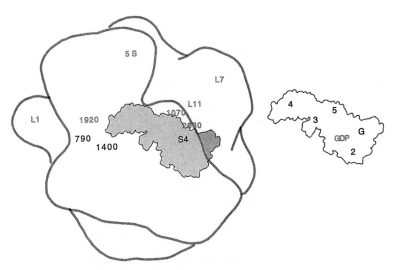

Figure 7 Position of translational elongation factor EF-G in the ribosome in the post-translocational state, as determined by directed hydroxyl radical probing (Wilson and Noller 1998). Protein and RNA landmarks within the small and large subunits are indicated in bold and open lettering, respectively. At right, the different structural domains of EF-G and the position of GDP are indicated (Aevarsson et al. 1994; Czworkowski et al. 1994). (Reprinted, with permission, from Wilson and Noller 1998 [copyright Cell Press].)

There are some interesting parallels between the sarcin and 530 loops. Both are large, highly conserved hairpin loops (among the most highly conserved elements in rRNA), containing additional higher-order structure that has been determined by NMR spectroscopy in the case of the α-sarcin loop (Szewczak et al. 1993). The 530 loop is involved in at least one pseudoknot interaction (Woese and Gutell 1989; Powers and Noller 1991), and the α-sarcin loop appears to exist as a tetraloop structure in at least one of its physiological states (Wool et al. 1992; Szewczak et al. 1993). Both loops, although themselves conserved, are imbedded in RNA domains that otherwise show rather unremarkable sequence conservation, in clear contrast to the decoding and peptidyl transferase sites. This, along with their structural and functional association with some of the most highly conserved proteins that are involved in protein synthesis, including S12, L7/L12, EF-Tu, and EF-G, suggests a somewhat different character for this interdomain region of the ribosome: that of a ribonucleoprotein (RNP) domain. This view is compatible with the possibility that the GTPase-related functions evolved after establishment of the more fundamentally RNA-based translational mechanisms. Finally, both loops ap-

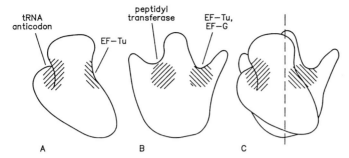

Figure 8 Location of the sites of interaction of the tRNA anticodon, its CCA end (peptidyl transferase), EF-Tu, and EF-G with (*A*) 30S subunits (*B*) 50S subunits, and (*C*) 70S ribosomes, in the electron microscopy model (Girshovich et al. 1986; Langer and Lake 1986; Oakes et al. 1986; Wagenknecht et al. 1988). The proposed RNA protodomains are located to the left of the dashed line, and the proposed RNP protodomains are to the right (see text for details).

pear to be located toward the right-hand side of the ribosome, as viewed in Figure 8, placing the RNP domain in a location that is physically distinct from that of the two primary RNA functional domains, which are located toward the left side of the ribosome.

EVOLUTION OF TRANSLATIONAL PROTODOMAINS FROM THE RNA WORLD

One of the main challenges in thinking about the evolution of the ribosome is its size and structural complexity. This is especially problematic for any RNA World scenario, because even modern RNA genomes are notorious for their low fidelity of replication. Because of this, it is likely that the RNA genes, and therefore the corresponding ribozymes of the RNA World, were constrained to be rather small—a few hundred nucleotides at best. Even so, their sequences would have shown much higher mutational variation than seen in DNA genes. An important question is then to explain how the large rRNA structures could have evolved from a world where the largest RNAs would have been at least an order of magnitude smaller.

Although the ribosome is large and complex, it seems increasingly evident that, for at least two primary translational functions, the regions of tRNA and rRNA that interact with each other are relatively compact. Codon recognition appears to involve no more than about 15 nucleotides of the tRNA structure, and a localized subdomain of 16S rRNA. The peptidyl transferase function involves only three nucleotides (the conserved CCA terminus) of tRNA, and a limited number of features of one domain of 23S rRNA. Thus, these two primarily functional domains, including their tRNA sub-fragment "substrates," may represent structures simple enough to have arisen from preexisting functional RNAs.

Here, one can appreciate an important evolutionary advantage of RNA over proteins. It is unlikely that the active site of a protein enzyme could exist on its own; this is because such structures are usually stabilized by an elaborate hydrophobic core, typically involving a much larger proportion of the total molecular mass than the active site itself. Individual β strands do not exist on their own, and even α helices are rarely stable in isolation. In contrast, individual structural features of larger RNAs often fold independently of the rest of the RNA molecule. Nucleic acids are not believed to have hydrophobic interiors; rather, stability is provided by local hydrogen-bonded and base-stacking interactions and by coordination with magnesium and other multivalent cations. These properties of RNA make possible the creation of very small functional RNAs—RNA "naked active sites." One notable example is the "hammerhead" ribozyme, containing as few as 19 nucleotides, which catalyzes a highly sequence-specific RNA-cleavage reaction (Uhlenbeck 1987). Another example is the model oligonucleotide analog of the decoding site of 16S rRNA, which faithfully binds aminoglycoside antibiotics, as mentioned above (Purohit and Stern 1994; Fourmy et al. 1996). Thus, the existence of compact, functional RNA protodomains provides a plausible solution to the dilemma of the spontaneous evolution of a large, complex macromolecular structure. Most significantly, the structural properties of RNA more readily suit it to such a role than do the properties of proteins.

Starting with small RNAs carries the further advantage that the daunting problem of correctly folding a very large RNA is avoided (the difficulty of correctly folding an RNA increases with something like the second power of its length). Given the existence of compact protodomains, larger complex assemblies resembling the ribosome could then arise first by noncovalent association of protodomains, followed by eventual ligation into larger, contiguous RNA molecules. Since in the RNA World, gene products are also templates for replication, any such ligation products become potential genes, which can then be carried forward in evolution.

A central question underlying the evolution of rRNA and tRNA protodomains is, What was the driving force (selective advantage) for their evolution? This leads to the further question, What were the preexisting RNA World molecules from which the protodomains evolved? As already pointed out, the RNA World could not have "known" that it was in the process of evolving protein synthesis until the whole process became functional. Most likely, the functions of protodomains evolved as improvements or "spinoffs" from existing ribozyme-catalyzed functions. What were they? An RNA World genetic system would most likely be based on two well-known functions: replication and recombination; the latter would be used in order to exploit to their full combinatorial potential any catalytic RNAs that might emerge. Here, there is no need to speculate on whether these two functions could have been RNA-catalyzed, since both have already been shown to be carried out by present-day ribozymes. Recombination in an RNA World amounts to what is currently referred to as *trans*-splicing. Both one step of RNA replication and *trans*-splicing can be catalyzed by group I introns (Cech and Bass 1986).

Whether group I introns are ancient or modern is not yet clear; however, their existence alone is sufficient for the purpose of arguing for the plausibility of basic RNA World functions as starting points for the evolution of translational protodomains. The discovery that certain ribosome-directed antibiotics such as neomycin, kanamycin, and gentamycin also inhibit the function of group I introns raises the intriguing possibility that there could even be an evolutionary link between present-day ribozymes and ribosomes (von Ahsen et al. 1991).

A possible mechanistic basis for such a link is that recognition of splice sites involves base-pairing, via short complementary sequences. An interesting consequence of this is that a splicing complex, at the instant of splicing, bears a striking resemblance to two tRNAs bound to mRNA at adjacent A- and P-site codons (Fig. 9). If the *trans*-esterification activity of the ribozyme could be adapted to catalysis of the chemically analogous acyl-transfer reaction of peptide-bond formation, the stage would be set for the evolution of a primitive ribosome. Such a possibility may not be out of the question. Piccirilli et al. (1992) engineered a group I intron so that its internal guide sequence could base-pair with the P-site peptidyl transferase substrate CAACCA(f-Met). This places the formyl-methionyl carbonyl group at the position of the phosphate group of the normal *trans*-esterification reaction. The modified ribozyme was shown to catalyze hydrolysis of the model peptidyl ester substrate, a reaction catalyzed by ribosomal peptidyl transferase.

Conversely, it has been shown that ribosomal peptidyl transferase can catalyze transfer of aminoacyl groups to a phosphate center. Tarrusova et al. (1981) have reported the formation of Ac-Met.GlyP-Phe from Ac-Met.GlyP and Phe-tRNA, where GlyP is a glycine analog in which the carboxyl group is replaced by phosphate. The reaction results in a phosphinoamide, in which a P–N bond is formed. This result is consistent with the possible evolution of peptidyl transferase from nucleic acid-based chemistry.

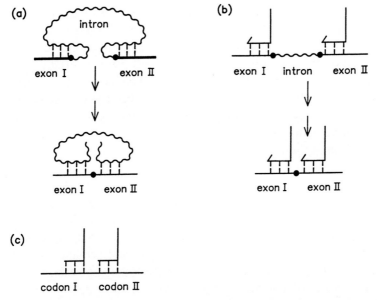

Figure 9 Base-pairing of short sequences flanking splice junctions in putative ancestral RNAs related to (*a*) group I introns, (*b*) small nuclear RNAs, which could have given rise to complexes that are structurally analogous to two tRNAs bound to adjacent codons on mRNA (*c*). (Adapted from Noller 1991b.)

Recently, the ability of an RNA molecule to catalyze peptide-bond formation has been demonstrated directly by in vitro selection experiments (Zhang and Cech 1997). A 196-nucleotide RNA was selected that performs the same reaction as that catalyzed by ribosomal peptidyl transferase, in which the α-amino group of a phenylalanine residue linked to the 5′ end of the RNA forms a peptide bond with the carbonyl group of an α-biotinylated methionine residue esterified to the 3′ hydroxyl of an AMP that is bound noncovalently to the RNA molecule. Although the rate of the reaction catalyzed by this particular ribozyme is several orders of magnitude slower than that of ribosomal peptidyl transferase, these findings clearly establish that RNA is itself capable of catalyzing peptide-bond formation. Equally important is the demonstration that an RNA with the properties of peptidyl transferase can readily be selected from a random pool of small RNAs.

FROM INDEPENDENT PROTODOMAINS TO PROTO-RIBOSOME

Independent evolution of the two extremities of tRNA (proto-tRNAs) in separate protodomains introduces a new paradox: uncoupled evolution of coding and peptidyl transferase. Coding without polypeptide synthesis, or polypeptide synthesis without coding, both seem pointless from an evolutionary standpoint. One solution to this paradox is that each of the proto-tRNAs had the ability to participate in both coding and polymerization, but that each became specialized for one of the two functions as the modern tRNA–ribosome system emerged. In this scenario, the anticodon stem-loop proto-tRNA would have amino acid acceptor activity, and the acceptor-end proto-tRNA would have recognition capability, in addition to their present-day properties. Both kinds of proto-tRNA could have emerged from a hypothetical precursor RNA consisting of an anticodon stem-loop containing a 3'-CCA acceptor tail. Obviously, such proto-tRNAs could function only if the geometrical relationship between codons and peptidyl transferase were significantly different from that of present-day ribosomes.

An alternative scenario is that a kind of nontemplated ordered polymerization preceded coupling of peptide-bond formation to codon recognition. Structural complementarity between adjacent proto-tRNAs could program a repeating polypeptide sequence; for example, complementarity of proto-tRNA A to B, B to C, and C to A would program the repeating amino acid sequence $(a-b-c)_n$. Coupling of codon recognition to such a primitive mechanism could begin with a fixed mRNA sequence embodied in the structure of the proto-16S rRNA, again programming a simple, repeating polypeptide. Diversity would then originate in the different coding sequences of divergent proto-16S rRNAs. The ability of all proto-23S rRNAs to associate with any proto-16S rRNA would give a strong selective advantage. An important consequence of this would be selection for independence of the two ribosomal subunits, in which the coding and polymerization functions were uncoupled, at an early stage of ribosome evolution. Polypeptides with more complex sequences could be generated, even at this stage, by dissociation and reassociation with different proto-16S rRNA subunits. The emergence of a *trans*-coding mRNA template is then a short step from its *cis*-coding precursor. For example, short autonomous mRNAs containing a few codons, associating and dissociating via Shine-Dalgarno-like pairing with proto-16S rRNA, could begin to program complex polypeptide sequences in a modular, combinatorial way.

The final stage, at which the use of longer, more modern mRNAs is adopted, would necessitate a translocation mechanism. The striking structural similarity between domain IV of elongation factor EF-G and the anticodon arm of tRNA as bound to EF-Tu (Aevarsson et al. 1994; Czworkowski et al. 1994; Nissen et al. 1995) provides potential clues to the origins of EF-G-catalyzed translocation. This domain of EF-G may mimic tRNA for the reason that that translocation may have originally been catalyzed by tRNA itself (Wilson and Noller 1998). Further speculation about the way in which translocation may have arisen will probably require a deeper understanding of its mechanism. However, the observation that translocation of the anticodon arm of tRNA can be uncoupled from that of its acceptor end suggests that evolution of the translocation mechanism is also rooted in the independent evolution of the two extremities of tRNA.

CONCLUSION

The point of departure for this discussion is the rapidly unfolding picture of ribosome structure and function, and of ribosomal RNA, in particular. Our current understanding is already beginning to place significant constraints on models for the evolution of translation, and, as suggested in this chapter, may provide substantial clues to the nature of early versions of the translational apparatus. It seems almost inevitable that ribosomes evolved from simpler pieces of RNA that were functionally independent, or nearly so. The absence of any detectable structural or sequence homology within or between 16S and 23S rRNA is consistent with independent evolutionary origins, as is the ability of the two extremities of tRNA to interact independently with their respective ribosomal subunits. However, some serious conceptual gaps remain in the scheme suggested here. Not the least of these is the merging of the two ends of tRNA. This, and other problems left unaddressed here, emphasize the importance of another longstanding question: How did tRNA identity evolve? Continuing progress in comparative molecular biology and in in vitro RNA selection methods provides hope for answers to questions that have, until recently, been considered to be inaccessible.

ACKNOWLEDGMENTS

Work in the author's laboratory was supported by National Institutes of Health grant GM-17129, National Science Foundation grant MCB-9406186, and by a grant to the Center for Molecular Biology of RNA from the Lucille P. Markey Charitable Trust. The previous version of this chapter was motivated by a Symposium on Molecular Biology and the Origin of Life, held at Berkeley, California in July, 1990, sponsored by the Institute for Advanced Studies in Biology. I thank J. Abelson, R. Green, H. Hartman, S. Joseph, T. Powers, J. Sampson, M. Saks, P. Schimmel, K. Wilson, C. Woese, and M. Yarus for criticism and discussions.

REFERENCES

Aevarsson A., Brazhnikov E., Garber M., Zheltonosova J., Chirgadze Y., al-Karadaghi S., Svensson L.A., and Liljas A. 1994. Three-dimensional structure of the ribosomal translocase: Elongation factor G from *Thermus thermophilus*. *EMBO J.* **13:** 3669–3677.

Agrawal R.K., Penczek P., Grassucci R.A., Li Y., Leith A., Nierhaus K.H., and Frank J. 1996. Direct visualization of A-, P-, and E-site transfer RNAs in the *Escherichia coli* ribosome. *Science* **271:** 1000–1002.

Allen P.N. and Noller H.F. 1991. A single base substitution in 16S ribosomal RNA suppresses streptomycin dependence and increases the frequency of translational errors. *Cell* **66:** 141–148.

Barta A., Kuechler E., and Steiner G. 1990. Photoaffinity labelling of the peptidyltransferase region. In *Ribosomes: Structure, function and evolution* (eds. W.E. Hill et al.), pp. 358–365. American Society Microbiology, Washington, D.C.

Barta A., Steiner G., Brosius J., Noller H.F., and Kuechler E. 1984. Identification of a site on 23S ribosomal RNA located at the peptidyl transferase center. *Proc. Natl. Acad. Sci.* **81:** 3607–3611.

Bowman C.M., Dahlberg J.E., Ikemura T., Konisky J., and Nomura M. 1971. Specific inactivation of 16S ribosomal RNA induced by colicin E3 in vivo. *Proc. Natl. Acad. Sci.* **68:** 964–968.

Breitmeyer J.B. and Noller H.F. 1976. Affinity labeling of specific regions of 23S rRNA by reaction of N-bromoacetyl-phenylalanyl-transfer RNA with *Escherichia coli* ribosomes. *J. Mol. Biol.* **101:** 297–306.

Cech T.R. and Bass B.L. 1986. Biological catalysis by RNA. *Annu. Rev. Biochem.* **55:** 599–629.

Crick F.H.C. 1968. The origin of the genetic code. *J. Mol. Biol.* **38:** 367–379.

Cundliffe E. 1990. Recognition sites for antibiotics within rRNA. In *Ribosomes: Structure, function and evolution* (eds. W.E. Hill et al.), pp. 479–490. American Society for Microbiology, Washington, D.C.

Czworkowski J., Wang J., Steitz T.A., and Moore P.B. 1994. The crystal structure of elongation factor G complexed with GDP, at 2.7 Å resolution. *EMBO J.* **13:** 3661–3668.

Dabbs E.R. 1986. Mutant studies on the prokaryotic ribosome. In *Structure, function and genetics of ribosomes* (ed. B. Hardesty and G. Kramer), pp. 733–748. Springer-Verlag, New York.

Endo Y. and Wool I.G. 1982. The site of action of α-sarcin on eukaryotic ribosomes. The sequence at the α-sarcin cleavage site in 28S ribosomal ribonucleic acid. *J. Biol. Chem.* **257:** 9094–9060.

Endo Y., Mitsui M., Motizuki M., and Tsurugi K. 1987. The mechanism of action of ricin and related toxic lectins on eukaryotic ribosomes. The site and the characteristics of the modification in 28S ribosomal RNA caused by the toxins. *J. Biol. Chem.* **262:** 5908–5912.

Fellner P. 1974. Structure of the 16S and 23S ribosomal RNAs. In *Ribosomes* (ed. M. Nomura et al.), pp. 169–191. Cold Spring Harbor Laboratory, Cold Spring Harbor, New York.

Fink D.L., Chen R.O., Noller H.F., and Altman R.B. 1996. Computational methods for defining the allowed conformational space of 16S rRNA based on chemical footprinting data. *RNA* **2:** 851–866.

Fourmy D., Recht M.I., Blanchard S.C., and Puglisi J.D. 1996. Structure of the A site of *Escherichia coli* 16S ribosomal RNA complexed with an aminoglycoside antibiotic. *Science* **274:** 1367–1371.

Gesteland R.F. and Atkins J.F., eds. 1993. *The RNA world.* Cold Spring Harbor Laboratory Press, Cold Spring Harbor, New York.

Girshovich A.S., Bochkareva E.S. and Vasiliev V.D. 1986. Localization of elongation factor Tu on the ribosome. *FEBS Lett.* **197:** 192–198.

Green R. and Noller H.F. 1997. Ribosomes and translation. *Annu. Rev. Biochem.* **66:** 679–716.

Green R., Switzer C., and Noller H.F. 1998. Ribosome-catalyzed peptide bond formation with an A-site substrate covalently linked to 23S rRNA. *Science* **280:** 286–289.

Guerrier-Takada C., Gardiner K., Marsh T., Pace N., and Altman S. 1983. The RNA moiety of ribonuclease P is the catalytic subunit of the enzyme. *Cell* **35:** 849–857.

Hill W.E., Dahlberg A.E., Garrett R.A., Moore P.B., Schlessinger D. and Warner J.R., eds. 1990. *The ribosome: Structure, function and evolution.* American Society for Microbiology, Washington, D.C.

Hüttenhoffer A. and Noller H.F. 1992. Hydroxyl radical cleavage of tRNA in the ribosomal P site. *Proc. Natl. Acad. Sci.* **89:** 7851–7855.

Kruger K., Grabowski P.J., Zaug A.J., Sands J., Gottschling D.E., and Cech T.R. 1982. Self-splicing RNA: Autoexcision and autocyclization of the ribosomal RNA intervening sequence of *Tetrahymena. Cell* **31:** 147–157.

Kurland C.G., Jorgensen F., Richter A., Ehrenberg M., Bilgin N., and Rojas A.M. 1990. Through the accuracy window. In *The ribosome: Structure, function and evolution* (ed. W.E. Hillet al.), pp. 513–526. American Society for Microbiology, Washington, D.C.

Langer J.A. and Lake J.A. 1986. Elongation factor Tu localized on the exterior surface of the small ribosomal subunit. *J. Mol. Biol.* **187:** 617–621.

Lodmell J.S. and Dahlberg A.E. 1997. A conformational switch in *Escherichia coli* 16S ribosomal RNA during decoding of messenger RNA. *Science* **277:** 1262–1267.

Malhotra A. and Harvey S.C. 1994. A quantitative model of the *Escherichia coli* 16 S RNA in the 30 S ribosomal subunit. *J. Mol. Biol.* **240:** 308–340.

Matheson A.T., Davies J.E., Dennis P.P., and Hill W.E. 1995. *Frontiers in translation.* National Research Council, Ottawa, Canada.

Moazed D. and Noller H.F. 1986. Transfer RNA shields specific nucleotides in 16S ribosomal RNA from attack by chemical probes. *Cell* **47:** 985–994.

———. 1987a. Interaction of antibiotics with functional sites in 16S ribosomal RNA. *Nature* **327:** 389–394.

———. 1987b. Chloramphenicol, erythromycin, carbomycin and vernamycin B protect overlapping sites in the peptidyl transferase region of 23S ribosomal RNA. *Biochimie* **69:** 879–884.

———. 1989. Interaction of tRNA with 23S rRNA in the ribosomal A, P and E sites. *Cell* **57:** 585–597.

———. 1990. Binding of tRNA to the ribosomal A and P site protects two distinct sets of nucleotides in 16S rRNA. *J. Mol. Biol.* **211:** 135–145.

———. 1991. Sites of interaction of the CCA end of peptidyl-tRNA with 23S rRNA. *Proc. Natl. Acad. Sci.* **88:** 3725–3728.

Moazed D., Robertson J.M., and Noller H.F. 1988. Interaction of elongation factors EF-G and EF-Tu with a conserved loop in 23S ribosomal RNA. *Nature* **334:** 362–364.

Moore P.B. 1990. Comments on the 1989 International Conference on Ribosomes. In *Ribosomes: Structure, function and evolution* (ed. W.E. Hill et al.), pp. xxi–xxiii. American Society for Microbiology, Washington, D.C.

Mueller F., Stark H., van Heel M., Rinke-Appel J., and Brimacombe R. 1997. A new model for the three-dimensional folding of *Escherichia coli* 16 S ribosomal RNA. III. The topography of the functional centre. *J. Mol. Biol.* **271:** 566–587.

Nissen P., Kjeldgaard M., Thirup S., Polekhina G., Reshetnikova L., Clark B.F., and Nyborg J. 1995. Crystal structure of the ternary complex of Phe-tRNAPhe, EF-Tu, and a GTP analog. *Science* **270:** 1464–1472.

Nitta I., Ueda T., and Watanabe K. 1998. Possible involvement of *Escherichia coli* 23S ribosomal RNA in peptide bond formation. *RNA* **4:** 257–267.

Noller H.F. 1991a. Ribosomal RNA and translation. *Annu. Rev. Biochem.* **60:** 191–227.

———. 1991b. Drugs and the RNA world. *Nature* **353:** 302–303.

———. 1993. On the origin of the ribosome: Coevolution of subdomains of tRNA and rRNA. In *The RNA world* (ed. R.F. Gesteland and J.F. Atkins), pp. 137–156. Cold Spring Harbor Laboratory Press, Cold Spring Harbor, New York.

Noller H.F. and Chaires J.B. 1972. Functional modification of 16S ribosomal RNA by kethoxal. *Proc. Natl. Acad. Sci.* **69:** 3115–3118.

Noller H.F., Hoffarth V., and Zimniak L. 1992. Unusual resistance of peptidyl transferase to protein extraction methods. *Science* **256:** 1416–1419.

Nomura M., Mizushima S., Ozaki M., Traub P., and Lowry C.V. 1970. Structure and function of ribosomes and their molecular components. *Cold Spring Harbor Symp. Quant. Biol.* **34:** 49–61.

Oakes M.I., Clark M.W., Henderson E., and Lake J.A. 1986. DNA hybridization electron microscopy: Ribosomal RNA nucleotides 1392–1407 are exposed in the cleft of the small subunit. *Proc. Natl. Acad. Sci.* **83:** 275–279.

Piccirilli J.A., McConnell T.S., Zaug A.J., Noller H.F., and Cech T.R. 1992. Aminoacyl esterase activity of the *Tetrahymena* ribozyme. *Science* **256:** 1420–1424.

Powers T. and Noller H.F. 1991. A functional pseudoknot in 16S ribosomal RNA. *EMBO J.* **10:** 2203–2214.

———. 1993. Evidence for functional interaction between elongation factor Tu and 16S ribosomal RNA. *Proc. Natl. Acad. Sci.* **90:** 1364–1368.

Prince J.B., Taylor G.N., Thurlow D., Ofengand J., and Zimmermann R.A. 1982. Covalent crosslinking of tRNAVal to 16S RNA at the ribosomal P site: Identification of crosslinked residues. *Proc. Natl. Acad. Sci.* **79:** 5450–5454.

Purohit P. and Stern S. 1994. Interactions of a small RNA with antibiotic and RNA ligands of the 30S subunit. *Nature* **370:** 659–662.

Rose S.J., Lowry P.T., and Uhlenbeck O.C. 1983. Binding of yeast tRNAPhe anticodon arm to *Escherichia coli* 30S ribosomes. *J. Mol. Biol.* **167:** 103–117.

Samaha R.R., Green R., and Noller H.F. 1995. A base pair between tRNA and 23S rRNA in the peptidyl transferase centre of the ribosome. *Nature* **377:** 309–314.

Santer M., Bennett-Guerrero E., Byahatti S., Czarnecki S., O'Connell D., Meyer M., Khoury J., Cheng X., Schwartz I., and McLaughlin J. 1990. Base changes at position 792 of *Escherichia coli* 16S rRNA affect assembly of 70S ribosomes. *Proc. Natl. Acad. Sci.* **87:** 3700–3704.

Schimmel P. and Henderson B. 1994. Possible role of aminoacyl-RNA complexes in noncoded peptide synthesis and origin of coded synthesis. *Proc. Natl. Acad. Sci.* **91:** 11283–11286.

Senior B.W. and Holland I.B. 1971. Effect of colicin E3 upon the 30S ribosomal subunit of *Escherichia coli*. *Proc. Natl. Acad. Sci.* **69:** 959–963.

Stark H., Orlova E.V., Rinke-Appel J., Junke N., Mueller F., Rodnina M., Wintermeyer W., Brimacombe R., and van Heel M. 1997. Arrangement of tRNAs in pre- and post-translocational ribosomes revealed by electron cryomicroscopy. *Cell* **88:** 19–28.

Steiner G., Kuechler E., and Barta A. 1988. Photoaffinity labelling at the peptidyl transferase center reveals two different positions for the A- and P-sites in domain V of 23S rRNA. *EMBO J.* **7:** 3949–3955.

Stern S., Weiser B., and Noller H.F. 1988. Model for the three-dimensional folding of 16S ribosomal RNA. *J. Mol. Biol.* **204:** 447–481.

Stern S., Powers T., Changchien L.M., and Noller H.F. 1989. RNA-protein interactions in the 30S ribosomal subunits: Folding and function of 16S rRNA. *Science* **244:** 783–790.

Szewczak A.A., Moore P.B., Chang Y.L., and Wool I.G. 1993. The conformation of the sarcin/ricin loop from 28S ribosomal RNA. *Proc. Natl. Acad. Sci.* **90:** 9581–9585.

Tapprich W.E. and Dahlberg A.E. 1990. A single base mutation at position 2661 in *E. coli* 23S RNA affects the binding of ternary complex to the ribosome. *EMBO J.* **9:** 2649–2655.

Tapprich W.E., Goss D.J., and Dahlberg A.E. 1989. Mutation at position 791 in *Escherichia coli* 16S ribosomal RNA affects processes involved in the initiation of protein synthesis. *Proc. Natl. Acad. Sci.* **86:** 4927–4931.

Tarussova N.B., Jacovleva G.M., Victorova L.S., Kukhanova M.K., and Khomutov R.M. 1981. Synthesis of an unnatural P-N bond catalyzed with *Escherichia coli* ribosomes. *FEBS Lett.* **130:** 85–87.

Thompson J., Schmidt F., and Cundliffe 1982. Site of action of a ribosomal RNA methylase conferring resistance to thiostrepton. *J. Biol. Chem.* **257:** 7915–7917.

Thompson R.C. 1988. EF-Tu provides an internal kinetic standard for translational accuracy. *Trends Biochem. Sci.* **13:** 91–93.

Triman K.L., Peister A., and Goel R.A. 1998. Expanded versions of the 16S and 23S ribosomal RNA mutation databases (16SMDBexp and 23SMDBexp). *Nucleic Acids Res.* **26:** 280–284.

Uhlenbeck O.C. 1987. A small catalytic oligoribonucleotide. *Nature* **328:** 596–600.

von Ahsen U. and Noller H.F. 1995. Identification of bases in 16S rRNA essential for tRNA binding at the 30S ribosomal P site. *Science* **267:** 234–237.

von Ahsen U., Davies J., and Schroeder R. 1991. Antibiotic inhibition of group I ribozyme function. *Nature* **353:** 368–370.

Wagenknecht T., Frank J., Boublik M., Nurse K., and Ofengand J. 1988. Direct localization of the tRNA-anticodon interaction site on the *Escherichia coli* 30S ribosomal subunit by electron microscopy and computerized image averaging. *J. Mol. Biol.* **203:** 753–760.

Wilson K.S. and Noller H.F. 1998. Mapping the position of translational elongation factor EF-G in the ribosome by directed hydroxyl radical probing. *Cell* **92:** 131–139.

Woese C.R. 1980. Just so stories and Rube Goldberg machines: Speculations on the origin of the protein synthetic machinery. In *Ribosomes: Structure, function and genetics* (ed. G. Chambliss et al.), pp. 357–373. University Park Press, Baltimore.

Woese C.R. and Gutell R.R. 1989. Evidence for several higher order structural elements in ribosomal RNA. *Proc. Natl. Acad. Sci.* **86:** 3119–3122.

Wool I.G., Gluck A., and Endo Y. 1992. Ribotoxin recognition of ribosomal RNA and a proposal for the mechanism of translocation. *Trends Biochem. Sci.* **17:** 266–269.

Woodcock J., Moazed D., Cannon M., Davies J., and Noller H.F. 1991. Interaction of antibiotics with A- and P-site-specific bases in 16S ribosomal RNA. *EMBO J.* **10:** 3099–3103.

Zhang B. and Cech T.R. 1997. Peptide bond formation by in vitro selected ribozymes. *Nature* **390:** 96–100.

9
Introns and the RNA World

Walter Gilbert and Sandro J. de Souza
Department of Molecular and Cellular Biology
The Biological Laboratories
Harvard University
Cambridge, Massachusetts 02138

The RNA World is a hypothesis about the origin of life based on the view that the most critical event is the emergence of a self-replicating molecule, a molecule that can both copy itself and mutate and, hence, evolve to more efficient copying (Gilbert 1986). Evolution works on variation and selection, and selection is always measured in terms of more efficient multiplication, the ability to make more of the entity in question. The concept of an RNA World is a way of answering the basic problem of what was the molecular biology involved at the beginning of life. Our understanding of the molecular basis of biology today is in terms of a genetic material, commonly DNA, translated through an apparatus involving RNA and the mechanism of protein synthesis to specify the positions of 20 amino acids in protein enzymes. That picture of life, in which the genetic material is of one chemical kind, DNA, made up of four bases, a second chemical, RNA, is used for structural and transfer purposes, and the enzymatic activities in the cell are a third chemical kind made up of 20 ingredient amino acids, creates a complex paradox in trying to formulate how life could have begun. This paradox was resolved by two realizations. One was that RNA is likely to be more primary than DNA, but the picture of an RNA–protein world, in which RNA is the genetic material specifying the positions of amino acids in proteins, still left one with a complex problem of beginnings. The second realization, however, was that there was no intrinsic reason that enzymatic activity must be limited to proteins. The discovery of the first two RNA enzymes showed that RNA molecules could carry out the phosphodiester bond transfers needed for RNA synthesis (Kruger et al. 1982; Guerrier-Takada et al. 1983).

Why RNA rather than DNA? The current biochemistry of these molecules suggests that RNA was antecedent to DNA. First, the synthesis of the deoxynucleotides is not primary, but secondary, to the synthesis of the ribonucleotides. The biochemical processes in all cells today create ribonucleotide precursors, and then at the ribonucleotide diphosphate level, convert the sugars into the deoxy form using ribonucleotide reductases. This produces three of the deoxyribonucleotide precursors directly, but the fourth is produced as deoxyUDP and only later is the uracil methylated to produce the thymine of DNA. Second, the mechanism of DNA synthesis is completely dependent on previous RNA synthesis. The synthesis of the lagging strand of DNA, made in short pieces which are then connected, has each such piece initiated by an RNA primer which is then elongated into DNA and finally removed before the DNA strands are ligated. In general, DNA-copying enzymes cannot initiate new strands de novo but must elongate some preexisting primer, usually RNA. RNA-synthesizing enzymes, on the other hand, can initiate with a ribonucleotide triphosphate. Furthermore, the ends of linear DNA chromosomes are constructed by a telomerase function, which uses an RNA template to extend the 3' end of the DNA chain. One last observation along these lines is that the RNA-synthesizing enzymes seem to be more primitive than that for DNA, in that they are less efficient and less rapid: The rate of RNA synthesis is about 50 bases per second, whereas DNA synthesis runs ten times faster, about 500–1,000 bases per second. For all of these reasons, RNA appears to be a biochemically primitive molecule that could have served as a precursor to a later DNA involvement. Thus, one would be led back from our current DNA–RNA–protein world to contemplate an RNA–protein world in which RNA would serve as the genetic material as well as the messengers translated by ribosomes into protein enzymes.

But which then came first? The classic chicken-and-egg problem, which needs a complicated protein machinery involving 20 amino acids in order to synthesize the enzymes necessary to synthesize new copies of the genetic material, which in turn dictates the structure of the protein-assembling machinery and the enzymes, would be simplified if either one or the other chemical entity served as the first structures. The suggestion that life begins with protein molecules creates a pattern of chemical reactions but provides no mechanism for genetic inheritance, since there is no form of protein–protein self-copying, replication, and mutation. The key aspect of evolution is the ability of molecules to grow; i.e., to replicate themselves, but in a form that embodies variation, mutation, and thus can provide the novel patterns on which natural selection will operate to improve the replication in a changing environment.

The ribonuclease P activity (Guerrier-Takada et al. 1983) and the self-splicing intron from *Tetrahymena* (Kruger et al. 1982) showed that RNA molecules could cleave and join other RNA molecules; this is a sufficient enzymology to establish that an RNA molecule should be able to catalyze RNA-dependent RNA synthesis. Furthermore, one should expect that RNA would be capable of catalyzing a whole range of reactions. Broadly speaking, an enzyme is any structure that can bind to, and hence stabilize, the transition state of a chemical process. In that view of enzymology, the issue is one of the shape and complexity of a binding site rather than the chemical nature of the binding site; the issue is simply whether one can find hydrogen bonds and hydrophobic surfaces to construct binding pockets. RNA can serve these functions. In another view of enzymology, the nature of the enzyme is to bring into apposition with the substrate the cofactors and other groups that will participate in the chemical reaction. Again RNA molecules could, in principle, bring charges or metal ions to bear on the substrate.

In fact, today's view of the ability of RNA to catalyze phosphodiester bond formation and cleavage is that the RNA binds two essential magnesium ions which are used to handle the phosphate intermediates in exactly the same way as the protein RNA polymerases carry out those same processes using metal ions (Steitz and Steitz 1993; Steitz 1998).

Although protein enzymes frequently use cofactors, White (1976) pointed out that a large number of these cofactors are related to RNA moieties and look like residual pieces of RNA enzymes held in a protein framework to catalyze chemical reactions.

THE RNA WORLD

The minimal enzymology that RNA can do, phosphodiester bond cleavage and transfer, is enough enzymology to show that it would be possible for the first self-replicating molecule to be RNA, in the sense that in principle an RNA enzyme could copy other RNA molecules, including copies of itself. Although no such ribozyme has yet been created in the laboratory, Bartel's group has come extremely close by constructing a ribozyme that is capable of adding up to six bases in a template-directed fashion using RNA triphosphate precursors (Ekland and Bartel 1996). This is the key activity for an RNA-copying enzyme. The issue remains of finding a way to do the copying so that the two strands do not become inevitably hydrogen-bonded, and finding a way for the enzyme to accept an arbitrary RNA molecule. These are more mechanical problems than chemical problems; the critical demonstration is that an RNA molecule can do the synthetic chemistry. Thus, the RNA World contemplates a self-replicating RNA molecule, arising in a puddle containing all the RNA precursors, catalyzing the formation of more molecules like itself, and in so doing, leading to mutant molecules and ever-better replication.

After this beginning, the RNA World picture considers a far more extensive use of RNA as enzymes: ribozymes that catalyze the synthesis of all the precursors needed to synthesize RNA; ribozymes to construct charge-neutralizing polyamines; and ribozymes to construct lipids. The full expression of the RNA World conceives of RNA-based organisms with RNA genetic material and RNA enzymes contained in lipid vesicles growing and multiplying. To have natural selection work to develop better RNA enzymes, one must effectively construct organisms, which can multiply and outgrow each other, in which a bounding membrane connects the mutated RNA genes with the better and more effective enzymatic ribozymes that are their products.

These views of the emergence of self-copying molecules imagine that one begins with pools filled with concentrated solutions of all the biochemical precursors. The appearance of a self-copying RNA would catalyze the formation of many molecules like itself in a pool. One could contemplate essentially pools of liquid as the first "organisms," each pool developing a selected "best" replicator; but to go much farther, one must have a way of enfolding the genetic material and the copying function in some boundary coat, some boundary membrane. This is needed if the genes and gene products are to be linked, as they must be if natural selection is to be able to identify a gene that makes a "better" product. Initially these membranes do not have to be impermeable to small molecules. They need only be able to hold together the macromolecular genes and gene products, because we have assumed a high concentration of precursors outside as well as inside these primordial cells. Only later, as one develops ribozymes to make precursors, does one need tighter membranes, with pores to let specific chemicals through and pumps to create and control gradients. Even at the beginning, however, a problem is posed by the negative charges on RNA and on lipids. Some supply of charge-neutralizing molecules, polyamines or oligo-lysines, would be necessary to permit RNA to be wrapped in lipid membranes and to aid in the wrapping (Jay and Gilbert 1987).

Although we can imagine a plethora of ribozymes to do all necessary reactions, the issue arises of how random processes could yield RNA enzymes of appreciable size. This is the classic size paradox, emerging for RNA. An RNA enzyme might be 300–600 bases long in order to function. Even for 300 bases, that is 4^{300} or 10^{180} molecules. Clearly, one could not get such a molecule by a random process. One way around this problem is to hypothesize that the enzymatic activities are carried by rather short

RNA pieces. We think a more likely possibility is that the RNA World had an intron–exon structure. The RNA genetic material consisted of RNA exons held together by self-splicing RNA introns, either group I or group II introns, both of which functions are catalyzed by RNA molecules acting as ribozymes. In this picture, the RNA genetic material would have an extended, presumably a more linear, structure adapted for copying. After the gene is copied, the introns would splice themselves out to leave a set of RNA exons tied together so that they could fold up and become a ribozyme. This use of introns at the RNA level does three things. First, it solves the size paradox by making complex RNA functions out of shorter, simpler pieces, RNA exons 30–40 bases long. Second, it provides a way of distinguishing genetic material in a form that could be copied from RNA folded up to form enzymes, and thus removes a requirement that the ribozyme be in a form that can be directly copied. Third, it provides a mechanism for enhanced illegitimate recombination. There may be a background of recombination in an RNA-copying world, just by the RNA synthetase occasionally jumping from molecule to molecule during the replication and thus creating recombinant products, but this would be primarily homologous recombination. The critically useful process created by the intron–exon structure at the RNA level is essentially transposition of exons. Two introns surrounding an RNA exon can cut out across the exon and make a transposon that can then enter an intron in some other molecule. Figure 1 shows this process. This concept provides a way to shuffle exons at the RNA level to create novel RNA molecules. Finally, as recently discussed by Jeffares et al. (1998), recombination tends to minimize the "Eigen limit" problem; i.e., the limits in genome size imposed by high rates of replication error (see Eigen 1993).

This picture of a fully developed RNA World uses RNA genetic material, molecules about a few thousand (1–10,000) bases long with an intron–exon structure. The splicing out of the introns ties the exons together to make functional ribozymes, which make the precursors, the charge neutralizers, and the membranes. There would be many copies of the genes in each cell so that the division does not have to be extremely accurate.

Is the genetic material single- or double-stranded? If the genetic material were single-stranded, with the ribozyme polymerase copying a plus-strand into a minus-strand and a minus-strand into a plus-strand, the intron–exon structure might be unstable. If the RNA genetic material were double-stranded, the intron–exon structure would be stable, and upon transcribing to make a plus-strand, that strand can either be copied back to make a new double-stranded molecule or can splice out its introns to become a ribozyme. In this picture, one has as many transcription units

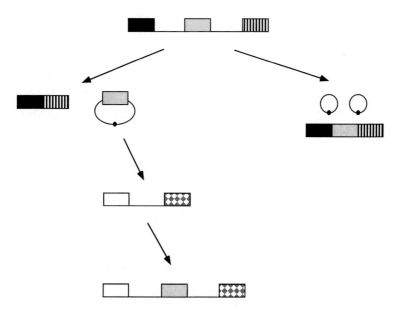

Figure 1 Schematic view of an intron–exon structure at the RNA level, spliced in two ways. The introns are shown as type I, although they could as easily be type II. Either both introns splice out to produce a ribozyme, or the extreme ends of the two introns splice together to carry an intervening exon to a new position in a novel gene.

as one has ribozymes, and about as many separate chromosomes as ribozymes.

THE RNA–PROTEIN WORLD

The concept that there is rather complete development of an RNA World accumulating in proto cells, with an RNA genetic material and RNA enzymology capable of making a variety of the biochemical structures needed, produces a simple way to develop an RNA–protein world. The reason for the simplicity is that one can introduce proteins first as short homo-oligo peptides. Their functions are to bind to RNA molecules to enhance the catalytic functions of the ribozymes. One would expect the process to begin with the ability to activate and charge a single amino acid onto an RNA molecule and to develop a precursor to the ribosome to catalyze the peptide bond formation. One must develop a way of encoding that process into mRNA in order to make short oligopeptides. One conjectures that the first amino acid used would be either lysine or arginine: lysine because of its chemical simplicity and codon simplicity; poly A, or arginine, because of its much more extensive codon usage in the ultimate code. The first product would be short, positively charged oligopeptides to bind to RNA molecules to aid in charge neutralization made on a primitive RNA ribosome (Maizels and Weiner 1987). Products and structures that increase the fidelity of the proto-translational system would be positively selected (see discussion on proto-ribosomes in Jeffares et al. 1998). The ability of a ribozyme to activate an amino acid has recently been demonstrated (Illangasekare et al. 1995; see Chapter 7). A link between ribosomes and ribozymes is also suggested by the fact that group I introns have their function affected by kanamycin, gentamycin, and neomycin, anti-ribosome antibiotics (von Ahsen et al. 1991). A further link is the observation that an engineered ribozyme could hydrolyze an aminoacyl ester substrate (Piccirilli et al. 1992), and peptide bonds have been formed by selected ribozymes (Zhang and Cech 1997). Beginning with a single amino acid and a single transfer/activating RNA, one can then easily imagine mutant forms developing the ability to activate, encode, and transfer other amino acids, building up first to some 5 amino acids to carry out most of the protein functions and ultimately extending to the 20 amino acids (21 including seleno-cysteine) that are currently used.

Such a picture of the gradual development of the protein-synthetic machinery, for the transition from an RNA World to an RNA–protein world, does not require the big bang of a beginning for protein enzymes that spring into action completely functional. The first use of proteinaceous material is as oligopeptides to support the RNA enzymes. Gradually the protein chains become more complex, the support they can offer to the ribozymes becomes more extensive, and ultimately, protein molecules emerge that themselves carry the full enzymatic action. These can then replace the ribozyme functions for a number of biochemical processes. This view would suggest a unitary origin for the transfer RNAs arising from one common ancestor. At the RNA level, the activating ribozymes would have a unitary origin, but at the protein level the replacement of the ribozyme function need not be unitary. In fact, the protein-activating enzymes belong to two major families of dissimilar structure and differing chemical detail (Eriane et al. 1990).

As the mechanism of protein synthesis develops using RNA messages to encode proteins, the intron–exon structure at the RNA World level means that one can develop an intron–exon structure for the genetic information at the protein world level. In this hypothesis, the exons would encode small functional or folding elements of the ultimate proteins which could, through shuffling, create novel protein structures. The first protein enzymes were probably aggregates of short polypeptide chains, each folding up as a small component of the final structure: These products as exons were linked ultimately into a single gene by introns and then moved into novel combinations by exon shuffling. Experimental evidence showing the plausibility of such aggregates exists for triosephosphate isomerase (Bertolaet and Knowles 1995) and tRNA synthetase (Shiba 1995). The end product of an RNA–protein world would be a cell with the ability to handle and to synthesize RNA and protein precursors, able to do a great deal of biochemistry. This cell would have a cell membrane and probably a cell wall to support the contents against osmotic pressure differentials. This cell would have pores and pumps to scavenge organic precursors from the environment. Such a cell has all the attributes of the last common ancestor except for DNA. The genetic material still is RNA, even with an intron–exon structure. The exon structure would match the one we have detected today in the intron positions in ancient conserved proteins (de Souza et al. 1996), in which we infer that there is a pattern of original exons which represent modules, compact units of protein structure, of the order of 21 Å, 28 Å, and 33 Å in diameter, corresponding roughly to oligopeptide units 15, 22, and 30 amino acids long. This exon–intron structure, we think, would have been fully developed at an RNA World level.

Two lines of argument further support this concept. The peptidyl transferase activity today still appears to reside in the ribosomal RNA: The basic mechanism of protein chain extension is an RNA-catalyzed one, not a protein-synthetic one (Noller et al. 1992). This is clearly an indication that an RNA enzymology preceded a protein enzymology. We described the intron–exon structure of the RNA World as being type I and type II introns. Even in the RNA World, the introns could be of type III, for which the catalysis is done not by *cis*-acting RNA structures, but by *trans*-acting ones. Today the RNA spliceosomes, the particles involved in the splicing of nuclear pre-messenger RNAs, are RNA–protein particles in which the small nuclear RNAs involved are most likely to carry out the details of the catalytic reactions (for review, see Sharp 1994). Both the ribosomes and the RNA-splicing mechanism look like remnants of an ancient RNA World.

THE DNA–RNA–PROTEIN WORLD

The genetic material of the RNA–protein world would have its genetic material in the form of RNA molecules about 10,000 bases long, RNA molecules the length of the molecules in RNA double-stranded or single-stranded viruses; the reason is that the error rate in copying RNA is likely to be too high to permit the creation of extremely long molecules. One expects the error rate to be high because there is no proofreading mechanism, and thus the errors are likely to be on the order of 10^{-3} or 10^{-4}, about down to the tautomerism rate for the bases. Thus, at the RNA–protein world level, one expects a large number of short chromosomes, each encoding one or at most a few genes, and the assortment of these chromosomes into daughter cells probably taking place at random because the parent cell has large numbers of duplicates. The introduction of DNA solves these problems. DNA-based enzymology can develop a full-fledged error-correcting mechanism that ultimately drives the error rate down to 10^{-9}. This permits long chromosomes, an ordered development of a mitotic process, and a well-defined segregation of chromosomes into daughter organisms. The first enzymes involved would be the enzymes that create the DNA precursors from the RNA precursors, ribonucleotide reductases, and a reverse transcriptase function that can convert an RNA chain into a DNA chain. Ultimately then, we need DNA-directed RNA polymerases and DNA-directed DNA polymerases, as well as a telomerase. This picture suggests that the reverse transcriptase and the telomerase functions are of equal age and very old. Since there is an intron–exon structure for protein genes at the RNA–protein world level, this structure is simply copied into the DNA. This leads us to an intron–exon structure for the DNA genes, the exons still being primarily units of protein folding and function. Of course, at both the RNA and the DNA levels, simple exons can be fused together and reused as more complicated exons.

At both the RNA and DNA levels, tying together of the exons that correspond to the short polypeptides that might as an aggregate form enzymes increases their genetic linkage, so that the entire complement of polypeptides needed to generate some enzymatic activity can be passed in a simple form from parent to daughter. However, genetic linkage is not a required concept. Pieces of protein need not be genetically linked for the organism to survive, but if they are linked, it is easier to pass the function as a whole to the offspring and that pattern, we think, would be quite valuable in the very early stages of evolution. The intron–exon structure of a gene provides a certain amount of genetic linkage in that the exons are held together, but the separation of the exons along the genetic material provides an enhanced recombination rate, over evolutionary time, that could lead to better novel combinations, and to the creation, by illegitimate recombination, of entirely new structures.

We have discussed the origins of life focusing on the first self-replicating event. By self-replicating we mean the appearance of a molecule that can copy molecules like itself to produce more of its own kind along with a notion that that copying could be occasionally inaccurate, hence able to introduce mutations that would permit a population of molecules varying about a norm to emerge. This variation in structure permits the emergence of molecules that copy more effectively under any specified conditions, and as those conditions change, the molecules can evolve to make more of themselves under the new conditions. We regard this as the crucial aspect of evolution: that it involves exponential replication, including both multiplication and variation, which can be worked upon by natural selection. Before the emergence of such a molecule, there was a period of prebiotic synthesis of material, which we have not discussed. That synthesis needs to include ways of forming the precursors for RNA synthesis, oligonucleotides or RNA triphosphates, or polyphosphates. The synthesis of these molecules could even be catalyzed by inorganic catalysts, such as clays or other material, which might produce high-molecular-weight precursors. But these issues of where the precursors are synthesized, or whether there is a background of inorganic or organic catalysts that can be used to produce complex molecules, to our mind are not the same as the question of where the evolution begins. The replication of a nucleic acid in the presence of a catalyst, even though the nucleic acid moiety under these conditions will modify and evolve (Spiegelman 1971; Eigen 1987), is not a full model of evolution because the multiplication or amplification is not self-sustaining. It is only the ability of a molecule effectively to copy itself, to make more of its own kind, that creates the exponential growth that is characteristic of life.

REFERENCES

Bertolaet B. and Knowles J. 1995. Complementation of fragments of triosephosphate isomerase defined by exon boundaries. *Biochemistry* **34:** 5736–5743.

de Souza S.J., Long M., Schoenbach L., Roy S.W., and Gilbert W. 1996. Intron positions correlate with module boundaries in ancient proteins. *Proc. Natl. Acad. Sci* **93:** 14632–14636.

Eigen, M. 1987. New concepts for dealing with the evolution of nucleic acids. *Cold Spring Harbor Symp. Quant. Biol.* **52:** 307–320.

———. 1993. The origin of genetic information: Viruses as a model. *Gene* **135:** 37–47.

Ekland E. and Bartel, D. 1996. RNA-catalysed RNA polymerization using nucleoside triphosphates. *Nature* **382:** 373–376.

Eriane G., Delarue M., Poch O., Gangloff J., and Moras, D. 1990. Partition of tRNA synthetase into two classes based on mutually exclusive sets of sequence motifs. *Nature* **347:** 203–206.

Gilbert W. 1986. Origin of life: The RNA world. *Nature* **319:** 618.

Guerrier-Takada C., Gardiner K., Marsh T., Pace N., and Altman S. 1983. The RNA moiety of ribonuclease P is the catalytic subunit of the enzyme. *Cell* **35:** 849–857.

Illangasekare M., Sanchez G., Nickles T., and Yarus M. 1995. Aminoacyl-RNA synthesis catalyzed by an RNA. *Science* **267**: 643–647.

Jay D. and Gilbert W. 1987. Basic protein enhances the incorporation of DNA into lipid vesicles: Model for the formation of primordial cells. *Proc. Natl. Acad. Sci.* **84**: 1978–1980.

Jeffares D., Poole A., and Penny D. 1998. Relics from the RNA world. *J. Mol. Evol.* **46**: 18–36.

Kruger K., Grabowski P., Zaug A., Sands J., Gottschling D., and Cech T. 1982. Self-splicing RNA: Autoexcision and autocyclization of the ribosomal RNA intervening sequence of *Tetrahymena*. *Cell* **31**: 147–157.

Maizels N. and Weiner A. 1987. Peptide-specific ribosomes, genomic tags, and the origin of the genetic code. *Cold Spring Harbor Symp. Quant. Biol.* **52**: 743–749.

Noller H., Hoffarth V., and Zimniak L. 1992. Unusual resistance of peptidyl transferase to protein extraction procedures. *Science* **256**: 1416–1419.

Piccirilli J.A., McConnell T.S., Zaug A.J., Noller H.F., and Cech T.R. 1992. Aminoacyl esterase activity of the Tetrahymena ribozyme. *Science* **256**: 1420–1424.

Sharp P.A. 1994. Split genes and RNA splicing. *Cell* **77**: 805–815.

Shiba K. 1995. Dissection of an enzyme into two fragments of intron-exon boundaries. In *Tracing biological evolution in protein and gene structures*. (ed. M. Go and P. Schimmel), pp. 11–21. Elsevier Press, The Netherlands.

Spiegelman S. 1971. Extracellular strategies of a replicating RNA genome. *Ciba Found. Symp.* **1971**: 45–73.

Steitz T. 1998. A mechanism for all polymerases. *Nature* **391**: 231–232.

Steitz T. and Steitz J. 1993. A general two-metal-ion mechanism for catalytic RNA. *Proc. Natl. Acad. Sci.* **90**: 6498–6502.

von Ahsen U., Davies J., and Schroeder R. 1991. Antibiotic inhibition of group I ribozyme function. *Nature* **353**: 368–370.

White H. III 1976. Coenzymes as fossils of an earlier metabolic state. *J. Mol. Evol.* **7**: 101–104.

Zhang B. and Cech T. 1997. Peptide bond formation by in vitro selected ribozymes. *Nature* **390**: 96–100.

10
The Interactions That Shape RNA Structure

Mark E. Burkard and Douglas H. Turner
Department of Chemistry
University of Rochester
Rochester, New York 14627-0216

Ignacio Tinoco, Jr.
Department of Chemistry
University of California, Berkeley and
Structural Biology Division
Lawrence Berkeley National Laboratory
Berkeley, California 94720-1460

This chapter describes the effects of noncovalent interactions on RNA structure and evolution. The building blocks of RNA are well suited for taking advantage of relatively strong noncovalent interactions such as stacking and hydrogen-bonding to form ordered structures. These ordered structures are able to protect RNA from chemical degradation and to allow the specific binding and catalysis required for further evolution.

The noncovalent interactions important for shaping RNA during evolution are revealed by the RNA structures that occur naturally, and by thermodynamic measurements on model systems. First, we discuss the fundamentals of the molecular interactions, then the contributions of stacking, hydrogen-bonding, and metal ions to formation of helices and other motifs. Examples are given of how these interactions shape RNA structures. Finally, some speculations are presented as to how these interactions directed evolution. Since understanding of noncovalent interactions, and knowledge of three-dimensional structures of RNA, are limited, this chapter represents an early stage in the evolution of our understanding of how the two are connected.

FUNDAMENTALS

The equilibrium constant, K, relating the concentrations of two conformations, C_1 and C_2, of an RNA strand is given by

$$K = [C_1]/[C_2] = \exp(-\Delta G°/RT) \qquad (1)$$

For an association of two non-self-complementary strands A and B to give A·B, the relevant $K = [A·B]/[A][B]$. Here $\Delta G°$ is the standard free-energy difference between the two conformations, $[C_1]/[C_2]$ and $[A·B]/[A][B]$ are the ratios of the equilibrium concentrations, R is the gas constant (1.987 cal K^{-1} mole^{-1}), and T is the temperature in Kelvins. Thus, changes in $\Delta G°$ of 1.4 and 2.8 kcal mole^{-1} at 37°C change equilibrium constants roughly 10- and 100-fold, respectively. The free-energy change can be obtained from the enthalpy change or heat associated with the conformational change at constant pressure, $\Delta H°$, and from the change in disorder of the system as measured by the entropy change of the system, $\Delta S°$

$$\Delta G° = \Delta H° - T\Delta S° \qquad (2)$$

Thus, the shape of an RNA is dependent on both $\Delta H°$ and $\Delta S°$. The $\Delta H°$ is determined by interactions between atoms, as described by potential energy functions, since enthalpy is related to energy by $H = E + PV$, and the PV term is almost always negligible.

The interactions that shape RNA are the same interactions that determine all molecular structures. For RNA molecules and their bases, nucleosides, and nucleotides in aqueous solution, semiempirical potential energy functions provide a useful approximation for understanding the interactions. Understanding leads to prediction, and eventually, to control of RNA structure and thus to control of function.

Energies

Bonded Interactions

Semiempirical potential energy functions relate the energy to a sum of bonded and nonbonded interactions. The bonded interactions are bond-length stretching (and compressing), bond-angle bending, and torsion angle—also called dihedral angle—rotation. The energies of bond stretching and bond bending are treated as harmonic potentials. Each energy E is a quadratic function of the difference in actual bond distance, r, and the equilibrium distance, r_{eq}, and of the difference in bond angles, θ and θ_{eq}

$$E = k_r (r - r_{eq})^2 + k_b (\theta - \theta_{eq})^2 \qquad (3)$$

Here k_r and k_b are force constants for bond stretching and bond-angle bending. The energies of torsion angle rotations are represented as cosine functions. For the torsion angle between two sp³ hybrid atoms (for example C4'-C5' in ribose) there are three minima corresponding to *gauche⁺*, *trans*, and *gauche⁻* conformations

$$E = (V_3/2)[1 + \cos(3\phi)] \qquad (4)$$

The value of V_3 is the barrier to rotation between the conformations corresponding to the minima at $\phi = +60°$ (g^+), 180° (t), and −60° or 300° (g^-). The maxima in the energy as the torsion angle varies occur at 0°, 120°, and 240°, with a typical barrier to rotation of about 3 kcal mole⁻¹. Other types of torsion angles, such as the glycosidic torsion angle, or the torsion angles in the ribose ring, have a similar form but a different dependence on ϕ. For example, the glycosidic torsion angle typically has minima that are approximately 180° apart at *syn* and *anti* conformations.

The folding of a molecule—the change in conformation—does not change bond lengths or bond angles very much; changes of bond lengths by more than 0.1 Å, or bond angles by more than 10° are rare. A single-bond stretching force constant k_r is about 300 kcal mole⁻¹ Å⁻² (Cornell et al. 1995), thus 0.1 Å change in bond length costs 3 kcal mole⁻¹. A bond angle bending force constant k_b for tetrahedral bonding is about 0.015 kcal mole⁻¹ degree⁻², thus a 10° change in bond angle costs 1.5 kcal mole⁻¹. The major differences between a folded state and an unfolded state, or between different folded states, are the torsion angles. The different minima for each torsion angle are nearly the same, and the barrier to rotation is small.

Nonbonded Interactions

Nonbonded interactions refer to interactions between atoms separated by three or more bonds; interactions between atoms connected by one or two bonds are incorporated into the bond-stretching and bond-angle-bending energies.

The longest range interaction is the electrostatic energy—Coulomb's law. The energy of interaction of two charges is directly proportional to the reciprocal of the distance (r^{-1}) between them. The energy also depends on the magnitude of each charge q_i and on the medium (solvent, ions, RNA) between them. The medium shields the interaction between the charges; it is represented by a dielectric constant, ε.

$$E = \frac{q_1 q_2}{\varepsilon r} \qquad (5)$$

The dielectric constant is 1 for a vacuum and 80 for water at room temperature. Clearly, the value of ε has a large effect on energy. A distance-dependent value is usually used for ε. The idea is that when two charges are in contact the medium provides minimal shielding, but as the charges separate, the medium becomes more significant. A reasonable approximation is to equate ε to r in Å. Thus, the values of ε typically range from about 2 to 80.

To apply Equation 5 to nucleic acids, we need net charges for all atoms of the nucleotides and of water. Figure 1 shows net charges for the

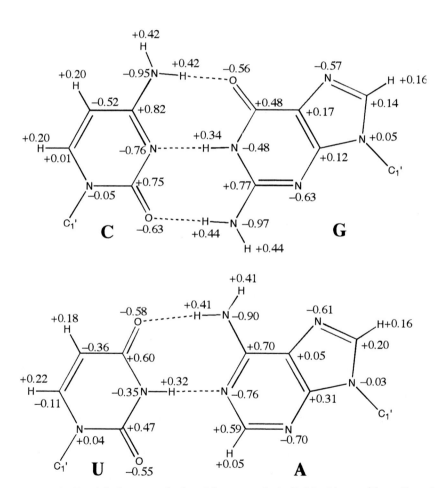

Figure 1 Partial charges calculated for atoms in individual bases (Cornell et al. 1995). Bases are shown in Watson-Crick pairings, which are isosteric, thus allowing formation of a regular helix independent of sequence. Other pairings are not isosteric as shown in Appendix 1.

four RNA bases; it is obvious that the bases are very polar and that electrostatics will be very important. Coulomb's law can account for most of the energy of a hydrogen bond. An atom with a net negative charge, such as a nitrogen or oxygen, attracts the positively charged hydrogen atom. For example, opposite charges of 0.2 electron each at a distance of 1.8 Å (a reasonable distance between a hydrogen-bonded heteroatom and hydrogen atom) have an attractive energy of about 4 kcal mole^{-1} for ε = 1.8. Of course, there is also the repulsive energy between the negative heteroatoms. The calculation just illustrates the magnitude of Coulombic interactions. Coulomb's law also describes a component of the interactions between stacked bases because there are large partial charges on many of the atoms (Fig. 1).

London attractions—named after the physicist Fritz London—are fluctuation dipole-induced dipole interactions dependent on r^{-6}. Whereas electrostatic interactions can be repulsive (for like charges) or attractive (for unlike charges), London interactions are always attractive. The attraction energy is proportional to the polarizability of each atom or group of atoms. The polarizability is a measure of how easy it is to move the electrons relative to the nuclei. More polarizable electrons cause larger fluctuation dipoles and larger induced dipoles. The London attraction will thus be large between bases with their highly polarizable pi electrons. Phosphate groups also are highly polarizable.

The van der Waals repulsions—named after Johannes van der Waals—are very short range repulsions that can be approximated as an r^{-12} interaction. The energy depends on the size of each atom, i.e., the sum of the van der Waals radii. Combining the London attraction with the van der Waals repulsion leads to the London–van der Waals interaction,

$$E = \frac{B_{ij}}{r_{ij}^{12}} - \frac{A_{ij}}{r_{ij}^{6}} \quad (6)$$

with A_{ij} and B_{ij} both positive and dependent on the interacting atoms i and j. The combined nonbonded interactions including the electrostatic interactions (Eqs. 5 and 6) are sometimes called van der Waals interactions.

Other interaction terms can be added, such as charge-induced dipole interactions, which depend on r^{-4}. Clearly, these equations are approximations to the interactions actually present in a solution of molecules. They can help us understand the forces that determine the conformations of RNAs, but they are not accurate enough yet to allow calculation of a correct conformation a priori. They can be, and are, used to improve conformations primarily obtained from NMR data, or from X-ray diffraction data. For example, the program X-PLOR (Brünger 1996) is routinely used to obtain optimal conformations for nucleic acids.

The parameters for Equations 3, 4, and 6 are obtained from semiempirical molecular mechanics force fields derived by fitting spectroscopic and thermodynamic data for small molecules. For example, parameters for bonded terms should give correct conformations for monomers. The nonbonded terms should give reasonable values for heats of vaporization of crystals of the small molecules. Interactions with solvent and with ions in solution are among the most difficult to calculate. A widely used force field is AMBER, developed at the University of California, San Francisco (Cornell et al. 1995). Louise-May et al. (1996) provide a useful review of methods for calculating conformations of nucleic acids. The journal *Current Opinion in Structural Biology* is a good source of up-to-date reviews on nucleic acid structure.

Entropies

The main contributions to entropy changes in various RNA reactions are translational entropy and conformational entropy. When two strands are brought together to form a duplex there is a loss of translational entropy, which depends on the concentration of each strand. Therefore, we must specify the concentration when quoting an entropy change. This is done by using a superscript $\Delta S°$, that specifies standard conditions—usually 1 M concentrations. Binding or release of water molecules and of ions also contributes to translational entropy changes.

Hydrophobic contributions to protein folding and to macromolecular conformation in general were originally defined as the aggregation of nonpolar molecules—such as phenylalanine or valine side chains—caused by release of bound water molecules (Tanford 1980). The gain in translational entropy on freeing water molecules was the main part of the favorable, negative free energy of the nonpolar solute aggregation reaction. A positive $\Delta S°$ and a small $\Delta H°$ for aggregation is an indicator of a hydrophobic-driven aggregation. The measured thermodynamics of folding RNA show that hydrophobic interactions are not dominant; the bases are all polar (Fig. 1), and ionic effects are very important.

The change in conformation when an RNA strand goes from a disordered single strand to an ordered strand in a duplex results in changes in torsion angles. The decrease in conformational entropy is dependent on the smaller range of torsion angles available to the ordered conformation compared to the disordered one. The statistical interpretation of entropy derived by Boltzmann relates the entropy to the logarithm of the number N of energetically equivalent states of the system.

$$S = R \ln N \qquad (7)$$

We can use Equation 7 to roughly estimate entropy changes for a disorder–order transition in a polynucleotide. There are seven torsion angles that determine the conformation of each mononucleotide in a polynucleotide. If we assume that in the disordered state—an unstacked single strand—each torsion angle has three equally probable values (*gauche*$^+$, *gauche*$^-$, *trans*), then the mononucleotide has 3^7 states. The ordered state—a duplex, for example—may have only one state. Thus, for this approximation the loss of conformational entropy (per mole of nucleotides) when a single strand forms a duplex is

$$\Delta S° = R \ln (1/3^7) = -7 R \ln 3 = -15.3 \text{ cal K}^{-1} \text{ mole}^{-1} \qquad (8)$$

This provides an unfavorable contribution to the free energy of folding at 37°C equivalent to $-T\Delta S° = + 4.7$ kcal mole^{-1}. The model is clearly too simple, but it does give a qualitative understanding of the entropy changes that occur in RNA transitions. Note that the conformational entropy change for folding would be considerably more unfavorable if the bases and sugars were not relatively rigid due to their cyclic structures.

The free-energy changes that characterize RNA folding involve both entropy and enthalpy (energy). The system will try to minimize energy and maximize entropy in order to minimize free energy as required by the laws of thermodynamics. How fast this will occur depends on kinetics, which is more difficult to predict.

INTERACTIONS IN SIMPLE HELICES

Base Stacking

In aqueous solution the polar groups on the periphery of each base (Fig. 1) hydrogen-bond to water molecules. The faces of the planar bases interact less than the edges with water, so the bases tend to stack. The equilibrium constants for stacking are small; at 25°C the equilibrium constants are about 1 M^{-1} for pyrimidine nucleosides, and 5–10 M^{-1} for purine nucleosides (Bloomfield et al. 1998). In dinucleoside monophosphates the thermodynamics for stacking of the bases was obtained from the temperature dependence of absorption and optical rotation. The standard enthalpy of stacking $\Delta H°$ is favorable, ranging from –5 to –8 kcal mole^{-1}, the standard entropy of stacking $\Delta S°$ is unfavorable at about –25 cal K^{-1} mole^{-1}, and the standard free energy of stacking at 25°C, $\Delta G°$, ranges from +0.9 to –1.7 kcal mole^{-1} corresponding to a range of roughly 20–95% stacked

(Davis and Tinoco 1968). The order of stacking of the bases is: guanine > adenine > cytosine > uracil, with uracil stacking significantly less than the other bases. The extent of stacking also depends on the base sequence. The bases in single-stranded RNA tend to form a right-handed helix with *anti* glycosidic torsion angles; this produces a larger overlap and increased stacking in 5′-purine-pyrimidine-3′ stacks than in 5′-pyrimidine-purine-3′ stacks. Although the differences in stacking of the bases are mainly caused by the enthalpy contributions, it is difficult to make quantitative calculations. All the energy terms contribute, but Coulombic electrostatic interactions among the net charges on each base atom (Fig. 1), plus London attractions of the polarizable electrons in the planar bases, are the main terms. The van der Waals repulsion of the stacked bases keeps the charges at least 3.4 Å apart. The electrostatic interaction energy can be repulsive or attractive, depending on the sequence and orientation of the bases. The London energy is always attractive, and is much less dependent on sequence or stacked orientation. The larger purines attract each other more than the pyrimidines. Because neither the stacked geometry nor the unstacked geometry is fixed, any energy or free energy calculations must be done for a range of structures. The solvent and ion interactions for each structure must also be taken into account.

The stacking in single strands decreases with increasing temperature—the stacking is not primarily caused by a hydrophobic interaction. The experimental value of $\Delta S° = -25$ cal K^{-1} mole^{-1} per stack is consistent with the loss of conformational entropy per nucleotide calculated in Equation 8. Increasing salt concentration favors stacking in single strands because the phosphate charges are closer together in a stacked conformation.

An unpaired base at the end of a double helix—a so-called dangling end—can stack on the helix and stabilize the duplex (Petersheim and Turner 1983). The thermodynamics of the dangling end is very sequence dependent for the 32 possible cases (Turner et al. 1988). A 3′-unpaired nucleotide, e.g., $\begin{smallmatrix}5'-CA3'\\G\end{smallmatrix}$, is usually more stabilizing than a 5′-unpaired nucleotide, e.g., $\begin{smallmatrix}5'-AC3'\\G\end{smallmatrix}$. For 3′-unpaired nucleotides, $\Delta H°$ of stacking ranges from –2 to –9 kcal mole^{-1}, with $\Delta S°$ from –5 to –23 cal K^{-1} mole^{-1}. Free-energy increments for 3′ stacking, $\Delta G°$, at 37°C range from –0.1 to –1.7 kcal mole^{-1}. For 5′-unpaired nucleotides, free-energy increments at 37°C range from 0 to –0.5 kcal mole^{-1}. About the same stabilization is induced by a 5′-terminal phosphate, probably because it restricts conformations available to the unpaired conformation. Thus, net stacking interactions of 5′-unpaired nucleotides are negligible. As shown in Table 1 and

Table 1 Comparison of free-energy increments ($\Delta G°_{37} < -1$ kcal/mole or $\Delta G°_{37} > -0.4$ kcal/mole) for stacking of dangling ends with structures determined by X-ray diffraction

Structure[a] and (motif)[b]	Sequence	$\Delta G°_{37}$[c] kcal/mole	Stacked or unstacked[d]	Structure[a] and (motif)[b]	Sequence	$\Delta G°_{37}$[c] kcal/mole	Stacked or unstacked[d]
tRNAi (h)	5'CA 14 3'GA	−1.7	S	tRNA^phe (j)	5'G 3'CC 48	−0.3	U
tRNAi (j)	5'CG 26 3'GG	−1.7	S	tRNAi (h)	5'CA 3'GA 21	−0.2	U
P4-P6 (j)	5'GG 176 3'CG	−1.3	S	tRNA^phe (h)	5'CA 3'GA 21	−0.2	U
P4-P6 (il)	5'CU 224 3'GU	−1.2	S	P4-P6 (j)	5'GG 3'CG 164	−0.2	U
tRNA^asp (h)	5'CU 32 3'GC	−1.2	S	tRNA^phe (j)	5'C 3'GA 9	−0.2	U
P4-P6 (il)	5'GA 113 3'CA	−1.1	S	tRNA^asp (h)	5'GU 3'CU 60	−0.1	U
tRNAi (h)	5'GA 54 3'CA	−1.1	S	hammerhd (j)	5'UC 3 3'AC	−0.1	S
tRNA^phe (h)	5'AC 3'UA 38	−0.3	S	hammerhd (j)	5'UC 17 3'AA	−0.1	U
tRNA^phe (h)	5'GU 3'CC 60	−0.3	U	tRNA^phe (j)	5'UU 8 3'A	−0.1	U
hammerhd (j)	5'UC 3'AC 17	−0.3	U	tRNA^asp (j)	5'G 3'CU 48	−0.1	U
hammerhd (j)	5'UC 3'AA 14	−0.3	S	P4-P6 (il)	5'CU 3'GU 249	0.0	U
tRNAi (j)	5'G 3'CC 48	−0.3	U	tRNAi (j)	5'CG 3'GG 9	0.0	U
tRNA^asp (j)	5'CU 3'GC 38	−0.3	S				

Bases adjacent to terminal A·U or G·C and not in A·U, G·C, G·A, or G·U pairs are considered. For details, see Burkard et al. 1999.

[a]References: tRNAi (Basavappa and Sigler 1991) P4-P6 (Cate et al. 1996); tRNA^asp (Westhof et al. 1988); tRNA^phe (Sussman et al. 1978; Westhof and Sundaralingam 1986); hammerhd (Scott et al. 1995).

[b](il) = Internal loop; (h) = hairpin; (j) = junction.

[c]The $\Delta G°_{37}$ is for the dangling end numbered and in bold as measured in oligoribonucleotide duplexes (Turner et al. 1988).

[d]A nucleotide is considered stacked (S) on the adjacent base pair shown in bold if the distance between closest non-hydrogen atoms of bases is less than 4.0 Å, the angle between bases is less than 30°, and there is overlap of base rings when the base pair is projected into the plane of the dangling end base or when the dangling end base is projected into the plane of either of the paired bases.

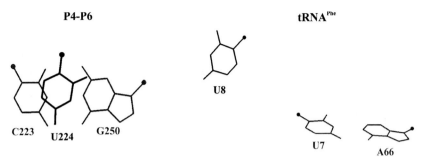

Figure 2 Examples of stacked and unstacked bases from Table 1. (*Left*) U224, in *black*, from P4-P6 (Cate et al. 1996) is stacked as a 3′ dangling end on the adjacent 223 C-G 250 pair, in *gray*; predicted stacking free energy is –1.2 kcal/mole (Turner et al. 1988). (*Right*) U8, in black, from tRNA^{Phe} (Sussman et al. 1978) is not stacked as a 3′ dangling end on the adjacent 7 U-A 66 pair; predicted free energy of stacking is –0.1 kcal/mole. Both views are from the plane of the unpaired base. Small circles represent ribose.

Figure 2, there is a correlation between free energy increments for dangling end stacking in oligonucleotides and structures at the ends of helices. As suggested previously based on the structure of phenylalanine tRNA (Turner et al. 1988), nucleotides with stacking free-energy increments more favorable than –1 kcal mole^{-1} are stacked in crystal structures. Nucleotides with free-energy increments less favorable than –0.4 kcal/mole tend to be unstacked unless they are in a G·A pair. Evidently, stacking interactions are an important determinant of 3D structure.

Hydrogen Bonding

Hydrogen bonds are formed between hydrogen atoms with partial positive charges and fluorine, nitrogen, or oxygen atoms with partial negative charges. The interaction is largely electrostatic with a small covalent component associated with sharing of electrons (Pauling 1960). The large charge separations in nucleic acid bases (Fig. 1) result in many potential hydrogen-bond donors and acceptors. As shown in Appendix 1, all pairwise combinations of A, C, G, and U can be oriented in multiple ways to form more than one hydrogen bond between bases if *syn* and *anti* glycosidic torsion angles or parallel and anti-parallel sugar-phosphate backbones are allowed (Saenger 1984; Tinoco 1993). Formation of multiple hydrogen bonds between rigid ring systems is particularly favorable because there is only a small conformational entropy penalty for forming hydrogen bonds beyond one. The geometries required to form hydrogen bonds are another important determinant of RNA shapes.

The net contribution of various hydrogen bonds to stabilizing folded structures is not well established. This is because all the hydrogen-bond donors and acceptors in unfolded RNA can be hydrogen-bonded to water. An experimental measure of the contribution of some hydrogen bonds to folding stability is the effect of replacing the NH_2 groups of G or A with hydrogen. Such substitutions in terminal G·C base pairs and in hydrogen-bonded G·A mismatches in different contexts indicate a single NH_2 group can contribute from -0.4 to -1.6 kcal mole^{-1} of favorable free-energy change for folding at 37°C (Turner et al. 1987; SantaLucia et al. 1991, 1992; Turner and Bevilacqua 1993). In DNA, substitution of the weakly hydrogen-bonding isostere, difluorotoluene, for thymine in a central A·T base pair reduces $\Delta G°_{25}$ for duplex formation by 3.6 kcal mole^{-1} (Moran et al. 1997), or 1.8 kcal mole^{-1} of hydrogen bond. Some carbon atoms are electronegative enough to be involved in C—H···O hydrogen bonds; these bonds may also be important in shaping RNA (Leonard et al. 1995; Wahl and Sundaralingam 1997).

Counterion Condensation

In the physiological pH range, RNA is a polyanion because each phosphate has a –1 charge. For the charge separation in polymeric double-helical or random-coil RNA, the sum of these repulsive electrostatic interactions would be hugely unfavorable in the absence of counterions (Manning 1978). Thus, this charge is partially neutralized by condensation of positively charged counterions into a small volume around the RNA chain, resulting in a local counterion concentration that is independent of the bulk counterion concentration (Manning 1978; Record et al. 1978). For solutions containing only univalent ions, this local concentration is about 1 M. Localization of a +2 or +3 counterion neutralizes more charge per cation than a +1 counterion. This provides an entropic advantage for condensing multivalent cations since fewer particles are localized. Thus, when enough multivalent cations are available, they will essentially exclude +1 cations from being condensed. The effects of counterion condensation can be important whenever the charge density of an RNA changes. For example, duplexes have a higher charge density than single strands and thus require a higher local concentration of counterions. The dependence of duplex stability on amount and type of ions in solution is a measure of the difference in ion binding by the single strands and the duplex. For a 1:1 electrolyte like NaCl, this leads to the following equation for the change in melting temperature with salt concentration:

$$T_m(2) - T_m(1) = -\frac{\Delta nR}{\Delta H^\circ} T_m(2) T_m(1) \ln \frac{M_2}{M_1} \qquad (9)$$

Here Δn is the difference in the number of ions bound by the duplex and by the single strands; ΔH° is the enthalpy change for formation of the duplex; and T_m (2), T_m (1) are the melting temperatures corresponding to salt concentrations M_2, M_1. Because of the dependence of ΔH° and Δn on sequence and chain length, salt effects can vary. For long RNAs when only NaCl is present, the melting temperature can increase 10–20°C for a tenfold increase in NaCl concentration. When saturating concentrations of +2 or +3 cations are present, however, increasing NaCl concentration decreases T_m. There are analogous salt effects for other folding transitions, such as intramolecular helix formation and assembly of helixes. The polymer length required for counterion condensation to be important is not known, but is probably on the order of 20 base pairs (Olmsted et al. 1991).

Metal Ion Coordination

In addition to the general effect of ionic strength on the folding of RNA molecules, there are specific binding sites for metal ions. The most biologically relevant metal ion is Mg(II), present in aqueous solution as the hexahydrate ion, $Mg(H_2O)_6^{++}$. Mg(II) can bind to RNA either through outer shell coordination involving water-mediated hydrogen bonds, or by direct coordination to the metal. The coordinated water molecules are in fast exchange with solvent water, so the reactions of magnesium to accessible sites on the RNA are essentially diffusion controlled. Of course, if the RNA must break base pairs, or otherwise rearrange before reacting with the metal, the rate of reaction can become very slow.

Likely locations for binding metal ions are sites with a high concentration of phosphate ions. These include small bulge loops, small hairpin loops, the junction of stems and loops in pseudoknots and kissing hairpins, and base triples. A useful method for identifying phosphates that are essential in binding Mg^{++} ions is to replace a phosphate oxygen atom by a sulfur atom; sulfur–magnesium binding is much weaker than oxygen–magnesium binding. Essential metal ions and the phosphates they bind have been identified by this phosphorothioate interference method for folding the P4-P6 domain of group I introns (Murphy and Cech 1993; Cate et al. 1997), and for catalytic activity in hammerhead ribozymes (Ruffner and Uhlenbeck 1990). Since Mn^{++} prefers sulfur to oxygen, activity can often be restored in phosphorothioate constructs by replacing Mg^{++} with Mn^{++}.

Metal ions can bind to the bases as well as to phosphate ions; all hydrogen-bonding acceptors on the bases are potential binding sites. Kazakov (1996) provides a useful review of metal-ion binding to nucleic acids. X-ray crystallography can identify metal-ion-binding sites with the greatest precision. Magnesium-binding sites have thus been identified in the hammerhead ribozyme (Pley et al. 1994a; Scott et al. 1995, 1996) and in the P4-P6 domain of the *Tetrahymena* group I intron (Cate and Doudna 1996; Cate et al. 1996, 1997). Metal ions that can mimic magnesium and compete for the same sites in RNA are useful for identifying binding sites in crystals and in solution. Cobalt(III) hexammine can replace magnesium(II) hexahydrate in outer shell complexes with RNA. The amino groups in $Co(NH_3)_6^{+++}$ are not in rapid exchange and the 18 protons are excellent NMR reporters of the location of the central metal ion. Adjacent G·U base pairs form an electronegative pocket (lined with N7 of guanine and the carbonyls of G and U) in the major groove of RNA that binds either $Mg(H_2O)_6^{++}$ or $Co(NH_3)_6^{+++}$ (Cate and Doudna 1996; Kieft and Tinoco 1997). Terbium(III) ion competes with Mg^{++} ion at one of four known ion-binding sites and inhibits activity of the hammerhead ribozyme although it binds 10 Å away from the cleavage site (Feig et al. 1998). Manganese(II) ions can bind at Mg^{++} ion-binding sites, and NMR can be used to monitor the location of Mn^{++} binding by the effect of its unpaired electron on neighboring protons (Allain and Varani 1995).

SECONDARY STRUCTURES

The number of RNA structures determined by NMR in solution and by X-ray diffraction in crystals is increasing rapidly. The only way to keep up with progress is to read the current literature such as *Biochemistry*, *Journal of Molecular Biology*, *Nature Structural Biology*, *Nucleic Acids Research*, and *RNA*. Reviews that periodically summarize RNA structure are very useful (Gesteland and Atkins 1993; Moore 1995; Varani 1995; Chang and Varani 1997; Ramos et al. 1997; Simons and Grunberg-Manago 1998). We only refer to a few structures that illustrate general characteristics.

A-Form Double Helix

The stacking of nucleosides and bases in aqueous solution is due to the interactions of the planar, aromatic bases. But why do two strands come together to form a double strand? What is the main driving force? Bringing two molecules together causes an unfavorable loss of translational entropy, and when the molecules are negatively charged, overcoming the charge repulsion requires energy of attraction. There is also an unfavorable loss of conformational entropy because the duplex structure is much more rigid than the partially stacked single strands. The favorable factor is the Watson-Crick complementarity that allows many base pairs to form in a uniform helix. The snug fit of the bases means that each additional base pair lowers the enthalpy through favorable London–van der Waals interactions: intrastrand stacking, plus interstrand stacking and pairing. The measured thermodynamics of double-strand formation can be approximately divided into an initiation step of forming the first base pair followed by successive propagation steps of adding base pairs (Borer et al. 1974). Thermodynamic data on many oligonucleotides were fit to this nearest-neighbor model to obtain parameters for initiation and for ten Watson-Crick nearest neighbors (Freier et al. 1986). The favorable enthalpy and unfavorable entropy terms for adding a base pair to a helix range from $\Delta H° = -6.6$ to -14.2 kcal mole^{-1} and $\Delta S° = -18.4$ to -34.9 cal K^{-1} mole^{-1}, respectively, resulting in favorable free-energy changes at 37°C ranging from -0.9 to -3.4 kcal mole^{-1}. Note that the enthalpy changes are similar to, but more favorable than, those for stacking the bases in a dinucleoside phosphate. The thermodynamic parameters provide a useful method to calculate equilibrium constants and melting temperatures for any sequence of complementary RNA strands (Serra and Turner 1995).

For RNA helixes composed completely of the isosteric Watson-Crick base pairs, the interactions discussed above result in a double helix with a relatively constant and uniform structure independent of sequence. An A-form helix has about 11 base pairs per turn with the base pairs displaced from the helix axis. This forms a deep and narrow major groove, and a very shallow minor groove. The ribose sugars all have C3′-endo conformations and the bases have an *anti* glycosidic torsion angle (Saenger 1984). Due to hydrogen bonding (Fig. 1) and their protected location inside the helix, imino protons on base-paired guanine and uracil exchange slowly with water and give sharp resonances in NMR. However, the terminal base pairs are less constrained so imino resonances of the end base pairs are broad or not observable because they can exchange more rapidly with water. The base pairs at the end of the helix are said to breathe. Dangling nucleotides that stack on the duplex stabilize the helix and sharpen the end imino resonances.

Coaxial Stacks

Coaxial stacking is the end-to-end stacking of double helixes (Appendix 2). This stacking interaction is a major determinant for the alignment of helixes in large RNAs, and provides a way to form long column structures without continuous backbones on each side. Examples are apparent in the structures of tRNA and group I self-splicing introns (Kim et al. 1973; Robertus et al. 1974; Michel and Westhof 1990; Cate et al. 1996). The strength of coaxial stacking interactions has been measured in model systems such as that shown in Figure 3 (Walter and Turner 1994; Walter et al. 1994; Kim et al. 1996). For Watson-Crick base pairs in coaxial stacks created by the absence of a phosphate in an otherwise continuous helix, the free-energy increment ranges from –2.2 to –4.3 kcal mole^{-1} at 37°C, and is about 1 kcal mole^{-1} more favorable than for propagation of a continuous helix by the same base pairs. Presumably, the enhancement over helix propagation is due to the flexibility inherent in a nicked backbone, which allows optimization of stacking interactions while retaining some conformational freedom. When the chains are extended with an unpaired nucleotide beyond the nick, the free-energy increment for coaxial stacking is approximately equal to that for propagation of an uninterrupted helix. Favorable coaxial stacking is also observed when a single mismatch is at the helix/helix interface (Kim et al. 1996). Surprisingly, the free-energy increments for mismatches appear relatively insensitive to sequence and to chain extension beyond the helix/helix interface. Both G·A and C·C mismatches at the interface provide roughly a –2 kcal mole^{-1} increment in binding free energy at 37°C. This contrasts with the 25-fold more abundant occurrence of G·A over C·C mismatches at helix/helix interfaces in

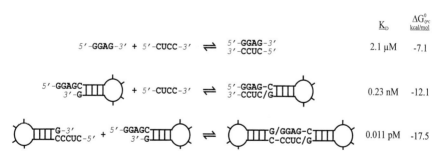

Figure 3 Comparison of the predicted thermodynamics for duplex formation at 0°C with 0, 1, or 2 coaxial stacks (Walter and Turner 1994; Walter et al. 1994). The / indicates the 3′ end of one chain and the 5′ end of the adjacent chain; there is no phosphate at this position.

a set of known RNA secondary structures (Kim et al. 1996). Apparently, evolution selected G·A mismatches in coaxial stacks for structural rather than thermodynamic reasons.

Coaxial stacking interactions may have facilitated association of RNA fragments during early stages of evolution. The interactions are both strong and highly temperature dependent. Thus, as illustrated in Figure 3, an association with 3 G·C and 1 A·U base pairs is predicted to have dissociation constants of 2.1 µM, 0.23 nM, and 0.011 pM at 0°C when there are 0, 1, or 2 coaxial stacks formed. Assuming a typical association rate of 1×10^7 M^{-1} s^{-1}, the half-lives of the complexes with one and two coaxial stacks would be roughly 5 minutes and 73 days, respectively. This could provide enough time for ligation of the fragments.

Internal Loops

Internal loops form when internal nucleotides on both sides of a double helix cannot pair to form Watson-Crick or G·U pairs (Appendix 2). Internal loops provide novel shapes for molecular recognition and flexibility for folding, and can decrease the folding stability of a long helical region. The thermodynamic stabilities and structures of internal loops are very sequence dependent. For example, $\Delta G°_{37}$ for forming the duplexes $\begin{array}{l}5'GAGA\underline{AG}GAG3'\\3'CUC\underline{GAC}UC5'\end{array}$ and $\begin{array}{l}5'GAGA\underline{AC}GAG3'\\3'CUC\underline{ACC}UC5'\end{array}$ are –6.0 and –3.0 kcal mole^{-1}, respectively (Xia et al. 1997). Both duplexes have internal loops with two mismatches but the dissociation constants differ by about 100-fold. NMR spectra indicate that each G·A mismatch in the $\begin{array}{l}AG\\GA\end{array}$ sequence forms two hydrogen bonds to make the imino hydrogen-bonded structure shown in Appendix 1. A·A and C·C mismatch structures with two hydrogen bonds do not easily fit into an A-form helix (see Appendix 1). This difference in cross-strand hydrogen-bonding potential may largely account for the difference in stability.

Cross-strand hydrogen-bonding potential between mismatches is not always expressed in internal loops (Xia et al. 1997). For example, $\Delta G°_{37}$ for forming the duplex $\begin{array}{l}5'GAGU\underline{U}GAG3'\\3'CUC\underline{UU}CUC5'\end{array}$ is –5.9 kcal mole^{-1}, similar to the –6.0 kcal mole^{-1} for $\begin{array}{l}5'GAGA\underline{GG}AG3'\\3'CUC\underline{GAC}UC5'\end{array}$. NMR spectra indicate two imino proton hydrogen bonds within each UU mismatch (see Appendix 1). A duplex with one G·A and one adjacent U·U mismatch,

5'GAG<u>AU</u>GAG3'
3'CUC<u>GU</u>CUC5', however, forms with $\Delta G°_{37} = -3.8$ kcal mole^{-1}, and NMR spectra provide no evidence for hydrogen bonding within the G·A mismatch. Evidently, an imino proton hydrogen-bonded G·A mismatch is not compatible with an adjacent imino proton hydrogen-bonded U·U mismatch in this motif. The different shapes for the two mismatches shown in Appendix 1 suggest that coexistence would require a large distortion of the backbone. Evidently, such steric effects are another factor important in shaping RNA.

An example of the sequence dependence of internal loop structure is shown in Figure 4 (SantaLucia and Turner 1993; Wu and Turner 1996; Wu et al. 1997). All three duplexes have 6 G·C pairs with 2 adjacent G·A mismatches in the center. The G·A mismatches in the duplexes (GCG<u>GA</u>CGC)$_2$ and (GGC<u>AG</u>GCC)$_2$ have the imino proton hydrogen-bonding conformation shown in Appendix 1. This results in a small expansion of the sugar phosphate backbone as shown by the yellow ribbons. The G·A mismatches in the duplex (GGC<u>GA</u>GCC)$_2$, however, have

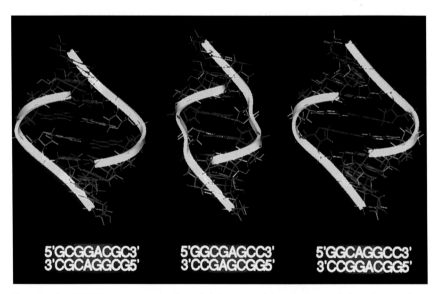

Figure 4 NMR structures for duplexes with tandem GA mismatches: (GCG<u>GA</u>CGC)$_2$ (Wu and Turner 1996), (GGC<u>GA</u>GCC)$_2$ (SantaLucia and Turner 1993), and (GGC<u>AG</u>GCC)$_2$ (Wu et al. 1997). The yellow ribbon follows the sugar-phosphate backbone. In mismatches, G is *red* and A is *green*. The GA mismatches in (GGC<u>GA</u>GCC)$_2$ have a sheared conformation and the GA mismatches in (GCG<u>GA</u>CGC)$_2$ and (GGC<u>AG</u>GCC)$_2$ have imino hydrogen-bonded conformations (see Appendix 1).

the sheared conformation shown in Appendix 1. This results in a marked contraction of the sugar phosphate backbone at the mismatches. Formation of a sheared conformation is favored by a hydrogen bond from the G amino to a nonbridging oxygen of the phosphate 5' to A (Heus and Pardi 1991; SantaLucia and Turner 1993) and possibly by a water-bridged hydrogen bond from the G imino hydrogen to the same oxygen (Santa-Lucia et al. 1992). These contacts cannot be made when the sequence is switched to (GGCAGGCC)$_2$ because the phosphate 5'of A must be in position to allow formation of the adjacent Watson-Crick C·G pair, another example of a steric, compatible fit effect (Gautheret et al. 1994; Wu et al. 1997). This restriction, however, does not prevent the sheared G·A conformation for the sequence (GCGGACGC)$_2$. In this case, it has been suggested that unfavorable electrostatic stacking interactions between a sheared G·A and the adjacent G·C base pair favor the imino hydrogen-bonded G·A conformation (Wu and Turner 1996). Evidently, the balance between hydrogen bonding and stacking can determine local conformation. It would not be surprising if this balance could be reversed by additional contacts with a ligand or protein. This would change the shape of the RNA duplex and thereby change the relative positions of the ends of the duplex.

The studies described above identify cross-strand hydrogen bonding, compatible fit of adjacent pairs, and stacking electrostatics as interactions that shape internal loops with two mismatches. Structural studies indicate these interactions are also important in shaping larger internal loops, and suggest some additional generalizations. In loops of known structure that have an equal number of nucleotides on both strands, each pair of opposing bases exhibits noncanonical hydrogen bonds. These non-Watson-Crick pairs distort the helix from A-form, affecting the helical twist and interphosphate distance, as illustrated in Figure 4 for tandem G·A mismatches. The distortion can also bend the helix; for example, Baeyens et al. (1996) determined the structure of an internal loop with tandem A·A mismatches flanked by G·A mismatches, which produced a 34° angle between the flanking helices. A crystal structure of a 5S rRNA domain contains seven consecutive noncanonical base pairs, yielding a slightly bent shape, stabilized in part by cross-strand stacking of purines (Correll et al. 1997). The known loops with equal numbers of nucleotides on both strands all have antiparallel backbones with anti-glycosidic torsion angles; this limits the possible non-Watson-Crick pairs to those in which the base faces are oriented in opposite directions, i.e., those labeled +/– in Appendix 1.

Loops with a different number of bases on the two sides have less predictable structures. It is not obvious which bases are likely to interact or even which direction will be taken by the backbone. Three asymmetric internal loops are in the P4-6 domain of the group I intron (Cate et al. 1996), and one is in the HIV Rev-responsive element (RRE) (Battiste et al. 1994; Peterson and Feigon 1996). Three of these have an unstacked U, consistent with the limited propensity of U to stack in most contexts. An unexpected feature in the P4-6 domain is an internal loop (J6a/6b) with hydrogen bonding between two consecutive As on the same strand to form an "A-platform" (Cate et al. 1996). This internal loop forms a tertiary interaction with a GAAA hairpin tetraloop. The A-platform is absent in an NMR structure of the isolated J6a/6b internal loop, illustrating that shapes of internal loops can depend on long-range as well as short-range interactions (Butcher et al. 1997). The J4/5 internal loop of the P4-6 domain contains two A·A pairs with one hydrogen bond each and an extraordinary helical twist between, so that the second pair stacks cross-strand on the first. The internal loop from the RRE contains an imino-hydrogen-bonded G·A, an unstacked U, and a G·G pair in which the guanines have parallel-oriented faces. When Rev protein is bound, the single unstacked U allows the backbone on this strand to become parallel for the adjacent G. Clearly, asymmetric internal loops can provide large distortions from A-form and a wide variety of possible base pairs, thus providing unique structures for protein or RNA binding.

Hairpin Loops

Hairpin loops reverse the direction of an RNA chain (Appendix 2). In principle, they can contain any number of unpaired bases. DNA hairpin loops have been studied in which only one (Zhu et al. 1995; Bhaumik 1996) or two (Davison and Leach 1994) nucleotides are unpaired. The C(UN)G family of hairpin loops occurs often in ribosomal RNA. The structure of the C(UU)G member determined by NMR shows that it is indeed a biloop closed by a C·G pair (Jucker and Pardi 1995a). The first U of the biloop folds down into the minor groove of the stem. An RNA triloop with the sequence C(UUU)G was studied by NMR (Davis et al. 1993); it is very dynamic with the bases out in solution. Both structures are consistent with the lack of stacking observed in single-stranded UU and polyU. Tetraloops are the most numerous hairpin loops in ribosomal RNAs, and C(GNRA)G and C(UNCG)G are the most common sequences (Woese et al. 1990). The UNCG family is particularly stable thermodynamically (Antao et al. 1991). These tetraloops are structurally biloops because the first and last bases of the tetraloop hydrogen bond to form G·A or U·G base pairs (Varani 1995). A pentaloop with the sequence

U(UUCUG)A is structurally a triloop closed by a U·G base pair (Puglisi et al. 1990). A conserved hexaloop from ribosomal RNAs, C(GUAAUA)G, is a tetraloop closed by a G·A pair (Fountain et al. 1996; Huang et al. 1996).

The tRNA anticodon loops of seven nucleotides have been the most extensively studied of all hairpin loops. A characteristic U-turn (also called a π turn) first seen in phenylalanine tRNA (Saenger 1984) also occurs in hairpin loops with fewer nucleotides (Fountain et al. 1996; Huang et al. 1996; Jucker et al. 1996; Stallings and Moore 1997) as well as in junctions (Pley et al. 1994a; Scott et al. 1995). The U-turn involves a change in backbone direction stabilized by a hydrogen bond to a non-bridging phosphate oxygen, and a hydrogen bond from a $2'$-OH to a purine N7 (Jucker and Pardi 1995b; SantaLucia et al. 1992). The bases are left free to form tertiary interactions. The importance of dynamics as well as structure is illustrated by a comparison between initiator and elongator tRNA anticodon loops in solution (Schweisguth and Moore 1997). Although both have structures very similar to the structures of elongator tRNAs in crystals, the elongator tRNA loop is significantly more dynamic. The central hairpin loop in the hepatitis delta virus ribozyme contains seven nucleotides (Kolk et al. 1997), including a *syn* cytidine (Lynch and Tinoco 1998). RNA stems and most of the loop nucleotides have the bases in an *anti* orientation around the glycosidic bond. A few *syn* purines have been found, but *syn* pyrimidines have not been identified in RNA previously. Clearly, the library of shapes is still incomplete.

Large hairpin loops can exist, but as illustrated by the loops that have already been studied, we expect the loop bases to hydrogen-bond to form Watson-Crick or mismatch base pairs. This means that hairpin loops with eight or more unpaired nucleotides are very unlikely.

Bulge Loops

Bulges form when internal nucleotides on one strand of a double helix are unpaired while those on the other strand are all in Watson-Crick or G·U pairs (Appendix 2). Single purine bulges tend to stack in the double-stranded helix and bend the helix, whereas single pyrimidine bulges are extrahelical (van den Hoogen et al. 1988; Borer et al. 1995). This behavior is expected from the stacking properties of the bases. Bulge loops with a larger number of nucleotides also bend the helix. The amount of bending, which can be measured by electric birefringence (Zacharias and

Hagerman 1995a,b) and by gel electrophoresis (Riordan et al. 1992; Luebke and Tinoco 1996), is dependent on the sequence and size of the bulge, and on the concentration of Mg^{++} ions.

The structure of the A-rich bulge in the P5a stem of the *Tetrahymena* group I intron was determined by X-ray crystallography of the P4-P6 domain to be a bulge of four nucleotides followed by a single base pair and a one-base bulge (Cate et al. 1996). This secondary structure is not thermodynamically stable by itself, but is stabilized by tertiary interactions involving both bulge loops (Cate et al. 1997). The structure of the A-rich bulge in solution is a bulge of five nucleotides (Luebke et al. 1997) as expected from thermodynamic measurements on other bulges. Thus tertiary interactions can sometimes modify base pairing in a secondary structure in order to minimize free energy.

Junctions

Junctions are formed when more than two double helices are connected in a closed structure (Appendix 2). For example, the four helical arms of tRNA emanate from a junction. Junctions are a major determinant of the overall shape of an RNA, and the available structures suggest several interactions important for determining these shapes (Kim et al. 1973; Robertus et al. 1974; Scott et al. 1995; Cate et al. 1996). One is coaxial stacking. For the one tRNA structure with two alternate possibilities for coaxial stacking (Biou et al. 1994), the observed structure corresponds to the thermodynamically most favorable coaxial stacking as measured in model systems such as those shown in Figure 3 (Walter and Turner 1994).

Alignment of helices in junctions is also affected by formation of nucleotide triples. This can be seen in structures of tRNA and the model of group I introns based on sequence comparison (Michel and Westhof 1990). Many base triple combinations are known, including U·(A·U), A·(U·A), C^+·(G·C), U·(C·G), G·(C·G), and A·(G·C) (Felsenfeld et al. 1957; Lipsett 1964; Lee et al. 1979; Broitman et al. 1987; Michel and Westhof 1990). Nucleotide triples can also involve base to 2′-OH contacts (Chastain and Tinoco 1992, 1993). Metal ions also can have a large effect on the shape of a junction (Bassi et al. 1997).

ASSEMBLY OF SECONDARY STRUCTURES

Large, functional RNAs must assemble their secondary structural elements into well defined three-dimensional shapes. The specific contacts responsible for assembly include coaxial stacking, base pairing to form pseudoknots (Banerjee et al. 1993; Harris et al. 1994; Lehnert et al. 1996), nucleotide triples (Kim et al. 1973; Robertus et al. 1974; Michel and Westhof 1990; Chastain and Tinoco 1992), stacking, hydrogen bonding, and metal ion coordination. Only limited information is available for the sequence dependence and energetics for some of the contacts, including those between tetraloop hairpins and tetraloop receptors (Michel and Westhof 1990; Pley et al. 1994b; Cate et al. 1996; Costa and Michel 1997), hydrogen bonds to 2′-OH groups (Sugimoto et al. 1989; Bevilacqua and Turner 1991; Pyle and Cech 1991; Pyle et al. 1992; Strobel and Cech 1993), hydrogen bonds to the exocyclic amino group of a G·U pair (Strobel and Cech 1995), hydrogen bonds to nonbridging phosphate oxygens (Quigley and Rich 1976; Li and Turner 1997; Profenno et al. 1997), and specific metal ion coordination (Christian and Yarus 1993; Piccirilli et al. 1993; Laing et al. 1994; Cate et al. 1997). Nonspecific cation interactions can also be important. For example, under certain conditions, counterion condensation theory predicts an attraction between helixes due to delocalization of the condensed counterions (Ray and Manning 1994). It has also been suggested that mobile counterions induce an attraction between helixes by a mechanism similar to dispersion forces between atoms and molecules (Rouzina and Bloomfield 1996). The net contribution of tertiary interactions can be substantial. For example, separate RNA domains can associate in solution in the absence of any Watson-Crick pairing (Szostak 1986; Doudna and Cech 1995). It is clear that many studies of the global folding of RNAs will be necessary to learn all the interactions that determine overall shape.

IMPLICATIONS FOR EVOLUTION

Due to their relatively large size, rigidity, and inherent charge separations, nucleic acid bases can form strong stacking and hydrogen-bonding interactions without large entropy costs. Thus, short oligonucleotides are able to form stable, folded structures and complexes. For example, the melting temperature of the hairpin formed by rGGC<u>GCAA</u>GCC is 71°C in 0.1 M NaCl (SantaLucia et al. 1992). The melting temperatures of $(GCCGGCp)_2$ at 10^{-6}, 10^{-9}, and 10^{-12} M are 51, 30, and 11°C, respectively, in 1 M NaCl (Freier et al. 1986). Thus, only a few coupling steps are required to form a sequence with a stable structure. In contrast, oligopeptides must be much longer to form stable structures because the interactions of individual monomers are weak. In the early stages of evolution, when covalent bond formation was difficult, oligonucleotides would have a competitive advantage over oligopeptides (Turner and Bevilacqua 1993).

RNA/RNA interactions are strong enough to allow association of different strands or domains. At intermediate stages of evolution, however, it would be a competitive advantage to covalently link all the domains required for a particular function. At this stage, there could be a competitive disadvantage to the strong interactions of RNA because they permit a plethora of kinetically trapped structures (Emerick and Woodson 1993; Uhlenbeck 1995). This is because rates of secondary structure formation are typically fast, but because of the strong local interactions formed, unfolding rates are typically slow and very temperature dependent. A large variety of relatively stable structures for a given sequence would be a particular disadvantage with irregular variations of temperature and other environmental factors, since the particular structures trapped kinetically would depend, for example, on how fast the temperature changed. Thus, there would be a competitive advantage for sequences that are not easily trapped kinetically, but rather are able to rearrange to find their lowest free-energy structure. These sequences could perform their function regardless of the latest "renaturation protocol." This implies an advantage for sequences that form structures with relatively short helixes interrupted by destabilizing loops. For example, the architecture of tRNA is based on two helical stems, each of which is interrupted roughly in the middle by a coaxial stacking interaction to form a junction. The melting temperatures of the TψC and acceptor stems are 61 and 77°C, respectively, in the native fMet tRNA structure (Crothers et al. 1974). If this helical motif were made from a single continuous helix, the T_m is predicted to be 112°C (Serra and Turner 1995). In principle, A is the best unpaired nucleotide for avoiding formation of multiple kinetically trapped species, since it can only pair with U to form the relatively weak A·U pair. A survey of small and large ribosomal RNA secondary structures (Gutell et al. 1993; Gutell 1994) shows that the percentages of unpaired nucleotides that are A, C, G, and U are 38%, 16%, 24%, and 22%, respectively. This contrasts with 15%, 28%, 35%, and 22%, respectively, for regions that are Watson-Crick and G·U paired. Thus adenine is favored in unpaired regions.

The strong, local interactions in RNA result from bonding interactions and are thus associated with large, favorable enthalpy changes. From Equations 1 and 2, there is therefore a large temperature dependence to the equilibrium constants associated with RNA/RNA interactions. This could be a competitive disadvantage for organisms required to live over a large temperature range. The large temperature dependence can be reduced by folding processes associated with unfavorable enthalpy terms, as has been observed for docking of the P1 helix into the catalytic core of the *Tetrahymena* ribozyme (Li et al. 1995). Formation of internal loops and processes involving uptake of metal ions is also often associated with unfavorable enthalpy changes, so there might be a selection pressure favoring them. Another way to reduce the temperature dependence of a process depending on RNA/RNA interactions is to require switching of base pairs or other interactions rather than completely new formation of structure (Li et al. 1995).

Strong RNA/RNA interactions also create a problem with replication in an RNA World. Helicase activity is required to separate the new copy from the RNA template. Initially, this helicase activity could be provided by increased temperature (Turner and Bevilacqua 1993). At some point, however, a single-stranded DNA template would have a competitive advantage. Single-stranded DNA is more chemically stable than RNA, and RNA/DNA hybrids usually have lower melting temperatures than RNA/RNA helices (Sugimoto et al. 1995). An added advantage of RNA/DNA hybrids is that the sequence could be selected to favor hydrolysis of the RNA strand at particular nucleotides. This would facilitate unwinding while assuring spatial localization of particular RNAs.

Nucleotides have strong interactions with each other, but their shapes are limited. Thus, it is possible that small molecules were recruited to allow RNA to fold in new ways (Davies et al. 1993). Although it is unlikely such interactions would release enough free energy to drastically affect secondary structure, they could easily affect three-dimensional folding, as has been observed when arginine binds to TAR RNA (Puglisi et al. 1992). Presumably, peptides were selected for this role because they could be polymerized into a wide variety of molecules.

As more RNAs were generated, the strong interactions of nucleotides and the limited repertoire of shapes presented a problem for specific recognition by RNA/RNA contacts. For example, when many sequences are present, it is more difficult to generate the unique complementarity required for specificity. If each contact is strong, as in base pairing, then tight binding will also be possible for sequences that are not perfect matches (Herschlag 1991; Roberts and Crothers 1991). Specific binding between linear polymers can be attained by associations involving many weak contacts, however. One strategy is to have adaptive binding between two macromolecules; the free-energy change associated with the final contacts is partially offset by the free energy required either to break up a preexisting structure (Roberts and Crothers 1991), or to turn a flexible structure into a rigid one. Thus, the net free energy per final contact is reduced. As a result, the 3D shape of RNA is expected to be affected often by specific binding of proteins.

Most biological functions are performed and regulated by linear polymers like RNA and proteins. Presumably, this is because it is relatively easy for a new RNA or protein to evolve since polymerization requires only one type of bond formation, and the machinery required for this bond formation is already available. In limited cases, however, organisms have found it necessary to synthesize totally new molecules, e.g., natural products like taxol. Because such compounds have low molecular weights and relatively rigid shapes that are complementary to binding sites, they can bind tightly and specifically to either RNA or protein by forming relatively few strong contacts, perhaps with minor effects on receptor shape. Organic chemists have the capability of generating many new compounds of this type that can alter RNA function. Perhaps this will play a role in the next stage of evolution.

CONCLUSIONS

The large charge separations, polarizable electrons, and relative rigidity of nucleotides allow strong, local RNA/RNA interactions. The sequence dependence of these interactions can be quantified from thermodynamic experiments on small model systems. These results are used to model secondary structures based on sequence (Walter et al. 1994; Mathews et al. 1997; Lück et al. 1996). Correlations with known 3D structures of internal loops, hairpins, and junctions indicate that they can also be useful for modeling local 3D structure (Table 1) (Turner et al. 1988). Much more information is required for this purpose, however (Major et al. 1991). These same interactions will also be important for global folding, but other interactions, for example with metal ions, are now being revealed. Discovery of new classes of molecules having strong, specific interactions with RNA will affect future stages of evolution. Understanding stacking, hydrogen bonding, and cation interactions with RNA should facilitate design of such compounds.

ACKNOWLEDGMENTS

We thank Dr. Ming Wu for Figure 4. This work was supported by National Institutes of Health grants GM-22939 (D.H.T.) and GM-10840 (I.T.), and Department of Energy grant DE-FG03-86R60406 (I.T.).

REFERENCES

Allain F. H.-T. and Varani G. 1995. Divalent metal ion binding to a conserved wobble pair defining the upstream site of cleavage of group I self-splicing introns. *Nucleic Acids Res.* **23:** 341–350.

Antao V.P., Lai S.Y., and Tinoco Jr., I. 1991. A thermodynamic study of unusually stable RNA and DNA hairpins. *Nucleic Acids Res.* **19:** 5901–5905.

Baeyens K.J., DeBondt H.L., Pardi A., and Holbrook S.R. 1996. A curved RNA helix incorporating an internal loop with G·A and A·A non-Watson-Crick base pairing. *Proc. Natl. Acad. Sci.* **93:** 12851–12855.

Banerjee A.R., Jaeger J.A., and Turner D.H. 1993. Thermal unfolding of a group I ribozyme: The low-temperature transition is primarily disruption of tertiary structure. *Biochemistry* **32:** 153–163.

Basavappa R. and Sigler P.B. 1991. The 3Å crystal structure of yeast initiator tRNA: Functional implications in initiator/elongator discrimination. *EMBO J.* **10:** 3105–3111.

Bassi G.S., Murchie A.I.H., Walter F., Clegg R.M., and Lilley D.M.J. 1997. Ion-induced folding of the hammerhead ribozyme: A fluorescence resonance energy transfer study. *EMBO J.* **16:** 7481–7489.

Battiste J.L., Tan R., Frankel A.D., and Williamson J.R. 1994. Binding of an HIV peptide to Rev responsive element RNA induces formation of purine-purine base pairs. *Biochemistry* **33:** 2741–2747.

Bevilacqua P.C. and Turner D.H. 1991. Comparison of binding of mixed ribose-deoxyribose analogues of CUCU to a ribozyme and to GGAGAA by equilibrium dialysis: Evidence for ribozyme specific interactions with $2'$ OH groups. *Biochemistry* **30:** 10632–10640.

Bhaumik S.R. 1996. Conformational feasibility of a DNA hairpin with one-base loop. *Biochem. Biophys. Res. Commun.* **220:** 853–857.

Biou V., Yaremchuk A., Tukalo M., and Cusack S. 1994. The 2.9 Å crystal structure of *T. thermophilus* seryl-tRNA synthetase complexed with tRNA^{Ser}. *Science* **263:** 1404–1410.

Bloomfield V.A., Crothers D.M., and Tinoco Jr., I. 1999. *Nucleic acids: Structures, properties, and functions.* University Science Books, Mill Valley, California.

Borer P.N., Dengler B., Tinoco Jr., I., and Uhlenbeck O.C. 1974. Stability of ribonucleic acid double-stranded helices. *J. Mol. Biol.* **86:** 843–853.

Borer P.N., Lin Y., Wang S., Roggenbuck M.W., Gott J.M., Uhlenbeck O.C., and Pelczer I. 1995. Proton NMR and structural features of a 24-nucleotide RNA hairpin. *Biochemistry* **34:** 6488–6503.

Broitman S.L., Im D.D., and Fresco J.R. 1987. Formation of the triple-stranded polynucleotide helix, poly(A·A·U). *Proc. Natl. Acad. Sci.* **84:** 5120–5124.

Brünger A.T. 1996. *X-PLOR: A system for X-ray crystallography and NMR.* Yale University, New Haven, Connecticut.

Burkard M.E., Kierzek R., and Turner D.H. 1999. Thermodynamics of unpaired terminal nucleotides on short RNA helixes correlates with stacking at helix termini in larger RNAs. *J. Mol. Biol.* **290:** 967–982.

Butcher S., Dieckmann T., and Feigon J. 1997. Solution structure of a GAAA tetraloop receptor RNA. *EMBO J.* **16:** 7490–7499.

Cate J.H. and Doudna J.A. 1996. Metal-binding sites in the major groove of a large

ribozyme domain. *Structure* **4:** 1221–1229.
Cate J.H., Hanna R.L., and Doudna J.A. 1997. A magnesium ion core at the heart of a ribozyme domain. *Nat. Struct. Biol.* **4:** 553–558.
Cate J.H., Gooding A.R., Podell E., Zhou K., Golden B.L., Kundrot C.E., Cech T.R., and Doudna J.A. 1996. Crystal structure of a group I ribozyme domain: Principles of RNA packing. *Science* **273:** 1678–1685.
Cate J.H., Gooding, A.R., Podell C., Zhou K., Golden B.L., Szewczak A.A., Kundrot C.E., Cech T.R., and Doudna J.A. 1996. RNA tertiary structure mediation by adenosine platforms. *Science* **273:** 1696–1699.
Chang K.-Y. and Varani G. 1997. Nucleic acids structure and recognition. *Nat. Struct. Biol.* **4:** 854–858.
Chastain, M. and Tinoco Jr., I. 1992. A base-triple structural domain in RNA. *Biochemistry* **31:** 12733–12741.
———. 1993. Nucleoside triples from the group I intron. *Biochemistry* **32:** 14220–14228.
Christian E.L. and Yarus M. 1993. Metal coordination sites that contribute to structure and catalysis in the group I intron from *Tetrahymena*. *Biochemistry* **32:** 4475–4480.
Cornell W.D., Cieplak P., Bayly C.I., Gould I.R., Merz Jr., K.M., Ferguson D.M., Spellmeyer D.C., Fox T., Caldwell J.W., and Kollman P.A. 1995. A second generation force field for the simulation of proteins, nucleric acids, and organic molecules. *J. Am. Chem. Soc.* **117:** 5179–5197.
Correll C.C., Freeborn B., Moore P.B., and Steitz T.A. 1997. Metals, motifs, and recognition in the crystal structure of a 5S rRNA domain. *Cell* **91:** 705–712.
Costa M. and Michel F. 1997. Rules for RNA recognition of GNRA tetraloops deduced by in vitro selection: Comparison with in vivo evolution. *EMBO J.* **16:** 3289–3302.
Crothers D.M., Cole P.E., Hilbers C.W., and Shulman R.G. 1974. The molecular mechanism of thermal unfolding of *Escherichia coli* formylmethionine transfer RNA. *J. Mol. Biol.* **87:** 63–88.
Davies J., von Ahsen U., and Schroeder R. 1993. Antibiotics and the RNA world: A role for low-molecular-weight effectors in biochemical evolution? In *The RNA world* (ed. R.F. Gesteland and J.F. Atkins), pp. 185–204. Cold Spring Harbor Laboratory Press, Cold Spring Harbor, New York.
Davis R.C. and Tinoco Jr., I. 1968. Temperature-dependent properties of dinucleoside phosphates. *Biopolymers* **6:** 223–242.
Davis P.W., Thurmes W., and Tinoco Jr., I. 1993. Structure of a small RNA hairpin. *Nucleic Acids Res.* **21:** 537–545.
Davison A. and Leach D.R. 1994. Two-base DNA hairpin-loop structures in vivo. *Nucleic Acids Res.* **22:** 4361–4363.
Doudna J. and Cech T.R. 1995. Self-assembly of a group I intron active site from its component tertiary structural domains. *RNA* **1:** 36–45.
Emerick V.L. and Woodson S.A. 1993. Self-splicing of the *Tetrahymena* pre-rRNA is decreased by misfolding during transcription. *Biochemistry* **32:** 14062–14067.
Feig A.L., Scott W.G., and Uhlenbeck O.C. 1998. Inhibition of the hammerhead ribozyme cleavage reaction by site-specific binding of Tb(III). *Science* **279:** 81–84.
Felsenfeld G., Davies D.R., and Rich A. 1957. Formation of a three-stranded polynucleotide molecule. *J. Am. Chem. Soc.* **79:** 2023–2024.
Fountain M.A., Serra M.J., Krugh T.R., and Turner D.H. 1996. Structural features of a six-nucleotide RNA hairpin loop found in ribosomal RNA. *Biochemistry* **35:** 6539–6548.

Freier S.M., Kierzek R., Jaeger J.A., Sugimoto N., Caruthers M.H., Neilson T., and Turner D.H. 1986. Improved free-energy parameters for predictions of RNA duplex stability. *Proc. Natl. Acad. Sci.* **83**: 9373–9377.

Gautheret D.F., Konings D., and Gutell R.R. 1994. A major family of motifs involving G·A mismatches in ribosomal RNA. *J. Mol. Biol.* **242**: 1–8.

Gesteland R.F. and Atkins J.F., eds. 1993. *The RNA world*. Cold Spring Harbor Laboratory Press, Cold Spring Harbor, New York.

Gutell R.R. 1994. Collection of small subunit (16S- and 16S-like) ribosomal RNA structures. *Nucleic Acids Res.* **22**: 3502–3507.

Gutell R.R., Gray M.W., and Schnare M.N. 1993. A compilation of large subunit (23S and 23S-like) ribosomal RNA structures. *Nucleic Acids Res.* **21**: 3055–3074.

Harris M.E., Nolan J.M., Malhotra A., Brown J.W., Harvey S.C., and Pace N.R. 1994. Use of photoaffinity crosslinking and molecular modeling to analyze the global architecture of ribonuclease P RNA. *EMBO J.* **13**: 3953–3963.

Herschlag D. 1991. Implications of ribozyme kinetics for targeting the cleavage of specific RNA molecules in vivo: More isn't always better. *Proc. Natl. Acad. Sci.* **88**: 6921–6925.

Heus H.A. and Pardi A. 1991. Structural features that give rise to the unusual stability of RNA hairpins containing GNRA loops. *Science* **253**: 191–194.

Huang S., Wang Y.X., and Draper D.E. 1996. Structure of a hexanucleotide RNA hairpin loop conserved in ribosomal RNAs. *J. Mol. Biol.* **258**: 308–321.

Jucker F.M. and Pardi A. 1995a. Solution structure of the CUUG hairpin loop: A novel RNA tetraloop motif. *Biochemistry* **34**: 14416–14427.

———. 1995b. GNRA tetraloops make a U-turn. *RNA* **1**: 219–222.

Jucker F.M., Heus H.A., Yip P.F., Moors E.H.M., and Pardi A. 1996. A network of heterogenous hydrogen bonds in GNRA tetraloops. *J. Mol. Biol.* **264**: 968–980.

Kazakov S.A. 1996. Nucleic acid binding and catalysis by metal ions. In *Bioorganic chemistry: Nucleic acids* (ed. S. Hecht), pp. 244–287. Oxford University Press, New York.

Kieft J.S. and Tinoco Jr., I. 1997. Solution structure of a metal-binding site in the major groove of RNA complexed with cobalt (III) hexammine. *Structure* **5**: 713–721.

Kim J., Walter A.E., and Turner D.H. 1996. Thermodynamics of coaxially stacked helixes with GA and CC mismatches. *Biochemistry* **35**: 13753–13761.

Kim S.-H., Quigley G.J., Suddath F.L., McPherson A., Sneden D., Kim J.J., Weinzierl J., and Rich A. 1973. Three-dimensional structure of yeast phenylalanine transfer RNA: Folding of the polynucleotide chain. *Science* **179**: 285–288.

Kolk M.H., Heus H.A., and Hilbers C.W. 1997. The structure of the isolated, central hairpin of the HDV antigenomic ribozyme: Novel structural features and similarity of the loop in the ribozyme and free in solution. *EMBO J.* **16**: 3685–3692.

Laing L.G., Gluick T.C., and Draper D.E. 1994. Stabilization of RNA structure by Mg ions. Specific and non-specific effects. *J. Mol. Biol.* **237**: 577–587.

Lee J.S., Johnson D.A., and Morgan A.R. 1979. Complexes formed by (pyrimidine)$_n$: (purine)$_n$ DNAs on lowering the pH are three-stranded. *Nucleic Acids Res.* **6**: 3073–3091.

Lehnert V., Jaeger L., Michel F., and Westhof E. 1996. New loop-loop tertiary interactions in self-splicing introns of subgroup IC and ID: A complete 3D model of the *Tetrahymena thermophila* ribozyme. *Chem. Biol.* **3**: 993–1009.

Leonard G.A., McAuley-Hecht K., Brown T., and Hunter W.N. 1995. Do C—H···O hydrogen bonds contribute to the stability of nucleic acid base pairs? *Acta Cryst.* **D51**: 136–139.

Li Y. and Turner D.H. 1997. Effects of Mg^{++} and the 2'OH of guanosine on steps required for substrate binding and reactivity with the *Tetrahymena* ribozyme reveal several local folding transitions. *Biochemistry* **36:** 11131–11139.

Li Y., Bevilacqua P.C., Mathews D., and Turner D.H. 1995. Thermodynamic and activation parameters for binding of a pyrene-labeled substrate by the *Tetrahymena* ribozyme: Docking is not diffusion-controlled and is driven by a favorable entropy change. *Biochemistry* **34:** 14394–14399.

Lipsett M.N. 1964. Complex formation between polycytidylic acid and guanine oligonucleotides. *J. Biol. Chem.* **239:** 1256–1260.

Louise-May S., Auffinger P., and Westhof E. 1996. Calculations of nucleic acid conformations. *Curr. Opin. Struct. Biol.* **6:** 289–298.

Lück R., Steger G., and Riesner D. 1996. Thermodynamic prediction of conserved secondary structure: Application to the RRE element of HIV, the tRNA-like element of CMV and the mRNA of prion protein. *J. Mol. Biol.* **258:** 813–826.

Luebke K.J. and Tinoco Jr., I. 1996. Sequence effects on RNA bulge-induced helix bending and a conserved five-nucleotide bulge from the group I introns. *Biochemistry* **35:** 11677–11684.

Luebke K.J., Landry S.M., and Tinoco Jr., I. 1997. Solution conformation of a five-nucleotide RNA bulge loop from a group I intron. *Biochemistry* **36:** 10246–10255.

Lynch S.R. and Tinoco Jr., I. 1998. The structure of the L3 loop from the hepatitis delta virus ribozyme: A *syn* cytidine. *Nucleic Acids Res.* **26:** 980–987.

Major F., Turcotte M., Gautheret D., Lapalme G., Fillion E., and Cedergren R. 1991. The combination of symbolic and numerical computation for three-dimensional modeling of RNA. *Science* **253:** 1255–1260.

Manning G.S. 1978. The molecular theory of polyelectrolyte solutions with applications to the electrostatic properties of polynucleotides. *Q. Rev. Biophys.* **11:** 179–246.

Mathews D.H., Banerjee A.R., Luan D.D., Eickbush T.H., and Turner D.H. 1997. Secondary structure model of the RNA recognized by the reverse transcriptase from the R2 retrotransposable element. *RNA* **3:** 1–16.

Michel F. and Westhof E. 1990. Modelling of the three-dimensional architecture of group I catalytic introns based on comparative sequence analysis. *J. Mol. Biol.* **216:** 585–610.

Moore P.B. 1995. Determination of RNA conformation by nuclear magnetic resonance. *Accts. Chem. Res.* **28:** 251–256.

Moran S., Ren R.X.F., and Kool E. T. 1997. A thymidine triphosphate shape analog lacking Watson-Crick pairing ability is replicated with high sequence selectivity. *Proc. Natl. Acad. Sci.* **94:** 10506–10511.

Murphy F.L. and Cech T.R. 1993. An independently folding domain of RNA tertiary structure within the *Tetrahymena* ribozyme. *Biochemistry* **32:** 5291–5300.

Olmsted M.C., Anderson C.F., and Record M.T. Jr. 1991. Importance of oligoelectrolyte end effects for the thermodynamics of conformational transitions of nucleic acid oligomers: A grand canonical Monte Carlo analysis. *Biopolymers* **31:** 1593–1604.

Pauling L. 1960. *The nature of the chemical bond*, 3rd edition, chapt. 12. Cornell University Press, Ithaca, New York.

Petersheim M. and Turner D.H. 1983. Base-stacking and base-pairing contributions to helix stability: Thermodynamics of double-helix formation with CCGG, CCGGp, CCGGAp, ACCGGp, CCGGUp, and ACCGGUp. *Biochemistry* **22:** 256–263.

Peterson R.D. and Feigon J. 1996. Structural change in Rev responsive element RNA of

HIV-1 on binding Rev peptide. *J. Mol. Biol.* **264:** 863–877.
Piccirilli J.A., Vyle J.S., Caruthers M.H., and Cech T.R. 1993. Metal ion catalysis in the *Tetrahymena* ribozyme reaction. *Nature* **361:** 85–88.
Pley H.W., Flaherty K.M., and McKay D.B. 1994a. Three-dimensional structure of a hammerhead ribozyme. *Nature* **372:** 68–74.
———. 1994b. Model for an RNA tertiary interaction from the structure of an intermolecular complex between a GAAA tetraloop and an RNA helix. *Nature* **372:** 111–113.
Profenno L.A., Kierzek R., Testa S.M., and Turner D.H. 1997. Guanosine binds to the *Tetrahymena* ribozyme in more than one step, and its $2'$-OH and the nonbridging *pro*-*S*p phosphoryl oxygen at the cleavage site are required for productive docking. *Biochemistry* **36:** 12477–12485.
Puglisi J.D., Wyatt J.R., and Tinoco Jr., I. 1990. Solution conformation of an RNA hairpin loop. *Biochemistry* **29:** 4215–4226.
Puglisi J.D., Tan R., Calnan B.J., Frankel A.D., and Williamson J.R. 1992. Conformation of the TAR RNA-arginine complex by NMR spectroscopy. *Science* **257:** 76–80.
Pyle A.M. and Cech T.R. 1991. Ribozyme recognition of RNA by tertiary interactions with specific ribose $2'$-OH groups. *Nature* **350:** 628–631.
Pyle A.M., Murphy F.L., and Cech T.R. 1992. RNA substrate binding site in the catalytic core of the *Tetrahymena* ribozyme. *Nature* **358:** 123–128.
Quigley G.J. and Rich A. 1976. Structural domains of transfer RNA molecules. The ribose $2'$ hydroxyl which distinguishes RNA from DNA plays a key role in stabilizing tRNA structure. *Science* **194:** 796–806.
Ramos A., Gubser C.C., and Varani G. 1997. Recent solution structures of RNA and its complexes with drugs, peptides and proteins. *Curr. Opin. Struct. Biol.* **7:** 317–323.
Ray J. and Manning G.S. 1994. An attractive force between two rodlike polyions mediated by the sharing of condensed counterions. *Langmuir* **10:** 2450–2461.
Record Jr., M.T., Anderson C.F., and Lohman T.M. 1978. Thermodynamic analysis of ion effects on the binding and conformational equilibra of proteins and nucleic acids: The roles of ion association or release, screening, and ion effects on water activity. *Q. Rev. Biophys.* **11:** 103–178.
Riordan F.A., Bhattacharyya A., McAteer S., and Lilley D.M. 1992. Kinking of RNA helices by bulged bases, and the structure of the human immunodeficiency virus transactivator response element. *J. Mol. Biol.* **226:** 305–310.
Roberts R.W. and Crothers D.M. 1991. Specificity and stringency in DNA triplex formation. *Proc. Natl. Acad. Sci.* **88:** 9397–9401.
Robertus J.D. Ladner J.E., Finch J.T., Rhodes D., Brown R.S., Clark B.F.C., and Klug A. 1974. Structure of yeast phenylalanine tRNA at 3Å resolution. *Nature* **250:** 546–551.
Rouzina I. and Bloomfield V.A. 1996. Macroion attraction due to electrostatic correlation between screening counterions. *J. Phys. Chem.* **100:** 9977–9989.
Ruffner D.E. and Uhlenbeck O.C. 1990. Thiophosphate interference experiments locate phosphates important for the hammerhead RNA self-cleavage reaction. *Nucleic Acids Res* **18:** 6025–6029.
Saenger W. 1984. *Principles of nucleic acid structure*. Springer-Verlag, New York.
SantaLucia Jr., J. and Turner D.H. 1993. Structure of (rGGCGAGCC)$_2$ in solution from NMR and restrained molecular dynamics. *Biochemistry* **32:** 12612–12623.
SantaLucia Jr., J., Kierzek R., and Turner D.H. 1991. Functional group substitutions as probes of hydrogen bonding between GA mismatches in RNA internal loops. *J. Am.*

Chem. Soc. **113:** 4313–4322.
———. 1992. Context dependence of hydrogen bond free energy revealed by substitutions in an RNA hairpin. *Science* **256:** 217–219.
Schweisguth D.C. and Moore P.B. 1997. On the conformation of the anticodon loops of initiator and elongator methionine tRNAs. *J. Mol. Biol.* **267:** 505–519.
Scott W.G., Finch J.T., and Klug A. 1995. The crystal structure of an all-RNA hammerhead ribozyme: A proposed mechanism for RNA catalytic cleavage. *Cell* **81:** 991–1002.
Scott W.G., Murray J.B., Arnold J.R.P., Stoddard B.L., and Klug A. 1996. Capturing the structure of a catalytic RNA intermediate: The hammerhead ribozyme. *Science* **274:** 2065–2069.
Serra M.J. and Turner D.H. 1995. Predicting thermodynamic properties of RNA. *Methods Enzymol.* **259:** 242–261.
Simons R.W. and Grunberg-Manago M., eds. 1998. *RNA structure and function.* Cold Spring Harbor Laboratory Press, Cold Spring Harbor, New York.
Stallings S.C. and Moore P.B. 1997. The structure of an essential splicing element: Stem loop IIa from yeast U2 snRNA. *Structure* **5:** 1173–1185.
Strobel S.A. and Cech T.R. 1993. Tertiary interactions with the internal guide sequence mediate docking of the P1 helix into the catalytic core of the *Tetrahymena* ribozyme. *Biochemistry* **32:** 13593–13604.
———. 1995. Minor groove recognition of the conserved G·U pair at the *Tetrahymena* ribozyme reaction site. *Science* **267:** 675–679.
Sugimoto N., Tomka M., Kierzek R., Bevilacqua P.C., and Turner D.H. 1989. Effects of substrate structure on the kinetics of circle opening reactions of the self-splicing intervening sequence from *Tetrahymena thermophila*: Evidence for substrate and Mg^{++} binding interactions. *Nucleic Acids Res.* **17:** 355–371.
Sugimoto N., Nakano S., Katoh M., Matsumura A., Nakamuta H., Ohmichi T., Yoneyama M., and Sasaki M. 1995. Thermodynamic parameters to predict stability of RNA/DNA hybrid duplexes. *Biochemistry* **34:** 11211–11216.
Sussman J.L., Holbrook S.R., Warrant R.W., Church G.M., and Kim S.H. 1978. Crystal structure of yeast phenylalanine transfer RNA. I. Crystallographic refinement. *J. Mol. Biol.* **123:** 607–630.
Szostak J.W. 1986. Enzymatic activity of the conserved core of a group I self-splicing intron. *Nature* **322:** 83–86.
Tanford C. 1980. *The hydrophobic effect: formation of micelles and biological membranes.* Wiley, New York.
Tinoco Jr., I. 1993. Appendix 1: Structures of base pairs involving at least two hydrogen bonds. In *The RNA world* (ed. R.F. Gesteland and J.F. Atkins), pp. 603–607. Cold Spring Harbor Laboratory Press, Cold Spring Harbor, New York.
Turner D.H. and Bevilacqua P.C. 1993. Thermodynamic considerations for evolution by RNA. In *The RNA world* (ed. R.F. Gesteland and J.F. Atkins), pp. 447–464. Cold Spring Harbor Laboratory Press, Cold Spring Harbor, New York.
Turner D.H., Sugimoto N., and Freier S.M. 1988. RNA structure prediction. *Annu. Rev. Biophys. Biophys. Chem.* **17:** 167–192.
Turner D.H., Sugimoto N., Kierzek R., and Dreiker S.D. 1987. Free energy increments for hydrogen bonds in nucleic acid base pairs. *J. Am. Chem. Soc.* **109:** 3783–3785.
Uhlenbeck O.C. 1995. Keeping RNA happy. *RNA* **1:** 4–6.
van den Hoogen Y.T., van Beuzekom A.A., de Vroom E., van der Marel G.A., van Boom

J.H., and Altona C. 1988. Bulge-out structures in the single-stranded trimer AUA and in the duplex (CUGGUGCGG)·(CCGCCCAG). A model-building and NMR study. *Nucleic Acids Res.* **16:** 5013–5030.

Varani G. 1995. Exceptionally stable nucleic acid hairpins. *Annu. Rev. Biophys. Biomol. Struct.* **24:** 379–404.

Wahl M.C. and Sundaralingam M. (1997). C-H···O hydrogen bonding in biology. *Trends Biochem. Sci.* **22:** 97–102.

Walter A.E. and Turner D.H. 1994. Sequence dependence of stability for coaxial stacking of RNA helixes with Watson-Crick base paired interfaces. *Biochemistry* **33:** 12715–12719.

Walter A.E., Turner D.H., Kim J., Lyttle M.H., Müller P., Mathews D.H., and Zuker M. 1994. Coaxial stacking of helices enhances binding of oligoribonucleotides and improves predictions of RNA folding. *Proc. Natl. Acad. Sci.* **91:** 9218–9222.

Westhof E. and Sundaralingam M. 1986. Restrained refinement of the monoclinic form of yeast phenylalanine transfer RNA. Temperature factors and dynamics, coordinated waters, and base-pair propeller twist angles. *Biochemistry* **25:** 4868–4878.

Westhof E., Dumas P., and Moras D. 1988. Restrained refinement of two crystalline forms of yeast aspartic acid and phenylalanine transfer RNA crystals. *Acta Crystallogr. A* **44:** 112–123.

Woese C.R., Winker S., and Gutell R.R. 1990. Architecture of ribosomal RNA: Constraints on the sequence of "tetra-loops." *Proc. Natl. Acad. Sci.* **87:** 8467–8471.

Wu M. and Turner D.H. 1996. Solution structure of (rGCGGACGC)$_2$ by two-dimensional NMR and the iterative relaxation matrix approach. *Biochemistry* **35:** 9677–9689.

Wu M., SantaLucia Jr., J., and Turner D.H. 1997. Solution structure of (rGGCAGGCC)$_2$ by two-dimensional NMR and the iterative relaxation matrix approach. *Biochemistry* **36:** 4449–4460.

Xia T., McDowell J.A., and Turner D.H. 1997. Thermodynamics of nonsymmetric tandem mismatches adjacent to G·C base pairs in RNA. *Biochemistry* **36:** 12486–12497.

Zacharias M. and Hagerman P.J. 1995a. The bend in RNA created by the *trans*-activation response element bulge of human immunodeficiency virus is straightened by arginine and by *tat*-derived peptide. *Proc. Natl. Acad. Sci.* **92:** 6052–6056.

———. 1995b. Bulge-induced bends in RNA: Quantification by transient electric birefringence. *J. Mol. Biol.* **247:** 486–500.

Zhu L., Chou S.H., Xu J., and Reid B.R. 1995. Structure of a single-cytidine hairpin loop formed by the DNA triplet GCA. *Nat. Struct. Biol.* **2:** 1012–1017.

11
Small Ribozymes

David B. McKay and Joseph E. Wedekind
Department of Structural Biology
Stanford University School of Medicine
Stanford, California 94305-5126

Often referred to as "small" ribozymes, the hammerhead (Forster and Symons 1987a), hairpin (Buzayan et al. 1986), hepatitis delta virus (HDV) (Kuo et al. 1988), and *Neurospora* VS ribozymes (Saville and Collins 1990) all cleave a RNA phosphodiester backbone to yield a 5' hydroxyl and a 2', 3' cyclic phosphodiester as product). Since the reaction produces a cyclic phosphodiester, but does not hydrolyze it (thereby undergoing only the first step of a classic ribonuclease reaction and approximately conserving the substrate bond energy in the products), these ribozymes can also ligate the products to re-form a phosphodiester. Each of these ribozymes has been derived from a self-cleavage activity of linear RNA intermediates in the rolling circle replication cycle of small satellite RNAs, or in the case of the *Neurospora* VS ribozyme, of the mitochondrial Varkud satellite plasmid. The native function of these catalytic motifs is to cleave a linear RNA having multiple tandem (plus or minus) copies of the satellite sequence into segments of unit length prior to their ligation into circles.

Given such a biological function, the activity is required to be precise with regard to site of cleavage and to yield a product whose ends can readily be ligated, but it is not required to be kinetically rapid. In the discussions below, it will become apparent that cleavage-site specificity is often achieved through Watson-Crick base-pairing between RNA sequences flanking the target phosphodiester bond and those flanking the catalytic "core" of the ribozyme (although the *Neurospora* VS ribozyme appears to be an exception to this trend). For all of these ribozymes, the time scale of the chemical step of bond cleavage is on the order of one minute; although this is slow compared to typical enzymatic activities, it is rapid compared to the time scale of a viral infection cycle. Consequently, in addition to being small in size when compared to group I and group II introns or RNAse P, members of this collection of ribozymes all catalyze the same single-step reaction, with relatively slow rate constants. In this context, it might be more suitable to refer to these as the "small, slow, and simple" ribozymes—Nature's most rudimentary enzymes.

This chapter is a general overview of what is currently known about the structure and mechanism of this group of small catalytic RNAs. Detailed discussions on the hammerhead (Wedekind and McKay 1998), hairpin (Burke 1996), and HDV (Been and Wickham 1997) ribozymes are available for more extensive information.

FROM SINGLE TURNOVER SELF-CLEAVING ACTIVITY TO MULTIPLE TURNOVER ENZYME

Linear RNAs that are several hundred nucleotides in length, albeit the native context of the self-cleaving activities, are unwieldy for biochemical and structural studies. Consequently, it has been necessary to extract, for each of these systems, a minimal essential core of nucleotides that retains the cleavage activity. Figure 1 shows representative core sequences and predicted secondary structures for each of the four ribozymes. Delineation of minimal sequence requirements was first accomplished with the hammerhead, where comparisons of several self-cleaving satellite RNA sequences revealed an approximate consensus for the requisite catalytic core, consisting of thirteen conserved nucleotides at the junction of three duplex stems (Forster and Symons 1987b). On the basis of this consensus, it was possible to design and synthesize small RNAs that included the proposed secondary structure and core. This was done with bipartate constructs, one of which had stem III of the hammerhead (as shown in Fig. 1a) closed with a loop and stems I and II open-ended (Uhlenbeck 1987), and a second of which had stem II closed and the other two open-ended (Haseloff and Gerlach 1988). Both constructs were active with the ribozyme and substrate strands present as separate molecules (in *trans*). Furthermore, under suitable conditions, both types of constructs catalyzed multiple turnover cleavage, a requisite characteristic of a bona fide enzymatic activity. Defining the RNA strand harboring the cleavage site as the "substrate" and the complementary strand of a bipartate construct as "enzyme," it was then possible to analyze the steady-state kinetics of these constructs with a standard Michaelis-Menten model, yielding the classic parameters K_m and k_{cat} for the reaction. (Although it has become colloquial to refer to such bipartate constructs as being enzyme and substrate, since one is cleaved and the other is not, it should be appreciated that this partitioning of function is inaccurate, since both RNA strands contribute elements of the catalytic core that is required for ribozyme activity.)

These small hammerhead ribozymes gave credibility to the proposal, based on sequence comparisons, that the activity is housed in a minimal core at the junction of three Watson-Crick base-paired stems. The activity tolerates substantial variation of the lengths, sequences, and connec-

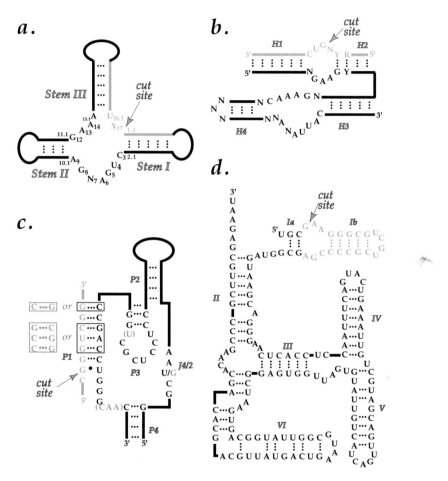

Figure 1 Schematic representations of the approximate minimal core sequences and predicted secondary structures of four small ribozymes. Triple dots represent Watson-Crick base-pair interactions. Stem labels are in *green* letters; the scissile bonds are shown by a *green arrow*; approximate minimal substrate strands are labeled in *blue*; *red* indicates positions of the hepatitis delta virus present in the genomic RNA that are either not present in the antigenomic RNA as indicated by parentheses (insertions) or take the place of a corresponding antigenomic sequence, labeled in *blue* or *black*. (U) Uracil; (A) adenine; (C) cytosine; (G) guanine. (*a*) The hammerhead, (*b*) hairpin, (*c*) hepatitis delta virus, and (*d*) *Neurospora* VS ribozymes.

tivity of the stems, as long as the consensus sequence within the core is retained. This concept—that ribozymes are constituted of small nonduplex segments of RNA of defined sequence connected by duplex stems—has subsequently been extended to delineate similar minimal sequence requirements for activity in the hairpin, HDV, and *Neurospora* VS ribozymes.

For the hairpin ribozyme, comparison of self-cleaving sequences from several virus satellite RNAs (tobacco ringspot virus [Buzayan et al. 1986], arabis mosaic virus [DeYoung et al. 1995], and chicory yellow mottle virus [Rubino et al. 1990]) suggested there were two cores in duplex stems, possibly with a flexible hinge region between them (Fig. 1b). The consensus sequence was refined further by in vitro selection methods (Berzal-Herranz et al. 1992, 1993). When a hairpin ribozyme is partitioned into two fragments, the first of which includes stems H1 and H2 with their concomitant nonduplex region, and the second of which includes stems H3 and H4 with their nonduplex region, the ribozyme still cleaves when sufficiently high concentrations of both fragments are present (Butcher et al. 1995).

In the cases of the HDV (Perrotta and Been 1990, 1991) and *Neurospora* VS (Guo et al. 1993) ribozymes, where only one or two naturally occurring representatives are known, a minimal core has been delimited by pruning away nonessential sequences by systematic deletion. Interestingly, the minimal substrate found for the *Neurospora* VS ribozyme is a stem-loop (Fig. 1d), rather than a linear segment of RNA, providing an example where tertiary interactions other than simple Watson-Crick base-pairing may be important for cleavage-site recognition (Guo and Collins 1995).

At this time, approximate minimal core sequences and anticipated secondary structures are known for all four ribozymes. The lack of any recognizable sequence or secondary-structure similarity between them suggests that these ribozymes also have different tertiary structures.

A Michaelis-Menten analysis of steady-state kinetics provides an overall parameterization of an enzymatic reaction, but suffers the disadvantage that the values of K_m and k_{cat} may include significant and indistinguishable contributions from several reversible reaction steps—substrate binding and release, bond cleavage and ligation, product release and re-binding. Delineation of a minimal kinetic scheme for a ribozyme activity, in which the rate constant of each individual step of the reaction is determined, separates the rate of the chemical cleavage activity from the rates of binding and release of substrate and product. Complete description of a minimal kinetic scheme was first accomplished for the hammerhead ribozyme, using a construct with the strands of stem II connected with a loop and duplexes of eight base pairs in stems I and III (Hertel et al. 1994). For such a construct, release of the products after cleavage of the substrate strand is one to two orders of magnitude slower than the rate of bond cleavage. With this system it was found that, to a good approximation, binding and release of substrate and product follow the kinetics expected for formation and dissociation of standard Watson-Crick duplexes in the stem regions. The rate of the chemical step, i.e., bond cleavage, was found to be essentially the same for different hammerheads under a given set of reaction conditions. For example, in conditions that have become relatively standard, i.e., 10 mM $MgCl_2$, 40–50 mM Tris-HCl, pH 7.5, 25°C, cleavage rates reported by multiple groups for at least 20 different hammerhead constructs span only a relatively narrow range of 0.25 min^{-1} to 2.8 min^{-1} (for tabulation, see McKay 1996; Clouet-d'Orval and Uhlenbeck 1997).

Recently, deviations from this trend have been found with variants of the hammerhead. In particular, the rate of cleavage is sensitive to the sequence of the first two base pairs adjacent to the core in stem I and can vary as much as tenfold, with the more rapid rates occurring with A-U base pairs at these positions and the slower rates occurring with G-C base pairs (Clouet-d'Orval and Uhlenbeck 1997). The effect of the stem I sequences illustrates that the historical perception of the hammerhead as having a distinct catalytic core at the junction of three stems is only partially accurate.

The chemical cleavage step of the hammerhead reaction is reversible; under the standard conditions described above, the equilibrium is 130-fold in favor of product (Table 1) (Hertel and Uhlenbeck 1995). For free substrate and product strands in solution, the bond-energy difference between product and substrate would favor substrate slightly, since the cyclic 2′,3′-phosphodiester bond of the product is somewhat strained relative to the 3′,5′-phosphodiester bond of the substrate. For substrate and product strands bound to the ribozyme, the shift of equilibrium in the direction of product is due to a large increase in entropy upon bond cleavage, interpreted qualitatively as the ends of the product RNA strands being "floppy" after cleavage. The specific value of this equilibrium constant depends on the particular divalent metal ion present in the reaction and has a hyperbolic dependence on ion concentration, suggesting that divalent ion dissociation contributes to the free-energy difference of the product/substrate equilibrium and drives the reaction toward product (Long et al. 1995).

It is notable that the cleavage rates for representatives of all four small ribozymes are of the same magnitude, ~1 min^{-1} (Table 1). It is equally

Table 1 Representative kinetic parameters for small ribozymes

Ribozyme	k_{cleave} (min^{-1})	K_{eq}^a	Phosphoro-thioate effect?[b]	Log-linear rate dependence on pH?	Assay	References
Hammerhead	1.0	0.008	yes	yes	50 mM Tris-HCl 10 mM MgCl$_2$ pH 7.5 25°C	Ruffner and Uhlenbeck (1990); Dahm et al. (1993); Hertel et al. (1994); Hertel and Uhlenbeck (1995)
Hairpin	0.45	4.4	no	no	50 mM Tris-HCl 10 mM MgCl$_2$ 0.1 mM EDTA pH 7.5 25°C	Nesbitt et al. (1997)
Hepatitis delta virus	2.1	NR[d]	no	yes (pH 4 to 6)	50 mM Tris-HCl 10 mM MgCl$_2$ pH 7.4 37°C	Fauzi et al. (1997)
Neurospora VS	0.7[c]	NR[d]	NR[d]	no	50 mM Tris-HCl 10 mM MgCl$_2$ pH 7.5 37°C	Collins and Olive (1993); Guo and Collins (1995)

[a]K_{eq}, the equilibrium constant for the bond-cleavage step, is defined as the ratio of ribozyme-bound substrate to product at equilibrium, and is equal to k_{cleave}/k_{ligate}.
[b]Phosphorothioate effect is defined as (i) a significant loss of activity when Mn^{++} is substituted for Mg^{++}; (ii) a restoration of activity in presence of Mg^{++} when phosphorothioates are introduced, coupled with
[c]Estimated from steady-state k_{cat}.
[d]NR indicates not reported.

notable that the equilibrium between product and substrate for the hairpin ribozyme is inverted relative to that of the hammerhead, with ligation being favored almost fivefold over cleavage (Hegg and Fedor 1995; Nesbitt et al. 1997). This suggests that the hairpin ribozyme constrains the RNA backbone at the cleavage site so that re-ligation is facile; presumably the ends of the product strands cannot flop around as they can in the hammerhead.

THE CHEMICAL ASPECTS OF CATALYSIS

Once the self-cleaving activity of a large RNA has been reduced to a minimal, well-defined ribozyme, experiments to delineate what is required for catalytic activity become feasible. The activity of small ribozymes is optimized by, and in most cases strictly dependent on, divalent cations such as Mg^{++}. For example, hammerhead activity is maximal at $[Mg^{++}] \geq$ 10 mM. In this context, significant effort has been focused on determining the interactions of divalent ions with the ribozymes. Historically, these interactions have been divided into two classes, "structural"—those in which an ion contributes to maintaining a ribozyme in an active conformation but does not participate directly in catalysis, and "catalytic"—ions which are thought to be directly involved in the chemistry of bond cleavage.

Biochemical studies using mimics of hydrated Mg^{++} ions (or more explicitly, octahedrally coordinated $Mg(H_2O)_6^{++}$ in solution) have given information on divalent ion interactions with small ribozymes. Replacing $Mg(H_2O)_6^{++}$ with $Co(NH_3)_6^{+++}$, which is thought to bind in a manner similar to $Mg(H_2O)_6^{++}$ but whose ligands are relatively substitution-inert, discriminates between "inner sphere" interactions, in which the metal ion ligates to the RNA directly, and "outer sphere" interactions, which are mediated through a bridging H_2O molecule (or in the case of $Co(NH_3)_6^{+++}$, through a bridging amine group). The hairpin ribozyme is equally active in 10 mM $MgCl_2$ and 0.1 mM $Co(NH_3)_6Cl_3$, which argues against inner-sphere-metal–RNA interactions being required for activity (Hampel and Cowan 1997; Nesbitt et al. 1997; Young et al. 1997).

In contrast, inner-sphere-metal interactions are important for hammerhead ribozyme activity (Ruffner and Uhlenbeck 1990; Knoll et al. 1997). Interactions of Mg^{++} ions with the nonbridging phosphate oxygens of RNAs have been defined with phosphorothioate substitution/Mn^{++} rescue experiments. In this method, nonbridging phosphate oxygens are replaced with sulfur, either by in vitro transcription (in which T7 RNA polymerase selectively utilizes the *pro*-Sp isomer of α-S-NTPs and, by inversion of configuration in the polymerization reaction, incorporates *pro*-Rp thiophosphates in a polynucleotide strand), or by sulfurization during an oxidation step of chemical synthesis. A loss of activity in the presence of Mg^{++} (due to the orders-of-magnitude lower affinity of Mg^{++} for sulfur than for oxygen), in combination with rescue of activity by Mn^{++} (which has approximately equal affinity for oxygen and sulfur), indicates a likely inner-sphere interaction with the divalent ion (see Feig and Uhlenbeck, this volume).

Thiophosphate substitution experiments with the hammerhead have indicated inner-sphere interactions of divalent ions with the *pro*-Rp nonbridging phosphate oxygen of the scissile bond (nucleotide $N_{1.1}$ of stem I) (van Tol et al. 1990; Koizumi and Ohtsuka 1991; Slim and Gait 1991). Substitution of the other nonbridging phosphate oxygen (*pro*-Sp) of the scissile bond with sulfur does not diminish cleavage activity. These experiments identify the *pro*-Rp nonbridging oxygen of the scissile phosphodiester bond as a ligand for a catalytic Mg^{++} ion, as shown schematically in Figure 2.

In contrast, substitution of either nonbridging oxygen of the scissile phosphodiester bond with sulfur for the hairpin (Chowrira and Burke 1992; Hampel and Cowan 1997; Nesbitt et al. 1997; Young et al. 1997) and HDV (Jeoung et al. 1994) ribozymes does not significantly reduce the rate of bond cleavage. If a catalytic metal ion is essential for the cleavage activity of these ribozymes, it must bind in the active site in a manner that is different from the hammerhead.

Further thiophosphate substitution experiments with the hammerhead have demonstrated binding of (presumably structural) divalent ions to the *pro*-Rp oxygen of the phosphates of nucleotides A_9, A_{13}, and A_{14} of the catalytic core (Ruffner and Uhlenbeck 1990; Knoll et al. 1997). The binding at position A_9 has been confirmed crystallographically (Pley et al. 1994; Scott et al. 1995).

The pH dependence of the ribozyme activity can suggest whether a hydroxide ion may participate as a proton acceptor in the bond-cleavage reaction. In the case of the hammerhead, the cleavage rate in the presence of Mg^{++} increases as a function of pH over the range 5.5–8.5, with a slope of ~1 in a log linear plot (Dahm and Uhlenbeck 1991; Dahm et al. 1993). When Mg^{++} is replaced with other transition-series metals in the reaction, there is a general correlation between the order of relative reaction rate under a given set of conditions ($Co^{++} > Mn^{++} > Mg^{++}$) and the pK_a for titration of a proton from a metal-bound H_2O molecule to form a hydroxide ion (pK_a = 10.2, 10.6, 11.4 for Co^{++}, Mn^{++}, and Mg^{++}, respectively). (Notably, this trend breaks down for divalent ions such as Ca^{++} that have

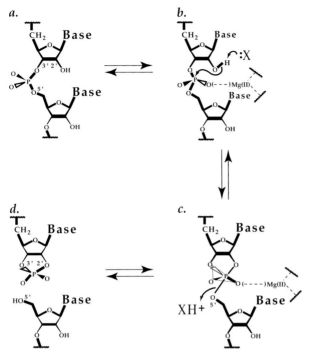

Figure 2 Schematic representation of the ribozyme chemical reaction. (*a*) the ground-state precatalytic enzyme–substrate complex. (*b*) activation of the 2′-OH group by a base, X (possibly a metal-bound hydroxide). This step is accompanied by metal binding to the *pro*-R_p nonbridging oxygen in the hamnmerhead reaction. (*c*) In-line attack leads to formation of a trigonal-bipyramidal intermediate or transition state. An acid "XH⁺" must be present to alleviate charge build-up on the 5′-leaving group. (*d*) products are a cyclic-2′, 3′-phosphodiester and a free 5′-OH group.

a different preferred coordination number). These data suggest a metal-bound hydroxide ion is a candidate for a proton acceptor (shown schematically as "X" in Fig. 2b) in the hammerhead cleavage reaction (Dahm and Uhlenbeck 1991; Dahm et al. 1993).

In contrast, the activities of the hairpin (Nesbitt et al. 1997) and *Neurospora* VS (Collins and Olive 1993) ribozymes are relatively pH-insensitive. The activity of the HDV ribozyme increases with pH over the range 4–6, but decreases thereafter (Fauzi et al. 1997), making it difficult to draw any firm conclusions about whether a metal-bound hydroxide ion might act as a proton acceptor in this reaction.

For the hammerhead, it has been demonstrated that the bond-cleavage reaction, starting with a 3′, 5′-phosphodiester and yielding a 2′, 3′-phosphodiester, results in inversion of the stereochemical configuration of the phosphate of nucleotide $N_{1.1}$ (van Tol et al. 1990; Koizumi and Ohtsuka 1991; Slim and Gait 1991). The most reasonable mechanism by which this would occur is by in-line attack of the ribose $2'$-OH on the phosphorus atom at the cleavage site, proceeding through a trigonal bipyramidal intermediate or transition state, with proton abstraction from the $2'$-OH and proton donation to the leaving group, as shown in Figure 2b. Although the configuration of substrate and product has not been determined for the other small ribozymes, it is likely that they undergo a similar reaction. An essential constraint of this mechanism is that the phosphodiester bond must be "twisted" into a conformation where it is poised for catalysis (Fig. 2a,b). This point is discussed further below in the context of the three-dimensional structures of the hammerhead ribozyme.

HAMMERHEAD RIBOZYME STRUCTURE

From the discussions above, it should be apparent that although the four small ribozymes all catalyze the same reaction with approximately equal rate constants, they must differ in both their overall tertiary structure and their catalytic mechanisms. From among this group of ribozymes, the hammerhead is the only one whose three-dimensional structure has been solved at this time. This was first accomplished with a complex between an all-DNA "inhibitor" and an all-RNA "ribozyme" (Pley et al. 1994), and subsequently with an all-RNA construct that had a $2'$-O-methyl ribonucleotide at the cleavage site to prevent bond cleavage (Table 2) (Scott et al. 1995). Despite being different constructs (the RNA–DNA complex had stems I and III open and stem II closed with a tetraloop, whereas the all-RNA complex had stem III closed and the other two open) and despite being crystallized under very different conditions (high salt for the RNA–DNA complex versus low ionic strength for the all-RNA hammerhead; see Table 2), the overall tertiary structures of these two hammerheads were essentially identical. These structures revealed three duplex stems, as predicted from the secondary structure (although the first base pair in stem III does not form Watson-Crick hydrogen bonds), and a conserved catalytic core (Fig. 3). This core has two distinct structural domains. The first domain is a sharp "uridine turn" comprising nucleotides C_3-A_6 at the junction of stem I and the core; it is identical in conformation to the uridine turns found in tRNAs. The second domain comprises U_7-A_9 and G_{12}-A_{14} and includes a tandem GA mismatch. Stems II and III are nearly collinear, whereas the sharp turn of domain I juxtaposes stem I and stem II, giving the hammerhead a wishbone conformation.

Table 2 Crystallographic structures reported for hammerhead ribozymes

Construct (ribozyme; substrate/inhibitor)	Crystal stabilization conditions[a]	Resolution (Å)	NDB entry[b]	Reference
1. 34-mer RNA; 13-mer DNA; GAAA tetraloop in stem II	2.4 M Lis_2SO_4, 1 mM spermine, 10 mM Na-cacodylate, pH 6.0	2.6	UHX026	Pley et al. (1994)
2. 16-mer RNA; 25-mer RNA with 2′-O-Me at cleavage site; GUAA tetraloop in stem III	23% PEG 6000, 100 mM NH_4OAc, 10 mM $Mg(OAc)_2$, 50 mM spermine, 5% glycerol, 10 mM NH_4-cacodylate pH 6.5	3.1	URX035	Scott et al. (1995)
3. 16-mer RNA; 25-mer RNA; GUAA tetralooop in stem III	1.8 M Li_2SO_4, 1.25 mM EDTA, 50 mM Na-cacodylate, pH 6.0	3.0	URX057	Scott et al. (1996)
4. Same as 3	1.8 M Li_2SO_4, Mn^{++}, buffer, pH 5.0	3.0	URX058	Scott et al. (1996)
5. Same as 3	1.8 M Li_2SO_4, 100 mM $MgSO_4$, buffer, pH 8.5	3.1	URX059	Scott et al. (1996)
6. Same as 3	1.8 M Li_2SO_4, with 2 mM $TbCL_3$	2.9	URX067	Feig et al. (1998)
7. Same construct as 3, but with 5′-C-methyl ribose at position 1.1	1.8 M Li_2SO_4, 50 mM Tris buffer, pH 8.5, with 50 mM $CoCl_2$	3.0	URX071	Murray et al. (1998)

[a]To flash-freeze crystals for data collection, crypoprotectant of 20% glycerol was added to 1 and 3–7; 25% glycerol to 2.
[b]Nucleic Acids Database.

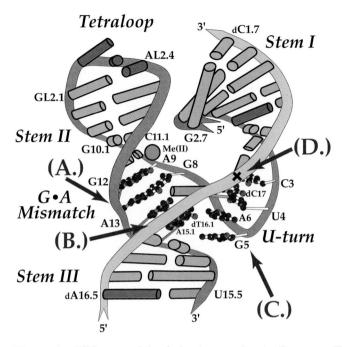

Figure 3 Ribbon model of the hammerhead ribozyme. The substrate and ribozyme strands are depicted at *yellow* and *magenta* ribbons, respectively. Bases are indicated as balls and sticks in the ribozyme core, or cylinders outside the core colored *red*, A; *orange*, G; *blue*, C; *green*, U and T. An "X" indicates the site of cleavage (*A–D* designate features of the ground-state X-ray structures that cannot be reconciled with the current chemical modification data: (*A*) Mg(II) binding to the *pro*-Rp oxygens at A_{13} and A_{14} is required, but no metal ions are observed in the structures. (*B*) No discernible reason for conserved core sequence at $A_{15.1}$ and $U_{16.1}$. (*C*) The G_5 base is solvent exposed, but essential for activity. (*D*) The active site is not in the correct conformation for in-line attack. (Derived from coordinate entry UHX026 in the Nucleic Acids Database.)

Thus, the catalytic core of the hammerhead has a simple tertiary fold. Subtle differences between the RNA–DNA structure and the all-RNA structure are seen in the vicinity of the cleavage site, where a 2′-OH of nucleotide $U_{16.1}$ 5′ to the cleavage site in the all-RNA structure forms specific hydrogen bonds that are necessarily absent in the RNA–DNA complex (for discussion, see Scott et al. 1995).

One divalent ion-binding site in domain 2 of the hammerhead core shows up unambiguously in all of the crystal structures (Fig. 3). The metal ion interacts with the *pro*-Rp phosphate oxygen of A_9, as anticipated from the results of phosphorothioate substitution/Mn^{++} rescue experiments

(Ruffner and Uhlenbeck 1990), and also with N7 of $G_{10.1}$, rationalizing the preference for a purine at this position (Ruffner et al. 1990; Tuschl and Eckstein 1993). It is notable, however, that no metal ions have been observed to bind to the *pro*-Rp phosphate oxygens of A13 or A14, although phosphorothioate substitution experiments also implicate these sites as inner-sphere ligands for divalent ions. Additionally, electron density peaks interpreted as hydrated Mg^{++} ions have been reported for hammerhead crystals grown at low ionic strength (Scott et al. 1995), and more recently for crystals in 1.8 M Li_2SO_4 (Scott et al. 1996). Tb^{+++} competes with a single Mg^{++}-binding site and inhibits the hammerhead; it has been shown to bind in crystals adjacent to the base of G_5 of the catalytic core (Feig et al. 1998).

Despite revealing an overall tertiary fold of the hammerhead, the crystallographic structures all introduce the same dilemma regarding the catalytic mechanism. As discussed earlier, the bond-cleavage reaction probably proceeds by in-line attack by the 2′-OH of the ribose hydroxyl on the scissile phosphate, as shown schematically in Figure 2b. However, in both structures, the phosphodiester backbone is in a conformation similar to that of A-form duplex RNA, similar to that shown in Figure 2a; it is not twisted into the requisite catalytic conformation. At a minimum, the phosphodiester bond must twist relative to the ribose hydroxyl before bond cleavage can occur. This can be accomplished either by the phosphate rotating inward, or by the base flipping out, as illustrated in greater detail elsewhere (Pley et al. 1994). Additionally, no divalent ion has been observed bound in the active site in a manner consistent with its proposed participation in the catalytic mechanism.

THE DILEMMA OF THE NONCATALYTIC CONFORMATION

This dilemma—that the overall tertiary structure is essentially the same in different crystal structures, but the active site is clearly not in a catalytic conformation—raises the question of the relevance of the structures to the catalytic mechanism. The observed structures can be rationalized within the reaction pathway of the hammerhead, drawn schematically as a free-energy profile in Figure 4. In this pathway, the substrate strand first binds to form a ground state complex E·S with the ribozyme. Then, it proceeds to the transition state E·S* of bond cleavage. The ground state to transition state activation barrier is approximately 22 kcal/mole (Hertel and Uhlenbeck 1995). In this context, it is reasonable to suggest that the crystal structures approximate the structure of the ground state E·S of the reaction, and that a significant conformational rearrangement must take place

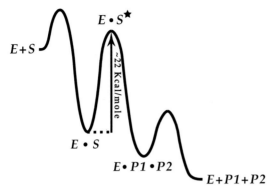

Figure 4 Schematic representation of the free-energy profile for the hammerhead ribozyme reaction. E·S represents the enzyme–substrate complex; E·P$_1$·P$_2$ are the products on the enzyme whereas P$_1$ and P$_2$ are the products after release into solution. Formation of the E·P$_1$·P$_2$ complex is described by K$_{eq}$, the ratio of rate constants k_{cleave}/k_{ligate} prior to product release, separated by a barrier of ~22 kcal/mole. (See footnote to Table 1.)

in order to twist the active site into a catalytic conformation, thereby allowing the ribozyme to cleave the scissile bond by a mechanism consistent with the known stereochemistry of the reaction.

Efforts have been made to determine structures of reaction pathway intermediates that bear closer resemblance to the elusive transition state. Using an active all-RNA construct (i.e., having a ribonucleotide rather than a modified nucleotide at the cleavage site), Scott and coworkers crystallized a hammerhead at low pH and triggered the reaction in the crystals by raising the pH to 8.5 (Scott et al. 1996). The crystals disintegrated during the course of the experiment, but by flash-freezing after a limited time at the higher pH, the investigators managed to trap and collect data on a conformation that showed a shift around the cleavage site. There were also new peaks in difference electron density maps, one of which was interpreted as a divalent magnesium ion bound to the *pro*-Rp oxygen of the scissile phosphodiester bond. More recently, they have used an RNA with a 5′-C-methyl ribose at position 1.1 adjacent to the cleavage site to trap a long-lived reaction intermediate conformation in the presence of Co^{++} at pH 8.5 (Murray et al. 1998). In the resulting structure, C$_{17}$ is displaced from its ground-state position, with its purine ring stacking on the pyrimidine ring of A$_6$. On the basis of these data, they proposed a cleavage mechanism that requires only a modest conformational shift within the molecule as a whole, coupled with an inward rotation of the phosphate at the cleavage site.

It is clear that some conformational rearrangement must take place in order for bond cleavage to occur. The magnitude of the change—whether it is localized around the cleavage site or whether there is a much more global rearrangement—is not known. The hammerhead is a flexible molecule; it follows a two-step folding pathway as a function of increasing divalent ion concentration to reach an active state (Fig. 5) (Bassi et al. 1995, 1996). The first step occurs at [Mg^{++}] ≤ 0.5 mM and puts stems II and III into a nearly collinear configuration (Fig. 5b). The second transition requires [Mg^{++}] > 1 mM and pulls stem I into proximity to stem II,

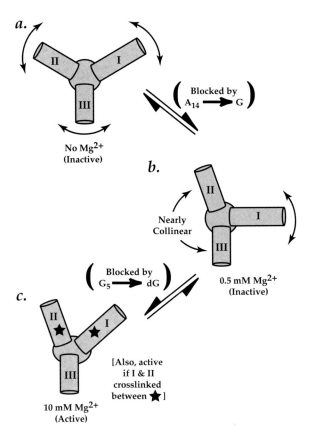

Figure 5 Schematic diagram of the hammerhead folding properties and stem conformations under various ionic conditions, based on gel electrophoresis measurements (Bassi et al. 1995, 1996). Double arrows indicate essentially random orientation of stem for a given condition. (*a*) no Mg^{++}, inactive. (*b*) 0.5 mM Mg^{++}, inactive. (*c*) 10 mM Mg^{++}, active. Stems I and II may be cross-linked between positions marked with a star and the hammerhead will remain active.

consistent with formation of the wishbone configuration observed in the crystal structures. When stems I and II are cross-linked in a manner that constrains them to the wishbone configuration of the crystal structures (Fig. 5c), the hammerhead is fully active (Sigurdson et al. 1995). Thus, although the molecule as a whole can sample a range of global conformations, the relative orientation of the three duplex stems that allows activity is well-determined.

In the context of this constraint on global conformation of the active hammerhead, it is instructive to consider information on what local interactions within the core influence catalytic activity. A picture for which functional groups are important for hammerhead activity has begun to emerge from site-specific alteration studies. Mutagenesis of one naturally occurring nucleotide to another changes several functional groups simultaneously; in most cases, this results in a decrease of catalytic activity of two to three orders of magnitude (Sheldon and Symons 1989; Ruffner et al. 1990). Although this approach is useful for determining which nucleotides must be conserved for activity, it is too "coarse" to identify specific interactions that are essential for catalysis. A methodology that allows a more specific delineation of these interactions is to incorporate modified nucleotides, in which only one or two functional groups are modified, and to determine the consequent effect on activity. For example, many of the ribonucleotides of the ribozyme core have been individually substituted with either a deoxyribonucleotide, thereby "deleting" the 2'-hydroxyl, or with 2'-amino, 2'-fluoro, or 2'-O-methyl nucleotides, thereby changing both the hydrogen-bonding pattern and the chemical reactivity of this site. In a similar vein, a spectrum of nucleotides with modified purine or pyrimidine rings has been incorporated to determine the consequence of deleting or altering specific exocyclic functional groups. The results from more than a hundred such modifications have been tabulated and mapped onto the three-dimensional structure of the hammerhead. These results are summarized in detail elsewhere (McKay 1996); only the general conclusions are reiterated here.

In some regions of the ribozyme core, and particularly in domain 2, many of the functional groups whose modification results in loss of activity are involved in specific tertiary interactions or in interactions with divalent metal ions. For example, replacing G_8 with deoxyguanosine, thereby deleting the 2'-OH group, reduces activity approximately three orders of magnitude. In the hammerhead structure, the 2'-OH of G_8 hydrogen bonds to the exocyclic amino group of G_{12}. Regrettably, it is not possible to distinguish whether this modification affects hammerhead folding or the chemical cleavage step from the steady-state rate constants that are reported—a difficulty encountered with interpreting much of the functional group modification data. However, this is a case where there is a clear correlation between the results of altering a specific functional group, demonstrating that this hydroxyl is essential for activity, and the crystallographic structures showing that the hydroxyl is involved in specific tertiary interactions.

In sharp contrast, there are other instances where the three-dimensional structure and the consequences of functional group alterations cannot be reconciled (Fig. 3). Possibly the most overt discrepancy involves nucleotide G_5 of domain 1; any alteration of the exocyclic groups of the guanosine base reduces hammerhead activity two to three orders of magnitude; this observation confounds the X-ray structures in which the base of G_5 is solvent-exposed and its exocyclic groups are completely devoid of intramolecular interactions. Additionally, the phosphates of A_{13} and A_{14} have been indicted as binding sites for essential divalent ions, yet as discussed earlier, no metal ion has been seen crystallographically at these positions. If the crystallographic structures of the hammerhead mimic a ground-state enzyme–substrate complex, then the functional group modification data, in conjunction with the observation that the active site is not poised for catalysis in the crystallographic structures, suggest a substantial conformational rearrangement involving both the cleavage site and the uridine turn must take place prior to reaching the transition state. By implication, significant intramolecular and/or metal-ion-binding interactions of the transition state must be absent, and hence not observed, in the crystallographic structures.

In this context, an intriguing suggestion has been made from data on the rescue of bond-cleavage activity by Cd^{++} for a hammerhead having a single *pro*-Rp thiophosphate at position A_9 (Peracchi et al. 1997). The apparent K_d for Cd^{++} binding to this site (i.e., to the divalent metal site shown in Fig. 3) in the ground state of the reaction is 25 µM. Measuring the chemical cleavage rates in the presence (1.2 min^{-1}) and absence (1.4 × 10^{-4} min^{-1}) of Cd^{++}, Perrachi et al. compute a K_d of 2.5 nM for Cd^{++} bound at this site in the transition state. The 10^4-fold increase in affinity in going from ground to transition state suggests that the Cd^{++} ion acquires at least one additional ligand in the transition state. The identity of this putative ligand is unknown. Notably, this divalent ion binding site is >20 Å away from the scissile bond in the hammerhead structures; a direct participation of this divalent ion at the cleavage site would require a dramatic rearrangement of the catalytic core.

The discussions above illustrate a key problem in understanding the hammerhead mechanism: We are unable to extrapolate from apparent ground-state structures to a catalytically competent transition-state structure. It appears that this rudimentary enzyme probably relies on flexibility and being able to dynamically sample a catalytic state transiently, rather than on a near-static structural complementarity between ribozyme and the transition state of the reaction, for its activity. This is in contrast to the large ribozymes discussed in Chapters 13 and 14.

CONCLUSIONS AND FUTURE PROSPECTS

We may reflect at this point on what generalities have emerged about the structure and mechanism of small ribozymes. First, all four of these ribozymes catalyze the same reaction, and they all enhance the cleavage over its uncatalyzed rate by approximately the same factor of $\sim 10^6$. However, the consensus sequences of the four ribozymes indicate that each has a unique structure (Fig. 1). Furthermore, the characteristics of the bond-cleavage activities (how Mg^{++} ions are involved in catalysis; substrate/product equilibrium, pH dependence of the reaction rate; Table 1) suggest that each ribozyme has a unique catalytic mechanism; the current evidence indicates that we should not anticipate convergent evolution in which different RNA structures employ identical catalytic mechanisms.

Among these four ribozymes, the hammerhead has been the most extensively characterized, and in many ways it has become a paradigm for studies on small ribozymes. Ironically, the current wealth of biochemical and structural data on this system has not translated into a richness of insight into the chemical mechanism of its activity; a picture of the catalytic transition state remains elusive.

Departing for a moment into the realm of speculation, we may ask what role small ribozymes, which have been relegated to a relatively obscure niche of modern biology, might have played in a hypothetical RNA World. It is fashionable to suggest a self-replicating ribozyme as the initial perpetrator of an RNA World; at the same time it is difficult to imagine plausible routes by which such a ribozyme could emerge in a single step from prebiotic pools of random mono- and polynucleotides. In this context, we may ask what capabilities small ribozymes might contribute to the success of this overall process. The diversity of structure and mechanism that we observe in a few representative ribozymes that have survived evolution suggests that many more small RNA motifs, capable of catalyzing simple cleavage and ligation of a diverse spectrum of polynucleotides, are possible and may have existed. (For example, the "leadzyme" is a small, 6 nucleotide motif that has been artificially selected for the ability to cleave a ribonucleotide backbone in the presence of Pb^{++} [Pan and Uhlenbeck 1992].) From such a diverse sampling of prebiotic chemical reactions, one can envision mechanisms by which a plethora of larger, multidomain RNA structures, capable of performing more complex catalytic tasks, could be constructed using a limited pool of mono- and poly-ligated nucleotides as starting material. We may wonder whether small ribozymes, by virtue of their diversity and simplicity, were obligatory progenitors to the more complex self-replicating ribozymes of the RNA World.

ACKNOWLEDGMENTS

This work was supported by National Institutes of Health grant AI-30606 to D.B.M. and a Burroughs Wellcome Fund fellowship of the Life Science Research Foundation to J.E.W.

REFERENCES

Bassi G.S., Murchie A.I., and Lilley D.M. 1996. The ion-induced folding of the hammerhead ribozyme: Core sequence changes that perturb folding into the active conformation. *RNA* **2:** 756–758.

Bassi G.S., Mollegaard N.E., Murchie A.I., von Kitzing E., and Lilley D.M. 1995. Ionic interactions and the global conformations of the hammerhead ribozyme. *Nat. Struct. Biol.* **2:** 45–55.

Been M.D. and Wickham G.S. 1997. Self-cleaving ribozymes of hepatitis delta virus RNA. *Eur. J. Biochem.* **247:** 741–753.

Berzal-Herranz A., Joseph S., and Burke J.M. 1992. In vitro selection of active hairpin ribozymes by sequential RNA-catalyzed cleavage and ligation reactions. *Genes Dev.* **6:** 129–134.

Berzal-Herranz A., Joseph S., Chowrira B.M., Butcher S.E., and Burke J.M. 1993. Essential nucleotide sequences and secondary structure elements of the hairpin ribozyme. *Embo J.* **12:** 2567–2573.

Burke J.M. 1996. Hairpin ribozyme: Current status and future prospects. *Biochem. Soc. Trans.* **24:** 608–615.

Butcher S.E., Heckman J.E., and Burke J.M. 1995. Reconstitution of hairpin ribozyme activity following separation of functional domains. *J. Biol Chem.* **270:** 29648–29651.

Buzayan J.M., Hampel A., and Bruening G. 1986. Nucleotide sequence and newly formed phosphodiester bond of spontaneously ligated satellite tobacco ringspot virus RNA. *Nucleic Acids Res.* **14:** 9729–9743.

Chowrira B.M. and Burke J.M. 1992. Extensive phosphorothioate substitution yields highly active and nuclease-resistant hairpin ribozymes. *Nucleic Acids Res.* **20:** 2835–2840.

Clouet-d'Orval B. and Uhlenbeck O.C. 1997. Hammerhead ribozymes with a faster cleavage rate. *Biochemistry* **36:** 9087–9092.

Collins R.A. and Olive J.E. 1993. Reaction conditions and kinetics of self-cleavage of a ribozyme derived from *Neurospora* VS RNA. *Biochemistry* **32:** 2795–2799.

Dahm S.C. and Uhlenbeck O.C. 1991. Role of divalent metal ions in the hammerhead RNA cleavage reaction. *Biochemistry* **30:** 9464–9469.

Dahm S.C., Derrick W.B., and Uhlenbeck O.C. 1993. Evidence for the role of solvated metal hydroxide in the hammerhead cleavage mechanism. *Biochemistry* **32:** 13040–13045.

DeYoung M., Siwkowski A.M., Lian Y., and Hampel A. 1995. Catalytic properties of hairpin ribozymes derived from chicory yellow mottle virus and arabis mosaic virus satellite RNAs. *Biochemistry* **34:** 15785–15791.

Fauzi H., Kawakami J., Nishikawa F., and Nishikawa S. 1997. Analysis of the cleavage reaction of a *trans*-acting human hepatitis delta virus ribozyme. *Nucleic Acids Res.* **25:** 3124–3130.

Feia A.L., Scott W.G., and Uhlenbeck O.C. 1998. Inhibition of the hammerhead ribozyme cleavage reaction by site-specific binding of Tb. *Science* **279:** 81–84.

Forster A.C. and Symons R.H. 1987a. Self-cleavage of plus and minus RNAs of a virusoid and a structural model for the active sites. *Cell* **49:** 211–220.

———. 1987b. Self-cleavage of virusoid RNA is performed by the proposed 55-nucleotide active site. *Cell* **50:** 9–16.

Guo H.C. and Collins R.A. 1995. Efficient *trans*-cleavage of a stem-loop RNA substrate by a ribozyme derived from *Neurospora* VS RNA. *Embo J.* **14:** 368–76.

Guo H.C., De Abreu D.M., Tillier E.R., Saville B.J., Olive J.E., and Collins R.A. 1993. Nucleotide sequence requirements for self-cleavage of *Neurospora* VS RNA. *J. Mol. Biol.* **232:** 351–361.

Hampel A. and Cowan J.A. 1997. A unique mechanism for RNA catalysis: The role of metal cofactors in hairpin ribozyme cleavage. *Chem. Biol.* **4:** 513–517.

Haseloff J. and Gerlach W.L. 1988. Simple RNA enzymes with new and highly specific endoribonuclease activities. *Nature* **334:** 585–591.

Hegg L.A. and Fedor M.J. 1995. Kinetics and thermodynamics of intermolecular catalysis by hairpin ribozymes. *Biochemistry* **34:** 15813–15828.

Hertel K.J. and Uhlenbeck O.C. 1995. The internal equilibrium of the hammerhead ribozyme reaction. *Biochemistry* **34:** 1744–1749.

Hertel K.J., Herschlag D., and Uhlenbeck O.C. 1994. A kinetic and thermodynamic framework for the hammerhead ribozyme reaction. *Biochemistry* **33:** 3374–3385.

Jeoung Y.H., Kumar P.K., Suh Y.A., Taira K., and Nishikawa S. 1994. Identification of phosphate oxygens that are important for self-cleavage activity of the HDV ribozyme by phosphorothioate substitution interference analysis. *Nucleic Acids Res.* **22:** 3722–3727.

Knoll R, Bald R., and Furste J.P. 1997. Complete identification of nonbridging phosphate oxygens involved in hammerhead cleavage. *RNA* **3:** 132–140.

Koizumi M. and Ohtsuka E. 1991. Effects of phosphorothioate and 2-amino groups in hammerhead ribozymes on cleavage rates and Mg^{++} binding. *Biochemistry* **30:** 5145–5150.

Kuo M.Y.-P., Sharmeen L., Dinter-Gottlieb G., and Taylor J. 1988. Characterization of self-cleaving RNA sequences on the genome and antigenome of human hepatitis delta virus. *J. Virol.* **62:** 4439–4444.

Long D.M., LaRiviere F.J., and Uhlenbeck O.C. 1995. Divalent metal ions and the internal equilibrium of the hammerhead ribozyme. *Biochemistry* **34:** 14435–14440.

McKay D.B. 1996. Structure and function of a hammerhead ribozyme: An unfinished story. *RNA* **2:** 395–403.

Murray J.B., Terwey D.P., Maloney L., Karpeisky A., Usman N., Beigelman L., and Scott W.G. 1998. The structural basis of hammerhead ribozyme self-cleavage. *Cell* **92:** 665–673.

Nesbitt S., Hegg L.A., and Fedor M.J. 1997. An unusual pH-independent and metal-ion-independent mechanism for hairpin ribozyme catalysis. *Chem. Biol.* **4:** 619–630.

Pan T. and Uhlenbeck O.C. 1992. A small metalloribozyme with a two-step mechanism. *Nature* **358:** 560–563.

Peracchi A., Beigelman L., Scott E.C., Uhlenbeck O.C., and Herschlag D. 1997. Involvement of a specific metal ion in the transition of the hammerhead ribozyme to its catalytic conformation. *J. Biol. Chem.* **272:** 26822–26826.

Perrotta A.T. and Been M.D. 1990. The self-cleaving domain from the genomic RNA of hepatitis delta virus: Sequence requirements and the effects of denaturant. *Nucleic Acids Res.* **18:** 6821–6827.

———. 1991. A pseudoknot-like structure required for efficient self-cleavage of hepatitis delta virus RNA. *Nature* **350:** 434–436.

Pley H.W., Flaherty K.M., and McKay D.B. 1994. Three-dimensional structure of a hammerhead ribozyme. *Nature* **372:** 68–74.

Rubino L., Tousignant M.E., Steger G., and Kaper J.M. 1990. Nucleotide sequence and structural analysis of two satellite RNAs associated with chicory yellow mottle virus. *J. Gen. Virol.* **71:** 1897–1903.

Ruffner D.E. and Uhlenbeck O.C. 1990. Thiophosphate interference experiments locate phosphates important for the hammerhead RNA self-cleavage reaction. *Nucleic Acids Res.* **18:** 6025–6029.

Ruffner D.E., Stormo G.D., and Uhlenbeck O.C. 1990. Sequence requirements of the hammerhead RNA self-cleavage reaction. *Biochemistry* **29:** 10695–10702.

Saville B.J. and Collins R.A. 1990. A site-specific self-cleavage reaction performed by a novel RNA in *Neurospora* mitochondria. *Cell* **61:** 685–696.

Scott W.G., Finch J.T., and Klug A. 1995. The crystal structure of an all-RNA hammerhead ribozyme: A proposed mechanism for RNA catalytic cleavage. *Cell* **81:** 991–1002.

Scott W.G., Murray J.B., Arnold J., Stoddard B.L., and Klug A. 1996. Capturing the structure of a catalytic RNA intermediate: The hammerhead ribozyme. *Science* **274:** 2065–2069.

Sheldon C.C. and Symons R.H. 1989. Mutagenesis analysis of a self-cleaving RNA. *Nucleic Acids Res.* **17:** 5679–5685.

Sigurdson S.T., Tuschl T., and Eckstein F. 1995. Probing RNA tertiary structure: Interhelical crosslinking of the hammerhead ribozyme. *RNA* **1:** 575–583.

Slim G. and Gait M.J. 1991. Configurationally defined phosphorothioate-containing oligoribonucleotides in the study of the mechanism of cleavage of hammerhead ribozymes. *Nucleic Acids Res.* **19:** 1183–1188.

Tuschl T. and Eckstein F. 1993. Hammerhead ribozymes: Importance of stem-loop II for activity. *Proc. Natl. Acad. Sci.* **90:** 6991–6994.

Uhlenbeck O.C. 1987. A small catalytic oligoribonucleotide. *Nature* **328:** 596–600.

van Tol H., Buzayan J.M., Feldstein P.A., Eckstein F., and Bruening G. 1990. Two autolytic processing reactions of a satellite RNA proceed with inversion of configuration. *Nucleic Acids Res.* **18:** 1971–1975.

Wedekind J.E. and McKay D.B. 1998. Crystallographic structures of the hammerhead ribozyme: Relationship to ribozyme folding and catalysis. *Annu. Rev. Biophys. Biomol. Struct.* **27:** 475–502.

Young K.J., Gill F., and Grasby J.A. 1997. Metal ions play a passive role in the hairpin ribozyme catalysed reaction. *Nucleic Acids Res.* **25:** 3760–3766.

12

The Role of Metal Ions in RNA Biochemistry

Andrew L. Feig and Olke C. Uhlenbeck
Department of Chemistry and Biochemistry
University of Colorado
Boulder, Colorado 80309

Approximately two-thirds of the elements in the periodic table can be categorized as metals. Besides luster, malleability, and conductivity, one of the fundamental characteristics of metals is their low ionization potential. As a result, the ionic forms of these elements predominate in the biosphere. Considering the diverse properties of these ions, it is not surprising that through the process of evolution, metal ions have been co-opted into numerous roles in biology. Metal ions are required for so many biochemical reactions that it is likely that they also had an important role in the RNA world. To understand both modern and prebiotic RNA biochemistry, it is therefore essential to have a basic understanding of these inorganic elements.

Metal ions were abundant in the primordial soup. It is believed that 3.8×10^9 years ago, the ocean was between 80°C and 100°C with a pH possibly as low as 6 (Bengston 1994). Table 1 shows the concentrations of the most common metal ions in today's seas and in blood plasma. Although the concentrations of most of these ions in the prebiotic ocean are not known, the higher temperature and lower pH relative to the current ocean would have solvated a variety of ions and leached metal ions from the mineral-rich ocean beds. Therefore, the concentrations would have been significantly higher than the current values. One important additional difference is the extremely low concentration of easily oxidized metal ions such as Fe(II). Ferrous ion has been predicted to have been very abundant in the primordial ocean, as high as 0.1 mM (Bengston 1994). Interestingly, this change is probably due to the fact that early life forms produced molecular oxygen from photosynthetic reactions that in turn caused a cascade of oxidation reactions to occur. The conversion of Fe(II) to the more acidic Fe(III) resulted in the formation of oxyhydroxide polymers that precipitated out of the ancient oceans, depleting the ion

Table 1 Metal ion abundance in the modern oceans and biological fluids

Ionic species	Modern ocean (mM)[a]	Blood plasma (mM)[a]	In mammalian cells[b]	Mammalian extracellular fluid[b]
Na^+	470	138	10	145
K^+	10	4	140	5
$Mg(^{++})$	50	1	30	1
$Ca(^{++})$	10	3	1	4
$Fe(^{++})$	1×10^{-4}	2×10^{-2}		
$Zn(^{++})$	1×10^{-4}	2×10^{-2}		
$Cu(^{++})$	1×10^{-3}	1.5×10^{-2}		
$Co(^{++})$	3.1×10^{-6}	2×10^{-3}		
$Ni(^{++})$	1×10^{-6}	0		

[a]Data from Pan et al. (1993).
[b]Data from Cowan (1995).

from the biosphere and giving the banded iron formations observed in sedimentary rocks. With respect to modern biochemical studies, this shift in metal ion availability justifies studying what are today considered "non-physiologically relevant" metal ions in RNA catalysis and structure in order to probe events related to the early RNA world. The more reactive metal ions may in fact have helped to drive the evolution from the RNA world to the modern protein world, where the polymer backbone is more stable to the potential side reactions induced by the metal cofactors.

A variety of physical properties can be used to characterize the behavior of metal ions. A number of textbooks have excellent discussions of metal ion behavior (Huheey 1983; Cotton and Wilkinson 1988; Cowan 1995; Richens 1997) and should be consulted for more detailed descriptions. Table 2 lists ionic radii, pK_a values (of the aqua ions), hydration numbers, water exchange rates and ΔH_{hyd} for a number of relevant metal ions. It should be remembered when dealing with tables such as these that many of the properties listed are for the most common forms of the ions. Parameters specific to a metal ion complex can have significant effects on the values for other physical properties, especially for the transition metals.

Another useful concept in consideration of metal ions is that of hardness and softness. This property summarizes the general affinity of a Lewis acid (the metal ion) for a Lewis base (the ligand) in a manner independent of the acidity or basicity of the species and reflects the degree of covalency in the metal–ligand bond. The general trend is that hard metal ions preferentially bind hard ligands and vice versa. In RNA, the hardest ligand is the anionic phosphate oxygen. In a recent quantitative description, the absolute hardness (η) is proportional to the difference between I, the ionization potential, and A, the electron affinity, of the species (Pearson 1988). Absolute softness is defined as η^{-1}. The interested reader should consult Pearson's paper for how η and χ (the absolute electronegativity) are applied quantitatively to any given acid–base reaction. It should be noted, however, that the ligands coordinated to a metal ion will influence its actual hardness or softness when in the context of a specific interaction. As an approximation, the presence of soft ligands in the coordination sphere of a metal tends to make the central ion softer, whereas hard ligands make the ion harder.

With these physical properties in mind, one can address the question of the role metal ions play in RNA biochemistry. Since each residue contains an anionic phosphodiester group, the principles of charge neutralization and electrostatic condensation dictate that cations must be closely associated with the polyanionic RNA molecule (Record et al. 1978; Manning 1979; Anderson and Record 1995). In principle, these can be any cationic species, but in general, the condensation layer consists of the abundant surrounding monovalent and divalent ions. In vivo, Mg(II) and K$^+$ are believed to dominate in this role. Charge neutralization becomes particularly important during the process of RNA folding, as the negatively charged backbones from two or more regions of the primary sequence most come close together in space. Without cations to screen these charges, the repulsive forces generated in the close-packed structure would overwhelm the energetically favorable interactions that dictate the proper three-dimensional structure. Since there is a formal charge of –1 for every residue, RNAs carry around a sufficient number of metal ions in a condensation layer to neutralize the charge. Studies measuring the number of Mg(II) ions bound to different RNAs have borne out this expectation (Table 3). The majority of these metal ions bind the RNA nonspecifically, solely dictated by electrostatic considerations. Each individual counterion is very weakly bound and in rapid exchange with more freely diffusing ions. Furthermore, they cannot be localized by most biophysical techniques because of the diversity of binding environments at any given instant.

Among the metal ions that bind an RNA, a subset interact specifically. Metal ions generally bind to these sites more tightly than to the nonspecific ones. They are better localized because of discrete interactions and they cannot be as easily substituted by other ions (Laing et al. 1994; Gluick et al. 1997). For that reason, these sites dictate the metal preferences of the RNA molecule as a whole. These specific sites can be further subdivided based on the role of the metal ion in the biochemistry. Metal ions can serve in structural roles, or as catalytic cofactors or potentially as

Table 2 Properties of selected metal ions

Metal	Oxidation state	Hydration number[a]	Ionic radius (Å)[a,b]	ΔH_{hyd} (kJ/mole)[a]	Water exchange rate (s^{-1})[a]	pKa of aqua ion[a]	Absolute hardness η^c	Absolute electro-negativity χ^c
Group 1A								
Li	1+	4	0.59	−514.1	~10^9	13.8[e]	35.12	40.52
Na	1+	6	1.02	−405.4	~10^9 [d]	14.48[e]	26.21	21.08
K	1+	6	1.38	−320.9	~10^9 [d]		17.99	13.64
Group 2A								
Mg	2+	6	0.72	−1922.1	~10^6	11.42[e]	47.59	32.55
Ca	2+	8	1.12	−1592.4	~10^8	12.70[e]	19.52	31.39
Sr	2+	6	1.18	−1444.7	~10^9 [d]	13.18[e]	27.3	16.3
Ba	2+	6	1.35	−1303.7	~10^9	13.82[e]		
1st row transition metals								
Cr	2+	6	0.80	−1849.7	~10^9		7.23	23.73
Cr	3+	6	0.62	−4401.6	2.4×10^{-6}	4.00	9.1	40.0
Mn	2+	6	0.67	−1845.6	2.1×10^7	10.6	9.02	24.66
Fe	2+	hs 6	0.78	−1920.0	4.4×10^6	9.5	7.24	23.42
		ls 6	0.61					
Fe	3+	hs 6	0.65	−4376.5	~10^{-3}	0.70	12.08	42.73
		ls 6	0.55			2.19[e]		
Co	2+	6	0.75	−2054.3	3.2×10^6	9.65	8.22	25.28
Co	3+	hs 6	0.61		~10^{-1} [e]		8.9	42.4
		ls 6	0.55					
Ni	2+	T_d 4	0.55	−2105.8	3.4×10^4	9.86	8.50	26.67
		sq 4	0.66					

(continues)

Table 2 (Continued)

Metal	Oxidation state	Hydration number[a]	Ionic radius (Å)[a,b]	ΔH_{hyd} (kJ/mole)[a]	Water exchange rate (s^{-1})[a]	pKa of aqua ion[a]	Absolute hardness η^c	Absolute electronegativity χ^c
Cu	1+	6	0.77	−594.1		7.97	6.28	14.01
Cu	2+	6	0.73	−2100.4	4.4×10^9	8.96	8.27	28.56
Zn	2+	6	0.74	−2044.3	$\sim 10^7$	9.60^e	10.88	28.84
2nd row transition metals								
Ru	3+	6	0.68		1.8×10^{-2}	2.3	10.7	39.2
Pd	2+	4	0.64		5.6×10^2	10.08	6.75	26.18
Cd	2+	6	0.95	−2384.9	$\sim 10^8$	11.7^e	10.29	27.20
3rd row transition metals								
Pt	2+	4	0.60		3.9×10^{-4}	~2.5	8.0	27.2
Hg	2+	6	1.02	−1853.5	$\sim 10^9$	3.4	7.7	26.5
Miscellaneous								
Pb[f]	2+	6	1.19	−1479.9		7.8	8.46	23.49
Tl	1+	6	1.50	−325.9		13.2	7.16	13.27

[a] Data from Richens (1997).
[b] Ionic radii are listed for the appropriate coordination number. Other coordination numbers are known for most of these ions and have different effective radii. Species with higher coordination numbers in general have larger ionic radii.
[c] Data taken from Pearson (1988). χ and η parameters for a number of potential ligands are also tabulated in this reference, however, exact data on biological ligands are not currently available.
[d] Data from Lincoln and Merbach (1995).
[e] Data from Huheey (1983).
[f] There is no evidence for a monomeric aqua ion of Pb(II). It is observed as an oligomer in aqueous solution (Richens 1997).

Table 3 Mg(II) binding and uptake studies on various RNAs

						K_d (M)		
RNA	Method	Conditions	Length (nt)	Metal ions	Nt/metal ion	strong sites (# of sites)	weak sites (# of sites)	Reference
E. coli tRNA^{Met}	equilibrium dialysis	0.17 M Na$^+$, pH 7.0, 4°C	76	27	2.8	3.4×10^{-5} (1)	2.4×10^{-3} (26)	Stein and Crothers (1976)
E. coli tRNA^{Glu}	equilibrium dialysis	0.10 M Na$^+$, pH 7.0, 4°C	76	37	2.1	1.3×10^{-5} (1)	1.2×10^{-3} (36)	Bina-Stein and Stein (1976)
Yeast tRNA^{Phe}	fluorescence titration	0.032 M Na$^+$, pH 6.0, 10°C	76	23.5 ± 8	3.2	1.1×10^{-5} (6.5 ± 3)	1.7×10^{-4} (17 ± 5)	Römer and Hach (1975)
Yeast tRNA^{Phe}	calorimetry	0.01 M Na$^+$, pH 7.2	76	24	3.2	1.0×10^{-6} (4)	9.1×10^{-5} (20)	Rialdi et al. (1972)
Yeast tRNA^{Phe}	^{25}Mg NMR	0.17 M Na$^+$, pH 7.0, 25°C	76	53 ± 8	1.4	$<10^{-4}$ (3–4)	4.5×10^{-3} (50 ± 8)	Reid and Cowan (1990)
B. subtilis RNase P	gel filtration	0.1 M NH$_4^+$, pH 8.0, 37°C	400	90–130	~4		$(0.39 \pm 0.08) \times 10^{-3}$ (95 ± 6)	Beebe et al. (1996)
Hammerhead ribozyme 16	fluorescence titration	0.40 M Li$^+$, pH 6.0, 25°C	55	29 ± 5	1.9	n.d.	n.d.	A. Feig and O. Uhlenbeck (unpubl.)

n.d. indicates not determined.

both simultaneously. To date, the designation of a site as functional generally has been related to the spatial proximity of the ion to the catalytic center rather than direct experimental evidence that the ion actively participates in the catalytic event.

The specificity of a binding pocket, defined as the relative affinity of different metal ions for the site, can result from a variety of factors, including the hardness of an ion, the identity of the coordinating ligands, the ionic radius, the preferred coordination geometry of the ion, and the metal's hydration number. Very few metal-binding sites have been probed sufficiently to fully define their specificities (Bukhman and Draper 1997; A.L. Feig et al., in prep.). The main problem is that to probe specificity, the site under consideration must remain independent even when ions bind at other places on the RNA. The site must also be uniquely identified by the biophysical technique being used. The specific metal ion interactions among catalytic RNAs are the best defined because the enzymatic activity can be used as a probe (Table 4). The metal ion specificities for these RNAs are composite parameters, however, simultaneously reflecting all of the binding sites required for activity. Whereas some of the ribozymes require a specific metal cofactor, others have less strict requirements. The larger RNAs tend to be more specific than the smaller ribozyme species. This increased specificity makes sense. Since the overall fold is more complicated, there are greater opportunities for specific interactions and there are more tertiary contacts that must be maintained by these ions. In addition to specificity, the apparent Michaelis constants for metal ion cofactors ($^{Metal}K_M$) are often used to describe the total contribution of the metal-binding sites on the catalytic activity (Clouet-d'Orval and Uhlenbeck 1996; McConnell et al. 1997). Although this parameter is an important characteristic of a catalytic RNA, it potentially reflects the effects of multiple metal-ion-binding events and may give little information regarding the properties of any individual metal-binding site.

There are several substantial experimental challenges for the RNA bioinorganic chemist. First and foremost, methods must be found that identify the metal ions bound to specific sites and separate them from the bulk ionic condensation events. As discussed below, any single technique is unlikely to locate all such sites, so multiple parallel approaches will almost certainly be required. Once a site has been identified, the next step is to relate it to a discrete property of the RNA. Any given site can be involved in maintaining local or global structure and can potentially participate in catalysis. Another complication is that RNA molecules and their associated metal ions are conformationally dynamic. Specific metal

Table 4 Metal ion specificity of various natural ribozymes

Ribozyme	Functional	Nonfunctional	Reference
hammerhead	Mg(II), Mn(II), Ca(II), Cd(II), Co(II)	Ba(II), Sr(II), [Cr(NH$_3$)$_6$]$^{+++}$, Pb(II), Zn(II), Tb(III), Eu(III)	Dahm and Uhlenbeck (1991); A.L. Feig et al. (in prep.)
hairpin	All tested, including [Cr(NH$_3$)$_6$]$^{+++}$		Hampel and Cowan (1997); Nesbitt et al. (1997); Young et al. (1997)
hepatitis δ virus	Mg(II), Mn(II), Ca(II), Sr(II)	Cd(II), Ba(II), Co(II), Pb(II), Zn(II)	Wu et al. (1989); Suh et al. (1993);
Neurospora VS	Mg(II), Mn(II), Ca(II)		Collins and Olive (1993)
RNase P	Mg(II), Mn(II), Ca(II)	Sr(II), Ba(II), Zn(II), Co(II), Cu(II), Fe(II), Ni(II)	Smith et al. (1992); Smith and Pace (1993)
Tetrahymena Group I	Mg(II), Mn(II)	Ca(II), Sr(II), Ba(II), Zn(II), Co(II), Cu(II)	Grosshans and Cech (1989); McConnell et al. (1997)
Group II	Mg(II)	Ca(II), Mn(II)	Chin and Pyle (1995)

ions generally exchange rapidly and need not be present in the same location in all of the conformational states. Therefore, if one is to accurately define the function of the metal ion, techniques must be available that can study metal–RNA interactions in different time frames. Finally, metal ion interactions may significantly affect RNA folding pathways, and controls should always be included to determine whether the conditions used resulted in the stabilization of an alternate conformation of the RNA (Uhlenbeck 1995).

One important issue addressed by this chapter is how metal-binding sites are identified. One can broadly group the techniques into three categories: (1) use of a biophysical technique (X-ray crystallography, NMR, etc.) with the native RNA and the native metal ions, (2) replacing the native metal ion with a nonnative ion more sensitive to a biophysical technique, and (3) specific synthetic modifications of the RNA that alter the local metal-binding properties. All three approaches have advantages and disadvantages and thus are best used in combination.

A few dozen specific RNA-binding sites have been identified. A number of the better-studied examples are presented in Table 5. We focus on a few of these sites that show either interesting structures or substantial specificity. We do not discuss the role of some ions in promoting catalysis. That topic is covered in Chapters 11 and 13 of this volume and has been the subject of a number of recent review articles (Pan et al. 1993; Yarus 1993; Smith 1995; McKay 1996; Pyle 1996). Instead, the central issue will be the binding sites themselves. What are the interactions between metal ions and RNA? Why does one site show specificity for a particular metal ion whereas another site does not? Does this specificity derive from the metal ion being used in the experiment, or is it a result of the RNA structure? Finally, we address the issue of whether the experimental approach being used to study a particular question biases the results toward identifying one type of metal-binding site over another.

CRYSTALLOGRAPHICALLY CHARACTERIZED METAL-ION-BINDING SITES

One useful way in which bioinorganic chemists classify metal-binding sites is by their nuclearity, or the number of metal ions that are held together in the structural/functional unit. The common divisions used for proteins, mononuclear, dinuclear, and polynuclear sites, are also suitable for RNA sites. The rationale behind this organization derives from the functional and spectroscopic differences between these classes. Although many fewer RNA sites are available, mono- and dinuclear clusters have both been characterized crystallographically. Polynuclear clusters of ions have also been observed and given names such as "metal zippers" (Correll et al. 1997), but it is still unclear whether the sites are cooperatively linked and thus act as a single element or just represent the clustering of mononuclear sites in a complex folded region of the RNA molecule. As more of these multinuclear motifs become available for detailed study, it will hopefully become clear whether the core ions act individually or as a unit to promote RNA structures.

The majority of the well-characterized metal-binding sites fall into the mononuclear category. These sites include all of the metal-binding sites observed on tRNAs, as well as most of the sites in the hammerhead ribozyme, P4-P6 domain of the *Tetrahymena* group I intron and the 5S rRNA fragment. The mononuclear sites tend to be quite variable with

Table 5 Selected well-characterized metal-ion-binding sites on RNA

RNA	Binding site	Metal ions that bind site	Closest contacts	Methodology[a]	Notes
tRNA					
	8–12 turn	Mg(II), Sm(III)	O_p-U8 O_p-A9		b–d
	D-stem	Sm(III)	O_p-U7		b
	D-loop-1	Mg(II), Sm(III)	O_p-A14 O_p-A20 O_p-A21		b–d
	D-loop-2	Co(II)	N7-G15		b
	D/TΨC-loop-1	Mg(II), Mn(II), Pb(II)	O_p-G19, N7-G20, O4-U59, N3-C60	Pb(II) cleavage, X-ray	b–d
	D/TΨC-loop-2	Mg(II), Sm(III)	O_p-G57, O_p-A14		b, d
	acceptor arm 1	Mn(II), Mg(II)	G3·U70	NMR	e
	acceptor arm 2	[Co(NH$_3$)$_6$]$^{+++}$	O4-U69		d
	anti-codon loop-1	Mg(II)	O_p-Y37		c
	anti-codon loop-2	Pt(II)	N7-^{m}G34		b
	anti-codon loop-3	Pb(II)	O2-^{m}C32 N7-Y37		f
	anti-codon stem	[Co(NH$_3$)$_6$]$^{+++}$	N7-G42		d
	variable loop-1	Sm(III), Pb(II)	O6-G45 N7-G45		b, f
	variable loop-2	Hg(II)	O4-U47		b
	variable loop-3	[Co(NH$_3$)$_6$]$^{+++}$	O_p-A44		d
Hammerhead ribozyme					
	domain II	Mg(II), Mn(II)	O_p-A9	X-ray, phosphoro- thioate	g, h

Metal Ions and RNA   297

uridine turn (G5)	Mg(II), Mn(II), Tb(III)	O6-G5	X-ray, inhibition, electrophoretic mobility	h–j
cleavage site	Mg(II)	O_p-A1.1	X-ray, phosphorothioate	i
P4-P6 domain				
P5-U·G	Mg(II), [Os(NH$_3$)$_6$]$^{+++}$	U120·G201 G121·U202		k
P5b-G·U	[Co(NH$_3$)$_6$]$^{+++}$, [Os(NH$_3$)$_6$]$^{+++}$	G147·U156 G148·U155	X-ray, NMR	k, l
P5c	[Os(NH$_3$)$_6$]$^{+++}$	O4-G174 N7/O6-G175 O_p-G176 O_p-U177		k
3 helix junction (P5a, P5b, P5c)	Mg(II)	O6-G164		k
J6/6a	Mg(II), [Os(NH$_3$)$_6$]$^{+++}$	C213 A256 G257		k
A-rich bulge di-Mg(II) site	Mg(II), Sm(III)	O_p-185,186		k
P5a-near A-rich bulge				
J5/5a	Mg(II)	O6-G188		k
J4/5	Mg(II)	O_p-C128		k
	Mg(II)	O_p-G112		k

(*continues*)

Table 5 (Continued)

RNA	Binding site	Metal ions that bind site	Closest contacts	Methodology[a]	Notes
5S rRNA					
	loop E di-Mg(II) site	Mg(II)	O_p-G100 O_p-A101		m
	loop E	Mg(II)	O6-G98 O_p-C97 O_p-A76		m
	stacked GG site	Mg(II)	O6-G105 O6-G106		m
IRE		$[Co(NH_3)_6]^{+++}$	U6-G26	NMR	n

[a] Data from x-ray crystallographic studies unless otherwise noted.
[b] Data from Jack et al. (1977).
[c] Data from Holbrook et al. (1977).
[d] Data from Hingerty et al. (1982).
[e] Data from Ott et al. (1993); Allain and Varani (1995).
[f] Data from Brown et al. (1985).
[g] Data from Pley et al. (1994).
[h] Data from Scott et al. (1995).
[i] Data from Scott et al. (1996).
[j] Data from A.L. Feig et al. (in prep.).
[k] Data from Cate and Doudna (1996).
[l] Data from Kieft and Tinoco (1997).
[m] Data from Correll et al. (1997).
[n] Data from Gdaniec et al. (1998).

respect to their ligand coordination sphere, and almost universally appear in regions of irregular secondary or tertiary structure. The nonbridging phosphate oxygens are the most common non-water ligands found in the inner-coordination sphere of Mg ions in RNA structures. Because they are very hard ligands and the sites of greatest charge density on the RNA, this finding is not unexpected. Other common and somewhat softer ligands include the 2'-hydroxyl groups, the N7 nitrogens of the purine bases, and the keto oxygens of G and U. The only consistent feature in RNA metal-binding sites is that the ions are significantly hydrated. Often only one or two inner-sphere contacts are made between the ion and the RNA to which it is specifically bound, and in some cases, the binding is entirely mediated through outer-sphere contacts. This hydration contrasts with the majority of the metal-binding sites observed in proteins where water ligands tend to occupy very few coordination sites at the metal center (Lippard and Berg 1995).

It remains a challenge in RNA crystallography to unambiguously identify Mg ions, because the electron density is similar to that of hydrated sodium ions or waters. Holbrook et al. (1977) proposed several criteria, including the size and the height of the electron density peaks and the coordination geometry, to distinguish Mg ions from its look-alikes. Hydrated magnesium ions are separated from sodium aqua ions based on the smaller diameter (0.8 Å smaller) of the latter. Another useful approach is to make use of the fact that in many cases, Mn(II) can compete with Mg(II) for specific binding sites. Due to their greater electron density and anomalous scattering properties, Mn(II) ions are often more easily identified in the electron density maps and provide additional support for the assignment of a peak as a hydrated Mg(II) ion (Holbrook and Kim 1997). One must remember, however, that Mg(II) and Mn(II) are not identical. This substitution is often advantageous, but the exact orientation of the metal ion within the overall binding site may shift as a function of this substitution.

Among several well-defined metal-ion-binding sites in tRNA^{Phe} (Fig. 1A), the first example of a mononuclear site is at the intersection of the D-loop and the TΨC-loops and was designated site 1 in the original orthorhombic crystal form (Holbrook et al. 1977) and site 3 in the monoclinic structure (Fig. 1B) (Jack et al. 1977). The closest contact between this ion and the RNA molecule is the 1.9 Å distance to the pro-S_P oxygen of phosphate 19 (Holbrook et al. 1977), which clearly represents an inner-sphere interaction. The remaining contacts to the RNA are 3.5–4.5 Å and thus are probably indicative of outer-sphere interactions mediated by hydrogen bonding of a bound water to the RNA. Although in the best

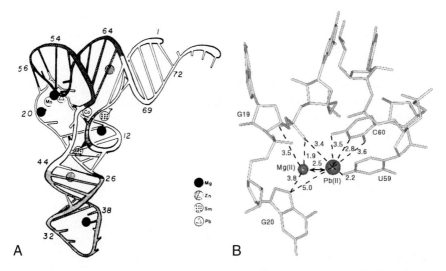

Figure 1 (A) Metal ions found in the crystal structure of yeast tRNA^{Phe} and the relationship between these sites and the overall structure. (Reprinted, with permission, from Pan et al. 1993.) (B) Superposition of the Pb(II) and Mg(II) binding sites from the intersection of the D and TΨC loops. Waters of hydration are not shown and contact distances are given in angstroms.

cases these water molecules can be observed directly in the electron density maps, this site is typical in that the ordering of the water molecules around the metal ion is not resolved.

This tRNA site is useful to illustrate how various ions bind to the same site, since lead and manganese ions have also been characterized crystallographically in this pocket. When Pb(II) binds at this site, the center of the electron density is shifted 2.5 Å with respect to Mg(II). Inner-sphere interactions are observed between the lead and the carbonyl oxygen (O4) of residue U59 (2.2 Å) and the heterocyclic nitrogen (N4) of C60 (2.8 Å), and the ion is positioned much further from p19 (3.4 Å) than was the Mg(II) (Fig. 1B). On the other hand, the position of the Mn(II) is closest to the N7 of G20, but maintains the inner-sphere interaction with the pro-S_P oxygen of p19 in a manner similar to Mg(II) when it occupies this site (Jack et al. 1977). Thus, all three ions occupy the same general site but show distinct coordination preferences. These differences make sense for the three ions. Mg(II) is an extremely hard metal ion, so its position is most strongly influenced by the hard phosphate oxygen. Lead, on the other hand, is much softer. The shift toward the carbonyl oxygen of U59 and ring nitrogen of C60 respects this difference. Furthermore, the larger size of lead and its greater hydration number allow it to fill the entire gap between the residues C60 and C20. The inner-sphere interactions with C60 and U59 do not preclude the outer-sphere coordination to G20 and p19. Mn(II) is too small to span this gap and thus is held to one side of the pocket by its interactions with the N7 and phosphate oxygen. All three ions are sufficiently close together, however, that their binding is mutually exclusive.

Several mononuclear metal-binding sites have been identified crystallographically on the hammerhead ribozyme (Fig. 2A) (Pley et al. 1994; Scott et al. 1995, 1996; Feig et al. 1998). One of these sites is located adjacent to position G5 and was identified based on the binding of Mn(II) and Tb(III), although a small amount of electron density was observed at this site in the presence of Mg(II) (Fig. 2B). This site is located approximately 10 Å from the scissile phosphate. Like the tRNA^{Phe} site discussed above, the different ions occupy overlapping positions that are about 2.1 Å apart. The closest contact in both cases is to the base-pairing face of G5, with longer contacts to the adenosine at residue 6 and the 2'-OH groups of positions 15.3 and 16.2. All three ions are nearly the same distance from G5 but sit either above or below the plane of the base. The resolutions of these structures are insufficient to localize the water molecules coordinated to these ions. It is important to note that these structures were determined from crystals, grown from a mother liquor containing 1.8 M Li$_2$SO$_4$. Because lithium would not be observed in the electron density maps, it is possible that the presence of these ions could perturb the position of another ion nearby such that it would not bind in exactly the same manner as it might in a typical solution study. If there is an inner-sphere contact with the N1 position of G5, it almost certainly implies that metal ion coordination at this site induces a tautomerization, because this nitrogen is protonated in the dominant tautomer. Metal binding at this site has been linked to a structural rearrangement at low Mg(II) concentrations by gel electrophoretic mobility studies (Bassi et al., 1995, 1996). It has also been shown that the binding of Tb(III) to this site results in inhibition of the cleavage reaction (Feig et al. 1998).

A second metal-binding site from the hammerhead ribozyme is located near the phosphate of residue A9 (Fig. 2C). This metal-binding site was identified in both X-ray crystal structures (Pley et al. 1994; Scott et al. 1995), but there is a discrepancy between the two with regard to the details observed. The Mg ion found in the all-RNA structure makes an inner-sphere contact to the pro-S$_P$ oxygen of p9 and outer-sphere interactions with the 2'-OH of G8 (3.8 Å), the N7 of G10.1 (3.9 Å) and the N2 of G12 (4.6 Å). In contrast, when Mn(II) was used to localize the metal-

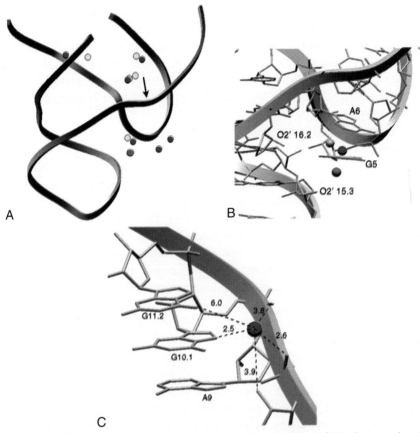

Figure 2 (A) Metal ions found in the crystal structures of the hammerhead ribozyme and the relationship between these sites and the overall structure. The location of the cleavage site is marked by the arrow. Mg(II) is shown in *yellow*, Mn(II) is shown in *blue*, and Tb(III) is shown in *red*. (B) Expanded view of the G5 metal-binding site of the hammerhead ribozyme. Waters of hydration have been omitted. (C) Expanded view of Mn(II) binding to the p9 site of the hammerhead ribozyme. Waters of hydration are not shown and distances are given in angstroms.

binding sites of the RNA/DNA chimeric hammerhead, the metal ion was observed binding directly to the pro-R_P oxygen of p9 as well as to the N7 position of G10.1 (Pley et al. 1994). The different coordination modes, as in the tRNA site discussed above, might result from the metal substitution used to better visualize the ion. Whereas metal binding to this site appears essential for cleavage, the role of this ion in the mechanism is still uncertain (McKay 1996; Peracchi et al. 1997).

The first dinuclear site identified in an RNA molecule is from the P4-6 domain of the group I intron from *Tetrahymena* (Cate and Doudna 1996; Cate et al. 1996, 1997). In total, 12 well-defined Mg(II) ions were identified in the native crystal structure (Fig. 3A). The dimagnesium site from P4-P6 is found as an integral component of the A-rich bulge (Fig. 3B). In this region, the phosphate backbone is highly distorted and the two magnesium ions are found bridged by two phosphate groups. One Mg(II) lies on either side of the backbone with a Mg–Mg separation of 5.3 Å. Each Mg ion makes three inner-sphere contacts with phosphate oxygens. Three unresolved water molecules probably fill out the rest of the coordination sphere. The metal binding helps to hold the backbone in an unusual conformation with a 4.2 Å phosphate-to-phosphate distance between A184 and A186. Residue 185 is involved in reversing the direction of the phosphate backbone. The inner-sphere Mg-O_P distances are unremarkable

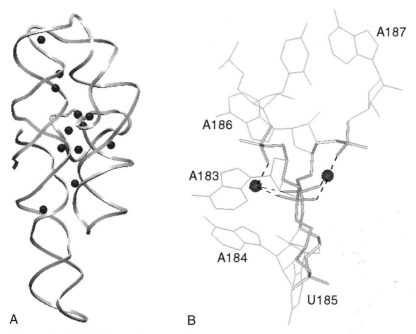

Figure 3 (A) Ribbon diagram of the P4-P6 domain of the group I intron from *Tetrahymena* showing the location of the bound Mg(II) ions. (B) Expanded view of the dimagnesium site from the A-rich bulge of the P4-P6 domain. The Mg(II) ions hold the backbone in a highly distorted conformation by coordination to the phosphate oxygens of A184 and A186. Phosphate oxygens from A183 and A187 also make inner sphere contacts with the Mg(II) ions. The Mg–Mg distance is 5.3 Å. The Mg(II)–O_P distances range from 2.0 to 2.3 Å. Waters of hydration are not shown.

at 2.0–2.3 Å. Both Mg ions lie in the phosphinyl plane of phosphate 186. This orientation has been shown to be less favorable for the binding of metal ions to phosphate oxygens than one in which the metal ion sits 0.9 ± 0.5 Å above or below that plane (Alexander et al. 1990). In the case of phosphate 184, one Mg ion is in-plane, whereas the other rests in the more favorable out-of-plane orientation.

A variety of other metal ions have been soaked into the crystals of P4-P6 and this site will accept Sm(III) ions. Although osmium hexammine and cobalt hexammine are observed to bind in a number of the mononuclear sites on this molecule, neither ion binds efficiently to the A-rich bulge, based on the crystallographic experiments. These two ions are exchange-inert, and the specificity may relate to their inability to make the close contacts to the phosphate oxygens observed in the native structure. Three other Mg ions also bind in this region of the RNA in mononuclear sites and presumably assist in the stabilization of the extensive interhelical packing.

The dinuclear magnesium center observed in the loop E fragment of the 5S rRNA is currently unique among RNA metal-binding sites (Fig. 4A) (Correll et al. 1997). These ions are two of the five that line the major groove of the helical fragment and help pull the backbones to a very narrow 6 Å separation between the phosphate atoms. The high resolution of the structure (1.5 Å) allows the Mg ions to be clearly visualized with their

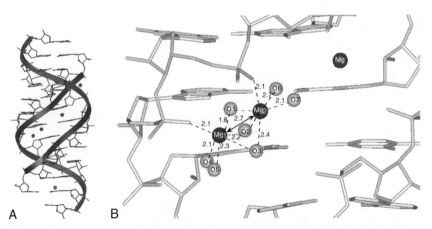

Figure 4 (A) Metal-binding sites from the loop E fragment of the *E. coli* 5S rRNA. (B) Expanded view of the dimagnesium center from the 5S rRNA determined at 1.5 Å resolution. Water molecules directly coordinated to the Mg ions of the dinuclear center are shown, but others have been omitted for clarity. Contact and bond distances are given in angstroms.

coordinated water molecules (Fig. 4B). The magnesiums lie just 2.7 Å apart and have been modeled as being bridged by three water molecules. The protonation state of these waters cannot be determined based on the electron density map, but it is unlikely that they are fully protonated. For the extremely short metal–metal distance, it is likely that at least one and possibly two of the bridging waters are partly or completely deprotonated. In protein and small molecule systems, short metal–metal interactions are often mediated by monatomic bridging ligands (Lippard and Berg 1995; Dismukes 1996; Yachandra et al. 1996). The manganese centers from the photosystem II reaction center and catalase make for good comparisons. The manganese core in PSII consists of a dimer of dimers with a Mn–Mn distance of ~2.7 Å derived from EXAFS and model studies (Yachandra et al. 1996). To maintain this short Mn–Mn distance, the core contains two μ-oxo (O^{2-}) bridges. For catalase, the Mn(II)–Mn(II) separations are much larger, separated by ~3.7 Å for the aquo bridged form (Dismukes 1996). The water labeled O1 in the 5S rRNA (Fig. 4B) is a likely candidate for deprotonation, as it is located just 1.8 and 1.9 Å from Mg_1 and Mg_2, respectively. This distance is 0.3 Å shorter than a typical Mg–H_2O bond and 0.4 Å shorter than the next shortest bridging ligand. The other two bridges have 2.2 Å and 2.4 Å contacts to the Mg ions and might reflect hydroxo and aquo bridges, respectively. Insufficient small molecule model chemistry is currently available for Mg(II) to predict the actual protonation state of this core based on the observed distances.

The coordination chemistry of the dimagnesium center in the 5S rRNA is interesting. In the other Mg(II) sites described above, the coordination spheres of the Mg ions are believed to be quite close to octahedral. This generality holds true for the mononuclear Mg sites in the high-resolution 5S rRNA fragment as well. However, deformation of the octahedral geometry is much more evident in this dinuclear center. The coordination sphere of Mg_1 is only slightly distorted such that one of the bridging waters (O2) is pulled out of the equatorial plane consisting of O1, O2, O4, and O5. The bridging water located *trans* to the axial phosphate is significantly elongated relative to the other bridging ligands and may reflect a *trans* influence (Huheey 1983) in the chemistry of Mg(II). The geometry around Mg_2 is much more distorted. If one defines the equatorial plane as containing O1, O2, O6, and O7, the O_p–Mg_2–O3 bond angle is only 160° instead of the ideal 180°. The elongation of the monatomic bridge *trans* to the phosphate oxygen is also observed at this metal center. Both sites of phosphate coordination occur in the optimal, anti, out-of-plane geometry. The source of the distortion is likely to be the necessity to satisfy the coordination requirements of both Mg ions simultaneously.

By looking at these sites, it is clear that we are beginning to see patterns in the coordination of metal ions to specific binding sites based on X-ray crystallographic studies. A variety of metal ions will often be able to occupy the same site. The manner in which they do so, however, depends on the properties of the particular ion. From this information, it is clear that metal substitution chemistry will allow the identification of general sites quite well, but exact information about the way Mg(II) coordinates to the RNA will not be obtained through these studies. As structural data become available on larger and more complicated RNAs, more variation in the types of sites will likely be observed. In all RNAs, the majority of the sites appear to be mononuclear, but higher nuclearity sites do exist. High-resolution data unfortunately are going to be required for the greatest understanding of these interactions so that the coordinated water molecules and the geometry around the metal center can be observed.

SPECTROSCOPIC METHODS TO STUDY METAL-ION-BINDING SITES IN RNA

NMR has great potential for the localization of metal-binding sites in relatively small RNAs. Depending on the metal ion involved, the binding event can be observed by changes in the chemical shift or through the line broadening of RNA resonances. This methodology unfortunately requires the assignment of resonances. Currently, only relatively few RNAs have been studied at this level of detail, so limited data on metal binding are available.

The more common NMR approach is the use of paramagnetic metals such as Mn(II) (Bertini and Luchinat 1986). When these ions bind to the RNA, nearby atoms experience enhanced relaxation and the line broadening with a distance dependence of r^{-6} from the metal-binding site. The potential to locate binding sites using this technique is pretty clear, but the lack of data at very short distances from tightly bound ions often makes it difficult to determine the exact binding mode within the site. This technique has been applied successfully to metal binding adjacent to GU and GG sites embedded within helices (Ott et al. 1993; Allain and Varani 1995). In these studies, Mn(II) was found to be in fast exchange between its free and bound states, since substoichiometric concentrations of Mn(II) affected the entire RNA signal.

A second NMR approach relies on the chemical shifts of RNA resonances that change upon metal binding. This technique has an advantage over paramagnetic relaxation experiments in that it allows Mg(II) itself to be used. Unfortunately, the results can be difficult to interpret. Upon addition of Mg(II) to a sample, many signals may change. The main problem is determining whether the altered spectra result from a specific binding event or a global conformational change. An example of the successful application of ^{31}P NMR in this manner involves a study of the hammerhead ribozyme where one of the Mg(II)-binding sites was localized by this method and the binding constant for the interaction determined (J.P. Simore and A. Pardi, unpubl.).

NMR has also been used to further probe the solution conformation of a metal-binding site that was observed in the P4-P6 crystal structure. In this example, the exchange inert compound $[Co(NH_3)_6]^{+++}$ was used to mimic a hydrated Mg ion (Kieft and Tinoco 1997). Observation of five NOE contacts between the ammine protons and the RNA allowed the modeling of the exact position of the metal ion in the major groove of the RNA fragment. In this case, the position and orientation of the cobalt within the binding site is much better determined than in the X-ray structure where the inner-sphere ligands were not located. The $[Co(NH_3)_6]^{+++}$ was found to be in fast exchange (on the NMR time scale) between the bound and unbound states, even though the interaction was specific and reasonably tight. Other NMR experiments that have not been fully utilized in RNA biochemistry employ metal ions that are NMR active. These metals include species like ^{113}Cd. The Cd(II) substitution has been used very successfully in the study of Zn(II)-binding proteins since NMR can provide information regarding the ligands that coordinate to the metal ion (Summers 1988).

Many other biophysical techniques can be used to extract important information regarding an established metal-binding site. One set of data that is currently of great interest is the specificity of a site as derived from the relative binding affinities in different ions. The easiest way to extract these data is to perform competition experiments. Luminescence spectroscopy recently has been used to probe the specificity of the G5 metal-binding site of the hammerhead ribozyme (Fig. 2B) (A.L. Feig et al., in prep.). Since Tb(III) binds specifically to this site and has a sensitized luminescence spectrum while bound, the extent of binding can be probed as a function of the concentration of a competitor. It is still unclear exactly what property of the ion the RNA recognizes, but it clearly does so with great specificity, as there is a 10^3 difference in the binding affinity at this site for the six different divalent ions tested.

The biophysical techniques that allow spatial localization are obviously important for the study of metal ions in RNA systems. The current focus in the field has been on finding the metal-binding sites. Once a sufficient number of examples are available, the next phase of these projects is going to turn to the more detailed probing of the coordination sphere and role of these ions. A host of spectroscopic techniques have been developed for the study of metal ions (Drago 1977). These techniques have been applied quite successfully to the protein systems, but with much more limited usage within the area of nucleic acid biochemistry (Lippard and Berg 1995). Some of these methods have a relatively narrow range of metal ions that are suitable for study. As we learn more about the specificity of RNA-binding sites, it is hoped that we will come to a point where we will be able to either predict the metal-binding sites that will be affected under a certain set of conditions, or to control the population of individual sites.

BIOCHEMICAL APPROACHES TO LOCATE METAL-ION-BINDING SITES

Many RNA biochemical experiments involve varying the concentration and identity of metal ions. Most, however, involve examining some bulk property of the RNA and only indirectly conclude information about the location of the metal ions. Very few methods actually attempt to locate the binding site.

One of the first methods used involves ion-specific cleavage of RNA. This method relies on the fact that when metal ions such as Pb(II) or Eu(III) with sufficiently low pK_a values bind to RNA, they sometimes induce cleavage by deprotonating a nearby $2'$-OH that subsequently attacks the adjacent phosphodiester bond. This method is best exemplified by cleavage at position 17 of tRNA^{Phe} by metal ions bound at the site shown in Figure 1B. Clearly, this cleavage is dependent on the coordination of the metal ion, and the site of cleavage can be slightly modulated by changing the metal ion used in the experiment (Ciesiolka et al. 1989; Michalowski et al. 1996a). Cleavage sites on other RNAs have been mapped, purportedly identifying the location of metal-ion-binding sites (Zito et al. 1993; Winter et al. 1997). However, this methodology is not ideal because (1) the site of cleavage may not be near the metal-ion-binding site in the primary sequence, and (2) not all metal-binding sites lead to cleavage. This method is better suited to assay for structural changes upon mutagenesis similar to other chemical modification reagents (Ciesiolka et al. 1989; Behlen et al. 1990; Michalowski et al. 1996a,b).

Currently, the most effective biochemical approach to locate metal-ion-binding sites in RNA involves the selective replacement of oxygen atoms suspected of being ligands to metal ions by sulfur atoms. The development of powerful chemical (Eckstein 1991) and enzymatic (Ruffner and Uhlenbeck 1990) synthesis methods has made this approach feasible. The most common substitution used for these experiments is the

R_P phosphorothioate, which can be prepared by in vitro transcription. Both the R_P and the S_P isomers are prepared by chemical synthesis and can often be separated by HPLC methods if the oligos are sufficiently short (Slim and Gait 1991). Other oxygen atoms that have been specifically substituted with sulfur include the 3'-bridging (Sun et al. 1997; Weinstein et al. 1997) and 5'-bridging (Piccirilli et al. 1993; Kuimelis and McLaughlin 1995, 1996; Zhou et al. 1996, 1997) phosphate oxygens, the keto oxygens of U and C (^{4S}U, ^{4S}C, ^{2S}C) (Wang and Ruffner 1997), and the 2'-OH group (Hamm and Piccirilli 1997). Varying degrees of efficacy have been observed with respect to the modulation of the catalytic activity and metal ion specificity with these modifications.

The basic chemistry of the experiment derives from hard and soft acid base theory. The idea is that the sulfur modification will reduce the affinity of a given oxygen ligand for a Mg(II) ion. Since Mg(II) is an extremely hard Lewis acid ($\eta = 47.59$), binding to the sulfur atom will be disfavored. If this ligand is critical for positioning a metal ion involved in the folding or catalysis of the ribozyme, activity will be lost. In experiments often referred to as rescue experiments, the addition of softer metals, such as Mn(II) ($\eta = 9.02$) or Cd(II) ($\eta = 10.29$), can restore activity by interacting with the sulfur atom and thus replace the Mg ion. The choice of a rescue metal should be governed by the hardness parameter η listed in Table 2, but several ions should be tested because the structural details of the binding site and a variety of experimental parameters will dictate whether any given ion will be able to successfully replace Mg(II) in a particular site. In early experiments, the rescue metal was used to the exclusion of other metal ions in solution (Dahm and Uhlenbeck 1991). More recent work has shown that better data result from the use of lower concentrations of the rescue metal in the presence of a significant excess of Mg(II) to assure that the other metal-ion-binding sites contain Mg(II) and are therefore not perturbed by the presence of the softer ion (Christian and Yarus 1993; Piccirilli et al. 1993; Peracchi et al. 1997). In model chemistry that used adenosine nucleotide di- and triphosphates and their thiophosphate analogs, where the individual binding constants for a variety of ions could be measured accurately, Mg(II) showed a 31,000-fold preference for binding the oxygen rather than the sulfur (Pecoraro et al. 1984). In the same experiments, Cd(II) exhibited a 55-fold preference for the sulfur and Mn(II) showed approximately equal affinity for oxygen and sulfur ligands. Qualitatively similar findings were found for a series of metal ions with respect to their affinity for AMP^{2-} and $AMPS^{2-}$ (Sigel et al. 1997).

Phosphorothioate rescue experiments have been used in a number of ribozyme systems, including the hammerhead (Dahm and Uhlenbeck

1991; Slim and Gait 1991; Peracchi et al. 1997), the hepatitis δ virus (Jeoung et al. 1994; Fauzi et al. 1997), hairpin (Chowrira and Burke 1991, 1992), group I intron (Waring 1989; Christian and Yarus 1993; Piccirilli et al. 1993), group II intron (Chanfreau and Jacquier 1994; Podar et al. 1995), and RNase P (Harris and Pace 1995; Warnecke et al. 1996). These experiments provide the most useful data when activity in Mg(II) alone is reduced but the addition of Mn(II) or Cd(II) restores catalysis. This combination of results is interpreted as strong evidence for a metal-binding site that utilizes the specific phosphate oxygen as an inner-sphere ligand. When one phosphorothioate isomer but not the other displays this behavior, the conclusions are even more strongly supported, as is the case with the cleavage site phosphorothioate substitution in the hammerhead ribozyme (Koizumi and Ohtsuka 1991; Slim and Gait 1991) because it controls for the structural deformations that might have occurred as a result of this substitution. In certain cases, activity is lost upon incorporation of the phosphorothioate into the ribozyme at a certain position, but addition of a rescue metal does not restore activity (Strobel and Shetty 1997). When this behavior is observed, a role for the phosphate oxygen in metal binding is not excluded, but it must be remembered that even though the change is a single oxygen atom to a sulfur, significant perturbation of local structure can result. The charge localization on a phosphorothioate residue is also different from a standard phosphodiester, residing primarily on the sulfur as opposed to being distributed evenly between the nonbridging oxygen atoms (Frey and Sammons 1985). Furthermore, the phosphorus–sulfur bond is longer than a typical phosphate–oxygen interaction. Together, these structural and electronic differences can lead to misalignment of functional groups required for catalytic activity irrespective of metal binding. In a few cases, data from phosphorothioate rescue experiments can be related to metal binding at the site observed crystallographically. One example of such corroboration comes from the p9 site in the hammerhead ribozyme (Fig. 2C) (Peracchi et al. 1997). Another example comes from the P4-P6 structure (Cate and Doudna 1996; Cate et al. 1997). Rescue experiments therefore appear to accurately predict metal-binding sites that contain phosphate–oxygen ligation.

There are a few caveats and potential biases in sulfur modification experiments. The main limitation with the use of phosphorothioate chemistry is that only metal ions that contain inner-sphere contacts to the phosphate backbone are sensitive to this class of experiments. This technique may find many of the important metal-binding sites, but it can never find all of them because some of the site(s) of interest may not contain these ligands. A further potential problem is referred to as "recruitment." As the name implies, there is the possibility that by making the modification to the RNA, a metal-binding site was created that did not exist in the parent molecule. Recruitment cannot formally be excluded as a possible side effect of these modifications; however, the incorporation of phosphorothioates in most positions in ribozymes does not affect the catalytic rate even when thiophilic metal ions are added. Thus, recruitment is not a particularly serious problem. Systematic probing of phosphorothioates in RNAs that are easily characterized will be required to determine the significance of this potential artifact. Small-molecule modeling studies can also be used to address this problem.

The use of competitive inhibition experiments is another versatile method to identify metal-binding events that are critical for activity in a catalytic ribozyme system. This technique answers questions about metal specificity in a critical site, but alone does not allow the identification or localization of the metal-binding site(s) involved in the inhibition event. The choice of inhibitors should therefore be made with the problem of localization in mind. Luminescent metal ions are generally relatively easy to localize based on energy-transfer experiments, the utility of which was originally demonstrated on tRNAs (Kayne and Cohn 1974; Wolfson and Kearns 1974; Draper 1985). In principle, localization of the sites of Cr(II) or Co(II) binding to RNA can also be accomplished by oxidizing them to Cr(III) or Co(III) in situ because the ligand exchange rates of the +3 ions are sufficiently slow that the complex remains intact during analysis (Danchin 1973, 1975).

METAL IONS IN THE RNA WORLD

By combining our limited knowledge of metal-ion-binding to contemporary RNAs and our more extensive knowledge of metal-ion-binding to proteins, it is possible to speculate on the role of metal ions in prebiotic molecular evolution. It seems clear that specifically bound metal ions co-evolved with RNA molecules. Many of the mononuclear sites in Table 5 are formed with, or can be engineered into, small RNA fragments. Since such sites are highly hydrated and contain limited direct contact with the RNA, the observed affinities are only moderate, in the 1–1000 μM range. These sites are also expected to show limited specificity, predominantly dictated by the chemical nature of the ligands. Furthermore, in these examples, the RNA structures themselves are likely to be quite flexible and can accommodate a variety of metal ions with only minor distortions to the overall RNA fold. These minimalist sites are sufficient to stabilize the secondary and tertiary structures observed in these motifs.

The metal ion sites generated on small RNAs appear to be capable of facilitating a variety of different types of chemistry. Activities range from the transesterification and hydrolytic reactions of small ribozymes (Pyle 1996; Sigurdsson et al. 1998) to the more exotic porphyrin metalation (Conn et al. 1996) and Diels-Alder condensation reactions (Tarasow et al. 1997) catalyzed by aptamers produced from in vitro selection experiments. These small RNAs have only limited amounts of structure and therefore are likely to position the catalytic metal ions by only a few points of contact. The relatively modest rate enhancements supported by catalytic RNAs such as these probably reflect the types of species that first evolved from random polymerization events. Very active metal ions might have assisted in this process but would have increased the danger of side reactions that would accidentally damage the catalyst.

A striking difference between most RNA metal-binding sites studied thus far and those seen in proteins is the degree of hydration. Both structural and catalytic metal-ion-binding sites in proteins are predominantly dehydrated (Lippard and Berg 1995). Water molecules occasionally appear in the coordination spheres of these metal ions, but in these cases, they are often believed either to be displaced by the substrate when it enters the active site or to take part in the catalytic mechanism of the enzyme. Such protein sites also bind their metal ions much more tightly than the RNA systems. In fact, tight binding is a requirement for dehydrated sites, since there is a characteristic energy (ΔH_{hyd}) associated with the hydration of any ion. The net binding energy upon coordination of the ion must account for the energetic cost of dehydration. The question arises, Why are such dehydrated sites not observed in RNAs?

One possibility is that metal-binding sites in RNAs are intrinsically different from those in proteins. RNA has a much more limited set of ligands to use in generating a specific metal-binding pocket. Amino acid side chains containing thiols and thioethers are well suited to binding a variety of softer metals. In addition, the carboxylate side chains provide anionic ligands with great versatility in their potential modes of coordination. They can act as either terminal or bridging ligands and bind in either monodentate or bidentate geometries. The nucleotides, on the other hand, are much larger and more rigid than the corresponding amino acids. The anionic ligand in RNA, the nonbridging phosphate oxygen, is an integral component of the backbone and therefore is more limited in its conformational freedom than the aspartate and glutamate carboxylate groups. The heterocyclic ring nitrogens and the keto oxygens from the bases are held in rigidly planar orientations by the aromatic rings. This geometric constraint severely limits the ability of an RNA to compactly encompass a metal ion and provide more than facial coordination and therefore complete dehydration. It also explains why the most specific metal-binding sites are not in the Watson-Crick base-paired regions of the structure where the conformation is too constrained. Instead, metal-ion-binding sites are clustered in regions of extensive distortion from the A-form RNA helices.

There is also the question of the folding of RNAs relative to that of proteins. It is possible that in RNAs there is insufficient energy in the folding and metal-binding process to completely displace the waters of hydration around a metal ion. It has been suggested that in contemporary RNAs, modified nucleotides might be present to assist in metal ion binding (Agris 1996). A more straightforward possibility, however, is that most RNAs studied to date are structurally too simple. In these RNAs, most residues involved in metal ion binding are solvent-exposed. Thus, the RNAs have no real inside comparable to the hydrophobic core of a protein. The largest RNA crystallographically characterized to date is the P4-P6 domain. On the basis of that structure, it was proposed that an ionic core may substitute in RNA folding for the hydrophobic core of proteins such that the 3° structure assembles around a fixed number of discrete metal-binding sites (Cate et al. 1997). Even in this structure, however, the most buried of the metal-binding sites are significantly hydrated.

It could be that all metal-ion-binding sites in RNA are at least partially hydrated. One can imagine several advantages to using hydrated ions within the ionic core of a large RNA. Hydrated ions would span larger voids than dehydrated ions and allow looser packing of secondary structure elements. The hydrated ion also can accommodate a wide range of structural interactions through its orientation of the water molecules as compared to direct coordination of metal ions at every site. In addition, the energy associated with deforming the outer-sphere interactions should be significantly less than what would be observed for distorting the inner-sphere coordination. A consequence of RNAs having a core of hydrated ions is that one might expect this core to be much more dynamic than the hydrophobic core of a protein.

In the modern protein world, metal cofactors are associated with a variety of reaction types, including electron transfer, redox chemistry, and hydrolysis reactions. Transesterification and hydrolytic activities, however, are the primary catalytic behaviors observed in ribozymes. Did these other catalytic activities not develop until the dawn of the protein world, or are there undiscovered natural catalytic RNAs that are the ancestors of the early redox enzymes? Through the use of in vitro selection experiments, the scope of RNA catalysis has been significantly broadened. RNA is almost certainly capable of catalyzing these other classes of reactions, but it is still unclear whether there are naturally occurring examples. Such an enzyme would likely use a metal ion cofactor other than Mg(II), so the search for RNA molecules that naturally use alternative ions is of significant interest. A recent selection experiment showed that a single base change results in an altered metal ion specificity for RNase P (Frank and Pace 1997). It is clear from this result that catalytic RNAs retain the ability to adapt to an everchanging environment, using the resources available to evolve and to overcome evolutionary pressures. Were RNAs to have evolved out of an environment devoid of metal ions, they probably would have found a way around the problems of folding and generating reactive functional groups. The primordial soup and all cellular environments that have evolved subsequently contained a variety of ions, however. Given the availability of metal ions, they will certainly play a significant role in the biology of current and future RNAs.

ACKNOWLEDGMENTS

This work was sponsored by a grant from the National Institutes of Health (GM-36944 to O.C.U.). A.L.F. acknowledges a postdoctoral fellowship from the National Science Foundation (CHE-9504698). We thank A. Pyle, C. Correll, and Y. Lu for their helpful comments.

REFERENCES

Agris P.F. 1996. The importance of being modified: Roles of modified nucleosides and Mg(II) in RNA structure and function. *Prog. Nucleic Acid Res. Mol. Biol.* **53:** 79–129.

Alexander R.S., Kanyo Z.F., Chirlian L.E., and Christianson D.W. 1990. Stereochemistry of phosphate-Lewis acid interactions: Implications for nucleic acid structure and recognition. *J. Am. Chem. Soc.* **112:** 933–937.

Allain F.H.T. and Varani G. 1995. Divalent metal ion binding to a conserved wobble pair defining the upstream site of cleavage of group I self-splicing introns. *Nucleic Acids Res.* **23:** 341–350.

Anderson C.F. and Record, Jr. M.T. 1995. Salt-nucleic acid interactions. *Annu. Rev. Phys. Chem.* **46:** 657–700.

Bassi G.S., Murchie A.I.H., and Lilley D.M.J. 1996. The ion-induced folding of the hammerhead ribozyme: Core sequence changes that perturb folding into the active conformation. *RNA* **2:** 756–768.

Bassi G.S., Møllegaard N.-E., Murchie A.I.H., von Kitzing E., and Lilley D.M.J. 1995. Ionic interactions and the global conformations of the hammerhead ribozyme. *Nat. Struct. Biol.* **2:** 45–55.

Beebe J.A., Kurz J.C., and Fierke C.A. 1996. Magnesium ions are required by *Bacillus subtilis* ribonuclease P RNA for both binding and cleaving precursor tRNA[Asp]. *Biochemistry* **35:** 10493–10505.

Behlen L.S., Sampson J.R., DiRenzo A.B., and Uhlenbeck O.C. 1990. Lead-catalyzed cleavage of yeast tRNA^{Phe} mutants. *Biochemistry* **29:** 2515–2523.

Bengston S., ed. 1994. *Early life on earth.* Columbia University Press, New York.

Bertini I. and Luchinat C. 1986. *NMR of paramagnetic molecules in biological systems.* Benjamin/Cummings, Menlo Park, California.

Bina-Stein M. and Stein A. 1976. Allosteric interpretation of Mg(II) binding to the denaturable *Escherichia coli* tRNA^{Glu}. *Biochemistry* **15:** 3912–3917.

Brown R.S., Dewan J.C., and Klug A. 1985. Crystallographic and biochemical investigation of the lead(II)-catalyzed hydrolysis of yeast phenylalanine tRNA. *Biochemistry* **24:** 4785–4801.

Bukhman Y.V. and Draper D.E. 1997. Affinities and selectivities of divalent cation binding sites within an RNA tertiary structure. *J. Mol. Biol.* **273:** 1020–1031.

Cate J.H. and Doudna J.A. 1996. Metal-binding sites in the major groove of a large ribozyme domain. *Structure* **4:** 1221–1229.

Cate J.H., Hanna R.L., and Doudna J.A. 1997. A magnesium ion core at the heart of a ribozyme domain. *Nat. Struct. Biol.* **4:** 553–558.

Cate J.H., Gooding A.R., Podell E., Zhou K., Golden B.L., Kundrot C.E., Cech T.R., and Doudna J.A. 1996. Crystal structure of a group I ribozyme domain: Principles of RNA packing. *Science* **273:** 1678–1685.

Chanfreau G. and Jacquier A. 1994. Catalytic site components common to both splicing steps of a group II intron. *Science* **266:** 1383–1387.

Chin K. and Pyle A.M. 1995. Branch-point attack in group II introns is a highly reversible transesterification, providing a potential proofreading mechanism for 5′-splice site selection. *RNA* **1:** 391–406.

Chowrira B.M. and Burke J.M. 1991. Binding and cleavage of nucleic acids by the hairpin ribozyme. *Biochemistry* **30:** 8515–8522.

———. 1992. Extensive phosphorothioate substitution yields highly active and nuclease-resistant hairpin ribozymes. *Nucleic Acids Res.* **20:** 2835–2840.

Christian E.L. and Yarus M. 1993. Metal coordination sites that contribute to structure and catalysis in the group I intron from *Tetrahymena*. *Biochemistry* **32:** 4475–4480.

Ciesiolka J., Wrzesinski J., Gornicki P., Podkowinski J., and Krzyzosiak W.J. 1989. Analysis of magnesium, europium and lead binding sites in methionine initiator and elongator tRNAs by specific metal-ion-induced cleavages. *Eur. J. Biochem.* **186:** 71–77.

Clouet-d'Orval B. and Uhlenbeck O.C. 1996. Kinetic characterization of two I/II format hammerhead ribozymes. *RNA* **2:** 483–491.

Collins R.A. and Olive J.E. 1993. Reaction conditions and kinetics of self-cleavage of a ribozyme derived from *Neurospora* VS RNA. *Biochemistry* **32:** 2795–2799.

Conn M.M., Prudent J.R., and Schultz P.G. 1996. Porphyrin metalation catalyzed by a small RNA molecule. *J. Am. Chem. Soc.* **118:** 7012–7013.

Correll C.C., Freeborn B., Moore P.B., and Steitz T.A. 1997. Metals, motifs, and recognition in the crystal structure of a 5S rRNA domain. *Cell* **91:** 705–712.

Cotton F.A. and Wilkinson G. 1988. *Advanced inorganic chemistry*, 5th edition. Wiley, New York.

Cowan J.A., ed. 1995. *The biological chemistry of magnesium.* VCH, New York.

Dahm S.C. and Uhlenbeck O.C. 1991. Role of divalent metal ions in the hammerhead RNA cleavage reaction. *Biochemistry* **30:** 9464–9469.

Danchin A. 1973. Biological macromolecules labeling with covalent complexes of magnesium analogs. I. The cobaltic Co(III) ion. *Biochimie* **55:** 17–27.

———. 1975. Labeling of biological macromolecules with covalent analogs of magnesium. II. Features of the chromic Cr(III) ion. *Biochimie* **57:** 875–880.

Dismukes G.C. 1996. Manganese enzymes with binuclear active sites. *Chem. Rev.* **96:** 2909–2926.

Drago R. 1977. *Physical methods in chemistry.* W.B. Saunders, Philadelphia, Pennsylvania.

Draper D.E. 1985. On the coordination properties of Eu(III) bound to tRNA. *Biophys. Chem.* **21:** 91–101.

Eckstein F. 1991. *Oligonucleotides and analogues: A practical approach.* IRL Press, Oxford, United Kingdom.

Fauzi H., Kawakami J., Nishikawa F., and Nishikawa S. 1997. Analysis of the cleavage reaction of a trans-acting human hepatitis delta virus ribozyme. *Nucleic Acids Res.* **25:** 3124–3130.

Feig A.L., Scott W.G., and Uhlenbeck O.C. 1998. Inhibition of the hammerhead ribozyme cleavage reaction by site-specific binding of Tb(III). *Science* **279:** 81–84.

Frank D.N. and Pace N.R. 1997. In vitro selection for altered divalent metal specificity in the RNase P RNA. *Proc. Natl. Acad. Sci.* **94:** 14355–14360.

Frey P.A. and Sammons R.D. 1985. Bond order and charge localization in nucleoside phosphorothioates. *Science* **228:** 541–545.

Gdaniec Z., Sierzputowska-Gracz H., and Theil E.C. 1998. Iron regulatory element and internal loop/bulge structure for ferritin mRNA studied by cobalt(III) hexammine binding, molecular modeling and NMR spectroscopy. *Biochemistry* **37:** 1505–1512.

Gluick T.C., Gerstner R.B., and Draper D.E. 1997. Effects of Mg^{2+}, K^+, and H^+ on an equilibrium between alternative conformations of an RNA pseudoknot. *J. Mol. Biol.* **270:** 451–463.

Grosshans C.A. and Cech T.R. 1989. Metal ion requirements for sequence-specific endoribonuclease activity of the *Tetrahymena* ribozyme. *Biochemistry* **28:** 6888–6894.

Hamm M.L. and Piccirilli J.A. 1997. Incorporation of 2'-deoxy-2'-mercaptocytidine into oligonucleotides via phosphoramidite chemistry. *J. Org. Chem.* **62:** 3415–3420.

Hampel A. and Cowan J.A. 1997. A unique mechanism for RNA catalysis: the role of metal cofactors in hairpin ribozyme cleavage. *Chem. Biol.* **4:** 513–517.

Harris M.E. and Pace N.R. 1995. Identification of phosphates involved in catalysis by the ribozyme RNase P RNA. *RNA* **1:** 210–218.

Hingerty B.E., Brown R.S., and Klug A. 1982. Stabilization of the tertiary structure of yeast phenylalanine tRNA by $[Co(NH_3)_6]^{3+}$. X-ray evidence for hydrogen bonding to pairs of guanine bases in the major groove. *Biochim. Biophys. Acta* **697:** 78–82.

Holbrook S.R. and Kim S.-H. 1997. RNA crystallography. *Biopolymers* **44:** 3–21.

Holbrook S.R., Sussman J.L., Warrant R.W., Church G.M., and Kim S.-H. 1977. RNA-ligand interactions. (I)Mg binding sites in yeast tRNA^{Phe}. *Nucleic Acids Res.* **4:** 2811–2820.

Huheey J.E. 1983. *Inorganic chemistry*, 3rd edition. Harper & Row, New York.

Jack A., Ladner J.E., Rhodes D., Brown R.S., and Klug A. 1977. A crystallographic study of metal-binding to yeast phenylalanine tRNA. *J. Mol. Biol.* **111:** 315–328.

Jeoung Y.-H., Kumar P.K.R., Suh Y.-A., Taira K., and Nishikawa S. 1994. Identification of phosphate oxygens that are important for self-cleavage activity of the HDV ribozyme by phosphorothioate substitution interference analysis. *Nucleic Acids Res.* **22:** 3722–3727.

Kayne M.S. and Cohn M. 1974. Enhancement of Tb(III) and Eu(III) fluorescence in complexes with *E. coli* tRNA. *Biochemistry* **13:** 4159–4165.

Kieft J.S. and Tinoco, Jr. I. 1997. Solution structure of a metal-binding site in the major groove of RNA complexed with cobalt (III) hexammine. *Structure* **5:** 713–721.

Koizumi M. and Ohtsuka E. 1991. Effects of phosphorothiolate and 2′ amino groups in hammerhead ribozymes on cleavage rates and Mg^{2+} binding. *Biochemistry* **30:** 5145–5150.

Kuimelis R.G. and McLaughlin L.W. 1995. Hammerhead ribozyme-mediated cleavage of a substrate analogue containing an internucleotidic bridging 5′-phosphorothioate: Implications for the cleavage mechanism and the catalytic role of the metal cofactor. *J. Am. Chem. Soc.* **117:** 11019–11020.

———. 1996. Ribozyme-mediated cleavage of a substrate analogue containing an internucleotide-bridging 5′-phosphorothioate: Evidence for the single-metal model. *Biochemistry* **35:** 5308–5317.

Laing L.G., Gluick T.C., and Draper D.E. 1994. Stabilization of RNA structure by Mg ions. Specific and non-specific effects. *J. Mol. Biol.* **237:** 577–587.

Lincoln S.F. and Merbach A.E. 1995. Substitution reactions of solvated metal ions. *Adv. Inorg. Chem.* **42:** 1–88.

Lippard S.J. and Berg J. (1995). *Principles of bioinorganic chemistry*. University Science Books, Mill Valley, California.

Manning G.S. 1979. Counterion binding in polyelectrolyte theory. *Accts. Chem. Res.* **12:** 443–449.

McConnell T.S., Herschlag D., and Cech T.R. 1997. Effects of divalent metal ions on individual steps of the *Tetrahymena* ribozyme reaction. *Biochemistry* **36:** 8293–8303.

McKay D.B. 1996. Structure and function of the hammerhead ribozyme: An unfinished story. *RNA* **2:** 395–403.

Michalowski D., Wrzesinski J., and Krzyzosiak W. 1996a. Cleavages induced by different metal ions in yeast tRNA^{Phe} U59C60 mutants. *Biochemistry* **35:** 10727–10734.

Michalowski D., Wrzesinski J., Ciesiolka J., and Krzyzosiak W.J. 1996b. Effect of modified nucleotides on structure of yeast tRNA^{Phe}. Comparative studies by metal ion-induced hydrolysis and nuclease mapping. *Biochimie* **78:** 131–138.

Nesbitt S., Hegg L.A., and Fedor M.J. 1997. An unusual pH-independent and metal-ion-independent mechanism for hairpin ribozyme catalysis. *Chem. Biol.* **4:** 619–630.

Ott G., Arnold L., and Limmer S. 1993. Proton NMR studies of manganese ion binding to tRNA derived acceptor arm duplexes. *Nucleic Acids Res.* **21:** 5859–5864.

Pan T., Long D.M., and Uhlenbeck O.C. 1993. Divalent metal ions in RNA folding and catalysis. In *The RNA world* (ed. R. Gesteland and J. Atkins), pp. 271–302. Cold Spring Harbor Laboratory Press, Cold Spring Harbor, New York.

Pearson R.G. 1988. Absolute electronegativity and hardness: Applications to inorganic chemistry. *Inorg. Chem.* **27:** 734–740.

Pecoraro V.L., Hermes J.D., and Cleland W.W. 1984. Stability constants of Mg(II) and Cd(II) complexes of adenine nucleotides and thionucleotides and rate constants for formation and dissociation of Mg-ATP and Mg-ADP. *Biochemistry* **23:** 5262–5271.

Peracchi A., Beigelman L., Scott E.C., Uhlenbeck O.C., and Herschlag D. 1997. Involvement of a specific metal ion in the transition of the hammerhead ribozyme to its catalytic conformation. *J. Biol. Chem.* **272:** 26822–26826.

Piccirilli J.A., Vyle J.S., Caruthers M.H., and Cech T.R. 1993. Metal ion catalysis in the *Tetrahymena* ribozyme reaction. *Nature* **361:** 85–88.

Pley H.W., Flaherty K.M., and McKay D.B. 1994. Three-dimensional structure of a hammerhead ribozyme. *Nature* **372:** 68–74.

Podar M., Perlman P.S., and Padgett R.A. 1995. Stereochemical selectivity of group II intron splicing, reverse splicing, and hydrolysis reactions. *Mol. Cell. Biol.* **15:** 4466–4478.

Pyle A.M. 1996. Role of metal ions in ribozymes. In *Metal ions in biological systems* (ed. A. Sigel and H. Sigel), pp. 479–520. Marcel Dekker, New York.

Record M.T., Jr., Anderson C.F., and Lohman T. 1978. Thermodynamic analysis of ion effects on the binding and conformational equilibria of proteins and nucleic acids: The roles of ion association or release, screening, and ion effects of water activity. *Q. Rev. Biophys.* **11:** 103–178.

Reid S.S. and Cowan J.A. 1990. Biostructural chemistry of magnesium ion: Characterization of the weak binding sites on tRNA^{Phe} (yeast). Implications for conformational change and activity. *Biochemistry* **29:** 6025–6032.

Rialdi G., Levi J., and Biltonen R. 1972. Thermodynamic studies of transfer ribonucleic acids. I. Magnesium binding to yeast phenylalanine transfer ribonucleic acid. *Biochemistry* **11:** 2472–2479.

Richens D.T. 1997. *The chemistry of aqua ions.* Wiley, New York.

Römer R. and Hach R. 1975. tRNA conformation and magnesium binding. A study of yeast phenylalanine-specific tRNA by a fluorescent indicator and differential melting curves. *Eur. J. Biochem.* **55:** 271–284.

Ruffner D.E. and Uhlenbeck O.C. 1990. Thiophosphate interference experiments locate phosphates important for the hammerhead RNA self-cleavage reaction. *Nucleic Acids Res.* **18:** 6025–6029.

Scott W.G., Finch J.T., and Klug A. 1995. The crystal structure of an all-RNA hammerhead ribozyme: A proposed mechanism for RNA catalytic cleavage. *Cell* **81:** 991–1002.

Scott W.G., Murray J.B., Arnold J.R.P., Stoddard B.L., and Klug A. 1996. Capturing the structure of a catalytic RNA intermediate: The hammerhead ribozyme. *Science* **274:** 2065–2069.

Sigel R.K.O., Song B., and Sigel H. 1997. Stabilities and structures of metal ion complexes of adenosine 5′-O-thiomonophosphate (AMPS^{2-}) in comparison with those of its parent nucleotide (AMP^{2-}) in aqueous solution. *J. Am. Chem. Soc.* **119:** 744–755.

Sigurdsson S.T., Thomson J.B., and Eckstein F. 1998. Small ribozymes. In *RNA structure and function* (ed. R.W. Simons and M. Grunberg-Mango), pp. 339–376. Cold Spring Harbor Laboratory Press, Cold Spring Harbor, New York.

Slim G. and Gait M.J. 1991. Configurationally defined phosphorothiolate-containing oligoribonucleotides in the study of the mechanism of cleavage by hammerhead ribozymes. *Nucleic Acids Res.* **19:** 1183–1188.

Smith D. 1995. Magnesium as the catalytic center of RNA enzymes. In *The biological chemistry of magnesium* (ed. J.A. Cowan), pp. 111–136. VCH, New York.

Smith D. and Pace N.R. 1993. Multiple magnesium ions in the ribonuclease P reaction mechanism. *Biochemistry* **32:** 5273–5281.

Smith D., Burgin A.B., Haas E.S., and Pace, N.R. 1992. Influence of metal ions on the ribonuclease P reaction. Distinguishing substrate binding from catalysis. *J. Biol. Chem.* **267:** 2429–2436.

Stein A. and Crothers D.M. 1976. Equilibrium binding of magnesium(II) by *Escherichia coli* tRNA^{fMet}. *Biochemistry* **15:** 157–160.

Strobel S.A. and Shetty K. 1997. Defining the chemical groups essential for *Tetrahymena* group I intron function by nucleotide analog interference mapping. *Proc. Natl. Acad. Sci.* **94:** 2903–2908.

Suh Y.-A., Kumar P.K.R., Taira K., and Nishikawa S. 1993. Self-cleavage activity of the genomic HDV ribozyme in the presence of various divalent metal ions. *Nucleic Acids Res.* **21:** 3277–3280.

Summers M.F. 1988. ^{113}Cd NMR spectroscopy of coordination compounds and proteins. *Coord. Chem. Rev.* **86:** 43–134.

Sun S., Yoshida A., and Piccirilli J.A. 1997. Synthesis of 3′-thioribonucleosides and their incorporation into oligoribonucleotides via phosphoramidite chemistry. *RNA* **3:** 1352–1363.

Tarasow T.M., Tarasow S.L., and Eaton B.E. 1997. RNA catalyzed carbon-carbon bond formation. *Nature* **389:** 54–57.

Uhlenbeck O.C. 1995. Keeping RNA happy. *RNA* **1:** 4–6.

Wang L. and Ruffner D.E. 1997. An ultraviolet crosslink in the hammerhead ribozyme dependent on 2-thiocytidine or 4-thiouridine substitution. *Nucleic Acids Res.* **25:** 4355–4361.

Waring R.B. 1989. Identification of phosphate groups important to self-splicing of the *Tetrahymena thermophila* rRNA intron as determined by phosphorothioate substitution. *Nucleic Acids Res.* **17:** 10281–10293.

Warnecke J.M., Fürste J.P., Hardt W.D., Erdmann V.A., and Hartmann R.K. 1996. Ribonuclease P (RNase P) RNA is converted to a Cd(II)-ribozyme by a single Rp-phosphorothioate modification in the precursor tRNA at the RNase P cleavage site. *Proc. Natl. Acad. Sci.* **93:** 8924–8928.

Weinstein L.B., Jones B.C.N.M., Cosstick R., and Cech T.R. 1997. A second catalytic metal ion in a group I ribozyme. *Nature* **388:** 805–808.

Winter D., Polacek N., Halama I., Streicher B., and Barta A. 1997. Lead-catalyzed specific cleavage of ribosomal RNAs. *Nucleic Acids Res.* **25:** 1817–1824.

Wolfson J.M. and Kearns D.R. 1974. Europium as a fluorescent probe of metal binding sites on transfer ribonucleic acid. I. Binding to *Escherichia coli* formylmethionine transfer ribonucleic acid. *J. Am. Chem. Soc.* **96:** 3653–3654.

Wu H.-N., Lin Y.-J., Lin F.-P., Makino S., Chang M.-F., and Lai M.M. 1989. Human hepatitis δ virus RNA subfragments contain an autocleavage activity. *Proc. Natl. Acad. Sci.* **86:** 1831–1835.

Yachandra V.K., Sauer K., and Klein M.P. 1996. Manganese cluster in photosynthesis: Where plants oxidize water to dioxygen. *Chem. Rev.* **96:** 2927–2950.

Yarus M. 1993. How many catalytic RNAs? Ions and the Cheshire cat conjecture. *FASEB J.* **7:** 31–39.

Young K.J., Gill F., and Grasby J.A. 1997. Metal ions play a passive role in the hairpin ribozyme catalyzed reaction. *Nucleic Acids Res.* **25:** 3760–3766.

Zhou D.-M., Zhang L.H., and Taira K. 1997. Explanation by the double-metal-ion mechanism of catalysis for the differential metal ion effects on the cleavage rates of 5′-oxy and 5′-thio substrates by a hammerhead ribozyme. *Proc. Natl. Acad. Sci.* **94:** 14343–14348.

Zhou D.-M., Usman N., Wincott F.E., Matulic-Adamic J., Orita M., Zhang L.-H., Komiyama M., Kumar P.K.R., and Taira K. (1996). Evidence for the rate-limiting departure of the 5′-oxygen in nonenzymatic and hammerhead ribozyme-catalyzed reactions. *J. Am. Chem. Soc.* **118:** 5862–5866.

Zito K., Hüttenhofer A., and Pace N.R. 1993. Lead-catalyzed cleavage of ribonuclease P RNA as a probe for integrity of tertiary structure. *Nucleic Acids Res.* **21:** 5916–5920.

13
Building a Catalytic Active Site Using Only RNA

Thomas R. Cech and Barbara L. Golden
Howard Hughes Medical Institute
Department of Chemistry and Biochemistry
University of Colorado
Boulder, Colorado 80309-0215

WHERE DO RIBOZYMES FIT IN THE RNA WORLD HYPOTHESIS?

Two of the most fundamental requirements for life are information storage and catalytic function. Without storage, transfer, and replication of information, a system cannot learn from its past and improve its viability; that is, there can be no natural selection. Equally essential to life is catalytic (enzymatic) function. At the very least, there must be machinery to catalyze the copying of the informational molecules. This process must proceed with considerable fidelity, yet some frequency of errors is also necessary to provide the diversity that allows adaptation and evolution. Beyond replication of the genome, additional catalytic functions would be highly advantageous to provide basic metabolism for even a primitive self-reproducing system.

In contemporary organisms information is stored in the form of DNA. However, the persistence of RNA genomes in many viruses shows us that this sister nucleic acid is competent for information storage, at least for small genomes. Biocatalytic function in the modern world is mostly the domain of protein enzymes, although ribonucleoproteins (RNPs) still catalyze the essential cellular reactions of protein synthesis and RNA splicing. By what evolutionary pathway did this DNA-RNA-protein solution to the problem of life come about? The finding that RNA, an informational molecule, can by itself catalyze biochemical reactions has rekindled enthusiasm for the possibility that a key intermediate stage was an RNA World, with RNA providing both information and function, genotype and phenotype.

One version of this RNA World hypothesis is diagrammed in Figure 1. RNA is usually considered to be too complex a polymer to arise by

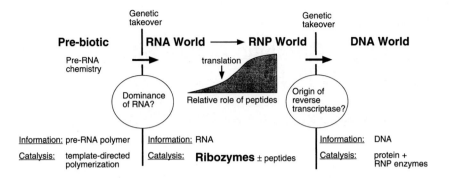

Figure 1 Role of ribozymes in an RNA World hypothesis.

random chemistry, thus the proposal of a simpler progenitor (see Chapters 2 and 6). The advantages of replicating a more homogeneous polymer with a single type of building block may have driven the emergence of an RNA World, with RNA catalyzing its own replication (see Chapter 5). At this stage ribozymes, the subject of this chapter, would speed up and provide specificity to biochemical reactions. Even at the very earliest stages, ribozymes would probably work in concert with whatever randomly assembled peptides and other molecules were in their environment. But once a system stumbled upon a means to direct the synthesis of specific peptide reproducibly—that is, to translate information from RNA to protein (see Chapters 7 and 8)—ribozymes would start working as RNP enzymes. Thus, the RNA and RNP worlds would overlap extensively. Many of the ribozymes that function in contemporary cells continue to work as RNPs, as discussed in Chapters 14 and 18.

Will we ever know whether such an RNA World really existed? Perhaps not, but we can test the chemical plausibility of ribozymes performing diverse catalytic roles. Is RNA catalysis fast enough, specific enough, and versatile enough to support complex metabolism? How does catalysis by RNA compare with that attainable by proteins, and what is the structural basis of the similarities and differences? These are the questions we address here.

RIBOZYME-CATALYZED REACTIONS

How Versatile is RNA Catalysis?

Natural ribozymes (Table 1) are all involved in RNA processing reactions. RNase P cleaves the 5' leader from primary transcripts of tRNAs. Chemically, this reaction is phosphodiester bond hydrolysis (Fig. 2a).

Table 1 Natural ribozymes

Category	Number sequenced	Biological sources	Reaction performed (reaction product)
Self-splicing RNAs			transesterification (3′-OH)
Group I	>500	Eukaryotes (nuclear and organellar), prokaryotes, bacteriophage	
Group II	>100	Eukaryotes (organellar), prokaryotes	
Self-cleaving			
Group I-like	6	*Didymium, Naeglaria*	Hydrolysis (3′-OH)
Small self-cleavers			transesterification (2′,3′>p)
hammerheads	11	Plant viroids and satellite RNAs, newt	
hairpin	1	Satellite RNA of tobacco ringspot virus	
HDV	2	Human hepatitis virus	
VS	1	*Neurospora* mitochondria	
RNase P RNAs	>100 bacterial	Eukaryotes (nuclear and organellar), prokaryotes	Hydrolysis (3′-OH)

Selected references. Group I: Gott et al. 1986; Reinhold-Hurek and Shub 1992; Saldanha et al. 1993; http://pundit.icmb.utexas.edu. Group II: Cupertino and Hallick 1991; Michel and Ferat 1995. Self-cleaving group I-like: Einvik et al. 1997. Small self-cleavers: Symons, 1992. RNase P: http://jwbrown.mbio.ncsu.edu/ RNaseP/home.html; see Chapter 14.

Group I and II introns mediate the splicing of the RNA in which they reside by a series of two concerted phosphodiester bond cleavage–ligation (transesterification) reactions (Fig. 2a). Group II RNAs also cleave and insert themselves into double-stranded DNA in a reaction that requires the cooperation of the ribozyme active site and a protein moiety (Zimmerly et al. 1995). Small ribozymes, such as the hammerhead, clip RNA replicative intermediates to make unit-size progeny molecules by a different transesterification mechanism (Fig. 2b) (see Chapter 11). All these reactions are quite similar, in that the substrates are themselves nucleic acids, and chemically they all involve attack of a nucleophile at a phosphate. This is only a very small segment of the constellation of reactions catalyzed by protein enzymes.

Ribozymologists have reengineered self-splicing and self-cleaving ribozymes to act as multiple-turnover catalysts by separating the portion

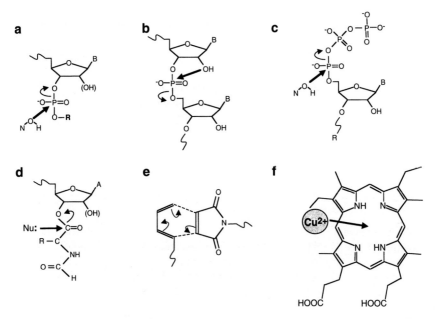

Figure 2 Reactions catalyzed by ribozymes, both natural (*a, b*) and selected in vitro (*c–f*). (*a*) Hydrolysis or transesterification of a phosphodiester linkage (R = polynucleotide) or a phosphomonoester linkage (R = H). NOH = water for RNase P, NOH = 3′-OH of guanosine for group I introns, and NOH = 2′-OH of an adenosine within the RNA for group II introns. (*b*) RNA cleavage by attack of the adjacent 2′-OH, the reaction catalyzed by the hammerhead and other small ribozymes. (*c*) RNA chain elongation using a nucleoside triphosphate substrate (R = H) or RNA ligation (R = polynucleotide). In either case, NOH = RNA chain with a free 3′-OH (or 2′-OH). (*d*) Amide-bond formation (Nu = 5′-NH$_2$ of a modified RNA), peptidyl transfer (Nu = NH$_2$ group of another amino acid), or ester hydrolysis (Nu = water). In all cases, R is an amino acid side chain. (*e*) Diels-Alder cycloaddition. (*f*) Porphyrin metalation.

of the molecule that contains the active site from the portion that undergoes reaction (Zaug and Cech 1986a; Zaug et al. 1986). This concept has been extended to so many catalytic RNA systems that it is already commonplace, yet it is critical for versatility: An "enzyme" restricted to a single round of reaction with a portion of itself would have extremely limited roles in an RNA World.

Ribozymologists have also coaxed natural ribozymes to catalyze alternative chemistries. Group I and group II ribozymes can cleave single-stranded DNA as well as RNA (Herschlag and Cech 1990a; Robertson and Joyce 1990; Mörl et al. 1992). The *Tetrahymena* ribozyme can transfer a 3'-terminal phosphate from one RNA strand to another, and it can hydrolyze a 3'-terminal phosphate (Zaug and Cech 1986b). These reactions involve phosphate monoester substrates instead of the usual diester. These DNA-cleavage and phosphomonoester reactions do not extend the ribozyme repertoire greatly, as the substrates are still nucleic acid and the reactions still involve phosphorus centers. Although the *Tetrahymena* ribozyme has aminoacyl ester hydrolysis activity, a reaction that requires attack of water at a carbon center (Fig. 2d), the amount of catalysis is modest (Piccirilli et al. 1992).

The explosion of types of ribozyme-catalyzed reactions came with the advent of in vitro selection/evolution technology, developed in the laboratories of L. Gold, G. Joyce, and J. Szostak in 1990 (see Chapters 2, 5 and 6). In brief, a large combinatorial library of diverse RNA sequences is challenged to perform some "task," like catalyzing a specific reaction, and the rare molecules that succeed are isolated and amplified to give a new population, enriched in competent molecules. The beauty of the method is that huge populations of sequences, such as 10^{15}, can be sampled. In vitro evolution has led to the discovery of RNA and DNA molecules that utilize nucleoside triphosphate substrates (Fig. 2c) (Lorsch and Szostak 1994; Ekland and Bartel 1996); make and break amide bonds (Dai et al. 1995; Lohse and Szostak 1996; Wiegand et al. 1997), including an amide bond between two amino acids (Zhang and Cech 1997); alkylate a nucleoside or a thiophosphate (Wilson and Szostak 1995; Wecker et al. 1996), and add an amino acid to a nucleotide via an ester linkage (Illangasekare et al. 1995). In all of these cases, the substrates still have an essential nucleic acid component, but the reactions involve carbon centers as well as phosphorus centers. Thus, the spectrum of reactions that can be catalyzed by RNA is far greater than could be inferred from looking at natural ribozymes.

Two recent examples push the envelope even further. In vitro selection has led to the discovery of modified RNA that catalyzes a classic organic chemistry reaction, Diels-Alder cycloaddition, in the presence of Cu^{++} ions (Fig. 2e) (Tarasow et al. 1997). In addition, both RNA and DNA molecules have been selected to catalyze the insertion of metal ions (Cu^{++}, Zn^{++}) into porphyrin rings (Fig. 2f), similar to a step in the biosynthesis of heme (Conn et al. 1996; Li and Sen 1996). The former reaction involves carbon–carbon bond formation, whereas the latter requires a conformational change in the substrate but no covalent bond formation. The other breakthrough in these studies was that the substrates were not directly attached to a nucleic acid component, so interactions other than substrate–ribozyme base-pairing are responsible for positioning the substrate within the ribozyme active site.

How Fast is RNA Catalysis?

Catalytic rate can be assessed in numerous ways, only two of which are discussed here. First, the turnover number of the enzyme (k_{cat}) is informative, because it gives the number of substrate molecules converted to product per minute by a single catalyst at saturating substrate concentration. Second, the rate constant for the chemical step of the catalyzed reaction can be compared to the rate constant for the same reaction uncatalyzed. In more technical terms, the first-order rate constant for the reaction of one enzyme-bound substrate (this is simply k_{cat} in those cases where the chemical step is rate-limiting overall) is divided by the first-order rate constant for the uncatalyzed reaction at the same temperature. Both of these parameters are listed in Table 2 for a few select ribozymes and for some protein enzymes that catalyze related reactions.

One main conclusion from Table 2 is ribozymes that evolved in nature to catalyze reactions with nucleic acid substrates (e.g., group I introns) increase the intrinsic rate of a reaction by huge factors, within the range achieved by protein enzymes. Thus, RNA is not intrinsically a poor catalyst compared to protein. On the other hand, ribozymes selected in the laboratory to catalyze reactions with non-nucleic acid substrates have been measured to have more modest rate enhancements, no more than ~1000-fold. It remains to be seen if this represents a fundamental inferi-

Table 2 Speed of ribozyme-catalyzed reactions

	Rate enhancement k_{cat}/k_{uncat}[a]	Turnover number k_{cat} (min^{-1})
Ribozyme		
Group I, *Tetrahymena*	10^{11}	0.1
Group I, *Anabaena*	10^{10}	4
hammerhead	10^{6}	≥1
RNA ligase	10^{9}	100
porphyrin metalation	460	0.9
Diels-Adler	800	—
Enzyme		
T7 RNA polymerase	3×10^{11}	14,000
DNA ligase, *E. coli*	10^{9}	28

Values are taken from Herschlag and Cech 1990b, Zaug et al. 1994, Hertel et al. 1997, Ekland and Bartel 1995, Conn et al. 1996, Tarasow et al. 1997 and Bartel and Szostak 1993.

[a] k_{cat} refers to the chemical step of the reaction (where known), and therefore may not be the same as k_{cat} (turnover number). For the Diels-Adler reaction, rate enhancement is instead based on ratio of second-order rate constants.

ority of ribozymes for such reactions, or a limitation of the in vitro selection experiments.

Considering now the turnover numbers, even the ribozymes with impressive catalytic rate enhancements are slow at processing substrates under multiple turnover conditions. How is this possible? A reaction can only proceed as fast as its slowest step. Ribozymes often bind nucleic acid substrates by Watson-Crick base-pairing (secondary structure) plus additional "tertiary" interactions, which add up to give very tight binding. The same interactions stabilize binding of the nucleic acid reaction product, slowing its dissociation from the active site. Thus, rate-limiting product dissociation makes some ribozymes such as the *Tetrahymena* ribozyme very sluggish multiple-turnover catalysts (Herschlag and Cech 1990b; Young et al. 1991). However, this should not be considered a deficiency, because these catalytic introns have evolved as single turnover catalysts. If these RNAs had a rapid turnover rate, they would be able to react with other cellular RNAs once excised from their host RNAs.

How Specific Is an RNA Catalyst?

Protein enzymes are notable for their exquisite specificity, their ability to distinguish between very closely related substrates. How do ribozymes compare? Group I introns use guanosine (G) as a nucleophile to cleave the 5' splice site (Fig. 3), and other nucleosides have much reduced or undetectable activity (Bass and Cech 1984). When Michel et al. (1989b) located the G-binding site, they also identified specific hydrogen-bonding interactions that explain the specificity (Fig. 4a). For the *Tetrahymena* ribozyme, the specificity for G over 2-aminopurine ribonucleoside is about 1000-fold (Legault et al. 1992), whereas single base-pair changes in the G-binding site reverse this specificity. An even higher specificity for G over deoxyG (Bass and Cech 1984) has been attributed to the favorable interaction of a metal ion with the 2'-OH (Sjögren et al. 1997).

Another remarkable illustration of RNA's potential to discriminate between closely related molecules is an in-vitro-selected aptamer that binds theophylline >10,000 times more tightly than caffeine, a molecule that is identical except for one methyl group (Jenison et al. 1994). In this case, NMR structural analysis supports the model that substitution of the N7 proton with an N7 methyl group in caffeine would remove two H bonds that the aptamer uses to bind theophylline and also disrupt stacking interactions above and below the plane shown in Figure 4b (Zimmermann et al. 1997).

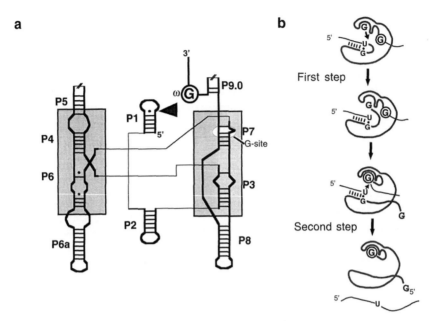

Figure 3 Group I intron structure and reactivity. (*a*) Expanded view of secondary structure of a minimal group I intron. P = paired region. Thick lines represent the RNA backbone. Thin lines show connectivity in this expanded view but in fact have zero length. Shaded regions = catalytic core. In the first step of self-splicing, exogenous guanosine or GTP binds in the G-site and cleaves the 5′ splice site (*large arrowhead*). The 3′-terminal (ω) guanosine of the intron (*circled*) then occupies the G-site for the second step of splicing. (*b*) Self-splicing. First step, 5′ splice-site cleavage by exogenous guanosine. Second step, 3′ splice-site cleavage by the 5′ exon at the bond following ωG (*circled*), resulting in exon ligation.

On the other hand, ribozymes that rely on base-pairing to bind nucleic acid substrates often have their specificity limited by the fact that mismatches within nucleic acid helices are not very destabilizing (see Chapter 10). Even worse, the slow rates of helix dissociation can give enough time for cleavage to occur before a more weakly bound mismatched substrate can dissociate (Herschlag and Cech 1990c; Hertel et al. 1996). For example, the *Tetrahymena* ribozyme shows a 200-fold specificity for cleavage of the "matched" substrate GGCCCUCUAAAAA over the "mismatched" substrate GGCCCGCUAAAAA at low concentration of G nucleophile where the reaction is slow, but the specificity decreases to 4.5-fold at high G concentration (Herschlag and Cech 1990c). Mutant ribozymes that bind substrate less tightly do show increased sequence specificity (Young et al.

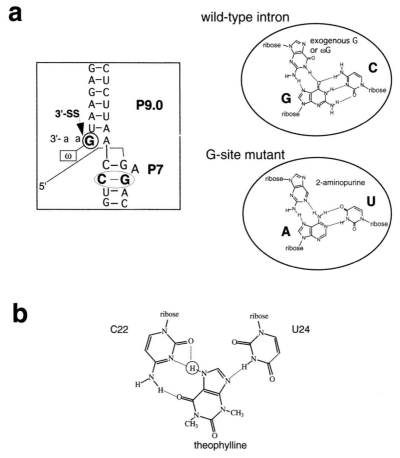

Figure 4 Specific binding of small molecules by RNA. (*a*) Guanosine binding by a conserved G-C base pair in the P7 duplex of group I introns, and change of specificity (G → 2-aminopurine ribonucleoside) upon switching to an A-U base pair in P7 (Michel et al. 1989b). (*b*) Theophylline-binding site in an in-vitro-selected aptamer. The circled H is substituted by a methyl group in caffeine, disrupting binding (Zimmermann et al. 1997).

1991). The hammerhead ribozyme also has some ability to distinguish mismatches in a ribozyme-substrate helix away from the catalytic core, but it has dramatic discrimination against mismatches that flank the core (Werner and Uhlenbeck 1995; Hertel et al. 1997). Presumably, these proximal mismatches destabilize the catalytically active structure. Thus, there are cases in which a ribozyme recognizes a nucleic acid substrate with more specificity than that intrinsic to base-pairing interactions. Such a feature seems critical to obtaining sufficient fidelity for RNA-catalyzed RNA replication in an RNA World.

Ribozymes discriminate easily between stereoisomers of the same substrate, as expected for any catalyst that provides a chiral three-dimensional active site. A dramatic example is provided by the *Tetrahymena* ribozyme, which cleaves the R_P and S_P diastereomers of an RNA substrate with a >1000-fold different rate, even though these differ only in the position of a subtle oxygen-to-sulfur substitution (Cech et al. 1992). As another example, an RNA selected for binding to D-tryptophan agarose bound >700 times more weakly to L-tryptophan agarose (Famulok and Szostak 1992). An exciting application of this stereospecificity is "mirror image" drug development: One can select a nucleic acid aptamer that binds to the mirror image (enantiomer) of a target protein, and then be assured that the enantiomer of the selected aptamer will be specific for the naturally occurring isomer of the target protein (Klussmann et al. 1996; Williams et al. 1997). The enantiomer of the aptamer, composed of unnatural L-ribose sugars, is not recognized by natural nucleases and therefore has the potential to be a long-lived drug in vivo.

MECHANISMS OF RNA CATALYSIS

How do ribozymes lower the activation energy and thereby speed up biochemical reactions? The much-studied group I ribozymes will be used to illustrate the catalytic strategies available to RNA, with other systems being compared and contrasted along the way (see also Narlikar and Herschlag 1997).

Binding and Orienting Reactive Groups

When enzymes bind substrates, they create a high local concentration and optimal orientation to increase the efficiency of reaction (Jencks 1969). Group I introns use this same strategy, binding guanosine in proximity to the 5' splice site and converting what would normally be an intermolecular reaction into an effectively intramolecular reaction (Bass and Cech 1984). Guanosine binding is entropically driven, perhaps indicative of release of solvent from the RNA upon G-binding (McConnell and Cech 1995). Elements of the ribozyme that contribute to G-binding (Fig. 4a) (Michel et al. 1989b; Yarus et al. 1991) and to binding of the 5'-splice site have been identified. (The latter set of interactions is described below in Building a Catalytic Center Using Only RNA, because it provides a good general model for specific helix-packing interactions in large RNAs.)

Although we know that these two reactants are bound in a manner that allows them to react, there is no information about whether they are held in anything close to an optimal orientation.

Group II introns, found in certain mitochondria, chloroplasts, and bacteria, have conserved structure distinct from that of group I (Michel et al. 1989a). Group II introns also accomplish self-splicing by consecutive transesterification reactions, but with a twist: In the first step, the 2′-OH of a specific intronic adenosine (A) cleaves the 5′ splice site to form a "lariat" intermediate, and in step two, the cleaved 5′ exon attacks the 3′-splice site, ligating the surrounding exons (Peebles et al. 1986; Schmelzer and Schweyen 1986; van der Veen et al. 1986). In the laboratory, group II introns can catalyze *trans* reactions, such as binding a free 5′-exon analog and ligating it to the 3′-exon (Jacquier and Rosbash 1986). Both the *cis* and *trans* reaction rely on two exon-binding sites that base-pair with complementary 5′-exon sequences (intron binding sites or IBS, Fig. 5; Jacquier and Michel 1987). These long-range interactions are similar to the internal guide sequence-5′-exon pairing of group I introns (Davies et al. 1982; Michel et al. 1982). In addition, a portion of the conserved GUGYG sequence at the 5′ end of group II introns pairs with a loop within domain I of the intron, further contributing to 5′-splice site recognition (Jacquier and Michel 1990). Presumably, additional, as-yet-unidentified tertiary interactions position the 2′-OH of the nucleophilic A in domain VI for attack.

Step 2 of group II intron self-splicing requires an alternate structural interaction (Chanfreau and Jacquier 1996; η/η′ in Fig. 5). This is especially noteworthy, because although there is much evidence for conformational changes in ribozyme catalysis (see, e.g., Been and Perrotta 1991; Wang et al. 1993; Golden and Cech 1996), there are few examples where a specific structure has been identified that orchestrates a conformational switch. In the group II intron, the branch-site A for step 1 of splicing must be displaced from the active site to allow entry of the last intron nucleotide and the 3′-splice site (Steitz and Steitz 1993; Jacquier 1996). Branch formation or 5′-splice site cleavage may result in a displacement that allows the η/η′ base-pairing to form, and its formation may in turn pull the branched A from the active site, allowing the last intron nucleotide to enter for step 2 of splicing.

Metal Ion Catalysis

Activity of most ribozymes requires, or is greatly stimulated by, divalent metal ions such as Mg^{++}, Mn^{++}, and in some cases Ca^{++}. These could be

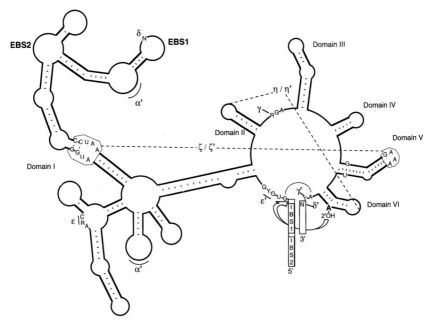

Figure 5 Group II intron structure. Intron-binding sites (IBS 1 and 2) located in the 5' exon are base-paired to two intronic exon-binding sites (EBS 1 and 2, respectively). Greek letters designate other tertiary interactions; those discussed in the text are connected by dashed lines. The first step of self-splicing, attack of the 2'-OH of an A in domain VI at the 5' splice site, is indicated by an open arrow.

involved in folding the polyanionic RNA catalyst, in active-site chemistry, or in both. Plausible chemical roles have been much discussed previously (Pyle 1993; Yarus 1993) and include

1. electrostatic stabilization of an anionic attacking or leaving group
2. electrostatic stabilization of an electrophilic group (Lewis acid catalysis)
3. electrostatic destabilization of a substrate in the ground state
4. specific orientation of substrates by metal ion coordination
5. acid-base catalysis by metal-bound water or hydroxide
6. oxidation-reduction chemistry

In favorable cases, the structural and catalytic roles of metal ions have been separated. The *Tetrahymena* ribozyme requires divalent metal ions in order to attain its active three-dimensional folded structure (Latham and Cech 1989; Celander and Cech 1991), a requirement that can be met by a variety of divalent cations and partially met by high concentrations of monovalent cation (Downs and Cech 1996). In addition, correct positioning of the 5′ splice site helix within its active site has a requirement for Mg^{++} that cannot be met by Ca^{++} (Wang and Cech 1994; McConnell et al. 1997).

Because numerous metal ions are required to fold the ribozyme, it was difficult to identify additional metal ions involved directly in active-site chemistry, especially since they are already bound at the 2 mM Mg^{++} concentration required for ribozyme folding (McConnell et al. 1997). This problem was addressed by site-specific substitution of atoms that would predictably perturb the binding and function of a coordinated metal ion. In the *Tetrahymena* ribozyme, substitution of bridging oxygen atoms by sulfur at the reactive bonds led to loss of activity with Mg^{++}. Restoration of cleavage with thiophilic metal ions such as Mn^{++} and Cd^{++} in both the forward and reverse reactions led to the proposal of metal ions 1 and 2 in Figure 6. Metal ion 1 destabilizes the substrate in the ground state but stabilizes the developing negative charge as the O–P bond is broken in the transition state, thereby providing roughly 10^6-fold catalysis (Piccirilli et al. 1993; Narlikar et al. 1995). Metal ion 2 acts to deprotonate the 3′-OH of the G nucleophile (Weinstein et al. 1997). These two metal ions may form the sort of "two metal ion center" proposed by Steitz and Steitz (1993), a catalytic strategy with clear connections to the world of protein enzymes (Beese and Steitz 1991; Kim and Wyckoff 1991). Substitution of the 2′-OH of the G nucleophile by an amino group also led to a metal specificity switch, leading to the proposal of metal ion 3 (Fig. 6) (Sjögren et al. 1997). Metal ion 3 could function to position or activate the nucleophile.

Divalent metal ions are required for the hammerhead ribozyme reaction under physiological conditions. There is evidence for their binding to the reaction-site phosphate, enhancing its ability to be attacked by the oxyanionic 2′ nucleophile, and also evidence for metal ion activation of the 2′-OH nucleophile (see Chapter 11). However, the hairpin ribozyme, which cleaves RNA by the same 2′, 3′-cyclic phosphate mechanism as the hammerhead, appears to use metal ions just for folding, not for chemistry. Cobalt hexammine, a metal complex inert to changes in its coordination sphere, supports full hairpin ribozyme cleavage, providing evidence against the importance of direct coordination of the metal to the RNA. Also unlike the hammerhead, phosphorothioate substitution at either non-bridging oxygen has small and metal-ion-independent effects on activity, evidence against any direct metal coordination (Hampel and Cowan 1997; Nesbitt et al. 1997; Young et al. 1997). This system serves as an impor-

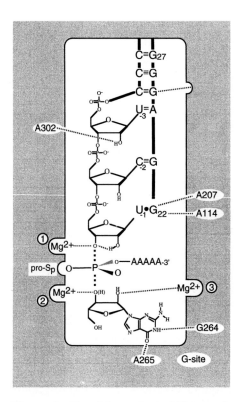

Figure 6 Transition state stabilization by the *Tetrahymena* group I ribozyme. In-line attack of the 3′-OH of guanosine at the phosphorus atom equivalent to the 5′ splice site is supported by the stereochemical course of the reaction (McSwiggen and Cech 1989; Rajagopal et al. 1989). Metal ion catalysis (sites labeled 1, 2, and 3) is discussed in the text.

tant reminder that RNA enzymes, like protein enzymes, will embrace diverse catalytic strategies.

Covalent Catalysis

Many protein enzymes use a nucleophilic amino acid (e.g., serine, lysine, cysteine, or histidine) to attack a substrate, forming a covalent intermediate with a portion of the substrate. In a second step, attack of the covalent intermediate by an external nucleophile (such as water) releases the product and restores the enzyme (e.g., serine proteases, or alkaline phosphatase in Fig. 7).

The *Tetrahymena* ribozyme performs reactions that are highly analogous to those carried out by *Escherichia coli* alkaline phosphatase (Fig. 7).

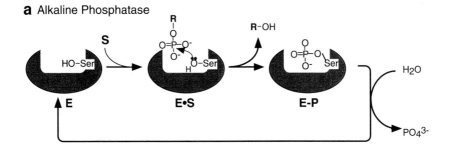

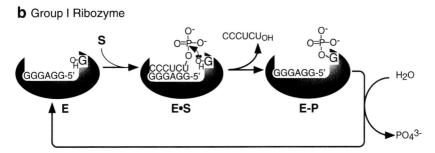

Figure 7 Covalent catalysis by the *Tetrahymena* group I ribozyme. (*a*) Reaction of the *E. coli* protein enzyme alkaline phosphatase (E) involves binding of a phosphorylated substrate (S) to form a noncovalent E·S complex, followed by transfer of the phosphate to an active-site serine to form a covalent E-P intermediate. Hydrolysis then restores the active enzyme. (*b*) The ribozyme has analogous activity, with the 3'-terminal G of the nucleic acid acting like the serine of alkaline phosphatase.

Because the biological function of this RNA involves self-splicing reactions at phosphodiesters, it was not obvious that it should have reactivity with phosphomonoester substrates, since the two have very different transition states (Zaug and Cech 1986b). Nevertheless, the ribozyme efficiently transfers a 3'-terminal phosphate from an oligonucleotide to its own 3'-terminal guanosine, forming a covalent E-p intermediate. The phosphate can then be transferred to a different oligonucleotide (one that can also bind in the active site) or it can be hydrolyzed, the latter reaction being enhanced at low pH. In either case, the free catalyst E is restored, ready for another catalytic cycle.

Why is covalent catalysis advantageous? It allows an active site to be occupied successively by two different substrates; in the example given above, the *Tetrahymena* ribozyme first binds the oligo CCCUCUp, stores the phosphate as an E-p complex, binds a second oligo UCU, and transfers the phosphate to it to form UCUp (Zaug and Cech 1986b). More generally, the immobilization gained by forming a covalent intermediate may be advantageous for catalysis.

Contemporary protein enzymes use covalent catalysis to act as proteases, esterases, phosphoglucomutases, transphosphorylases, and ligases, often with a serine OH group as a nucleophile. In an RNA World, it would be feasible for ribozymes to perform similar reactions using a ribose 2′- or 3′-OH group as a nucleophile.

General Acid–Base Catalysis

Many chemical reactions require transfer of a proton to or from a reactant. The proximity of a proton donating or accepting group, that is, a general acid or base, can greatly speed such reactions. RNA does not have histidine, but functional groups on the RNA bases or backbone could perhaps serve as proton donors or acceptors (Pace and Marsh 1985). However, with the exception of the 5′-terminal phosphate and modified nucleotides such as 7-methylguanosine, the pK_a values of these groups are normally far from pH 7. Could an RNA structure perturb a pK_a in an active site to provide efficient general acid–base catalysis? In the absence of a confirmed example, it seems possible that ribozymes will in general use metal ions and metal-ion-bound solvent in lieu of general acid–base catalysis (Cech et al. 1992).

A DNAzyme that catalyzes the insertion of metal ions into mesoporphyrin IX works by enhancing the basicity of the bound mesoporphyrin substrate by 3–4 pH units, thereby enhancing its ease of metallation (Li and Sen 1998). This may occur by distortion of the planar mesoporphyrin structure by the DNAzyme, although involvement of a catalytic group such as a negatively charged phosphate is another possibility.

Use of Binding Energy away from the Site of Reaction

To decrease the energy barrier for a reaction, an enzyme must stabilize the reaction's transition state more than it stabilizes the bound substrate(s) in the ground state. One strategy for doing so is to use binding energy away from the site of reaction to "force" an interaction at the reaction site, an interaction that induces the substrate to react (Jencks 1975). The enzyme can force entropic fixation of a substrate with respect to another substrate or with respect to active-site groups, or force electrostatic destabilization by juxtaposing a (partially) charged substrate atom with a like charge on the enzyme. In either case, if the strained interaction is relieved upon approach to the transition state, catalysis will result. When enzymes use this trick, the observed binding of the substrate(s) is weaker than it would otherwise be, because part of the intrinsic binding energy is used to pay the price of forcing the unfavorable ground-state interaction.

Recent studies with the *Tetrahymena* ribozyme have shown that this sophisticated strategy is used by RNA as well as protein enzymes. The *Tetrahymena* ribozyme binds its oligonucleotide substrate (CCCUCUpA) and product (CCCUCU-OH) by base-pairing with the CCCUCU portion plus additional "tertiary" interactions. The tertiary interactions are much weaker with the substrate than with the reaction product, a difference due to the phosphate rather than the 3′-A residue (Narlikar et al. 1995). The authors suggest that the ribozyme uses binding interactions to force the bridging oxygen atom, partially positive in the ground state, next to the positively charged magnesium ion 1 (Fig. 6); this substrate destabilization is relieved in the transition state as negative charge accumulates on the oxygen atom. They estimate 280-fold rate enhancement from the ground state destabilization and another 60-fold from the additional positive interactions in the transition state, for a combined catalytic effect that accounts for 10^4-fold of the total 10^{11}-fold catalysis achieved by this ribozyme (Narlikar et al. 1995).

BUILDING A CATALYTIC CENTER USING ONLY RNA

Solvent-inaccessible Core

Protein enzymes form a densely packed, hydrophobic core to support a concave active site. Given the absence of hydrophobic side chains and the polyanionic backbone of nucleic acids, how is it possible for them to form a functionally equivalent structure? Single-stranded nucleic acids certainly fold into more compact structures than their double-helical counterparts. Transfer RNA and guanine quadruplexes provide well-studied examples, but even the tRNA structure does not provide much solvent inaccessibility to its backbone; most of the backbone is still on the outside of the molecule, bathed in solvent, very different from a folded protein.

Large ribozymes, such as the *Tetrahymena* group I intron, do form a relatively solvent-inaccessible core as judged by protection from free-radical bombardment (Latham and Cech 1989; Celander and Cech 1991; Sclavi et al. 1997). This suggested that nucleic acid building blocks can be packed together with the help of divalent cations to form something roughly equivalent to the core of a globular protein. In the following section, we explore the types of interactions that contribute to formation of such a structure.

Interactions that Stabilize a Close-packed RNA Core

The crystal structure of one domain of the *Tetrahymena* ribozyme at 2.6 Å resolution provided the first atomic view of an RNA large enough to have a relatively solvent-inaccessible core (Cate et al. 1996). This 160-nucleotide domain, called P4-P6, comprises about half of the active site of the ribozyme. When synthesized as a separate RNA molecule, it assumes its native secondary structure as well as a higher order tertiary structure that appears to be a simple subset of its structure within the context of the whole, active ribozyme (Murphy and Cech 1993, 1994). Examination of the structure (Fig. 8a) reveals both the expected nucleic acid features and

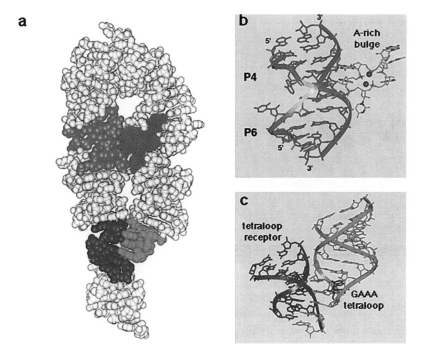

Figure 8 An RNA domain. (*a*) The P4-P6 domain of the *Tetrahymena* group I intron rendered as CPK atoms, showing the side-by-side arrangement of two helical subdomains. The two major sites of interaction between the two subdomains are highlighted: the junction of P4 and P6 (*blue*), the A-rich bulge (*red*), the GAAA tetraloop (*green*), and the tetraloop receptor (*violet*). (*b*) Coordination of two magnesium ions (*violet*) by the backbone of the A-rich bulge displays the nucleotide bases for tertiary interactions with the rest of the domain. (*c*) The GAAA tetraloop-receptor is stabilized both by stacking (highlighted in *red*) and specific hydrogen bonds between the two modules. For orientation purposes, the G of the GAAA tetraloop is drawn in *blue*.

also features that appear protein-like, at least superficially. The molecule has a sharp bend at the top, which allows two punctuated double-helical regions to be aligned side by side. The protein-like feature is the tightly packed core produced by sandwiching together the two halves of the molecule. Thus, the P4-P6 structure provides the first opportunity to evaluate the molecular basis for formation of such a higher order RNA structure, as discussed below.

Long-range Base Pairs and Triples

The two halves of the P4-P6 RNA domain are brought together mainly through two tertiary interactions: The A-rich bulge docks into the minor groove of P4 (Flor et al. 1989) and also into the P5abc three-helix junction (Fig. 8b), whereas the L5b tetraloop is bound by a receptor sequence (Costa and Michel 1995) located between P6a and P6b in the other half of the molecule (Fig. 8c). Each of these interactions involves base triples between a base pair within a duplex region and a third base that is far away in the secondary structure of the molecule. However, thinking of these as simply base triples oversimplifies the interactions, as they involve multiple H–bonds between 2′-hydroxyl groups and phosphate oxygens, and also extensive base-stacking interactions (Cate et al. 1996).

A similar lesson about base triples can be gleaned from an intermolecular tetraloop/minor groove interaction seen in the crystal structure of the hammerhead ribozyme (Pley et al. 1994). Again base triples provide the specificity of the interaction (often because substitution of a different base would cause a steric clash), but a larger number of long-range H–bonds involve ribose hydroxyls and phosphate oxygen atoms.

Comparative sequence analysis and site-specific mutagenesis studies have identified base triples and long-range base pairs in many other ribozymes whose detailed structures have not yet been solved. Examples in group I ribozymes include the triple-helical scaffold involving P4 and P6 (Michel and Westhof 1990; Michel et al. 1990), base-pairing between peripheral loops that stabilize the *sunY* and *Tetrahymena* introns (Michel et al. 1992; Jaeger et al. 1993; Lehnert et al. 1996), and a long-range triple between P4 and J8/7 that brings together the domains of the *Tetrahymena* and cyanobacterial introns (Tanner and Cech 1997). Examples in group II ribozymes so far include long-range base pairs between two hairpin loops or between a hairpin loop and an internal loop (Michel et al. 1989a) and a GAAA tetraloop-receptor interaction that joins domains I and V (Costa and Michel 1995), as summarized in Figure 5. RNase P RNAs are compacted by long-range pseudoknots, one of which involves a hairpin loop (James et al. 1988; Haas et al. 1991), and other interactions (see

Chapter 14). As in the interactions seen in crystal structures, it seems likely that each of these biochemically verified tertiary interactions will be buttressed by a network of additional H–bonds involving the RNA backbone.

The 2′-Hydroxyl Group

The 2′-OH of the ribose sugar, the group that distinguishes RNA from DNA, is exploited frequently for structural interactions in large RNAs. This should be expected, because once RNA engages in base-paired secondary structure, its backbone is its most available element, and the 2′-OH groups lining the minor groove are particularly available for higher-order interactions. As detailed above, H-bonds involving 2′-OH groups buttress base triple interactions. In addition, helical regions can be held together by ribose zippers—the ribose and attached base of one nucleotide donate and accept H-bonds from the same 2′-OH of a second nucleotide, and this interaction is repeated at the adjacent level in a "ladder" of interactions.

Ribose 2′-hydroxyls also have a key role in 5′-splice-site recognition in group I introns. This system is of general interest because it provides a model for specific recognition of base-paired helices within folded RNA molecules. The last few nucleotides of the 5′-exon are first recognized by base-pairing to an intronic internal guide sequence (IGS; Davies et al. 1982; Michel et al. 1982). The resulting duplex is called P1 (Fig. 3a). In a kinetically separable step, this P1 helix is then positioned within the ribozyme active site (Bevilacqua et al. 1992; Herschlag 1992). As shown in the model in Figure 9, certain 2′-OH groups on both strands of P1 are recognized by the intron core (see, e.g., Pyle et al. 1992; Strobel and Cech 1993; Strobel et al. 1998). Additionally, the exocyclic amino group of the G·U wobble base pair at the cleavage site is also important for splice-site recognition and specificity (see, e.g., Strobel and Cech 1995). This G·U wobble pair is conserved among most of the 500 known group I introns, and the two that have instead a G-C become more reactive when a G·U is substituted at this position (Hur and Waring 1995). The docking of the P1 helix within the group I active site illustrates how backbone and minor groove interactions can be used to recognize an RNA element that has its base sequence information "hidden" in a helix.

In group II self-splicing introns (Fig. 5), domain V is a phylogenically conserved element that is essential for catalysis (Jarrell et al. 1988). As in the P1–group I intron case, specific 2′-OH groups on one side of the helix mediate binding of domain V to the remainder of the intron (Abramovitz et al. 1996). On the opposite (major groove) face of the domain V helix, specific 2′-OH groups, phosphates, and bases (including the G of a G·U

Mechanisms of RNA Catalysis   **341**

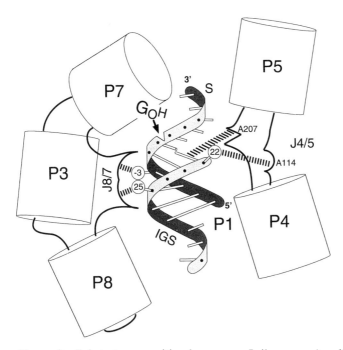

Figure 9 Substrate recognition by a group I ribozyme. An oligonucleotide substrate (S) is first recognized by base-pairing to a short complementary sequence called the internal guide sequence (IGS), forming the helix P1. The cleavage site is defined by a G·U wobble pair at the top of the helix. The P1 helix then docks into the active site of the intron (including elements P3, P4, P5, P7, P8, J4/5, and J8/7). It is correctly positioned by interactions with 2′-OH groups (*circled*) and the exocyclic amine group of the G·U wobble at the cleavage site. (Figure template courtesy of Scott Strobel.)

pair) participate in transition-state stabilization and therefore appear to form part of the active site for 5′ splice-site cleavage (Chanfreau and Jacquier 1994; Peebles et al. 1995; Abramovitz et al. 1996).

Metal Ions

The role of divalent metal ions in stabilizing sites of close backbone–backbone interaction has been studied in transfer RNA, both thermodynamically (Cole et al. 1972) and structurally (see Chapter 12). However, tRNA only provided details for the shorter-range interactions, not those which mediate helix packing on globular domains. The crystal structure of the P4-P6 ribozyme domain revealed metal-ion-mediated long-range interactions. The close packing of helices brings their phosphates into proximity; the resulting electrostatic repulsion is ameliorated by binding of hydrated magnesium ions. These outer-sphere magnesium-ion complexes bridge P5 and P5a as well as P5b and P6a (Cate et al. 1996). The crystal structure brought atomic detail to a picture that had been inferred from studies of ribozyme folding as a function of Mg^{++} concentration (Celander and Cech 1991; Banerjee et al. 1993; Jaeger et al. 1993; Murphy and Cech 1994; Szewczak and Cech 1997).

A striking role of divalent cations is seen in the A-rich bulge of P4-P6. The backbone wraps around two divalents, each of which makes an inner-sphere coordination complex with three phosphate oxygens (Cate et al. 1996). As a result, the backbone is on the inside of this local structure and the bases protrude to make tertiary interactions with distant portions of the molecule (Fig. 8b). This structure provides an answer to the paradox of Sigler (1975), who pointed out that proteins form sensible secondary structure, in that the side chains protrude to allow higher-order interactions, whereas nucleic acids form helices that tuck the side chains (bases) inside, not optimal for higher-order interactions. In this A-rich bulge, the metal ions turn a portion of the RNA "inside out," constraining the backbone to be inside and displaying the adenosine "side chains" for tertiary interactions. Phosphorothioate substitution at the metal-coordinating sites interferes both with local folding and with folding of the P4-P6 domain, confirming the essentiality of this metal binding (Cate et al. 1997).

In going from the isolated P4-P6 domain to the whole ribozyme, much more surface area is buried. Therefore, metal ions are likely to be even more important. Consistent with this idea, a long string of functionally significant metal-binding sites in the ribozyme core has been mapped by Christian and Yarus (1993) using phosphorothioate interference and Mn^{++} rescue.

In summary, both inner-sphere and outer-sphere coordination of RNA functional groups by divalent cations contribute to the specificity and stability of RNA tertiary structure.

Small Ribozymes Are Different

At this point we must introduce a disclaimer. We have been developing a picture of a globular RNA with a concave active site that maintains a similar form with or without substrates bound. That this is a reasonable model for the *Tetrahymena* ribozyme is supported by the small amount of energetic coupling between the binding of the guanosine substrate and the P1 helix, indicative of a site that is only slightly reorganized upon binding either single substrate (Bevilacqua et al. 1993; McConnell et al. 1993). Similarly, chemical modification or cleavage studies have seen only minor differences between the free ribozyme and substrate-bound versions. However, small ribozymes such as the hammerhead are much different. Here the binding of substrate is necessary for formation of the catalytic center (Hertel et al. 1997), and even after the substrate–ribozyme complex is formed, the catalytically active conformation may be formed only occasionally (see Chapter 11). Thus, the small ribozymes that were useful in the early stages of the RNA World were probably relatively floppy, and only after RNA self-replication became efficient enough to copy a larger RNA (>100 nucleotides) could ribozymes with permanent active sites come on the scene.

Ribozyme Dynamics

An atomic-resolution crystal structure or NMR structure of a macromolecular catalyst is the single most valuable set of structural information one can obtain. It provides a framework for planning and interpreting experiments that probe the relationships between structure and catalytic function. Yet it gives largely a static picture, and even if a completely rigid macromolecule could bind its substrates, it could never promote their reaction. Formation of a transition state requires movement. One type of movement—switches between distinct conformers—has been described above. We now consider another type of dynamics—thermal motions. These can be thought of as the excursions that a molecule undergoes from its most stable ground-state structure.

The magnitudes of thermal motions between two domains (the P1 substrate domain and the P3-P9 domain) of the *Tetrahymena* ribozyme have been investigated by disulfide cross-linking (Cohen and Cech 1997). Reactive groups were appended to the RNA backbone by an adaptation of the 2′-modification method pioneered by Sigurdsson et al. (1995). Upon collision, these groups can form a disulfide cross-link. Sites separated by 50 Å in the structure were found to be cross-linked at rates only 3 to 15 times slower than the rates for proximal sites, indicative of an unexpectedly high degree of flexibility. The motions were estimated to occur on the microsecond time scale at 30°C (Cohen and Cech 1997).

Presumably, there will be fewer degrees of freedom and therefore less flexibility within a single RNA domain than between two domains. Nevertheless, these measurements suggest that large, globular RNAs are much more dynamic than proteins. It remains to be tested whether one of the functions of RNA-binding proteins is to damp out these large-scale motions, thereby conferring a selective advantage to the RNP World over the RNA World (Fig. 1).

REFERENCES

Abramovitz D.L., Friedman R.A., and Pyle A.M. 1996. Catalytic role of 2′-hydroxyl groups within a group II intron active site. *Science* **271:** 1410–1413.

Banerjee A.R., Jaeger J.A., and Turner D.H. 1993. Thermal unfolding of a group I ribozyme: The low-temperature transition is primarily disruption of tertiary structure. *Biochemistry* **32:** 153–163.

Bartel D.P. and Szostak J.W. 1993. Isolation of new ribozymes from a large pool of random sequences. *Science* **261:** 1411–1418.

Bass B.L. and Cech T.R. 1984. Specific interaction between the self-splicing RNA of *Tetrahymena* and its guanosine substrate: Implications for biological catalysis by RNA. *Nature* **308:** 820–826.

Been M.D. and Perrotta A.T. 1991. Group I intron self-splicing with adenosine: Evidence for a single nucleoside-binding site. *Science* **252:** 434–437.

Beese L.S. and Steitz T.A. 1991. Structural basis for the 3′-5′ exonuclease activity of *Escherichia coli* DNA polymerase I: A two metal ion mechanism. *EMBO J.* **10:** 25–33.

Bevilacqua P.C., Johnson K.A., and Turner D.H. 1993. Cooperative and anticooperative binding to a ribozyme. *Proc. Nat. Acad. Sci.* **90:** 8357–8361.

Bevilacqua P.C., Kierzek R., Johnson K.A., and Turner D.H. 1992. Dynamics of ribozyme binding of substrate revealed by fluorescence detected stopped-flow. *Science* **258:** 1355–1358.

Cate J.H., Hanna R.L., and Doudna J.A. 1997. A magnesium ion core at the heart of a ribozyme domain. *Nat. Struct. Biol.* **4:** 553–558.

Cate J.H., Gooding A.R., Podell E., Zhou K., Golden B.L., Kundrot C.E., Cech T.R., and Doudna J.A. 1996. Crystal structure of a group I ribozyme domain: Principles of RNA packing. *Science* **273:** 1678–1685.

Cech T.R., Herschlag D., Piccirilli J.A., and Pyle A.M. 1992. RNA catalysis by a group I ribozyme: Developing a model for transition state stabilization. *J. Biol. Chem.* **267:** 17479–17482.

Celander D.W. and Cech T.R. 1991. Visualizing the higher order folding of a catalytic RNA molecule. *Science* **251:** 401–407.

Chanfreau G. and Jacquier A. 1994. Catalytic site components common to both splicing steps of a group II intron. *Science* **266:** 1383–1387.

———. 1996. An RNA conformational change between the two chemical steps of group II self-splicing. *EMBO J.* **15:** 3466–3476.

Christian E.L. and Yarus M. 1993. Metal coordination sites that contribute to structure and catalysis in the group I intron from *Tetrahymena*. *Biochemistry* **32:** 4475–4480.

Cohen S.B. and Cech T.R. 1997. Dynamics of thermal motions within a large catalytic RNA investigated by cross-linking with thiol-disulfide interchange. *J. Am. Chem. Soc.* **119:** 6259–6268.

Cole P.E., Yang S.K., and Crothers D.M. 1972. Conformational changes of transfer ribonucleic acid. Equilibrium phase diagrams. *Biochemistry* **11:** 4358–4368.

Conn M.M., Prudent J.R., and Schultz P.G. 1996. Porphyrin metalation catalyzed by a small RNA molecule. *J. Am. Chem. Soc.* **118:** 7012–7013.

Costa M. and Michel F. 1995. Frequent use of the same tertiary motif by self-folding RNAs. *EMBO J.* **14:** 1276–1285.

Copertino D.W. and Hallick R.B. 1991. Group II twintron: An intron within an intron in a cytochrome b-559 gene. *EMBO J.* **10:** 433–442.

Dai X.-C., de Mesmaeker A., and Joyce G.F. 1995. Cleavage of an amide bond by a ribozyme. *Science* **267:** 237–240.

Davies R.W., Waring R.B., Ray J.A., Brown T.A., and Scazzocchio C. 1982. Making ends meet: A model for RNA splicing in fungal mitochondria. *Nature* **300:** 719–724.

Downs W.D. and Cech T.R. 1996. Kinetic pathway for folding of the *Tetrahymena* ribozyme revealed by three ultraviolet-inducible cross-links. *RNA* **2:** 718–732.

Einvik C., Decatur W.A., Embley T.M., Vogt V.M., and Johansen S. 1997. *Naegleria* nucleolar introns contain two group I ribozymes with different functions in RNA splicing and processing. *RNA* **3:** 710–720.

Ekland E.H. and Bartel D.P. 1995. The secondary structure and sequence optimization of an RNA ligase ribozyme. *Nucleic Acids Res.* **23:** 3231–3238.

―――. 1996. RNA-catalyzed RNA polymerization using nucleoside triphosphates. *Nature* **382:** 373–376.

Famulok M. and Szostak J.W. 1992. Stereospecific recognition of tryptophan agarose by *in vitro* selected RNA. *J. Am. Chem. Soc.* **114:** 3990–3991.

Flor P., Flanegan J.B., and Cech T.R. 1989. A conserved base-pair within helix P4 of the *Tetrahymena* ribozyme helps to form the teritary structure required for self-splicing. *EMBO J.* **8:** 3391–3399.

Golden B.L. and Cech T.R. 1996. Conformational switches involved in orchestrating the successive steps of group I RNA splicing. *Biochemistry* **35:** 3754–3763.

Gott J.M., Shub D.A., and Belfort M. 1986. Multiple self-splicing introns in bacteriophage T4: Evidence from autocatalytic GTP labeling of RNA *in vitro*. *Cell* **47:** 81–87.

Haas E.S., Morse D.P., Brown J.W., Schmidt F.J., and Pace N.R. 1991. Long-range structure in ribonuclease P RNA. *Science* **254:** 853–856.

Hampel A. and Cowan J.A. 1997. A unique mechanism for RNA catalysis: The role of metal cofactors in hairpin ribozyme cleavage. *Chem. Biol.* **4:** 516–517.

Herschlag D. 1992. Evidence for processivity and two-step binding of the RNA substrate from studies of J1/2 mutants of the *Tetrahymena* ribozyme. *Biochemistry* **31:** 1386–1399.

Herschlag D. and Cech T.R. 1990a. DNA cleavage catalysed by the ribozyme from *Tetrahymena*. *Nature* **344:** 405–409.

―――. 1990b. Catalysis of RNA cleavage by the *Tetrahymena thermophila* ribozyme. 1. Kinetic description of the reaction of an RNA substrate complementary to the active site. *Biochemistry* **29:** 10159–10171.

―――. 1990c. Catalysis of RNA cleavage by the *Tetrahymena thermophila* ribozyme. 2. Kinetic description of the reaction of an RNA substrate that forms a mismatch at the active site. *Biochemistry* **29:** 10172–10180.

Hertel K.J., Herschlag D., and Uhlenbeck O.C. 1996. Specificity of hammerhead ribozyme cleavage. *EMBO J.* **15:** 3751–3757.

Hertel K.J., Peracchi A., Uhlenbeck O.C., and Herschlag D. 1997. Use of intrinsic binding energy for catalysis by an RNA enzyme. *Proc. Natl. Acad. Sci.* **94:** 8497–8502.

Hur M. and Waring R.B. 1995. Two group I introns with a C·G basepair at the 5′ splice-site instead of the very highly conserved U·G basepair: Is selection post-translational? *Nucleic Acids Res.* **23:** 4466–4470.

Illangasekare M., Sanchez G., Nickles T., and Yarus M. 1995. Aminoacyl-RNA synthesis catalyzed by an RNA. *Science* **267:** 643–647.

Jacquier A. 1996. Group II introns: Elaborate ribozymes. *Biochimie* **78:** 474–487.

Jacquier A. and Michel F. 1987. Multiple exon-binding sites in class II self-splicing introns. *Cell* **50:** 17–29.

———. 1990. Base-pairing interactions involving the 5′ and 3′-terminal nucleotides of group II self-splicing introns. *J. Mol. Biol.* **213:** 437–447.

Jacquier A. and Rosbash M. 1986. Efficient *trans*-splicing of a yeast mitochondrial RNA group II intron implicates a strong 5′ exon-intron interaction. *Science* **234:** 1099–1104.

Jaeger L., Westhof E., and Michel F. 1993. Monitoring of the cooperative unfolding of the *sunY* group I intron of bacteriophage T4. *J. Mol. Biol.* **234:** 331–346.

James B.D., Olsen F.J., Liu J., and Pace N.R. 1988. The secondary structure of ribonuclease P RNA, the catalytic element of a ribonucleoprotein enzyme. *Cell* **52:** 19–26.

Jarrell K.A., Dietrich R.C., and Perlman P.S. 1988. Group II intron domain 5 facilitates a *trans*-splicing reaction. *Mol. Cell Biol.* **8:** 2361–2366.

Jencks W.P. 1969. *Catalysis in chemistry and enzymology* McGraw Hill, New York.

———. 1975. Binding energy, specificity and enzymic catalysis: The Circe effect. *Adv. Enzymol.* **43:** 219–410.

Jenison R.D., Gill S.C., Pardi A., and Polisky B. 1994. High-resolution molecular discrimination by RNA. *Science* **263:** 1425–1429.

Kim E.E. and Wyckoff H.W. 1991. Reaction mechanism of alkaline phosphatase based on crystal structures. Two metal ion catalysis. *J. Mol. Biol.* **218:** 449–464.

Klussmann S., Nolte A., Bold R., Erdmann V.A, and Fürste J.P. 1996. Mirror-image RNA that binds D-adenosine. *Nat. Biotechnol* **14:** 1112–1115.

Latham J.A. and Cech T.R. 1989. Defining the inside and outside of a catalytic RNA molecule. *Science* **245:** 276–282.

Legault P., Herschlag D., Celander D.W., and Cech T.R. 1992. Mutations at the guanosine-binding site of the *Tetrahymena* ribozyme also affect site-specific hydrolysis. *Nucleic Acids Res.* **20:** 6613–6619.

Lehnert V., Jaeger L., Michel F., and Westhof E. 1996. New loop-loop tertiary interactions in self-splicing introns of subgroup IC and ID: A complete 3D model of the *Tetrahymena thermophila* ribozyme. *Chem. Biol.* **3:** 993–1009.

Li Y. and Sen D. 1996. A catalytic DNA for porphyrin metallation. *Nat. Struct. Biol.* **3:** 743–747.

———. 1998. The *modus operandi* of a DNA enzyme: Enhancement of substrate basicity. *Chem. Biol.* **5:** 1–12.

Lohse P.A. and Szostak J.W. 1996. Ribozyme-catalyzed amino-acid transfer reactions. *Nature* **381:** 442–444.

Lorsch J.R. and Szostak J.W. 1994. *In vitro* evolution of new ribozymes with polynucleotide kinase activity. *Nature* **371:** 31–36.

McConnell T.S. and Cech T.R. 1995. A positive entropy change for guanosine binding and for the chemical step in the *Tetrahymena* ribozyme reaction. *Biochemistry* **34:** 4056–4067.

McConnell T.S., Cech T.R., and Herschlag D. 1993. Guanosine binding to the *Tetrahymena* ribozyme: Thermodynamic coupling with oligonucleotide binding. *Proc. Natl. Acad. Sci.* **90:** 8362–8366.

McConnell T.S., Herschlag D., and Cech T.R. 1997. Effects of divalent metal ions on individual steps of the *Tetrahymena* ribozyme reaction. *Biochemistry* **36:** 8293–8303.

McSwiggen J.A. and Cech T.R. 1989. Stereochemistry of RNA cleavage by the *Tetrahymena* ribozyme and evidence that the chemical step is not rate-limiting. *Science* **244:** 679–683.

Michel F. and Ferat J.-L. 1995. Structure and activities of group II introns. *Annu. Rev. Biochem.* **64:** 435–461.

Michel F. and Westhof E. 1990. Modelling of the three-dimensional architecture of group I catalytic introns based on comparative sequence analysis. *J. Mol. Biol.* **216:** 585–610.

Michel F., Jacquier A., and Dujon B. 1982. Comparison of fungal mitochondrial introns reveals extensive homologies in RNA secondary structure. *Biochimie* **64:** 867–881.

Michel F., Umesono K., and Ozeki H. 1989a. Comparative and functional anatomy of group II catalytic introns. *Gene* **82:** 5–30.

Michel F., Ellington A.D., Couture S., and Szostak J.W. 1990. Phylogenetic and genetic evidence for base-triples in the catalytic domain of group I introns. *Nature* **347:** 578–580.

Michel F., Hanna M., Green R., Bartel D.P., and Szostak J.W. 1989b. The guanosine binding site of the *Tetrahymena* ribozyme. *Nature* **342:** 391–395.

Michel F., Jaeger L., Westhof E., Kuras R., Tihy F., Xu M.Q., and Shub D.A. 1992. Activation of the catalytic core of a group I intron by a remote 3′ splice junction. *Genes Dev.* **6:** 1373–1385.

Mörl M., Niemer I., and Schmelzer C. 1992. New reactions catalyzed by a group II intron ribozyme with RNA and DNA substrates. *Cell* **70:** 803–810.

Murphy F.L. and Cech T.R. 1993. An independently folding domain of RNA tertiary structure within the *Tetrahymena* ribozyme. *Biochemistry* **32:** 5291–5300.

———. 1994. GAAA tetraloop and conserved bulge stabilize tertiary structure of a group I intron domain. *J. Mol. Biol.* **236:** 49–63.

Narlikar G.J. and Herschlag. D. 1997. Mechanistic aspects of enzymatic catalysis: Lessons from comparison of RNA and protein enzymes. *Annu. Rev. Biochem.* **66:** 19–59.

Narlikar G.J., Gopalakrishnan J.V., McConnell T.S., Usman N., and Herschlag D. 1995. Use of binding energy by an RNA enzyme for catalysis by positioning and substrate destabilization. *Proc. Natl. Acad. Sci.* **92:** 3668–3672.

Nesbitt S., Hegg L.A., and Fedor M.J. 1997. An unusual pH-independent and metal-ion-independent mechanism for hairpin ribozyme catalysis. *Chem. Biol.* **4:** 619–630.

Pace N.R. and Marsh T.L. 1985. RNA catalysis and the origin of life. *Origins Life* **16:** 97–116.

Peebles C.L., Zhang M., Perlman P.S., and Franzen J.S. 1995. Catalytically critical nucleotide in domain 5 of a group II intron. *Proc. Natl. Acad. Sci.* **92:** 4422–4426.

Peebles C.L., Perlman P.S., Mecklenburg K.L., Petrillo M.L., Tabor J.H., Jarrell K.A., and Cheng H.-L. 1986. A self-splicing RNA excises an intron lariat. *Cell* **44:** 213–223.

Piccirilli J.A., Vyle J.S., Caruthers M.H., and Cech T.R. 1993. Metal ion catalysis by the *Tetrahymena* ribozyme. *Nature* **361:** 85–88.

Piccirilli J.A., McConnell T.S., Zaug A.J., Noller H.F., and Cech T.R. 1992. Aminoacyl esterase activity of the *Tetrahymena* ribozyme. *Science* **256:** 1420–1424.

Pley H.W., Flaherty K.M., and McKay D.B. 1994. Model for an RNA tertiary interaction from the structure of an intermolecular complex between a GAAA tetraloop and an RNA helix. *Nature* **372:** 111–113.

Pyle A.M. 1993. Ribozymes: A distinct class of metalloenzymes. *Science* **261:** 709–714.

Pyle A.M., Murphy F.L., and Cech T.R. 1992. RNA substrate binding site in the catalytic core of the *Tetrahymena* ribozyme. *Nature* **358:** 123–128.

Rajagopal J., Doudna J.A., and Szostak J.W. 1989. Stereochemical course of catalysis by the *Tetrahymena* ribozyme. *Science* **244:** 692–694.

Reinhold-Hurek B. and Shub D.A. 1992. Self-splicing introns in tRNA genes of widely divergent bacteria. *Nature* **357:** 173–176.

Robertson D.L. and Joyce G.F. 1990. Selection *in vitro* of an RNA enzyme that specifically cleaves single-stranded DNA. *Nature* **344:** 467–468.

Saldanha R., Mohr G., Belfort M., and Lambowitz A.M. 1993. Group I and group II introns. *FASEB J.* **7:** 15–24.

Schmelzer C. and Schweyen R.J. 1986. Self-splicing of group II introns *in vitro*: Mapping of the branch point and mutational inhibition of lariat formation. *Cell* **46:** 557–565.

Sclavi B., Woodson S., Sullivan M., Chance M.R., and Brenowitz M. 1997. Time-resolved synchrotron X-ray "footprinting," a new approach to the study of nucleic acid structure and function: Application to protein-DNA interactions and RNA folding. *J. Mol. Biol.* **266:** 144–159.

Sigler P.B. 1975. An analysis of the structure of tRNA. *Annu. Rev. Biophys. Bioeng.* **4:** 447–527.

Sigurdsson, S.T., Tuschl T., and Eckstein F. 1995. Probing RNA tertiary structure: Interhelical crosslinking of the hammerhead ribozyme. *RNA* **1:** 575–583.

Sjögren A.-S., Pettersson E., Sjöberg B.-M., and Strömberg R. 1997. Metal interaction with cosubstrate in self-splicing of group I introns. *Nucleic Acids Res* **25:** 648–653.

Steitz T.A. and Steitz J.A. 1993. A general two-metal ion mechanism for catalytic RNA. *Proc. Natl. Acad. Sci.* **90:** 6498–6502.

Strobel S.A. and Cech T.R. 1993. Tertiary interactions with the internal guide sequence mediate docking of the P1 helix into the catalytic core of the *Tetrahymena* ribozyme. *Biochemistry* **32:** 13593–13604.

———. 1995. Minor groove recognition of the conserved G·U pair at the *Tetrahymena* ribozyme reaction site. *Science* **267:** 675–679.

Strobel S.A., Ortoleva-Donnelly L., Ryder S.P., Cate J.H., and Moncoeur E. 1998. Complementary sets of noncanonical base pairs mediate RNA helix packing in the group I intron active site. *Nat. Struct. Biol.* **5:** 60–66.

Symons R.H. 1992. Small catalytic RNAs. *Annu. Rev. Biochem.* **61:** 641–671.

Szewczak A.A. and Cech T.R. 1997. An internal loop acts as a hinge to facilitate ribozyme folding and catalysis. *RNA* **3:** 838–849.

Tanner M.A. and Cech T.R. 1997. Joining the two domains of a group I ribozyme to form the catalytic core. *Science* **275:** 847–849.

Tarasow T.M., Tarasow S.L., and Eaton B.E. 1997. RNA-catalysed carbon-carbon bond formation. *Nature* **389:** 54–57.

van der Veen R., Arnberg A.C., van der Horst G., Bonen L., Tabak H.F., and Grivell L.A. 1986. Excised group II introns in yeast mitochondria are lariats and can be formed by self-splicing in vitro. *Cell* **44:** 225–234.

Wang J.-F. and Cech T.R. 1994. Metal ion dependence of active-site structure of the *Tetrahymena* ribozyme revealed by site-specific photocrosslinking. *J. Am. Chem. Soc.* **116:** 4178–4182.

Wang J.F., Downs W.D., and Cech T.R. 1993. Movement of the guide sequence during RNA catalysis by a group I ribozyme. *Science* **260:** 504–508.

Wecker M., Smith D., and Gold L. 1996. *In vitro* selection of a novel catalytic RNA: Characterization of a sulfur alkylation reaction and interaction with a small peptide. *RNA* **2:** 982–994.

Weinstein L.B., Jones B.C.N.M., Cosstick R., and Cech T.R. 1997. A second catalytic metal ion in a group I ribozyme. *Nature* **388:** 805–807.

Werner M. and Uhlenbeck O.C. 1995. The effect of base mismatches in the substrate recognition helices of hammerhead ribozymes on binding and catalysis. *Nucleic Acids Res.* **23:** 2092–2096.

Wiegand T.W., Janssen R.C., and Eaton B.E. 1997. Selection of RNA amide synthases. *Chem. Biol.* **4:** 675–683.

Williams K.P., Liu X.H., Schumacher T.N., Lin H.Y., Ausiello D.A., Kim P.S., and Bartel D.P. 1997. Bioactive and nuclease-resistant L-DNA ligand of vasopressin. *Proc. Natl. Acad. Sci.* **94:** 11285–11290.

Wilson C. and Szostak J.W. 1995. *In vitro* evolution of a self-alkylating ribozyme. *Nature* **374:** 777–782.

Yarus M. 1993. How many catalytic RNAs? Ions and the Cheshire cat conjecture. *FASEB J.* **7:** 31–39.

Yarus M., Illangesekare M., and Christian E. 1991. An axial binding site in the *Tetrahymena* precursor RNA. *J. Mol. Biol.* **222:** 995–1012.

Young B., Herschlag D., and Cech T.R. 1991. Mutations in a nonconserved sequence of the *Tetrahymena* ribozyme increase activity and specificity. *Cell* **67:** 1007–1019.

Young K.J., Gill F., and Grasby J.A. 1997. Metal ions play a passive role in the hairpin ribozyme catalyzed reaction. *Nucleic Acids Res.* **25:** 3760–3766.

Zaug A.J. and Cech T.R. 1986a. The intervening sequence RNA of *Tetrahymena* is an enzyme. *Science* **231:** 470–475.

———. 1986b. The *Tetrahymena* intervening sequence ribonucleic acid enzyme is a phosphotransferase and an acid phosphatase. *Biochemistry* **25:** 4478–4482.

Zaug A.J., Been M.D., and Cech T.R. 1986. The *Tetrahymena* ribozyme acts like an RNA restriction endonuclease. *Nature* **324:** 429–433.

Zaug A.J., Dávila-Aponte J.A., and Cech T.R. 1994. Catalysis of RNA cleavage by a ribozyme derived from the group I intron of *Anabaena* pre-tRNA^{Leu}. *Biochemistry* **33:** 14935–14947.

Zhang B. and Cech T.R. 1997. Peptide bond formation by *in vitro* selected ribozymes. *Nature* **390:** 96–100.

Zimmerly S., Guo H., Eskes R., Yang J., Perlman P.S., and Lambowitz A.M. 1995. A group II intron RNA is a catalytic component of a DNA endonuclease involved in intron mobility. *Cell* **83:** 529–538.

Zimmermann G.R., Jenison R.D., Wick C.L., Simone J.-P., and Pardi A. 1997. Interlocking structural motifs mediate molecular discrimination by a theophylline-binding RNA. *Nat. Struct. Biol.* **4:** 644–649.

14
Ribonuclease P

Sidney Altman
Department of Molecular, Cellular and Developmental Biology
Yale University
New Haven, Connecticut 06520

Leif Kirsebom
Department of Microbiology
Biomedical Center
Uppsala University
Uppsala S-751 23, Sweden

RNase P was first characterized in *Escherichia coli* as an enzyme that is required for the processing of the 5′ termini of tRNA in the pathway of biosynthesis of tRNA from precursor tRNAs (ptRNAs; Altman and Smith 1971; for review, see Altman 1989). Subsequently, it was discovered that this enzyme consisted of one protein and one RNA subunit, the latter being the catalytic subunit (Guerrier-Takada et al. 1983). In retrospect, it should not be surprising that an enzyme with the chemical composition of RNase P exists. Ribosomes are also ribonucleoproteins (RNPs), but they are much more complicated than RNase P. It seems unreasonable that ribosomes, with three RNAs and about 50 proteins, would exist without less complex RNPs having come into existence first and having persisted throughout evolution. Indeed, relatively simple RNPs with and without catalytic activity have been discovered with regularity during the past 20 years.

What important questions about RNase P have not yet been answered? Certainly, for the biochemist, problems not yet solved for *E. coli* RNase P include a full understanding of enzyme–substrate interactions, RNA–protein subunit interactions, and the chemical details of the hydrolysis reaction. In every case, much progress has been made, but much has yet to be learned.

A second issue of interest to biochemists and "evolutionists" is the puzzle presented by the different compositions of RNase P from eubacteria (one catalytic RNA and one protein subunit) and from eukaryotes (one RNA subunit and several protein subunits: no understanding yet of which subunit is responsible for catalysis). The complexity of eukaryotic

RNase P presents the questions of why there are so many subunits and what relation this complexity might have to the intracellular localization of the enzyme, possible isoforms (Li and Williams 1995), and its relation to other enzymes with which it might interact in vivo. RNase P in organelles, e.g., chloroplasts and mitochondria, present their own complexities in terms of the origin and function of their subunits (Wang et al. 1988; Dang and Martin 1993; Baum et al. 1996; Lang et al. 1997; Martin and Lang 1997). More generally, the role of RNase P in the regulation of tRNA and rRNA biosynthesis in both prokaryotes and eukaryotes has not yet been fully explored. Indeed, new roles for RNase P in cell physiology are still being uncovered (Stolc and Altman 1997).

This chapter focuses on enzyme–substrate interactions and summarizes briefly what is known about RNase P from bacteria (predominantly *E. coli*) that is pertinent to the other questions presented above and contrasts the characteristics of *E. coli* RNase P with eukaryotic RNase P (predominantly human RNase P). The discussion is not meant to be encyclopedic: We apologize to those whose work we have not mentioned.

FUNCTION OF RNASE P

In *E. coli*, RNase P processes not only ptRNAs (Fig. 1) but also, at the very least, the precursors to 4.5S RNA (Bothwell et al. 1976a), an analog of the eukaryotic SRP RNA (Poritz et al. 1990), tmRNA (10Sa RNA; Komine et al. 1994; Keiler et al. 1996; Muto et al. 1996; Felden et al. 1997), and some small phage RNAs (Hartmann et al. 1995). Formally, this enzyme also has another substrate, 30S prRNA, which it cleaves at the 5' termini of tRNA sequences embedded in the long prRNA (for reviews, see Apirion and Miczak 1993; Deutscher 1993). There may very well be other "hidden" substrates, e.g., ones that resemble the polycistronic *his* operon mRNA, which have not yet been identified in the complexity of RNA processing events in vivo. In this latter case, a cleavage site by RNase P is accessible only after a previous cleavage of the mRNA by RNase E (Alifano et al. 1994). Only after the RNase E cleavage can the remaining 3' terminal fragment of the polycistronic mRNA adopt a conformation that presents the recognition features of substrates for RNase P.

To date, only one set of substrates for purified human RNase P has been identified: These are ptRNAs. However, in crude extracts of both yeast and HeLa cells, RNase P appears to be part of a complex that includes RNase MRP and that is involved in the cleavage of prRNAs (Lygerou et al. 1996; Jarrous et al. 1998). Highly purified RNase P, however, cannot carry out this latter cleavage reaction (Jarrous et al. 1998).

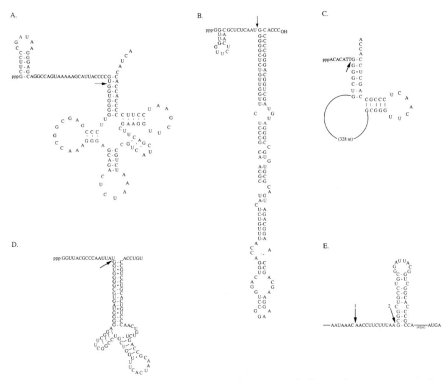

Figure 1 Natural substrates for RNase P from *E. coli*. Secondary structures of several substrates are shown. (*A*) Precursor to tRNA[Tyr]. (*B*) Precursor to 4.5S RNA. (*C*) Precursor to tmRNA. (*D*) Precursor to phages P1 and P7 antisense C4 RNA. (*E*) *his* operon mRNA at the junction of the *his* C and *his* B genes of *Salmonella typhimurium* (see text for explanation and references).

Identification of new substrates for eukaryotic RNase P should involve the examination of cell extracts depleted in RNase P activity.

The fact that nuclear RNase P from both human and yeast cells have several protein subunits points to the possibility that these subunits, alone or together in the RNase P holoenzyme complex, have a function other than that of RNA cleavage. (RNase P from *Saccharomyces cerevisiae* and HeLa cells [Eder et al. 1997; Chamberlain et al. 1998] have been sufficiently highly purified to enable the previous statement regarding the number of subunits to be made. In *S. cerevisiae*, there is also genetic evidence for the identity of at least five subunits [for references, see Stolc and Altman 1997; Stolc et al. 1998].) In fact, the sequences of at least two of the human subunits have motifs resembling those found in proteins that are known to have nuclear localization signals (Jarrous et al. 1998).

In human cells, RNase P must also be transported to the mitochondria (Doersen et al. 1985). (Unlike mitochondria in human cells, in several yeasts and other lower eukaryotes, the mitochondria appear to encode their own RNase P [Hollingsworth and Martin 1986; Martin and Lang 1997].) Accordingly, one of the protein subunits of the human enzyme may be important for the transport process. One candidate is the protein called p38 or Rpp38 (Eder et al. 1997), a nuclear antigen that is also found in the RNP enzyme known as RNase MRP (Yuan et al. 1991). As with RNase P, this enzyme is thought to have both nuclear (nucleolar) and mitochondrial function. The p38-binding domains in the RNA subunits of each of these enzymes are similar in structure (see Fig. 2) (Forster and Altman 1990b). Indeed, there are significant similarities in the overall proposed secondary structure of these RNAs. We note also that the two enzymatic functions are physically associated with each other in crude extracts of both human and yeast cells (Tollervey 1995; Stolc and Altman 1997), an indication of function in similar, or the same, biosynthetic pathways (e.g., rRNA biosynthesis) and perhaps a common origin during evolution (see below).

What are the common features of the various substrates that are recognized by RNase P? Extensive genetic and biochemical analyses of both natural and model substrates indicate that the minimal features recognized by *E. coli* RNase P are a hydrogen-bonded stem of >3 bp and at least one unpaired nucleotide at the 5' end of the sequence (Liu and Altman 1996). The sequence –CCA at the 3' end of this stem (common to all mature tRNAs) enhances cleavage efficiency (Seidman and McClain 1975; Guerrier-Takada et al. 1984; Green and Vold 1988). The variety of known model substrates is shown in Figure 3. In contrast, human RNase P requires a somewhat more complex minimal structure for recognition by the enzyme (Fig. 3), one in which there is a bulge of one or more nucleotides at position 8 (Yuan and Altman 1995; see below).

SOME DETAILS OF THE REACTION WITH DIFFERENT SUBSTRATES

The simplest way of studying an enzymatic reaction is to perform an analysis of its kinetics and to quantitate the well-known parameters, K_m and k_{cat}. The values of these parameters for several substrates for *E. coli* RNase P are listed in Table 1. Note that whereas some model substrates have a value of k_{cat}/K_m (a measure of the efficiency of the reaction) close to that of the enzyme with a natural substrate, the K_m with ptRNA is the lowest, an indication that the interaction between ptRNA and the holoenzyme is stronger than with other substrates. We conclude, not

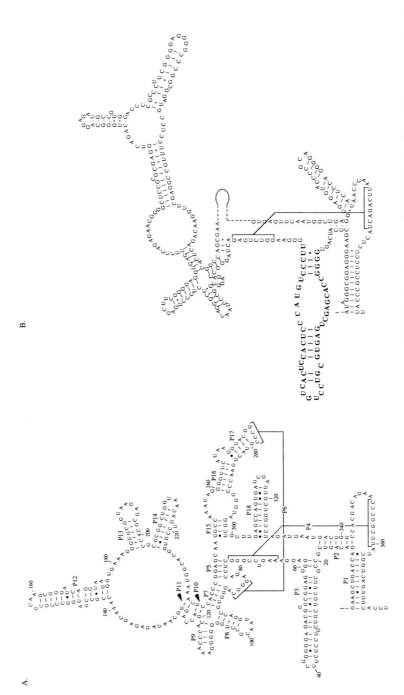

Figure 2 Secondary structure schemes of (*A*) M1 RNA and (*B*) H1 RNA. The domain of H1 RNA that is analogous to a domain to which the Th/To antigen, p38, binds (Yuan et al. 1991) in MRP RNA is shown in bold letters (Forster and Altman 1990b). (*A*, Reprinted, with permission, from Chen and Pace 1997, *B* modified from Chen and Pace 1997.)

unexpectedly, that a natural substrate has more contacts with the enzyme than do model substrates. Indeed, mutagenesis studies show that any mutation that disturbs the secondary and tertiary structure of the tRNA moiety of the substrate in any part of the molecule alters the K_m of the

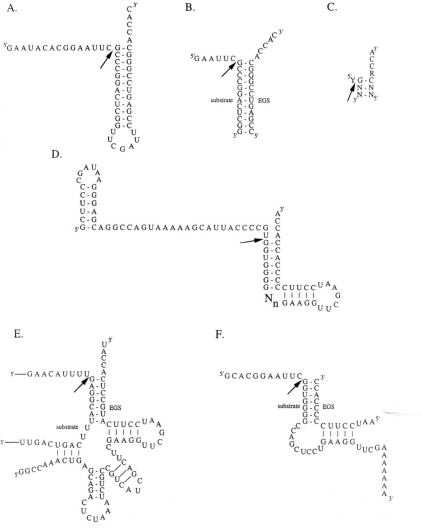

Figure 3 Model substrates for *E. coli* (*A–C*) and human RNase P (*D–F*). (*A*) pAT1, (*B*) Target RNA-EGS (external guide sequence) complex. (*C*) Minimal substrate with base preferences shown. (*D*) Bulged hairpin (analog of pAT1) recognized by human (Yuan and Altman 1995), *X. laevis* (Carrara et al. 1989), and *Drosophila melanogaster* (Levinger et al. 1997). (*E*) Target RNA-EGS complex (3/4 EGS). (*F*) Mini-EGS and target complex (Werner et al. 1997).

Table 1 Kinetic parameters of the reaction of M1 RNA and RNase P with various substrates

Substrate	Enzyme	K_m (nM)	k_{cat} (min^{-1})	k_{cat}/K_m (min^{-1}/μM)
pTyr[a]	M1	30	0.4	13
	M1 + C5[b]	33	29	879
	Δ65		0	
	Δ65 + C5	200	3	15
	Δ[273-281]	97	1	10
	Δ[94-204]		0	
	Δ[94-204] + C5		0	
pTyr-CCA[c]	M1	960	30	31
	M1 + C5	110	101	920
	Δ[273-281]	250	11	44
p4.5S[d]	M1	11,500	1	0.1
	M1 + C5	150	60	400
	Δ65		0	
	Δ65 + C5		0	
	Δ[273-281]	84,700	11	0.13
	Δ[94-204]		0	
	Δ[94-204] + C5	1,170	44	37
p10AT1[e]	M1	1,460	7	4.8
	M1 + C5	1,010	32	32
	Δ[273-281]	1,086	0.7	0.6

[a]This natural substrate terminates in the sequence ACCAUCA. The presence of the extra three nucleotides has little effect on the kinetics of the cleavage reaction.
[b]M1 RNA and C5 protein are in molar ratios of 1:10 in the reaction mixtures.
[c]This substrate terminates in A: The terminal CCAUCA sequence is missing.
[d]The precursor to 4.5S RNA of *E. coli*.
[e]The model substrate pAT1 (McClain et al 1987) with ten nucleotides in its 5′ leader sequence.

reaction (Kirsebom and Altman 1989; Svärd and Kirsebom 1992; Hardt et al. 1993; Kufel and Kirsebom 1994). Note also that the k_{cat} of the reaction is enhanced by the presence of the protein cofactor, C5 protein. In certain cases, C5 protein also greatly reduces the K_m, e.g., with p4.5S RNA (Peck-Miller and Altman 1991). C5 protein also enables the enzyme to discriminate among several tRNA substrates by differentially altering the relative rates of the cleavage reaction with different substrates (Kirsebom and Altman 1989; Gopalan et al. 1997).

When ptRNAs have the common 3'-terminal RCCA sequence, the rate-limiting step of the RNA alone reaction, as determined under multiple turnover conditions, is product release: Addition of the protein cofactor facilitates the release of the product (Reich et al. 1988; Tallsjö and Kirsebom 1993). Recently, a thorough analysis of the kinetics of the reaction with *Bacillus subtilis* RNase P RNA (a type B RNase P RNA; see Haas et al. 1996) has established a model that attempts to assign a kinetic parameter to every stage of the reaction, from substrate and ion binding to product release (Beebe and Fierke 1994). The model should be applicable, in most respects, to all bacterial RNase P RNAs, but some of its steps remain to be characterized more completely both in theory and in relation to experimental data.

The 3' terminal sequence, –CCA, is found in mature tRNAs in all organisms. In eukaryotes, this sequence is absent from the tRNA gene transcripts but is added posttranscriptionally. It is present in all tRNA gene transcripts in *E. coli* and in almost all transcripts in other bacteria, too (Fleischmann et al. 1995; Fraser et al. 1995, 1997; Himmelreich et al. 1996; Blattner et al. 1997; Kunst et al. 1997; Tomb et al. 1997). The presence of the CCA sequence in all ptRNAs in *E. coli* raises the question of whether this sequence plays a role in recognition of substrates by RNase P. (Note that p4.5S RNA ends in –CC and that the terminal A turns over in tRNAs and is not essential for recognition [Deutscher 1993 and references therein; see below].) The CCA sequence is important for recognition by M1 RNA but plays a lesser role when C5 protein is also present (see Tables 1 and 2). When the 5' C of the terminal two Cs is substituted by dC, the rate of the reaction is reduced about tenfold (Perreault and Altman 1992). The absence of the second C of the CCA sequence also decreases the rate of the reaction considerably (Table 2). Curiously, if the entire CCA sequence is missing, the rate of cleavage is not so severely affected, an indication, perhaps, that the other important features of the measuring and cleavage mechanisms are no longer affected by the identity or positioning of the 3'-terminal sequence. These results have been derived from studies of both model and natural substrates from which terminal nucleotides have been individually and sequentially removed. Genetic studies have also determined that the CCA sequence hydrogen-bonds to a specific, complementary sequence in M1 RNA (Kirsebom and Svärd 1994; Svärd et al. 1996; see below). This interaction also involves the discriminator base at position +73 in the ptRNA, since it has been shown that the absence of an interaction between the discriminator base and M1 RNA facilitates product release (Tallsjö et al. 1996). Accordingly, we can assign this sequence, the unpaired nucleotide at the 5' side of the mature tRNA sequence (the

Table 2 The effect of changes in the 3'-terminal CCA sequence on the kinetic parameters of the M1 RNA and RNase P cleavage reactions

Substrate	Enzyme	K_m (nM)	k_{cat} (min^{-1})	k_{cat}/K_m (min^{-1}/μM)
pTyr[a]	M1	33	0.4	12
	M1 + C5[b]	29	76	260
pTyr-CA[c]	M1	200	0.15	0.75
	M1 + C5	125	18	140
pTyr-CCA[d]	M1	500	3	6
	M1 + C5	200	27	140
Yeast pSupS1[e]	M1	100	0.44	4.4
	M1 + C5	200	7.6	38

The experiments reported in Tables 1 and 2 were performed with different preparations of M1 RNA.
[a]This natural substrate terminates in the sequence ACCAUCA. The presence of the extra three nucleotides has little effect on the kinetics of the cleavage reactions.
[b]M1 RNA and C5 protein are in molar ratios of 1:10 in the reaction mixtures.
[c]Terminates in AC: The terminal CAUCA is missing.
[d]Terminates in A: The terminal CCAUCA is missing.
[e]Derived from *S. cerevisiae* tRNA^{Ser}. This substrate does not have a 3'-terminal CCA sequence.

"–1" nucleotide), and a short double-stranded region in the aminoacyl stem as part of the "minimal" recognition features (see Fig. 3C).

Since RNase P requires a divalent metal ion, of which Mg^{++} is the most efficient as a cofactor (Gardiner et al. 1985; Guerrier-Takada et al. 1986; Kazakov and Altman 1991), there has been a great deal of interest in the number of divalent metal ions that are required for the hydrolysis reaction to proceed. The results of various studies suggest that two or three metal ions are required for catalysis (Kahle et al. 1993; Smith and Pace 1993; Warnecke et al. 1996; Chen et al. 1997). There are, apparently, several strong metal-ion-binding sites in M1 RNA but whether, and how many of, the ions bound to these sites are involved directly in the cleavage event is not clear (Kazakov and Altman 1991; Zito et al. 1993; Ciesiolka et al. 1994). However, recent data suggest that a Mg^{++} ion is coordinated directly to the pro-Rp oxygen at the cleavage site and that this ion participates in the chemistry of cleavage (Warnecke et al. 1996; Chen et al. 1997). These recent data are consistent with another study which indicates that one of the required metal ions is bound to the substrate near the site of cleavage and is carried to the active center of the enzyme in this manner (Perreault and Altman 1993).

The substitution with chemical analogs of the usual nucleotides found in RNA has proved to be useful in determining which nucleotides are essential for catalysis in M1 RNA, its substrates, and in other ribozymes. Many studies of E-S complexes have been performed in which each of these RNAs, randomly substituted with phosphorothioate backbones, deoxynucleotides or 4-thiouracil, are probed in E-S complexes with chemical agents that react with the RNAs (see below). Those phosphorothioate bonds or substituted nucleotides that do not appear in the reaction products or that fail to react with the chemical agents are judged to be protected by E-S contacts. Experiments can also be carried out directly on unsubstituted RNAs. Finally, cross-linking agents can be attached to specific positions in "cyclized" M1 RNA or its substrate: Cross-links formed between RNAs must be in areas of close contact (~15 Å). The results of all these experiments, and others carried out with deoxy-substituted model substrates, can be summarized as follows:

1. The aminoacyl stem, or its equivalent in a substrate, is denatured during catalysis by M1 RNA, probably after binding of the substrate but before hydrolysis occurs. These data are derived from protection from chemical reagents of a ptRNA substrate in E-S complexes (Knap et al. 1990).
2. Deoxynucleotide substitutions at various positions in model and ptRNA substrates have been used to show that positions –1 and –2, as well as the 5'-terminal C of the CCA sequence (Perreault and Altman 1992; Kleineidam et al. 1993; Kufel and Kirsebom 1996a), severely affect the k_{cat} of hydrolysis. Similar substitutions in other parts of tRNA substrates show that binding to M1 RNA is critically affected by the loss of the 2'-OH at G + 1, and in the T stem and T loop (Gaur and Krupp 1993; Conrad et al. 1995; Loria and Pan 1997).
3. Phosphorothioate substitutions have been used to show that the aminoacyl and T stems of substrates unfold during the reaction, as well as to demonstrate that ptRNAs with either a short or a long variable arm interact differently with M1 RNA (Gaur et al. 1996). This approach has also been used to identify phosphates and 2'-OH in RNase P RNA important for tRNA binding (Hardt et al. 1995, 1996) and catalysis (Harris and Pace 1995). Here, 2'-OH and phosphates important for tRNA binding have been localized at and near the P4 helix as well as to residue C247 in M1 RNA (see Fig. 2). Phosphates in P4 are also thought to be important for Mg^{++} binding and catalysis.
4. Chemical cross-linking agents have been used in various ways to show that the 3'-terminal CCA sequence, the aminoacyl stem, the T

stem (positions 49–53 and 61–65), and loop are critical for binding of a tRNA precursor to M1 RNA (Nolan et al. 1993; Harris et al. 1994; Oh and Pace 1994) as well as to identify nucleotides in M1 RNA that are in close contact with its substrate (see below; Guerrier-Takada et al. 1989; Burgin and Pace 1990; Harris et al. 1994; Kufel and Kirsebom 1996a; Massire et al. 1998). However, some of these experiments have been carried out with tRNAs rather than ptRNAs. It is not yet clear that both ptRNAs and tRNAs bind in precisely the same manner to M1 RNA. Other data indicate that product inhibition of the reaction is noncompetitive (Guerrier-Takada and Altman 1993), but there are some differences in experiments of this kind from different laboratories. Through the use of various substrates, differences in the way in which M1 RNA interacts with particular substrates have been observed both by cross-linking and chemical protection studies (Guerrier-Takada et al. 1989; Kufel and Kirsebom 1996a,b; Pan and Jakacka 1996; see also section 3 above). Nevertheless, many of the data agree with results cited in sections 2 and 3 above and are compatible, to a large extent, with three-dimensional models of M1 RNA–ptRNA interactions.

Human RNase P, as mentioned above, has more constraints in recognizing its substrates than does *E. coli* RNase P, with one exception: The 3′-terminal–CCA sequence is unimportant in determining enzyme efficiency. We note also that eukaryotic RNase P RNAs lack the region in the internal loop of P15 in the secondary structure scheme depicted in Figure 2. These RNAs, as well as their analogs from *Chlamydia spp.* and some cyanobacteria, do not need the specific hydrogen-bonded interaction with the –CCA sequence found in *E. coli* M1 RNA (Herrmann et al. 1996; Vioque 1997). Hardt et al. (1993) have also shown that, unlike hairpin substrate for RNase P from *E. coli*, RNase P from HeLa cells requires some tertiary structure at the junction between the equivalents of the aminoacyl and the T stems in its substrates for recognition to occur. These results were confirmed by Yuan and Altman (1995) and Werner et al. (1997).

Both substrate affinity and efficient catalysis by *S. cerevisiae* RNase P appear to be influenced by the presence of a mismatch of the nucleotide at -1 (Lee et al. 1997). Furthermore, in the absence of the 3′-terminal CCA sequence, if the nucleotide at -1 is base-paired with a nucleotide at the 3′ end of the ptRNA, denaturation of the aminoacyl stem to facilitate positioning and cleavage by RNase P becomes energetically more difficult (see below).

Since no eukaryotic RNase P RNA has been shown to be catalytic in vitro, it is likely that the protein subunits of these RNase Ps play a greater role in substrate recognition than does the protein (subunit) of *E. coli* RNase P.

SEQUENCE AND CLEAVAGE-SITE RECOGNITION IN SUBSTRATES

RNase P does not recognize specific sequences in its substrates. From studies of RNase P from *E. coli*, we conclude that there are, however, some general preferences of the enzyme for certain nucleotides in the minimal recognition domain. Not surprisingly, these preferences reflect the naturally occurring composition of tRNAs. For example, the vast majority of mature tRNAs have a G in the first position of their sequences, a high G-C content in the aminoacyl stem, U in position 8 and, as discused above, a 3′-terminal CCA sequence. Some data suggest that G at –2 also influences both cleavage efficiency and cleavage-site recognition (Kirsebom and Svärd 1993; Meinnel and Blanquet 1995; M. Bränvall et al., unpubl.). The efficiency of the RNase P cleavage reaction in ptRNAs is frequently decreased considerably if any of these features of substrates are altered. Another interesting aspect of the reaction, namely miscleavage, is revealed when G1 is changed, or the base-pairing near the ends of the aminoacyl stem is destroyed (see, e.g., Kirsebom 1995 and references therein).

One of the early questions posed in studies of RNase P concerns the accuracy of the cleavage reaction; i.e., how does the enzyme always find the right cleavage site in a large number of substrates that differ in sequence? Put another way, we can ask if RNase P has a measuring device that senses the tRNA domain of ptRNAs and, accordingly, cleaves a certain distance from particular features in this domain. These features include the structural integrity of the T loop, the length of the T stem and the length of the aminoacyl stem, commonly 7 bp (with two exceptions; see below). The enzyme appears to measure the distance, in almost every case 12 bp between the T loop and the cleavage site (Kahle et al. 1990; Thurlow et al. 1991; Svärd and Kirsebom 1992, 1993; Kufel and Kirsebom 1994, 1996b). Additionally, the RCCA at the 3′ termini of ptRNAs has a specific role in this process (see below). The physical attributes of these various features of M1 RNA have to operate in concert in order to accomplish cleavage of the substrate at the correct position. We propose, specifically, that the enzyme interacts with the T loop (P11 and A118 in M1 RNA; G64 in the substrate with A118 and C100 in M1 RNA;

see Figs. 4 and 5) as proposed by Massire et al. (1998) a certain distance in the substrate, equivalent to 12 bp, from where the actual cleavage event takes place. This has to occur in concert with the RCCA–RNase P RNA interaction (with the conserved GGU in the internal loop of P15; interacting residues as shown in Figure 6 as suggested by Kirsebom and Svärd 1994; see also LeGrandeur et al. 1994; Oh and Pace 1994; Svärd et al. 1996; Tallsjö et al. 1996). A third contact point between the RNase P RNA (A249; see Fig. 4) and its substrate would be an interaction with, in particular, a G at the cleavage site (+1) and perhaps also with the nucleotides at positions –1 and –2. The outcome of these interactions would be to expose the cleavage site and consequently correctly position the scissile bond in the active site (A351, A352, J18/2, P4; see Fig. 4). This model can also account for the correct cleavage of various model substrates, since these substrates still exhibit at least two of the features important for cleavage-site recognition. RNase P RNA folding would generate a pocket that can accommodate the T loop and the coaxially stacked T stem and aminoacyl stem as shown in Figure 4. The stem region, P5, is a likely candidate for part of the "measuring" apparatus.

The exact contacts between substrates and RNase P RNA from different species might change according to the nature of the RNA from that species. We would expect that RNase P RNA and its substrates would have co-evolved to preserve the necessary interactions (see below). (Sequence analysis reveals that the tRNA genes in *Borrelia burgdorferei* [Fleischmann et al. 1995] do not encode CCA despite the presence of the P15 loop in its RNase P RNA. This may be an example of a branch point in the evolution of RNase P RNA.)

Evidence supporting our picture of a measuring mechanism comes from studies with model substrates in which extra base pairs have been added to or deleted from the aminoacyl and/or the T stem in combination, in some cases with removal of the terminal RCCA sequence (Svärd and Kirsebom 1992, 1993; Kufel and Kirsebom 1994, 1996b). Addition of 1 or 2 bp in the analogs of the aminoacyl and T stems has no effect on precision of cleavage (Svärd and Kirsebom 1992). In these cases, the precision of cleavage depends, we assume, mainly on the RCCA–RNase P RNA interaction as well as the nucleotide at the cleavage site. When 3 bp are inserted (Svärd and Kirsebom 1993), cleavage occurs at the 5′ side of positions +1 (the 5′ side of the first base pair in the altered aminoacyl stem) and +4, an indication that both aspects of the measuring device are operative. If the 3′-terminal RCCA is deleted, cleavage occurs almost exclusively at the +4 site (Kufel and Kirsebom 1994). Thus, there is some

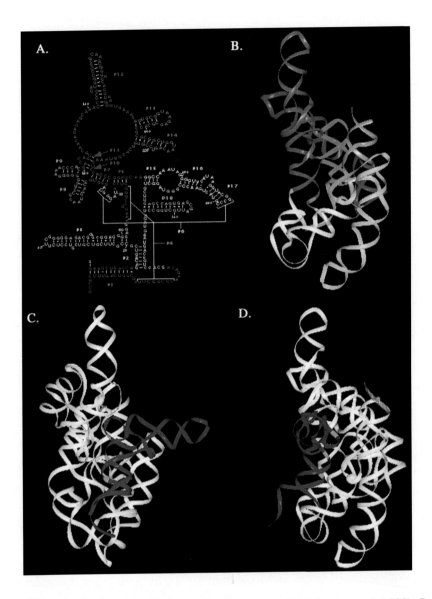

Figure 4 Computer-assisted modeling of M1 RNA (Massire et al. 1998). Proposed secondary structure (*A*) and three-dimensional ribbon model of M1 RNA (*B*). Color coding of domains is the same for each domain shown in (*A*) and in (*B*). (Reprinted, with permission from Massire et al. 1998 [copyright Academic Press].) (*C*, *D*) Two views of a three-dimensional ribbon model of M1 RNA (*white*) and bound precursor tRNA^{Asp} (*red*). Positions +1 (cleavage site), 64 and 53 (T stem), from bottom to top, in the tRNA domain are fixed accurately (*cyan blue*) as reference points for the interaction with the enzyme (M1 RNA, in this case).

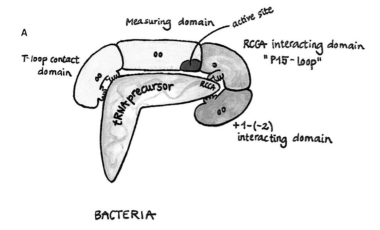

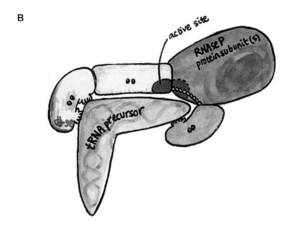

Figure 5 Scheme for the "measuring" device in, and selection of cleavage site in a tRNA precursor by, RNase P RNAs. (*A*) Bacterial RNase P RNA. (*B*) Eukaryotic RNase P RNA. The common features of each scheme are the domains that interact with the T loop, that "measure" the distance between the T loop and the cleavage site, that encompass the active center of the enzyme, and that interact with nucleotides in the vicinity of the cleavage site (+1 to –2). The active center also includes part of the substrate, since a Mg^{++} ion is coordinated to the Rp oxygen at the site of cleavage (Perreault and Altman 1993; Warnecke et al. 1996; Chen et al. 1997). The RCCA sequence in substrates interacts with a P15 domain found only in bacterial RNase P RNAs and, in particular, in those bacteria in which the RCCA sequence is encoded in the genomic DNA of tRNA genes. Additional data on which this model is based are given in the text. Bothwell et al. (1976b) offered a much simpler version of this scheme.

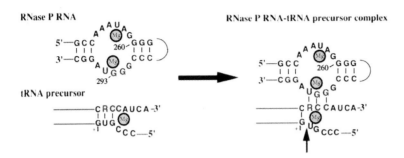

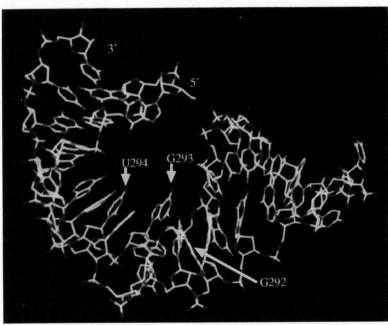

Figure 6 P15-RCCA interactions between M1 RNA and precursor tRNA. (*A*) R in the substrate denotes an A or a G and results in an A-U or a G-U base pair in the enzyme–substrate complex. The figure also shows three Mg^{++} ions (*yellow*), one carried in by the substrate and two bound to the P15 loop in M1 RNA. (*B*) Structure of a model RNA that contains the P15 loop (Glemarec et al. 1996). The arrows denote the highly conserved GGU-motif in RNase P RNAs that base-pairs with the 3′-terminal RCCA sequence of a tRNA precursor.

play in the correlation between the length of the aminoacyl and T stems and the point at which the measuring device stops relying solely on the RCCA and T-loop reference points for positioning of substrates. (We note that p4.5S RNA, which has a very long stem, is lacking a T-loop equivalent. Thus, the precision of cleavage must depend primarily on recognition features near the site of cleavage on the 5′ strand and the 3′-terminal CC sequence.)

In contrast to the situation with *E. coli* RNase P, changing the number of base pairs in substrates for human or *Xenopus Laevis* RNase P by even 1 bp does affect cleavage precision in the sense that the enzyme always cleaves at a position 7 bp from the junction with the T stem (Carrara et al. 1989; Yuan and Altman 1995). It appears that the bacterial enzyme has sufficient information in a model substrate to determine where G1 is, but the human enzyme needs additional contacts in the T stem-loop to make this determination. It follows that this phenomenon might be explained in part by the absence of the RCCA–RNase P RNA interaction in eukaryotes.

It is, indeed, possible to make RNase P alter the site of cleavage in bacterial substrates other than by adding 3 bp to the combined length of the aminoacyl and T stems (see below). The substitution of only one nucleotide within one of the minimal substrate recognition domains in certain substrates can also achieve this effect (Kirsebom and Svärd 1994; see below). The introduction of these structural features into a substrate overrides the measuring mechanism of even the minimal substrate recognition features.

Changes of G1, or base-pairing at the end of the aminoacyl stem, can, in some cases, affect the choice of cleavage site. This phenomenon is dramatically illustrated with *E. coli* ptRNA^{His}. tRNA^{His} is the one of two tRNAs that has 8 bp in its aminoacyl stem: the other is tRNA^{SeCys}. In bacteria, the 8-bp stem in these tRNAs results from RNase P cleavage at a position that would normally be position –1 in other ptRNAs. Replacement of the guanosine at the ptRNA^{His} cleavage site or substitution of C73 at the 3′ end results in miscleavage by M1 RNA alone or by reconstituted holoenzyme such that 5′ matured tRNAs with 7 bp forming an aminoacyl stem are produced. However, addition of C5 protein to the reaction mixture results in an increased frequency of cleavage at the correct position to yield an aminoacyl stem of 8 bp (Burkard et al. 1988; Green and Vold 1988; Kirsebom and Svärd 1992). This is consistent with the fact that details of the 5′ and 3′ termini of the tRNA domain affect the cleavage-site recognition and the measuring process. In contrast, eukaryotic RNase P cleavage of ptRNA^{His} generates a 7-bp aminoacyl stem and the extra G is added after RNase P cleavage (Cooley et al. 1982). This phenomenon is again in keeping with our speculations concerning measuring and the absence of the CCA at the 3′ termini of eukaryotic tRNA precursors, as well as the lack of a P15 loop equivalent, the site of RCCA interaction, in eukaryotic RNase P RNA.

3-D MODELS AND SUBSTRATE RECOGNITION

Any successful model of RNase P must be able to account for the data gathered so far on substrate binding and the detailed mechanism of the cleavage reaction. The latter, unfortunately, is still largely unknown. In the absence of a crystal structure of the enzyme, theoretical, computer-assisted models (Harris et al. 1994, 1997; Westhof and Altman 1994) derived from phylogenetic and experimental data have been useful in leading the way to further experimentation. The most recent model (Fig. 4) (Massire et al. 1998) is derived primarily from phylogenetic analysis but is also consistent with a large fraction, but not all, of the data available from chemical, cross-linking studies and genetic studies. It is more compatible with Fe-EDTA protection data for M1 RNA (Westhof et al. 1996) than are previous models (D. Wesolowski and S. Altman, unpubl.).

A compact, catalytic center for M1 RNA is a feature of the new model. In its organization of RNA domains, it shares some features with the catalytic center of group I introns. Nevertheless, there are still some discrepancies between the proposal in the model that certain structures (e.g., that are encompassed by nucleotides 94–204) are essential for catalysis and experimental data which show they can be deleted from M1 RNA molecules that still retains function in vitro. Although the precise details of the orientation of a ptRNA substrate on the surface of M1 RNA remain to be worked out, it is clear that the model generally agrees with what is known about contacts with the T stem and loop and the RCCA domain of substrates (see above). These constraints, themselves, give credence to the model and also allow it to be considered as a platform on which to dock and test experimental data regarding the conformation and binding to M1 RNA of the C5 protein cofactor (Gopalan et al. 1998).

NATURAL HISTORY OF RNASE P

Several groups have shown that M1 RNA can be linked to other RNAs and still retain its catalytic function. Three kinds of constructs have been made: (1) M1 RNA covalently linked at either terminus to a ptRNA or model substrate (Kikuchi et al. 1993; Frank et al. 1994; Kikuchi and

Suzuki-Fujita 1995); (2) M1 RNA linked at either terminus to an external guide sequence (EGS: i.e., the 3'-terminal segment of the aminoacyl stem of a tRNA [Li et al. 1992; Yuan et al. 1992; Guerrier-Takada et al. 1995, 1997; Liu and Altman 1995, 1996; Li and Altman 1996]); and (3) M1 RNA covalently linked to a ptRNA followed by a portion of the bacteriophage T4 gene, thymidylate synthase (*td*), which contains two exons separated by an intron (A. Kaplin et al., unpubl. cited in Altman 1989). In these cases, individually, cleavage was observed at the correct site in (1) the ptRNA segment; (2) the target sequence of an EGS-target complex that resembles part of a ptRNA (i.e., the 5' segment of a ptRNA aminoacyl stem and the upstream leader sequence; and (3) in the ptRNA sequence concomitantly with self-splicing of the partial *td* gene. We note that the M1 RNA-ptRNA or M1 RNA-EGS RNA construct reactions are quite efficient. Presumably, binding of the potential substrate to the active center of M1 RNA is facilitated by the tether linking it to the enzyme (the tether must be longer than a minimum length, about 7 nucleotides [Liu and Altman 1996]). These experiments suggest that complex RNAs can exist, and could have existed eons ago in an RNA World, which contain two or more independent catalytic activities and substrates (see Fig. 7A).

Contemporary M1 RNA and its substrates must have been somewhat simpler at one point in time than they are today. The reasoning for this statement is presented below, but there is an implicit assumption that should be noted first: We equate "simpler" with "earlier" in time of evolution. Such an assumption depends in a not very well-defined manner on the meaning of the word "simpler."

Working with purely phylogenetic analyses, Pace and coworkers have defined what they call the "minimal" RNase P RNA, i.e., the smallest molecule (211 nucleotides; Siegel et al. 1996) needed to carry out the hydrolysis reaction associated with RNase P. Furthermore, an even smaller RNase P RNA (140 nucleotides) has been identified in the mitochondria of *Saccharomyces fibuligera* (Wise and Martin 1991), although this RNA is not catalytic by itself. One can also demonstrate in another fashion that M1 RNA must have been smaller at a point in time earlier than today. Contemporary M1 RNA has two basic classes of substrates, exemplified by precursor tRNA and precursor 4.5S RNA, respectively. The latter substrate is structurally the simpler of the two, since it contains only a hairpin and an upstream leader sequence. The K_m of the reaction of M1 RNA (not including C5 protein) with each substrate is an indication of the complexity of each substrate: The K_m with pTyr is about 100-fold smaller than with p4.5S RNA, a reflection of the larger number of contacts made with the former substrate. Furthermore, the entire "upper" domain of M1 RNA (nucleotides 94–204; see Fig. 2) can be deleted without a major effect on the reaction with p4.5S RNA, whereas the reaction with various ptRNAs is less efficient when this domain is lost. We conclude, therefore, that a simpler form of today's M1 RNA, dictated by either Pace's criterion or the one outlined immediately above, must have existed in which the upper

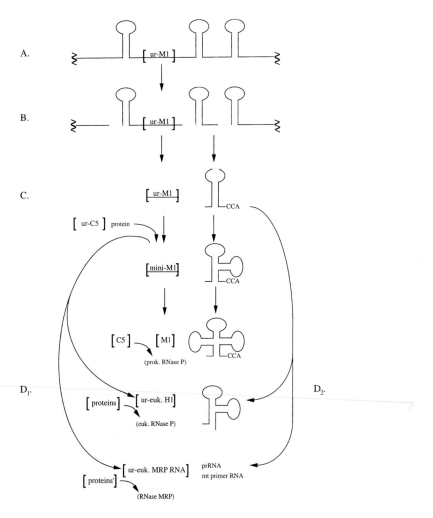

Figure 7 Scheme for the early evolution of RNA and RNPs. (*A*) Long RNA molecule with occasional hairpins and a sequence that contains an ancient M1 RNA-like (ur-M1) cleavage activity. (*B*) Molecule in *A* is self-cleaved at 5′ side of hairpins by ur-M1. (*C*) Beginning of the progression of coevolution of M1 RNA, substrates, and protein cofactors. (D_1, D_2) Branch points leading to eukaryotic RNase P and RNase MRP.

domain was absent, since it is not needed for the binding of "hairpin" substrates (the simpler class of substrates) or catalysis. We recall that the simplest model substrates for M1 RNA consist only of a small hairpin and a few extra nucleotides at the 5' terminus.

It is not irrelevant to mention that the simplest substrates for tRNA synthases are also small hairpins (Francklyn et al. 1992). In both cases, i.e., aminoacylation and RNase P cleavage, the K_m values for the reactions with the model substrates are very high compared to reactions with today's more complex, natural substrates. We speculate, accordingly, that the original role of an ancient M1 RNA may have been to cleave long RNAs at the 5' side of hairpin structures to generate a variety of smaller molecules (Fig. 7A, B). (Maizels and Weiner [1993] also recognized a possible important early role of hairpin structures in their development of the "genome tagging" theory of RNA replication.) The population of more numerous smaller molecules should be endowed with a larger number of functional capabilities than might be possible with the original, parent, larger RNA molecule. One can imagine that greater functional variation comes in part from discrete structural domains of RNA interacting with each other, not necessarily through covalent bonds, to create larger complexes with active functions and to present even more possibilities for assembly of an ancient M1 RNA with cleavage capability. Indeed, complexes with M1 RNA activity have been constructed in vitro from subdomains with no catalytic activity (Guerrier-Takada and Altman 1992; Pan 1995).

We do not assign an absolute time to the origins of M1 RNA and its substrates. The possibility that the first self-replicating genetic system was not made of RNA, or the specific place of M1 RNA in an evolutionary time series in the RNA World is not particularly relevant. The only important assertion we make in a discussion of the evolution of M1 RNA and other catalytic RNAs and their attendant proteins is that there was an RNA World at some time prior to the appearance of proteins as entities encoded in genetic material. Furthermore, given the uncertainties in "replication times" of the inhabitants of the RNA World, the nature of the then-contemporary selective pressures and nucleotide substitution rates, it seems of little value to attempt to assign rough, absolute time intervals to different stages of complexity of the RNA World.

Notwithstanding the uncertainties cited above, it is still possible to consider the nature and fate of individual molecules in, and their progression during, evolution to a resemblance of those that we see today. For example, an evolutionary scheme can be described in which we consider a set of RNA–protein complexes that represent a progression in an evolutionary sense, and that might contain an ancient M1-like catalytic RNA (a "complex" with one RNA and no protein, R_1P_0), M1-like catalytic RNA plus a protein cofactor (R_1P_1), and ultimately, 16S rRNA plus 20 proteins (R_2P_{2-21}) if 16S rRNA is a direct evolutionary descendant of M1 RNA and its proteins are the immediate descendants of P_1. We note that this particular progression is unlikely in reality: It is apparent that there must be more intermediates in the series, as noted above, in the evolution of M1 RNA itself (i.e., $R_1P_0 \ldots R_NP_0$; see Fig. 7) as well as branch points in which the original R_1 or its predecessor evolved into another molecule, R_2, that might have been the evolutionary precursor of the self-splicing intron (or possibly, R_3, a molecule that generated different end groups during cleavage and was the precursor of the class of catalytic RNAs known as viroids). Each of these three progenitor RNA molecules must have evolved further in terms of catalytic specificity and, ultimately, in their ability to work together with one or more proteins (see Fig. 7).

In addition to these catalytic RNAs, a parallel evolution of noncatalytic RNAs, e.g., tRNA, tmRNA, rRNA, is required, and these RNAs might (since some were substrates for the catalytic RNAs) or might not be co-varying with the catalytic RNAs. In fact, today we can see an example of coevolution of an RNA enzyme and its substrates in the ability of RNase P RNAs from two different sources, the so-called type A or type B RNase P RNAs, to cleave or not to cleave, respectively, p4.5S RNA from *E. coli* (Guerrier-Takada et al. 1983).

Once proteins were on the scene, coevolution of catalytic RNAs, proteins, and substrate RNAs (Liu and Altman 1994) had to have been the general case. A clearer example of evolutionary progression is, perhaps, more obvious if we consider the lineage of RNase P from an early progenitor organism to *E. coli* RNase P to human RNase P with a relatively early branch point (perhaps at the same juncture in time as the branch between eubacteria and eukaryotes) that leads to RNase MRP (Chang and Clayton 1987), an enzyme found only in eukaryotes but which has a close relationship to RNase P, as noted previously (Fig. 7C, D). The thread of evolution can also be discerned in the observation that M1 RNA functions in mammalian cells in tissue culture, possibly by co-opting a protein subunit of host cell RNase P (Liu and Altman 1995).

Is it possible to make a statement concerning the "time" at which proteins began to interact with RNA and infer something about the state of the RNA World when proteins first appeared? We use as examples the facts that tRNA synthases can aminoacylate micro-hairpin substrates (Francklyn et al. 1992), and that RNase P can cleave similar, model ptRNAs (McClain et al. 1987; Forster and Altman 1990a). If these examples bear any relation to today's natural world, then in the cases of tRNA synthases and of RNase P, and very likely of other RNPs as well, RNA evolution, catalytic and otherwise, was far from finished when proteins arrived on the scene. One prediction from these ideas is that fragments of tRNA synthases and C5 protein will be able to carry out, or participate in as a cofactor, respectively, enzymatic reactions on micro-helical, model substrates.

ACKNOWLEDGMENTS

We thank our colleagues for comments on the manuscript, and we thank D. Wesolowski and D. Pomeranz-Krummel for help with the figures. Work in the laboratory of S.A. is funded by U.S. Public Health Service grant GM-19422 and a Human Frontiers of Science program grant RG-0291. Work in the laboratory of L.K. is supported by the Swedish Natural Science Research Council.

REFERENCES

Alifano P., Rivellini F., Piscitelli C., Arraiano C.M., Bruni C.B., and Carlomagno M.S. 1994. Ribonuclease E provides substrates for ribonuclease P-dependent processing of a polycistronic mRNA. *Genes Dev.* **8:** 3021–3031.

Altman S. 1989. Ribonuclease P: An enzyme with a catalytic RNA subunit. *Adv. Enzymol. Relat. Areas Mol. Biol.* **62:** 1–36.

Altman S. and Smith J.D. 1971. Tyrosine tRNA precursor molecule polynucleotide sequence. *Nat. New Biol.* **233:** 35–39.

Apirion D. and Miczak A. 1993. RNA processing in prokaryotic cells. *BioEssays* **15:**113–120.

Baum M., Cordier A., and Schön A. 1996. RNase P from a photosynthetic organelle contains an RNA homologous to the cyanobacterial counterpart. *J. Mol. Biol.* **257:** 43–52.

Beebe J.A. and Fierke C.A. 1994. A kinetic mechanism for cleavage of precursor tRNA^{Asp} catalysed by the RNA component of *Bacillus subtilis* ribonuclease P. *Biochemistry* **33:** 10294–10304.

Blattner F.R., Plunkett G., III, Bloch C.A., Perna N.T., Burland V., Riley M., Collado-Vides J., Glasner J.D., Rode C.K., Mayhew G.F., Gregor J., Davis N.W., Kirkpatrick H.A., Goeden M.A., Rose D.J., Mau B., and Shao Y. 1997. The complete genome sequence of *Escherichia coli* K-12. *Science* **277:** 1453–1462.

Bothwell A.L.M., Garber R.L., and Altman S. 1976a. Nucleotide sequence and *in vitro* processing of a precursor molecule to *Escherichia coli* 4.5 S RNA. *J. Biol. Chem.* **251:** 7709–7716.

Bothwell A.L.M., Stark B.C., and Altman S. 1976b. Ribonuclease P substrate specificity: Cleavage of a bacteriophage f80-induced RNA. *Proc. Natl. Acad. Sci.* **73:** 1912–1916.

Burgin A.B. and Pace N.R. 1990. Mapping the active site of ribonuclease P RNA using a substrate containing a photoaffinity agent. *EMBO J.* **9:** 4111–4118.

Burkard U., Willis I., and Söll D. 1988. Processing of histidine transfer RNA precursors. *J. Biol. Chem.* **263:** 2447–2451.

Carrara G., Calandra P., Fruscoloni P., Doria M., and Tocchini-Valentini G. 1989. Site selection by *Xenopus laevis* RNase P. *Cell* **58:** 37–45.

Chamberlain J.R., Lee Y., Lane W.S., and Engelke D.R. 1998. Purification and characterization of the nuclear RNase P holoenzyme complex reveals extensive subunit overlap with RNase MRP. *Genes Dev.* **12:** 1678–1690.

Chang D. and Clayton D.A. 1987. A novel endoribonuclease cleaves at a priming site of mouse mitochondrial DNA replication. *EMBO J.* **6:** 409–417.

Chen J.-L. and Pace N.R. 1997. Identification of the universally conserved core of ribonuclease P RNAs. *RNA* **3:** 557–560.

Chen Y., Li X., and Gegenheimer P. 1997. Ribonuclease P catalysis requires Mg^{2+} coordinated to the pro-Rp oxygen of the scissile bond. *Biochemistry* **36:** 2425–2438.

Ciesiolka J., Hardt W.-D., Schlegl J., Erdmann V.A., and Hartmann R.K. 1994. Lead-ion-induced cleavage of RNase P RNA. *Eur. J. Biochem.* **219:** 49–56.

Conrad F., Hanne A., Gaur R.K., and Krupp G. 1995. Enzymatic synthesis of 2′-modified nucleic acids: Identification of important phosphate and ribose moieties in RNase P substrates. *Nucleic Acids Res.* **23:** 1845–1853.

Cooley L., Appel B., and Söll D. 1982. Post-transcriptional nucleotide addition is responsible for the formation of the 5′ terminus of histidine tRNA. *Proc. Natl. Acad. Sci.* **79:** 6475–6479.

Dang Y.L. and Martin N.C. 1993. Yeast mitochondrial RNase P: Sequence of the RPM2 gene and demonstration that its product is a protein subunit of the enzyme. *J. Biol. Chem.* **268:** 19791–19796.

Deutscher M.P. 1993. RNA maturation nucleases. In *Nucleases*, 2nd edition (ed. S.M. Linn et al.), pp. 377–406. Cold Spring Harbor Laboratory Press, Cold Spring Harbor, New York.

Doersen C.J., Guerrier-Takada C., Altman S., and Attardi G. 1985. Characterization of an RNase P activity from HeLa cell mitochondria. Comparison with the cytosol RNase P activity. *J. Biol. Chem.* **260:** 5942–5949.

Eder P.S., Kekuda R., Stolc, V., and Altman S. 1997. Characterization of two scleroderma autoimmune antigens that copurify with human ribonuclease P. *Proc. Natl. Acad. Sci.* **94:** 1101–1106.

Felden B., Himeno H., Muto A., McCutcheon J.P., Atkins J.F., and Gesteland R.F. 1997. Probing the structure of the *E. coli* 10Sa RNA (tmRNA). *RNA* **3:** 89–103.

Fleischmann R.D., Adams M.D., White O., Clayton R.A., Kirkness E.F., Kerlavage A.R., Bult C.J., Tomb J.F., Dougherty B.A., Merrick J.M., et al. 1995. Whole-genome random sequencing and assembly of *Haemophilus influenzae* Rd. *Science* **269:** 496–512.

Forster A.C. and Altman S. 1990a. External guide sequences for an RNA enzyme. *Science* **249:** 783–786.

———. 1990b. Similar cage-shaped structures for the RNA components of all ribonuclease P and ribonuclease MRP enzymes. *Cell* **62:** 407–409.

Francklyn C., Shi J.P., and Schimmel P. 1992. Overlapping nucleotide determinants for specific aminoacylation of RNA microhelices. *Science* **255:** 1121–1125.

Frank D.N., Harris M.E., and Pace N.R. 1994. Rational design of self-cleaving pre-tRNA-ribonuclease P RNA conjugates. *Biochemistry* **33:** 10800–10808.

Fraser C.M., Casjens S., Huang W.M., Sutton G.G., Clayton R., Lathigra R., White O., Ketchum K.A., Dodson R., Hickey E.K., Gwinn M., Dougherty B., Tomb J.F., Fleischmann R.D., Richardson D., Peterson J., Kerlavage A.R., Quackenbush J., Salzberg S., Hanson M., van Vugt R., Palmer N., Adams M.D., Gocayne J., Venter J.C., et al. 1997. Genomic sequence of a Lyme disease spirochaete, *Borrelia burgdorferi. Nature* **390:** 580–586.

Fraser C.M., Gocayne J.D., White O., Adams M.D., Clayton R.A., Fleischmann R.D., Bult C.J., Kerlavage A.R., Sutton G., Kelley J.M., Fritchman J.L., Weidman J.F., Small K.V., Sandusky M., Fuhrmann J., Nguyen D., Utterback T.R., Saudek D.M., Phillips C.A., Merrick J.M., Tomb J.-F., Dougherty B.A., Bott K.F., Hu P.-C., and Lucier T.S. 1995. The minimal gene complement of *Mycoplasma genitalium. Science* **270:** 397–403.

Gardiner K.J., Marsh T.L., and Pace N.R. 1985. Ion dependence of the *Bacillus subtilis* RNase P reaction. *J. Biol. Chem.* **260:** 5415–5419.

Gaur R.K. and Krupp G. 1993. Modification interference approach to detect ribose moieties important for the optimal activity of a ribozyme. *Nucleic Acids Res.* **21:** 21–26.

Gaur R.K., Hahne A., Conrad F., Kahle D., and Krupp G. 1996. Differences in the interaction of *Escherichia coli* RNase P RNA with tRNA containing a short or a long extra arm. *RNA* **2:** 674–681.

Glemarec C., Kufel J., Foldesi A., Sandstrom A., Kirsebom L.A., and Chattopadhyaya J. 1996. The NMR structure of 31 mer RNA domain of *Escherichia coli* RNase P RNA using its non-uniformly deuterium labelled counterpart [the 'NMR-window' concept]. *Nucleic Acids Res.* **24:** 2022–2035.

Gopalan V., Baxevanis A., Landsman D., and Altman S. 1997. Analysis of the functional role of conserved residues in the protein subunit of ribonuclease P from *Escherichia coli. J. Mol. Biol.* **267:** 818–829.

Gopalan V., Ledman D.W., Fox R.O., and Altman S. 1998. Mapping RNA-protein interactions in ribonuclease P from *Escherichia coli* using disulfide linked EDTA-Fe. *J. Mol. Biol.* (in press).

Green C.J. and Vold B.S. 1988. Structural requirements for processing of synthetic tRNAHis precursors by the catalytic RNA components of RNase P. *J. Biol. Chem.* **263:** 652–657.

Guerrier-Takada C. and Altman S. 1992. Reconstitution of enzymatic activity from fragments of M1 RNA. *Proc. Natl. Acad. Sci.* **89:** 1266–1270.

———. 1993. A physical assay for and kinetic analysis of the interactions between M1 RNA and tRNA precursor substrates. *Biochemistry* **32:** 7152–7161.

Guerrier-Takada C., Li Y., and Altman S. 1995. Artificial regulation of gene expression in *Escherichia coli* by RNase P. *Proc. Natl. Acad. Sci.* **92:** 11115–11119.

Guerrier-Takada C., Lumelsky N., and Altman S. 1989. Specific interactions in RNA enzyme-substrate complexes. *Science* **286:** 1578–1584.

Guerrier-Takada C., McClain W.H., and Altman S. 1984. Cleavage of tRNA precursors by the RNA subunit of *E. coli* ribonuclease P (M1 RNA) is influenced by 3′-proximal CCA in the substrates. *Cell* **38:** 219–224.

Guerrier-Takada C., Salavati R., and Altman S. 1997. Phenotypic conversion of drug-resistant bacteria to drug sensitivity. *Proc. Natl. Acad. Sci.* **94:** 8468–8472.

Guerrier-Takada C., Haydock K., Allen L., and Altman S. 1986. Metal ion requirements and other aspects of the reaction catalyzed by M1 RNA, the RNA subunit of ribonuclease P from *Escherichia coli. Biochemistry* **25:** 1509–1515.

Guerrier-Takada C., Gardiner K., Marsh T., Pace N., and Altman S. 1983. The RNA moiety of ribonuclease P is the catalytic subunit of the enzyme. *Cell* **35:** 849–857.

Haas E.S., Banta A.B. Harris J.K., Pace N.R., and Brown J.W. 1996. Structure and evolution of ribonuclease P RNA in gram-positive bacteria. *Nucleic Acids Res.* **24:** 4775–4782.

Hardt, W.-D., Erdmann V.A., and Hartmann R.K. 1996. Rp-deoxy-phosphorothioate modification interference experiments identify 2′-OH groups in RNase P RNA that are crucial to tRNA binding. *RNA* **2:** 1189–1198.

Hardt W.-D., Schlegl J., Erdmann V.A., and Hartmann R.K. 1993. Role of the D arm and the anticodon arm in tRNA recognition by eubacterial and eukaryotic RNase P enzymes. *Biochemistry* **32:** 13046–13053.

Hardt W.-D., Warnecke J.M., Erdmann V.A., and Hartmann R.K. 1995. Rp-phosphorothioate modifications in RNase P RNA that interfere with tRNA binding. *EMBO J.* **14:** 2935–2944.

Harris M.E. and Pace N.R. 1995. Identification of phosphates involved in catalysis by the ribozyme RNase P RNA. *RNA* **1:** 210–218.

Harris M.E., Kazantsev A.V., Chen J.-L., and Pace N.R. 1997. Analysis of the tertiary structure of the ribonuclease P ribozyme-substrate complex by site-specific photoaffinity crosslinking. *RNA* **3:** 561–576.

Harris M.E., Nolan J.M., Malhotra A., Brown J.W., Harvey S.C., and Pace N.R. 1994. Use of photoaffinity crosslinking and molecular modeling to analyze the global structure of ribonuclease P RNA. *EMBO J.* **13:** 3953–3963.

Hartmann R.K., Heinrich J., Schlegl J., and Schuster H. 1995. Precursor of C4 antisense RNA of bacteriophages P1 and P7 is a substrate for RNase P of *Escherichia coli. Proc. Natl. Acad. Sci.* **92:** 5822–5826.

Herrmann B., Winqvist O., Mattsson J.G., and Kirsebom L.A. 1996. Differentiation of Chlamydia spp. by sequence determination and restriction endonuclease cleavage of RNase P RNA genes. *J. Clin. Microbiol.* **34:** 1897–1902.

Himmelreich R., Hilbert H., Plagens H., Pirkl E., Li B.C., and Herrmann R. 1996. Complete sequence analysis of the genome of the bacterium *Mycoplasma pneumoniae. Nucleic Acids Res.* **24:** 4420–4449.

Hollingsworth M.J. and Martin N.C. 1986. RNase P activity in the mitochondria of *Saccharomyces cerevisiae* depend on both mitochrondrion and nucleus-encoded components. *Mol. Cell. Biol.* **6:** 1058–1064.

Jarrous N., Eder P.S., Guerrier-Takada C., Hoog C., and Altman S. 1998. Autoantigenic properties of some protein subunits of catalytically active complexes of human ribonuclease P. *RNA* **4:** 407–417.

Kahle D., Küst B., and Krupp G. 1993. Phosphorothioates in pre-tRNAs can change the specificities of RNases P or reduce the cleavage efficiencies. *Biochimie* **75:** 955–962.

Kahle D., Wehmeyer U., and Krupp G. 1990. Substrate recognition by RNase P and by the catalytic M1 RNA: Identification of possible contact points in pre-tRNAs. *EMBO J.* **9:** 1929–1937.

Kazakov S. and Altman S. 1991. Site-specific cleavage by metal ion cofactors and inhibitors of M1 RNA, the catalytic subunit of RNase P from *Escherichia coli. Proc. Natl. Acad. Sci.* **88:** 9193–9197.

Keiler K.C., Waller R.H., and Sauer R.T. 1996. Role of a peptide tagging system in degradation of proteins synthesized from damaged messenger RNA. *Science* **271:** 990–993.

Kikuchi Y. and Suzuki-Fujita K. 1995. Synthesis and self-cleavage reaction of a chimeric molecule between RNase P-RNA and its model substrate. *J. Biochem.* **117:** 197–200.

Kikuchi Y., Sasaki-Tozawa N., and Suzuki K. 1993. Artificial self-cleaving molecules consisting of tRNA precursor and the catalytic RNA of RNase P. *Nucleic Acids Res.* **21:** 4685–4689.

Kirsebom L.A. 1995. RNase P—A 'Scarlet Pimpernel.' *Mol. Microbiol.* **17:** 411–420.

Kirsebom L.A. and Altman S. 1989. Reaction in vitro of some mutants of RNase P with wild-type and temperature-sensitive substrates. *J. Mol. Biol.* **207:** 837–840.

Kirsebom L.A. and Svärd S.G. 1992. The kinetics and specificity of cleavage by RNase P is mainly dependent on the structure of the amino acid acceptor stem. *Nucleic Acids Res.* **20:** 425–432.

———. 1993. Identification of a region within M1 RNA of *Escherichia coli* RNase P important for the location of the cleavage site on a wild-type tRNA precursor. *J. Mol. Biol.* **231:** 594–604.

———. 1994. Base pairing between *Escherichia coli* RNase P RNA and its substrate. *EMBO J.* **13:** 4870–4876.

Kleineidam R.G., Pitulle C., Sproat B., and Krupp G. 1993. Efficient cleavage of pre-tRNAs by *E. coli* RNase P RNA requires the 2′-hydroxyl of the ribose at the cleavage site. *Nucleic Acids Res.* **21:** 1097–1101.

Knap A.K., Wesolowski D., and Altman S. 1990. Protection from chemical modification of nucleotides in complexes of M1 RNA, the catalytic subunit of RNase P from *E. coli*, and tRNA precursors. *Biochimie* **72:** 779–790.

Komine Y., Kitabatake M., Yokogawa T., and Nishikawa K. 1994. A tRNA-like structure is present in 10Sa RNA, a small stable RNA from *E. coli. Proc. Natl. Acad. Sci.* **91:** 9223–9277.

Kufel J. and Kirsebom L.A. 1994. Cleavage site selection by M1 RNA, the catalytic subunit of *Escherichia coli* RNase P, is influenced by pH. *J. Mol. Biol.* **244:** 511–521.

———. 1996a. Different cleavage sites are aligned differently in the active site of M1 RNA, the catalytic subunit of *Escherichia coli* RNase P. *Proc. Natl. Acad. Sci.* **93:** 6085–6090.

———. 1996b. Residues in *Escherichia coli* RNase P RNA important for cleavage site selection and divalent metal ion binding. *J. Mol. Biol.* **263:** 685–698.

Kunst F., Ogasawara N., Moszer I., Albertini A.M., Alloni G., Azevedo V., Bertero M.G., Bessieres P., Bolotin A., Borchert S., Borriss R., Boursier L., Brans A., Braun M., Brignell S.C., Bron S., Brouillet S., Bruschi C.V., Caldwell B., Capuano V., Carter N.M., Choi S.K., Codani J.J., Connerton I.F., Danchin A., et al. 1997. The complete genome sequence of the gram-positive bacterium *Bacillus subtilis. Nature* **390:** 249–256.

Lang B.F., Burger G., O'Kelly C.J., Cedergren R., Golding G.B., Lemieux C., Sankoff D., Turmel M., and Grey M.W. 1997. An ancestral mitochondrial DNA resembling a eubacterial genome in miniature. *Nature* **387:** 493–496.

Lee Y., Kindelberger D.W., Lee J-Y., McClennen S., Chamberlain J., and Engelke D.R. 1997. Nuclear pre-tRNA terminal structure and RNase P recognition. *RNA* **3:** 175–185.

LeGrandeur T.E., Huttenhofer A., Noller N.F., and Pace N.R. 1994. Phylogenetic comparative chemical footprint analysis of the interaction between ribonuclease P RNA and tRNA. *EMBO J.* **13:** 3945–3952.

Levinger L., Bourne R., Kolla S., Cylin E., Russell K., Wang X., and Mohan A. 1997. Processing kinetics and minimal substrates for *Drosophila* RNase P and 3′-tRNase. *Nucleic Acids Symp. Ser.* **36:** 78–82.

Li K. and Williams R.S. 1995. Cloning and characterization of three new murine genes encoding short homologues of RNase P RNA. *J. Biol. Chem.* **42:** 25281–25285.

Li Y. and Altman S. 1996. Cleavage by RNase P of gene N mRNA reduces bacteriophage 1 burst size. *Nucleic Acids Res.* **24:** 835–842.

Li Y., Guerrier-Takada C., and Altman S. 1992. Targeted cleavage of mRNA *in vitro* by RNase P from *Escherichia coli. Proc. Natl. Acad. Sci.* **89:** 3185–3189.

Liu F. and Altman S. 1994. Differential evolution of substrates for an RNA enzyme in the presence and absence of its protein cofactor. *Cell* **77:** 1093–1100.

———. 1995. Inhibition of viral gene expression by the catalytic RNA subunit of RNase P from *Escherichia coli. Genes Dev.* **9:** 471–480.

———. 1996. Requirements for cleavage by a modified RNase P of a small model substrate. *Nucleic Acids Res.* **24:** 2690–2696.

Loria A. and Pan T. 1997. Recognition of the T stem-loop of a pre-tRNA substrate by the ribozyme from *Bacillus subtilis* ribonuclease P. *Biochemistry* **36:** 6317–6325.

Lygerou Z., Pluk H., van Venrooij W.J., and Seraphin B. 1996. hPop1: An autoantigenic protein subunit shared by the human RNase P and RNase MRP ribonucleoproteins. *EMBO J.* **15:** 5936–5948.

Maizels N. and Weiner A.M. 1993. The genomic tag hypothesis: Modern viruses as molecular fossils of ancient strategies for genomic replication. In *The RNA world* (ed. R.F. Gesteland and J.F. Atkins), pp. 577–602. Cold Spring Harbor Laboratory Press, Cold Spring Harbor, New York.

Martin N.C. and Lang B.F. 1997. Mitochondrial RNase P: The RNA family grows. *Nucleic Acids Symp. Ser.* **36:** 42–44.

Massire C., Jaeger L., and Westhof E. 1998. Derivation of the three-dimensional architecture of bacterial ribonuclease P RNAs from comparative sequence analysis. *J. Mol. Biol.* **279:** 773–793.

McClain W.H., Guerrier-Takada C., and Altman S. 1987. Model substrates for an RNA enzyme. *Science* **238:** 527–530.

Meinnel T. and Blanquet S. 1995. Maturation of pre-tRNA$_f^{Met}$ by *Escherichia coli* RNase P is specified by a guanosine of the 5'-flanking sequence. *J. Biol. Chem.* **270:** 15908–15914.

Muto A., Sato M., Tadaki T., Fukushima M., Ushida C., and Himeno H. 1996. Structure and function of 10Sa RNA: *trans*-translation system. *Biochimie* **78:** 985–991.

Nolan J.M., Burke D.H., and Pace N.R. 1993. Circulary permutated tRNAs as specific photoaffinity probes of ribonuclease P RNA structure. *Science* **261:** 762–765.

Oh B-K. and Pace N.R. 1994. Interaction of the 3'-end of tRNA with ribonuclease P RNA. *Nucleic Acids Res.* **22:** 4087–4094.

Pan T. 1995. Higher order folding and domain analysis of the ribozyme *B. subtilis* ribonuclease P. *Biochemistry* **34:** 902–909.

Pan T. and Jakacka M. 1996. Multiple substrate binding sites in the ribozyme from *Bacillus subtilis* RNase P. *EMBO J.* **15:** 2249–2255.

Peck-Miller K.A. and Altman S. 1991. Kinetics of the processing of the precursor to 4.5 S RNA, a naturally occurring substrate for RNase P from *Escherichia coli. J. Mol. Biol.* **221:** 1–5.

Perreault J-P. and Altman S. 1992. Important 2'-hydroxyl groups in model substrates for M1 RNA, the catalytic subunit of RNase P from *Escherichia coli. J. Mol. Biol.* **226:** 399–409.

———. 1993. Pathway of activation by magnesium ions of substrates for the catalytic subunit of RNase P from *Escherichia coli*. *J. Mol. Biol.* **230:** 750–756.

Poritz M.A., Bernstein H.D., Strub K., Zopf D., Wilhelm H., and Walter P. 1990. An *E. coli* ribonucleoprotein containing 4.5S RNA resembles mammalian signal recognition particle. *Science* **250:** 1111–1117.

Reich C., Olsen G.J., Pace B., and Pace N.R. 1988. Role of the protein moiety of RNase P, a ribonucleoprotein enzyme. *Science* **239:** 178–181.

Seidman J.G. and McClain W.H. 1975. Three steps in conversion of large precursor RNA into serine and proline transfer RNAs. *Proc. Natl. Acad. Sci.* **72:** 1491–1495.

Siegal R.W., Banta A.B., Haas E.S., Brown J.W., and Pace N.R. 1996. *Mycoplasma fermentans* simplifies our view of the catalytic core of ribonuclease P RNA. *RNA* **2:** 452–462.

Smith D. and Pace N.R. 1993. Multiple magnesium ions in the ribonuclease P reaction mechanism. *Biochemistry* **32:** 5273–5281.

Stolc V. and Altman S. 1997. Rpp1, an essential protein subunit of nuclear RNase P required for processing of precursor tRNA and 35S precursor rRNA in *Saccharomyces cerevisiae*. *Genes Dev.* **11:** 2926–2937.

Stolc V., Katz A., and Altman S. 1998. Rpp2, an essential protein subunit of nuclear RNase P is required for processing of precursor tRNAs and 35S precursor rRNA in *Saccharomyces cerevisiae*. *Proc. Natl. Acad. Sci.* **95:** 6716–6721.

Svärd S.G. and Kirsebom L.A. 1992. Several regions of a tRNA precursor determine the *Escherichia coli* RNase P cleavage site. *J. Mol. Biol.* **227:** 1019–1031.

———. 1993. Determinants of *Escherichia coli* RNase P cleavage site selection: A detailed *in vitro* and *in vivo* analysis. *Nucleic Acids Res.* **21:** 427–434.

Svärd S.G., Kagardt U., and Kirsebom L.A. 1996. Phylogenetic comparative mutational analysis of the base-pairing between RNase P RNA and its substrate. *RNA* **2:** 463–472.

Tallsjö A. and Kirsebom L.A. 1993. Product release is a rate-limiting step during cleavage by the catalytic RNA subunit of *Escherichia coli* RNase P. *Nucleic Acids Res.* **21:** 51–57.

Tallsjö A., Kufel J., and Kirsebom L.A. 1996. Interaction between *Escherichia coli* RNase P RNA and the discriminator base results in slow product release. *RNA* **2:** 299–307.

Thurlow D.L., Shilowski D., and Marsh T.L. 1991. Nucleotides in precursor tRNAs that are required intact for catalysis by RNase P RNAs. *Nucleic Acids Res.* **19:** 885–891.

Tollervey D. 1995. Genetic and biochemical analyses of yeast RNase MRP. *Mol. Biol. Rep.* **22:** 75–79.

Tomb J.F., White O., Kerlavage A.R., Clayton R.A., Sutton G.G., Fleischmann R.D., Ketchum K.A., Klenk H.P., Gill S., Dougherty B.A., Nelson K., Quackenbush J., Zhou L., Kirkness E.F., Peterson S., Loftus B., Richardson D., Dodson R., Khalak H.G., Glodek A., McKenney K., Fitzegerald L.M., Lee N., Adams M.D., Venter J.C., et al. 1997. The complete genome sequence of the gastric pathogen *Helicobacter pylori*. *Nature* **388:** 539–547.

Vioque A. 1997. The RNase P RNA from cyanobacteria: Short tandemly repeated repetitive (STRR) sequences are present within the RNase P RNA gene in heterocyst-forming cyanobacteria. *Nucleic Acids Res.* **25:** 3471–3477.

Wang M.J., Davis N.W., and Gegenheimer P. 1988. Novel mechanisms for maturation of chloroplast transfer RNA precursors. *EMBO J.* **7:** 1567–1574.

Warnecke J.M., Fürste J.P., Hardt W.-D., Erdmann V.A., and Hartmann R.K. 1996. Ribonuclease P (RNase P) RNA is converted to a Cd^{2+}-ribozyme by a single Rp-phosphorothioate modification in the precursor tRNA at the RNase P cleavage site. *Proc. Natl. Acad. Sci.* **93:** 8924–8928.

Werner M., Rosa E., and George S. 1997. Design of short external guide sequences (EGSs) for cleavage of target molecules with RNase P. *Nucleic Acids. Symp. Ser.* **36:** 19–21.

Westhof E. and Altman S. 1994. Three dimensional working model of M1 RNA, the catalytic RNA subunit of ribonuclease P from *Escherichia coli*. *Proc. Nat. Acad. Sci.* **91:** 5133–5137.

Westhof E., Wesolowski D., and Altman S. 1996. Mapping in three dimensions regions in a catalytic RNA protected from attack by an Fe(II)-EDTA reagent. *J. Mol. Biol.* **258:** 600–613.

Wise C. and Martin N.C. 1991. Dramatic size variation of yeast mitochondrial RNAs suggests that RNase P RNAs can be quite small. *J. Biol. Chem.* **266:** 19154–19157.

Yuan Y. and Altman S. 1995. Substrate recognition by human RNase P identification of small, model substrates for the enzyme. *EMBO J.* **14:** 159–168.

Yuan Y., Hwang E.-S., and Altman S. 1992. Targeted cleavage of mRNA by human RNase P. *Proc. Natl. Acad. Sci.* **89:** 8006–8010.

Yuan Y., Tan E., and Reddy R. 1991. The 40-kilodalton to autoantigen associates with nucleotides 21 to 64 of human mitochondrial RNA processing/7-2 RNA *in vitro*. *Mol. Cell. Biol.* **11:** 5266–5274.

Zito K., Huttenhofer A., and Pace N.R. 1993. Lead-catalyzed cleavage of ribonuclease P RNA as a probe for integrity of tertiary structure. *Nucleic Acids Res.* **21:** 5916–5920.

15

The RNA Folding Problem

Peter B. Moore
Departments of Chemistry, and Molecular Biophysics, and Biochemistry
Yale University
New Haven, Connecticut 06520-8107

Because the chemical and biological properties of many RNAs are determined by their conformations, an RNA equivalent exists to the protein folding problem. The RNA folding problem has both a practical side and an academic side. If the practical problem were solved, it would be possible to predict the conformations of RNAs of known sequence. The academic problem has to do with determining what happens when denatured RNAs fold in vitro, and nascent RNA chains fold in vivo. The focus of this essay is structure prediction, but it does include some commentary on what has been learned about RNA folding pathways.

Structure prediction is a matter of some urgency today because RNAs are being discovered much faster than their three-dimensional structures are being solved. Furthermore, the techniques available for determining structures experimentally, X-ray crystallography and NMR, are so time-consuming, applied to RNAs by communities so small, and so uncertain of success when applied to specific RNAs that, absent some technological and/or sociological revolution, the mismatch between supply and demand is certain to persist indefinitely.

In principle, biochemists ought to be able to do without the services of crystallographers and NMR spectroscopists. Most (all?) denatured RNAs renature spontaneously in vitro, which means that the structures of RNAs are determined by their sequences. Furthermore, the physical interactions that drive RNA folding are largely understood (see Chapter 10). Why can't the structures of RNAs of known sequence be predicted, and failing that, why can't procedures be developed that determine conformations by combining theory with nonphysical data?

RNA ARCHITECTURE

Before launching into a discussion of RNA structure prediction, it would be wise to remind ourselves exactly what it is that needs to be predicted.

Double Helices

Like the rest of modern biology, the science of RNA conformation begins with the structure of DNA, and, as everyone knows, the typical DNA molecule consists of two polynucleotide strands arranged in an antiparallel double helix, in which As, Gs, Cs, and Ts on one strand pair with Ts, Cs, Gs, and As, respectively, on the other. Except for the (almost) trivial difference that in RNA, As pair with Us, RNA forms antiparallel double helices the same way. Under all conditions, the conformation of RNA helices resembles that adopted by DNA helices under low-humidity conditions; i.e., RNA helices are A-form, never B-form. In A-form helices, the planes of base pairs are tilted with respect to the helix axis, the major groove is narrow and deep, and the minor groove is broad and shallow (see Saenger 1984). Furthermore, the sugar pucker in A-form helices is C3'-endo, or N(orth), whereas in B-form helices, it is C2'-endo, or S(outh). The reason all RNA helices are A-form is that B-form helix formation is sterically hindered by the placement of their 2'-OH groups when their riboses are C2'-endo. Note, however, that A-form is not the only helical geometry possible for RNA. Three- and even four-strand helices exist, but they are possible only for specific sequences, and hence are seldom encountered.

Single-stranded Polynucleotides: Stem-loops

Most RNAs are single-stranded molecules that have sequences incompatible with the formation of long double helices, but, nevertheless, many are more than 60% as hyperchromic as they would be if they were perfect duplexes. In the late 1950s, Doty and coworkers explained why (Doty et al. 1959; Fresco and Alberts 1960; Fresco et al. 1960). Except for molecules with bizarre sequences, like polyuridylic acid or polycytidylic acid, RNAs are full of short sequences that are "accidentally" complementary, and RNA chains fold back on themselves to form hairpin loops or stem-loops, so that these sequences can form helices.

Many of the helical stems in RNA hairpins are interrupted by internal loops, which is to say regions where sequences are juxtaposed that cannot form GCs, AUs, or wobble GUs, which most regard as "honorary" Watson-Crick base pairs. Some form irregularly helical structures that consist of a succession of non-Watson-Crick base pairs (see, e.g., Correll et al. 1997; Dallas and Moore 1997). Others, especially those in which the number of bases contributed by the two strands is different, form more irregular structures (Puglisi et al. 1992, 1995; Battiste et al. 1994, 1996; Aboul-ela et al. 1995; Ye et al. 1995; Fan et al. 1996; Fourmy et al. 1996; Yang et al. 1996; Jiang et al. 1997; Kalurachchi et al. 1997). In some cases, the trajectory of the backbone becomes so distorted that bases project into solution, away from the body of the stem (bulged bases). In others, "extra" bases form base triples by hydrogen bonding edge-on with other base pairs, or intercalate between base pairs.

Necessarily, all stem-loops include a terminal loop where the trajectory of the RNA backbone changes direction by 180° so that the 3' side of the sequence can pair with the 5' side. Terminal loops vary in length from 2 nucleotides (Butcher et al. 1997), to hundreds or even thousands of bases. Some 4-base loops, or tetraloops, are unusually common and play the same role in RNA as β turns in proteins (Tuerk et al. 1988; Cheong et al. 1990; Woese et al. 1990; Allain and Varani 1995; Jucker and Pardi 1995; Jucker et al. 1996). Loops bigger than 20 or 30 nucleotides may contain internal stem-loops.

Ever since the hairpin model for RNAs was advanced, identification of the helical segments in RNAs has been a major activity, and Figure 1 shows what has been learned about the stem-loops in *Escherichia coli* 16S rRNA (Gutell 1996). The RNA literature is full of similar diagrams. Note that sequences for which conformational information is lacking for any reason are always shown as single-stranded in these diagrams.

Describing RNA Architecture

By tradition, RNA chemists describe the architecture of RNAs using terminology devised to describe proteins. It isn't very appropriate. The use of the term "secondary structure" to describe the helices in Figure 1 is a case in point. The backbone of any polymer is helical wherever there are runs of residues that have the same backbone torsion angles, and that helicity is secondary structure. Both strands of a nucleic acid double helix have secondary structure because taken in isolation, both are helical. Their antiparallel, side-by-side association, on the other hand, is not secondary structure. It is tertiary structure if two strands are part of the same sequence, and quaternary structure if they are not, because in both cases, sequences are being juxtaposed that are distantly associated, if associated at all. Thus, Figure 1 is better described as a helix diagram or a stem-loop diagram than as a "secondary structure diagram," which is the terminology commonly used. Note that the same confusion arises when the β-sheets in proteins are discussed.

The potential of most RNAs for intramolecular interaction is not exhausted by its helices. Additional interactions involving helices, the nonhelical segments, terminal loops, metal ions, water molecules, etc., can bring stem-loops together, and force RNAs to adopt globular conformations. The arrangements of helical segments that result are what most RNA chemists mean by "tertiary structure," but it is better described as suprahelical structure.

The last protein-related nomenclature issue that deserves comment is the application of the term "domain" to RNA structures. A domain is a region of a protein's structure, the conformation of which is stabilized en-

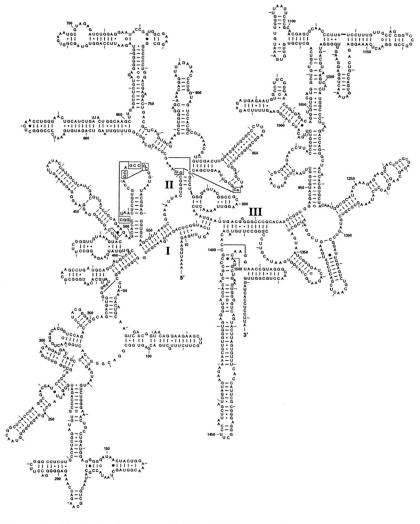

Figure 1 The helix diagram of 16S rRNA from *Escherichia coli*. This diagram was provided by Dr. Robin Gutell. (Reprinted, with permission, from Gutell 1996 [copyright CRC Press].)

tirely by internal interactions; it is a structurally autonomous unit. Domains are interesting because they retain their conformations and (often) their biological properties when they are excised from the structures to which they belong. In RNAs, there two kinds of domains. The simplest is a stem-loop that has a small terminal loop, i.e., a local hairpin loop. Many of these structures (most of them?) retain their conformations in isolation. The second type of domain is an assembly of stem-loops capable of maintaining its proper suprahelical organization in isolation, e.g., the P4-P6 domain from the *Tetrahymena* self-splicing intron (Cate et al. 1996). There are strong reasons for believing that ribosomal RNAs, for example, also contain suprahelical domains (Weitzmann et al. 1993; Samaha et al. 1994).

PREDICTING STRUCTURES FROM SEQUENCES

The essence of the conformation prediction problem was identified 30 years ago by Levinthal, who pointed out that the number of conformations possible for a random coil protein is so huge that if renaturation depended on a denatured protein finding its lowest energy conformation by a thermally driven, random walk through conformation space, renaturation would take an eon of eons (Levinthal 1968). Levinthal's argument is valid for RNA, and it follows that RNA structures cannot be predicted using techniques that rely on exhaustive searches to find lowest free-energy conformations. Thus, there are only two general strategies available that could solve the RNA folding problem. Either approaches must be found that reduce the number of conformations to be compared by astronomical factors, or methods must be developed for simulating what actually happens when RNAs refold in vitro. Both strategies have been pursued by protein chemists, and an immense literature now exists that those interested in RNA structure prediction ignore at their peril.

Ab Initio Computation of RNA Conformation

Over the past 20 years, a number of computer programs have been developed for simulating the thermal motions of biological macromolecules (e.g., CHARMM [Brooks et al. 1983], and AMBER [Weiner and Kollman 1981]). At the outset of the typical dynamics simulation, a structure is assumed for the molecule of interest, and both the energy of the molecule and the net force experienced by each of its atoms are computed. Harmonic potentials are (usually) used to represent the energetic cost of distorting bond lengths and angles away from their crystallographically determined, normal values, and classic approximations are used to represent nonbonded interactions. The trajectory of each atom is then computed for about a femtosecond, assuming force equals mass times acceleration. At that point, both the energy of the system and the forces on its atoms are re-evaluated, so that trajectories can be corrected for changes that have taken place, and another femtosecond's worth of trajectory is computed, and so on, for as many iterations as desired. The reason trajectories have to be updated so often is that molecules vibrate and collide in solution at inverse picosecond rates, and unless trajectories are updated much more rapidly than that, these motions will not be represented faithfully. Since the computer equivalent of temperature can be controlled during molecular dynamics simulations, one could imagine using these programs to simulate the thermal denaturation and renaturation of a single RNA molecule, and if that can be done, why can't RNA conformations be computed ab initio?

One reason RNA conformations cannot be computed this way is time. When RNAs renature, their stem-loops form in microseconds, but it takes seconds to minutes for suprahelical structures to develop (see below). Unfortunately, it strains the capacity of even the most powerful computers to simulate a single microsecond of the history of a modest sized macromolecule. Efforts are being made to lengthen the time interval simulated by each dynamics iteration (see Schlick et al. 1997), but still, until computer time gets many orders of magnitude cheaper than it is today, this avenue is closed.

Another reason RNA conformations cannot be computed today is accuracy. Contrary to what the chemically naive believe, the total nonbonded energy of a molecular system cannot be represented as a sum of independent atom i/atom j interaction energies. The reason is that in general, both the magnitude and direction of the force atom i exerts on atom j depends on the locations and identities of the atoms that surround them (Maitland et al. 1981). Nevertheless, because the computational cost of taking neighboring group effects into account would be prohibitive, all macromolecular dynamics programs assume pairwise additivity. At best, the pairwise potentials used are potentials of mean force, which is to say approximations that correct for the average effect of neighboring groups in solution. At worst, they are potentials that are approximately correct for groups interacting in the gas phase. It is not clear that conformations can be predicted accurately at this level of approximation.

Finally, it must be acknowledged that the RNA folding problem is intrinsically less tractable than its protein counterpart. The typical protein domain is stabilized primarily by interactions occurring in a hydrophobic core that is composed of densely packed alkyl and aromatic side chains, and polar groups, virtually all of which are involved in hydrogen bonding or other neutralizing interactions. Since the influence of solvent on intramolecular interactions in proteins can be dealt with by treating it as a continuous, dielectric medium (Sharp and Honig 1990), it is conceivable that computations based solely on the covalent structures of proteins could lead to reliable predictions of conformation.

RNAs are more complicated. The suprahelical structure of the P4-P6 domain, for example, depends on bound Mg^{++} ions, as does the conformation of an internal loop in prokaryotic 5S rRNAs, and hydrogen bonds involving water are commonplace in RNA (Cate et al. 1996; Correll et al. 1997). Since no one I know of can predict the numbers and locations of water molecules and ions tightly bound to an RNA on the basis of its sequence, ab initio computations of RNA structure must model the surrounding solvent in molecular detail, and that increases the cost of computations considerably. The only solace is that solvent effects are better evaluated this way than they are using continuum dielectric models.

RNAs are also more complicated because they have a very high charge density (one anionic charge per residue at neutral pH). As has been long understood, ionic interactions play a big role in determining RNA conformation, and unless they are dealt with properly, RNA simulations are guaranteed to fail. Ionic interactions are hard to cope with computationally because their range is much longer than that of any other kind of nonbonded interaction, but a solution to this problem may have just been found. Recent investigations suggest that the computational problems posed by Coulombic interactions can be solved using Ewald particle mesh algorithms (see Louise-May et al. 1995).

It does not follow from the above that today's molecular dynamics programs should never be applied to RNA. If the volume of conformation space to be explored is restricted, they can be very useful. For example, experimentalists routinely use molecular dynamics programs to compute and refine RNA structures (see, e.g., Brünger 1992). The X-ray and NMR data included in their computations restrict the range of conformations available and reduce the impact that force-field inaccuracies might have. Under these circumstances, current molecular dynamics codes routinely find lowest-energy molecular conformations. Molecular dynamics computations are also providing insight into the dynamics of RNAs of known conformation, RNA ligand-binding properties, etc. (see Louise-May et al. 1996).

For completeness, I note that most issues that can be addressed by molecular dynamics methods can also be approached using Metropolis

Monte Carlo algorithms. However, it is not obvious that they are any less limited by force-field inaccuracies or the large number of states available to be explored.

Semi-empirical Approaches to the Determination of Stem-loop Diagrams

There is a lot of conformational information in a diagram like Figure 1. Suppose a certain RNA had 100,000,000 possible conformations in the Levinthal sense, and suppose its stem-loops could be identified. Since the backbone torsion angles of nucleotides in helices are fixed, if half of its nucleotides were involved in helices, the number of conformations that remained possible for the molecule would be reduced to 10,000, and if three-quarters of its bases were tied up in helices, the number of possible conformations would be 100. Obviously, it would be a good thing if helix diagrams could be deduced from sequences.

In the 1970s, it was discovered that the thermal stabilities of RNA helices can be predicted using thermodynamic data obtained from melting studies done on small, synthetic RNA oligonucleotides (Tinoco et al. 1973). The reason this works is that the stabilizing effect of a base pair is determined primarily by its identity and the identities of the bases that flank it on either side. Thus, if Watson-Crick and wobble G·U pairs are the only ones considered, $\Delta G°$'s, $\Delta H°$'s, and $\Delta S°$'s for just 21 combinations of base pairs and flanking sequences should suffice to make predictions (Serra and Turner 1995). In addition, the entropic cost of forming duplexes from separate strands has been worked out, as have thermodynamic parameters for terminal loop formation, internal loop formation, etc. Thus, the information required to predict the energetics of simple RNA hairpins exists, and their melting temperatures can be predicted with reasonable accuracy (see Chapter 10).

A number of programs have been devised that are capable of identifying all combinations of stem-loops possible for a sequence, and then estimating their free energies from thermodynamic data. MFOLD, which is a well-known example, produces the lowest-energy helix diagrams for a sequence, rank-ordered by free energy (Zuker 1989). It works well for small RNAs like tRNAs and 5S rRNAs, but its performance degrades as the size of the RNAs increases (Zuker and Jacobson 1995). The larger an RNA, the more likely it is to include structural elements, the thermodynamic properties of which are unknown. Although it is clear that the accuracy of predictions improves in parallel with the thermodynamic database, there is a practical limit to what can be done. The labor required to fill the "catalog" of thermodynamic data grows geometrically with the number of substructures it includes. In addition, it is not clear that a complete catalog would do the job. Non-nearest neighbor effects do affect conformational stability (Turner et al. 1988). This is the reason programs like SAPSSARN are interesting (Gaspin and Westhof 1995). SAPSSARN is an interactive program for constructing helix diagrams that, in addition to taking base-pairing energies into account, can be guided by nonthermodynamic constraints.

It should be noted that secondary-structure determination and prediction have long been a part of the protein field. Just as the amount of helix in an RNA can be estimated from experiments that measure hyperchromicity, the amount of helix and β-sheet in proteins can be estimated from CD/ORD spectra. In addition, in the 1970s, as the number of experimentally determined protein structures increased, it became evident that the amino acid compositions of β-sheets, on average, are not the same as those of α-helices. From this insight, it was but a short step to rules for identifying regions in protein sequences that are likely to be helix or β-sheet. The best known are the ones formulated by Chou and Fasman almost 25 years ago (Chou and Fasman 1974). Secondary-structure prediction remains a significant activity in the protein field, and the information sought is equivalent to that contained in an RNA's helix diagram.

BIOCHEMICAL METHODS FOR DETERMINING RNA CONFORMATION

Given how poorly RNA biochemists and molecular biologists have been served by their biophysical colleagues over the years, it is little wonder that many of them have devoted immense effort to the characterization of RNA structures by nonphysical means. The only protein biochemists making similar investments in the nonphysical analysis of conformation today are those interested in membrane proteins, and they also have reason to complain about the small number of crystal structures and NMR structures available to them.

The Phylogenetic Approach

In 1975, Fox and Woese proposed the three-stem helix diagram for 5S rRNA now known to be correct (Fox and Woese 1975). The reason their paper is remembered is not that their model was novel—others had already proposed similar models (Madison 1968; DuBuy and Weissman 1971; Nishikawa and Takemura 1974)—but the nature of the argument made to support it. Since 5S rRNAs from different species perform the same function, they must have similar conformations. Therefore, the correct stem-loop structure for 5S rRNAs is that which is compatible with the sequences of all of them. The same argument had been made in support of the familiar tRNA cruciform almost a decade earlier, but there was a difference. The cruciform was identified initially because it is one of three maximally base-paired structures that could be proposed for the first tRNA sequence (Holley et al. 1965). As soon as a few more tRNAs were sequenced, it was obvious that the cruciform was right. From the point of view of Fox and Woese, the number of base pairs in an RNA helix diagram is irrelevant, and when they began, it was not obvious that a unique, maximally paired structure exists for 5S rRNA. Six years later, the phylogenetic approach was applied with spectacular results to the sequence of 16S rRNA (Noller and Woese 1981).

A phylogenetic analysis has three steps, the first of which is the accumulation of the sequences of as many members of a family of homologous RNAs as possible. The second step is sequence alignment, which is important because deletions and insertions are often encountered. Alignment usually depends on the identification of sequences that are conserved within families. The third step is analysis of sequence covariations. For example, if residues x and y form a Watson-Crick pair in an RNA, what you expect is that y will be a G whenever x is a C, and y will be a U whenever x is an A, etc. Not only can regular helices be identified this way, but if the number of sequences available is large enough, noncanonical pairings can be worked out, and even some of the long-range interactions on which suprahelical structure depends (Gutell and Woese 1990). The idea that an RNA's conformation will be preserved if a base change at one position is matched by an appropriate base change made at another is the basis for the compensating mutation method for identifying RNA–RNA interactions experimentally. Indeed, one could take the view that the sequences the phylogenetic analyst compares are the products of innumerable compensating mutation experiments carried out by Nature over the course of evolution.

Biochemical Methods for Probing RNA Conformation

For over 30 years, RNA chemists have been doing chemical and enzymatic experiments to validate helix diagrams, to determine the conformation of "unstructured" sequences in helix diagrams, and to work out suprahelical structures. A remarkable variety of strategies have been developed, and examples of the application of almost every one of them can be found in the 5S rRNA literature (Moore 1995).

In the 1960s and 1970s, most of these investigations exploited the fact that the susceptibility of RNA sequences to enzymatic digestion depends on their conformations. Many sequence-specific nucleases cleave single-stranded RNA more rapidly than double-helical RNA, and less specific nucleases exist that cleave either single-stranded RNA or double-helical RNA, but not both. Nevertheless, enzymatic approaches to RNA conformation gradually fell out of favor. Enzymes are so big compared to the structural details they were being used to examine that one had to be concerned that reactivities might depend on more than just local conformation. In addition, no one could be sure how any of these enzymes would respond to RNA structures other than A-form helix. For this reason, the RNA community turned increasingly to low-molecular-weight, group-specific reagents as conformational probes.

The number of group-specific reagents that have been used to probe RNA conformation is very large (for descriptions, see Noller and Moldave 1988). The chemical probes popular today share the property that products of their reactions with RNA nucleotides do not function as templates for reverse transcriptases. Thus, the nucleotides that have reacted in a lightly modified RNA can easily be identified by primer extension. The variations in reactivity observed are attributed to differences in the degree of reagent access allowed by the conformation of the parent molecule. This type of reactivity mapping has been used to identify Watson-Crick helices, to distinguish Watson-Crick-paired bases from those paired in other ways, and has even provided the basis for detailed models of entire RNAs (see, e.g., Brunel et al. 1991).

Any protein chemist who interpreted reactivity data this way would be a laughing stock. The reactivity of protein groups can be modulated by interactions with their neighbors in ways that have nothing to do with accessibility. Indeed, the unusual reactivity of many enzymes is explained by effects of just that kind. Nevertheless, when RNAs of known conformation are probed chemically, the reactivity patterns observed are those anticipated if conformation were in fact the dominant determinant. It must be the case that RNAs seldom adopt conformations that result in significant, nongeometric alterations in reactivities, and this may be one of the reasons that protein enzymes won out over ribozymes during the course of evolution.

The most important generality to emerge from this body of work is that isolated RNA molecules usually have stem-loop structures close to those predicted phylogenetically. This is not trivial, because most RNAs perform their functions in association with proteins, and as Fox and Woese pointed out, it is their conformations in those contexts that count, not their conformations as free molecules. Association with protein must not alter the stem-loop organization of RNAs very much.

Two types of chemical probes are used to obtain information about the suprahelical organization of RNAs: cross-linking reagents and short-range labeling reagents. Cross-linking agents covalently link nucleotides in RNAs that are neighbors in space, but not necessarily neighbors in sequence (see, e.g., Harris et al. 1994), and here too the number of reagents available is large. The short-range labeling reagents favored today all contain Fe(II) complexed with EDTA. When ascorbic acid, H_2O_2, and Fe(II)-EDTA are combined, short-lived hydroxyl radicals are generated that react with almost anything. When ribose rings are attacked, chain scission often results, and that blocks reverse transcriptase (Pogozelski et al. 1995). Thus, if an Fe(II)-EDTA group is bound to a specific location in an RNA, and ascorbate and H_2O_2 are added, the RNA residues that react will be those closer to the Fe(II)-EDTA group than the distance a hydroxyl radical can diffuse before it decays (~20 Å) (see, e.g., Joseph and Noller 1996; Joseph et al. 1997).

Model-building

It is difficult to understand the conformational implications of biochemical information of the type we have been discussing without computer assistance. Programs that use the distance-geometry/simulated annealing algorithms, which NMR spectroscopists favor for their structure determinations, lend themselves to this kind of data interpretation, and YAMMP, which was developed by Harvey and his colleagues, is an example of a program of that kind that can use chemical information to construct three-dimensional RNA models (Tan and Harvey 1993).

The model-building program, MC-SYM, takes a different approach (Major et al. 1991). The conformation of any nucleotide in an RNA can be specified by defining the rotational orientations, or torsion angles, associated with 6 backbone single bonds and its glycosidic bond. Although in principle 360° rotation is possible around single bonds, in all molecules certain rotational orientations are always favored, and often heavily so: the ones in which the groups joined by the bond in question interfere the least sterically. The values for any torsion angle that have low steric energy are its "rotamers." The maximum number of rotamers available to any one of the seven torsion angles that determine the conformation of a nucleotide is three, and the number of rotamer combinations possible for a nucleotide is about 400. Nevertheless, only 20–30 of these combinations have ever been observed in nucleic acid crystal structures (Gautheret et al. 1993).

MC-SYM contains a library of these preferred nucleotide conformations, which it uses to build RNA models sequentially. For example, if a three-nucleotide RNA is being modeled, and there are 20 conformations possible for each nucleotide, a priori, the universe of possible models has 8000 members. MC-SYM never considers that entire universe. Instead, each of the 400 two-nucleotide models produced when the second nucleotide is added to the first is tested for its compatibility with everything that is known about the conformation of the molecule being modeled. The information taken into account includes the molecule's stem-loop diagram and might also include chemical cross-linking and protection data and even distances deduced from NMR experiments. The only three-nucleotide models it produces are those that can be built by adding a nucleotide to one of the acceptable two-nucleotide structures, and they too are screened for compatibility. The pruning of incompatible conformations that occurs at each step in the construction of an RNA model in MC-SYM usually reduces the number of structures that emerge by orders of magnitude. Only the survivors need be analyzed further.

What Is a Structure?

Over the years, many RNAs have been modeled in many different ways, including tRNA (Levitt 1969), 5S rRNA (Brunel et al. 1991), M1 RNA (Harris et al. 1994; Westhof et al. 1996), the self-splicing intron from *Tetrahymena* (Michel and Westhof 1990), and 16S rRNA (Malhotra and Harvey 1994; Mueller and Brimacombe 1997; Stern et al. 1988), to name a few. The richness of the data that underlie these models, and the methods used to arrive at them, vary enormously, of course. However, if the amount of biochemical data taken into account in one of these exercises were large enough, and if the method used to analyze them was rigorous enough, there is no reason why a model might not emerge that is the truth, for all intents and purposes. A biochemical model that good would be equivalent to an X-ray or NMR structure. The problem nonspecialists confront is that whether a nonphysical RNA model is that good or not, it tends to look that good when it is displayed using an all-atom representation. How is one to know the difference? Are any of the biochemical models published to date that good?

One way to answer this question is to ask what the properties of physically determined structures are that scientists prize. I submit that there are just two. First, physical structures are reproducible, which means to say that if two groups independently determine the conformation of the same molecule by physical means, the structures they produce will be the same, to within experimental error. Second, physical structures have high explanatory power. Not only do they rationalize the data used to obtain them, they also rationalize large bodies of data that were not used. Typically they explain most of the physical, chemical, and biological properties of the molecules they represent.

A biochemical model that is reproducible and has high explanatory power would be a good model, no matter how crude the representation of the molecule it portrays, but in my estimation, none of the biochemical RNA models produced so far is "good." When different laboratories analyze the same RNA by nonphysical means, the models that emerge are different, especially in their suprahelical organization. For example, compare Altmann and Westhof's models for M1 RNA with that of Pace and his colleagues (Harris et al. 1994; Westhof et al. 1996) or the 16S rRNA models of Noller, Harvey, and Brimacombe (Stern et al. 1988; Malhotra and Harvey 1994; Mueller and Brimacombe 1997). This is not surprising because in almost every case, the data available are demonstrably insufficient to narrow the number of suprahelical conformations possible to just one. There is no harm in that, provided the reader is given some idea of the range of models the data permit, which has seldom been the case. Happily, methods have been developed recently for evaluating the uncertainty of such models quantitatively, and that is a huge step in the right direction (Malhotra and Harvey 1994).

The biochemical RNA models in the literature faithfully represent the stem-loop structures of their RNAs, insofar as they were known at the time they were built, and at worst, can be considered three-dimensional summaries of the data available. Most of them contain important amounts of truth, and not uncommonly they suggest explanations for some of the properties of the molecules they represent. The problem is distinguishing the wheat from the chaff.

WHAT HAPPENS WHEN RNAs FOLD?

There can be little argument that local stem-loop structures are simpler than suprahelical structure, and implicit in all chemical and semiempirical approaches to RNA structure determination is the notion that an RNA's stem-loop structure must be determined before its suprahelical structure is addressed; i.e., the simple must be dealt with before the complex. Is that the logic that underlies the folding of real RNA molecules, or is it merely an ordering humans find attractive? This question was first addressed experimentally in the late 1960s and early 1970s using tRNAs as model systems (see Crothers and Cole 1978; Crothers 1979). The field then went into an eclipse from which it is now emerging, thanks to the development of methods for producing RNAs of defined sequence in large quantities, and thanks also to the interest in catalytic RNAs. It is worth taking a look at what has emerged (see Brion and Westhof 1997).

Early Events in RNA Folding

T-jump experiments done in the early 1970s demonstrated that in the presence of physiological concentrations of monovalent ions, the time constants for local stem-loop formation are 10s to 100s of μsec (Crothers et al. 1974). Furthermore, consistent with polymer statistics, the time constants for stem-loop formation are (roughly) proportional to the lengths of their loops raised to the 3/2 power (Crothers et al. 1974). Thus, if under some set of conditions a stem with a seven-base loop takes 80 μsec to form, a stem with a 1000-nucleotide loop will take about 135 msec. These same pioneering studies demonstrated that the suprahelical structure of tRNA is considerably less stable than the structures of its helices, and that the suprahelical structure of most tRNAs is destabilized by the absence of divalent cations, which local stem-loops generally are not.

There are parallels between local stem-loop formation in RNAs and the early events in protein folding (see Friesner and Gunn 1996; Levitt et al. 1996). When small, globular proteins are denatured, and then transferred from denaturing solvents to physiological solvents, they are almost instantaneously transformed from random coils into globular structures that are not much bigger than their final, fully folded conformations. This occurs on about the same time scale as the formation of local stem-loops in RNAs. The condensed state that results is referred to as a "molten globule," and molten globules form for the same reason that lipids form micelles. They are conformations that minimize the contact of hydrophobic groups with water while maximizing the contact of hydrophilic groups with water.

A first-year graduate student might claim that the two processes are fundamentally different; that stem-loop formation is driven by hydrogen bonding, and that molten globule formation is driven by the hydrophobic effect. A second-year graduate student would not be so sure. The most hydrophobic parts of an RNA are its bases, and significant amounts of base surface become protected from solvent when helix forms; the middle of a helix is the hydrophobic core of a nucleic acid. Furthermore, in aqueous environments, the energetic value of intramolecular hydrogen bonds is only about one kcal/mole because water molecules can always substitute for macromolecular hydrogen-bond donors and acceptors. The driving force for base-pairing in nucleic acids is thus the free-energy penalty exacted if a base donor or acceptor group fails to find a mate in the middle of a helical stem. The same thermodynamic drive is at work on the hydrogen-bond donors and acceptors buried in the cores of molten globules, and not surprisingly, molten globules contain significant amounts of α-helix and β-sheet.

Late Events in RNA Folding

Since suprahelical structure formation occurs on a seconds-to-minutes time scale, it can be studied using stopped-flow methods, and the folding pathways of many catalytic RNAs are now being addressed that way. Of the many RNAs larger than tRNAs that have been investigated recently, the one that has been most thoroughly scrutinized is the self-splicing intron from *Tetrahymena* (see Zarrinkar and Williamson 1996 and references therein). Surprisingly, significant progress has been made with this system even though the three-dimensional structure of only one of its domains is known (Cate et al. 1996).

At least for the self-splicing intron, the formation of suprahelical structures is a process, in which the nth step depends on the prior completion of $(n-1)^{th}$ step, and folding intermediates accumulate because some steps are slow. It is interesting that some of the late steps in this RNA's pathway involve the formation of helix by sequences that are well-separated in the molecule. This observation validates a point made earlier, namely that helices should not be regarded as merely secondary structure elements.

The existence of folding pathways has been a matter of hot dispute in the protein field for many years. They may exist for some proteins, but almost certainly do not for all proteins. The slow step in the folding of small, globular protein domains, which is the transformation of molten globules into folded proteins, does not seem to be a strictly sequential process. The fundamental difference between a molten globule and a mature protein is that the hydrophobic core of the latter is about as densely packed as it conceivably could be, whereas the core of the former is not. The rearrangement of side chains required to achieve the densely packed, final state is slow because correlated motions are required to achieve it, and as it occurs, the helices and sheets in the molecule stabilize in their final forms.

The folding pathway of the self-splicing intron may prove to be the paradigm for all RNAs, but it would be foolish to assume that this is so. RNA folding may prove as sequence-specific and idiosyncratic as protein folding is, and the number of RNAs whose folding properties are well-characterized is so small that it would be wise not to attempt to generalize at this point.

CONCLUSION

Although significant progress has been made on the RNA folding problem, it is hard to imagine that we will be sophisticated enough to predict RNA structures ab initio any time soon, and it will be a while before biochemically derived conformational models are produced that are accurate enough to pass for structures. Those responsible for staffing academic departments and research institutes will have to keep hiring structural biologists for at least a little while longer.

ACKNOWLEDGMENTS

I thank the many colleagues who helped me locate references and answered my questions. I am particularly grateful to Professors I. Tinoco and D. Turner for their comments on the first draft of this manuscript. This work was supported by a grant from the National Institutes of Health (GM-41651).

REFERENCES

Aboul-ela F., Karn J., and Varani G. 1995. The structure of the human immunodeficiency virus type 1 TAR RNA reveals principles of RNA recognition by TAT protein. *J. Mol. Biol.* **253:** 313–332.

Allain F.H.-T. and Varani G. 1995. Structure of the P1 helix from group I self-splicing introns. *J. Mol. Biol.* **250:** 333–353.

Battiste J.L., Tan R., Fraenkel A., and Williamson J.R. 1994. Binding of an HIV Rev peptide to Rev responsive element RNA induces formation of purine-purine base pairs. *Biochemistry* **33:** 2741–2747.

Battiste J.L., Hongyuan M., Rao N.S., Tan R., Muhandiram D.R., Kay L.E., Frankel A.D., and Williamson J.R. 1996. Alpha helix-RNA major groove recognition in an HIV-1 Rev peptide-RRE RNA complex. *Science* **273:** 1547–1551.

Brion P. and Westhof E. 1997. Hierarchy and dynamics of RNA folding. *Annu. Rev. Biophys. Biomol. Struct.* **26:** 113–137.

Brooks B., Bruccoleri R., Olafson B., States D., Swaminathan S., and Karplus M. 1983. CHARMM: A program for macromolecular energy minimization and molecular dynamics calculations. *J. Computat. Chem.* **4:** 187–217.

Brunel C., Romby P., Westhof E., Ehresmann C., and Ehresmann B. 1991. Three-dimensional model of *Escherichia coli* ribosomal 5S RNAs as deduced from structure probing in solution and computer modelling. *J. Mol. Biol.* **221:** 293–308.

Brünger A. 1992. *X-PLOR Version 3.1: A System for X-ray crystallography and NMR.* Yale University Press, New Haven, Connecticut.

Butcher S.E., Dieckmann T., and Feigon J. 1997. Solution structure of the conserved 16 S-like ribosomal RNA UGAA tetraloop. *J. Mol. Biol.* **268:** 348–358.

Cate J., Gooding A.R., Podell E., Zhou K., Golden B.L., Kundrot C.E., Cech T.R., and Doudna J.A. 1996. Crystal structure of a group I ribozyme domain: Principles of RNA packing. *Science* **273:** 1678–1685.

Cheong C., Varani G., and Tinoco I., Jr. 1990. Solution structure of an unusually stable RNA hairpin, 5'GGAC(UUCG)GUCC. *Nature* **346:** 680–682.

Chou P.Y. and Fasman G.D. 1974. Conformational parameters for amino acids in helical, beta sheet and random coil regions calculated from proteins. *Biochemistry* **13:** 211–221.

Correll C.C., Freeborn B., Moore P.B., and Steitz T.A. 1997. Metals, motifs and recognition in the crystal structure of a 5S rRNA domain. *Cell* **91:** 705–712.

Crothers D.M. 1979. Physical studies of tRNA in solution. In *Transfer RNA: Structure, properties and recognition* (ed. P.R. Schimmel et al.), pp. 163–176. Cold Spring Harbor Laboratory, Cold Spring Harbor, New York.

Crothers D.M. and Cole P.E. 1978. Conformational changes in tRNA. In *Transfer RNA* (ed. S. Altman), pp. 196–247. MIT Press, Cambridge, Massachusetts.

Crothers D.M., Cole P.E., Hilbers C.W., and Shulman R.G. 1974. The molecular mechanism of thermal unfolding of *Escherichia coli* formyl methionine transfer RNA. *J. Mol. Biol.* **87:** 63–88.

Dallas A. and Moore P.B. 1997. The loop E-loop D region of *Escherichia coli* 5S rRNA: the solution structure reveals an unusual loop that may be important for binding ribosomal proteins. *Structure* **5:** 1639–1653.

Doty P., Boedtker H., Fresco J.R., Haselkorn R., and Litt M. 1959. Secondary structure in ribonucleic acids. *Proc. Natl. Acad. Sci.* **45:** 482–499.

DuBuy B. and Weissman S.M. 1971. Nucleotide sequence of *Pseudomonas fluorescens* 5S ribonucleic acid. *J. Biol. Chem.* **246:** 747–761.

Fan P., Suri A.K., Fiala R., Live D., and Patel D. 1996. Molecular recognition in the FMN-RNA aptamer complex. *J. Mol. Biol.* **258:** 480–500.

Fourmy D., Recht M.I., Blanchard S.C., and Puglisi J.D. 1996. Structure of the A site of *Escherichia coli* 16S ribosomal RNA complexed with an aminoglycoside antibiotic. *Science* **274:** 1367–1371.

Fox G.E. and Woese C.R. 1975. 5S RNA secondary structure. *Nature* **256:** 505–507.

Fresco J.R. and Alberts B.M. 1960. The accommodation of non-complementary bases in helical polyribonucleotides and deoxyribonucleic acids. *Proc. Natl. Acad. Sci.* **46:** 311–321.

Fresco J.R., Alberts B.M., and Doty P. 1960. Some molecular details of the secondary structure of ribonucleic acid. *Nature* **188:** 98–101.

Friesner R.A. and Gunn J.R. 1996. Computational studies of protein folding. *Annu. Rev. Biophys. Biomol. Struct.* **25:** 315–342.

Gaspin C. and Westhof E. 1995. An interactive framework for RNA secondary structure prediction with a dynamical treatment of constraints. *J. Mol. Biol.* **254:** 163–174.

Gautheret D., Major F., and Cedergren R. 1993. Modeling the three-dimensional structure of RNA using discrete nucleotide conformation sets. *J. Mol. Biol.* **229:** 1049–1064.

Gutell R.R. 1996. Comparative sequence analysis and the structure of 16S and 23S rRNA. In *Ribosomal RNA. Structure, evolution, processing and function in protein biosynthesis* (ed. A. Dahlberg and R. Zimmerman), pp. 111–128. CRC Press, Boca Raton, Florida.

Gutell R.R. and Woese C.R. 1990. Higher order structural elements in ribosomal RNAs: Pseudo-knots and the use of non-canonical pairs. *Proc. Natl. Acad. Sci.* **87:** 663–667.

Harris M.E., Nolan J.M., Malhotra A., Brown J.W., Harvey S.C., and Pace N.R. 1994. Use of photoaffinity crosslinks and molecular modeling to analyze the global architecture of ribonuclease P RNA. *EMBO J.* **13:** 3953–3963.

Holley R.W., Apgar J., Everett G.A., Madison J.T., Marquisee M., Merrill S.H., Penswick J.R., and Zamir A. 1965. Structure of a ribonucleic acid. *Science* **147:** 1462–1465.

Jiang L., Suri A.K., Fiala R., and Patel D.J. 1997. Saccharide-RNA recognition in an aminoglycoside antibiotic-RNA aptamer complex. *Chem. Biol.* **4:** 35–50.

Joseph S. and Noller H.F. 1996. Mapping the rRNA neighborhood of the acceptor end of tRNA in the ribosome. *EMBO J.* **15:** 910–916.

Joseph S., Weiser B., and Noller H.F. 1997. Mapping the inside of the ribosome with an RNA helical ruler. *Science* **278:** 1093–1098.

Jucker F.M. and Pardi A. 1995. Solution structure of the CUUG hairpin loop: A novel RNA tetraloop motif. *Biochemistry* **34:** 14416–14427.

Jucker F.M., Heus H.A., Yip P.F., Moors E.H.M., and Pardi A. 1996. A network of heterogeneous hydrogen bonds in GNRA tetraloops. *J. Mol. Biol.* **264:** 968–980.

Kalurachchi K., Uma K., Zimmermann R.A., and Nikonowicz E.P. 1997. Structural features of the binding site for ribosomal protein S8 in *Escherichia coli* 16S rRNA defined using NMR spectroscopy. *Proc. Nat. Acad. Sci.* **94:** 2139–2144.

Levinthal C. 1968. Are there pathways for protein folding? *J. Chem. Phys.* **65:** 44–47.

Levitt M. 1969. Detailed molecular model for transfer ribonucleic acid. *Nature* **224:** 759–763.

Levitt M., Gerstein M., Huang E., Subbiah S., and Tsai J. 1996. Protein folding: The endgame. *Annu. Rev. Biochem.* **66:** 549–579.

Louise-May S., Auffinger P., and Westhof E. 1995. RNA structure from molecular dynamics simulations. In *Biological structure and dynamics. Proceedings of the 9th Conversation, State University of New York* (ed. R.H. Sarma and M.H. Sarma), pp. 1–18. Adenine Press, Albany, New York.

———. 1996. Calculations of nucleic acid conformations. *Curr. Opin. Struct. Biol.* **6:** 298–298.

Madison J.T. 1968. Primary structure of RNA. *Annu. Rev. Biochem.* **37:** 131–148.

Maitland G.C., Rigby M., Smith E.B., and Wakeham, W.A. 1981. Intermolecular forces. Their origin and determination. *Int. Ser. Monogr. Chem.*, vol. 3. Oxford University Press, Oxford, United Kingdom.

Major F., Turcotte M., Gautheret D., Laplame G., Fillion E., and Cedergren R. 1991. The combination of symbolic and numerical computation for three-dimensional modeling of RNA. *Science* **253:** 1255–1260.

Malhotra A. and Harvey S.C. 1994. A quantitative model of the *Escherichia coli* 16S RNA in the 30S ribosomal subunit. *J. Mol. Biol.* **240:** 308–340.

Michel F. and Westhof E. 1990. Modelling of the three-dimensional architecture of group I catalytic introns based on comparative sequence analysis. *J. Mol. Biol.* **216:** 585–610.

Moore P.B. 1995. Structure and function of 5S RNA. In *Ribosomal RNA: Structure, evolution, processing and function in protein synthesis* (ed. R.A. Zimmermann and A.E. Dahlberg), pp. 199–236. CRC Press, Boca Raton, Florida.

Mueller F. and Brimacombe R. 1997. A new model for the three-dimensional folding of *Escherichia coli* 16S ribosomal RNA. I. Fitting the RNA to a 3D electron microscopic map at 20 Å. *J. Mol. Biol.* **271:** 524–544.

Nishikawa K. and Takemura S. 1974. Nucleotide sequence of 5S RNA from *Torulopsis utilis*. *FEBS Lett.* **40:** 106–109.

Noller H.F., Jr., and Moldave K., eds. 1988. Ribosomes. *Methods Enzymol.* vol. 164. Academic Press, San Diego.

Noller H.F. and Woese C.R. 1981. Secondary structure of 16S ribosomal RNA. *Science* **212:** 403–411.

Pogozelski W.K., McNeese T.J., and Tullius T.D. 1995. What species is responsible for strand scission in the reaction of $[Fe^{II}EDTA]^{2-}$ and H_2O_2 with DNA? *J. Am. Chem. Soc.* **117:** 6428–6433.

Puglisi J.D., Chen L., Blanchard S., and Frankel A.D. 1995. Solution structure of a bovine immunodeficiency virus Tat-TAR peptide-RNA complex. *Science* **270:** 1200–1203.

Puglisi J.D., Tan R., Calnan B.J., Frankel A.D., and Williamson J.R. 1992. Conformation of the TAR-Arginine complex by NMR spectroscopy. *Science* **257:** 76–80.

Saenger W. 1984. *Principles of nucleic acid structure.* (Springer Advanced Texts in Chemistry series). Springer-Verlag, New York.

Samaha R.R., O'Brien B., O'Brien T.W., and Noller H.F. 1994. Independent in vitro assembly of a ribonucleoprotein containing the 3' domain of 16S rRNA. *Proc. Natl. Acad. Sci.* **91:** 7884–7888.

Schlick T., Barth E., and Mandziuk M. 1997. Biomolecular dynamics at long timesteps: Bridging the timescale gap between simulation and experimentation. *Annu. Rev. Biophys. Biomol. Struct.* **26:** 181–222.

Serra M.J. and Turner D.H. 1995. Predicting thermodynamic properties of RNA. *Methods Enzymol.* **259:** 242–261.

Sharp K.A. and Honig B. 1990. Electrostatic interactions in macromolecules: Theory and applications. *Annu. Rev. Biophys. Biophys. Chem.* **19:** 301–332.

Stern S., Weiser B., and Noller H.F. 1988. Model for the 3-dimensional folding of 16S ribosomal RNA. *J. Mol. Biol.* **204:** 447–481.

Tan R.K.Z. and Harvey S.C. 1993. YAMMP: Development of a molecular mechanics program using the modular programming method. *J. Computat. Chem.* **14:** 455–470.

Tinoco I., Jr., Borer P.N., Dengler B., Levine M.D., Uhlenbeck O.C., Crothers D.M., and Gralla J. 1973. Improved estimation of secondary structure in ribonucleic acids. *Nat. New Biol.* **246:** 40–41.

Tuerk C., Gauss P., Thermes C., Groebe D.R., Gayle M., Guild N., Stormo G., d'Aubenton-Carafa Y., Uhlenbeck O.C., Tinoco I., Jr., Brody E.N., and Gold L. 1988. CUUCGG hairpins: Extraordinarily stable RNA secondary structures associated with various biochemical processes. *Proc. Natl. Acad. Sci.* **85:** 1364–1368.

Turner D., Sugimoto N., and Freier S.M. 1988. RNA structure prediction. *Annu. Rev. Biophys. Biophys. Chem.* **17:** 167–192.

Weiner P. and Kollman P. 1981. AMBER: Assisted model building with energy refinement. A general program for modeling molecules and their interactions. *J. Computat. Chem.* **2:** 287–303.

Weitzmann C.J., Cunningham P.R., Nurse K., and Ofengand J. 1993. Chemical evidence for domain assembly of the *Escherichia coli* 30S ribosome. *FASEB J.* **7:** 177–180.

Westhof E., Wesolowski D., and Altman S. 1996. Mapping in three-dimensions of regions in catalytic RNA protected from attack by an Fe(II)-EDTA reagent. *J. Mol. Biol.* **258:** 600–613.

Woese C.R., Winker S., and Gutell R.R. 1990. Architecture of ribosomal RNA: constraints on the sequence of "tetra-loops." *Proc. Natl. Acad. Sci.* **87:** 8467–8471.

Yang Y., Kochpyan M., Burgstaller P., Westhof E., and Famulok M. 1996. Structural basis of ligand discrimination by two related RNA aptamers resolved by NMR spectroscopy. *Science* **272:** 1343–1347.

Ye X., Kumar R.A., and Patel D.J. 1995. Molecular recognition in the bovine immunodeficiency virus Tat peptide-TAR RNA complex. *Chem. Biol.* **2:** 827–840.

Zarrinkar P.P. and Williamson J.R. 1996. The kinetic folding pathway of the *Tetrahymena* ribozyme reveals possible similarities between RNA and protein folding. *Nat. Struct. Biol.* **3:** 432–438.

Zuker M. 1989. On finding all suboptimal foldings of an RNA molecule. *Science* **244:** 48–52.

Zuker M. and Jacobson A.B. 1995. Well-determined regions in RNA secondary structure prediction: Analysis of small subunit ribosomal RNA. *Nucleic Acids Res.* **23:** 2791–2798.

16

RNA Interaction with Small Ligands and Peptides

Joseph D. Puglisi
Department of Structural Biology
Stanford University School of Medicine
Stanford, California 94305-5400

James R. Williamson
Department of Molecular Biology and
The Skaggs Institute of Chemical Biology
The Scripps Research Institute
La Jolla, California 92037

RNA is able to bind small molecule (MW <2000) ligands. These ligands can be drugs that bind to sites in biological RNAs, peptide fragments of larger proteins, or molecules for which RNA-binding sites (aptamers) have been selected by in vitro evolution. The RNAs that bind small molecules are modular, and the high local thermodynamic stability of RNA often assures stable folding of RNA domain fragments. Therefore, RNA–ligand interactions can often be studied using drastically reduced systems. RNA oligonucleotides can be produced in large quantities, and advances in NMR spectroscopy have allowed structure determination of RNA by NMR (Varani and Tinoco 1991; Chang and Varani 1997; Puglisi and Puglisi 1998). RNA–small molecule ligand complexes are particularly amenable to NMR structure determination.

This review focuses on the large number of NMR structures of RNA–ligand complexes that have been determined in recent years. These structures have revealed general themes for ligand recognition and have provided insights into the biological functions of RNAs. The binding and manipulation of small-molecule substrates was probably a central feature of the RNA World.

STRUCTURAL STUDIES OF RNA APTAMERS

Three RNA aptamers whose structures have been determined at high resolution using multidimensional heteronuclear NMR (Feigon et al.

1996): ATP (Dieckmann et al. 1996; Jiang et al. 1996), FMN (Fan et al. 1996), and theophylline (Zimmermann et al. 1997), are shown in Figure 1. For comparison, the complex of HIV TAR with argininamide (Puglisi et al. 1992; Aboul-ela et al. 1995) is included as a small ligand–RNA complex, although this structure is not an aptamer. The structure of two

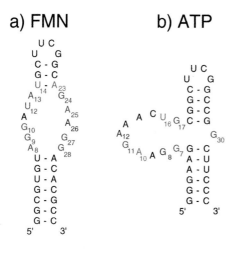

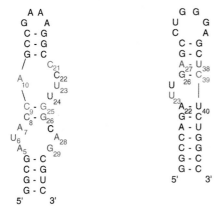

Figure 1 Secondary structures of RNA aptamers discussed in the text. (*a*) FMN aptamer; (*b*) ATP aptamer; (*c*) theophylline aptamer; (*d*) HIV TAR RNA. Nucleotides that are conserved in the selected aptamer are colored: *blue* are conserved nucleotides without non-Watson-Crick hydrogen binding, *green* are conserved nucleotides with non-Watson-Crick hydrogen bonding, and *magenta* are conserved nucleotides with direct hydrogen bonds to the ligand.

other aptamers that bind arginine and citrulline have been studied by NMR (Yang et al. 1996), but the role of many of the conserved sequence elements in these two structures is not yet clear, and these are not discussed here.

Aptamer sequences are selected from pools of random-sequence RNAs by repeated passage over a column derivatized with the desired ligand (Gold et al. 1995). After each passage, the RNAs bound to the column are eluted and amplified by a reverse transcription-PCR-transcription procedure (see Chapter 5). After the last round, the RNA pools are cloned, and individual clones are sequenced. Analysis of the sequences in all cases yielded core sequences for the aptamer that were highly conserved (Fig. 1). Each core sequence was flanked by helical regions composed of Watson-Crick base pairs whose sequence is not critical for small-molecule recognition. For all three of these aptamer sequences, it was possible to construct a minimal aptamer that was sufficiently small to permit NMR structure determination and retained high-affinity ligand interaction ($K_d < 10^{-5}$ M).

On the basis of our present knowledge and understanding of RNA structure, it is not possible to predict from these secondary structures what the three-dimensional structures might be. Fortunately, it is now possible to determine these structures by NMR spectroscopy, as described below.

The FMN Aptamer

FMN (Flavin mononucleotide) is a nucleotide cofactor found in enzymes. The secondary structure of the FMN aptamer (Burgstaller and Famulok 1994) (Fig. 1a) is composed of an asymmetric internal loop of six nucleotides opposite five nucleotides. All of the nucleotides in the internal loop except one are purines. The schematic structure of the bases in the conserved region is shown in Figure 2a. A continuous stack of purines is formed on one strand from A23 to G28, and all of these purines are involved in hydrogen-bonding interactions, primarily with the opposite strand of the internal loop. G24, A25, G27, and G28 all form non-Watson-Crick base pairs across the helix, and A26 forms hydrogen bonds to the flavin heterocycle. A base triple structure is formed between A25, U12, and G10, and this forms a platform on which the flavin heterocycle is stacked; the stacking continues through G9 and A8. The FMN ligand is effectively sandwiched between stacked purines, as shown in Figure 3a, and specific hydrogen bonds supplied by A26 provide additional stability.

The ATP Aptamer

RNA aptamers that are highly specific for binding ATP (adenosine triphosphate) over GTP have been selected (Sassanfar and Szostak 1993). The ATP aptamer, which also binds AMP with high affinity, contains an asymmetric internal loop of 11 nucleotides opposite a single nucleotide (Fig. 1b). In contrast to the FMN aptamer, several of the nucleotides in this loop are not conserved in the selected pools. In the structure of the RNA–AMP complex (Dieckmann et al. 1996; Jiang et al. 1996), it is revealed that these nucleotides either provide a stacking scaffold or they act as spacers to connect the essential structural elements (Fig. 2b). The adenine base of AMP is stacked between two conserved purines and is hydrogen-bonded by two other conserved purines in the asymmetric internal loop, as shown in Figure 3b. The two A-form helical segments in the ATP aptamer are oriented nearly perpendicular to each other, and each helix is capped by a G-G base pair. The structure formed by G8, A9, A10, and the bound adenine base is very similar to the structure found in GNRA tetraloops (Heus and Pardi 1991). This U-turn motif is combined with other hydrogen-bonding and stacking elements to form the essential recognition element.

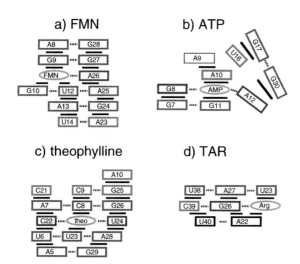

Figure 2 Schematics of the tertiary structures for RNA–aptamer complexes. (*a*) FMN aptamer; (*b*) ATP aptamer; (*c*) theophylline aptamer; (*d*) HIV TAR RNA. Nucleotides are colored as in Fig. 1; the ligand is shown in *orange*. Stacking is shown by *black bars*, hydrogen bonding by *dashed lines*.

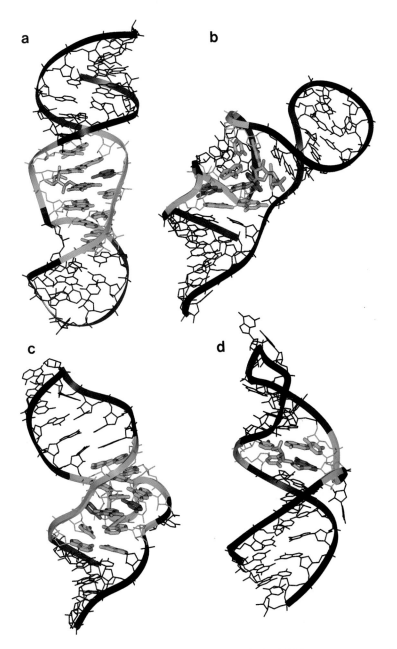

Figure 3 Three-dimensional structures of the RNA–aptamer complexes discussed in the text. (*a*) FMN aptamer; (*b*) ATP aptamer; (*c*) theophylline aptamer; (*d*) HIV TAR–arginine complex. The same coloring scheme as in Figs. 1 and 2 is used.

The Theophylline Aptamer

RNA aptamers have been selected that discriminate between theophylline and caffeine, two planar aromatic purine derivatives that differ by a single methyl group (Jenison et al. 1994). The secondary structure of the theophylline aptamer is shown in Figure 1c, and the schematic of the base arrangements in the structure (Zimmermann et al. 1997) is shown in Figure 2c. The theophylline is bound by a complex network of base-stacking and hydrogen-bonding interactions. The central element of the binding site is a stack of three base-triple interactions. The central triple of this stack is formed by hydrogen bonding between the theophylline base and two conserved nucleotides, and the outer triples are formed by three of the conserved nucleotides. These interactions completely surround the ligand by stacking interactions from above and below, and with hydrogen-bonding interactions at the sides (Fig. 3c). The central triple core is supported by other non-Watson-Crick interactions. However, the stacking interactions follow an unusual nonadjacent pattern, with bases U6 and A7 forming an intercalated stack with C21 and C22. The entire conserved core structure is an intricate laminar network of stacking and hydrogen-bonding interactions.

The TAR–Arginine Complex

The TAR element is the binding site for the HIV Tat protein. The portion of the Tat protein involved in binding contains a number of arginines, and it has been shown that the guanidinium group of an arginine derivative, argininamide, will specifically bind to the TAR element (Tao and Frankel 1992). The secondary structure of TAR is shown in Figure 1d, and the bases important for Tat and argininamide binding are indicated. In contrast to the high affinities of the aptamer-ligand complexes, argininamide has a weak affinity for TAR RNA ($K_d = \sim 10^{-3}$ M). In the HIV TAR–argininamide complex (Puglisi et al. 1992, 1993; Aboul-ela et al. 1995; Brodsky and Williamson 1997), the side-chain guanidinium group makes a pair of hydrogen bonds with the N7 and O6 of G26 (Fig. 2d), as well as electrostatic contacts with phosphate groups. Argininamide binding induces a conformational change in TAR RNA (Aboul-ela et al. 1995). In the free RNA, the pyrimidine bulge is stacked between the two helical stems, with distortion induced in the width of the helical major groove. On binding, the bulge nucleotides are extruded from between the stems, and U23 is positioned in the major groove near A27 (Fig. 3d). A large body of evidence (Puglisi et al. 1992, 1993; Brodsky and Williamson 1997) suggests that U23 forms a base-triple interaction with A27-U38, with the arginine side chain positioned below the U23 base. The base-triple interaction stabilizes the RNA-binding pocket for the arginine side chain. Interaction of the delocalized electrons in the guanidinium group with the uracil base may also stabilize the RNA–amino acid interaction through van der Waals interactions.

Evolutionary Pressure on Aptamers

The three aptamer–ligand complex structures described above provide a tantalizing view of the versatility of RNA structure. However, these RNA sequences are completely nonbiological, and there is no clear role for such RNA–small molecule complexes in cellular processes. The aptamer sequences were usually identified as rare members from very large random pools of sequences. In addition, a biological selection pressure for these motifs does not clearly exist. Although such RNA motifs may have been functional in ancient RNA machines, their presence in current organisms is unsubstantiated.

Interestingly, all of the aptamer sequences are not well-structured in the absence of their ligands, and formation of the complicated RNA structure is induced by ligand binding. In contrast, TAR RNA adopts a similar secondary structure in the presence and absence of argininamide, and structural rearrangements only involve bulge-stem base-triple formation. The aptamer RNA sequences are under no pressure during the in vitro selection process to form a particular structure in the absence of ligand. The only requirements for propagation during the selection are binding to the affinity matrix, elution from the affinity matrix, and the ability to be enzymatically amplified. It may be that flexible RNA structures provide an advantage during the in vitro selection, and that extremely stable structures are selectively depleted from the pools due to poor amplification. The selection pressures in vitro and in vivo are obviously distinct.

The generation of aptamer sequences naturally results in a phylogenetic-like family of sequences. Some nucleotides, as shown in Figure 1a–c, are completely conserved in every individual aptamer clone that is identified. Some nucleotides are quite variable, and their identity is unimportant, but their presence is required for function. These nucleotides are often involved in stacking or packing interactions, or as structural linkers. In all of the four ligand-RNA complexes, the conserved nucleotides in the RNA sequence were either directly involved in formation of the unique RNA architecture of the binding site, or were directly involved in contacts to the ligand. Thus, the structure of these RNAs provides a rationale for the functional role of the conserved nucleotides.

DRUG–RNA INTERACTIONS: AMINOGLYCOSIDES

Paromomycin-A-site Ribosomal RNA Complex

Aminoglycoside antibiotics bind to the ribosome and inhibit translation (Davies et al. 1965). Beyond the obvious pharmacological importance of these compounds, they have provided fundamental insights into the mechanism of protein synthesis. Aminoglycosides bind to 16S ribosomal RNA in the 30S ribosomal subunit (Moazed and Noller 1987) and cause misreading of the genetic code. A large body of biochemical and genetic data supports a highly conserved binding site in ribosomal RNA that is near the aminoacyl-tRNA-binding site (A site). RNA oligonucleotides (Fig. 4c) that correspond to this region of 16S rRNA (Purohit and Stern 1994; Recht et al. 1996) bind aminoglycosides in the same manner as the ribosome.

Aminoglycoside antibiotics contain common chemical features that are required for drug action (Fig. 4a,b). All active aminoglycosides contain a nonsugar deoxystreptamine ring (ring II), and a sugar aminoglucose ring (ring I) is always attached at position 4 of ring II, although the substitution pattern of this ring can vary. The number of additional rings and position of attachment to ring II vary as well. The 4,5 disubstituted compounds have an additional 5-membered sugar ring attached at position 5 of ring II. Neomycin and paromomycin are members of this class that have a fourth ring attached to the ribose sugar. The 4,6 disubstituted compounds have an additional 6-membered sugar ring attached at position 6 of ring II. The aminoglycosides are positively charged at biological pH, with charges of +2 to +5, yet there is no strong correlation between antibiotic activity and total charge.

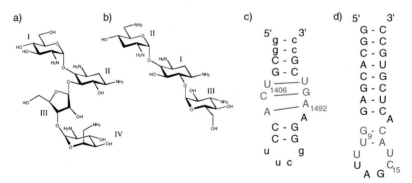

Figure 4 Aminoglycoside antibiotics and their RNA-binding sites. (*a*) Paromomycin; (*b*) tobramycin; (*c*) *E. coli* 16S ribosomal RNA A-site oligonucleotide; (*d*) tobramycin aptamer. Nucleotides involved in ligand contacts are shown in green.

High-resolution structures of aminoglycoside-rRNA complexes (Fourmy et al. 1996, 1998a,b) provide insights into how aminoglycoside antibiotics recognize ribosomal RNA and how they might interfere with translation. As suggested by biochemical data, the antibiotic binds in the RNA major groove, in the asymmetric internal loop (Fig. 5a). The structure of the complex in the region of the internal loop is well defined by the NMR data, as are rings I and II of paromomycin, whereas rings III and IV are more disordered. Paromomycin adopts a specific L-shaped structure, with rings II, III, and IV in a line along the major groove, and ring I positioned approximately 90° from the other three rings. Chemical groups that are common among all aminoglycosides make specific hydrogen bonds to bases and phosphates in the major groove.

Distortions in RNA structure allow formation of the antibiotic binding site. Two noncanonical base pairs are formed in the asymmetric internal loop, resulting in a closed, stable structure for this loop. A U1406-U1495 pair, which involves two hydrogen bonds, is formed. The asymmetric loop is closed by an A1408-A1493 base pair, with the Watson-Crick face of A1408 contacting the N7, N6 face of A1493. The conformation of this base pair is buckled, and long N-N distances suggest water-mediated

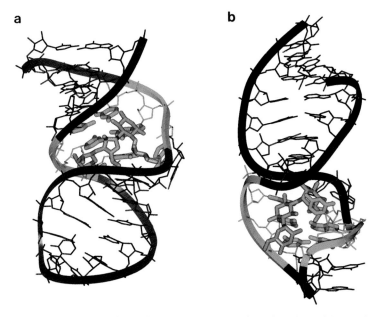

Figure 5 Three-dimensional structures of aminoglycoside-RNA complexes solved by NMR spectroscopy. (*a*) Paromomycin-A-site RNA complex; (*b*) tobramycin–RNA aptamer complex. The same coloring scheme as in Fig. 4 is used.

hydrogen bonding. A1492 is stacked below A1493 and is displaced toward the minor groove. The phosphodiester backbone is distorted at the junction of the asymmetric loop (A1492/A1493) and the lower helical stem (G1491). Ring I of paromomycin is located within a binding pocket formed by this distorted RNA structure. Mutations that disrupt the shape of the binding pocket lead to reduced affinity for aminoglycosides and drug resistance (Recht et al. 1996).

The structure of the A-site RNA–paromomycin complex explains many previous observations of aminoglycoside activity. Resistance enzymes modify aminoglycosides on rings I and II (Shaw et al. 1993); these enzymes include acetyl, phospho, and adenylyl transferases, and such modifications introduce electrostatic and steric penalties to aminoglycoside–RNA interaction. Interestingly, the 1-amino group of ring II, which makes a critical hydrogen bond to U1495, is not a target of resistance acetylases, as the acetyl group can readily exit from the top of the major groove, above U1495, with little steric penalty. Furthermore, semisynthetic aminoglycosides, such as amikacin, contain modifications at position 1 that can also exit the major groove.

The A-site RNA undergoes only a minor conformational change upon binding of aminoglycoside antibiotics (Fourmy et al. 1998a). In the free RNA, the noncanonical base pairs in the asymmetric internal loop are formed, although there is evidence for conformational dynamics in the loop. Upon drug binding, the major conformational change occurs at A1492 and A1493, whose base moieties are displaced by 2–3 Å toward the minor groove by ring I of paromomycin. A1492 and A1493 are universally conserved nucleotides that have been implicated in the interaction of the ribosome with the tRNA anticodon–mRNA codon pair. The effect of the aminoglycoside binding on their conformation may be related to aminoglycoside-induced misreading of mRNA codons.

Tobramycin–RNA Aptamer Complex

Aptamers that bind tightly to aminoglycosides have been selected (Lato et al. 1995; Wallis et al. 1995; Wang and Rando 1995) (Fig. 4d), and the structure of a tobramycin–RNA aptamer complex has been solved (Jiang et al. 1997). As with the paromomycin–ribosomal RNA complex, tobramycin interacts with the aptamer in the major groove at the junction between helical stem and nonhelical regions (Fig. 5b). Ring II of tobramycin interacts with a $^{5'}GU^{3'}$ helical step, as it does in the aminoglycoside–rRNA complexes. In contrast with the interaction with ribosomal RNA, tobramycin ring I is not in contact with RNA, but juts into solution. The

6′-NH$_2$ on ring I was the site of derivatization of tobramycin for attachment to the column matrix, so that it can not contact the RNA as in the paromomycin–rRNA complex. Ring III is positioned relative to ring II in an almost parallel orientation near the G9-C18 and U10-A17 base pairs and makes close contact with the RNA. A residue in the hairpin loop, C15, is flipped out of the stacked hairpin and forms a lower flap on ring III. The interactions between C17 and ring III are apparently dipole-dipole and hydrophobic.

The high-affinity tobramycin–aptamer complex shows similar details of interaction as the aminoglycoside–ribosomal RNA complex. Ring II, the 2-deoxystreptamine ring, interacts along the edge of the major groove; in both structures a 6-membered sugar ring, whose identity is different in the two interactions, makes more structure-specific recognition with the RNA within a binding pocket. In both structures, the bottom of the binding pocket is an RNA base, although in the tobramycin structure, the ring interacts edgewise with the RNA base, whereas in the paromomycin structure, the sugar ring I is "flat" above the base. The difference in interaction of the aptamer and ribosomal RNA is that aminoglycosides must interfere with ribosomal function in addition to binding, whereas the aptamer was selected solely on the basis of binding. This reflects the difference between high-affinity binding and functional binding of small ligands.

PEPTIDE–RNA INTERACTIONS

RNA-binding proteins in retroviruses and phages play a central role in regulation of replication (Karn et al. 1994). These RNA-binding proteins contain a functional domain and an arginine-rich RNA-binding domain. Peptides that correspond to the arginine-rich domains of these proteins interact specifically with their RNA targets. Therefore, both the protein and RNA components in this system are modular. Despite the implication that the arginine-rich domain is a conserved structural motif, the NMR structures described below reveal a variety of different folds and modes of interaction of these peptides with RNA.

HIV Tat–TAR Complex

The retroviral Tat proteins bind to an RNA stem-loop (TAR) at the 5′ end of viral transcripts, and activate transcription (Frankel 1992). The Tat proteins are essential for viral replication and have been the focus of intense study. The Tat proteins contain a conserved, cysteine-rich core that is required for transcriptional activation and an arginine-rich RNA-binding domain at the carboxyl-terminus. Peptides that correspond to the arginine-rich domain bind to TAR RNA with affinity and specificity similar to those of the intact protein (Weeks et al. 1990). As discussed above, free arginine also binds specifically to HIV TAR and drives a conformational change in the RNA bulge. Peptide interactions with HIV TAR induce a similar RNA conformational change. The RNA-binding domain of Tat does not form a regular structure either in the absence of RNA or when bound.

BIV Tat-TAR Complex

The biochemistry (Chen and Frankel 1994, 1995) and structure (Puglisi et al. 1995; Ye et al. 1995) of the bovine immunodeficiency virus Tat-TAR interaction have been delineated. BIV Tat protein has a domain structure related to that of HIV Tat, with sequence differences in the RNA-binding domain. Peptide fragments that correspond to the RNA-binding domain bind with high affinity to BIV TAR (Fig. 6b). Unlike HIV Tat protein, the

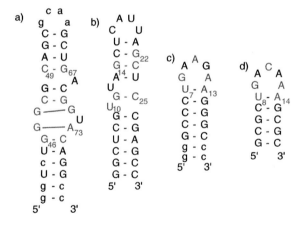

Figure 6 RNAs and arginine-rich peptides whose complexes are discussed in the text. (*a*) HIV RRE; (*b*) BIV TAR; (*c*) phage λ boxB RNA; (*d*) phage P22 boxB RNA. Nucleotides involved in RNA–peptide contacts are highlighted in *green*. Lowercase nucleotides were changed from the biological sequence in the oligonucleotide construct. (*e*) Peptides corresponding to the arginine-rich RNA-binding motifs.

sequence of the arginine-rich domain is very important for specific recognition. Eight of 14 residues in the RNA-binding peptide cannot be changed to other related amino acids without a significant loss of affinity. The sequence and secondary structure of BIV TAR RNA is distinct from that of HIV TAR. BIV TAR RNA contains two single uracil bulges, separated by a single G-C base pair. The nucleotide requirements for Tat peptide binding and function have been mapped and are clustered in the helical stems near the uracil bulges.

Two structures of BIV Tat peptide bound to TAR RNA have been solved by NMR, and the structures are essentially similar (Puglisi et al. 1995; Ye et al. 1995). BIV Tat peptide is unstructured in the absence of RNA, and adopts a β-strand conformation upon interaction with BIV TAR RNA. The peptide lies in the major groove of BIV TAR, near the uracil bulges (Fig. 7b). These two bulged nucleotides have different roles in peptide interaction. U12 is extruded from the RNA helix and is disordered on the minor groove side of the helix. In fact, this bulge can be deleted with little effect on binding affinity. In contrast, U10 interacts with an isoleucine side chain in the major groove, and this interaction between a hydrophobic side chain and bulged nucleotide is required for specific RNA–peptide interaction. In both structures, U10 is positioned near an A-U pair in the major groove, and there is some evidence for a base triple at low temperature, as observed in the HIV TAR–arginine complex; however, in the BIV peptide interaction, this triple is not required for binding. The RNA undergoes a minor conformational change upon binding, and the peptide recognizes the distortions in helical structure near the U10 bulge. In contrast to the HIV TAR–peptide interaction, there are multiple specific amino acid–RNA contacts that guide complex formation.

The RNA-binding domain of BIV Tat contains three arginines, of which two make well-defined contacts with helical guanosines. The guanidinium groups of these arginines interact with the guanosine O6 and N7 groups, as discussed above for the arginine–HIV TAR interaction. The third arginine was less well defined in the structures, but makes essential electrostatic contacts in the major groove. A threonine side chain forms hydrogen bonds with the phosphate oxygen of a helical base pair. The BIV peptide contains three essential glycine residues. One is required for β-turn formation, and the other two allow deep penetration of the peptide into the relatively narrow major groove of the RNA helix: The lack of side-chain steric bulk allows deep penetration. One glycine makes a main-chain hydrogen bond to a guanosine N7. The variety of specific RNA–peptide contacts may explain the relatively high specificity of this interaction compared to other RNA-protein interactions.

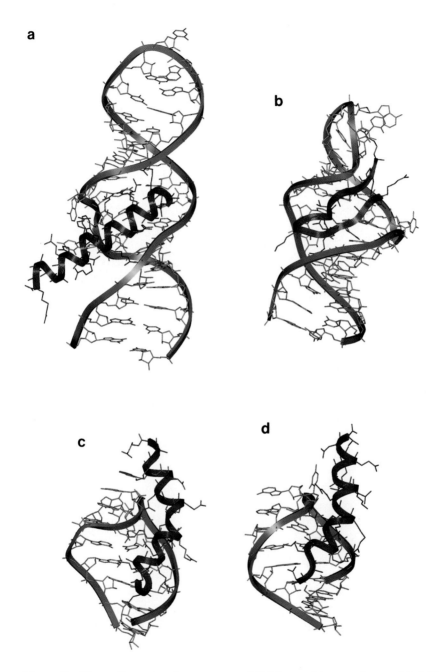

Figure 7 Three-dimensional structures of RNA–peptide complexes. (*a*) HIV Rev-RRE complex; (*b*) BIV Tat-TAR complex; (*c*) phage λ N-boxB–RNA complex; (*d*) phage P22 N-boxB RNA. RNA is *blue* and peptide is *red*.

Rev–RRE Complex

The HIV Rev protein is a small basic protein containing an arginine-rich motif that binds to the RNA regulatory sequence called the Rev response element (RRE). Binding of Rev to RRE results in a change of the relative ratios of unspliced and singly spliced to completely spliced mRNAs that are exported from the nucleus to the cytoplasm, thus altering the pattern of viral gene expression (Fritz and Green 1996). The Rev protein bears a nuclear export signal as the effector domain and the basic region is both a nuclear localization signal and the RNA-binding domain. Short peptides from the basic region of Rev bind specifically to the RRE. The full RRE is a large, complex structure that binds multiple Rev molecules, but a high-affinity Rev-binding site has been identified that binds a single Rev molecule. The minimal RRE is shown in Figure 6a.

The structure of the high-affinity Rev-binding site in the RRE in complex with Rev peptide has been determined using NMR (Battiste et al. 1996). A similar aptamer–Rev peptide complex structure has also been solved (Ye et al. 1996). The peptide forms an α helix that binds in the widened major groove of the RRE (Fig. 7a). The RRE forms a continuous helical structure that contains a G-A and a G-G base pair at the center of the widened major groove (Battiste et al. 1994). Two single nucleotides are bulged out of the helix, but the stacking in the helix is otherwise continuous on both strands. The geometry of the G-G base pair induces a pronounced kink in the backbone of one strand of the RNA, and this kink results in a widening of the major groove by 5 Å compared to a standard A-form helix. This widening permits the α helix of Rev to penetrate deeply into the otherwise restricted groove of the RNA.

The overall shape of the groove is determined by the purine-purine base pairs, but the specific recognition is mediated by side-chain contacts to bases in the major groove and to the backbone. Three arginine residues contact guanine bases in a manner very similar to that observed in DNA–protein complexes. An asparagine residue forms hydrogen bonds to the N6 and the O6 of the G-A base pair. One arginine makes contacts to particular phosphates on the backbone. Finally, a threonine residue makes an N-cap structure that stabilizes the amino terminus of the helical peptide and simultaneously contacts a backbone phosphate.

N-Protein–Box B Complexes

Bacteriophage lambda and P22 N-proteins are transcriptional antiterminators that bind to RNA hairpin loops (Box B) in their respective mRNAs (Das 1993). The N-proteins contain amino-terminal arginine-rich RNA-binding domains (Fig. 6e), and peptides that correspond to the arginine-rich domain interact specifically with boxB RNA hairpin loops (Fig. 6c,d) (Tan and Frankel 1995; Cilley and Williamson 1997). NMR studies have shown that N-proteins are unfolded, and that only the RNA-binding domain folds into an α helix upon complex formation (Mogridge et al. 1998). The remainder of the N-protein presumably folds upon interaction with other proteins involved in transcriptional antitermination. Flexible structures are an apparent hallmark of both arginine-rich domains and the viral/phage proteins that contain them (Frankel and Smith 1998).

The detailed structure of the lambda N-peptide-boxB RNA interaction (Legault et al. 1998) confirms the general features of RNA recognition discussed above. The peptide adopts a bent α helix structure (Su et al. 1997) upon binding the major groove of the RNA stem-loop (Fig. 7c). The 5-nucleotide GAAGA loop forms a GNRA tetraloop fold with a sheared G8-A12 pair and stacking of loop nucleotides. The 4th guanine of the loop, G11, is extruded from stacking, and this guanine is required for binding of the nusA factor to the N-protein–RNA complex. Arginine and glutamine side chains likely form specific hydrogen bonds with the sheared G-A pair, and a required tryptophan side chain stacks on the end of the tetraloop to further stabilize the loop fold. Other arginine and lysine side chains make electrostatic interactions with phosphate groups of helical nucleotides. No base-specific hydrogen bonds are formed to helical nucleotides, but a network of hydrophobic and electrostatic interactions, plus the specific bend of the α helix, lead to intimate RNA-protein recognition.

The structure of the P22 N-peptide-boxB RNA complex (Cai et al. 1998) is similar to the λ N-peptide complex. The peptide forms an α helix, which interacts with a GNRA-type tetraloop and the major groove of the helical duplex (Fig. 7d). The α helix has a slight bend, which allows more intimate interaction with RNA. Basic amino acids, which are required for specific complex formation, contact the phosphodiester backbone of the RNA helix. No base-specific contacts are observed to helical nucleotides. The boxB loop consists of 5 nucleotides, 4 of which (G9, A10, A12, and A13) form a -GNRA- fold with a sheared G9-A13 base pair. C11 is looped out and makes extensive hydrophobic contacts with peptide side chains (alanine and isoleucine). Base-specific hydrogen bonds are formed between a critical arginine (Arg 6) and the O6 and N7 positions of G9 in the tetraloop. The sheared G-A base-pair geometry leaves these positions available for protein recognition. The structure explains the limited biochemical data on the interaction on P22 N-protein–RNA interaction.

Structural Variety of Arginine-rich Peptides

The striking result of these studies of arginine-rich peptides interacting with RNA is their structural diversity. All arginine-rich peptides interact in the major groove of an RNA target, all are highly positively charged, and all are apparently unfolded in the absence of RNA. Upon RNA interaction, the peptides form very different secondary structure folds to achieve recognition: an extended structure in HIV Tat, a β hairpin in BIV Tat, and α helices in Rev and N-peptides. The different protein scaffolds for RNA recognition each achieve different detailed modes of recognition. The β hairpin of BIV Tat is able to penetrate deep into the major groove, which is only slightly distorted, because it contains glycines, whereas α-helical peptides in Rev-RRE interact with a major groove that is widened by purine-purine pairs. The structural diversity of arginine-rich motifs is a reflection of RNA structural diversity.

RNA STRUCTURAL PRINCIPLES

RNA structures and protein structures are both assembled from a combination of secondary structures held together by tertiary interactions. However, the stabilities of the secondary and tertiary interactions are very different for the two classes of macromolecules. Protein secondary structures tend to be very weak on their own, and they are stabilized by strong tertiary packing forces. In contrast, the secondary structures of RNAs are extremely stable in isolation, and the tertiary interactions are much weaker. Many RNAs are stabilized by binding of monovalent or divalent ions. In fact, the ligands can be an intimate part of the RNA fold, as in the AMP aptamer structure.

All of the structures discussed here, and in fact most other RNA structures, contain large amounts of A-form helical secondary structures. Typically, these very stable helices flank irregular structures that are usually less stable. Although the nonstandard regions of RNA structures provide a varied context for ligand discrimination, it appears to be important to hold things together with A-form helices. Recognition by ligands often occurs at the junctions between helices and nonstandard structures. Many of the binding sites use the A-form helix as a platform on which the more elaborate structures rest.

Base stacking is a defining element of the A-form helical segment, and it also appears that stacking is an important contribution to the stabilization of RNA tertiary structures. The amount of stacking of the bases appears to be maximized in the observed RNA structures. The exceptions to this rule are bulged nucleotides that are completely unstacked, but in these cases, there is usually a requirement for an unusual backbone geometry that can only be accommodated by presence of a true bulge.

The global structures of the RNAs in these complexes are highly stacked. The planar nature of the bases apparently exerts a profound influence on the geometries that RNAs can adopt. Even the much larger tRNA (Saenger 1984) and P4-P6 domain of the group I intron (Cate et al. 1996) are formed from two stacked helical segments, as are most of the other RNAs and RNA–ligand complexes whose structures are known (see Chapter 11). Just as protein structures are compacted to form a hydrophobic core, RNA structures seem to form compact structures by maximization of base-stacking interactions. This tends to result in RNA structures that are more linear and less globular than is typical for proteins.

Hydrogen-bonding interactions also appear to be optimized in these RNA structures. Simple base-pairing by hydrogen bonding is a commonly observed interaction that stabilizes both the A-form and nonstandard regions of structure. Although hydrogen-bonding interactions tend to be dominated by the bases, many hydrogen bonds occur to the phosphate backbone and to the ribose sugar. Although the bases are a more obvious device to provide specific hydrogen bonds for a particular structure, the sugar and backbone can also provide specific contacts or elements of stabilization when placed in an unusual geometry.

Other common features in these complexes are the prevalence of purines in the conserved regions and the prevalence of non-Watson-Crick interactions. The purines offer two opportunities for increased interactions compared to pyrimidines. First, they have a larger surface area available for stacking interactions that predominate in these structures. Second, they have two distinct edges that can form multiple hydrogen bonds, and thus they are well suited to form a variety of non-Watson-Crick interactions.

Electrostatic interactions also play a large role in shaping RNA structure. In the nonhelical parts of RNA structures, phosphates may be brought into close proximity. This unfavorable effect may be offset by other favorable interactions, such as ligand binding in the case of the aptamers, or by binding of mono- or divalent ions. Magnesium ions are required for binding of ligands to the FMN and theophylline aptamers, but not for the ATP aptamer or the HIV TAR-argininamide complex. In peptide and aminoglycoside–RNA complexes, Mg^{++} is not required for complex formation, reflecting the high positive charge density of these ligands.

LIGAND RECOGNITION

Ligand recognition of RNA makes use of the above structural principles. Ligand–RNA stacking plays a major stabilizing role in these complexes.

Theophylline, ATP, and FMN all contain a polycyclic aromatic ring, whereas argininamide bears a planar guanidinium group. These planar groups naturally lend themselves to favorable stacking interactions with the planar aromatic bases present in RNA, and it is perhaps no surprise that stacking is a prevalent feature in these RNA–ligand complexes. These ligands use the free energy of stacking to drive fairly large conformational rearrangements of the RNA upon binding. Stacking interactions are also used by the Rev and N-peptides, which contain aromatic side chains that stack on RNA bases.

Stacking does not appear to provide extensive specificity in RNA–ligand interaction. The aromatic rings on the aptamer ligands contain exocyclic chemical groups that can be recognized by hydrogen bonding. The specificity of the AMP–RNA and theophylline–RNA recognition arise in large part due to a collection of hydrogen bonds to the aromatic chemical groups.

Aliphatic ligands also pack tightly against the RNA. The arginine groups in the Tat–TAR interaction make hydrogen bonds to a G-C pair, but also pack against the uracil of the base triple. In the aminoglycoside–RNA complexes, the aminoglucose ring is packed directly above a G-C pair. These interactions between polar aliphatic ligands and the RNA bases may be stabilized by dipolar contacts. Hydrophobic ligand–RNA interaction has also been observed. In the BIV Tat-TAR complex, an isoleucine side chain packs against a bulged uracil.

Electrostatics are a major contribution to RNA–ligand interaction. Many of the ligands discussed here are positively charged, and these ligands (aminoglycosides and arginine-rich peptides) interact in the major groove of the RNA. Obviously, RNA is negatively charged, but this charge density is focused in the major groove, where the phosphate oxygens point. Magnesium binding in RNA major grooves has been shown to stabilize unfavorable RNA–RNA interactions. Likewise, the binding of positively charged ligands in the major groove also stabilizes unstable RNA backbone conformations. The long-range effects of electrostatic interactions can also be a driving force for conformational changes.

Whereas the small aromatic and aliphatic ligands fit tightly into well-packed RNA-binding pockets, the larger ligands—aminoglycosides and peptides—have more extended interaction surfaces with their target RNAs. These ligands are all positively charged and bind in the major groove. Electrostatic and hydrogen-bonding contacts are made with a number of base and backbone positions along the length of a helical stretch. Specificity is achieved by specific hydrogen-bonding interactions with RNA bases and by recognition of some structural distortion: an internal loop, bulge or hairpin loop. For bulky peptides to fit in the major groove, widening of the groove by these distortions is often required. Because of the variety in their side-chain composition, the peptide can use an ensemble of electrostatic, hydrogen-bonding, hydrophobic, and stacking interactions to achieve recognition.

The large conformational changes observed in aptamer–RNA complexes are more subtle in biological RNA–ligand complexes. The groove recognition in the aminoglycoside and peptide complexes leads to less rearrangement than the stacking interactions of aromatic ligands. Nonetheless, these subtle manipulations of RNA structure may be critical to modulating RNA function. In the case of aminoglycoside antibiotics, the minor conformational change observed on binding may be the origin of antibiotic action.

CONCLUSIONS

Structural studies on RNA–ligand complexes have provided insights into how ligands recognize RNA. The ligands utilize the same range of interactions that stabilize RNA itself: stacking, hydrogen bonding, and electrostatics. RNA is a hospitable site for ligand recognition, and RNA clearly can form binding pockets for small molecules. The basic chemistry of enzymes requires small molecule substrate and cofactor binding. The small ligand–RNA complexes discussed here highlight how primitive RNA machines may have bound small-molecule substrates in the RNA World.

ACKNOWLEDGMENTS

The authors thank D. Patel and L. Kay for providing unpublished coordinates, M. Recht and S. Blanchard for critical reading of the manuscript, and the members of the Puglisi and Williamson groups for stimulating discussions.

REFERENCES

Aboul-ela F., Karn J., and Varani G. 1995. The structure of the human immunodeficiency virus type-1 TAR RNA reveals principles of RNA recognition by Tat protein. *J. Mol. Biol.* **253:** 313–332.

Battiste J.L., Tan R., Frankel A.D., and Williamson J.R. 1994. Binding of an HIV Rev Peptide to Rev responsive element RNA induces formation of purine-purine base pairs. *Biochemistry.* **33:** 2741–2747.

Battiste J.L., Mao H., Rao N.S., Tan R., Muhandiram D.R., Kay L.E., Frankel A.D., and Williamson J.R. 1996. α helix-RNA major groove recognition in an HIV-1 Rev peptide-RRE RNA complex. *Science* **273**: 1547–1551.

Brodsky A.S. and Williamson J.R. 1997. Solution structure of the HIV-2 TAR-argininamide complex. *J. Mol. Biol.* **267**: 624–639.

Burgstaller P. and Famulok M. 1994. Isolation of RNA aptamers for biological cofactors by in vitro selection. *Angew. Chem. Int. Ed. Engl.* **33**: 1084–1087.

Cai Z., Gorin A., Frederick R., Ye X., Hu W., Majumdar A., Kettani A., and Patel D. J. 1998. Solution structure of P22 transcriptional antitermination N peptide-box B RNA complex. *Nat. Struct. Biol.* **5**: 203–212.

Cate J.H., Gooding A.R., Podell E., Zhou K., Golden B.L., Kundrot C.E., Cech T.R., and Doudna J.A. 1996. Crystal structure of a group I ribozyme domain: Principles of RNA packing. *Science* **273**: 1678–1685.

Chang K. and Varani G. 1997. Nucleic acids structure and recognition. *Nat. Struct. Biol. (suppl.)* **4**: 854–858.

Chen L. and Frankel A.D. 1994. An RNA-binding peptide from bovine immunodeficiency virus Tat protein recognizes an unusual RNA structure. *Biochemistry* **33**: 2708–2715.

———. 1995. A peptide interaction in the major groove of RNA resembles protein interactions in the minor groove of DNA. *Proc. Natl. Acad. Sci.* **92**: 5077–5081.

Cilley C.D. and Williamson J.R. 1997. Analysis of bacteriophage N protein and peptide binding to boxB RNA using polyacrylamide gel coelectrophoresis (PACE). *RNA* **3**: 57–67.

Das A. 1993. Control of transcription termination by RNA binding proteins. *Annu. Rev. Biochem.* **62**: 893–930.

Davies J., Gorini L., and Davis B.D. (1965). Misreading of RNA codewords induced by aminoglycoside antibiotics. *Mol. Pharmacol.* **1**: 93–106.

Dieckmann T., Suzuki E., Nakamura G.K., and Feigon J. 1996. Solution structure of an ATP-binding RNA aptamer reveals a novel fold. *RNA* **2**: 628–640.

Fan P., Suri A.K., Fiala R., Live D., and Patel D. J. 1996. Molecular recognition in the FMN-RNA aptamer complex. *J. Mol. Biol.* **258**: 480–500.

Feigon J., Dieckmann T., and Smith F.W. 1996. Aptamer structures form A to ζ. *Chem. Biol.* **3**: 611–617.

Fourmy D., Recht M.I., and Puglisi J.D. 1998a. Binding of neomycin-class aminoglycoside antibiotics to the A-site of 16 S rRNA. *J. Mol. Biol.* **277**: 347–362.

Fourmy D., Yoshizawa S., and Puglisi J.D. 1998b. Paromomycin binding induces a local conformational change in the A-site of 16 S rRNA. *J. Mol. Biol.* **277**: 333–345.

Fourmy D., Recht M.I., Blanchard S.C., and Puglisi J.D. 1996. Structure of the A site of *E. coli* 16 S rRNA complexed with an aminoglycoside antibiotic. *Science* **274**: 1367–1371.

Frankel A.D. 1992. Activation of HIV transcription by Tat. *Curr. Opin. Genet. Dev.* **2**: 293–298.

Frankel A.D. and Smith C.A. 1998. Induced folding in RNA-protein recognition: More than a simple molecular handshake. *Cell* **92**: 149–151.

Fritz C.C. and Green M.R. 1996. HIV Rev uses a conserved cellular protein export pathway for the nucleocytoplasmic transport of viral RNAs. *Curr. Biol.* **6**: 848–854.

Gold L., Polisky B., Uhlenbeck O., and Yarus M. 1995. Diversity of oligonucleotide functions. *Annu. Rev. Biochem.* **64**: 763–797.

Heus H.A. and Pardi. A. 1991. Structural features that give rise to unusual stability of RNA hairpins containing GNRA loops. *Science* **253:** 191–194.

Jenison R.D., Gill S.C., Pardi A., and Polisky B. 1994. High-resolution molecular discrimination by RNA. *Science* **263:** 1425–1429.

Jiang F., Kumar R.A., Jones R.A., and Patel D.J. 1996. Structural basis of RNA folding and recognition in an AMP-RNA aptamer complex. *Nature* **382:** 183–186.

Jiang L., Suri A.K., Fiala R., and Patel D.J. 1997. Saccharide-RNA recognition in an aminoglycoside-RNA aptamer complex. *Chem. Biol.* **4:** 35–50.

Karn J., Gait M.J., Churcher M.J., Mann D.A., Mikaelian I., and Pritchard C. 1994. Control of human immunodeficiency virus gene expression by the RNA binding proteins Tat and Rev. In *RNA-protein interactions* (ed. K. Nagai and I.W. Mattaj), pp. 193–220. IRL Press, New York.

Lato S.M., Boles A.R., and Ellington A.D. 1995. In vitro selection of RNA lectins: Using combinatorial chemistry to interpret ribozyme evolution. *Chem. Biol.* **2:** 291–303.

Legault P., Li J., Mogridge J., Kay L.E., and Greenblatt J. 1998. NMR structure of the bacteriophage l N peptide/boxB RNA complex: Recognition of a GNRA fold by an arginine-rich motif. *Cell* **93:** 289–299.

Moazed D. and Noller H.F. 1987. Interaction of antibiotics with functional sites in 16S ribosomal RNA. *Nature* **327:** 389–394.

Mogridge J., Legault P., Li J., Van Oene M.D., Kay L.E., and Greenblatt J. 1998. Independent ligand-induced folding of the RNA binding domain and two functionally distinct activating regions for transcriptional antitermination in the N protein of bacteriophage l. *Mol. Cell.* **1:** 265–275.

Puglisi E.V. and Puglisi J.D. 1998. Nuclear magnetic resonance spectroscopy of RNA. In *RNA structure and function*, (ed. R.W. Simons and M. Grunberg-Manago), pp. 117–146, Cold Spring Harbor Laboratory Press, Cold Spring Harbor, New York.

Puglisi J.D., Chen L., Blanchard S., and Frankel A.D. 1995. Solution structure of a bovine immunodeficiency virus Tat-TAR peptide-RNA complex. *Science* **270:** 1200–1203.

Puglisi J.D., Chen L., Frankel A.D., and Williamson J.R. 1993. Role of RNA structure in arginine recognition of TAR RNA. *Proc. Natl. Acad. Sci.* **90:** 3680–3684.

Puglisi J.D., Tan R., Calnan B.J., Frankel A.D., and Williamson J.R. 1992. Conformation of the TAR RNA-arginine complex by NMR spectroscopy. *Science* **257:** 76–80.

Purohit P. and Stern S. 1994. Interactions of a small RNA with antibiotic and RNA ligands of the 30S subunit. *Nature* **370:** 659–662.

Recht M.I., Fourmy D., Blanchard S.C., Dahlquist K.D., and Puglisi J.D. 1996. RNA sequence determinants for aminoglycoside binding to an A-site rRNA model oligonucleotide. *J. Mol. Biol.* **262:** 421–436.

Saenger W. (1984) *Principles of nucleic acid structure*. Springer-Verlag, Berlin.

Sassanfar M. and Szostak J.W. 1993. An RNA motif that binds ATP. *Nature* **364:** 550–553.

Shaw K.J., Rather P.N., Hare R.S., and Miller G.H. 1993. Molecular genetics of aminoglycoside resistance genes and familial relationships of the aminoglycoside-modifying enzymes. *Microbiol. Rev.* **57:** 138–163.

Su L., Radek J.T., Hallenga K., Hermanto P., Chan G., Labeots L.A., and Weiss M.A. 1997. RNA recognition by a bent α-helix regulates transcriptional antitermination in phage l. *Biochemistry* **36:** 12722–12732.

Tan R. and Frankel A.D. 1995. Structural variety of arginine-rich RNA-binding peptides. *Proc. Natl. Acad. Sci.* **92:** 5282–5286.

Tao J. and Frankel A.D. 1992. Specific binding of arginine to TAR RNA. *Proc. Natl. Acad. Sci.* **89:** 2723–2726.

Varani G. and Tinoco I., Jr. 1991. RNA structure and NMR spectroscopy. *Q. Rev. Biophys.* **24:** 479–532.

Wallis M.G., Vonahsen U., Schroeder R., and Famulok M. 1995. A novel RNA motif for neomycin recognition. *Chem. Biol.* **2:** 543–552.

Wang Y. and Rando R.R. 1995. Specific binding of aminoglycoside antibiotics to RNA. *Chem. Biol.* **2:** 281–290.

Weeks K.M., Ampe C., Schultz S.C., Steitz T.A., and Crothers D.M. 1990. Fragments of the HIV-1 Tat protein specifically bind TAR RNA. *Science* **249:** 1281–1285.

Yang Y., Kochoyan M., Burgstaller P., Westhof E., and Famulok M. 1996. Structural basis for ligand discrimination by two related RNA aptamers resolved by NMR. *Science* **272:** 1343–1347.

Ye X.M., Kumar R.A., and Patel D.J. 1995. Molecular recognition in the bovine immunodeficiency virus Tat peptide TAR RNA complex. *Chem. Biol.* **2:** 827–840.

Ye X., Gorin A., Ellington A.D., and Patel D.J. 1996. Deep penetration of an α-helix into a widened RNA major groove in the HIV-1 rev peptide-RNA aptamer complex. *Nat. Struct. Biol.* **2:** 827–840.

Zimmermann G.R., Jenison R.D., Wick C.L., Simorre J.P., and Pardi A. 1997. Interlocking structural motifs mediate molecular discrimination by a theophylline-binding RNA. *Nat. Struct. Biol.* **4:** 644–649.

17
RNA Recognition by Proteins

Thomas A. Steitz
Department of Molecular Biophysics and Biochemistry and Department of Chemistry
Yale University and Howard Hughes Medical Institute
New Haven, Connecticut 06520-8114

Many RNA molecules are recognized by proteins that interact preferentially with a particular RNA molecule. We address here the structural principles by which these proteins recognize their target nucleic acid. RNA molecules invariably consist of duplex regions that are A-form, often stacked one on another, as well as regions of single-stranded loops and bulges giving rise in general to the possibility of a more complex and richly varied three-dimensional shape than can be assumed by duplex DNA. Although the structural database for RNA-binding proteins and their complexes with RNA is still smaller than for DNA-binding proteins, it has expanded rapidly in the past few years. As a consequence, some patterns of similarities and differences in the structural basis of recognition by proteins can be seen at this time.

Structural, biochemical, and molecular genetic studies of protein–RNA complexes have established at least four important sources of sequence specificity in protein–RNA interactions: (1) Direct hydrogen bonding and van der Waals interactions between protein side chains and the exposed edges of base pairs provide structural complementarity to the correct but not to the incorrect sequences. The interactions in the major groove can distinguish between all four Watson-Crick base pairs, whereas interactions in the minor groove can only distinguish between G-C and A-T. However, the formation of duplex RNA in loop regions from noncanonical base pairs presents a far more varied pattern of hydrogen-bond donors and acceptors in the minor groove which then forms an ideal target for protein recognition; (2) the sequence-dependent ability of some RNA sequences to adapt a particular structure required for binding to a protein at a lower free-energy cost than other sequences can provide sequence selectivity; (3) RNA molecules can and do form unique structures, in contrast with simple B-form DNA. Furthermore, the non-Watson-Crick base pairs formed in "loop" regions, as well as motifs like the "cross-strand purine stack," result in non-regular RNA backbone structures, which provide recognition opportunities. Finally, large RNA molecules such as ribosomal RNA can form branched structures; (4) bases of RNA that are in single-stranded regions or in bulges can be directly recognized by pockets on the protein that are complementary to these bases in shape and hydrogen-bonding capabilities.

THE PROBLEM THAT IS SET: WHAT IS BEING RECOGNIZED?

Let us first consider the problem confronting proteins interacting with the portion of an RNA molecule that forms a regular Watson-Crick duplex and contrast it with that confronted by DNA recognition proteins. The three-dimensional structure of double-stranded DNA is highly polymorphic (Kennard and Hunter 1989), but variations of two forms, A-form and B-form, are of relevance to the proteins that interact with DNA and RNA. Figure 1 shows an important difference between A-form RNA and B-form DNA. In B-DNA the major groove is wide enough to accommodate either an α-helix or an antiparallel β ribbon and the functional groups on the exposed edges of the base pairs can be directly contacted by side chains of the protein. The minor groove, on the other hand, is deep and narrow (5.8 Å wide) and thus less accessible to secondary structures such as an α-helix. For RNA, which is always A-form, the opposite is true. The minor groove is shallow and broad (10–11 Å wide) whereas the major groove is very deep and narrow (4 Å) (Delarue and Moras 1989).

Although on the basis of RNA structure alone one might expect proteins to discriminate among duplex RNA molecules by interactions with sequences via the minor groove, examples of interactions between protein in RNA in both the major and minor groove are now known (Rould et al. 1989; Ruff et al. 1991). It is true that the edges of base pairs are inaccessible in the major groove of A-form RNA in the central portion of a long duplex, but most naturally occurring RNA molecules contain relatively

---

Figure 1 Structures of A- and B-form DNA in space-filling representation showing differences in major and minor groove widths and shapes. On the top is shown B-DNA and on the bottom A-DNA. In the models on the left the helix axes are parallel to the page, whereas on the right the helix axes have been tilted up by 32° to show the groove shapes. The bases are colored blue, the phosphorous atoms are green, and all other atoms are white. The edges of the bases are easily accessible from the major groove of B-DNA and the minor or shallow groove of A-DNA (or RNA). The minor groove is designated by m and the major groove by M. (Reprinted, with permission, from Steitz 1990.)

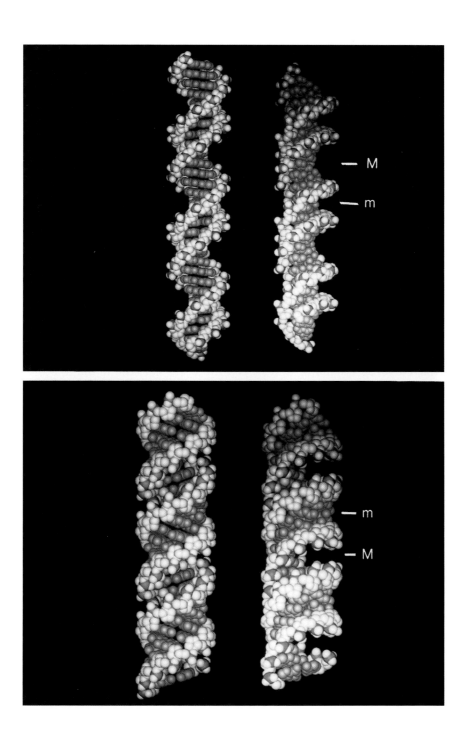

short duplex regions interrupted by bulges or loops. Although these short duplex regions may be expected to stack one on another as occurs in tRNA, the P4-P6 fragment of *Tetrahymena* group I intron (Cate et al. 1996) and the ribosome (Correll et al. 1997; Ban et al. 1998), the edges of base pairs exposed in the major groove are accessible at the ends of these RNA helices. Furthermore, features widespread in RNA structures, such as loop nucleotides and cross-strand purine stacks, provide local distortions to the RNA duplex that serve to greatly widen the major groove.

A second important consideration in the suitability of the major and minor grooves for direct sequence recognition is the degree of structural variation of the four base pairs as viewed from the two grooves. Seeman et al. (1976) pointed out that the base pairs present a more richly varied set of hydrogen-bond donors and acceptors to the major groove as compared to the minor groove. Figure 2 shows that the minor groove side of base pairs is a veritable recognition desert with only the N2 of guanine distinguishing AT from GC. The patterns of donors and acceptors on the major groove side, however, can distinguish all four base pairs. In duplex regions, RNA has an opportunity available that does not exist in duplex DNA sequences: non-Watson-Crick base pairs can exist within RNA helices (e.g., Figs. 3 and 7) and thus present to both the major and the minor groove hydrogen bonding and shape differences not seen in the four orientations of the two Watson-Crick base pairs. Although G-U base pairs are perhaps the most common non-Watson-Crick base pairs seen in RNA, A-G, A-A, G-G, A-C, and various kinds of U-U base pairs have been seen, and others may exist.

Three other recognition opportunities that occur in RNA and not in duplex DNA and appear to be utilized by proteins binding specifically to RNA are the single-stranded loop regions at the ends of helices, single-stranded bulges within helices, and modified bases. Finally, RNA secondary structures imply that it can form branched and richly varied tertiary structures that offer recognition opportunities not afforded by DNA.

THE ROLE OF THE MAJOR GROOVE IN RNA RECOGNITION

As pointed out initially by Seeman et al. (1976) and extensively documented from high-resolution crystal structures and a few NMR structures of DNA complexes (for reviews, see Steitz 1990; Harrison 1991; Pabo and Sauer 1992), the hydrogen-bond donors and acceptors presented by the base pairs to the major groove are important to DNA recognition by proteins. From the now rather large structural database of protein–DNA complexes, it appears that the structural complementarity between a pro-

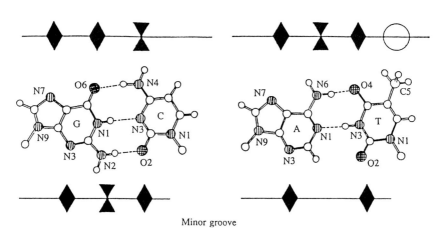

Figure 2 The hydrogen-bond donors and acceptors presented by Watson-Crick pairs to the major groove and the minor groove (adapted from Lewis et al. 1985). The symbols for hydrogen-bond donors (▼) and acceptors (♦) (Woodbury et al. 1980) show a varied pattern presented by the base pairs into the major groove and a poor information array presented into the minor groove. Although it is possible to distinguish among AT, TA, GC, and CB in the major groove, functional groups in the minor groove allow only easy discrimination between AT- and GC-containing base pairs. O, methyl group. (Reprinted, with permission, from Steitz 1990.)

tein and a specific DNA sequence is achieved in idiosyncratic manners: There does not appear to be a code for nucleic acid sequence recognition (Pabo 1983; Matthews 1988). Although the number of known examples of proteins interacting in the major groove of RNA is far smaller, one would expect the same principles to apply. Particular amino acid side chains do not always recognize the same base pair; however, there are some apparent preferences as suggested by Seeman et al. (1976). The guanidinium group of arginine very often makes a bidentate interaction with the N7 and O6 of guanine, although other interactions are also seen. Similarly, the hydrogen-bond donors and acceptors of the glutamine side chains are observed frequently to interact with the corresponding hydrogen-bond donors and acceptors of adenine. The ability of these side chains to make bidentate interactions with DNA greatly enhances their suitability for sequence-specific recognition (Seeman et al. 1976).

Information concerning protein interaction in the major groove of RNA is provided by a protein loop at the end of a β-hairpin in aspartyl tRNA synthetase observed to interact with the terminal base pairs via the major groove of the acceptor stem of tRNA^{Asp} (Ruff et al. 1991). Interactions between HIV *tat* and its target RNA TAR were initially hypothesized (Weeks et al. 1990, 1991) and later demonstrated to occur in the region of a three-nucleotide bulge and on the major groove side (see, e.g.,

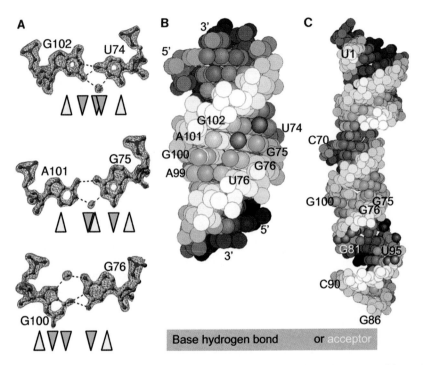

Figure 3 The role of non-Watson-Crick base pairs in RNA recognition. (*A*) Three non-Watson-Crick base pairs seen in the structure of loop E in the 5S rRNA fragment. These have only one direct inter-base hydrogen bond, but have bridging water molecules between them. The hydrogen-bond donors and acceptors exposed in the minor groove are indicated by green and yellow arrowheads, respectively. Superimposed on the base pairs is a 1.5 Å resolution F_o-F_c electron density map obtained using phases calculated with the atoms shown omitted. (*B*) A van der Waals surface representation of the dodecamer showing a complex H-bonding array in the minor groove, including the bridging structural water molecules colored blue. (*C*) A van der Waals surface representation of fragment I showing the nearly A-form helix I on top, loop E in the middle, and helix IV with its widened major groove on the bottom. The regions of fragment I that the binding of L25 protects from ethylation (Toukifimpa et al. 1989) are blue, and from RNase IV hydrolysis (Douthwaite et al. 1982) are purple. (Reprinted, with permission, from Correll et al. 1997 [copyright Cell Press].)

Puglisi et al. 1992). This bulged structure widens the major groove sufficiently to allow access to proteins.

The potential accessibility of an RNA major groove to protein side chains has been probed by Weeks and Crothers (1993), using diethyl pyrocarbonate (DEPC). DEPC carbethoxylates purines primarily at the N7 position in a reaction that is sensitive to the solvent exposure of the base (Vincze et al. 1973; Peattie and Gilbert 1980). The reagent is comparable in size to those protein side chains such as arginine that mediate RNA–protein interactions, suggesting that the rates of reactivity of this probe are likely to reflect the steric accessibility of purines to protein interaction. Although the major groove of an uninterrupted RNA duplex is relatively inaccessible to this reagent, as expected, the major groove at helix termini is accessible to modification, with the effect extending further on the 3' strand (Fig. 4). Furthermore, bulges in RNA helices larger than one nucleotide greatly increase the accessibility of flanking duplexes to reaction with DEPC.

The structure of a portion of HIV TAR RNA containing a three-nucleotide bulge and bound to arginine has been deduced from NMR data, showing one example of how a bulge can make the major groove of RNA accessible to a protein side chain (Puglisi et al. 1992). A cytosine from the three-nucleotide bulge makes a triple base pair with an adjacent G-C forming a binding site for the guanidinium group of arginine and opening the major groove.

It appears that other RNA motifs can serve to widen the major groove sufficiently to allow protein access for recognition. A "cross-strand G-stack" seen in the structure of a 62-nucleotide 5S rRNA fragment struc-

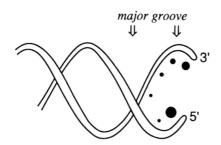

Figure 4 Schematic representation of an RNA duplex with the reactivity of purines to DEPC shown by filled circles whose diameter is proportional to reactivity (accessibility). Although the 5' base is most accessible, the accessibility to DEPC extends further into the duplex on the 3' strand. (Reprinted, with permission, from Weeks and Crothers 1993 [copyright American Association for the Advancement of Science].)

ture results in a major groove that is wide enough to accommodate an α-helix, although it is not yet known whether this in fact occurs (Correll et al. 1997).

ROLE OF SEQUENCE-DEPENDENT ALTERNATIVE CONFORMATIONS

The sequence-dependent RNA distortability, or more generally, the sequence-dependent ability of RNA to adapt alternative structures, is a very important source of specificity in many protein–RNA interactions. Nucleic acid distortability as a more indirect source of sequence specificity arises from two facts: (1) Proteins often bind a conformation of RNA that is altered from its uncomplexed solution conformation; (2) the free-energy cost for various nucleic acid sequences to assume the conformation that is required for its binding to the protein is not the same for different sequences.

The sequence-dependent deformability of duplex RNA that provides specificity for sequences being recognized by a protein can include the melting of base pairs. If binding to a protein requires melting of one or more base pairs, then the binding of mismatched base pairs should be favored over A-T pairs that in turn should bind better than G-C base pairs. The order of binding should reflect the thermodynamic stability of base pairs. An example of the role of duplex meltability in sequence specificity that can be cited is the binding of tRNA^{Gln} to its cognate synthetase, which results in the breaking of the terminal base pair of the acceptor stem between nucleotides U_1 and A_{72} (Rould et al. 1989). For glutaminyl tRNA synthetase (GlnRS) recognition and charging of tRNA, it is important that this base pair be not G-C (Yarus et al. 1977). The added free-energy cost of breaking the G-C base pair makes tRNAs containing a G-C at 1-72 less suitable for proper binding to the enzyme reducing k_{cat}/K_m by about tenfold (Jahn et al. 1991). The breaking of this terminal base pair in the acceptor stem is necessary to allow the 3′ terminus to hairpin into the active site, and thus a G-C base-pair will make formation of this essential structure less energetically favorable.

Sequence recognition in RNA also arises from the sequence-dependent ability of single-stranded RNA to take up the conformation required for protein binding as occurs in the single-stranded acceptor end of tRNA^{Gln} (Fig. 5). The observed interaction between the N^2 of G73 and the backbone phosphate of A72 is not possible for the other three bases (Rould et al. 1989), consistent with the observation that changing G73 to A, C, or U reduces the k_{cat}/K_m for charging by one, three, and four orders of magnitude, respectively (Jahn et al. 1991).

Two other examples of sequence specific structures that are assumed for protein recognition are provided by the anticodon loops of tRNA^{Gln} and tRNA^{Ile} when complexed to their cognate synthetases. Although the structures of their anticodon loops must be largely the same when these charged tRNAs are serving as substrates on the ribosome, they both form different and unique, sequence-dependent structures when complexed with their cognate synthetases. Two non-Watson-Crick base pairs are formed by four nucleotides in the anticodon loop of tRNA^{Gln} (Fig. 6), producing a noncanonical structure that is recognized by the synthetase

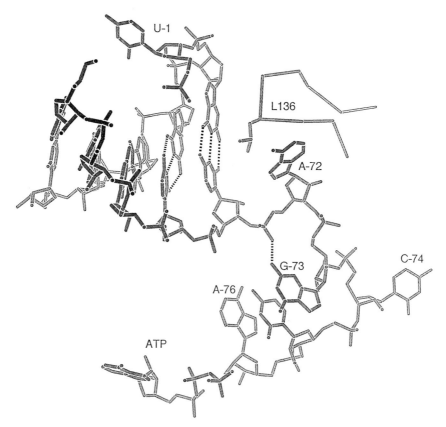

Figure 5 Conformation of the end of the acceptor stem and the 3' strand in tRNA^{Gln} bound to GlnRS. The expected base pair between U1 and A72 is broken by leucine 136, which packs against the guanine of the G2-C71 base pair. The 2-amino group of guanine 73 hydrogen-bonds to the phosphate backbone, stabilizing the hairpin conformation of the 3' strand into the active site. Cytosine 74 binds into a tight pocket in the protein, allowing the bases of nucleotides 73, 75, and 76 to stack. (Reprinted, with permission, from Rould and Steitz 1992.)

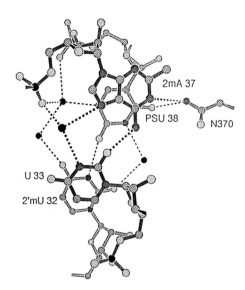

Figure 6 View of the two novel non-Watson-Crick base pairs that extend the anticodon stem of tRNA^{Gln} when complexed to GlnRS, showing the water network between these bases and the sugar-phosphate backbone. The U-U base pair has only one direct inter-base hydrogen bond, but has bridging water molecules. Asparagine 370 directly contacts both base pairs via the minor groove. (Reprinted, with permission, from Rould and Steitz 1992.)

(Rould et al. 1991). Other bases are unable to make these non-Watson-Crick base pairs and thus would not allow formation of the structure being recognized and bound by this enzyme. In this case of tRNA^{Gln}, the non-Watson-Crick interactions are formed by the four non-anticodon bases in the loop. In contrast, U36, the third base of the tRNA^{Ile} anticodon, is seen to be making a hydrogen-binding interaction with U33 in the anticodon loop, stabilizing a novel structure that is formed in its complex with the IleRS enzyme. A non-Watson-Crick A38-C32 base pair is formed while A37 and G34 are looped out to make base-specific interactions with IleRS (Fig. 7).

ROLE OF WATER MOLECULES IN SEQUENCE RECOGNITION

Buried water molecules appear to be playing a very important role in both DNA and RNA sequence recognition. Ascertaining the role of water molecules in sequence recognition requires crystal structures at sufficiently high resolution (usually 2.5 Å or better) and well enough refined

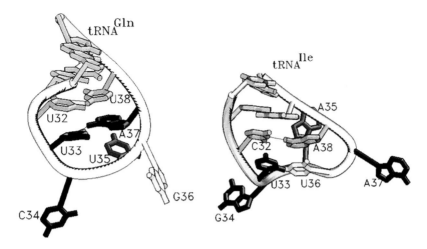

Figure 7 Schematic drawing showing the interactions between anticodon loop bases in tRNA^{Gln} (*left*) and tRNA^{Ile} (*right*) when bound to their respective synthetases (L. Silvian et al., unpubl.). Some sequence specificity for binding to the synthetases arises from the ability of these anticodon bases to stabilize unique structures for these loops that differ from their common uncomplexed structure.

so that water molecules can be reliably located. Water (or a protein hydroxyl group) can only make a base-specific hydrogen bond if it is also making at least two other hydrogen bonds with obligate donors or acceptors on the protein and is sequestered from bulk solvent. In this circumstance, the two unsatisfied water hydrogen-bond donors/acceptors directed toward the nucleic acid become obligate donors/acceptors and consequently become part of the hydrogen-bonding template surface of the protein to which the nucleic acid must be complementary for optimal binding (Fig. 8). Furthermore, it would appear that water molecules bridging non-Watson-Crick base pairs in loop regions (as seen in the structure of RNA Loop E) are not only integral parts of the RNA structure, but provide additional recognition opportunities to the protein (Correll et al. 1997). In this case, the specifically oriented water molecule is presenting obligate hydrogen-bond donors/acceptors to the protein. In the GlnRS complex with tRNA, two buried water molecules are an integral part of the hydrogen-bonding matrix presented in the shallow groove of the tRNA acceptor stem (Rould et al. 1989; Rould and Steitz 1992). Hydrogen bonds between these two water molecules as well as both a buried carboxylate of Asp-235 and a backbone amide of residue 183 serve to orient one hydrogen-bond donor of water toward the O2 of cytosine 71 and acceptor toward the N2 of guanine (Fig. 6).

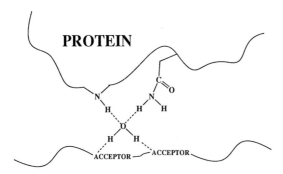

Figure 8 Schematic drawing showing how a water molecule can be specifically oriented by interactions with the protein turning it into a surrogate side chain. For example, here two obligate proton donors from the protein bind a water molecule such that it requires H-bond acceptors on the nucleic acid.

The water molecules that bridge between many non-Watson-Crick base pairs may be thought of as an integral part of the RNA structure and present unique and specific recognition opportunities; however, the involvement of such bridging in direct recognition by proteins has not yet been reported.

THE ROLE OF MINOR GROOVE IN RNA RECOGNITION

There are well-established examples of specific recognition of RNA in the minor groove (Rould et al. 1989; Musier-Forsyth and Schimmel 1992), where only the exocyclic of N2 of guanine distinguishes AT from GC and perhaps GC from CG, and recognition is restricted to a binary code—AT versus GC. Several sequence-specific interactions between GlnRS in the minor groove of tRNA^{Gln} have been observed (Rould et al. 1989, 1991; Rould and Steitz 1992). Base pairs G2-C71 and G3-C70 are recognized in a base-specific manner by two protein "fingers," one an α-helix and the other a turn of an antiparallel β-loop. In both cases, recognition involves contact between the 2-amino group of the guanine and hydrogen-bond acceptors of the protein. The carboxylate of an aspartic acid side chain 235 emanating from the amino end of α-helix H interacts with both the N2 of guanine three and a buried water molecule (Fig. 9). The peptide carboxyl group of proline 181 interacts with the N2 of guanine 2. Substantiating the hypothesis that these two base pairs are among the recognition elements

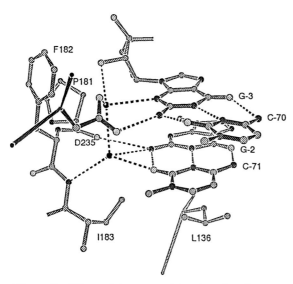

Figure 9 View of the recognition interface between GlnRS and base pairs G2-C71 and G3-C70 of tRNA^{Gln}. Aspartate 235 directly bonds to the 2-amino group of guanine 3 via the minor groove. The backbone carbonyl of proline 181 is rigidly directed to hydrogen-bond to the 2-amino group of guanine 2. A network of water molecules between the proteins and minor groove of the tRNA, only two of which are shown here, appear to enforce a requirement for G-C base pairs at these positions. The hydrophobic environment formed by the proline, phenylalanine, isoleucine, and the underside of the ribose sugars enhances the strength and specificity of these direct and water-mediated hydrogen bonds. (Reprinted, with permission, from Rould and Steitz 1992.)

of tRNA^{Gln}, replacement of either by A-U reduces the k_{cat}/K_m for charging by two to three orders of magnitude (Jahn et al. 1991). Furthermore, mutations in Gln RS that have increased rates of mischarging of noncognate tRNAs are changes of Asp-235 to asparagine or glycine (Conley et al. 1988; Perona et al. 1989), showing the importance of this interaction for discrimination. An additional interaction in the minor groove that is important for discrimination is between the carboxylate of Glu323 and the N2 of G10.

The importance of protein interaction with the two-amino group of guanine in the minor groove of RNA has also been established in the case of alanine tRNA synthetase recognition of tRNA^{Ala} (Hou and Schimmel 1988; McClain and Foss 1988; Hou et al. 1989; Musier-Forsyth and Schimmel 1992). The alanine synthetase has been clearly shown to recognize base pair 3-70, which is G-U in tRNA^{Ala}. Replacing G3 by an inosine, which lacks the N2, reduces charging of a minihelix dramatically (Musier-Forsyth and Schimmel 1992).

The loop regions of RNA can and do form duplexes using noncanonical base pairs, thereby expanding enormously the hydrogen-bond donor and acceptor diversity that can be the target of recognition (Correll et al. 1997). A co-crystal structure between ribosomal protein L25 and 5S rRNA loop E has not yet been determined; however, various types of chemical protection and interference studies suggest that L25 binds to 5S rRNA on the minor groove side of the loop E region, perhaps extending to the adjacent and much widened major groove (Fig. 3). The pattern of hydrogen-bonding groups on the RNA, including water base bridging molecules accessible in the loop E minor groove, is very different and more complex than that presented by normal Watson-Crick base-paired RNA.

ROLE OF SINGLE-STRANDED REGIONS IN RECOGNITION

Since RNA molecules have single-stranded regions in loops at the ends of helical stems and in bulges (or loops) between helical stems, these regions are potential targets for recognition by proteins that are not available in duplex DNA. Recently, the structures of numerous proteins and protein domains complexed with RNAs containing single-stranded regions being recognized have been determined by X-ray crystallography and by NMR. These include complexes between five aminoacyl-tRNA synthetases and their cognate tRNAs that show base-specific interactions with the anticodon loops. Furthermore, two structures of U1A with different RNA molecules containing recognized loops show how the conserved and ubiquitous RNP motif can recognize single-stranded RNAs with a large degree of base unstacking (Oubridge et al. 1994; Allain et al. 1996, 1997). A recently established structure of U2B shows how U1A and U2B″, which share nearly all of the residues involved in recognition of the single-stranded nucleotides, can have different substrate specificities (Price et al. 1998).

The recognition of anticodon loops in tRNAs by cognate synthetases provides extremely well characterized examples of protein recognition of single-stranded regions. Earlier molecular genetic and biochemical studies have established that the anticodon bases of tRNA serve as recognition elements for many of the aminoacyl-tRNA synthetases (Schulman and Pelka 1985; Normanly and Abelson 1989; Sampson et al. 1989). The co-crystal structures of synthetases responsible for charging tRNAs specific for glutamine, aspartic acid, proline, lysine, and isoleucine complexed with their cognate tRNAs all show that upon forming a complex, the anticodon bases become unstacked, allowing them to bind in separate base recognition pockets (Rould et al. 1991; Cavarelli et al. 1993; Cusack et al. 1996, 1998; L. Silvian et al., unpubl.). The energy required to unstack the anticodon bases (as they exist in the uncomplexed tRNA) is provided by interactions with the protein. Since bases in loop regions of uncomplexed RNAs will tend to be stacked on each other, and since optimal recognition of bases by a protein requires their unstacking in order for them to interact in separate recognition pockets, it may be more often the case than not that protein recognition of an RNA single-stranded region will be accompanied by a significant conformational change in the RNA.

The three anticodon bases of tRNA^{Gln} are splayed out and interacting with GlnRS (Fig. 10) with each anticodon nucleotide being recognized primarily by a polypeptide segment of five or six amino acids (Rould et al. 1991; Rould and Steitz 1992). In all three cases, at least one positively charged amino acid from this segment forms a salt link with an adjacent negatively charged phosphate. The aliphatic portion of this residue generally packs against either the base or the hydrophobic "underside" of ribose.

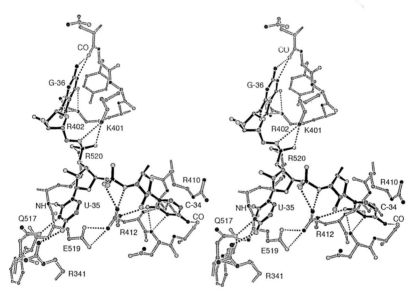

Figure 10 Stereo view of the binding pockets for the anticodon bases C34, U35, and G36 (*dark*) in the GlnRS·tRNA^{Gln} complex. Each nucleotide is recognized primarily by a single short polypeptide segment in the enzyme (*light*). In each case, an arginine or lysine from the polypeptide anchors the nucleotide by its phosphate group, allowing peptide backbone and side chains of the segment to specifically recognize the base. (Reprinted, with permission, from Rould and Steitz 1992.)

With all three anticodon bases, recognition is achieved by direct hydrogen bonding between the backbone and side chains of a short recognition peptide and the Watson-Crick hydrogen-bonding groups of the bases. Furthermore, several of the interactions presumed to be discriminating involve hydrogen bonds with charged side chains that are buried from solvent in the complex. Although there is no conserved sequence or structural similarity among these segments, they are predominantly in extended β-type conformation.

Four structures of class II aminoacyl-tRNA synthetases complexed to their cognate tRNAs are known (Asp, Pro, Lys, Ser), and in the first three cases the anticodon loop is interacting with a β-sheet (Fig. 11). The

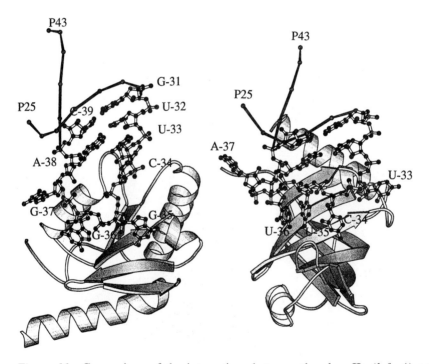

Figure 11 Comparison of the interactions between the class IIa (*left*, *A*) and class IIb (*right*, *B*) aminoacyl-tRNA synthetase anticodon-binding domains interacting with their cognate tRNA anticodon loops. (*Left*) Three anticodon loop bases of tRNA^{Pro} are unstacked and looped out (G-35, G-36, G-37), two remain stacked (U-33 and C-34), and two are specifically recognized by the ProRS protein (anticodon bases G-35 and G-36). (*Right*) Five anticodon loop bases of tRNA^{Lys} are looped out, and the three anticodon bases are recognized by the LysRS protein. (Reprinted, with permission, from Cusack et al. 1998 [copyright Current Biology Ltd.].)

largely β-sheet domain in the class IIb synthetases belongs to the oligonucleotide (or *o*ligosaccharide) *b*inding (OB) family of protein domains (Murzin 1993). Five of the seven bases of the anticodon loop are looped out and the three anticodon bases (34–36) are specifically recognized by the OB domain in both the AspRS and LysRS complexes with their cognate tRNAs. In these cases, the protein recognizes the U35 that occurs in both anticodons in a structurally homologous manner (Cavarelli et al. 1993; Cusack et al. 1996). However, the interactions between these two synthetases and anticodon bases 34 and 36 differ, as required by the differing anticodon sequences.

Whereas anticodon bases G35 and G36 of tRNA^{Pro} are specifically recognized by the class IIa synthetase, ProRS, the first anticodon base, C34, makes no interactions with the synthetase and is stacked under base 33, which in turn is stacked under an additional anticodon stem base pair formed between bases 32 and 38 (Fig. 11a) (Cusack et al. 1998). This lack of interaction between the protein and position 34 is required by the variability of this position in different proline anticodons.

There appears to be a general similarity in the interaction between the single-stranded RNA loops being recognized and the largely β-sheet structures of the RNA-binding domains of the U1A spliceosomal protein, the anticodon binding domain of AspRS, and the bacteriophage MS2 coat protein (Figs. 12–14). Although the structures of these three protein domains are not homologous, there are some structural analogies in the interactions observed. It remains to be evaluated as to whether there is some structural feature of antiparallel β-sheets that make them particularly suitable for interaction with the splayed-out bases of single-stranded RNA.

ROLE OF MODIFIED BASES

Many of the RNA molecules that are recognized by proteins contain modified bases, whose role in specific recognition remains largely unknown. There are at least two ways that modified bases might enhance recognition: by stabilizing RNA conformations that otherwise would be less favored and by changing the shape of a recognition site. The N at position 1 of pseudo uridine is observed in tRNA^{Gln} to interact with a water molecule that in turn is interacting with the phosphate of the pseudo uridine and the preceding phosphate, an interaction and conformation seen in all but one of the pseudo Us in tRNAs of known structure (Arnez and Steitz 1996). This structure so stabilized may be significant in protein recognition.

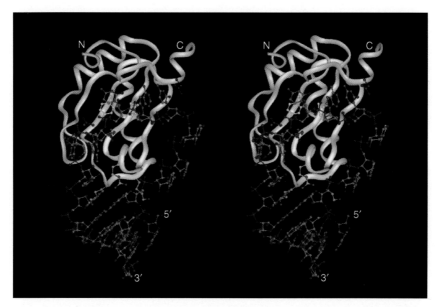

Figure 12 Interaction between a U1 snRNA hairpin loop and the β-sheet of the RNA-binding domain of the U1A spliceosomal protein. The aromatic protein side chains (Tyr-13 and Phe-56, yellow) that are highly conserved in all related RNP motifs are seen to stack on two looped-out bases. (Reprinted, with permission, from Oubridge et al. 1994.)

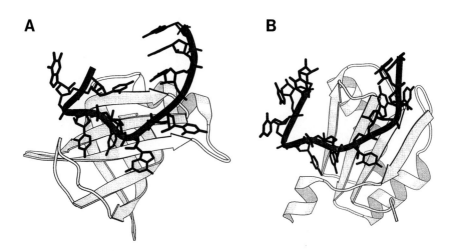

Figure 13 Schematic comparison between (*A*) the anticodon loop of tRNA^{Asp} interacting with the OB domain β-sheet of Asp-RS with (*B*) the U1 snRNA loop interacting with the β-sheet of the U1A RNP motif. (Reprinted, with permission, from Arnez and Moras 1998).

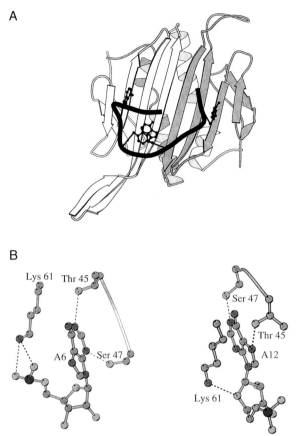

Figure 14 The interaction between single-stranded RNA and the bacteriophage MS2 coat protein (Valegard et al. 1994). (*A*) Schematic ribbon drawing of the structure of the coat protein dimer with the single-stranded RNA backbone and three splayed-out bases. (*B*) Hydrogen-bonding interactions in two quasi-symmetric contacts. (Reprinted, with permission, from Arnez and Cavarelli 1997 [copyright Cambridge University Press].)

Base modifications of noncognate tRNAs at nucleotides 34 and 37 may act as negative determinants of aminoacylation by GlnRS. The tightly packed interface between A37 and the protein (Fig. 10) suggests that bases with bulky modifications of A37, for example, 6-carbamoyl-threonyl adenine or 2-methylthio-N6-isopentenyl adenine, may provide an additional source of discrimination against the many noncognate tRNA molecules bearing these modifications. A role for modified bases at position 37 in tRNA discrimination has already been suggested by biochemical studies showing that ArgRS misacylates tRNA^{Asp} lacking modified bases (Perret et al. 1990). Likewise, C34 is tightly packed into a pocket that is covered by a loop of protein; this pocket may not accommodate certain modified bases at position 34, such as queuosine.

The 2'-ribosylated adenosine 64 in the initiator methionine tRNA from yeast appears to play an important role in assuring that this tRNA is only used in the initiation of protein synthesis and not in elongation (Kiesewetter et al. 1990). Again, the modification appears to function as a negative effector, since the modified $tRNA_i^{Met}$ will not bind to a EF-Tu or get inserted in elongation, whereas the $tRNA_i^{Met}$ that is unmodified at position 64 will both bind to EF-Tu and participate in elongation.

IleRS can recognize either a lysidine or a guanine at the wobble position 34, which can be explained by the co-crystal structure of IleRS with a $tRNA^{Ile}$ containing a G34 and by model building (L. Silvian et al., unpubl.). A G at position 34 is observed to make one set of specific interactions with the protein and model-building lysidine 34 suggests that it makes a different set of interactions (Fig. 15).

SUMMARY

The challenge of sequence-specific RNA recognition is greatest for those proteins that must discriminate among RNA molecules having nearly identical overall structures, such as tRNAs, which are specifically charged with correct amino acids by aminoacyl-tRNA synthetases. More generally, however, the overall structures that RNA molecules form are highly diverse. Initital inspection of a 6 Å resolution electron density map of the ribosomal large subunit suggests that very little of its nearly 3000 nucleotides form stretches of duplex RNA that look like Figure 1 (see Appendix 5, this volume; N. Ban et al., unpubl.). Rather, it consists of distorted, bent, and branched duplex with some regions of single-stranded RNA. Although little is known concerning ribosomal protein recognition of rRNA, these varied shapes provide recognition opportunities, as do the bases that form them.

For those families of similarly shaped RNA molecules whose members must be correctly identified by proteins, some principles of sequence-specific RNA recognition and discrimination have now emerged. Direct hydrogen bonding to the edges of base pairs exposed in either the minor or major grooves of duplex RNA is effective for either Watson-Crick or noncanonically base-paired regions. The minor groove of RNA is more accessible to proteins but provides a binary code (A,T or G,C) only, except in non-Watson-Crick base-paired regions. The major groove, with its more varied hydrogen-bonding features, is accessible to protein recogni-

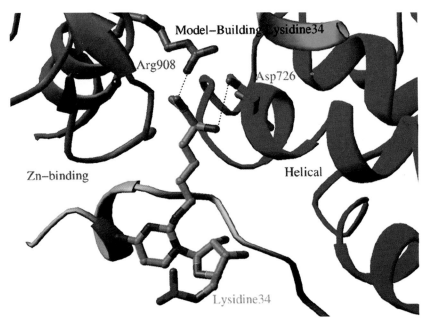

Figure 15 Model for the interaction of lysidine 34 of tRNA^{Ile} with IleRS (L. Silvian et al., unpubl.). The G34 observed in the co-crystal structure of IleRS with tRNA^{Lys} and mupirocin was replaced by lysidine in a model for the interaction. The carboxylate of the modification can interact with an arginine, whereas the amino may interact with an aspartic.

tion at the ends of helices or at bulges. The sequence-dependent ability of an RNA to assume a particular structure that may differ from its uncomplexed structure provides a second source of recognition. Finally, individual bases in the single-stranded regions can bind into specific protein recognition pockets. In many of the examples currently known, the splayed-out bases in a single-stranded RNA interact specifically with protein side chains emanating from one side of a largely antiparallel β-sheet structure, as exemplified by the RNA interactions of the anticodon-binding domains of GlnRS, AspRS, LysRS, and ProRS, the amino-terminal RNA-binding domains of the U1A and U2B" snRNP, and the coat protein of phage MS2.

ACKNOWLEDGMENTS

I thank G. Varani, L. Liljas, D. Moras, K. Nagai, S. Cusack, J. Greenblatt, D. Patel, and R. Giegé for preprints of unpublished manuscripts. This work was supported in part by National Institutes of Health grant GM-22778.

REFERENCES

Allain F.H.-T., Howe P.W.A., Neuhaus D., and Varani G. 1997. Structural basis of RNA-binding specificity of human U1A protein. *EMBO J.* **16:** 5764–5774.
Allain F.H.-T., Gubser C.C., Howe P.W.A., Nagai K., Neuhaus D., and Varani G. 1996. Specificity of ribonucleoprotein interaction determined by RNA folding during complex formation. *Nature* **380:** 646–650.
Arnez J.G. and Cavarelli J. 1997. Structure of RNA-binding proteins. *Q. Rev. Biophys.* **30:** 195–240.
Arnez J.G. and Moras D. 1998. tRNA/aminoacyl-tRNA synthetase interactions. In *RNA structure and function* (ed. R. W. Simons and M. Grunberg-Manago), pp. 465–494. Cold Spring Harbor Laboratory Press, Cold Spring Harbor, New York.
Arnez J.G. and Steitz T.A. 1996. Crystal structures of three misacylating mutants of *E. coli* glutaminyl-tRNA synthetase complexed with tRNA^{Gln} and ATP. *Biochemistry* **35:** 14725–14733.
Ban N., Freeborn B., Nissen P., Penczek P., Grassucii R.A., Sweet R., Frank J., Moore P.B., and Steitz T.A. 1998. A 9 Å resolution X-ray crystallographic map of the large ribosomal subunit. *Cell* **93:** 1105–1115.
Cavarelli J., Rees B., Ruff M., Thierry J.-C., and Moras D. 1993. Yeast tRNAAsp recognition by its class II aminoacyl-tRNA synthetase. *Nature* **362:** 181–184.
Cate J.H., Gooding A.R., Podell E., Zhou K., Golden B.L., Kundrot C.E., Cech T.R., and Doudna J.A. 1996. Crystal structure of a group I ribozyme domain: Principles of RNA packing. *Science* **273:** 1678–1685.
Conley J., Uemura H., Yamao F., Rogers J., and Söll D. 1988. *E. coli* glutaminyl tRNA synthetase: A single amino acid replacement relaxes tRNA specificity. *Protein Seq. Data Anal.* **1:** 479–485.
Correll C.C., Freeborn B., Moore P.B., and Steitz T.A. 1997. Metals, motifs and recognition in the crystal structure of a 5S rRNA domain. *Cell* **91:** 705–712.
Cusack S., Yaremchuk A., and Tukalo M. 1996. The crystal structures of *T. thermophilus* lysyl-tRNA synthetase complexed with *E. coli* tRNA^{Lys} and a *T. thermophilus* tRNA^{Lys} transcript: Anticodon recognition and conformational changes upon binding of a lysyl-adenylate analogue. *EMBO J.* **15:** 6321–6334.
Cusack S., Yaremchuk A., Krikliviy I., and Tukalo M. 1998. tRNA^{Pro} anticodon recognition by *Thermus thermophilus* prolyl-tRNA synthetase. *Structure* **6:** 101–108.
Delarue M. and Moras D. 1989. RNA structure. *Nucleic Acids Mol. Biol.* **3:** 182–196.
Douthwaite S., Christensen A., and Garrett R.A. 1982. Binding sites of ribosomal proteins on prokaryotic 5S rRNAs: A study with ribonucleases. *Biochemistry* **21:** 2313–2320.
Harrison S.C. 1991. A structural taxonomy of DNA-binding domains. *Nature* **353:** 715–719.
Hou Y.-M. and Schimmel P. 1988. A simple structural feature is a major determinant of the identity of a transfer RNA. *Nature* **333:** 140–145.
Hou Y.-M., Francklyn C., and Schimmel P. 1989. Molecular dissection of a transfer RNA and the basis for its identity. *Trends Biochem. Sci.* **14:** 233–237.
Jahn M., Rogers J., and Söll D. 1991. Anticodon and acceptor stem nucleotides in tRNA^{Gln} are major recognition elements for *E. coli* glutaminyl-tRNA synthetase. *Nature* **352:** 258–260.
Kennard O. and Hunter W.N. 1989. Oligonucleotide structure: A decade of results from single crystal X-ray diffraction studies. *Q. Rev. Biophys.* **22:** 327–379.

Kiesewetter S., Ott G., and Sprinzl M. 1990. The role of modified purine 64 in initiator/elongator discrimination of tRNA$_i^{Met}$ from yeast and wheat germ. *Nucleic Acids Res.* **18:** 4677–4682.

Lewis M., Wang J., and Pabo C. 1985. Structure of the operator binding domain of lambda repressor. In *Biological macromolecules and assemblies*, vol. 2 pp. 266–287 (ed. F.A. Jurnak and A. McPherson). Wiley, New York.

Matthews B.W. 1988. No code for recognition. *Nature* **335:** 294–295.

McClain W.C. and Foss K. 1988. Changing the identity of a tRNA by introducing a G-U wobble pair near the 3' acceptor end. *Science* **240:** 793–796.

Murzin A.G. 1993. OB (oligonucleotide/oligosaccharide binding)-fold: Common structural and functional solution for non-homologous sequences. *EMBO J.* **12:** 861–867.

Musier-Forsyth K. and Schimmel P. 1992. Functional contact of a transfer RNA synthetase with 2'-hydroxyl groups in the RNA minor groove. *Nature* **357:** 513–515.

Normanly J. and Abelson J. 1984. tRNA identity. *Annu. Rev. Biochem.* **58:** 1029–1049.

Oubridge C., Ito N., Evans P.R., Teo C.-H., and Nagai K. 1994. Crystal structure at 1.92 Å resolution of the RNA-binding domain of the U1A spliceosomal protein complexed with an RNA hairpin. *Nature* **372:** 432–438.

Pabo C.O. 1983. DNA-protein interactions. In *Proceedings of the 27th Robert A. Welch Foundation Conference on Chemical Research*, Stereospecificity in Chemistry and Biochemistry, ch. 7, pp. 223–255. Houston, Texas.

Pabo C.O. and Sauer R.T. 1992. Transcription factors: Structural families and principles of DNA recognition. *Annu. Rev. Biochem.* **61:** 1053–1095.

Peattie D.A. and Gilbert W. 1980. Chemical probes of higher-order structure in RNA. *Proc. Natl. Acad. Sci.* **77:** 4679–4682.

Perona J.J., Swanson R.N., Rould M.A., Steitz T.A., and Söll D. 1989. Structural basis for misaminoacylation by mutant *E. coli* glutaminyl-tRNA synthetase enzymes. *Science* **246:** 1152–1154.

Perret V., Garcia A., Grosjean H., Ebel J.-P., Florentz C., and Giegé R. 1990. Relaxation of a transfer RNA specificity by removal of modified nucleotides. *Nature* **344:** 783–789.

Price S.R., Evans P.R., and Nagai K. 1998. Crystal structure of the spliceosomal U2B''-U2A' protein complex bound to a fragment of U2 small nuclear RNA. *Nature* **394:** 645–650.

Puglisi J.D., Tan R., Calnan B.J., Frankel A.D., and Williamson J.R. 1992. Conformation of the TAR RNA-arginine complex by NMR spectroscopy. *Science* **257:** 76–80.

Rould M.A. and Steitz T.A. 1992. Structure of the glutaminyl tRNA synthetase-tRNA^{Glu}-ATP. *Mol. Biol.* **6:** 225–245.

Rould M.A., Perona J.J., and Steitz T.A. 1991. Structural basis of anticodon loop discrimination by glutaminyl-tRNA synthetase. *Nature* **352:** 213–218.

Rould M.A., Perona J.J., Söll D., and Steitz T.A. 1989. Structure of *E. coli* glutaminyl-tRNA synthetase complexed with tRNA^{Gln} and ATP at 2.8 Å resolution: Implications for tRNA discrimination. *Science* **246:** 1135–1142.

Ruff M., Krishnaswarmy S., Boeglin M., Poterszman A., Mitschler A., Podjarny A., Rees B., Thierry J.C., and Moras D. 1991. Class II aminoacyl tRNA synthetase: Crystal structure of yeast aspartyl-tRNA synthetase complexed with tRNA^{Asp}. *Science* **252:** 1682–1689.

Sampson J.R., DiRenzo A.B., Behlen L.S., and Uhlenbeck O.C. 1989. Nucleotides in yeast tRNA^{Phe} required for the specific recognition by its cognate synthetase. *Science* **243:** 1363–1366.

Schulman L.H. and Pelka H. 1985. *In vitro* conversion of a methionine to a glutamine-acceptor tRNA. *Biochemistry* **24:** 7309–7314.

Seeman N.C., Rosenberg J.M., and Rich A. 1976. Sequence-specific recognition of double helical nucleic acids by proteins. *Proc. Natl. Acad. Sci.* **73:** 804–808.

Steitz T.A. 1990. Structural studies of protein-nucleic acid interaction: The sources of sequence specific binding. *Q. Rev. Biophys.* **23:** 205–280.

Toukifimpa R., Romby P., Rozier C., Ehresmann C., Ehresmann B., and Mache R. 1989. Characterization and footprint analysis of two 5S rRNA binding proteins from spinach chloroplast ribosomes. *Biochemistry* **28:** 5840–5846.

Valegård K., Murray J.B., Stockley P.G., Stonehouse N.J., and Liljas L. 1994. Crystal structure of an RNA bacteriophage coat protein-operator complex. *Nature* **371:** 623–626.

Vincze A., Henderson R.E.L., McDonald J.J., and Leonard N.J. 1973. Reaction of diethyl pyrocarbonate with nucleic acid components. Bases and nucleosides derived from guanine, cytosine, and uracil. *J. Am. Chem. Soc.* **95:** 2677–2682.

Weeks K.M. and Crothers D.M. 1991. RNA recognition by tat-derived peptides: Interaction in the major groove? *Cell* **66:** 577–588.

———. 1993. Major groove accessibility and interhelix coupling in RNA. *Science* **261:** 1574–1577.

Weeks K.M., Ampe C., Schultz S.C., Steitz T.A., and Crothers D.M. 1990. Fragments of the HIV-1 *tat* protein specifically bind TAR RNA: Peptide recognition of bulged RNA. *Science* **249:** 1281–1285.

Woodbury C.P., Hagenbüchle O., and von Hippel P.H. 1980. DNA site recognition and reduced specificity of the EcoRI endonuclease. *J. Biol. Chem.* **255:** 11534–11546.

Yarus M., Knowlton R., and Soll L. 1977. Aminoacylation of the ambivalent Su+7 amber suppressor tRNA. In *Nucleic acid-protein recognition* (ed. H.J. Vogel), pp. 391–409. Academic Press, New York.

18

Group I and Group II Ribozymes as RNPs: Clues to the Past and Guides to the Future

Alan M. Lambowitz and Mark G. Caprara
Institute for Cellular and Molecular Biology
Departments of Chemistry and Biochemistry, and Microbiology
University of Texas at Austin, MBB 1.220
Austin, Texas 78712

Steven Zimmerly
Department of Biological Sciences
University of Calgary
Calgary, T2N 1N4, Canada

Philip S. Perlman
Department of Molecular Biology and Oncology
University of Texas Southwestern Medical Center
Dallas, Texas 75235

Group I and group II introns are not only catalytic RNAs, but also mobile genetic elements. The success of these introns as mobile elements almost certainly relates to their innate self-splicing capability, which enables them to propagate by inserting into host genes while only minimally impairing gene expression. Nevertheless, both types of introns have become dependent on proteins for efficient splicing in vivo to help fold the intron RNA into the catalytically active structure.

Although group I and group II introns have very different structures and splicing mechanisms (Chapter 13), there are striking parallels in the evolution of their protein-assisted splicing reactions. For example, the splicing factors for both types of introns include intron-encoded as well as cellular proteins, and the intron-encoded proteins, DNA endonucleases for group I introns and reverse transcriptases (RTs) for group II introns, also function in intron mobility. In addition, excised group I and group II intron RNAs remain associated with splicing factors in RNP particles, which can then cleave and insert into cellular RNA or DNA target sites by reverse splicing. The need to control this deleterious ribozyme activity may have been an evolutionary driving force favoring mutations that impaired self-splicing activity and resulted in dependence on protein factors (Nikolcheva and Woodson 1997).

In this chapter, we review protein-assisted reactions of group I and group II introns. These studies illustrate how proteins facilitate RNA folding and catalysis and provide unique insights into how splicing mechanisms evolve. A recurring theme, first developed in a previous review (Lambowitz and Perlman 1990), is that splicing factors evolved from preexisting cellular proteins, such as aminoacyl-tRNA synthetase, with independent functions. Such recent adaptation is supported by strong evidence that group I intron splicing factors differ even in closely related organisms. A similar adaptation of preexisting cellular proteins may have occurred for other types of introns, including spliceosomal introns. Finally, the integration of a ribozyme into the complex process of group II intron mobility may provide a model for the evolution of more complicated RNPs, including the spliceosome and ribosome.

GROUP I INTRON-ENCODED PROTEINS

Group I introns encode site-specific DNA endonucleases that function in intron mobility and in some cases have adapted to function in RNA splicing (Lambowitz and Belfort 1993; Belfort and Roberts 1997; Johansen et al. 1997). Open reading frames (ORFs) encoding endonucleases are found in about 30% of group I introns. The locations of the ORFs vary within the conserved RNA secondary structure, but they are generally found in loops where they would not interfere with folding of the catalytic core. Some ORFs are free-standing within the introns, whereas others are expressed as fusion proteins with the upstream exon and then proteolytically processed to mature form (Guo et al. 1995).

The group I intron endonucleases promote the site-specific insertion of the intron at the identical location in an intronless allele, a process referred to as "homing" (Dujon et al. 1989). As shown in Figure 1a, homing is initiated by the endonuclease, which makes a double-strand break at or near the intron insertion site in the recipient DNA. After DNA cleavage, mobility occurs by a double-strand break repair (DSBR) mechanism (i.e., gene conversion), leading to the transfer of the intron with flanking exon sequences. Homing is highly efficient, occurring at frequencies approaching 100% for some fungal mtDNA introns.

In addition to homing, phylogenetic evidence indicates that group I introns transpose to ectopic sites and undergo horizontal transmission between organisms (Lambowitz and Belfort 1993). Ectopic transposition has been proposed to occur by reverse splicing of the intron into an unre-

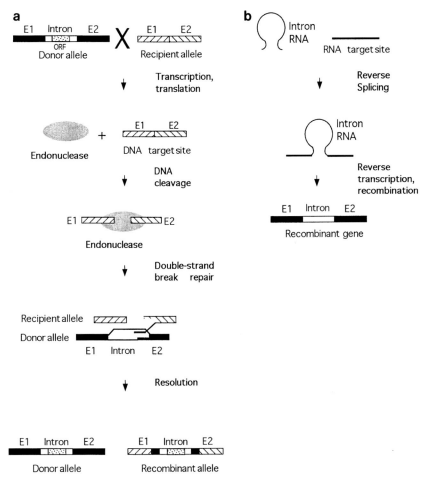

Figure 1 Group I intron mobility mechanisms. (*a*) Homing. The mobile intron in the donor allele encodes a site-specific DNA endonuclease that recognizes the exon junction sequence in the recipient allele and makes a double-strand break at or near the intron insertion site. After DNA cleavage, the intron is copied from the donor allele into the recipient allele by a DSBR mechanism involving cellular enzymes. Homing leads to insertion of the intron with co-conversion of flanking 5′ and 3′ exon sequences (*black*). (*b*) Ectopic transposition. Transposition of group I introns to new genomic sites may occur by reverse-splicing of the excised intron into a new RNA site, followed by reverse transcription of the resulting intron-containing RNA, and integration of the cDNA into the genome. The RNA target site for reverse splicing is selected by short (4–6 nt) base-pairing interactions with the intron's IGS (see text), ensuring that the new intron insertion site will be compatible with forward splicing. Reverse splicing can be assisted by splicing factors that remain bound to the excised intron RNA (Mohr and Lambowitz 1991).

lated RNA, followed by reverse transcription of the recombined RNA, and genomic integration of the resulting cDNA (see Fig. 1b) (Cech 1985; Roman and Woodson 1998). The reverse splicing reaction can be assisted by splicing factors that are bound to the excised intron RNA (Mohr and Lambowitz 1991). Since reverse splicing requires only short (4–6 nucleotides) base-pairing interactions between the intron's internal guide sequence (IGS) and the RNA target, there are many potential insertion sites in cellular RNAs.

The endonucleases that promote homing appear to be independent entities whose coding sequences inserted into preexisting group I introns (Lambowitz and Belfort 1993; Loizos et al. 1994; Sellem and Belcour 1997). Four different types of endonucleases have been found encoded in group I introns, and these are distinguished by the conserved amino acid motifs LAGLIDADG, GIY-YIG, H-N-H, and His-Cys (Fig. 2) (Belfort and Roberts 1997). Endonucleases of the first three types have also been found to be encoded by free-standing ORFs not associated with introns. The composite elements, in which a DNA endonuclease is inserted in a self-splicing intron, combine two cardinal features of a successful genomic parasite—an efficient transmission mechanism and minimal interference with host gene expression. A similar combination is found in inteins, which appear to have been formed by association of LAGLIDADG or H-N-H endonucleases with protein-splicing activities (Dalgaard et al. 1997; Duan et al. 1997).

Although they are primarily DNA endonucleases, several LAGLIDADG proteins have been found to function in RNA splicing. This splicing function was first demonstrated genetically for the proteins encoded by introns bI2, bI3, and bI4 of the yeast mtDNA cytochrome *b* gene (for review, see Lambowitz and Perlman 1990). In each case, mutations in the intron ORF resulted in defective splicing of the intron, which could then be rescued (complemented) by providing the intron-encoded protein in *trans*. Such intron-encoded splicing factors have been termed "maturases," a name adopted before the mobility function of related proteins was known (Lazowska et al. 1980). The bI4 maturase was also shown to splice the homologous intron aI4α, found in the yeast mtDNA *cox1* gene, which encodes cytochrome oxidase subunit I. Since these yeast maturases are intron-specific splicing factors, they presumably recognize unique structural features of their own unspliced precursor RNAs.

Recently, an in vitro splicing system was developed for the maturase encoded by the *Aspergillus nidulans cob* (Ancob) intron, a LAGLIDADG protein homologous to the yeast bI3 maturase (Ho et al. 1997). The Ancob protein was expressed in *Escherichia coli*, and both the splicing and en-

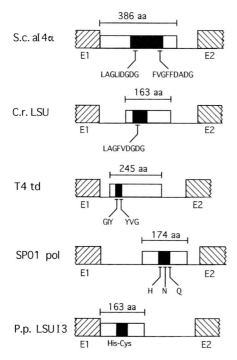

Figure 2 Group I introns encode different types of DNA endonucleases. The introns shown are the *S. cerevisiae* (S.c.) mtDNA intron aI4α, which encodes a LAGLIDADG endonuclease with two copies of the motif (Bonitz et al. 1980); the *C. reinhardtii* (C.r.) ctDNA large subunit rRNA (LSU) intron, which encodes a LAGLIDADG endonuclease (I-*Cre*I) with one copy of the motif (Rochaix et al. 1985); the phage T4 thymidylate synthase (*td*) intron, which encodes a GIY-YIG endonuclease (Chu et al. 1984); the bacteriophage SPO1 DNA polymerase I (*pol*) intron, which encodes an H-N-H endonuclease (Goodrich-Blair et al. 1990); and *Physarum polycephalum* (P.p.) LSU intron 3, which encodes a His-Cys endonuclease (Muscarella and Vogt 1989). Exons (E) are shown as hatched boxes, and ORFs are shown as rectangles within the intron. The conserved regions for each class of proteins are demarcated by gray boxes, with distinguishing amino acid motifs shown below.

donuclease activities were assayed biochemically. The purified protein binds specifically to the intron as a monomer with an apparent K_d of 3 nM, and splices stoichiometrically, with a very slow turnover rate. This in vitro splicing system should permit detailed biochemical analysis of the maturase activity.

Individual LAGLIDADG proteins have either splicing or endonuclease activity or both. Some LAGLIDADG proteins contain only one copy of the motif, but most are larger proteins with two copies separated by ~100 amino acids (Fig. 2) (Hensgens et al. 1983; Dalgaard et al. 1997). Thus far, only proteins with two motifs are known to have maturase activity (Lambowitz and Belfort 1993; Schafer et al. 1994; Ho et al. 1997). The protein encoded by the yeast mtDNA intron aI4α was the first shown to have both activities (Delahodde et al. 1989; Wenzlau et al. 1989). In wild-type yeast strains, the aI4α protein has endonuclease but not maturase activity. However, the protein has a latent maturase activity that can be activated either by a mutation in the intron ORF (Dujardin et al. 1982) or by dominant suppressors of an interacting nuclear gene *NAM2* (see below). Experiments with the activated maturase showed that mutations in the first LAGLIDADG motif inhibit endonuclease but not maturase activity, whereas identical mutations in the second motif have the opposite effect, suggesting that the two motifs function differentially in splicing and endonuclease activity (Henke et al. 1995).

Since LAGLIDADG endonucleases were likely acquired by preexisting group I introns, it seems probable that the endonuclease function existed first and the splicing function evolved subsequently (Lambowitz and Perlman 1990). A key question is whether the same nucleic acid-binding site is used for both DNA and RNA substrates. X-ray crystal structures for the *Chlamydomonas reinhardtii* I-*Cre*I homing endonuclease (Heath et al. 1997) and the yeast PI-*Sce*I intein endonuclease (Duan et al. 1997) revealed very similar active sites, consisting of two α-helical LAGLIDADG motifs packed together, resulting in twofold symmetry. The DNA-binding site corresponds to a platform or saddle formed by elements adjacent to the LAGLIDADG motifs. PI-*Sce*I has two LAGLIDADG motifs and is a monomer, whereas I-*Cre*I (Fig. 2) has only one motif and is a homodimer with the active site formed by structural elements from each subunit. Footprinting and molecular modeling based on the X-ray crystal structures indicated that LAGLIDADG endonucleases bind 5′ and 3′ exon DNA via the major groove and phosphodiester backbone (Duan et al. 1997; Heath et al. 1997). Because diverse DNA sequences are recognized by different LAGLIDADG endonucleases, the DNA-binding site may be highly adaptable, a characteristic that could facilitate dispersal of group I introns to different sites.

Given the nature of the DNA-binding site, an intriguing possibility is that LAGLIDADG proteins adapted to function in splicing by using analogous interactions to recognize similar sequences in the major groove of the DNA substrate and the major grooves of the intron's P1 and/or P10 helices, which contain the same 5′ and 3′ exon sequences base-paired to the IGS. The major groove of the P1 helix remains accessible after docking to the catalytic core (Strobel and Doudna 1997) and could be widened by RNA or protein interactions. However, the finding that mutations in the two LAGLIDADG motifs of the yeast aI4α protein differentially affect the splicing and endonuclease activities suggests that the binding of DNA and RNA substrates differ, at least in detail. The aI4α protein appears to be a dimer containing a total of four LAGLIDADG motifs and could thus form two active sites, one associated with splicing and the other with endonuclease activity (see above; Wernette et al. 1990; Henke et al. 1995).

Thus far, maturase activity has not been demonstrated for any other class of group I intron endonuclease. Negative evidence exists for the GIY-YIG and the H-N-H endonucleases in phage T4, where inactivation of the intron ORF does not inhibit intron splicing in vivo (Belfort 1990; Eddy and Gold 1991). A broader survey is needed to learn whether the ability to facilitate splicing is limited to LAGLIDADG proteins.

HOST-ENCODED SPLICING FACTORS FOR GROUP I INTRONS

Neurospora mt Tyrosyl-tRNA Synthetase

The *Neurospora crassa* mt tyrosyl-tRNA synthetase (mt TyrRS), encoded by the *cyt-18* gene, is the best-studied example of a host protein adapted to function in group I intron splicing. The CYT-18 protein functions in splicing three of the ten *N. crassa* mtDNA group I introns in vivo (LSU, *cob*-I2, and *ND1*; Mannella et al. 1979; Collins and Lambowitz 1985; Wallweber et al. 1997). The remaining *Neurospora* mtDNA introns contain large peripheral RNA structures that impede CYT-18 binding to the catalytic core (Wallweber et al. 1997; and see below). The three CYT-18-dependent introns show no detectable self-splicing in vitro, but all splice efficiently at physiological Mg^{++} concentrations in the presence of purified CYT-18 protein (Garriga and Lambowitz 1986; Wallweber et al. 1997). In addition to *Neurospora* mtDNA introns, CYT-18 binds to and facilitates the splicing of group I introns from a variety of other organisms (Guo and Lambowitz 1992; Mohr et al. 1992, 1994). The introns spliced by CYT-18 have relatively little sequence similarity, indicating that the protein primarily recognizes conserved secondary and tertiary structural features of the intron RNAs (Guo and Lambowitz 1992; Mohr et al. 1992).

The CYT-18 protein is homologous to well-studied bacterial TyrRSs. These enzymes, class I aaRSs, consist of a conserved nucleotide-binding fold region, which functions in the formation of tyrosyl-adenylate and binds the tRNA's acceptor stem, and a less conserved carboxy-terminal domain, which binds the tRNA's anticodon domain (Nair et al. 1997). The bacterial TyrRSs are homodimers with one tRNA^{Tyr} bound asymmetrically across the surface of the two subunits (Bedoulelle 1990), and CYT-18 is likewise a homodimer, which binds one molecule of intron RNA (Saldanha et al. 1995). Despite these structural and functional similarities, only the *N. crassa* mt TyrRS and that of the related fungus *Podospora anserina* have been found to function in splicing group I introns, whereas the *E. coli* and yeast mt TyrRS lack splicing activity (Cherniack et al. 1990; Kämper et al. 1992). The inference is that splicing activity resulted from adaptation of the basic TyrRS structure after the common ancestor of *N. crassa* and *P. anserina* diverged from yeast (Cherniack et al. 1990; Kämper et al. 1992).

CYT-18 and other group I intron splicing factors may have adapted to function in splicing by recognizing structural features of group I introns that resemble their normal cellular RNA substrates (Akins and Lambowitz 1987). Biochemical analysis of mutant CYT-18 proteins synthesized in *E. coli* showed that regions required for splicing overlap those required for binding tRNA^{Tyr} and include parts of the nucleotide-binding fold region, the carboxy-terminal tRNA-binding domain, and a small idiosyncratic amino-terminal domain that is absent from bacterial and yeast mt TyrRSs (Kittle et al. 1991). The idiosyncratic amino-terminal domain appears to be required for splicing and for aminoacylation of the *N. crassa* mt tRNA^{Tyr}, but not for the aminoacylation of *E. coli* tRNA^{Tyr} (Cherniack et al. 1990; Cherniack 1991). Hence, this domain may be a recent acquisition that contributes to the recognition of specific structural features of *N. crassa* mt tRNA^{Tyr} and group I introns.

CYT-18 promotes the splicing of group I introns by stabilizing the catalytically active structure of the intron core (Guo and Lambowitz 1992; Mohr et al. 1992). The purified protein binds group I introns stoichiometrically with K_d values in the fM range, among the tightest known RNA–protein interactions (Saldanha et al. 1995). RNA footprinting experiments revealed how CYT-18 binds to the group I intron catalytic core (Fig. 3a). The catalytic core consists of two extended helical domains, P4-P6 and P3-P9, which form a cleft containing the intron's active site (Michel and Westhof 1990; Chapter 13). Putative CYT-18 protection sites on the phosphodiester backbone (spheres in Fig. 3a) are located primarily on the side of the core opposite the active-site cleft (Caprara et al. 1996a,b). Many of the CYT-18-protected sites are in the P4-P6 domain, clustered around the junction of the P4-P6 stacked helix, but additional protected sites were found in regions of the P3-P9 domain, including P7 and P9. Considered with other studies, these findings support a model in which CYT-18 binds

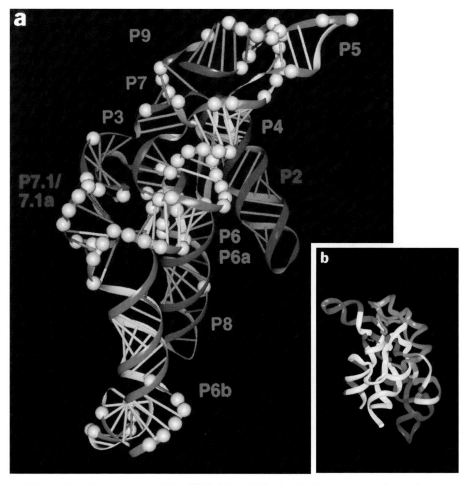

Figure 3 Comparison of the CYT-18 and Cbp2p binding sites on three-dimensional model stuctures of group I introns. For both proteins, the binding sites were inferred from RNA footprinting experiments combined with computer modeling (Weeks and Cech 1995b; Caprara et al. 1996a,b). (*a*) Putative CYT-18 protein protection sites (*gray spheres*) on the phosphodiester backbone of the *Neurospora* mt LSU intron. (*b*) Putative Cbp2p protection sites (*gray coloring*) on the phosphodiester backbone of the yeast mt LSU intron. CYT-18 binds primarily to the side of the catalytic core opposite the active-site cleft and interacts with both the P4-P6 and P3-P9 domains, as well as the peripheral RNA structures P6b and P7.1. Cbp2p binds on the side opposite CYT-18 and interacts with the P1 helix, which contains the 5′ splice site, and with P8 and the peripheral RNA structures P2 and P7.1. (Reprinted, with permission, from Weeks and Cech 1995b; Caprara et al. 1996b [copyright Cell Press].)

first to the P4-P6 domain to stabilize the correct geometry at the junction of the P4-P6 stacked helix. The assembled protein-bound P4-P6 domain then forms a scaffold for the assembly of the P3-P9 domain, and the protein makes additional contacts with the P3-P9 domain, stabilizing the two domains in the correct relative orientation to form the active site (Caprara et al. 1996a,b; Myers et al. 1996; Saldanha et al. 1996).

Exemplifying the ability of proteins to replace RNA structures, CYT-18 functionally substitutes for a peripheral RNA structure P5abc found in the *Tetrahymena* LSU intron and related group I introns (Mohr et al. 1994). Such peripheral RNA structures facilitate folding of the intron but are not required for catalytic activity. Derivatives of the *Tetrahymena* LSU intron that lack P5abc no longer self-splice at physiological Mg^{++}, but splicing is restored either by binding P5abc as a separate RNA molecule or by binding CYT-18. CYT-18 and P5abc RNA interact with overlapping sites in the P4-P6 domain of the catalytic core and induce similar conformational changes required for splicing activity (van der Horst et al. 1991; Mohr et al. 1994; Murphy and Cech 1994; Cate et al. 1996). Both CYT-18 and P5abc are thought to act by nucleating the folding of the P4-P6 domain, and it will be interesting to see if this is accomplished by similar interactions with the domain in both cases.

Does CYT-18 recognize a tRNA-like structure in group I introns? To address this question, the CYT-18-binding site in group I introns was compared with that in the *N. crassa* mt $tRNA^{Tyr}$ by three-dimensional graphic modeling based on RNA footprinting data for each of the RNA species (Caprara et al. 1996b). Remarkably, alignment of the binding sites revealed that the tRNA could be superimposed over an extended region of the group I intron catalytic core. In this alignment, the P4-P6 stacked helix corresponds to the D-anticodon arm of the tRNA, P7 to the long variable arm, and P9 to the acceptor stem. In principle, these structural similarities could reflect either convergent evolution or an evolutionary relationship between group I introns and tRNAs. One possibility is that tRNAs evolved from group I introns during the evolution of protein synthesis. However, the structural similarities between group I introns and modern tRNAs favor an alternate scenario: that group I introns evolved from tRNAs, which had the preexisting structural complexity required to develop an active site (Caprara et al. 1996b).

Yeast mt Leucyl-tRNA Synthetase

The yeast mt leucyl-tRNA synthetase (mt LeuRS), encoded by the nuclear gene *NAM2*, is another class I aaRS that functions in splicing group I introns. Genetic evidence indicates that the Nam2p assists the splicing of the closely related mtDNA introns bI4 and aI4α by acting in concert with a maturase (Dujardin et al. 1983; Labouesse et al. 1987; Herbert et al. 1988; Labouesse 1990; Li et al. 1996a). *NAM2* was identified by dominant nuclear mutations that suppressed the splicing defects of strains lacking the bI4 maturase (NAM = *n*uclear *a*ccommodation of *m*itochondria) (Dujardin et al. 1980; Groudinsky et al. 1981). Surprisingly, a suppressor allele, *NAM2-1*, which has a single missense mutation, was found to alleviate the need for the bI4 maturase by activating the latent maturase activity of the aI4α protein (Dujardin et al. 1983; Labouesse et al. 1987). The mechanism of activation is not known, but it could involve protein–protein interaction.

In the absence of an in vitro splicing system, additional functional information about Nam2p comes primarily from in vivo studies of site-directed mutations. Because Nam2p is essential for mt protein synthesis, the most cogent experiments employed a system in which the bI4 maturase is expressed in the nucleus and imported into mitochondria. A key result was that a strain with the imported maturase could not splice unless an active wild-type *NAM2* gene was expressed, indicating that both proteins are required for splicing activity (Labouesse 1990). Various *NAM2* mutations affect the splicing and/or synthetase functions, indicating that these activities are at least partially separable (Labouesse 1990; Li et al. 1992, 1996a). Based on the assumption that Nam2p functions similarly to the *Neurospora* mt TyrRS, Li et al. (1996a) suggested that the synthetase helps structure the core of the ribozyme, whereas the maturase aligns the splice sites. However, it remains possible that the two proteins act as a complex at the same splicing step.

Cbp2p

The yeast *CBP2* gene, which is required for splicing the yeast mtDNA intron bI5, encodes a 630-amino-acid protein lacking clear homology with other proteins (McGraw and Tzagoloff 1983; Hill et al. 1985; Li et al. 1996b). Disruption of *CBP2* specifically inhibited the splicing of bI5 but caused no discernible defect in a strain lacking bI5. The splicing function was demonstrated directly by experiments in which purified Cbp2p stimulated the splicing of bI5 at low (5 mM) Mg^{++}, an inefficient condition for self-splicing (Gampel et al. 1989). Cbp2p binds specifically to bI5 RNA with a K_d of ~50 pM, but also has moderate affinity for the *Tetrahymena* LSU and phage T4 *nrdB* group I introns (K_d values = 10–100 nM) (Weeks and Cech 1995a, 1996). Dissociation of the protein from the excised intron RNA is rate-limiting for splicing in vitro (Weeks and Cech 1996).

RNA footprinting experiments showed that Cbp2p binds to the catalytic core of bI5 on the same surface that binds P1, the side opposite that which binds CYT-18 (Fig. 3b) (Weeks and Cech 1995b). Kinetic studies indicated that Cbp2p promotes splicing by a mechanism termed tertiary-structure capture (Weeks and Cech 1995a,b, 1996). In this mechanism, Cbp2p does not induce folding of the intron, but rather binds to and stabilizes the active tertiary structure after it is formed. Once associated with the core, the protein makes additional contacts with the P1 helix, thereby promoting the association of the 5' splice site with the core (Weeks and Cech 1995b, 1996).

Does *CBP2* fit the hypothesis that group I intron splicing factors evolved by adaptation of preexisting cellular proteins? The initial genetic studies suggested that *CBP2* is required only for splicing bI5 (McGraw and Tzagoloff 1983; Hill et al. 1985). There is still no indication of a second nonsplicing function, but recent findings indicate that Cbp2p influences the splicing of several other mt introns (Li et al. 1996b; Shaw and Lewin 1997). Interestingly, homologs of *CBP2* or bI5 have not been found outside of *Saccharomyces*, consistent with the possibility that Cbp2p is a dedicated splicing factor that evolved from another gene by duplication and divergence.

RNA Chaperones

RNA chaperones are proteins that facilitate RNA folding but do not remain associated with the folded RNA (Herschlag 1995). The concept is supported by in vitro experiments in which nonspecific RNA-binding proteins, such as the HIV-1 NC and hnRNP A1, facilitated folding of the hammerhead ribozyme by increasing the rate of dissociation of improper RNA secondary structures. Two *E. coli* proteins, ribosomal protein S12 and StpA, an analog of the nucleoid protein H-NS, stimulate the splicing of the phage T4 *td* intron in vitro, perhaps via RNA chaperone activity (Coetzee et al. 1994; A. Zhang et al. 1995). StpA was identified initially by its ability to suppress mutations in the P7 region of the *td* intron, suggesting that it also stimulates splicing in vivo (A. Zhang et al. 1995), and recent studies indicate this is also the case for S12 (Semrad and Schroeder 1998). Both S12 and StpA bind RNA nonspecifically and stimulate the association and/or dissociation of model RNA duplexes in vitro, properties expected for RNA chaperones (Herschlag 1995). However, a direct test of whether this chaperone activity is biologically relevant has not been possible. In yeast mitochondria, a putative RNA helicase, *MSS116*, appears to function in splicing both group I and group II introns, perhaps by acting as an RNA chaperone (see below).

Other Splicing Factors for Group I Introns

Several other yeast nuclear genes that influence the splicing of mt group I introns have been identified genetically but not studied in detail. A potentially interesting example, where multiple proteins may function in splicing the same intron, involves the yeast mtDNA intron aI5β, another intron of the *cox1* gene. The splicing of this intron is inhibited by mutations in four different nuclear genes, *PET54*, *MSS18*, *MRS1/PET157*, and *suv3* (Johnson and McEwen 1997). The best-characterized, *PET54*, encodes a translation factor that interacts with the 5′ end of *cox3* pre-mRNA and may recognize a similar sequence in the intron (Costanzo and Fox 1988; Valencik et al. 1989). Mss18p and Mrs1p are not homologous to known proteins. The former appears to have an additional nonsplicing function (Séraphin et al. 1988), whereas the latter may be required only for splicing aI5β and a second intron, bI3, since mutations have no phenotype in strains lacking these introns (Kreike et al. 1986; Bousquet et al. 1990). The remaining gene, *suv3*, encodes a putative RNA helicase involved in group I intron RNA turnover (see below). *suv3* mutations result in the accumulation of excised group I intron RNAs and may thus affect the splicing of aI5β indirectly by preventing the recycling of splicing factors tightly bound to the excised intron.

A recent example of regulated group I intron splicing has been found in *C. reinhardtii* chloroplasts (Deshpande et al. 1997). The *psbA* gene, which encodes an essential photosystem II protein, contains four group I introns, which self-splice only under nonphysiological conditions in vitro. Unspliced transcripts accumulate in dark-grown cells, and the splicing of all four introns is rapidly stimulated by light, suggesting that the splicing factors are regulated by the cellular machinery for light adaptation.

GROUP II INTRONS

Group II introns have been found in the genomes of fungal and plant mitochondria, chloroplasts, and eubacteria (Michel and Ferat 1995; Chapter 13). Those in fungi and bacteria generally have the canonical secondary structure, consisting of six conserved domains (denoted I–VI), whereas some of the plant introns lack substructures, and sometimes entire domains. The most degenerate group II introns, the so-called group III introns of *Euglena* ctDNA, have a length of only 91–119 nucleotides and contain little more than recognizable domains V and VI, and sometimes a domain I-like structure (Copertino and Hallick 1993). In another evolutionary variation, some plant group II introns have been fragmented by DNA recombination and consist of two or more separately encoded segments that are *trans*-spliced (Bonen 1993). Fragmentation of the introns into segments that reassociate to form the catalytic core may parallel the evolution of snRNAs in spliceosomal introns (Guthrie 1991; Sharp 1991).

Relatively few group II introns self-splice, and those that do generally require nonphysiological conditions, suggesting dependence on proteins in vivo. The best-characterized group II intron splicing factors are the intron-encoded reverse transcriptases (RTs) that also function in intron mobility. Several putative host-encoded splicing factors have been identified genetically, but have not yet been studied biochemically.

GROUP II INTRON-ENCODED PROTEINS

Mobile group II introns, which encode RTs, use an efficient retrotransposition mechanism for homing to intronless alleles (Zimmerly et al. 1995a, b; Curcio and Belfort 1996) and also transpose to ectopic sites at low frequency (Mueller et al. 1993; Sellem et al. 1993) (see below). The best-studied mobile group II introns are the yeast mtDNA introns aI1 and aI2, found in the *cox1* gene, and the *Lactococcus lactis ltrB* (Ll.LtrB) intron, found in a putative relaxase gene of a conjugative element (Fig. 4). All three introns contain lengthy ORFs that encode proteins shown biochemically to have RT activity (Kennell et al. 1993; Matsuura et al. 1997). Although the conserved regions of the ORF are always encoded in the loop of intron domain IV, the bacterial intron ORF is free-standing, whereas the yeast intron ORFs are translated as extensions of the upstream exons.

The RT domains of group II intron proteins contain conserved amino acid sequences (blocks I to VII) characteristic of retroviral and other RTs, including the highly conserved YXDD motif, which forms part of the RT active site (Fig. 4) (Michel and Lang 1985; Kohlstaedt et al. 1992). Phylogenetic analyses indicate that the group II intron-encoded RTs belong to the same subclass as those of non-LTR retrotransposons, a diverse group that includes the abundant human LINE elements (Eickbush 1994). The group II intron RTs are also related both by sequence and functional characteristics to the recently discovered telomerase RT (Eickbush 1997; Lingner et al. 1997) and may be evolutionary progenitors of telomerase (Zimmerly et al. 1995a; see below).

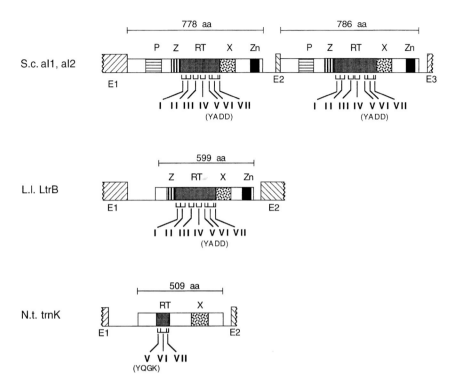

Figure 4 Group II intron-encoded proteins. Schematics of the *S. cerevisiae* (S.c.) mtDNA introns aI1 and aI2, which encode RTs that are translated in-frame with the upstream exon (Bonitz et al. 1980); the *L. lactis* (L.l.) *ltrB* intron, which encodes an RT translated from a free-standing ORF within the intron (Mills et al. 1996; Shearman et al. 1996); and the *Nicotiana tabacum* ctDNA tRNA^{Lys} (*trnK*) intron, which encodes a matK protein (Sugita et al. 1985). Exons (E) are shown as hatched boxes, and ORFs are shown as rectangles within the intron. Protein domains demarcated within the ORF are (P) retroviral protease-like domain; (Z) conserved region found at the amino terminus of the RT domain in non-LTR retroelements; (RT) reverse transcriptase domain; (X) putative RNA-binding domain involved in maturase activity; (Zn) Zn^{++}-finger-like domain. Roman numerals below the RT domain indicate conserved amino acid sequence blocks characteristic of RTs (Michel and Lang 1985; Xiong and Eickbush 1990). The highly conserved YXDD motif, which is part of conserved sequence block V, forms part of the RT active site. The matK protein contains a degenerate RT domain in which only sequence blocks V, VI, and VII are recognizable and the YADD motif has diverged to YQGK (Mohr et al. 1993).

In addition to the RT domain, group II intron-encoded proteins contain other conserved regions associated with additional activities (Fig. 4). Domain Z is a conserved region found at the amino terminus of the RT domain in non-LTR retroelements (McClure 1991). Domain X is a putative RNA-binding domain associated with maturase activity (Mohr et al. 1993; see below). The Zn domain possesses DNA endonuclease activity (Zimmerly et al. 1995a) and contains a Zn-finger-like motif interspersed with H-N-H DNA endonuclease motifs similar to those in one type of group I intron endonuclease (Gorbalenya 1994; Shub et al. 1994). The aI1 and aI2 proteins, which undergo proteolytic processing, contain an additional amino-terminal domain (P) that weakly resembles a retroviral protease domain (McClure 1991).

In addition to their role in intron mobility, the group II intron-encoded RTs also function in RNA splicing. The splicing function of the yeast aI1- and aI2-encoded proteins was established genetically by showing that mutations in the intron ORF result in splicing defects, which could be complemented by the respective intron-encoded protein in *trans* (Carignani et al. 1983; Moran et al. 1994). The aI1 and aI2 maturases are required only for splicing the intron that encodes them, and, like group I intron maturases, presumably recognize unique structural features of their own unspliced precursor RNA. Recently, the *Lactococcus* Ll.LtrB intron was expressed in *E. coli*, and the splicing activity of the intron-encoded protein (LtrA) was demonstrated both genetically and biochemically (Matsuura et al. 1997). The protein-assisted splicing reaction does not require ATP, and splicing appears to be accomplished by the protein promoting or stabilizing formation of the catalytically active RNA structure, analogous to the mechanism of group I intron splicing factors.

As for group I introns, it seems likely that the group II intron-encoded proteins functioned initially in intron mobility, and that the splicing function evolved subsequently (Kennell et al. 1993). If so, similar regions of the protein may be used to bind the intron RNA both as a template for reverse transcription and as a substrate for RNA splicing. The HIV RT consists of "finger," "palm," and "thumb" subdomains, with the RT active site in the palm (Kohlstaedt et al. 1992; Jacobo-Molina et al. 1993). It is thought that the RNA template interacts primarily with the finger subdomain, whereas the template–primer complex interacts with the thumb, and the remainder of the primer interacts with the other RT subunit. Assuming group II intron RTs are structurally analogous to HIV RT, domain Z and the amino-terminal region of the RT domain would comprise the fingers, and the remainder of the RT domain would comprise the palm. The HIV RT thumb has no sequence homolog in group II intron proteins, but its location coincides with domain X. Thus, the RT domain and domain X are expected to bind group II intron RNAs for reverse transcription, and the prediction is that one or both of these regions may also bind group II intron RNAs for splicing.

A splicing function for domain X is supported by genetic analysis, which shows that missense mutations in this region of the yeast aI2 and lactococcal LtrA proteins strongly inhibit splicing activity in vivo (Moran et al. 1994; Matsuura et al. 1997). In contrast, the Zn domain, which contributes to DNA cleavage during retrotransposition, can be deleted from the aI2 protein without affecting splicing in vivo (Zimmerly et al. 1995b). The catalytic proficiency of the RT domain is also not required, since missense mutations in the palm region, including mutations of the conserved YADD motif, abolish RT activity but do not inhibit splicing (Moran et al. 1995; Zimmerly et al. 1995a). Additional mutations must be analyzed to determine if the finger or other regions of the RT domain participate in intron binding.

Like group I introns, only a minority (~25%) of group II introns encode proteins. Virtually all of the group II intron-encoded proteins are related, but many appear to be degenerate proteins that lack some of the conserved domains (Mohr et al. 1993). The most degenerate are the matK proteins, encoded in the introns of ctDNA tRNA^{Lys} genes (Fig. 4). These proteins contain a conserved domain X, but have only remnants of the RT domain and lack the Zn domain, both of which are required for efficient mobility. The exclusive retention of domain X presumably reflects that it has an essential splicing function, whereas regions required for mobility are dispensable after the intron has inserted (Mohr et al. 1993). Direct evidence for the suspected splicing function remains fragmentary, however. The mustard ctDNA matK protein synthesized in *E. coli* binds with some specificity to its precursor RNA, but this binding does not result in in vitro splicing (Liere and Link 1995). Additionally, conserved sequences in domain X of certain matK, as well as mat-r, proteins encoded by plant mtDNA group II introns are restored by RNA editing, consistent with an essential function for this region (Thomson et al. 1994; Vogel et al. 1997).

Although the yeast aI1 and aI2 proteins are intron-specific splicing factors, several investigators have suggested that maturases in some organisms function in splicing multiple group II introns (Ems et al. 1995; L. Zhang et al. 1995). In one case, a maturase-like protein encoded by a *Euglena* ctDNA group III intron has been proposed to function as part of a general splicing apparatus for other chloroplast group III introns (Copertino et al. 1994; Doetsch et al. 1998). Although it is plausible that maturases could recognize structural features conserved in many intron

RNAs, the yeast maturases do not splice other group II introns in the same cellular compartment. At present, the only suggestive evidence for a generalized role of group II intron maturases comes from experiments showing that inhibition of chloroplast protein synthesis in barley and maize blocks the splicing of multiple group II introns, including some that lack ORFs (Hess et al. 1994; Hübschmann et al. 1996; Jenkins et al. 1997; Vogel et al. 1997). If true, the adaptation of an intron-specific maturase to function in splicing multiple group II introns might parallel a step in the evolution of a common cellular splicing machinery for spliceosomal introns.

HOST-ENCODED PROTEINS IN GROUP II INTRON SPLICING

Host genes affecting the splicing of group II introns have been identified genetically in several organisms, but in most cases, it is not yet clear whether the genes affect splicing directly. The two candidates that have received the most attention are the yeast nuclear genes *MSS116* and *MRS2*. The *MSS116* gene, which encodes a protein with DEAD motifs characteristic of RNA helicases, was identified in a screen for nuclear mutations that cause respiratory deficiency only in strains whose mtDNA contains introns (Séraphin et al. 1987). The mutant *MSS116* allele caused defective splicing of subsets of group I and group II introns in yeast mitochondria (Séraphin et al. 1989). Disruption of *MSS116* resulted in impaired mitochondrial function, even in strains that lack mtDNA introns, indicating that the protein has an additional nonsplicing function (Séraphin et al. 1989). Supporting the group II intron splicing function, overexpression of *MSS116* stimulated ATP-dependent splicing of bI1 in yeast mitochondrial lysates (Niemer et al. 1995). The involvement of a putative RNA helicase in group II intron splicing is a satisfying parallel to nuclear pre-mRNA splicing, which involves numerous RNA helicases (Staley and Guthrie 1998).

The second putative group II intron splicing factor, *MRS2* (MRS = *m*itochondrial *R*NA *s*plicing), was identified in a screen for multicopy nuclear suppressors of a splicing-defective bI1 intron with a 1-nucleotide deletion in domain III (Koll et al. 1987). Disruption of *MRS2* blocks splicing of all four group II introns in yeast mitochondria, but has at most modest effects on group I intron splicing (Wiesenberger et al. 1992). Like *MSS116*, *MRS2* is inferred to have a second function apart from splicing, since its disruption results in respiratory deficiency even in strains that lack mtDNA introns (Wiesenberger et al. 1992). The amino acid sequence of *MRS2* suggests that it may be a membrane-anchored protein with a soluble amino terminus that could face the mitochondrial matrix to interact with group II introns (U. Schmidt et al., in prep.; R.J. Schweyen, pers. comm.).

Recently, three dominant suppressor alleles of *MRS2* were isolated in a screen for chromosomal mutations that suppress point mutations in domain V of another group II intron, aI5γ (Boulanger et al. 1995; Schmidt et al. 1996 and in prep.). Those suppressors restore much higher levels of splicing than are obtained by overexpressing the wild-type gene. An additional screen for suppressors of an MRS2 disruption identified ten interacting genes, a number of which encode membrane proteins (Waldherr et al. 1993; van Dyck et al. 1995; Jarosch et al. 1996). It remains unclear whether MRS2 and its suppressors affect splicing directly. If so, they could be pointing to a membrane complex involved in group II intron splicing.

Several studies have identified nuclear genes that influence group II intron splicing in chloroplasts. A mutation in the maize nuclear gene *crs1* (*c*hloroplast *R*NA *s*plicing) inhibits splicing of only the chloroplast *atpF* group II intron, whereas a mutation in *crs2* inhibits the splicing of many chloroplast introns (Jenkins et al. 1997). In *Chlamydomonas* chloroplasts, the *psaA* gene, which encodes a photosystem I subunit, contains two *trans*-spliced group II introns, whose splicing is influenced by at least fourteen nuclear genes (Rochaix 1996). These genes fall into three classes, depending on whether they affect splicing of either or both introns. To date, the identities of these genes have not been reported, and it remains possible that they affect splicing indirectly, e.g., by affecting synthesis of a ctDNA component required for splicing.

GROUP II INTRON RNP PARTICLES FUNCTION IN INTRON MOBILITY

Group II intron mobility provides an unprecedented example of a ribozyme RNP particle that carries out a coordinated series of reactions involving both protein and RNA catalytic activity. The major mobility pathway, retrohoming, occurs by a target DNA-primed reverse transcription mechanism (Zimmerly et al. 1995a,b; Yang et al. 1996; Eskes et al. 1997). The process, outlined in Figure 5, is initiated by an intron-encoded DNA endonuclease, which is associated with RNP particles containing the intron-encoded RT and the excised intron RNA. Remarkably, the intron RNA integrates into the sense strand of the recipient DNA by a partial or complete reverse splicing reaction at the intron-insertion site. The intron-encoded protein then cleaves the antisense strand at position +9 or +10 of the 3′ exon, with the position differing for the bacterial and yeast intron endonucleases. After DNA cleavage, the 3′ end of the cleaved antisense strand is used as a primer for reverse transcription of the reverse spliced

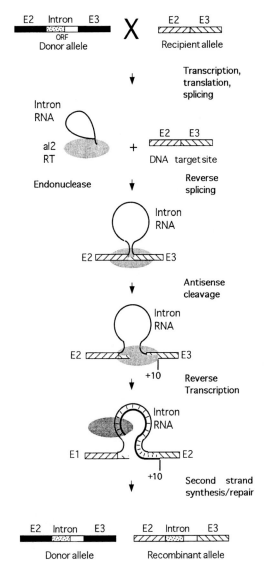

Figure 5 Group II intron mobility mechanism. Mobile group II introns insert site-specifically into intronless alleles by a process termed retrohoming (Curcio and Belfort 1996). Retrohoming is initiated by the intron-encoded DNA endonuclease, which is an RNP complex containing the intron-encoded protein and the excised intron lariat RNA. The endonuclease cleaves the recipient DNA in two steps. First, the intron RNA cleaves the sense strand of the DNA at the intron insertion site by a reverse-splicing reaction that leads to insertion of linear intron RNA between the two DNA exons. Then, the intron-encoded protein cleaves the antisense strand after positions +9 or +10 of the 3′ exon depending on the

intron RNA. In an alternative mechanism (not shown), the unspliced precursor RNA containing the intron may be used as the template (see Fig. 5). The cDNA copy of the intron may be inserted into genomic DNA by any of several recombination or repair pathways (Eskes et al. 1997). The infrequent ectopic transposition of group II introns could occur similarly via reverse splicing of the same group II intron RNP particle into a new DNA site (Yang et al. 1996, 1998) or, as proposed for group I introns, by reverse splicing into a new RNA site, followed by reverse transcription of the recombined RNA and integration of the resulting cDNA into the genome (Mueller et al. 1993; Sellem et al. 1993).

The splicing and mobility reactions of group II introns are intimately related. In the retrohoming pathway, the intron-encoded protein promotes splicing and remains associated with the excised intron RNA to constitute the DNA endonuclease that initiates mobility. The RNA and protein components of the endonuclease then function as an RNP particle. The intron-encoded protein stabilizes the active structure of the intron RNA, as expected for its maturase activity, and also contributes to DNA binding, enabling the intron RNA to reverse-splice efficiently into double-stranded DNA substrates (Guo et al. 1997). Without the protein, the intron reverse-splices into RNA substrates, but is unable to utilize double-stranded DNA substrates (Zimmerly et al. 1995b; Guo et al. 1997). The protein, whose active structure is stabilized by the intron RNA, also cleaves the antisense strand and reverse transcribes the intron RNA. Thus, the RNA and protein components of the RNP particle have coevolved to carry out a coordinated series of reactions leading to intron mobility.

Recent studies with the yeast aI2 endonuclease show that both the RNA and protein components contribute to recognition of the DNA target site (Guo et al. 1997). The DNA target for aI2 spans 31 bp, with the reverse splicing reaction requiring positions –21 in the 5' exon through +1 in the 3' exon, and antisense-strand cleavage requiring additional sequences extending to +10 in the 3' exon (positions are numbered from the intron-insertion site). A model for target-site recognition by the aI2 en-

---

endonuclease. After the double-strand break, the 3' end of the cleaved antisense strand is used as a primer for reverse transcription of the inserted intron RNA. Integration is accomplished by recombination or repair mechanisms using either a complete cDNA copy of the intron or a partial cDNA copy, which initiates DSBR by invading an intron-containing allele. Other variations of the retrohoming mechanism may involve sense-strand cleavage by partial reverse splicing and use of the intron in unspliced precursor RNA as the template for reverse transcription (for details, see Zimmerly et al. 1995a,b; Eskes et al. 1997).

donuclease is shown in Figure 6. The protein first recognizes a few key nucleotides in the distal 5′ exon region (positions E2-21 to −1) of the double-stranded DNA. This interaction leads to DNA unwinding, allowing the intron RNA to base-pair to a 13-nucleotide region (positions E2-12 through E3+1) that includes short sequence elements denoted IBS1, IBS2, and δ′. These base-pair with complementary intron sequences denoted EBS1, EBS2, and δ for the reverse splicing reaction, the same interactions involved in forward and reverse splicing of group II introns with RNA substrates (Fig. 6b) (Michel and Ferat 1995). Finally, protein recognition of target-site positions between +1 and +10 is required for antisense-strand cleavage.

The finding that a 13-nucleotide region of the DNA target site is recognized primarily by base-pairing with the intron RNA suggests that it might be possible to design group II intron endonucleases to target desired 13-nucleotide DNA sequences for site-specific cleavage and insertion, taking into account the additional requirements of the protein recognition region. Since additional genetic markers can be inserted into group II introns without compromising the mobility reactions (Matsuura et al. 1997), this approach might be used to introduce new genetic information at preselected sites in double-stranded DNA genomes.

TURNOVER OF EXCISED GROUP I AND GROUP II INTRON RNAs

A growing body of information suggests that cells have evolved mechanisms to degrade excised group I and group II intron RNAs. The degradation of the excised intron RNA is presumably necessary to recycle tightly bound splicing factors, such as CYT-18 and Cbp2p (see above). In addition, the excised intron RNAs can have deleterious effects. Both group I and group II intron RNAs can cleave short RNA target sites by exon-reopening reactions (Zaug et al. 1986; Jarrell et al. 1988), and the *Tetrahymena* LSU intron overexpressed in *E. coli* appears to impede protein synthesis directly by base-pairing to a target site in the LSU rRNA (Nikolcheva and Woodson 1997).

In fungal mitochondria, most excised group I intron RNAs are degraded efficiently (see, e.g., Conrad-Webb et al. 1990). The yeast mt enzyme responsible for that degradation was purified as a 3′ to 5′ exoribonuclease (mtEXO) that has 75-, 90-, and 110-kD subunits (Min et al. 1993). The 90-kD subunit is encoded by the nuclear gene *suv3* and is related to DExH box NTP-dependent RNA helicases (Margossian et al. 1996). *suv3* mutations lead to accumulation of excised group I intron, as expected, but also affect the metabolism of other mt RNAs, suggesting additional functions (Conrad-Webb et al. 1990).

Mutants of the *N. crassa cyt-4* gene have complex mt RNA phenotypes resembling those of the *suv3* mutants (Garriga et al. 1984; Dobinson et al. 1989). The *cyt-4* gene encodes a 120-kD protein resembling *E. coli* RNase II, which is involved in RNA turnover (Turcq et al. 1992; Saldanha

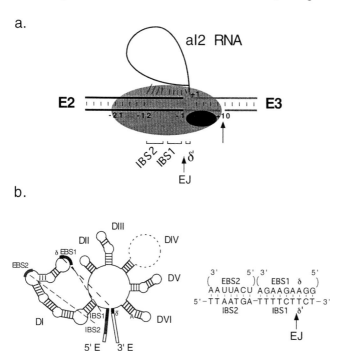

Figure 6 DNA target-site recognition by group II endonucleases. (*a*) Model showing how the yeast aI2 intron uses both its RNA and protein components to recognize its DNA target site. The target site extends from position –21 in the 5′ exon (E2-21) to +10 in the 3′ exon (E3+10), with positions E2-21 to E3+1 required for reverse splicing, and additional interactions extending to E3+10 required only for antisense-strand cleavage. The protein recognizes the upstream region of the DNA target site (E2-21 to -11) and unwinds the double-stranded DNA allowing the intron to base-pair to positions E2-12 to E3+1. The latter region includes short sequence elements IBS1, IBS2, and δ′ that base-pair to EBS1, EBS2, and δ sequences in the intron RNA. The conserved region of the Zn domain interacts with sequences between E3+1 and E3+10 for antisense-strand cleavage (Guo et al. 1997). The arrow indicates the antisense-strand cleavage site between positions +10 and +11. (*b*) Group II intron secondary-structure model showing the location of the EBS1, EBS2, and δ sequences in domain I of the intron RNA and their base-pairing with IBS1, IBS2, and δ′ sequences in the 5′ and 3′ exons. Only pertinent regions of the domain I RNA structure are shown. To the bottom right are shown the base-pairing interactions of the yeast aI2 intron with IBS1, IBS2, and δ′ sequences in the DNA target site (Guo et al. 1997). EJ indicates the exon junction.

et al. 1995). *DSS1*, a recently identified yeast homolog of CYT-4, appears to encode the 110-kD subunit of mtEXO (Dmochowska et al. 1995). Thus, the mtEXO complex may use an NTP-dependent helicase activity of Suv3p to facilitate degradation by the 3′ to 5′ exonuclease activity of Dss1p/CYT-4 (Margossian et al. 1996). The Dss1p/CYT-4 subunit also resembles fungal proteins involved in cell-cycle regulation via protein phosphatase activity and may be regulated by phosphorylation (Turcq et al. 1992; Uesono et al. 1997).

In contrast to group I introns, relatively little is known about the turnover of excised group II introns. In yeast mitochondria and *Euglena* chloroplasts, excised group II introns accumulate detectably as lariat RNAs (see, e.g., Hensgens et al. 1983; Copertino et al. 1994). The levels of group II lariats are not altered in *suv3* mutants, implying a different turnover pathway than that used by group I introns (Conrad-Webb et al. 1990). The lactococcal group II intron expressed in *E. coli* is cleaved efficiently by *E. coli* endonucleases (Matsuura et al. 1997). The turnover of excised group II introns may be more pressing in bacteria, which have larger genomes that are more vulnerable to the DNA endonuclease activity of the excised intron RNPs.

EVOLUTION OF PROTEIN-ASSISTED SPLICING REACTIONS

It is not known when group I and group II introns first encountered proteins. Although it is often suggested that group I introns are remnants of the RNA World, the only evidence that group I introns are ancient is the finding that a group I intron in the tRNA^{Leu} gene of cyanobacteria is conserved at the same location in chloroplast tRNA^{Leu} genes (Kuhsel et al. 1990; Xu et al. 1990; Paquin et al. 1997). Even at face value, however, this finding indicates only that group I introns existed during the evolution of prokaryotes, prior to the origin of chloroplasts as endosymbionts. This time period is also consistent with the possibility that group I introns evolved from structurally similar tRNAs, which had sufficient complexity to develop an active site (see above; Caprara et al. 1996b).

Group II introns are present in eubacteria and in mitochondria and chloroplasts, which evolved from eubacteria, but they have not been found in archaebacteria, based on the complete genome sequences available thus far. Although intron loss is always possible, this phylogenetic distribution suggests that group II introns evolved in eubacteria after their divergence from archaebacteria. A eubacterial origin of group II introns is consistent with the hypothesis that mobile group II introns entered eukaryotic cells with endosymbiotic eubacteria that became organelles, and then migrated to the nucleus where they evolved into spliceosomal introns (Cavalier-Smith 1991; Palmer and Logsdon 1991) and possibly non-LTR retrotransposons and telomerase (Zimmerly et al. 1995a). The group II intron ribozyme activity seems uniquely adapted for insertion into DNA, consistent with the possibility that group II introns evolved as DNA parasites (Griffin et al. 1995).

If group I and group II introns originated in the RNA World, then they may have undergone substantial evolution, possibly as self-replicating entities, before interacting with proteins. However, if they evolved in eubacteria, then proteins were present from their beginnings. In either case, dispersal in the DNA World, where competing processes lead to intron removal (see, e.g., Hill et al. 1985; Levra-Juillet et al. 1989), likely required the acquisition of proteins to promote efficient mobility. Group I introns acquired the coding sequences for several different types of DNA endonucleases, and at least one type adapted to function in splicing. Group II introns may have likewise acquired RTs (Lambowitz and Belfort 1993) or alternatively evolved from retrotransposable elements that acquired self-splicing capability (Curcio and Belfort 1996). The RT appears to have adapted secondarily to function in splicing, but unlike group I introns, the splicing function and mobility mechanism are intimately related. For both group I and group II introns, the acquisition of a splicing function by the intron-encoded proteins may help ensure that mobile introns retain splicing capability despite the larger size imposed by the intron ORF and the increased frequency of mutations that may result from mobility processes. The intron-encoded splicing factor could also facilitate the transmission of the introns to different hosts.

In the DNA World, the introns also encountered host-encoded proteins. The diversity of cellular proteins involved in group I intron splicing suggests that the introns were dispersed relatively recently in evolution, with different proteins adapting to function in splicing in different organisms (Lambowitz and Perlman 1990). This scenario is consistent with a post-RNA World origin of self-splicing group I introns in a single species of prokaryotes. Selective pressure for the involvement of proteins may have been the need to regulate the ribozyme activities of the intron RNAs, which otherwise would be deleterious to the host cell (Nikolcheva and Woodson 1997). Once established, protein-dependent splicing reactions could be adapted for regulating gene expression; for example, the light-regulated expression of chloroplast group I introns in *C. reinhardtii* (Deshpande et al. 1997).

Genes for putative group II intron splicing factors, such as *MRS2* and *MSS116*, have additional cellular functions, again suggesting adaptation of preexisting host proteins. If group II introns were dispersed relatively recently, the expectation is that different proteins will have adapted for splicing in different organisms, as has been found for group I introns. In contrast, if the protein-dependent splicing of group II introns was fixed early, homologous proteins might be involved in splicing the introns in different organisms. Information about the diversity of group II intron splicing factors is insufficient to make a reasonable prediction at this point. The question is fairly pressing, since if group II intron–protein interactions were fixed long ago, we might expect to find similar proteins involved with spliceosomal introns.

The adaptation of preexisting cellular proteins to function in splicing group I and II introns may be a guide to the evolution of other types of introns, including spliceosomal introns. Principles that may be generally relevant include the recruitment of proteins to compensate for defects in RNA function, the adaptation of proteins with related cellular functions, the incorporation of protein-dependent splicing reactions into cellular regulatory processes, and the development of intron turnover mechanisms to recycle splicing factors and minimize reverse splicing.

Finally, the example of group II intron mobility shows how the association with protein cofactors can expand the catalytic repertoire of an RNA, allowing it access to new substrates, in this case, double-stranded DNA. The coevolution of protein and RNA thus produced a different reaction than either component could achieve independently. Pointing toward the future, group II introns might be a paradigm for the evolution of other complex RNP particles, such as telomerase, ribosomes, and spliceosomes, in which coevolution has produced an RNP machinery in which RNA and proteins share catalysis.

ACKNOWLEDGMENTS

We thank Drs. Marlene Belfort (New York State Department of Health, Albany, NY) and Ellen Gottlieb (University of Texas at Austin) for comments on the manuscript. Research in the authors' laboratories was supported by National Institutes of Health grants GM-37949 and GM-37951 to A.M.L. and GM-31480 to P.S.P., and Robert A. Welch Foundation grant I-1211 to P.S.P.

REFERENCES

Akins R.A. and Lambowitz A.M. 1987. A protein required for splicing group I introns in *Neurospora* mitochondria is a mitochondrial tyrosyl-tRNA synthetase or a derivative thereof. *Cell* **50:** 331–345.

Bedoulelle H. 1990. Recognition of tRNA^{Tyr} by tyrosyl-tRNA synthetase. *Biochimie* **72:** 589–598.
Belfort M. 1990. Phage T4 introns: Self-splicing and mobility. *Annu. Rev. Genet.* **24:** 363–385.
Belfort M. and Roberts R.J. 1997. Homing endonucleases: Keeping the house in order. *Nucleic Acids Res.* **25:** 3379–3388.
Bonen L. 1993. *Trans*-splicing of pre-mRNA in plants, animals, and protists. *FASEB J.* **7:** 40–46.
Bonitz S.G., Coruzzi G., Thalenfeld B.E., and Tzagoloff A. 1980. Assembly of the mitochondrial membrane system. Structure and nucleotide sequence of the gene coding for subunit 1 of yeast cytochrome oxidase. *J. Biol. Chem.* **255:** 11927–11941.
Boulanger S.C., Belcher S.M., Schmidt U., Dib-Hajj S.D., Schmidt T., and Perlman P.S. 1995. Studies of point mutants define three essential paired nucleotides in the domain 5 substructure of a group II intron. *Mol. Cell. Biol.* **15:** 4479–4488.
Bousquet I., Dujardin G., Poyton R.O., and Slonimski P.P. 1990. Two group I mitochondrial introns in the *cob-box* and *coxI* genes require the same *MRS1/PET157* nuclear gene product for splicing. *Curr. Genet.* **18:** 117–124.
Caprara M.G., Mohr G., and Lambowitz A.M. 1996a. A tyrosyl-tRNA synthetase protein induces tertiary folding of the group I catalytic core. *J. Mol. Biol.* **257:** 512–531.
Caprara M.G., Lehnert V., Lambowitz A.M., and Westhof E. 1996b. A tyrosyl-tRNA synthetase recognizes a conserved tRNA-like structural motif in the group I intron catalytic core. *Cell* **87:** 1135–1145.
Carignani G., Groudinsky O., Frezza D., Schiavon E., Bergantino E., and Slonimski P.P. 1983. An mRNA maturase is encoded by the first intron of the mitochondrial gene for the subunit I of cytochrome oxidase in *S. cerevisiae*. *Cell* **35:** 733–742.
Cate J.H., Gooding A.R., Podell E., Zhou K., Golden B.L., Kundrot C.E., Cech T.R., and Doudna J.A. 1996. Crystal structure of a group I ribozyme domain: Principles of RNA folding. *Science* **273:** 1678–1685.
Cavalier-Smith T. 1991. Intron phylogeny: A new hypothesis. *Trends Genet.* **7:** 145–148.
Cech T.R. 1985. Self-splicing RNA: Implications for evolution. *Int. Rev. Cytol.* **93:** 3–22.
Cherniack A.D. 1991. "Involvement of mitochondrial tyrosyl-tRNA synthetase in splicing: Identification of an N-terminal domain that functions in splicing." Ph.D. thesis, The Ohio State University, Columbus, Ohio.
Cherniack A.D., Garriga G., Kittle J.D., Jr., Akins R.A., and Lambowitz A.M. 1990. Function of *Neurospora* mitochondrial tyrosyl-tRNA synthetase in RNA splicing requires an idiosyncratic domain not found in other synthetases. *Cell* **62:** 745–755.
Chu F.K., Maley G.F., and Belfort M. 1984. An intervening sequence in the thymidylate synthase gene of bacteriophage T4. *Proc. Natl. Acad. Sci.* **81:** 3049–3053.
Coetzee T., Herschlag D., and Belfort M. 1994. *Escherichia coli* proteins, including ribosomal protein S12, facilitate in vitro splicing of phage T4 introns by acting as RNA chaperones. *Genes Dev.* **8:** 1575–1588.
Collins R.A. and Lambowitz A.M. 1985. RNA splicing in *Neurospora* mitochondria. Defective splicing of mitochondrial mRNA precursors in the nuclear mutant *cyt-18-1*. *J. Mol. Biol.* **184:** 413–428.
Conrad-Webb H., Perlman P.S., Zhu H., and Butow R.A. 1990. The nuclear *SUV3-1* mutation affects a variety of post-transcriptional processes in yeast mitochondria. *Nucleic Acids Res.* **18:** 1369–1376.

Copertino D.W. and Hallick R.B. 1993. Group II and group III introns of twintrons: Potential relationships with nuclear pre-mRNA introns. *Trends Biochem. Sci.* **18:** 467–471.

Copertino D.W., Hall E.T., Van Hook F.W., Jenkins K.P., and Hallick R.B. 1994. A group III twintron encoding a maturase-like gene excises through lariat intermediates. *Nucleic Acids Res.* **22:** 1029–1036.

Costanzo M.C. and Fox T.D. 1988. Specific translational activation by nuclear gene products occurs in the 5' untranslated leader of a yeast mitochondrial mRNA. *Proc. Natl. Acad. Sci.* **85:** 2677–2681.

Curcio M.J. and Belfort M. 1996. Retrohoming: cDNA-mediated mobility of group II introns requires a catalytic RNA. *Cell* **84:** 9–12.

Dalgaard J.Z., Klar A.J., Moser M.J., Holley W.R., Chatterjee A., and Mian I.S. 1997. Statistical modeling and analysis of the LAGLIDADG family of site-specific endonucleases and identification of an intein that encodes a site-specific endonuclease of the HNH family. *Nucleic Acids Res.* **25:** 4626–4638.

Delahodde A., Goguel V., Becam A.M., Cruesot F., Banroques J., and Jacq C. 1989. Site-specific DNA endonuclease and RNA maturase activities of two homologous intron-encoded proteins from yeast mitochondria. *Cell* **56:** 431–441.

Deshpande N.N., Bao Y., and Herrin D.L. 1997. Evidence for light/redox-regulated splicing of *psbA* pre-RNAs in *Chlamydomonas* chloroplasts. *RNA* **3:** 37–48.

Dmochowska A., Golik P., and Stepien P.P. 1995. The novel nuclear gene *DSS-1* of *Saccharomyces cerevisiae* is necessary for mitochondrial biogenesis. *Curr. Genet.* **28:** 108–112.

Dobinson K.F., Henderson M., Kelley R.L., Collins R.A., and Lambowitz A.M. 1989. Mutations in nuclear gene *cyt-4* of *Neurospora crassa* result in pleiotropic defects in processing and splicing of mitochondrial RNAs. *Genetics* **123:** 97–108.

Doetsch N.A., Thompson M.D., and Hallick R.B. 1998. A maturase-encoding group III twintron is conserved in deeply rooted Euglenoid species: Are group III introns the chicken or the egg? *Mol. Biol. Evol.* **15:**76–86.

Duan X., Gimble F.S., and Quiocho F.A. 1997. Crystal structure of PI-SceI, a homing endonuclease with protein splicing activity. *Cell* **89:** 555–564.

Dujardin G., Jacq C., and Slonimski P.P. 1982. Single base substitution in an intron of oxidase gene compensates splicing defects of the cytochrome *b* gene. *Nature* **298:** 628–632.

Dujardin G., Labouesse M., Netter P., and Slonimski P.P. 1983. Genetic and biochemical studies of the nuclear suppressor *NAM2*: Extraneous activation of a latent pleiotropic maturase. In *Mitochondria 1983: Nucleo-mitochondrial interactions* (ed. R.J. Schweyen et al.), pp. 233–250. Walter de Gruyter, Berlin.

Dujardin G., Pavot P., Groudinsky O., and Slonimski P.P. 1980. Long range control circuit within mitochondria and between nucleus and mitochondria. I. Methodology and phenomenology of suppressors. *Mol. Gen. Genet.* **179:** 469–482.

Dujon B., Belfort M., Butow R.A., Jacq C., Lemieux C., Perlman P.S., and Vogt V.M. 1989. Mobile introns: Definition of terms and recommended nomenclature. *Gene* **82:** 115–118.

Eddy S.R. and Gold L. 1991. The phage T4 *nrdB* intron: A deletion mutant of a version found in the wild. *Genes Dev.* **5:** 1032–1041.

Eickbush T.H. 1994. Origin and evolutionary relationships of retroelements. In *The evolutionary biology of viruses* (ed. S.S. Morse), pp. 121–157. Raven Press, New York.

———. 1997. Telomerase and retrotransposons: Which came first? *Science* **277:** 911–912.
Ems S.C., Morden C.W., Dixon C.K., Wolfe K.H., dePamphilis C.W., and Palmer J.D. 1995. Transcription, splicing and editing of plastid RNAs in the nonphotosynthetic plant *Epifagus virginiana*. *Plant Mol. Biol.* **29:** 721–733.
Eskes R., Yang J., Lambowitz A.M., and Perlman P.S. 1997. Mobility of yeast mitochondrial group II introns: Engineering a new site specificity and retrohoming via full reverse splicing. *Cell* **88:** 865–874.
Gampel A., Nishikimi M., and Tzagoloff A. 1989. CBP2 protein promotes in vitro excision of a yeast mitochondrial group I intron. *Mol. Cell. Biol.* **9:** 5424–5433.
Garriga G. and Lambowitz A.M. 1986. Protein-dependent splicing of a group I intron in ribonucleoprotein particles and soluble fractions. *Cell* **46:** 669–680.
Garriga G., Bertrand H., and Lambowitz A.M. 1984. RNA splicing in *Neurospora* mitochondria: Nuclear mutants defective in both splicing and 3′ end synthesis of the large rRNA. *Cell* **36:** 623–634.
Goodrich-Blair H., Scarlato V., Gott J.M., Xu M.-Q., and Shub D.A. 1990. A self-splicing group I intron in the DNA polymerase gene of *Bacillus subtilis* bacteriophage SPO1. *Cell* **63:** 417–424.
Gorbalenya A.E. 1994. Self-splicing group I and group II introns encode homologous (putative) DNA endonucleases of a new family. *Protein Sci.* **3:** 1117–1120.
Griffin E.A., Qin Z.F., Michels W.J., and Pyle A.M. 1995. Group II intron ribozymes that cleave DNA and RNA linkages with similar efficiency, and lack contacts with substrate 2′-hydroxyl groups. *Chem. Biol.* **2:** 761–770.
Groudinsky O., Dujardin G., and Slonimski P.P. 1981. Long range control circuits within mitochondria and between nucleus and mitochondria. II. Genetic and biochemical analyses of suppressors which selectively alleviate the mitochondrial intron mutations. *Mol. Gen. Genet.* **184:** 493–503.
Guo H., Zimmerly S., Perlman P.S., and Lambowitz A.M. 1997. Group II intron endonucleases use both RNA and protein subunits for recognition of specific sequences in double-stranded DNA. *EMBO J.* **16:** 6835–6848.
Guo Q. and Lambowitz A.M. 1992. A tyrosyl-tRNA synthetase binds specifically to the group I intron catalytic core. *Genes Dev.* **6:** 1357–1372.
Guo W.W., Moran J.V., Hoffman P.W., Henke R.M., Butow R.A., and Perlman P.S. 1995. The mobile group I intron 3α of the yeast mitochondrial *COXI* gene encodes a 35-kDa processed protein that is an endonuclease but not a maturase. *J. Biol. Chem.* **270:** 15563–15570.
Guthrie C. 1991. Messenger RNA splicing in yeast: Clues to why the spliceosome is a ribonucleoprotein. *Science* **253:** 157–163.
Heath P.J., Stephens K.M., Monnat R.J., and Stoddard B.L. 1997. The structure of I-*Cre*I, a group I intron-encoded homing endonuclease. *Nat. Struct. Biol.* **4:** 468–476.
Henke R.M., Butow R.A., and Perlman P.S. 1995. Maturase and endonuclease functions depend on separate conserved domains of the bifunctional protein encoded by the group I intron aI4α of yeast mitochondrial DNA. *EMBO J.* **14:** 5094–5099.
Hensgens L.A., Arnberg A.C., Roosendaal E., van der Horst G., van der Veen R., van Ommen G.J., and Grivell L.A. 1983. Variation, transcription and circular RNAs of the mitochondrial gene for subunit I of cytochrome c oxidase. *J. Mol. Biol.* **164:** 35–58.
Herbert C.J., Labouesse M., Dujardin G., and Slonimski P.P. 1988. The *NAM2* proteins from *S. cerevisiae* and *S. douglasii* are mitochondrial leucyl-tRNA synthetases, and are involved in mRNA splicing. *EMBO J.* **7:** 473–483.

Herschlag D. 1995. RNA chaperones and the RNA folding problem. *J. Biol. Chem.* **270:** 20871–20874.

Hess W.R., Hoch B., Zeltz P., Hübschmann T., Kössel H., and Börner T. 1994. Inefficient *rpl2* splicing in barley mutants with ribosome-deficient plastids. *Plant Cell* **6:** 1455–1465.

Hill J., McGraw P., and Tzagoloff A. 1985. A mutation in yeast mitochondrial DNA results in a precise excision of the terminal intron in the cytochrome *b* gene. *J. Biol. Chem.* **260:** 3235–3238.

Ho Y., Kim S.-J., and Waring R.B. 1997. A protein encoded by a group I intron in *Aspergillus nidulans* directly assists RNA splicing and is a DNA endonuclease. *Proc. Natl. Acad. Sci.* **94:** 8994–8999.

Hübschmann T., Hess W.R., and Börner T. 1996. Impaired splicing of the *rps12* transcript in ribosome-deficient plastids. *Plant Mol. Biol.* **30:** 109–123.

Jacobo-Molina A., Ding J., Nanni R.G., Clark A.D., Jr., Lu X., Tantillo C., Williams R.L., Kamer G., Ferris A.L., Clarke P., Hizi A., Hughes S.H., and Arnold E. 1993. Crystal structure of human immunodeficiency virus type 1 reverse transcriptase complexed with double-stranded DNA at 3.0 Å resolution shows bent DNA. *Proc. Natl. Acad. Sci.* **90:** 6320–6324.

Jarosch E., Tuller G., Daum G., Waldherr M., Voskova A., and Schweyen R.J. 1996. Mrs5p, an essential protein of the mitochondrial intermembrane space, affects protein import into yeast mitochondria. *J. Biol. Chem.* **271:** 17219–17225.

Jarrell K.A., Peebles C.L., Dietrich R.C., Romiti S.L., and Perlman P.S. 1988. Group II intron self-splicing: Alternative reaction conditions yield novel products. *J. Biol. Chem.* **263:** 3432–3439.

Jenkins B.D., Kulhanek D.J., and Barkan A. 1997. Nuclear mutations that block group II RNA splicing in maize chloroplast reveal several intron classes with distinct requirements for splicing factors. *Plant Cell* **9:** 283–296.

Johansen S., Einvik C., Elde M., Haugen P., Vader A., and Haugli F. 1997. Group I introns in biotechnology: Prospects of application of ribozymes and rare-cutting homing endonucleases. In *Biotechnology annual review* (ed. M.R. El-Gewely), vol. 3, pp. 111–150. Elsevier, The Netherlands.

Johnson C.H. and McEwen J.E. 1997. Mitochondrial protein synthesis is not required for efficient excision of intron aI5β from *COX1* pre-mRNA in *Saccharomyces cerevisiae*. *Mol. Gen. Genet.* **256:** 88–91.

Kämper U., Kück U., Cherniack A.D., and Lambowitz A.M. 1992. The mitochondrial tyrosyl-tRNA synthetase of *Podospora anserina* is a bifunctional enzyme active in protein synthesis and RNA splicing. *Mol. Cell. Biol.* **12:** 499–511.

Kennell J.C., Moran J.V., Perlman P.S., Butow R.A., and Lambowitz A.M. 1993. Reverse transcriptase activity associated with maturase-encoding group II introns in yeast mitochondria. *Cell* **73:** 133–146.

Kittle J.D., Jr., Mohr G., Gianelos J.A., Wang H., and Lambowitz A.M. 1991. The *Neurospora* mitochondrial tyrosyl-tRNA synthetase is sufficient for group I intron splicing *in vitro* and uses the carboxy-terminal tRNA-binding domain along with other regions. *Genes Dev.* **5:** 1009–1021.

Kohlstaedt L.A., Wang J., Friedman J.M., Rice P.A., and Steitz T.A. 1992. Crystal structure at 3.5 Å resolution of HIV-1 reverse transcriptase complexed with an inhibitor. *Science* **256:** 1783–1790.

Koll H., Schmidt C., Wiesenberger G., and Schmelzer C. 1987. Three nuclear genes suppress a yeast mitochondrial splice defect when present in high copy number. *Curr. Genet.* **12:** 503–509.

Kreike J., Schulze M., Pillar T., Korte A., and Rodel G. 1986. Cloning of a nuclear gene MRS1 involved in the excision of a single group I intron (bI3) from the mitochondrial *COB* transcript in *S. cerevisiae*. *Curr. Genet.* **11:** 185–191.

Kuhsel M.G., Strickland R., and Palmer J.D. 1990. An ancient group I intron shared by eubacteria and chloroplasts. *Science* **250:** 1570–1573.

Labouesse M. 1990. The yeast mitochondrial leucyl-tRNA synthetase is a splicing factor for the excision of several group I introns. *Mol. Gen. Genet.* **224:** 209–221.

Labouesse M., Herbert C.J., Dujardin G., and Slonimski P.P. 1987. Three suppressor mutations which cure a mitochondrial RNA maturase deficiency occur at the same codon in the open reading frame of the nuclear *NAM2* gene. *EMBO J.* **6:** 713–721.

Lambowitz A.M. and Belfort M. 1993. Introns as mobile genetic elements. *Annu. Rev. Biochem.* **62:** 587–622.

Lambowitz A.M. and Perlman P.S. 1990. Involvement of aminoacyl-tRNA synthetases and other proteins in group I and group II intron splicing. *Trends Biochem. Sci.* **15:** 440–444.

Lazowska J., Claude J., and Slonimski P.P. 1980. Sequence of introns and flanking exons in wild-type and *box3* mutants of cytochrome *b* reveals an interlaced splicing protein coded by an intron. *Cell* **22:** 333–348.

Levra-Juillet E., Boulet A., Séraphin B., Simon M., and Faye G. 1989. Mitochondrial introns aI1 and/or aI2 are needed for the in vivo deletion of intervening sequences. *Mol. Gen. Genet.* **217:** 168–171.

Li G.-Y., Becam A.-M., Slonimski P.P., and Herbert C.J. 1996a. In vitro mutagenesis of the mitochondrial leucyl tRNA synthetase of *Saccharomyces cerevisiae* shows that the suppressor activity of the mutant proteins is related to the splicing function of the wild-type protein. *Mol. Gen. Genet.* **252:** 667–675.

Li G.-Y., Herbert C.J., Labouesse M., and Slonimski P.P. 1992. In vitro mutagenesis of the mitochondrial leucyl-tRNA synthetase of *S. cerevisiae* reveals residues critical for its in vivo activities. *Curr. Genet.* **22:** 69–74.

Li G.-Y., Tian G.L., Slonimski P.P., and Herbert C.J. 1996b. The *CBP2* protein from *Saccharomyces douglasii* is a functional homologue of the *Saccharomyces cerevisiae* gene and is essential for respiratory growth in the presence of a wild-type (intron-containing) mitochondrial genome. *Mol. Gen. Genet.* **250:** 316–322.

Liere K. and Link G. 1995. RNA-binding activity of the *matK* protein encoded by the chloroplast trnK intron from mustard (*Sinapis alba* L.). *Nucleic Acids Res.* **23:** 917–921.

Lingner J., Hughes T.R., Shevchenko A., Mann M., Lundblad V., and Cech T.R. 1997. Reverse transcriptase motifs in the catalytic subunit of telomerase. *Science* **276:** 561–567.

Loizos N., Tillier E.R.M., and Belfort M. 1994. Evolution of mobile group I introns: Recognition of intron sequences by an intron-encoded endonuclease. *Proc. Natl. Acad. Sci.* **91:** 11983–11987.

Mannella C.A., Collins R.A., Green M.R., and Lambowitz A.M. 1979. Defective splicing of mitochondrial rRNA in cytochrome-deficient nuclear mutants of *Neurospora crassa*. *Proc. Natl. Acad. Sci.* **76:** 2635–2639.

Margossian S.P., Li H., Zassenhaus H.P., and Butow R.A. 1996. The DExH box protein Suv3p is a component of a yeast mitochondrial 3′- to-5′ exoribonuclease that suppresses group I intron toxicity. *Cell* **84:** 199–209.

Matsuura M., Saldanha R., Ma H., Wank H., Yang J., Mohr G., Cavanagh S., Dunny G.M., Belfort M., and Lambowitz A.M. 1997. A bacterial group II intron encoding reverse transcriptase, maturase, and DNA endonuclease activities: Biochemical demonstration of maturase activity and insertion of new genetic information within the intron. *Genes Dev.* **11:** 2910–2924.

McClure M. A. 1991. Evolution of retrotransposons by acquisition or deletion of retrovirus-like genes. *Mol. Biol. Evol.* **8:** 835–856.

McGraw P. and Tzagoloff A. 1983. Assembly of the mitochondrial membrane system. Characterization of a yeast nuclear gene involved in the processing of the cytochrome *b* pre-mRNA. *J. Biol. Chem.* **258:** 9459–9468.

Michel F. and Ferat J.-L. 1995. Structure and activities of group II introns. *Annu. Rev. Biochem.* **64:** 435–461.

Michel F. and Lang B.F. 1985. Mitochondrial class II introns encode proteins related to the reverse transcriptases of retroviruses. *Nature* **316:** 641–643.

Michel F. and Westhof E. 1990. Modelling of the three-dimensional architecture of group I catalytic introns based on comparative sequence analysis. *J. Mol. Biol.* **216:** 585–610.

Mills D.A., McKay L.L., and Dunny G.M. 1996. Splicing of a group II intron involved in the conjugative transfer of pRSO1 in Lactococci. *J. Bacteriol.* **178:** 3531–3538.

Min J.J., Heuertz R.M., and Zassenhaus H.P. 1993. Isolation and characterization of an NTP-dependent 3′ exoribonuclease from mitochondria of *Saccharomyces cerevisiae*. *J. Biol. Chem.* **268:** 7350–7357.

Mohr G. and Lambowitz A.M. 1991. Integration of a group I intron into a ribosomal RNA sequence promoted by a tyrosyl-tRNA synthetase. *Nature* **354:** 164–167.

Mohr G., Perlman P.S., and Lambowitz A.M. 1993. Evolutionary relationships among group II intron-encoded proteins and identification of a conserved domain that may be related to maturase function. *Nucleic Acids Res.* **21:** 4991–4997.

Mohr G., Caprara M.G., Guo Q., and Lambowitz A.M. 1994. A tyrosyl-tRNA synthetase can function similarly to an RNA structure in the *Tetrahymena* ribozyme. *Nature* **37:** 147–150.

Mohr G., Zhang A., Gianelos J.A., Belfort M., and Lambowitz A.M. 1992. The *Neurospora* CYT-18 protein suppresses defects in the phage T4 *td* intron by stabilizing the catalytically active structure of the intron core. *Cell* **69:** 483–494.

Moran J.V., Mecklenburg K.L., Sass P., Belcher S.M., Mahnke D., Lewin A., and Perlman P.S. 1994. Splicing defective mutants of the *COXI* gene of yeast mitochondrial DNA: Initial definition of the maturase domain of the group II intron AI2. *Nucleic Acids Res.* **22:** 2057–2064.

Moran J.V., Zimmerly S., Eskes R., Kennell J.C., Lambowitz A.M., Butow R.A., and Perlman P.S. 1995. Mobile group II introns of yeast mitochondrial DNA are novel site-specific retroelements. *Mol. Cell. Biol.* **15:** 2828–2838.

Mueller M.W., Allmaier M., Eskes R., and Schweyen R.J. 1993. Transposition of group II intron *al1* in yeast and invasion of mitochondrial genes at new locations. *Nature* **366:** 174–176.

Murphy F.L. and Cech T.R. 1994. GAAA tetraloop and conserved bulge stabilize tertiary structure of a group I intron domain. *J. Mol. Biol.* **236:** 49–63.

Muscarella D.E. and Vogt V.M. 1989. A mobile group I intron in the nuclear rDNA of *Physarium polycephalum*. *Cell* **56:** 443–454.

Myers C.A., Wallweber G.J., Rennard R., Kemel Y., Caprara M.G., Mohr G., and Lambowitz A.M. 1996. A tyrosyl-tRNA synthetase suppresses structural defects in the two major helical domains of the group I intron catalytic core. *J. Mol. Biol.* **262:** 87–104.

Nair S., de Pouplana L.R., Houman F., Auruch A., Shen X., and Schimmel P. 1997. Species-specific tRNA recognition in relation to tRNA synthetase contact residues. *J. Mol. Biol.* **269:** 1–9.

Niemer I., Schmelzer C., and Börner G.V. 1995. Overexpression of DEAD box protein pMSS116 promotes ATP-dependent splicing of a yeast group II intron in vitro. *Nucleic Acids Res.* **23:** 2966–2672.

Nikolcheva T. and Woodson S.A. 1997. Association of a group I intron with its splice junction in 50S ribosomes: Implications for intron toxicity. *RNA* **3:** 1–12.

Palmer J.D. and Logsdon J.M., Jr. 1991. The recent origins of introns. *Curr. Opin. Genet. Dev.* **1:** 470–477.

Paquin B., Kathe S.D., Nierzwicki-Bauer S.A., and Shub D.A. 1997. Origin and evolution of group I introns in cyanobacterial tRNA genes. *J. Bacteriol.* **179:** 6798–6806.

Rochaix J.-D. 1996. Post-transcriptional regulation of chloroplast gene expression in *Chlamydomonas reinhardtii*. *Plant Mol. Biol.* **32:** 327–341.

Rochaix J.-D., Rahire M., and Michel F. 1985. The chloroplast ribosomal intron of *Chlamydomonas reinhardtii* codes for a polypeptide related to mitochondrial maturases. *Nucleic Acids Res.* **13:** 975–984.

Roman J. and Woodson S.A. 1997. Integration of the *Tetrahymena* group I intron into bacterial rRNA by reverse splicing in vivo. *Proc. Natl. Acad. Sci.* **95:** 2134–2139.

Saldanha R., Ellington A., and Lambowitz A.M. 1996. Analysis of the CYT-18 protein binding site at the junction of stacked helices in a group I intron RNA by quantitative binding assays and *in vitro* selection. *J. Mol. Biol.* **261:** 23–42.

Saldanha R.J., Patel S.S., Surendran R., Lee J.C., and Lambowitz A.M. 1995. Involvement of *Neurospora* mitochondrial tyrosyl-tRNA synthetase in RNA splicing. A new method for purifying the protein and characterization of physical and enzymatic properties pertinent to splicing. *Biochemistry* **34:** 1275–1287.

Schafer B., Wilde B., Massardo D.R., Manna F., Giudice L.D., and Wolf K. 1994. A mitochondrial group I intron in fission yeast encodes a maturase and is mobile in crosses. *Curr. Genet.* **25:** 336–341.

Schmidt U., Podar M., Stahl U., and Perlman P.S. 1996. Mutations of the two-nucleotide bulge of D5 of a group II intron block splicing in vitro and in vivo: Phenotypes and suppressor mutations. *RNA* **2:** 1161–1182.

Sellem C.H. and Belcour L. 1997. Intron open reading frames as mobile elements and evolution of a group I intron. *Mol. Biol. Evol.* **14:** 518–526.

Sellem C.H., Lecellier G., and Belcour L. 1993. Transposition of a group II intron. *Nature* **366:** 176–178.

Semrad K. and Schroeder R. 1998. A ribosomal function is necessary for efficient splicing of the T4 phage thymidylate synthase intron in vivo. *Genes Dev.* **12:** 1327–1337.

Séraphin B., Simon M., and Faye G. 1988. *MSS18*, a yeast nuclear gene involved in the splicing of intron aI5β of the mitochondrial *cox1* transcript. *EMBO J.* **7:** 1455–1464.

Séraphin B., Boulet A., Simon M., and Faye G. 1987. Construction of a yeast strain devoid of mitochondrial introns and its use to screen nuclear genes involved in mitochondrial splicing. *Proc. Natl. Acad. Sci.* **84:** 6810–6814.

Séraphin B., Simon M., Boulet A., and Faye G. 1989. Mitochondrial splicing requires a protein from a novel helicase family. *Nature* **337:** 84–87.

Sharp P.A. 1991. Five easy pieces. *Science* **254:** 663.

Shaw L.C. and Lewin A.S. 1997. The Cbp2 protein stimulates the splicing of the ω intron of yeast mitochondria. *Nucleic Acids Res.* **25:** 1597–1604.

Shearman C., Godon J.-J., and Gasson M. 1996. Splicing of a group II intron in a functional transfer gene of *Lactococcus lactis*. *Mol. Microbiol.* **21:** 45–53.

Shub D.A., Goodrich-Blair H., and Eddy S.R. 1994. Amino acid sequence motif of group I intron endonucleases is conserved in open reading frames of group II introns. *Trends Biochem. Sci.* **19:** 402–404.

Staley J.P. and Guthrie C. 1998. Mechanical devices of the spliceosome: Motors, clocks, springs and things. *Cell* **92:** 315–326.

Strobel S.A. and Doudna J.A. 1997. RNA seeing double: Close packing of helices in RNA tertiary structure. *Trends Biochem. Sci.* **22:** 262–266.

Sugita M., Shinozaki K., and Sugiura M. 1985. Tobacco chloroplast tRNA^{Lys} (UUU) gene contains a 2.5-kilobase-pair intron: An open reading frame and a conserved boundary sequence in the intron. *Proc. Natl. Acad. Sci.* **82:** 3557–3561.

Thomson M.C., Macfarlane J.L., Beagley C.T., and Wolstenholme D.R. 1994. RNA editing of mat-r transcripts in maize and soybean increases similarity of the encoded protein to fungal and bryophyte group II intron maturases: Evidence that mat-r encodes a functional protein. *Nucleic Acids Res.* **22:** 5745–5752.

Turcq B., Dobinson K.F., Serizawa N., and Lambowitz A.M. 1992. A protein required for RNA processing and splicing in *Neurospora* mitochondria is related to gene products involved in cell cycle protein phosphatase functions. *Proc. Natl. Acad. Sci.* **89:** 1676–1680.

Uesono Y., Akio T., and Kikuchi Y. 1997. Ssd1p of *Saccharomyces cerevisiae* associates with RNA. *J. Biol. Chem.* **272:** 16103–16109.

Valencik M.L., Kloeckener-Gruissem B., Poyton R.O., and McEwen J.E. 1989. Disruption of the yeast nuclear *PET54* gene blocks excision of mitochondrial intron aI5β from pre-mRNA for cytochrome *c* oxidase subunit I. *EMBO J.* **8:** 3899–3904.

van der Horst G., Christian A., and Inoue T. 1991. Reconstitution of a group I intron self-splicing reaction with an activator RNA. *Proc. Natl. Acad. Sci.* **88:** 184–188.

van Dyck E., Jank B., Ragnini A., Schweyen R.J., Duyckaerts C., Sluse F., and Foury F. 1995. Overexpression of a novel member of the mitochondrial carrier family rescues defects in both DNA and RNA metabolism in yeast mitochondria. *Mol. Gen. Genet.* **246:** 426–436.

Vogel J., Hübschmann T., Börner T., and Hess W.R. 1997. Splicing and intron-internal RNA editing of *trnK-matK* transcripts in barley plastids: Support for MatK as an essential splice factor. *J. Mol. Biol.* **270:** 179–187.

Waldherr M., Ragnini A., Jank B., Teply R., Wiesenberger G., and Schweyen R.J. 1993. A multitude of suppressors of group II intron-splicing defects in yeast. *Curr. Genet.* **24:** 301–306.

Wallweber G.J., Mohr S., Rennard R., Caprara M.G., and Lambowitz A.M. 1997. Characterization of *Neurospora* mitochondrial group I introns reveals different CYT-18 dependent and independent splicing strategies and an alternative 3' splice site for an intron ORF. *RNA* **3:** 114–131.

Weeks K.M. and Cech T.R. 1995a. Efficient protein-facilitated splicing of the yeast mitochondrial *bI5* intron. *Biochemistry* **34:** 7728–7738.

———. 1995b. Protein facilitation of group I intron splicing by assembly of the catalytic core and the 5' splice site domain. *Cell* **82:** 221–230.

———. 1996. Assembly of a ribonucleoprotein catalyst by tertiary structure capture. *Science* **271:** 345–348.

Wenzlau J.M., Saldanha R.J., Butow R.A., and Perlman P.S. 1989. A latent intron-encoded maturase is also an endonuclease needed for intron mobility. *Cell* **56:** 421–430.

Wernette C., Saldanha R.J., Butow R.A., and Perlman P.S. 1990. Purification of a site-specific endonuclease, I-*Sce*II, encoded by intron 4α of the mitochondrial *coxI* gene of *Saccharomyces cerevisiae. J. Biol. Chem.* **265:** 18976–18982.

Wiesenberger G., Waldherr M., and Schweyen R.J. 1992. The nuclear gene *MRS2* is essential for the excision of group II introns from yeast mitochondrial transcripts *in vivo. J. Biol. Chem.* **267:** 6963–6969.

Xiong Y. and Eickbush T.H. 1990. Origin and evolution of retroelements based upon their reverse transcriptase sequences. *EMBO J.* **9:** 3353–3362.

Xu M.Q., Kathe S.D., Goodrich-Blair H., Nierzwicki-Bauer S.A., and Shub D.A. 1990. Bacterial origin of a chloroplast intron: Conserved self-splicing group I introns in cyanobacteria. *Science* **250:** 1566–1570.

Yang J., Mohr G., Perlman P.S., and Lambowitz A.M. 1998. Group II intron mobility in yeast mitochondria: Target DNA-primed reverse transcription activity of aI1 and reverse splicing into DNA transposition sites *in vitro. J. Mol. Biol.* (in press).

Yang J., Zimmerly S., Perlman P.S., and Lambowitz A.M. 1996. Efficient integration of an intron RNA into double-stranded DNA by reverse splicing. *Nature* **381:** 332–335.

Zaug A.J., Been M.D., and Cech T.R. 1986. The *Tetrahymena* ribozyme acts like an RNA restriction endonuclease. *Nature* **324:** 429–433.

Zhang A., Derbyshire V., Salvo J.L.G., and Belfort M. 1995. *Escherichia coli* protein StpA stimulates self-splicing by promoting RNA assembly in vitro. *RNA* **1:** 783–793.

Zhang L., Jenkins K.P., Stutz E., and Hallick R.B. 1995. The *Euglena gracilis* intron-encoded mat2 locus is interrupted by three additional group II introns. *RNA* **1:** 1079–1088.

Zimmerly S., Guo H., Perlman P.S., and Lambowitz A.M. 1995a. Group II intron mobility occurs by target DNA-primed reverse transcription. *Cell* **82:** 545–554.

Zimmerly S., Guo H., Eskes R., Yang J., Perlman P.S., and Lambowitz A.M. 1995b. A group II intron RNA is a catalytic component of a DNA endonuclease involved in intron mobility. *Cell* **83:** 529–538.

19
The Growing World of Small Nuclear Ribonucleoproteins

Yi-Tao Yu, Elizabeth C. Scharl, Christine M. Smith, and Joan A. Steitz
Department of Molecular Biophysics and Biochemistry
Howard Hughes Medical Institute
Yale University School of Medicine
New Haven, Connecticut 06536-0812

Small *ribonucleoproteins* (RNPs)—defined as tight complexes of one or more proteins with a short RNA molecule (usually 60–300 nucleotides)—inhabit every compartment of eukaryotic cells. Those that reside in the nucleus, the *small nuclear* RNPs (snRNPs), can themselves be divided into two families. There are snRNPs of the nucleoplasm, whose major business is the generation of messenger RNAs for export to the cytoplasm. A different set of snRNPs, called snoRNPs, reside in the cell nucleolus, the subnuclear locale responsible for the synthesis, maturation, and assembly of rRNAs into ribosomal subunits, which are then exported to function in cytoplasmic protein synthesis. The last two years have produced the extraordinary realization that vertebrate cells contain about 200 distinct kinds of snRNPs with abundances between 10^4 (for snoRNPs directing rRNA modification) to over 10^6 (for snRNPs of the major spliceosome). All of those whose functions have been assigned play roles in gene expression, underscoring the pivotal participation of RNA molecules in the evolution of the gene expression apparatus. The one exception is the telomerase snRNP, essential for genome maintenance (see Chapter 23).

Curiously, snRNPs are often the target of autoantibodies present in the sera of patients suffering from rheumatic disease (as well as in other mammals afflicted with autoimmunity). These autoantibodies almost always target epitopes on the protein component(s) of the particles, and since an antigenic protein often associates with multiple snRNAs (small nuclear RNAs), they define families of snRNPs related in structure (and usually also in function) (see Table 1). For instance, U1, U2, U4, U5, U11, U12, U4atac, and U7 are all Sm snRNPs and participate in mRNA biogenesis within the nucleoplasm. Similarly, all known fibrillarin-associated *small nucleolar* RNPs (snoRNPs) are involved in pre-rRNA cleavage or

Table 1 Small nuclear ribonucleoproteins

(a) *Metazoan snRNPs*

RNP	Function	Abundance (copies/cell)	RNA polymerase	5' end	Common antigen
Nucleoplasm					
U1	major splicing	1×10^6	II	m$_3$GpppA	Sm
U2	major/*trans* splicing	5×10^5	II	m$_3$GpppA	Sm
U4	major/*trans* splicing	2×10^5	II	m$_3$GpppA	Sm
U5	major/AT-AC/*trans*[a] splicing	2×10^5	II	m$_3$GpppA	Sm
U6	major/*trans* splicing	4×10^5	III	mpppG	—
U7	histone mRNA 3'-end formation	4×10^3	II	m$_3$GpppA	Sm
U11	AT-AC splicing	1×10^4	II	m$_3$GpppA	Sm
U12	AT-AC splicing	5×10^3	II	m$_3$GpppA	Sm
U4atac	AT-AC splicing	2×10^3	II	m$_3$GpppA	Sm
U6atac	AT-AC splicing	2×10^3	III	mpppG	—
7SK	unknown	2×10^5	III	mpppG	—
RNase P	pre-tRNA cleavage	2×10^5	III	m$_3$GpppA	Th
SL RNA	*trans* splicing	5×10^5	II	m$_3$GpppG	Sm
Nucleolus					
U3	pre-rRNA cleavage	2×10^5	II[b]	m$_3$GpppA[b]	Fb
U8	pre-rRNA cleavage	4×10^4	II	m$_3$GpppA	Fb
U13	unknown	4×10^4	II	m$_3$GpppN	Fb
U14[c]	pre-rRNA 2'-O-methylation	1×10^4	II	pN	Fb
U22	pre-rRNA cleavage	1×10^4	II	pN	Fb
MRP	pre-rRNA cleavage	1×10^5	III	pppG	Th
U15, 16, U18, 20, 21, U24-63, U73-81	pre-rRNA 2'-O-methylation	1×10^4	II	pN	Fb
E2, 3, U19, U23, U64-72	pre-rRNA pseudo-uridylation	1×10^4	II	pN	Gar1[d]

(continued)

Table 1 (Continued)

		(b) Viral snRNPs			
RNP	Virus	Abundance (copies/cell)	RNA polymerase	5' end	Common antigen
EBER1, 2	Epstein-Barr	5×10^6	III	pppA	La
HSUR1-7	Herpes saimiri	$10^3 - 10^4$	II	m$_3$GpppA/G	Sm

[a]In *Ascaris*, U5 participates in both *cis* and *trans*-splicing (Maroney et al. 1996); putative U5 homologs have been identified in trypanosomes (Dungan et al. 1996; Xu et al. 1997).

[b]In plants, U3 is transcribed by RNA polymerase III and is capped with mpppG (see Maxwell and Fournier 1995).

[c]Yeast U14 is involved in cleavage of pre-rRNA, whereas *Xenopus* U14 is required only for 2'-O-methylation (Liang and Fournier 1995; Dunbar and Baserga 1998).

[d]Gar1 is a yeast protein whose vertebrate homolog has not been reported.

modification. The La snRNPs include all nascent transcripts of RNA polymerase III, which are either stably or transiently bound by the La autoantigen before trimming of their 3' ends. Even vertebrate RNase P, the snRNP responsible for 5' end maturation of all tRNAs, contains a protein component that can be targeted by autoantibodies (see Chapter 14).

As in studies of the ribosome, which is a large RNP, a most challenging question for snRNP investigators has been the contribution of the RNA component of each snRNP to function. To what extent can the snRNA be viewed as a catalytic component, whose base-pairing to pre-mRNA or pre-rRNA enables protein-assisted RNA catalysis? Do RNA–RNA interactions between the snRNP and its substrate serve only to specifically orient a protein enzyme for action? Alternatively, since snRNP–substrate interactions sometimes displace intramolecular base-pairing, do some snRNPs function as chaperones, critically molding the architecture of a much larger pre-RNA substrate? Why do certain viruses encode small RNAs that associate with some of the same proteins as host-cell small RNAs to form viral snRNPs? Progress on these questions, as well as the recent unanticipated identification of many novel low-abundance snRNPs, is summarized below.

snRNPs INVOLVED IN PRE-mRNA PROCESSING

Messenger RNA precursors (pre-mRNAs) must be efficiently and accurately processed (via splicing and 3' end processing) before the resulting mRNAs are transported to the cytoplasm where they direct the translation of proteins. A number of snRNPs, as well as protein factors, participate. All snRNPs (except for U6 and U6atac) involved in pre-mRNA processing belong to the Sm snRNP family (Table 1). U1, U2, U4, U5, U6, U11, U12, U4atac, and U6atac snRNPs orchestrate the splicing of pre-mRNAs (Fig. 1A, B); the SL snRNP serves as a donor in *trans*-splicing (Fig. 1C) in certain divergent eukaryotic organisms; and the U7 snRNP is essential for the maturation of the 3' ends of vertebrate histone mRNAs (see below).

The snRNA components of Sm snRNPs are synthesized by RNA polymerase II in response to special promoter and termination signals (Dahlberg and Lund 1988) and then immediately transported to the cytoplasm. There, a common set of 8 proteins, which are recognized by anti-Sm autoantibodies, assemble onto the conserved Sm-binding sequence

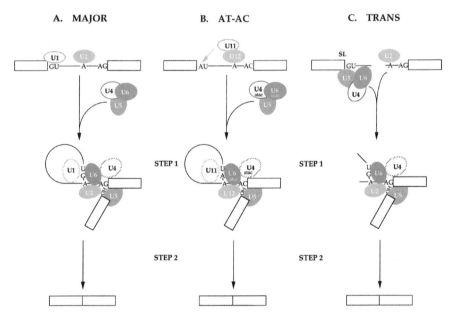

Figure 1 Spliceosome assembly and action. The major spliceosome (*A*), the AT-AC spliceosome (*B*), and the *trans*-spliceosome (*C*) are pictured at an early stage of spliceosome assembly and after the first reaction step (formation of the lariat intermediate) has occurred. U11 and U12 are pictured as assembling onto the pre-mRNA as a di-snRNP particle (*B*). Joining of the U4-U6-U5 tri-snRNP (*A*), the U4atac-U6atac-U5 tri-snRNP (*B*), or the SL RNA-U4-U6-U5 quadruple snRNP (*C*) forms the spliceosome. The newly assembled spliceosome then undergoes a conformational change that loosens the association of U1 (*A*) or U11 (*B*) and U4 (*A*, *C*) or U4atac (*B*) and leads to the first catalytic step, lariat formation. After a second conformational change, step 2—exon ligation—occurs.

($PuAU_{n\geq3}GPu$) present in each of these snRNAs (gray boxes in Fig. 2) (Lührmann 1988). After binding the Sm proteins, the 5′-monomethyl guanosine cap becomes 2,2,7-trimethylated (TMG) and the assembled particles are transported back into the nucleus (Mattaj 1988). Besides the common Sm proteins, each mature Sm snRNP contains its own specific proteins. For instance, the human U2 snRNP appears to contain as many as 11 specific proteins (Behrens et al. 1993); and most recently, 4 or possibly 5 more polypeptides in yeast were added to the list of 5 U1-specific proteins (Gottschalk et al. 1998). Because many specific snRNP proteins are loosely bound and thus dissociated by standard isolation procedures, the real number of specific proteins for each of the well-characterized Sm snRNPs may be quite large.

U6 and U6atac snRNAs are synthesized by RNA polymerase III and acquire a 5′ γ-methyl guanosine cap (Dahlberg and Lund 1988). Unlike the Sm snRNAs, U6 and U6atac snRNAs do not possess an Sm-binding site and may never leave the nucleus (Hamm and Mattaj 1989; Terns et al. 1993). U6 and U6atac associate with U4 and U4atac, respectively, to form an snRNP containing two snRNAs (see Fig. 2). It is this di-snRNP particle that actively participates in pre-mRNA splicing.

snRNPs of the Major Spliceosome

The most extensively studied snRNPs are those involved in excising the major class of introns from pre-mRNAs. These are the U1, U2, U4, U5, and U6 snRNPs. The secondary structures of these snRNAs (Fig. 2) are highly conserved from yeast to human, as are short stretches of sequence (colored boxes in Fig. 2). These regions participate in an intricate RNA–RNA interaction network that specifies, at least in part, the roles of snRNPs during spliceosome assembly and function. Curiously, at least seven U5 variants have been identified in human cells; all of them participate in splicing (Sontheimer and Steitz 1992).

It has long been known that spliceosome assembly is a stepwise event (at least in the test tube) (for review, see Guthrie 1991; Moore et al. 1993). At early times, the U1 snRNP recognizes, in an ATP-independent manner, the 5′ splice site to form a complex that commits the pre-mRNA to spliceosome assembly. This recognition involves base-pairing between the extremely conserved 10 nucleotides at the 5′ end of U1 (blue box in U1 in Fig. 2) and the 5′ splice site. Subsequent to U1 binding, the U2 snRNP recognizes the branch site of the pre-mRNA to form a pre-splicing complex. Again, this recognition involves base-pairing between a highly conserved sequence in U2 (green box in U2 in Fig. 2) and the pre-

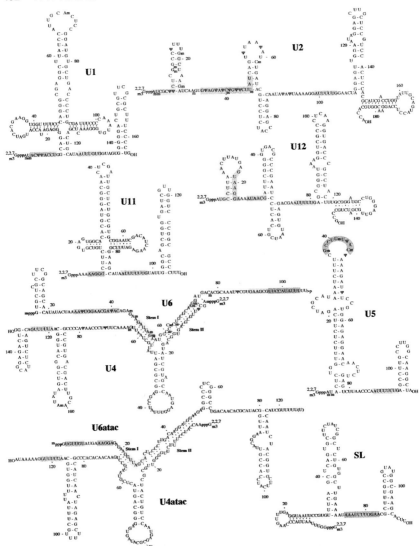

Figure 2 Primary and secondary structures of human spliceosomal snRNAs and nematode (*Ascaris*) SL RNA. The Sm-binding sites are indicated by *gray* boxes. Nucleotides known or predicted to be involved in intermolecular interactions with the pre-mRNA or other snRNP RNAs are colored: *blue* for interactions with intron sequences at the 5′ splice site; *green* for base-pairing with the branch site; *purple* for contacts with the exon sequences at the 5′ and 3′ splice sites; *light yellow*, *bright yellow*, and *dark yellow* for U2-U6 (or U12-U6atac) helix I, U2-U6 helix II and U2-U6 (or putative U12-U6atac) helix III, respectively; *pink* for SL RNA-U6 base-pairing. m indicates 2′-O-methylation except for U6 residue G72, which is a base methylation; other base methylations occur at U6 residue A43 and U2 residue A30.

mRNA. The U4/U5/U6 tri-snRNP, in which U4 and U6 snRNPs are involved in extensive intermolecular base-pairing (see Figs. 1A and 2), then joins this complex, converting the pre-splicing complex into a fully assembled spliceosome.

In the early 1990s, recognition of additional RNA–RNA interactions began to emerge with the use of powerful biochemical cross-linking and sensitive genetic suppression approaches. An evolutionarily invariant sequence in the first loop of U5 snRNA (purple in Fig. 2) contacts the exon sequence at the 5' splice site (Newman and Norman 1991; Wyatt et al. 1992; Sontheimer and Steitz 1993) and, probably later, the exon sequence at the 3' splice site (Newman and Norman 1992; Sontheimer and Steitz 1993). These contacts, together with the base-pairing interactions between the conserved 5' end of U1 and intron sequences at the 5' (see above) and 3' (Reich et al. 1992) splice sites, may form a four-armed Holliday-like structure, which juxtaposes the two splice sites in the newly assembled spliceosome (Steitz 1992). Subsequently, a complex dynamic rearrangement of RNA–RNA interactions occurs. U6 dissociates from U4, forms a new intramolecular stem-loop (Fig. 3), produces new base-paired duplexes with U2 (helices I, II, and III in Fig. 3; yellow boxes in U2 and U6 structures in Fig. 2), and displaces U1 in interacting with the 5' end of the intron (blue box in U6 in Fig. 2) (for review, see Nilsen 1994). Genetic studies in yeast have shown that the U2-U6 helix I (Fig. 3) is indispensable for splicing, arguing that it contributes to the catalytic core of the spliceosome (see below). Interestingly, although U2-U6 helix II is required for splicing in the mammalian system, it is redundant in yeast; only when U2-U6 helix Ib is disrupted does helix II become necessary (Field and Friesen 1996). Following these rearrangements, the first catalytic transesterification step of splicing takes place, generating the 2/3 lariat intermediate and the cut-off 5' exon (see Fig. 1A).

Before the second transesterification step, the spliceosome undergoes additional conformational changes, creating novel RNA–RNA interactions. Yeast genetic studies have suggested formation of a tertiary contact between the bulged nucleotide of U2 in U2-U6 helix I and a highly conserved sequence in U6 (purple in Fig. 3) (Madhani and Guthrie 1994), as well as a possible non-Watson-Crick interaction between the 5'- and the 3'-terminal nucleotides of the intron (orange in Fig. 3) (Parker and Siliciano 1993). Additionally, site-specific cross-linking with ^{4S}U in the second position of the intron has captured another contact between the 5' splice site in the 2/3 lariat intermediate and a conserved sequence in U6 (Sontheimer and Steitz 1993). Strikingly, this interaction can be chased to the completion of the splicing; i.e., the same cross-link can be identified between the lariat intron product and U6, implying functional significance for the second catalytic step. The new interactions, together with those preserved from the first catalytic conformation (such as the U2-U6 helices, the U6 intramolecular stem-loop, and the U5-exon contacts), then generate an altered catalytic site in which the free 5′ exon is properly

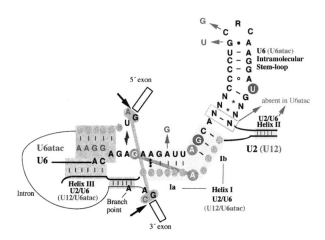

Figure 3 Model for the catalytic core of the spliceosome. Conserved sequences of U6 and U6atac are shown. Changes in U6atac RNA relative to U6 are indicated in *green*, as are terminal nucleotide changes in AT-AC introns relative to major class introns. Some AT-AC introns contain terminal nucleotides identical to those of major class introns (Sharp and Burge 1997). R and N stand for purines and variable nucleotides, respectively. Also shown are RNA–RNA interactions in the spliceosome (except for U5-exon contacts) known or predicted to be important for the catalytic reactions. The colored boxes represent the same interactions as in Fig. 2. *Red* and *blue spheres* indicate the nucleotides whose 5′ phosphate is important for the first or the second step of splicing as defined in yeast (Fabrizio and Abelson 1992) and/or *Ascaris lumbricoides* (Yu et al. 1995), respectively. The *purple sphere-bar* indicates a proposed tertiary interaction between the bulge of helix I and the second G of the highly conserved ACAGAGA sequence of U6, which contributes to the second step of splicing in yeast (Madhani and Guthrie 1994). The *orange oval-bar* indicates a possible non-Watson-Crick interaction between the 5′- and 3′-terminal nucleotides of major-class introns (Parker and Siliciano 1993) or AT-AC introns (Dietrich et al. 1997). This interaction may be important for the second step of splicing as well. The "lightning bolt" indicates a cross-link established between U6 and the 5′ splice site of a major-class intron (Sontheimer and Steitz 1993). *Black arrows* denote splice sites. The *dotted line* represents a Watson-Crick interaction in the yeast spliceosome and the human AT-AC spliceosome; *asterisks* stand for Watson-Crick base-pairing; the *closed* and *open circles* represent non-Watson-Crick interactions. (Modified from Tarn and Steitz 1997.)

aligned with the 3' exon, allowing nucleophilic attack of the 3'-hydroxyl group of the 5' exon on the 3' splice site. The invariant U5 loop sequence, which plays a pivotal role as guide, preserves its contact with the 5' exon through both catalytic steps (Sontheimer and Steitz 1993).

The dynamic rearrangements of RNA–RNA interactions in the spliceosome involve the formation of new RNA–RNA pairings accompanied by the disruption of prior RNA–RNA duplexes. To achieve this well-orchestrated coordination, RNA helicase-like activities (unwindases and rewindases) must be present in the spliceosome. The best candidates are proteins characterized as ATP-hydrolyzing non-snRNP splicing factors. Recent studies indicate that some of these proteins indeed have RNA unwindase and rewindase activities, which can disrupt the initial RNA–RNA interactions and replace them with new intermolecular base-pairing arrangements (for review, see Staley and Guthrie 1998).

A notable common feature of the spliceosomal snRNAs is their posttranscriptional modification. Aside from hypermethylation of the 5' cap, numerous internal nucleotides are also modified by pseudouridylation, 2'-O-methylation or base methylation (Fig. 2). Only recently has the function of these modifications begun to emerge. Both the 5' TMG cap and internal modifications of U2 snRNA are required for full assembly of a 17S snRNP, which contains all 11 U2-specific proteins (in addition to the common core proteins) and is essential for splicing (Y.-T. Yu et al., in prep.). By analogy with U2, highly conserved modifications in the other spliceosomal snRNAs are likely to be functionally important in snRNP biogenesis and/or splicing. Recent studies have suggested that, like the modifications of rRNAs (see below), spliceosomal snRNA modifications are guided by other snRNAs (K.T. Tycowski et al., in prep.). The modifying machinery has been localized to the nucleus (Y.-T. Yu et al., unpubl.) and could potentially be subcompartmentalized, such as in coiled bodies and/or the nucleolus (Lyon et al. 1997).

Despite significant progress on the mechanism of splicing of major class introns, a fundamental question remains: Are the catalytic transesterifications RNA- or protein-based? Since the discovery that group II self-splicing introns share a splicing pathway with nuclear pre-mRNAs, it has been fashionable to consider spliceosomal catalysis as RNA-based: The five snRNAs can be viewed as descendants of a self-splicing progenitor RNA that has been divided into pieces that act *in trans* on introns that have co-evolved to contain only limited conserved sequence elements (Sharp 1991). Each discovery of an RNA–RNA interaction in the spliceosome that mimics a critical base-pairing interaction in group II introns serves to reinforce this notion. The U5-5' exon sequence and U5-3' exon sequence interactions have direct analogies in group II introns. Also, the U2-U6 helix I together with the U6 intramolecular stem-loop (Fig. 3) appears analogous to domain V of the group II intron, a structure believed to play a catalytic role in self-splicing; the critical bases and backbone phosphates and their positions are nearly identical in the two structures (Chanfreau and Jacquier 1994; Yu et al. 1995). A recent study has demonstrated that the first step of nuclear pre-mRNA splicing is metal-ion-mediated, whereas the second step appears to occur by a different mechanism (Sontheimer et al. 1997). The metal-ion-mediated transesterification agrees with a model for the catalytic site that argues that the geometry of the spliceosomal catalytic center (at least for the first step) is similar, if not identical, to that of RNA-based ribozymes (Piccirilli et al. 1993; Steitz and Steitz 1993; Weinstein et al. 1997). Further support for the hypothesis that nuclear pre-mRNA splicing is RNA-centered comes from the studies of Hetzer et al. (1997), who have shown that an essential subdomain (ID3) of group II introns, previously proposed to be a U5 analog, can indeed be substituted *in trans* by human U5 snRNA.

Yet, all support for RNA-based nuclear pre-mRNA splicing can be regarded as indirect. In the past several years, unexpected observations have created more uncertainties. For instance, in the presence of excess SR proteins, U1—which normally forms essential base-pairing interactions with the 5' splice site (and with the 3' splice site of some yeast [*Schizosaccharomyces pombe*] pre-mRNAs)—is dispensable for splicing (Crispino et al. 1994; Tarn and Steitz 1994). Even more surprisingly, deletion of the invariant loop of U5, considered essential for aligning the 5' and 3' exons for the second (ligation) step of splicing, has no effect on the first step of yeast pre-mRNA splicing (O'Keefe et al. 1996). These results undermine the case for RNA-based catalysis by weakening the importance of some well-documented RNA–RNA interactions. However, a large multicomponent machine like the spliceosome must have evolved to ensure that disruption of any particular molecular contact could be compensated by other interactions.

Aside from the link to autoimmunity, until very recently no direct connections between human genetic disease and spliceosomal snRNP biogenesis or the pre-mRNA splicing machinery had been uncovered. Dreyfuss and coworkers (Fischer et al. 1997) have discovered that the protein product of the survival of motor neurons (SMN) gene—the spinal muscular atrophy (SMA) disease gene—and its associated novel protein (SIP1) contribute importantly to the biogenesis of spliceosomal snRNPs. This spectacular finding argues that more such connections may soon emerge.

The AT-AC Spliceosome—Divergent Yet Homologous snRNPs

Recently, a rare class of introns with sequence elements (at the 5' and 3' splice sites and the branch site) that are distinct from the consensus sequences of most pre-mRNA introns was identified in metazoan genes (Fig. 4). An intensive search for the splicing machinery that removes this type of intron led to the remarkable discovery of a new low-abundance spliceosome. The components of this spliceosome include three previously identified snRNPs, U11, U12, and U5, and two novel snRNPs termed U4atac and U6atac (Fig. 2 and Table 1) (Tarn and Steitz 1996a,b). Thus, with the exception of U5, the snRNPs (U11, U12, U4atac, and U6atac) in this spliceosome are distinct from those in the major spliceosome.

This new spliceosome was dubbed the AT-AC spliceosome, based on the fact that the divergent introns initially identified all have AU-AC (AT-AC in the DNA) at their termini (Fig. 4) (Jackson 1991; Hall and Padgett 1994). However, database analyses and mutational studies have now revealed that some introns with AU-AC at their termini are spliced by the major spliceosome; conversely, some introns with GU-AG at their termini

Pre-mRNA introns

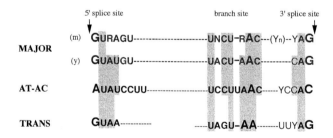

Group II introns

Figure 4 Conserved sequences in nuclear pre-mRNA introns and self-splicing group II introns (Tarn and Steitz 1997). The 5' and 3' splice-site and branch-site consensus sequences are shown (R, purines; Y, pyrimidines). The *trans* branch-site sequence is from a nematode *trans*-splicing substrate (Hannon et al. 1990). Enlarged letters in each intron indicate highly conserved nucleotides. Yn represents a long stretch of pyrimidines that is present in mammalian (m) introns; group II introns and most yeast (y) introns do not contain such a sequence; nematode *trans*-splicing substrates usually have a short stretch of pyrimidine residues. Identical nucleotides in major-class introns, AT-AC introns, and *trans*-splicing introns are shaded.

are spliced by the AT-AC spliceosome (Dietrich et al. 1997; Wu and Krainer 1997). What distinguishes the two classes of introns, rather than simply the terminal dinucleotides, are essential sequences near the 5' end and at the branch point. In AT-AC introns, the five nucleotides immediately downstream from the 5' AU are recognized by U11 (blue in U11 in Fig. 2; see below), whereas the branch site base-pairs with U12 (green in U12 in Fig. 2), leading to the suggestion that "U12-type" (or "U12-dependent") spliceosome might be a more suitable nomenclature (Dietrich et al. 1997; Sharp and Burge 1997).

Despite their distinct snRNP constituents, the major and minor spliceosomes active in metazoan cells exhibit many common features. Each snRNP in one spliceosome has a counterpart in the other (U11 and U1, U12 and U2, U4atac and U4, and U6atac and U6) (Tarn and Steitz 1996b; Yu et al. 1996). Although the sequences of the two sets of snRNAs show little identity, computer-predicted secondary structures reveal striking commonalities between each pair (Fig. 2). The Sm-binding sites (gray in Fig. 2) of U11, U12, and U4atac are all positioned in single-stranded regions and are almost identical to those of U1, U2, and U4, respectively. Although the 5' splice-site recognition sequence of U11 and the branch-site pairing sequence of U12 are pictured as partially double-stranded, both biochemical cross-linking and genetic studies have indicated that these regions are available for base pairing (see below). The overall secondary structure fold of human U6atac interestingly resembles that of yeast U6 more than human U6. However, when the structures in the di-snRNPs are compared, U4-U6 (human or yeast) and U4atac-U6atac are remarkably similar. More importantly, the placement of essential sequences (colored boxes in Fig. 2) within the structures is nearly identical. For instance, a U6atac sequence (blue box in U6atac) predicted to base-pair with the 5' splice site is positioned immediately upstream of stem I of U4atac-U6atac structure, precisely where the U6 5' splice site interacting sequence (blue box in U6) is positioned in the U4-U6 structure. Likewise, the U6atac sequences (light yellow in U6atac in Fig. 2) predicted to form base-pairing interactions with U12 (U12-U6atac helix I, Fig. 3) and to form an intramolecular stem-loop are buried in stems I and II of U4atac-U6atac, exactly as the U6 sequences that pair with U2 (U2-U6 helix I) and form an intramolecular stem-loop are located in the U4-U6 structure (Tarn and Steitz 1996b, 1997).

These predicted structures and intermolecular RNA interactions in the AT-AC spliceosome are bolstered by strong experimental support. At early times during the splicing of an AT-AC intron in HeLa nuclear extract, ATP-independent ^{4S}U site-specific cross-links between the 5' splice site and U11 were detected (Yu and Steitz 1997), presumably representing the base-pairing interactions deduced from genetic suppression experiments (Kolossova and Padgett 1997). As spliceosome assembly proceeds, the U11-5' splice site cross-links diminish, and a new ATP-dependent cross-link between the 5' splice site and U6atac is visualized, suggesting that U6atac may replace U11 in interacting with the 5' splice site, just as U6 replaces U1 in the major spliceosome (Yu and Steitz 1997). The interaction between U12 and the branch site of the AT-AC intron has also been confirmed both by in vivo analyses, where compensatory changes in U12 reversed the deleterious effects of branch-site mutations (Hall and Padgett 1996), and by in vitro psoralen cross-linking experiments (Tarn and Steitz 1996a). Thus, the proposed branch-site recognition sequence of U12 indeed base-pairs with the branch site of the intron, parallel to the role of U2 in splicing a major-class intron. Psoralen cross-linking analyses have also verified the existence of the U4atac-U6atac structure and the U12-U6atac helix I (Fig. 3) (Tarn and Steitz 1996b). In addition, a possible non-Watson-Crick interaction between the 5'- and the 3'-terminal nucleotides of an AT-AC intron can be postulated to mimic that established for major-class introns (orange in Fig. 3) (Dietrich et al. 1997). Finally, the role of U5, the only snRNP shared by both spliceosomes, is predicted to be the same as in the major spliceosome (see above). Interestingly, all U5 variants that participate in the splicing of major-class introns were also detected in the AT-AC spliceosome (Y.-T. Yu et al., unpubl.).

Although direct comparison of protein components (snRNP proteins and splicing factors) in the two metazoan spliceosomes is still lacking, the remarkable parallels in snRNA structures and RNA–RNA interactions have strongly reinforced our current picture of the spliceosomal active center (Fig. 3) and how it functions in intron removal (Nilsen 1996; Tarn and Steitz 1997). These parallels also support the notion that nuclear pre-mRNA splicing is RNA-centered.

Why are there two distinct spliceosomes in metazoan cells? Do the rare minor-class introns serve some regulatory function in those genes that have such an intron? So far, genes that contain a single AT-AC intron (the only exception so far is an *Arabidopsis* gene that contains two AT-AC introns) also possess multiple major-class introns (Sharp and Burge 1997). There are several significant human disease genes that have an AT-AC intron: e.g., Hermansky-Pudlak syndrome (Oh et al. 1996); Huntington's disease: the AT-AC intron is in the same position in *Fugu* as in human huntingtin! (Sharp and Burge 1997). In vitro, the rate of intron excision by the AT-AC spliceosome is much slower than that of major-class introns (Tarn and Steitz 1996a); yet, cooperative interspliceosome communication between an AT-AC and neighboring major-class spliceosome has been observed in vitro (Wu and Krainer 1996). Only one gene is known where alternative excision of a major-class versus an AT-AC intron occurs. This is the *Drosophila prospero* gene, which contains a so-called twintron with a major-class intron residing within an AT-AC intron (C. Doe, pers. comm.). Here, competition rather than cooperation between the two types of spliceosomes may decide the fate of neural development.

The evolutionary questions considered above must be reexamined in light of the existence of two independent spliceosomes in higher eukaryotic cells. Since the catalytic pathway of both spliceosomes is identical to that of group II self-splicing introns, relationships among all these splicing machines are suggested. Like the proposal that the major spliceosomal snRNAs have evolved from a group II self-splicing intron, the U11, U12, U4atac, and U6atac snRNAs can be reasonably viewed as descendants of a group II-like intron. Did the two spliceosomes evolve independently from a common ancestor, or has one arisen from the other? U6atac and U6 in human cells exhibit only approximately 40% sequence similarity, whereas human U6 shares approximately 60% identity with yeast U6 (Tarn and Steitz 1996b). Structural comparisons (see above) indicate that human U6atac may be more closely related to yeast U6 than to human U6. On the other hand, AT-AC introns share more features with group II introns and most yeast introns (the 3′ splice site sequence and the lack of a polypyrimidine tract between the branch site and the 3′ splice site) than does the metazoan major class of introns (Fig. 4). Recent detailed phylogenetic studies (Sharp and Burge 1997) have greatly reinforced the notion of a close relationship between the AT-AC introns and group II introns. Sharp and Burge (1997) have speculated that the evolutionary path led from an AT-AC intron, which initially evolved from a group II self-splicing intron, to a major-class intron.

Trans-splicing—Yet Another Spliceosomal Configuration

Another distinct kind of spliceosome catalyzes the addition of a specific spliced leader (SL) sequence (5′ exon, see below) to the 5′ end of pre-mRNA transcripts in a number of lower eukaryotic organisms, including trypanosomatid protozoans, nematodes, trematodes, and euglenoids (for review, see Blumenthal 1995; Nilsen 1997). In *trans*-splicing, exons from separately transcribed RNAs are spliced intermolecularly, rather than from within a single precursor RNA as in conventional *cis*-splicing (Fig. 1C).

U2, U4, and U6 are all required for *trans*-splicing, as demonstrated by deoxyoligonucleotide-directed RNase H digestion analysis in *Ascaris* extracts (Hannon et al. 1991) and in permeabilized trypanosome cells (Tschudi and Ullu 1990). In contrast, the nematode U1 snRNP is not necessary for *trans*-splicing (Hannon et al. 1991), consistent with the fact that a U1 homolog has not been identified in trypanosomes where only *trans*-splicing occurs. Since the U5 snRNP is not readily targeted for RNase H digestion by complementary deoxyoligonucleotides, its role in nematode *trans*-splicing remained unclear until the demonstration that U5 is present in the isolated *trans*-spliceosome (Maroney et al. 1996). Two U5 snRNA homologs have recently been reported for trypanosomes, but they are unusually short and contain no TMG cap at their 5′ termini (Dungan et al. 1996; Xu et al. 1997).

Trans-splicing requires—in addition to U2, U4, U5, and U6—a unique snRNP called the spliced-leader snRNP (SL RNP) (Fig. 2). The RNA component (SL RNA) of the SL RNP is composed of two domains: the 5′ SL sequence domain (5′ exon, 22 nucleotides in nematodes and 39 nucleotides in trypanosomes) and the 3′ snRNA-like domain (the intron portion) (Agabian 1990; Nilsen 1993). The 5′ SL sequence is highly conserved among nematodes (Nilsen et al. 1989) (except for the SL2 RNA in *Caenorhabditis elegans* [Huang and Hirsh 1989]) and among trypanosomes (Bruzik et al. 1988). In contrast, with the exception of one stringently conserved short sequence (the Sm-binding site in nematodes, and the Sm-like-binding site in trypanosomes), the SL RNA intron portions have diverged. Yet, the proposed secondary structures of all SL RNAs are virtually identical (Fig. 2) (Bruzik et al. 1988; Nilsen et al. 1989).

The SL RNA can be viewed as a specialized form of a spliceosomal snRNA (Nilsen 1993). Besides its Sm (or Sm-like)-binding site (gray box in Fig. 2), SL RNA shares another feature with spliceosomal Sm snRNAs: its 5′ TMG cap structure. However, the SL snRNA donates its 5′ SL sequence (mimicking a 5′ exon) to an acceptor RNA and therefore is consumed during the splicing process, whereas the other spliceosomal snRNPs are reusable cofactors for splicing. In vitro studies in *Ascaris* cell-free extracts indicate that participation of the nematode SL RNA in *trans*-splicing is independent of the SL portion of the sequence (Maroney et al. 1991). In contrast, in trypanosomes, the SL sequence appears to be required for SL RNA function (Lucke et al. 1996). This apparent discrepancy may reflect an important difference between these two systems: The trypanosome SL sequence is longer than the nematode SL sequence, arguing that essential U5 functions may be performed by the trypanosome

SL sequence (perhaps in cooperation with the trypanosome U5 analog, see above) but not by the nematode SL RNA (Steitz 1992).

Both in vivo and in vitro studies have demonstrated that many essential RNA–RNA interactions characterized in *cis*-splicing (see above) occur also in the *trans*-spliceosome (Nilsen 1997). In addition, a U6-SL RNA base-pairing interaction unique to *trans*-splicing has been documented in nematode *trans*-splicing extracts (pink in SL and U6 RNAs in Fig. 2) (Hannon et al. 1992). This duplex, together with the interaction between the U5 invariant loop sequence and the exon sequence at the 5' splice site (contained in the SL RNA) and U4/U5/U6 interactions, predicts formation of a SL/U4/U5/U6 quadruple snRNP, allowing the separate 5' exon to efficiently engage in *trans*-spliceosome assembly (Fig. 1C) (Maroney et al. 1996).

The existence of a distinct *trans*-spliceosome raises more evolutionary questions. Since *trans*- and *cis*-splicing are so closely related (both the snRNP components and the splicing mechanism are nearly identical), it is very likely that one appeared earlier and gave rise to the other. However, given the existence of both *trans*-splicing-only and *cis*-splicing-only primitive eukaryotes, it remains unclear whether the *trans*- or *cis*-spliceosome evolved first.

The U7 snRNP Measures Histone Pre-mRNAs

Maturation of the nonpolyadenylated 3' ends of cell-cycle-regulated metazoan histone mRNAs occurs by endonucleolytic cleavage (for review, see Muller and Schümperli 1998). Two *cis*-acting conserved sequence elements in the pre-mRNA are required: a stem-loop structure upstream of the site of cleavage and a purine-rich region downstream (called the HDE). The latter is recognized by the low-abundance U7 snRNP; the 5' end of U7 RNA has been demonstrated by biochemical and genetic suppression experiments to base-pair with the HDE (Fig. 5). A second well-characterized *trans*-acting factor is a 31-kD protein, similar in humans, mouse, *Xenopus laevis*, and *C. elegans* (Wang et al. 1996; Martin et al. 1997), that binds to the stem-loop element. This hairpin binding protein (HBP) is not absolutely required for in vitro processing, its effect being dependent on the histone pre-mRNA and type of nuclear extract used (Streit et al. 1993). HBP has also been implicated in post-transcriptional regulatory events such as stability, nuclear-cytoplasmic transport, translational efficiency, and cell-cycle-specific degradation of histone messages (for review, see Marzluff 1992; Muller and Schümperli 1998).

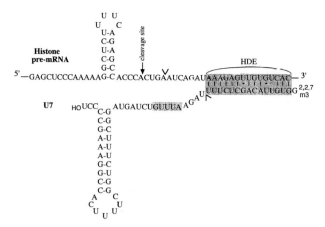

Figure 5 RNA–RNA interactions of the histone pre-mRNA processing complex. Shown are sequences of the mouse H2A-614 pre-mRNA (Hurt et al. 1989) and the human U7 snRNA (Mowry and Steitz 1987), with residues involved in base-pairing indicated by the larger shaded box. The U7 Sm-binding site is marked with a smaller shaded box. The site of insertions and deletions in the substrate is indicated by a caret. The caret in U7 RNA marks the site of insertions that result in length suppression.

Analysis of the U7 snRNP reveals several features not shared by the spliceosomal Sm snRNPs. The U7 Sm site (smaller gray box in Fig. 5), which differs from the consensus, is critical for proper functioning since replacement with the U1 Sm site produces a U7 snRNP that cannot support histone pre-mRNA processing (Stefanovic et al. 1995). The U7 Sm site also dictates its low cellular abundance (Grimm et al. 1993) and is necessary and sufficient for proper U7 snRNP localization to the sphere organelles of *X. laevis* oocyte nuclei (Wu et al. 1996). Affinity selection of U7 particles has revealed the core Sm proteins and two additional proteins of 14 kD and 50 kD (Smith et al. 1991).

In contrast to the spliceosomal snRNPs, whose base-pairing interactions with the substrate include the nucleotides involved in the trans-esterification reactions of splicing, base-pairing by the U7 snRNP acts at a distance (larger shaded box in Fig. 5). Stepwise insertions between the stem-loop and the HDE of a histone pre-mRNA (caret in the pre-mRNA in Fig. 5) showed that the site of cleavage moves downstream in concert with the HDE, but processing efficiencies decline (Scharl and Steitz 1994). Satisfyingly, length suppression could be achieved through compensatory insertions into the U7 RNA (caret in the U7 in Fig. 5), which restored both the efficiency and site of cleavage (Scharl and Steitz 1996). Thus, the U7 snRNP appears to act as a molecular ruler, somehow measuring 11 nucleotides along the pre-mRNA from its site of base-pairing (the HDE) to the cut site. Since this region of the pre-mRNA is nonconserved and apparently single-stranded, some snRNP protein(s) or associated factor(s) must rigidify the substrate, allowing accurate measurement by the processing complex. The architecture of this molecular measuring device is unknown.

The exact contribution of the U7 snRNP to catalytic cleavage of the histone pre-mRNAs is also unclear. Since the site of U7 snRNP base-pairing is distant from the cut site, U7 RNA could function to orient a protein factor as opposed to being directly involved in catalysis. Deletion analyses of the pre-mRNA substrate showed that cutting ceases when the HDE is moved closer than 5 nucleotides to the base of the stem-loop, suggesting that sufficient access to the cleavage site is required (Cho et al. 1995). Furthermore, cleavage occurs preferentially after A residues, whereas Gs following the cleavage site are inhibitory (Furger et al. 1998). Could the same cleavage factor be employed for histone pre-mRNA processing and polyadenylation? (for review, see Muller and Schümperli 1998). Both reactions require recognition elements upstream and downstream of the cleavage site. Likewise, both cleavages occur preferentially after A residues, leave 3'-OH and 5'-phosphates, and can take place without ATP or in the presence of EDTA. Identification of components interacting directly at the cleavage site of histone pre-mRNAs will be needed to define the exact role of the U7 snRNP in the cleavage process.

There are several other RNA processing events where catalysis occurs at a fixed distance from a recognition element. Many of these involve measurement along double-helical RNA: Examples are tRNA splicing (Baldi et al. 1992), RNase P cleavage of pre-tRNA (Altman et al. 1993; Pace and Brown 1995), and ribose methylation and pseudouridylation of pre-rRNA (see below). In a few situations, however, distance along a single-stranded RNA is determined: apolipoprotein B editing (Smith et al. 1997) and restriction of the length of the poly(A) tail (Wahle 1995; Colgan and Manley 1997). Elucidation of the architecture of the molecular measuring device assembled by the U7 snRNP can therefore be expected to shed light on other RNA processing contexts as well.

MYRIAD snoRNPs PARTICIPATE IN RIBOSOME BIOGENESIS

In the nucleolus of eukaryotic cells, more than 100 tandemly repeated units of ribosomal DNA are transcribed by RNA polymerase I into long precursor transcripts (Fig. 6) (for review, see Eichler and Craig 1995.) Following transcription, numerous residues of the ribosomal RNA

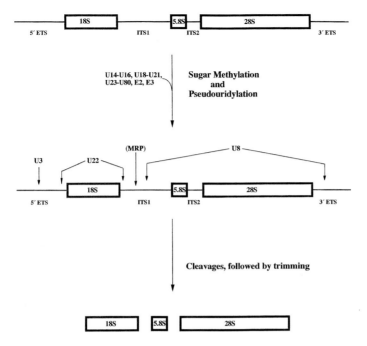

Figure 6 The majority of snoRNPs participate in sugar methylation and pseudouridylation of the nascent pre-rRNA transcript. Following modification, other snoRNPs participate in cleavage of the pre-rRNA to generate immediate precursors that are trimmed to the mature rRNAs. Internal and external transcribed spacer (ITS and ETS) regions are indicated. Cleavages involving yeast U3, U14, and snR30 are not shown, but are discussed in the text.

(rRNA) are modified by either sugar methylation or pseudouridylation. The resulting pre-rRNA is subsequently cleaved to form the mature 18S, 5.8S, and 28S rRNAs and assembled with 5S rRNA and approximately 80 proteins to form the large and small ribosomal subunits prior to their export to the cytoplasm. Over the past few years, it has become clear that a multitude of small nucleolar RNAs (snoRNAs) participate in both the modification and cleavage events that occur during ribosome biogenesis.

Most snoRNPs Target Pre-rRNA for Modification

In vertebrates, shortly after—or perhaps concomitant with—transcription, approximately 200 rRNA residues are modified (Eichler and Craig 1995). Over 95% of the modified sites have been reported: there are approximately 110 ribose residues methylated at the $2'$ position and approximately 95 uridine residues that are isomerized to pseudouridine, all located within highly conserved regions of the mature rRNAs (Maden 1990). The function of rRNA modifications is not yet clear, although it has been postulated that they may contribute in subtle ways to ribosome function. Ribose methylation may stabilize rRNA by increasing hydrophobic interaction surfaces; isomerization of uridine into pseudouridine creates the potential for an additional hydrogen bond at the N-1 position, which may contribute to rRNA folding.

The snoRNAs involved in modification of the precursor transcript can be categorized according to certain sequence motifs (for review, see Smith and Steitz 1997; Tollervey and Kiss 1997). Some snoRNAs contain boxes C (RUGAUG) and D (CUGA) whereas others contain boxes H (ANANNA) and ACA (Fig. 7). Additional box C and D sequences, termed C' and D', are located in internal positions of box C/D snoRNAs and do not strictly adhere to the box C and D consensus sequences. Both classes of snoRNAs exhibit bimodal structures: Box C/D snoRNAs possess terminal and often internal stems that juxtapose box C with D and box C' with D', respectively. Box H/ACA snoRNAs consist of two distinct stem-bulge-hairpin motifs that are followed by either box H (internally) or box ACA (at the 3' terminus).

A unique feature of these two classes of snoRNAs is that each exhibits extensive regions of complementarity to highly conserved regions of rRNA (Fig. 7). For box C/D snoRNAs, these stretches of complementarity to rRNA are uninterrupted, range from 10 to 21 nucleotides, and are located upstream of either box D or D'. For box H/ACA snoRNAs, two shorter regions of complementarity to rRNA are found within the bulged portions of either of the two stem-bulge-hairpin structures (Ganot et al. 1997b). It is important to note that although most snoRNAs target a single rRNA residue for modification, there are a few snoRNAs capable of targeting two rRNA residues for modification; examples of this latter type are depicted in Figure 7.

Analysis of the base-pairing regions of box C/D and box H/ACA snoRNAs indicated that sites of rRNA modification are precisely positioned with respect to the conserved box elements. Specifically, a sugar-methylated rRNA residue can base-pair to a box C/D snoRNA an invariant 5 nucleotides upstream of either box D or D' (Kiss-Laszlo et al. 1996; Nicoloso et al. 1996). Similarly, a reported site of pseudouridylation is located, unpaired, at the base of either of the two hairpin structures of a box H/ACA snoRNA (Fig. 7) (Ganot et al. 1997a; Ni et al. 1997). Since these correlations have been observed for over 70 species, the antisense snoRNAs are believed to serve as guide RNAs for rRNA modification. In-

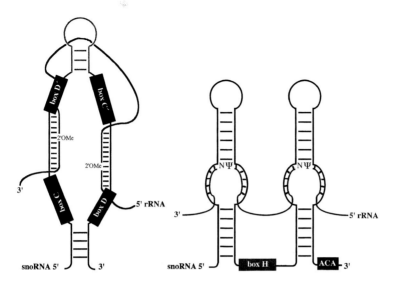

Figure 7 Box C/D and box H/ACA snoRNPs guide 2'-O-methylation and pseudouridylation of rRNA, respectively.

deed, deletion/rescue experiments in yeast, *Xenopus* oocytes and HeLa cells have confirmed that box C/D and box H/ACA snoRNAs function in 2'-O-methylation and in pseudouridylation of pre-rRNA, respectively (Kiss-Laszlo et al. 1996; Tycowski et al. 1996b; Ganot et al. 1997a; Ni et al. 1997). Furthermore, normally unmodified sites of rRNA can be methylated by transfecting into HeLa cells an artificial snoRNA with complementarity to an unmodified region of rRNA (Cavaille et al. 1996) or by moving the box D sequence within the snoRNA such that a normally unmodified rRNA residue base-pairs 5 nucleotides upstream of box D (Kiss-Laszlo et al. 1996). Curiously, depletion of one or more modification guide snoRNAs from either *Xenopus* oocytes or yeast does not appear to affect either the appearance of mature rRNAs or viability.

Much remains to be learned about the protein components of the modification guide snoRNPs. Box C/D snoRNAs can be immunoprecipitated with antibodies to the nucleolar antigen fibrillarin, which has been shown to associate via boxes C and D. Not surprisingly, mutations in the highly conserved yeast fibrillarin homolog (NOP1) affect rRNA ribose methylation (Tollervey et al. 1993). Box H/ACA snoRNAs in yeast, on the other hand, have been demonstrated to bind a protein called Gar1 (Balakin et al. 1996; Kiss et al. 1996; Bousquet-Antonelli et al. 1997; Ganot et al. 1997b). In addition, recent studies indicate that a protein having significant homology with bacterial and eukaryotic pseudouridylases associates with box

H/ACA snoRNAs, suggesting that the enzymes responsible for the modification reactions are components of these RNPs (Lafontaine et al. 1998).

Interestingly, all vertebrate members of the box C/D and box H/ACA classes of snoRNAs contain unmodified 5' end structures (Table 1) and have been found to be intron-encoded (see below). It seems likely that the presence of strong terminal secondary structures and proteins binding to the conserved sequence elements (boxes C and D and boxes H and ACA) protect these snoRNAs from degradation.

Some snoRNPs Are Required for Cleavage of Pre-rRNA

The release of 18S, 5.8S, and 28S rRNAs from the pre-rRNA transcript is a highly orchestrated process in which a handful of snoRNPs—namely, U3, U8, U22, and MRP in vertebrates, or U3, U14, MRP, and snR30 in yeast—facilitate endonucleolytic cleavages within the 5' ETS, ITS1, ITS2, and 3' ETS, followed by exonucleolytic trimming of the cleavage products (Fig. 6). Evidence of snoRNP involvement comes primarily from two types of experiments: disruption of pre-rRNA cleavage by (1) injection of antisense deoxyoligonucleotides complementary to a specific snoRNA into *Xenopus* oocytes where endogenous RNase H degrades the targeted snoRNA, or (2) genetic depletion of a specific snoRNA in yeast. Although the snoRNAs involved in rRNA cleavage were the first to be characterized, details of their mechanisms of action remain unclear. With the exception of MRP, each of these snoRNAs participates in cleavage at two sites, flanking both ends of a mature rRNA sequence.

The snoRNAs required for pre-rRNA cleavage, except for MRP, share sequence homology with the snoRNAs involved in pre-rRNA modification. Specifically, U3, U8, U14, and U22 possess boxes C and D and associate with fibrillarin, whereas snR30 contains boxes H and ACA and associates with Gar1 (see Table 1). MRP (also known as 7-2) possesses neither boxes C/D nor boxes H/ACA. Instead, it associates with the Th/To antigen and exhibits homology with RNase P RNA (or 8-2) (Forster and Altman 1990). Whereas U3 and U8 (and U13) are capped at their 5' ends by TMG, U14 and U22 possess unmodified 5' monophosphates: The capped snoRNAs are transcribed by RNA polymerase II from their own promoters, whereas the uncapped snoRNAs are encoded within the introns of other genes (see below). Compared to the modification guide snoRNAs, these RNAs are larger (>120 nucleotides) and more abundant (10^4–10^5) (see Table 1). U3 and U14 have been characterized from a variety of organisms; however, yeast homologs of U22 and U8 and a vertebrate homolog of snR30 have not yet been identified. Genetic depletion of the snoRNAs for which there are yeast homologs has indicated that these species, unlike the modification guide snoRNAs, are required for viability (for review, see Maxwell and Fournier 1995).

Co-purification of U3 with pre-rRNA and cross-linking of U3 to the 5' ETS region provided initial evidence that U3 participates in pre-rRNA processing (for review, see Maxwell and Fournier 1995; Sollner-Webb et al. 1996; Tollervey and Kiss 1997). Electron micrographs of rDNA being transcribed by RNA polymerase I, known as Miller Christmas trees, show complexes (ornaments) decorating the 5' end of each transcript (branch). Biochemical or genetic depletion of U3 from mouse cell extracts or yeast, respectively, inhibits processing within the 5' ETS and ITS1 (only in yeast), resulting in no 18S accumulation (Fig. 6) (Kass et al. 1990; Hughes and Ares 1991; Beltrame and Tollervey 1995); experiments in *Xenopus* oocytes have suggested that U3 is also involved in the later maturation of 5.8S (Savino and Gerbi 1990). Intriguingly, the ornaments on Miller Christmas trees are absent when the U3-binding site in pre-rRNA is deleted (Mougey et al. 1993), consistent with the finding that the U3 snoRNP stays complexed with the downstream product after 5' ETS cleavage (Kass et al. 1990). Base-pairing interactions between U3 and the pre-rRNA substrate at the 5' ETS and within the 18S sequence have been shown, through compensatory base changes in yeast, to be required for the cleavage reactions (Beltrame and Tollervey 1995; see Tollervey and Kiss 1997). It is important to note that the U3- rRNA duplexes, however, are not similar to those observed for the modification guide snoRNAs in that they are not located directly upstream of the U3 box D sequence.

U14 is unusual in that it possesses two regions, termed domains A and B, with complementarity to two sites that are in close proximity in the secondary structure of 18S rRNA (see Tollervey and Kiss 1997). Recently in *Xenopus* oocytes, domain B, which is positioned just upstream of box D, was shown to be required for methylation of 18S, whereas domain A was found to be nonessential for 18S maturation (Dunbar and Baserga 1998). In contrast, domain A of yeast U14 appears to be required for cleavage of pre-rRNA at sites within the 5' ETS and ITS1 (Liang and Fournier 1995). Morrissey and Tollervey (1997) have proposed that U14 originated as a modification guide snoRNA and only later acquired its role in 18S cleavage. Indeed, a third domain (called domain Y), which is present in yeast but absent in vertebrates, has been found to cross-link to pre-rRNA, suggesting it may play a role in 18S processing in yeast (see Morrissey and Tollervey 1997).

The other snoRNPs required for 18S maturation are U22, in vertebrates, and snR30, in yeast. Depletion of U22 from *Xenopus* oocytes reveals that it is involved in cleavages flanking the 5' and 3' ends of 18S, some 2000 nucleotides apart (Tycowski et al. 1994). Somewhat surprisingly, no complementarity between U22 and the pre-rRNA substrate is apparent. Cross-linking experiments have demonstrated that snR30 interacts with the pre-rRNA substrate; however, despite containing boxes H and ACA, it does not appear to be a pseudouridylation guide RNA (Morrissey and Tollervey 1993; Ni et al. 1997). In vertebrates, the U17 snoRNA appears to be similar to snR30 in that it contains boxes H and ACA and no apparent complementarity to rRNA; U17 has been implicated in 5′ ETS processing and has been found to cross-link to the pre-rRNA substrate (Enright et al. 1996; Mishra and Eliceiri 1997).

U8 and MRP are required for generation of 5.8S and 28S rRNAs. The U8 snoRNP is essential for cleavage at the 3′ end of 28S and upstream of 5.8S rRNA, approximately 6000 nucleotides apart (Peculis and Steitz 1993). Mutational and phylogenetic analyses indicate that the 5′ end of U8 may make important interactions with the 5′ end of 28S rRNA (Peculis and Steitz 1994; Peculis 1997). Yeast MRP functions as an endoribonuclease, cutting the pre-rRNA within ITS1 (for review, see Maxwell and Fournier 1995; Tollervey and Kiss 1997). The high degree of homology between the yeast and vertebrate MRP RNAs suggests that MRP is likely to perform an evolutionarily conserved function. Similar to U8 and U22, no cross-links between MRP RNA and pre-rRNA have been reported.

In summary, it appears that shortly following transcription, the pre-rRNA becomes coated with numerous snoRNA molecules. Based on the presence of approximately 200 modified nucleotides in vertebrate rRNAs, over one-third of the mature rRNA sequences at this stage are predicted to be engaged in duplexes with the snoRNAs since, in general, one snoRNP targets a single rRNA site for modification and each exhibits >10 nucleotides of complementarity to the rRNA. More analogous to the roles of snRNPs involved in splicing, the snoRNPs involved in cleavage of the modified pre-rRNA may assist in bringing distant sites into close proximity, enabling coordinated release of immediate precursors to the mature rRNAs (see Fig. 6). The numerous base-pairing interactions with snoRNPs must be disrupted prior to completion of ribosome assembly, a process likely to involve the action of RNA helicases. That such an elaborate RNA machine has evolved to facilitate the biogenesis of rRNAs bespeaks origins in the RNA World.

Intron-encoded snoRNAs

A unique feature of snoRNAs is that most are encoded within the introns of protein coding genes (for review, see Maxwell and Fournier 1995).

This economic use of intronic RNA is commonplace among intron-rich organisms, such as vertebrates, where the modification guide snoRNAs are exclusively intron-encoded. Only U3, U8, U13 (the function of which is currently unknown), and MRP snoRNAs are produced from independent transcription units in vertebrates. In contrast, in *Saccharoymyces cerevisiae* many snoRNAs are transcribed from their own promoters. A vertebrate host gene intron encodes only a single snoRNA, whereas in yeast and plants some snoRNA genes are located in polycistronic arrays without exons separating the snoRNA sequences. In all species investigated, the intron-encoded snoRNAs are transcribed from their host genes by RNA polymerase II as portions of the pre-mRNA (see Table 1). The functional snoRNAs are then produced by exonucleolytic trimming that follows either splicing or endonucleolytic cleavage of intron sequences. Surprisingly, the host gene for a particular snoRNA can differ even among closely related organisms; moreover, variant snoRNA sequences are often located in adjacent introns. These observations suggest that snoRNA genes are highly mobile elements. How they became inserted into introns remains an intriguing question.

Although many snoRNAs are encoded within introns of ribosomal protein genes, others are found within introns of genes specifying translation factors or nucleolar proteins. A ribosome-related function is not, however, obvious for some snoRNA host gene protein products (e.g., ATP synthase β). Furthermore, a subset of snoRNA host genes do not appear to produce any protein product: UHG, gas5, U17HG, and U19HG (Tycowski et al. 1996a; C.M. Smith and J.A. Steitz; P. Pelczar and W. Filipowicz; T. Kiss, all in prep.). The exons of each of these genes are processed to yield polyadenylated RNAs; however, no long open reading frames are apparent and the spliced transcripts are short-lived. Strikingly, inspection of known snoRNA host genes reveals that they all exhibit pyrimidine-rich transcription start sequences (C.M. Smith and J.A. Steitz, in prep.) and therefore share features with the 5' TOP (*t*erminal *o*ligo*p*yrimidine) family of genes (for review, see Meyuhas et al. 1996). Whereas 5' TOP genes have previously been scrutinized with respect to the effects of growth arrest on their mRNAs, their identification as snoRNA host genes suggests that the 5' TOP sequence may have some important nuclear function that coordinates or facilitates the production of ribosome components.

OTHER RNPs OF THE NUCLEUS

In addition to the well-characterized snRNPs of known function described above, there are snRNPs whose roles continue to be elusive. Included in this category is the 7SK snRNP, the most abundant (~10^5 copies) RNP of unknown function in vertebrate cells (Wassarman and Steitz 1991). The viral snRNPs found in lymphocytes infected with various herpesviruses have likewise not yet yielded their functional secrets. Finally, a growing roster of nuclear-localized polyadenylated RNAs form a family of intriguing larger RNPs, whose contributions to the life of the nucleus all remain to be deciphered.

Viral snRNPs Bind Important Host Proteins

Epstein-Barr virus (EBV) infection of human B lymphocytes often produces a transformed state (for review, see Kieff and Liebowitz 1990; Miller 1990). During latency, most viral genes are not expressed. Among the exceptions are those for two abundant small RNAs, EBER1 and 2 (for *E*pstein *B*arr-*e*ncoded *R*NA). The EBERs (each about 170 nucleotides) are transcribed by RNA polymerase III, are localized in the nucleoplasm, and quantitatively bind the host La protein (Table 1). Homologous small RNAs are found in cells infected by the related primate viruses *Herpesvirus papio* and *pan*.

EBER1 has been discovered to bind a host protein, ribosomal protein L22, about half of which becomes relocated to the nucleoplasm in EBV-positive cells (Toczyski et al. 1994). What advantage this provides for the virus is not yet clear, since EBER1, as well as EBER2, can be deleted from the viral genome with no apparent harmful effects on viral replication, transformation of host cells, or conversion from latency to lytic growth (Swaminathan et al. 1991), nor is the virus more sensitive to interferon (Swaminathan et al. 1992).

Herpesvirus saimiri targets the T lymphocytes of new-world primates to produce either lytic infection or malignant transformation. In transformed T cells, the most abundant viral transcripts are small RNAs (75–143 nucleotides) (Albrecht and Fleckenstein 1992) that assemble into particles of the Sm class; that is, they bind host Sm proteins and acquire TMG caps (Lee et al. 1988). There are at least seven different HSURs (for *H*erpesvirus *s*aimiri *U R*NAs), whose abundances range from moderate to low compared to the splicing snRNPs (Table 1). HSURs are synthesized by RNA polymerase II in response to the same unusual promoter and terminator signals as cellular U RNAs.

The sequences at the 5' ends of three of the HSURs mimic AU-rich mRNA destabilization signals found in the 3'-untranslated regions of messages encoding cellular oncoproteins, cytokines, and lymphokines (Shaw and Kamen 1986). The idea that HSURs might therefore act to manipulate the host's mRNA degradation machinery has growing experimental support. Upon transfection into mammalian cells, HSUR1 has been found to exhibit a short half-life, dependent on its AU-rich signal (Fan et al. 1997). A cellular protein that binds this sequence, HuR, is a member of the Elav family of RNA-binding proteins and shuttles between nucleus and cytoplasm; when overexpressed, HuR stabilizes short-lived mRNAs (Myer et al. 1997; X.C. Fan and J.A. Steitz, in prep.). Perhaps the HSURs containing AU-rich sequences sacrifice themselves, thereby sequestering some important component(s) of the decay machinery; this could lengthen the half-lives of unstable mRNAs, leading to increased production of a number of growth-related cellular proteins. This idea for how HSURs contribute to viral transformation has not yet been proven, nor have plausible functions for the other HSURs that do not contain AU-rich elements been proposed.

Most intriguing is a newly discovered virus-encoded nuclear RNA that combines the properties of both snRNAs and mRNAs (Sun et al. 1996; Zhong and Ganem 1997). In tumor cells and lymphoma cells infected with Kaposi sarcoma-associated herpesvirus (KSHV), a polyadenylated transcript of about 1 kb (called PAN or nut-1) is generated from a promoter that contains U snRNA-like proximal and distal sequence elements. Yet, this RNA is not capped with TMG as are U snRNAs, nor does it contain a long open reading frame like messages. It is highly abundant in the KSHV lytic cycle ($\sim 10^5$ copies per cell) and forms an RNP whose protein content has not yet been characterized. Hints of its association with Sm snRNPs suggests that this KSHV RNA might somehow manipulate the host's RNA processing machinery to the advantage of the virus.

Yet other intriguing nuclear RNAs—not all small—have been reported in latent or persistent viral infection. The LATs (latency associated transcripts) are the only viral RNAs abundantly expressed in the nuclei of neurons latently infected by *Herpes simplex* virus (for review, see Fraser et al. 1992). The LATs comprise at least three RNAs that are transcribed from repeat sequences in the viral genome; the smaller LATs (1.5 and 2.0 kb) lack 5' caps and are not polyadenylated, suggesting they represent stable introns. Recent reports (Chen et al. 1997; Garber et al. 1997) show that the LAT region exerts its influence in the early stage of neuronal infection to reduce productive-cycle viral gene expression. Both antisense effects and silencing of the viral genome have been proposed as mechanisms whereby LATs function. Similarly, in the insect virus Hz-1, only a 2.9-kb viral transcript is detected during viral persistency (Chao et al. 1998). It is apparently noncoding, exclusively nuclear, and has been implicated in the establishment of persistent Hz-1 viral infection.

The Enigma of Polyadenylated Nuclear RNAs

Although most RNAs that are spliced and polyadenylated are exported to the cytoplasm as mRNAs, a growing number of such transcripts have recently been identified as predominantly nuclear, even in nonviral infected cells. These RNAs appear to be noncoding: They contain multiple stop codons and only short potential open reading frames with poor translational initiation contexts, and they lack detectable translation products. Such transcripts, which undoubtedly exist as RNPs in the nucleus, are postulated to have important roles in gene expression or chromatin architecture, but in no case is the detailed mechanism understood. Numerous cytoplasmic noncoding RNAs also have been described, but are not reviewed in detail here.

Xist (*X i*nactive *s*pecific *t*ranscript), a 15–17-kb spliced and polyadenylated RNA required for X chromosome inactivation in mammals (Brockdorff et al. 1992; Brown et al. 1992; for review, see Kuroda and Meller 1997; Solter and Wei 1997; Willard and Salz 1997), functions *in cis* by associating exclusively with the chromosome from which it is transcribed. *Xist* contains two regions of direct repeats—one is nine copies of approximately 50 nucleotides, the other eight copies of approximately 300 nucleotides; within the 5′-most repeat regions are 10- and 11-nucleotide core regions conserved between mouse and human that may act as protein-binding sites. *Xist* is selectively stabilized on one X chromosome, thereby establishing an inactive chromatin state. *Xist* transcription from the active X chromosome is then silenced. In polyploid tissue-culture cells with more than two X chromosomes, all but one X is inactive. Exactly how all but one X chromosome are selected for silencing and how the stability of *Xist* RNA is regulated is not well understood (Penny et al. 1996; Marahrens et al. 1997, 1998).

In *Drosophila,* the single X chromosome in male cells, whose transcription is up-regulated when compared to the two X chromosomes in female cells, is coated by one or more polyadenylated RNAs of 1.1–1.3 kb (Amrein and Axel 1997; Meller et al. 1997). The exact mechanism of action of roX RNAs (*R*NA *o*n the *X*) is not understood, but the synthesis or accumulation of roX RNAs is known to be regulated by MSL proteins, which mediate dosage compensation and themselves assemble onto the X chromosome. roX1 can function *in trans* to bind the X chromosome and is able to recognize a translocated fragment of the X chromosome. Disruption of roX1 shows no obvious phenotype, suggesting that nonhomologous redundant genes may compensate for the missing roX1 RNA. Both roX and *Xist* RNAs are involved in dosage compensation, but with opposite results on overall gene expression. How these RNPs modulate chromatin structure and what role they have as important mediators in establishing and/or maintaining dosage compensation await further studies.

Drosophila hsr-omega is a constitutively active gene that is essential for viability in flies. It produces RNAs that are rapidly induced upon heat shock or in response to other agents (for reviews, see Pardue et al. 1990; Lakhotia and Sharma 1996). *Omega-n* is an exclusively nuclear unspliced but polyadenylated transcript of approximately 10 kb with a 5' nonrepetitive region that contains a 60-nucleotide conserved sequence and a 3' region of >5 kb consisting of tandem repeats of 115–284 nucleotides (depending on the species) (Hogan et al. 1994). The other gene transcript utilizes the same transcription start site but an alternative termination site to produce a shorter RNA (*omega-pre-c*) that is spliced, polyadenylated, and transported to the cytoplasm (omega-c). After the cessation of heat shock, the high levels of nuclear *omega-n* quickly return to lower levels, but if transcription is inhibited by actinomycin D, *omega-n* RNA is stabilized. Because of its rapid response to changes in cellular conditions, *omega-n* is proposed to have a regulatory role in the nucleus and possibly to coordinate nuclear-cytoplasmic activities (Hogan et al. 1994).

The *S. pombe* Mei2 protein, a 720-amino acid protein with three RNA recognition motifs (RRMs), is required for premeiotic DNA synthesis and for meiosis I (for review, see Yamamoto 1996). Genetic studies identified a 500-nucleotide polyadenylated RNA, called meiRNA, which specifically binds Mei2 in meiotic cells and is essential for meiosis I (Watanabe and Yamamoto 1994). The exact mechanism whereby meiRNA (and probably at least one other RNA that binds Mei2 to initiate premeiotic DNA synthesis) performs an essential role in the regulation of meiosis is unclear.

The 17-kb NTT (*n*oncoding *t*ranscript in *T* cells) RNA is expressed exclusively in activated human $CD4^+$ T cells in vivo and in a subset of activated human $CD4^+$ T-cell clones (Liu et al. 1997). Although polyadenylated, NTT is not spliced and is localized in the nucleus. Besides regions of unique sequence, the NTT gene contains two 74 bp tandem repeats, human Alu and L1 interspersed repetitive sequence homologs, and di-, tetra-, and penta-nucleotide repetitive sequences. Its function is not known.

PROSPECTS

In addition to marked progress in understanding snRNP function in the past several years, most stunning is the realization that a much larger number of low-abundance snRNPs populate eukaryotic cells than previously appreciated. A novel low-abundance spliceosome that excises a specific subset of introns from metazoan pre-mRNAs has been discovered, raising intriguing evolutionary questions. In all eukaryotes, virtually hundreds of nucleolar snRNPs comprise an enormous machinery devoted to introducing modified nucleotides (still of unknown function) into rRNAs and probably snRNAs as well. The world of snRNPs has therefore reached 200+, but even more may be revealed as we continue to delve into the RNA-based workings of the living cell.

ACKNOWLEDGMENTS

We are indebted to our colleagues in the Steitz lab for discussions and inspiration. Our work was supported by grants GM-26154 and CA-16038 from the National Institutes of Health. All authors contributed equally to the preparation of this chapter.

REFERENCES

Agabian N. 1990. *Trans* splicing of nuclear pre-mRNAs. *Cell* **61:** 1157–1160.

Albrecht J.C. and Fleckenstein B. 1992. Nucleotide sequence of HSUR 6 and HSUR 7, two small RNAs of herpesvirus saimiri. *Nucleic Acids Res.* **20:** 1810.

Altman S., Kirsebom L., and Talbot S. 1993. Recent studies of ribonuclease P. *FASEB J.* **7:** 7–14.

Amrein H. and Axel R. 1997. Genes expressed in neurons of adult male *Drosophila. Cell* **88:** 459–469.

Balakin A.G., Smith L., and Fournier M.J. 1996. The RNA world of the nucleolus: Two major families of small RNAs defined by different box elements with related functions. *Cell* **86:** 823–834.

Baldi M.I., Mattoccia E., Bufardeci E., Fabbri S., and Tocchini-Valentini G.P. 1992. Participation of the intron in the reaction catalyzed by the *Xenopus* tRNA splicing endonuclease. *Science* **255:** 1404–1408.

Behrens S.-E., Tyc K., Kastner B., Reichelt J., and Lührmann R. 1993. Small nuclear ribonucleoprotein (RNP) U2 contains numerous additional proteins and has a bipartite RNP structure under splicing conditions. *Mol. Cell. Biol.* **13:** 307–319.

Beltrame M. and Tollervey D. 1995. Base pairing between U3 and the pre-ribosomal RNA is required for 18S rRNA synthesis. *EMBO J.* **14:** 4350–4356.

Blumenthal T. 1995. *Trans*-splicing and polycistronic transcription in *Caenorhabditis elegans. Trends Genet.* **11:** 132–136.

Bousquet-Antonelli C., Henry Y., Gelunge J.-P., Caizergues-Ferrer M., and Kiss T. 1997. A small nucleolar RNP protein is required for pseudouridylation of eukaryotic ribosomal RNAs. *EMBO J.* **16:** 4769–4775.

Brockdorff N., Ashworth A., Kay G.F., McCage V.M., Norris D.P., Cooper P.J., Swift S., and Rastan S. 1992. The product of the mouse *Xist* gene is a 15 kb inactive X-specific transcript containing no conserved ORF and located in the nucleus. *Cell* **71:** 515–526.

Brown C.J., Hendrich B.D., Rupert J.L., Laferniere R.G., Xing Y., Lawrence H., and Willard H.F. 1992. The human *XIST* gene: Analysis of a 17 kb inactive X-specific

RNA that contains conserved repeats and is highly localized within the nucleus. *Cell* **71:** 527–542.
Bruzik J.P., Van Doren K., Hirsh D., and Steitz J.A. 1988. *Trans* splicing involves a novel form of small nuclear ribonucleoprotein particles. *Nature* **335:** 559–562.
Cavaille J., Nicoloso M., and Bachellerie J.P. 1996. Targeted ribose methylation of RNA *in vivo* directed by tailored antisense RNA guides. *Nature* **383:** 732–735.
Chanfreau G. and Jacquier A. 1994. Catalytic site components common to both splicing steps of a group II intron. *Science* **266:** 1383–1387.
Chao Y.-C., Lee S.-T., Chang M.-C., Chen H.-H., Chen S.-S., Wu T.-Y., Liu F.H., Hsu E.-L., and Hou R.F. 1998. A 2.9 kilobase noncoding nuclear RNA functions in the establishment of persistent Hz-1 viral infection. *J. Virol.* **72:** 2233–2245.
Chen S.-H., Kramer M.F., Schaffer P.A., and Coen D.M. 1997. A viral function represses accumulation of transcripts from productive-cycle genes in mouse ganglia latently infected with herpes simplex virus. *J. Virol.* **71:** 5878–5884.
Cho D.C., Scharl E.C., and Steitz J.A. 1995. Decreasing the distance between the two conserved sequence elements of histone pre-messenger RNA interferes with 3′ processing *in vitro*. *RNA* **1:** 905–914.
Colgan D.F. and Manley J.L. 1997. Mechanism and regulation of mRNA polyadenylation. *Genes Dev.* **11:** 2755–2766.
Crispino J.D., Blencowe B.J., and Sharp P.A. 1994. Complementation by SR proteins of pre-mRNA splicing reactions depleted of U1 snRNP. *Science* **265:** 1866–1869.
Dahlberg J.E. and Lund E. 1988. The genes and transcription of the major small nuclear RNAs. In *Small nuclear ribonucleoprotein particles* (ed. M.L. Birnstiel), pp. 38–70. Springer-Verlag, Berlin.
Dietrich R.C., Incorvaia R., and Padgett R.A. 1997. Terminal intron dinucleotide sequences do not distinguish between U2– and U12–dependent introns. *Mol. Cell* **1:** 151–160.
Dunbar D.A. and Baserga S.J. 1998. The U14 snoRNA is required for 2′-O-methylation of the pre-18S rRNA in *Xenopus* oocytes. *RNA* **4:** 195–204.
Dungan J.M., Watkins K.P., and Agabian N. 1996. Evidence for the presence of a small U5–like RNA in active *trans*-spliceosomes of *Trypanosoma brucei*. *EMBO J.* **15:** 4016–4029.
Eichler D.C. and Craig N. 1995. Processing of eukaryotic ribosomal RNA. *Prog. Nucleic Acid Res. Mol. Biol.* **49:** 197–239.
Enright C.A., Maxwell M.J., Eliceiri G.L., and Sollner-Webb B. 1996. 5′ ETS rRNA processing facilitated by four small RNAs: U14, E3, U17 and U3. *RNA* **2:** 1094–1099.
Fabrizio P. and Abelson J. 1992. Thiophosphates in yeast U6 snRNA specifically affect pre-mRNA splicing *in vitro*. *Nucleic Acids Res.* **20:** 3659–3664.
Fan X.C., Myer V.E., and Steitz J.A. 1997. AU-rich elements target small nuclear RNAs as well as mRNAs for rapid degradation. *Genes Dev.* **11:** 2557–2568.
Field D.J. and Friesen J.D. 1996. Functionally redundant interactions between U2 and U6 spliceosomal snRNAs. *Genes Dev.* **10:** 489–501.
Fischer U., Liu Q., and Greyfuss G. 1997. The SMN-SIP1 complex has an essential role in spliceosomal snRNP biogenesis. *Cell* **90:** 1023–1029.
Forster A.C. and Altman S. 1990. Similar cage-shaped structures for the RNA components of all ribonuclease P and ribonuclease MRP enzymes. *Cell* **62:** 407–409.
Fraser N.W., Block T.M., and Spivack J.G. 1992. The latency-associated transcripts of herpes simplex virus: RNA in search of function. *Virology* **191:** 1–8.

Furger A., Schaller A., and Schümperli D. 1998. Functional importance of conserved nucleotides at the histone RNA 3′ processing site. *RNA* **4:** 246–256.

Ganot P., Bortolin M.-L., and Kiss T. 1997a. Site-specific pseudouridine formation in pre-ribosomal RNA is guided by small nucleolar RNAs. *Cell* **89:** 799–809.

Ganot P., Caizergues-Ferrer M., and Kiss T. 1997b. The family of box ACA small nucleolar RNAs is defined by an evolutionarily conserved secondary structure and ubiquitous sequence elements essential for RNA accumulation. *Genes Dev.* **11:** 941–956.

Garber D.A., Schaffer P.A., and Knipe D.M. 1997. The latency-associated transcripts of herpes simplex virus: RNA in search of function. *Virology* **191:** 1–8.

Gottschalk A., Tang J., Puig O., Salgado J., Neubauer G., Colot H.V., Mann M., Séraphin B., Rosbash M., Lührmann R., and Fabrizio P. 1998. A comprehensive biochemical and genetic analysis of the yeast U1 snRNP reveals five novel proteins. *RNA* **4:** 374–393.

Grimm C., Stefanovic B., and Schümperli D. 1993. The low abundance of U7 snRNA is partly determined by its Sm binding site. *EMBO J.* **12:** 1229–1238.

Guthrie C. 1991. Messenger RNA splicing in yeast: Clues to why the spliceosome is a ribonucleoprotein. *Science* **253:** 157–163.

Hall S.L. and Padgett R.A. 1994. Conserved sequences in a class of rare eukaryotic nuclear introns with non-consensus splice sites. *J. Mol. Biol.* **271:** 1716–1718.

———. 1996. Requirement of U12 snRNA for *in vivo* splicing of a minor class of eukaryotic nuclear pre-mRNA introns. *Science* **271:** 1716–1718.

Hamm J. and Mattaj I.W. 1989. An abundant U6 snRNP found in germ cells and embryos of *Xenopus laevis*. *EMBO J.* **8:** 4179–4187.

Hannon G.J., Maroney P.A., and Nilsen T.W. 1991. U small nuclear ribonucleoprotein requirements for nematode *cis*- and *trans*-splicing *in vitro*. *J. Biol. Chem.* **266:** 22792–22695.

Hannon G.J., Maroney P.A., Denker J.A., and Nilsen T.W. 1990. *Trans*-splicing of nematode pre-mRNA *in vitro*. *Cell* **61:** 1247–1255.

Hannon G.J., Maroney P.A., Yu Y.-T., Hannon G.E., and Nilsen T.W. 1992. Interaction of U6 snRNA with a sequence required for function of the nematode SL RNA in *trans*-splicing. *Science* **258:** 1775–1780.

Hetzer M., Wurzer G., Schweyen R.J., and Mueller M.W. 1997. *Trans*-activation of group II intron splicing by nuclear U5 snRNA. *Nature* **386:** 417–420.

Hogan N.C., Traverse K.L., Sullivan D.E., and Pardue M.-L. 1994. The nucleus-limited Hsr-omega-n transcript is a polyadenylated RNA with a regulated intranuclear turnover. *J. Cell Biol.* **125:** 21–30.

Huang X.Y. and Hirsh D. 1989. A second *trans*-spliced RNA leader sequence in the nematode *Caenorhabditis elegans*. *Proc. Natl. Acad. Sci.* **86:** 8640–8644.

Hughes J.M.X. and Ares M.J. 1991. Depletion of U3 small nucleolar RNA inhibits cleavage in the 5′ external spacer of yeast pre-ribosomal RNA and prevents formation of 18S ribosomal RNA. *EMBO J.* **10:** 4231–4239.

Hurt M.M., Chodchoy N., and Marzluff W.F. 1989. The mouse histone H2a.2 gene from chromosome 3. *Nucleic Acids Res.* **17:** 8876.

Jackson I.J. 1991. A reappraisal of non-consensus mRNA splice sites. *Nucleic Acids Res.* **19:** 3795–3798.

Kass S., Tyc K., Steitz J.A., and Sollner-Webb B. 1990. The U3 small nucleolar ribonucleoprotein functions in the first step of preribosomal RNA processing. *Cell* **60:** 897–908.

Kieff E. and Liebowitz D. 1990. Epstein-Barr virus and its replication. *Virology* **2:** 1889–1920.

Kiss T., Bortolin M.-L., and Filipowicz W. 1996. Characterization of the intron-encoded U19 RNA, a new mammalian small nucleolar RNA that is not associated with fibrillarin. *Mol. Cell. Biol.* **16:** 1391–1400.

Kiss-Laszlo Z., Henry Y., Bachellerie J.-P., Caizergues-Ferrer M., and Kiss T. 1996. Site-specific ribose methylation of pre-ribosomal RNA: A novel function for small nucleolar RNAs. *Cell* **85:** 1077–1088.

Kolossova I. and Padgett R.A. 1997. U11 snRNA interacts *in vivo* with the 5′ splice site of U12–dependent (AU-AC) pre-mRNA introns. *RNA* **3:** 227–233.

Kuroda M.I. and Meller V.H. 1997. Transient *Xist*-ence. *Cell* **91:** 9–11.

Lafontaine D.L.J., Bousquet-Antonelli C., Henry Y., Caizergues-Ferrer M., and Tollervey D. 1998. The box H + ACA snoRNAs carry Cbf5p, the putative rRNA pseudouridine synthase. *Genes Dev.* **12:** 527–537.

Lakhotia S.C., and Sharma A. 1996. The 93D (*hsr-omega*) locus of *Drosophila*: Noncoding gene with house-keeping functions. *Genetica* **97:** 339–348.

Lee S.I., Murthy S.C.S., Trimble J.J., Desrosiers R.C., and Steitz J.A. 1988. Four novel U RNAs are encoded by a herpesvirus. *Cell* **54:** 599–607.

Liang W.Q. and Fournier M.J. 1995. U14 base-pairs with 18S rRNA: A novel snoRNA interaction required for rRNA processing. *Genes Dev.* **9:** 2433–2443.

Liu A.Y., Torchia B.S., Migeon B.R., and Siliciano R.F. 1997. The human *NTT* gene: Identification of a novel 17-kb noncoding nuclear RNA expressed in activated CD4$^+$ T cells. *Genomics* **39:** 171–184.

Lucke S., Xu G.-L., Palfi Z., Cross M., Bellofatto V., and Bindereif A. 1996. Spliced leader RNA of trypanosomes: *In vivo* mutational analysis reveals extensive and distinct requirements for *trans* splicing and cap4 formation. *EMBO J.* **15:** 4380–4391.

Lührmann R. 1988. snRNP proteins. In *Small nuclear ribonucleoprotein particles* (ed. M.L. Birnstiel), pp. 71–99. Springer-Verlag, Berlin.

Lyon C.E., Bohmann K., Sleeman J., and Lamond A.I. 1997. Inhibition of protein dephosphorylation results in the accumulation of splicing snRNPs and coiled bodies within the nucleolus. *Exp. Cell Res.* **230:** 84–93.

Maden B.E.H. 1990. The numerous modified nucleotides in eukaryotic ribosomal RNA. *Prog. Nucleic Acids Res.* **39:** 241–303.

Madhani H.D. and Guthrie C. 1994. Randomization-selection analysis of snRNAs *in vivo*: Evidence for a tertiary interaction in the spliceosome. *Genes Dev.* **8:** 1071–1986.

Marahrens Y., Loring J., and Jaenisch R. 1998. Role of the *Xist* gene is X chromosome choosing. *Cell* **92:** 657–664.

Marahrens Y., Panning B., Dausman J., Strauss W., and Jaenisch R. 1997. *Xist*-deficient mice are defective in dosage compensation but not spermatogenesis. *Genes Dev.* **11:** 156–166.

Maroney P.A., Hannon G.J., Shambaugh J.D., and Nilsen T.W. 1991. Intramolecular basepairing between the nematode spliced leader and its 5′ splice site is not essential for *trans*-splicing. *EMBO J.* **9:** 3667–3673.

Maroney P.A., Yu Y.-T., Jankowska M., and Nilsen T.W. 1996. Direct analysis of nematode *cis*- and *trans*-spliceosomes: A functional role for U5 snRNA in spliced leader addition *trans*-splicing and the identification of novel Sm snRNPs. *RNA* **2:** 735–745.

Martin F., Schaller A., Eglite S., Schümperli D., and Muller B. 1997. The gene for histone RNA hairpin binding protein is located on human chromosome 4 and encodes a novel type of RNA binding protein. *EMBO J.* **16:** 769–778.

Marzluff W. F. 1992. Histone 3' ends: Essential and regulatory functions. *Gene Expr.* **2:** 93–97.

Mattaj I.W. 1988. UsnRNP assembly and transport. In *Small nuclear ribonucleoprotein particles* (ed. M.L. Birnstiel), pp. 100–114. Springer-Verlag, Berlin.

Maxwell E.S. and Fournier M. 1995. The small nucleolar RNAs. *Annu. Rev. Biochem.* **64:** 897–934.

Meller V.H., Wu K.H., Roman G., Kuroda M.I., and Davis R.L. 1997. *RoX1* RNA paints the X chromosome of male *Drosophila* and is regulated by the dosage compensation system. *Cell* **88:** 445–457.

Meyuhas O., Avni D., and Shama S. 1996. Translational control of ribosomal protein mRNAs in eukaryotes. In *Translational control* (ed. J.W.B. Hershey et al.), pp. 363–388. Cold Spring Harbor Laboratory Press, Cold Spring Harbor, New York.

Miller G. 1990. Biology, pathogenesis, and medical aspects. In *Virology*, 2nd edition (ed. B.N. Fields, et al.), vol. 2, pp. 1921–1958. Raven Press, New York.

Mishra R.K. and Eliceiri G.L. 1997. Three small nucleolar RNAs that are involved in ribosomal RNA precursor processing. *Proc. Natl. Acad. Sci.* **94:** 4972–4977.

Moore M.J., Query C.C., and Sharp P.A. 1993. Splicing of precursors to mRNAs by the spliceosome. In *The RNA world* (ed. R.F. Gesteland and J.F. Atkins), pp. 303–357. Cold Spring Harbor Laboratory Press, Cold Spring Harbor, New York.

Morrissey J.P. and Tollervey D. 1993. Yeast snR30 is a small nucleolar RNA required for 18S rRNA synthesis. *Mol. Cell. Biol.* **13:** 2469–2477.

———. 1997. U14 small nucleolar RNA makes multiple contacts with the pre-ribosomal RNA. *Chromosoma* **105:** 515–522.

Mougey E.B., O'Reilly M., Osheim Y., Miller O.L., Jr., Beyer A., and Sollner-Webb B. 1993. The terminal balls characteristic of eucaryotic rRNA transcription units in chromatin are rRNA processing complexes. *Genes Dev.* **7:** 1609–1619.

Mowry K.L. and Steitz J.A. 1987. Identification of the human U7 snRNP as one of several factors involved in the 3' end maturation of histone premessenger RNA's. *Science* **238:** 1682–1687.

Muller B. and Schümperli D. 1998. The U7 snRNP and the hairpin binding protein: Key players in histone mRNA metabolism. *Semin. Cell Dev. Biol.* (in press).

Myer V.E. and Steitz J.A. 1997. Isolation and characterization of a novel, low abundance hnRNPprotein: AO. *RNA* **1:** 171–182.

Newman A. and Norman C. 1991. Mutations in yeast U5 snRNA alter the specificity of 5' splice-site cleavage. *Cell* **65:** 115–123.

———. 1992. U5 snRNA interacts with exon sequences at 5' and 3' splice sites. *Cell* **68:** 1–20.

Ni J., Tien A.L., and Fournier M.J. 1997. Small nucleolar RNAs direct site-specific synthesis of pseudouridine in ribosomal RNA. *Cell* **89:** 565–573.

Nicoloso M., Qu L.-H., Michot B., and Bachellerie J.-P. 1996. Intron-encoded, antisense small nucleolar RNAs: The characterization of nine novel species points to their direct role as guides for the 2'-O-ribose methylation of rRNAs. *J. Mol. Biol.* **260:** 178–195.

Nilsen T.W. 1993. *Trans*-splicing of nematode pre-mRNA. *Annu. Rev. Microbiol.* **47:** 413–440.

———. 1994. RNA–RNA interactions in the spliceosome: Unraveling the ties that bind. *Cell* **78:** 1–4.

———. 1996. A parallel spliceosome. *Science* **273:** 1813.

———. 1997. *Trans*-splicing. In *Eukaryotic RNA processing* (ed. A. Krainer), pp. 310–334. IRL Press, Oxford, United Kingdom.
Nilsen T.W., Shambaugh J., Denker J., Chubb G., Faser C., Putnam L., and Bennett K. 1989. Characterization and expression of a spliced leader RNA in the parasitic nematode *Ascaris Lumbricoides* var. *Suum*. *Mol. Cell. Biol.* **9:** 3543–3547.
Oh J., Bailin T., Fukai K., Feng G.H., Ho L., Mao J., Frenk E., Tamura N., and Spritz R.A. 1996. Positional cloning of a gene for Hermansky-Pudlak syndrome, a disorder of cytoplasmic organelles. *Nat. Genet.* **14:** 300–306.
O'Keefe R.T., Norman C., and Newman A.J. 1996. The invariant U5 snRNA loop 1 sequence is dispensable for the first catalytic step of pre-mRNA splicing in yeast. *Cell* **86:** 679–689.
Pace N.R. and Brown J.W. 1995. Evolutionary perspective on the structure and function of ribonuclease P, a ribozyme. *J. Bacteriol.* **177:** 1919–1928.
Pardue M.-L., Bendena W.G., Fini M.E., Garbe J.C., Hogan N.C., and Traverse K.L. 1990. *Hsr-omega*, a novel gene encoded by a *Drosophila* heat shock puff. *Biol. Bull.* **179:** 77–86.
Parker R. and Siliciano P.G. 1993. Evidence for an essential non-Watson-Crick interaction between the first and last nucleotides of a nuclear pre-mRNA intron. *Nature* **361:** 660–662.
Peculis B.A. 1997. The sequence of the 5' end of the U8 small nucleolar RNA is critical for 5.8S and 28S rRNA maturation. *Mol. Cell. Biol.* **17:** 3702–3713.
Peculis B.A. and Steitz J.A. 1993. Disruption of U8 nucleolar snRNA inhibits 5.8S and 28S rRNA processing in the *Xenopus* oocyte. *Cell* **73:** 1233–1245.
———. 1994. Sequence and structural elements critical for U8 snRNP function in *Xenopus* oocytes are evolutionarily conserved. *Genes Dev.* **8:** 2241–2255.
Penny G.D., Kay G.F., Sheardown S.A., Rastan S., and Brockdorff N. 1996. Requirement for *Xist* in X chromosome inactivation. *Nature* **379:** 131–137.
Piccirilli J.A., Vyle J.S., Caruthers M.H., and Cech T.R. 1993. Metal ion catalysis in the *tetrahymena* ribozyme reaction. *Nature* **361:** 85–88.
Reich C.I., VanHoy R.W., Porter G.L., and Wise J.A. 1992. Mutations at the 3' splice site can be suppressed by compensatory base changes in U1 snRNA in fission yeast. *Cell* **69:** 1159–1169.
Savino R. and Gerbi S.A. 1990. *In vivo* disruption of *Xenopus* U3 snRNA affects ribosomal RNA processing. *EMBO J.* **9:** 2299–2308.
Scharl E.C. and Steitz J.A. 1994. The site of 3' end formation of histone messenger RNA is a fixed distance from the downstream element recognized by the U7 snRNP. *EMBO J.* **13:** 2432–2440.
———. 1996. Length suppression in histone mRNA 3' end maturation: Processing defects of insertion mutant pre-mRNAs can be compensated by insertions into the U7 snRNA. *Proc. Natl. Acad. Sci.* **93:** 14659–14664.
Sharp P.A. 1991. Five easy pieces. *Science* **254:** 663.
Sharp P.A. and Burge C.B. 1997. Classification of introns: U2-type or U12-type. *Cell* **91:** 875–879.
Shaw G. and Kamen R. 1986. A conserved AU sequence from the 3' untranslated region of GM-CSF mRNA mediates selective mRNA degradation. *Cell* **46:** 659–667.
Smith C.M. and Steitz J.A. 1997. Sno storm in the nucleolus: New roles for myriad small RNPs. *Cell* **89:** 669–672.
Smith H.C., Gott J.M., and Hanson M.R. 1997. A guide to RNA editing. *RNA* **3:** 1105–1123.

Smith H.O., Tabiti K., Schaffner G., Soldati D., Albrecht U., and Birnstiel M.L. 1991. Two-step affinity purification of U7 small nuclear ribonucleoprotein particles using complementary biotinylated 2'-O-methyl oligoribonucleotides. *Proc. Natl. Acad. Sci.* **88:** 9784–9788.

Sollner-Webb B., Tycowski K.T., and Steitz J.A. 1996. Ribosomal RNA processing in eukaryotes. In *Ribosomal RNA. Structure, evolution, processing and function in protein biosynthesis* (ed. R.A. Zimmerman and A.E. Dahlberg), pp. 469–490. CRC Press, Boca Raton, Florida.

Solter D. and Wei G. 1997. Ends *Xist*, but where are the beginnings? *Genes Dev.* **11:** 153–155.

Sontheimer E.J. and Steitz J.A. 1992. Three novel functional variants of human U5 small nuclear RNA. *Mol. Cell. Biol.* **12:** 734–746.

———. 1993. Identification of U5 and U6 small nuclear RNA's as active site components of the spliceosome. *Science* **262:** 1989–1996.

Sontheimer E.J., Sun S., and Piccirilli J.A. 1997. Metal ion catalysis during premessenger RNA splicing. *Nature* **388:** 801–805.

Staley J.P. and Guthrie C. 1998. Mechanical devices of the spliceosome: Motors, clocks, springs, and things. *Cell* **92:** 315–326.

Stefanovic B., Hackl W., Lührmann R., and Schümperli D. 1995. Assembly, nuclear import and function of U7 snRNPs studied by microinjection of synthetic U7 RNA into *Xenopus* oocytes. *Nucleic Acids Res.* **23:** 3141–3151.

Steitz J.A. 1992. Splicing takes a Holliday. *Science* **257:** 888–889.

Steitz T.A. and Steitz J.A. 1993. A general two-metal-ion mechanism for catalytic RNA. *Proc. Natl. Acad. Sci.* **90:** 6498–6502.

Streit A., Koning T.W., Soldati D., Melin L., and Schümperli D. 1993. Variable effects of the conserved RNA hairpin element upon 3' end processing of histone pre-mRNA *in vitro*. *Nucleic Acids Res.* **21:** 1569–1575.

Sun R., Lin S.-F., Gradoville L., and Miller G. 1996. Polyadenylated nuclear RNA encoded by Kaposi sarcoma-associated herpesvirus. *Proc. Natl. Acad. Sci.* **93:** 11883–11888.

Swaminathan S., Tomkinson B., Kieff E. 1991. Recombinant Epstein-Barr virus with small RNA (EBER) genes deleted transforms lymphocytes and replicates *in vitro*. *Proc. Natl. Acad. Sci.* **88:** 1546–1550.

Swaminathan S., Huneycutt B.S., Reiss C.S., and Kieff E. 1992. Epstein-Barr virus-encoded small RNAs (EBERs) do not modulate interferon effects in infected lymphocytes. *J. Virol.* **66:** 5133–5136.

Tarn W.-Y. and Steitz J.A. 1994. SR proteins can compensate for the loss of U1 snRNP function *in vitro*. *Genes Dev.* **8:** 2704–2717.

———. 1996a. A novel spliceosome containing U11, U12 and U5 snRNPs excises a minor class (AT-AC) intron *in vitro*. *Cell* **84:** 801–811.

———. 1996b. Highly diverged U4 and U6 small nuclear RNAs required for splicing rare AT-AC introns. *Science* **273:** 1824–1832.

———. 1997. Pre-mRNA splicing: The discovery of a new spliceosome doubles the challenge. *Trends Biochem. Sci.* **22:** 132–137.

Terns M.P., Dahlberg J.E., and Lund E. 1993. Multiple *cis*-acting signals for export of pre-U1 snRNA from the nucleus. *Genes Dev.* **7:** 1898–1908.

Toczyski D.P., Matera A.G., Ward D.C., and Steitz J.A. 1994. Epstein-Barr virus (EBV) small RNA EBER1 binds and relocalizes ribosomal protein L22 in EBV Infected B-lymphocytes. *Proc. Natl. Acad. Sci.* **91:** 3463–3467.

Tollervey D. and Kiss T. 1997. Function and synthesis of small nucleolar RNAs. *Curr. Opin. Cell Biol.* **3:** 337–342.

Tollervey D., Lehtonen H., Jansen R., Kern H., and Hurt E. 1993. Temperature-sensitive mutations demonstrate roles for yeast fibrillarin in pre-rRNA processing, pre-rRNA methylation, and ribosome assembly. *Cell* **72:** 443–457.

Tschudi C. and Ullu E. 1990. Destruction of U2, U4, or U6 small nuclear RNAs blocks *trans*-splicing in trypanosome cells. *Cell* **61:** 459–466.

Tycowski K.T., Shu M.-D., and Steitz J.A. 1994. Requirement for intron-encoded U22 small nucleolar RNA in 18S ribosomal RNA maturation. *Science* **266:** 1558–1561.

———. 1996a. A mammalian gene with introns instead of exons generating stable RNA products. *Nature* **379:** 464–466.

Tycowski K.T., Smith C.M., Shu M.-D., and Steitz J.A. 1996b. A small nucleolar RNA required for site-specific ribose methylation of rRNA in *Xenopus. Proc. Natl. Acad. Sci.* **93:** 14480–14485.

Wahle E. 1995. 3'-end cleavage and polyadenylation of mRNA precursors. *Biochim. Biophys. Acta* **1261:** 183–194.

Wang Z.F., Whitfield M.L., Ingledue T.C. III, Dominski Z., and Marzluff W.F. 1996. The protein that binds the 3' end of histone mRNA: A novel RNA-binding protein required for histone pre-mRNA processing. *Genes Dev.* **10:** 3028–3040.

Wassarman D.A. and Steitz J.A. 1991. Structural analyses of the 7SK ribonucleoprotein: The most abundant human small RNP of unknown function. *Mol. Cell. Biol.* **11:** 3432–3445.

Watanabe Y. and Yamamoto M. 1994. *S. pombe mei2+* encodes an RNA-binding protein essential for premeiotic DNA synthesis and meiosis I, which cooperates with a novel RNA species meiRNA. *Cell* **78:** 487–498.

Weinstein L.B., Jones B.C.N.M., Cosstick R., and Cech T.R. 1997. A second catalytic metal ion in a group I ribozyme. *Nature* **388:** 805–808.

Willard H.F. and Salz H.K. 1997. Remodelling chromatin with RNA. *Nature* **386:** 228–229.

Wu C.-H., Murphy C., and Gall J.G. 1996. The Sm binding site targets U7 snRNA to coiled bodies (spheres) of amphibian oocytes. *RNA* **2:** 811–823.

Wu Q. and Krainer A.R. 1996. U1-mediated exon definition interactions between AT-AC and GT-AG introns. *Science* **274:** 1005–1008.

———. 1997. Splicing of a divergent subclass of AT-AC introns requires the major spliceosomal snRNAs. *RNA* **3:** 586–601.

Wyatt J.R., Sontheimer E.J., and Steitz J.A. 1992. Site-specific cross-linking of mammalian U5 snRNP to the 5' splice site prior to the first step of premessenger RNA splicing. *Genes Dev.* **6:** 2554–2568.

Xu Y., Ben-Shlomo H., and Michaeli S. 1997. The U5 RNA of trypanosomes deviates from the canonical U5 RNA: The *Leptomonas collosoma* U5 RNA and its coding gene. *Proc. Natl. Acad. Sci.* **94:** 8473–8478.

Yamamoto M. 1996. The molecular control mechanisms of meiosis in fission yeast. *Trends Biochem. Sci.* **21:** 18–22.

Yu Y.-T. and Steitz J.A. 1997. Site-specific crosslinking of mammalian U11 and U6atac to the 5' splice site of an AT-AC intron. *Proc. Natl. Acad. Sci.* **94:** 6030–6035.

Yu Y.-T., Maroney P.A., Darzynkiewicz E., and Nilsen T.W. 1995. U6 snRNA function in nuclear pre-mRNA splicing: A phosphorothioate interference analysis of the U6 phosphate backbone. *RNA* **1:** 46–54.

Yu Y.-T., Tarn W.-Y., Yario T.A., and Steitz J.A. 1996. More Sm snRNAs from vertebrate cells. *Exp. Cell Res.* **229:** 276–281.

Zhong W. and Ganem D. 1997. Characterization of ribonucleoprotein complexes containing an abundant polyadenylated nuclear RNA encoded by Kaposi's sarcoma-associated herpesvirus (human herpesvirus 8). *J. Virol.* **71:** 1207–1212.

20
Splicing of Precursors to mRNAs by the Spliceosomes

Christopher B. Burge,* Thomas Tuschl,* and Phillip A. Sharp
Center for Cancer Research and Department of Biology
Massachusetts Institute of Technology
Cambridge, Massachusetts 02139-4307

The splicing of precursors to mRNAs occurs in two steps, both involving single transesterification reactions (Fig. 1). The first step generates a 2'-5' bond at the branch site upstream of the 3' splice site and a free 3' hydroxyl group on the 5' exon. The resulting lariat RNA intermediate, with its slow migration in gels, is the most common assay for splicing in vitro. In the second step, attack of the 3' hydroxyl on the phosphodiester bond at the 3' splice site displaces the lariat intron with a 3' hydroxyl group and results in joining of the two exons. The bimolecular nature of the intermediate in splicing indicated that the reaction must occur within a stable splicing body or spliceosome. Surprisingly, assembly and functioning of the spliceosome requires approximately 100 polypeptides and five small nuclear RNAs (snRNAs), not considering gene-specific RNA-binding factors. There are two distinct types of spliceosomes in most cells. The major class or U2-type spliceosome is universal in eukaryotes, whereas the minor class or U12-type spliceosome may not be present in some organisms. The evolutionary relationship of these two spliceosomes is uncertain.

The sequence specificity for the splicing of introns must be encoded within the gene. In vertebrate genes, particularly for U2-type introns, the sequence specificity for splicing is not determined solely by the consensus sequences at the intron boundaries but is more broadly distributed within the gene. In contrast, the consensus sequences of the introns in the yeast *Saccharomyces cerevisiae* are generally adequate to specify their excision. The consensus sequences of U12-type introns are more highly conserved than those of vertebrate U2-type introns.

This review focuses on certain aspects of pre-mRNA splicing, including recent results in spliceosome formation, the U12-type spliceo-

*Both authors contributed equally to the writing of this review.

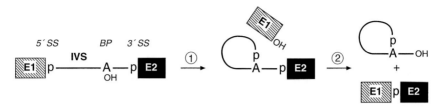

Figure 1 Two-step chemical mechanism for pre-mRNA splicing. A pre-mRNA with a single intron is shown at left, with exons (E1, E2) shown as boxes and the intron (IVS) shown as a line. The phosphodiester linkages that are broken or formed during the reaction are represented by the letter *p*: The branch adenosine (A) and 2′ and 3′ hydroxyl groups (OH) are also indicated. The ligated exon product and released lariat intron are shown at right.

some, and the sequence specificity of pre-mRNA splicing. Many subjects are discussed assuming an understanding of previous literature in the field. In particular, this review updates some aspects of the earlier article from this laboratory (Moore et al. 1993). In the interim, a number of excellent and provocative reviews have been written (Nilsen 1994, 1998; Krämer 1996; Hertel et al. 1997; Newman 1997; Wang and Manley 1997; Will and Lührmann 1997; Staley and Guthrie 1998).

SPLICEOSOME

The highly conserved nature of spliceosome composition and function deduced from studies of the human and yeast systems is now well established, although some fundamental differences exist. The snRNP particles containing the core Sm proteins (Table 1) associate with other factors to form the spliceosome. Discussions of the spliceosome assembly process below focus on the more established yeast system; the human homologs, if known, are given in parentheses. Information on individual yeast proteins and genes, mammalian homologs, and additional factors is listed in Tables 1–10 (see page 545). Many of these factors are not mentioned below, but the transition in the spliceosome cycle for which their activity is required can be deduced from Figure 2.

Assembly of a spliceosome for excision of an intron requires the recognition of sequences at the 5′ splice site as well as the branch site and nearby 3′ splice site (Reed 1996). As discussed below, each of these sites is located within consensus sequences that are strongly conserved in yeast and weakly conserved in vertebrates. There are at least two mechanisms for recognition of the 5′ consensus sequence, both based in part on RNA

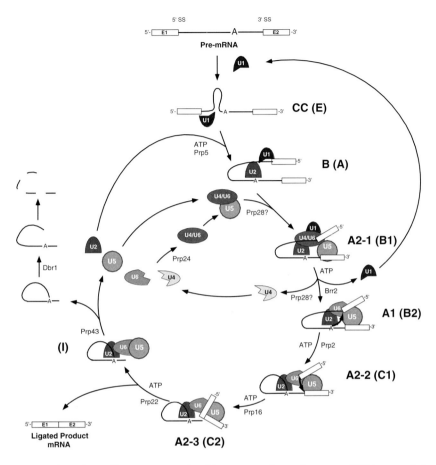

Figure 2 The spliceosome cycle. The processing of a pre-mRNA containing two exons and one intron (*top*, see legend to Fig. 1) into the ligated exon product (*lower left*) and lariat intron (*middle left*) is shown, emphasizing the involvement of the small nuclear ribonucleoprotein (snRNP) particles (*shaded*) at distinct steps in spliceosome formation and catalysis. Macromolecular complexes that have been distinguished biochemically and genetically are labeled using the designations suggested for yeast with the mammalian designations indicated in parentheses: CC (E), B (A), A2-1 (B1), A1 (B2), A2-2 (C1), A2-3 (C2), and (I). Arrows represent transitions between complexes and/or recycling of snRNP components. The debranching enzyme Dbr1 and several yeast RNA helicase motif-containing proteins implicated in the various ATP-dependent conformational changes (Brr2, Prp2, Prp5, Prp16, Prp22, Prp24, and Prp42) are also indicated; other non-snRNP factors are omitted. The debranching and subsequent degradation of the lariat intron is shown at left. (Modified from Moore et al. 1993.)

sequence complementarity: U1 snRNA binding to bases +1 to +6 of the intron and U6 snRNA binding to positions +4 to +6 (Kandels-Lewis and Seraphin 1993; Lesser and Guthrie 1993; Madhani and Guthrie 1994; Hwang and Cohen 1996). For the typical intron, U1 snRNP association with the 5′ sequence is critical for splicing in vitro, but there are introns that are processed very efficiently in the absence of U1 snRNP (Crispino et al. 1994, 1996; Tarn and Steitz 1994). In this case, the 5′ sequence is recognized by U6 snRNA and other spliceosome components before the first step in splicing.

The primary function of U1 snRNP (Table 2) is to promote the association of a U2 snRNP complex (17S form; Table 3) with the branch region (Fig. 2) (Barabino et al. 1990; Seraphin and Rosbash 1990). Although U2 snRNP will specifically bind pre-mRNA in the absence of U1 snRNP, for most RNAs this reaction is not efficient (Query et al. 1997). Early snRNP/pre-mRNA complexes are preferentially committed to splicing as compared to free RNA and thus are considered commitment complexes (CC; Tables 1–7). Four proteins, Prp5, Prp9 (SF3a60), Prp11 (SF3a66), and Prp21 (SF3a120), are critical for the ATP-dependent process of U2 snRNP association. They were originally identified through synthetic lethal interactions (Chapon and Legrain 1992; Legrain and Chapon 1993; Ruby et al. 1993; Wells and Ares 1994), and their function is now confirmed biochemically (Wiest et al. 1996). Prp5 is a putative ATP-dependent RNA helicase (O'Day and Abelson 1996) and Prp9, 11, and 21 were shown to correspond to the U2 snRNP-associated human ternary splicing factor complex SF3a (Table 3), which is required for stable association of U2 snRNP with the branch region (Bennet and Reed 1993; Brosi et al. 1993a,b; Krämer et al. 1995). Subsequent to the binding of U2 snRNP (complex B; Fig. 2), a tri-snRNP complex containing U4/U6 snRNP and U5 snRNP (Tables 4 and 5) associates in an ATP-dependent manner to form complex A2-1 (B1; Cheng and Abelson 1987; Konarska and Sharp 1987). It is likely that the U1 snRNA/pre-mRNA duplex is dissociated at this stage, through the action of the putative RNA helicase Prp28 (U5-100 kD; Table 4; Staley and Guthrie 1998). At this stage, the 5′ splice site sequence is probably bound on the intron side to U6 snRNA (Li and Brow 1996) and on the exon side by U5 snRNA (O'Keefe et al. 1996; Long et al. 1997). The U5 snRNA interaction is not essential for the first step but is required for the second step (Newman 1997; O'Keefe and Newman 1998).

The transition between complex A2-1 (B1) and A1 (B2) requires destabilization of at least U4 snRNA, which initially enters the spliceosome paired with U6 snRNA (see Figs. 2 and 4B). This transition again requires

ATP and possibly the putative RNA helicase Brr2 (U5-200 kD; Table 4; Staley and Guthrie 1998). Subsequently, another ATP-dependent transition occurs that is promoted by the putative RNA helicase Prp2 (Table 8). This is the most proximal transition known to date before the first catalytic step in splicing (Teigelkamp et al. 1994; Roy et al. 1995), and it is possible that this transition creates the catalytic site for the first step. Since only three snRNAs, U2, U6 and U5, are associated with the spliceosome at the moment of catalysis, and U5 snRNA pairing with exon sequences is not essential, the catalytic site is either created by U6 snRNA, U2 snRNA (or both), and/or associated proteins.

The actions of Prp16, Prp17, Prp18, and Slu7 (Table 4) are required for the transition to the second step in splicing (Schwer and Guthrie 1992; Jones et al. 1995; Umen and Guthrie 1995). The human homologs of these splicing factors have been identified, and they function similarly (Horowitz and Krainer 1997; Zhou and Reed 1998). Not surprisingly, these conformational rearrangements also require ATP, and Prp16 has a subdomain homologous to other ATPases. The catalytic site for the second step is again probably created by either U6 snRNA, U2 snRNA, or associated proteins.

The spliced exon RNA is released from the spliceosome through the action of Prp22 (Table 9) and perhaps other factors (Company et al. 1991; Schwer and Gross 1998). Prp22 is the first helicase motif-containing protein required for splicing for which ATP-driven RNA-unwinding activity has been demonstrated (Schwer and Gross 1998). Prp43 (Table 9) is another putative RNA helicase that is important for releasing the lariat intron and U2, U5, and U6 snRNPs (Arenas and Abelson 1997; Gee et al. 1997). The reannealing of released U4 and U6 snRNP and association with U5 forms the U4/U6.U5 tri-snRNP complex, which is then ready to reassemble on another commitment complex. The identification of other state-switching proteins such as the human tri-snRNP-specific 20-kD protein (Table 5), a potential peptidyl-prolyl *cis/trans* isomerase (Horowitz et al. 1997; Teigelkamp et al. 1998), or Snu114 (U5-116 kD; Table 4), a homolog of a ribosomal GTPase EF-2 (Fabrizio et al. 1997), suggests that many more conformational rearrangements remain to be described in the spliceosome.

A U5 snRNP-specific protein, Prp8 (U5-220 kD, p220 or hPrp8; Table 4), has attracted great interest because it is closely associated with sequences at the 5′ splice site, branch site, and 3′ splice site, implying a possibly direct role in catalysis or a function as a major scaffolding protein for aligning the splice sites. Prp8 is evolutionarily highly conserved (Hodges et al. 1995; Lücke et al. 1997) and required for spliceosome assembly before the first step. UV cross-linking studies show that Prp8 can be cross-linked to positions –8, –3, –2, –1, +1, +2, +3 of the 5′ splice site (Wyatt et al. 1992; Teigelkamp et al. 1995a,b; Chiara et al. 1996; Reyes et al. 1996), to sequences flanking the branch site (MacMillan et al. 1994), and to positions –1, +1, +7, +13 of the 3′ splice site (Teigelkamp et al. 1995a,b; Umen and Guthrie 1995; Chiara et al. 1996). Interestingly, substitution of U+2 at the 5′ splice site by 5-iodo-U or 5-methyl-U reduces both the rate of splicing and the efficiency of cross-linking to hPrp8, suggesting that the invariant GU dinucleotide at the 5′ end of the intron might be specifically recognized by Prp8 (Reyes et al. 1996). The time course of formation of these cross-links indicates that they form in complexes A2-1 (B1) through A2-3 (C2) of the spliceosome cycle (Chiara et al. 1996; Reyes et al. 1996). At this moment, there is no recognized subdomain structure in Prp8 and no indication of an explicit role in RNA splicing.

Formation of a catalytically competent spliceosome requires recognition of both the 5′ splice site and the branch site, but not necessarily the 3′ splice site. Early studies suggested that for pre-mRNAs with more than 20 nucleotides separating the branch site from the 3′ splice site, a scanning mechanism was responsible for 3′ splice site identification (Smith et al. 1989, 1993). This scanning process required Slu7 and other factors (Brys and Schwer 1996). Recently, the development of a bimolecular (*trans*-splicing) reaction, where a short RNA containing a 3′ splice site sequence could be utilized by a spliceosome arrested after the first step, has resulted in an analysis of this scanning process (Anderson and Moore 1997). Surprisingly, the only sequence in the short mRNA required for activity is the prototype YAG/ (Y=C or U) 3′ splice site: Neither a pyrimidine tract nor a specific exon sequence is important. The short *trans*-splicing RNA appears to access the spliceosome in a 5′ to 3′ polarity.

U2-TYPE AND U12-TYPE INTRONS AND SPLICEOSOMES

Some 20 years after the sequencing of the first introns and recognition of the "classic" consensus sequences at the 5′ splice site, branch site, and 3′ splice site, a new group of introns has been identified. These introns have a distinct set of consensus sequences (Fig. 3) and are spliced by a compositionally distinct spliceosome. The first inkling that there might be a distinct class of introns came from the observation of an unusual pair of introns with AT-AC termini, breaking the almost universal GT-AG rule (Jackson 1991). Later, through careful sequence analysis of the few known AT-AC introns, Hall and Padgett (1994) observed that unusual 5′ splice site sequences as well as a novel sequence upstream of the 3′ splice

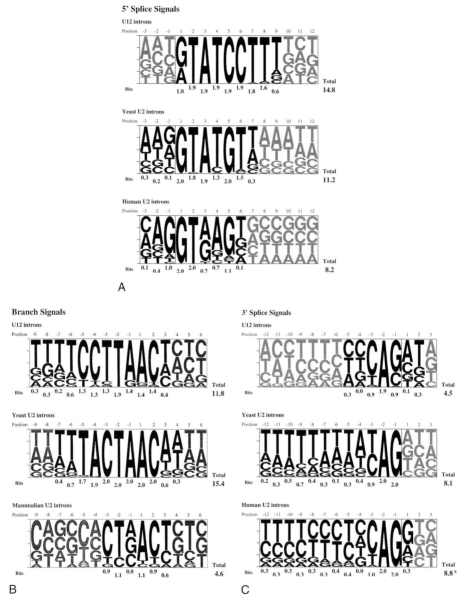

Figure 3 The sequence patterns at the 5' splice site (*A*), branch site (*B*), and 3' splice site (*C*) derived from sets of U12-type introns, *S. cerevisiae* U2-type introns, and human U2-type introns are represented using the Pictogram program (C.B. Burge and P.A. Sharp, unpubl.). At each sequence position, the frequencies f_1, f_2, f_3, and f_4 of the four DNA nucleotides A, C, G, and T are represented by the heights of the corresponding letters, with the letters shown in decreasing order of

site were very strongly conserved in this class of introns (Fig. 3). By comparing these conserved sequence elements to known snRNA sequences, they observed significant sequence complementarity with two low-abundance snRNAs of unknown function, U11 and U12. They proposed that these two snRNAs might be involved in recognition of the 5' splice site and branch site, respectively, in the splicing of these introns. Recently, it has been confirmed that this class of introns is spliced by a novel type of spliceosome involving U11 and U12 and U5 snRNAs, and evidence for the proposed interactions has been obtained (Hall and Padgett 1996; Tarn and Steitz 1996a). A more complete picture of this novel spliceosome emerged through the subsequent identification of two other novel snRNA components, termed U4atac and U6atac, which appear to be functionally analogous to U4 and U6 (Tarn and Steitz 1996b).

Members of this novel group of introns have strongly conserved 5' and 3' splice site sequences, either /ATATCCTTT and YAC/ or /GTATCCTTT and YAG/, respectively (where / represents the splice junction).

---

frequency from top to bottom. For each position j, the information content I_j (in bits) is defined by the formula $I_j = 2 + \sum_i f_i \log_2(f_i)$. To correct for the slight upward bias due to finite sample size n, the term $3/2\ln(n)$ was subtracted from each value of I_j (see appendix to Stephens and Schneider 1992). For each type of signal, positions are classified into those thought to be involved in recognition (letters shown in *black*) and those which are not (shown in *gray*): Only the former are included in the total information content values shown at right. For human 3' splice sites, the asterisk indicates that the total information content includes three positions (–13, –14, and –15) not shown in the figure. The set of U12 introns used is an updated version of that described by Sharp and Burge (1997) and contains 42 nonredundant introns of both GT-AG and AT-AC subtypes from a variety of animal and plant species [http://ccr-081.mit.edu/introns/U12.html]. The set of 205 yeast introns used [http://ccr-081.mit.edu/introns/Yeast.html] was derived from the annotated complete genomic sequence (Mewes et al. 1997), removing certain redundant entries corresponding to subtelomeric encoded proteins and correcting two apparent annotation errors. All but two introns contained a consensus branch site (TACTAAC), allowing at most one error. The set of 1683 human U2-type introns used was based on a set of nonredundant human complete gene sequences from GenBank Release 95 constructed by M.G. Reese and D. Kulp [http://www-hgc.lbl.gov/inf/genesets.html]. From this set, a few sequences were removed on the basis of their sequence annotation, e.g., mitochondrial encoded genes, incomplete genes. Details are given at [http://ccr-081.mit.edu/introns/Human.html]. Mammalian branch site data are from a set of 14 experimentally determined branch sites in wild-type (nonmutant) mammalian introns collected by Nelson and Green (1989).

For both the AT-AC and GT-AG subtypes of this rare intron class, the 5′splice site sequence is very tightly constrained, with variations apparently tolerated only at positions +7, +8, and +9 relative to the splice junction (Fig. 3A). These introns also have a conserved branch site signal with the consensus TCCTTA<u>A</u>C (branch site underlined), located between 10 and 20 nucleotides upstream of the 3′ splice site, and typically lack a polypyrimidine tract between these two elements (Fig. 3B,C). The presence of much stronger constraints on splice site and branch site sequences in this group of introns suggests that these sequences may play a greater role in specifying the sites for spliceosome formation than do the corresponding, more degenerate, sequences of major class introns (see below).

Since recognition of the branch site by an snRNP results in the formation of the first stable spliceosome complex, this new class of introns has been named "U12-type," as contrasted to the major class of "U2-type" introns (Dietrich et al. 1997). This nomenclature is preferable to that initially proposed where introns were classified according to the dinucleotides at their boundaries as either GT-AG or AT-AC, since this property is not diagnostic of the type of spliceosome used for intron excision. Specifically, although the vast majority (>99.9%) of U2-type introns have GT-AG (or, rarely, GC-AG) boundaries, a handful have recently been found that have AT-AC boundaries (Wu and Krainer 1997; Dietrich et al. 1997). Likewise, although the first U12-type introns identified had AT-AC boundaries, it was subsequently shown that some U12 introns have GT-AG boundaries (Dietrich et al. 1997), and it now appears that the GT-AG subtype of U12 introns is actually significantly more common than the AT-AC subtype (Dietrich et al. 1997; Sharp and Burge 1997). Including both the GT-AG and AT-AC subtypes, the proportion of all nuclear introns that are spliced by the U12 spliceosome is still very small, perhaps on the order of one in a thousand.

The extent of the phylogenetic distribution of U12-type introns and spliceosomes is not completely known, although they are clearly present in multiple vertebrate, insect, and plant species, including human and the major model organisms mouse, *Xenopus*, *Fugu*, *Drosophila*, and *Arabidopsis*. On the other hand, the yeast *S. cerevisiae* lacks U12-type introns and the associated machinery. Similarly, no U12-type introns have yet been identified in the nematode *Caenorhabditis elegans* despite the nearly complete genomic sequence, suggesting that the U12-type system has been lost in certain lineages. In those organisms where this system is present, it appears that the U2-type and U12-type spliceosomes are active within the same regions of the nucleus since the genes containing U12-type introns also contain U2-type introns. In fact, components from the

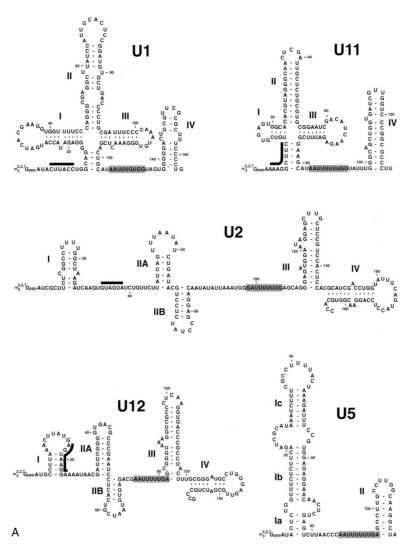

Figure 4 (See next page for legend.)

two spliceosomes may interact, since recognition of a 5′ splice site by U1 snRNP can promote the splicing of an upstream U12-type intron (Wu and Krainer 1996).

The sequences of U11, U12, U4atac, and U6atac snRNAs are quite different from their U2-type counterparts U1, U2, U4, and U6, respectively (Fig. 4A,B). However, when examined more closely, the most conserved core sequences of U6 snRNA are also present in U6atac (Fig. 5)

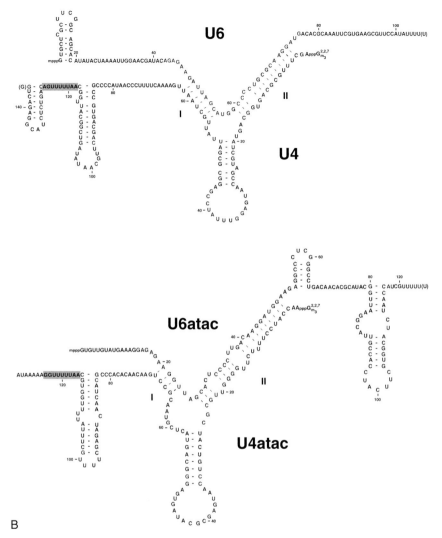

Figure 4 Sequences and conserved secondary structures of the small nuclear RNA (snRNA) components of the two spliceosomes. Positions in the snRNA are indicated by arabic numerals, binding sites for Sm proteins are shaded, and segments that form base-pairing interactions with the pre-mRNA are indicated by dark lines. Roman numerals indicate intrastrand or interstrand RNA helices. With the exception of the methylation of the guanosine cap structures, nucleotide modifications are not illustrated. (*A*) The human U1, U11, U2, U12, and U5 snRNAs are shown in 5′ to 3′ orientation. (Modified from Baserga and Steitz 1993.) (*B*) Sequences, secondary structures, and base-pairing interactions between human U4 and U6 snRNAs (*top*) and U4atac and U6atac snRNAs (*bottom*). The 5′ to 3′ orientation of U4 and U4atac reads from right to left. (Modified, with permission, from Tarn and Steitz 1996b.)

(Tarn and Steitz 1996b). In addition, the secondary structures of the cognate snRNAs are quite similar between the two spliceosomes (Fig. 4A,B) and form similar interactions with other snRNAs and with the pre-mRNA (Fig. 5). This suggests that the two types of spliceosomes may assemble and disassemble in similar cycles and probably share some common pro-

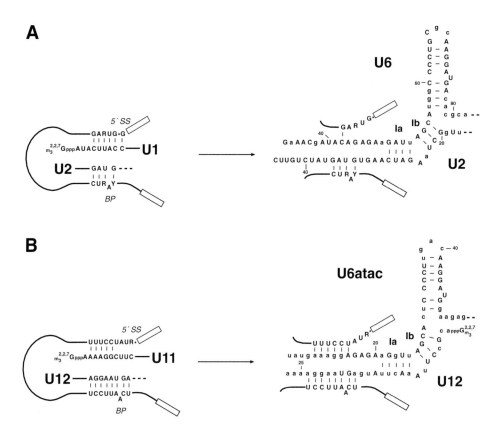

Figure 5 Base-pairing interactions at the 5' splice site and branch site for the U2-type (*A*) and U12-type (*B*) spliceosomes. The interactions of U1 and U11 snRNPs with the 5' splice site that occur early in spliceosome formation (*left*) are later displaced by U6 and U6atac, which associate with U2 and U12 snRNAs, respectively (*right*). Exons are shown as boxes, introns as curved lines, and the 5' splice site is represented by a star. Only residues involved in critical snRNA:snRNA or snRNA:pre-mRNA interactions are shown. The branch adenosine is shown bulged from an RNA duplex in both spliceosomes. Uppercase nucleotides in panel *A* (*right*) indicate phylogenetically conserved residues of the core sequences of U2 and U6 snRNAs (not considering *trans*-splicing organisms); uppercase nucleotides in panel *B* indicate residues that are conserved between U2-type and U12-type sequences. (Adapted from Tarn and Steitz 1996b.)

tein components. This proposal is consistent with the role of U5 snRNP as a component of both spliceosomes, and the observation that a protein which cross-reacts with antisera to the human homolog of Prp8 can be cross-linked to pre-mRNAs containing U12-type introns (M. J. Moore, pers. comm.). More specifically, some of the interactions which suggest that the transitions involved in the U12-type spliceosome cycle are similar to those described for the U2-type spliceosome are: (1) U11, U12, and U4atac snRNAs bind the same core Sm proteins as the U2-type snRNAs (Montzka and Steitz 1988; Tarn and Steitz 1996b); (2) analogous complementarity between the sequences at the 5' end of U11 snRNA and the 5' splice site (Hall and Padgett 1994, 1996; Tarn and Steitz 1996a); (3) analogous complementarity between a portion of U12 snRNA and the branch site with the potential to form a duplex with a bulged adenosine (Hall and Padgett 1994; 1996; Tarn and Steitz 1996a); (4) analogous complementarity between U4atac and U6atac (which likely dissociate within the mature spliceosome; Tarn and Steitz 1996b); (5) analogous complementarity suggesting base-pairing between U6atac and U12 snRNAs (direct cross-linking evidence for this interaction; Tarn and Steitz 1996b); and (6) conservation of sequences in U6atac which are highly conserved in all known U6 snRNAs.

The evolutionary origins of the two spliceosome types remain mysterious. At present, three possible scenarios for their (co)evolution seem most plausible, which we distinguish as the "fission-fusion" (FF) model, the "codivergence" (CD) model, and the "parasitic invasion" (PI) model. The FF and CD models assume that the U12 and U2 systems are homologous (derived from a common ancestor), and the many similarities in structures and functions of the five snRNAs support this point of view. However, the amount of direct sequence similarity between the analogous snRNA components of the two spliceosomes is not sufficient to strongly confirm a common ancestry (Tarn and Steitz 1996b). In the FF model, the divergence between the two systems is envisioned to have occurred in separate lineages (each with a single spliceosome type), followed by a merging of the genetic material, perhaps through endosymbiosis, in a progenitor of animals and plants. This model explains why the two types of spliceosomes were retained (to splice out the introns from the genes acquired through endosymbiosis), but does not readily explain why U12-type introns appear within genes containing multiple U2-type introns unless these introns or genes were formed after rejoining of the two systems or virtually all of the introns in these genes converted from U12-type to U2-type. Dietrich et al. (1997) conjectured that conversion of U12-type introns to U2-type introns was possible, and comparisons of homologous introns suggest that such conversion has occurred and is not uncommon (C. Burge, unpubl.).

The codivergence, or CD, model proposes that the entire set of snRNAs (with the possible exception of U5) was duplicated in a primordial organism, followed by coordinated divergence of two subsets of introns together with the two sets of snRNA components. For example, one of the duplicated snRNAs might have acquired a mutation that reduced its ability to recognize most introns, but increased the efficiency of splicing of a subset of introns with variant splice sites. Subsequent events are conjectured to lead to separation of the two splicing systems in this model. The observed similarities in snRNA structure could be explained if the two systems used the same or similar cellular proteins in formation of the spliceosome. The CD model appears to provide a simpler explanation for the common presence of U5 snRNA in the two spliceosomes and for the co-occurrence of U2 and U12 introns in the same genes than does the FF model.

A third possible explanation for the similarities observed between the two spliceosomes is that they evolved convergently, the PI model. In this scenario, a parasitic group II self-splicing intron entered the genome of a common precursor of animals and plants, which already had a developed splicing system, and was inserted into a number of genes. In time, the self-splicing RNA structure of this intron fragmented to give rise to a set of new (U12-type) snRNAs that could excise these introns, as may have occurred in the evolution of the U2-specific snRNAs. In this model, the generation of the U12-type snRNAs was facilitated by the presence of the other U2-type splicing factors such as the snRNP proteins and Prp factors. Since these factors work within a specific spliceosome cycle, the new snRNAs convergently evolved toward a similar set of secondary structures and interactions. Thus, in this model, the core spliceosomal proteins and cycle were conserved, except that now the machinery could splice out a new type of intron with distinctive sequence properties using a distinct set of cognate snRNAs.

The presence of parallel splicing systems could have conferred certain advantages to an organism if differences between the two systems were used to regulate gene expression levels at different times in the cell cycle or for alternative splicing, for example. However, as yet no significant evidence for regulation of U12-type splicing has been found.

These models highlight the unique role of U5 snRNA in pre-mRNA splicing in that it is a component of both spliceosomes. Although U5 snRNA can interact with the 5' and 3' exon sequences flanking U2-type introns (for review, see Newman 1997), these interactions are not essential for the first step in splicing (O'Keefe et al. 1996). The flanking exonic sequences of U12-type introns show few distinct base preferences (see Fig. 3A,C). This suggests that the essential function of U5 snRNP may be as a carrier of splicing factors common to both spliceosomes, and that interaction of U5 snRNA with exonic sequences may be specific to the U2-type spliceosome.

The catalytic core of the spliceosome has yet to be defined. In fact, it is not clear whether the spliceosome has two distinct catalytic sites or one predominant site that undergoes structural changes between the two transesterification steps. The first and second steps in the splicing reaction are single transesterifications where a 2'-OH displaces a 3'-OH in the former and a 3'-OH displaces a 3'-OH in the latter. This obvious chemical difference suggests two distinct catalytic sites. This idea is also supported by the stereoselectivity of the transesterification reactions observed when the phosphates at the splice sites are substituted by chiral phosphorothioates (Moore and Sharp 1993). Other evidence supporting two reaction sites comes from differences in the requirement for a metal ion between the two catalytic steps: Substitution of 3' sulfur at the 5' and 3' splice sites demonstrates that catalysis at the former depends on metal ion, whereas reaction at the latter may not (Sontheimer et al. 1997).

If RNA sequences are integral to the catalytic core of the spliceosome, then the evolutionarily highly conserved regions of U6 snRNA (also found in U6atac) are the most likely constituents. Both cross-linking studies and phylogenetic comparisons suggest that this region of U6 snRNA is at least transiently paired with sequences from U2 snRNA and also in a hairpin stem structure with a uridine bulge (Fig. 5A,B). There are 21 positions in the core of U6atac snRNA that are conserved with all known U6 snRNAs. Mutations in these portions of U6 snRNA inhibit splicing, whereas mutations of the complementary portions of U2 snRNA have a less pronounced phenotype (Madhani and Guthrie 1994; Sun and Manley 1995). It is also possible that these U6 snRNA sequences are conserved for a direct role in catalysis and that the two spliceosome processes form a common catalytic structure. Alternatively, these sequences could be a lattice upon which one or more proteins must interact to promote splicing.

SPECIFICITY OF RNA SPLICING

The sequences of human U2-type 5' and 3' splice signals are quite degenerate (Fig. 3), with only two positions in each signal, the /GT at the 5' splice site and the AG/ at the 3' splice site, which are more or less universally conserved. Using the terminology of information theory these two consensus sequences contain about 8 bits and 9 bits of information, respectively, which is almost certainly inadequate to reliably identify the splice junctions in long human pre-mRNA sequences, as discussed below. Since the application of concepts from information theory to the description of molecular specificity is not widely known, this subject is briefly reviewed in the legend to Figure 3. Readers interested in a more thorough treatment of this subject are referred to the classic text by Cover and Thomas (1991) and to the papers of Schneider and coworkers (see, e.g., Stephens and Schneider 1992).

Assuming that the nucleotides at different sequence positions are independent, the amount of information I in the entire 5' splice signal can be calculated by summing over all the positions: $I = \sum_j I_j$. The value $I \approx 8.2$ bits that is obtained (see also Stephens and Schneider 1992) implies that potential human U2-type 5' splice signals will occur on average about once every $2^{8.2} \approx 290$ bases in a tract of random RNA sequences. Since human genes typically have multiple introns, the 5' splice site must generally be located by the splicing machinery between flanking 3' splice junctions. This distance, the sum of the length of an exon and the downstream intron, averages about 1300 bases in human genes (derived from the sum of the mean exon length, 150 nucleotides, and the mean intron length, 1150 nucleotides, in the data described in the legend to Fig. 3). This mean value may well be an underestimate due to biases in the set of available human genes and does not capture the variability in this distance, which ranges from about 100 nucleotides to over 100 kb in some extreme examples. Nevertheless, it is fairly clear that the amount of information at the 5' splice signal is inadequate to accurately specify the locations of 5' splice junctions in most human primary transcripts.

The human U2-type 3' splice signal contains about 8.8 bits of information (see also Stephens and Schneider 1992), contributed in roughly equal parts by the pyrimidine-rich region upstream of the 3' splice junction and the YAG/ at the 3' splice site. This corresponds to the occurrence of such a site in random RNA sequences about once every $2^{8.8} \approx 450$ nucleotides, again insufficient to reliably localize the 3' splice junction between flanking 5' splice sites spaced an average of 1300 nucleotides apart. The branch signal, consensus CTRAY (the branch nucleophile is underlined), is even more degenerate in vertebrate U2-type introns, with only the T and A which are consistently present (Fig. 3B). This signal, because it is located at a highly variable position relative to the 3' splice junction, and contains only about 4.6 bits of information, is unlikely to contribute significantly to the specificity of vertebrate U2-type splicing. This analysis, although simplistic in some of its assumptions (e.g., the independence between nucleotides at different positions in splice signals and the independence between the locations of 5′ and 3′ splice sites), suggests that vertebrate cells depend critically on recognition of sequences other than the splice junctions and branch site for specification of the splice junctions in most transcripts.

The specificity problem for the yeast U2 splicing system is quite different from that in human cells. First, yeast transcripts, averaging about 1900 nucleotides (derived from data of Richard et al. 1997), are much shorter than human primary transcripts (averaging several kilobases) and contain shorter introns, averaging about 270 nucleotides (see Fig. 3) versus an average of over 1 kb in humans. On the other hand, most yeast genes do not contain introns, so that the splicing machinery must be able to pinpoint a few introns in a large excess of primarily intronless transcripts. The total transcribed length of the yeast genome is probably about 12 Mb (6,300 genes × 1,900 bp / gene), from which about 240 introns must be excised (approximate numbers from Mewes et al. 1997). Thus, introns occur only about once per 50 kb of transcribed RNA in yeast, implying that a signal with roughly $\log_2(50,000) \approx 15.6$ bits of specificity is required to distinguish intron termini without conferring significant constraints on other sequences. Interestingly, the strongly conserved yeast branch site sequence, TACTA$\underline{A}$C with only slight variations tolerated, contains essentially the needed information: about 15.4 bits (Fig. 3B). Although the location of yeast branch sites relative to the 3′ splice junction is somewhat variable, the 3′ splice signal information is about 8.1 bits (Fig. 3C), which, together with the branch site signal, appears adequate to accurately pinpoint the 3′ ends of yeast introns from among the large pool of transcripts.

Once the 3′ end of the intron has been identified, the specificity problem becomes much simpler, requiring only that the upstream portion of the transcript (typically 1 kb or less) be searched for the 5′ splice site. The amount of information present in the yeast 5′ splice signal (consensus GTATGT) is about 11.2 bits, sufficient to identify a site in roughly $2^{11.2} \approx 2350$ nucleotides, which should be adequate. Thus, it appears that yeast introns contain most or all of the information required to pinpoint the splice signals, in stark contrast to the situation in human U2-type splicing. A more complicated analysis accounting for the biased (A+T-rich) composition of the yeast genome gives somewhat lower estimates of the information content of the branch site and splice site signals, but still supports this general conclusion.

The specificity of splicing by the U12 system is like yeast in that the signals contain large amounts of information, but is reversed in that the 5′ splice signal, consensus ^A_GTATCCTTT, contains substantially more information (14.8 bits) than the branch signal, consensus TCCTTA<u>A</u>C (11.8 bits). The 3′ splice signal contains comparatively little information (4.5 bits) owing to the absence of a polypyrimidine tract, but this is compensated in a sense by the relatively restricted location of U12 branch sites, typically 10–16 nucleotides upstream of the 3′ splice junction. Is this enough information to specify the splice sites of U12 introns? At first glance, it would appear to be, since the 5′ splice signal alone contains enough information to specify one site per $2^{14.8} \approx 28{,}500$ nucleotides, longer than the average human transcript.

Once the 5′ splice junction has been located, the system would need to search only about a kilobase or so on average (assuming that U12 introns have a similar length distribution to U2 introns, which appears to be the case) to identify the branch site and 3′ splice junction, for which 11.8 (+ 4.5) bits should suffice. But since only one out of an estimated 200 transcripts contains a U12 intron, such introns probably occur only about once per megabase of transcribed RNA. If every nascent transcript is accessible to the U12 machinery, significantly more information should be necessary (on the order of 20 bits) to accurately identify U12 introns. Thus, the amount of information present in the U12 signals does not appear to be sufficient to precisely identify the splice sites of all U12 introns, suggesting that recognition of other sequences such as exonic or intronic splicing enhancers might also be involved in U12 splicing. In support of this idea, "exon bridging" interactions have been described in the splicing of a U12 intron (Wu and Krainer 1996).

The anticipation of massive amounts of vertebrate genomic sequences has spurred the development of computational methods that attempt to identify exons and introns in these sequences. Using a wide variety of information sources, including compositional differences between exons and introns as well as open reading frame information and properties of the signals involved in transcription, polyadenylation, splicing, and translation, recent methods are able to precisely identify approximately three-quarters of all coding exons in vertebrate genomic sequences (Burset and Guigo 1996; Burge and Karlin 1997). To achieve this level of accuracy appears to require that all aspects of gene structure be incorporated, including features such as the reading frame compatibility of adjacent exons that are unlikely to be used by the nuclear RNA processing machinery. However, inclusion of multiple types of sequence patterns such as those important in transcription and polyadenylation, as well as splicing, in combination to predict intron/exon structure is not mechanistically unreasonable; recent experiments suggest that the processes of cap formation, splicing, and polyadenylation interact with one another both in vivo and in vitro.

It is difficult to measure quantitatively how much of the accuracy of gene finding algorithms such as FGENEH (Solovyev et al. 1994), Genie (Reese et al. 1997) or GENSCAN (Burge and Karlin 1997) derives from splice signal information as compared to other sources. Qualitatively, the answer is that compositional differences between exons and introns are most important, followed by splice signals, with transcriptional, polyadenylation, and translational signals a distant third (Burge and Karlin 1998). These compositional differences between exons and introns are reflected in differences in the frequencies of specific hexamer sequences (Fickett and Tung 1992). If these computational studies have documented the critical sequences for splicing of introns, then these general compositional differences between exons and introns must be recognized in some way during formation of a spliceosome.

SUMMARY

The discovery of a new spliceosome and corresponding subclass of introns has re-emphasized some general and important questions. Particularly, the evolutionary origin of introns and nuclear RNA splicing will be reconsidered in view of the relationship between the major (U2-type) and minor (U12-type) classes of introns. It is possible that, with the impending flood of genome sequence data, a third subclass of introns might be discovered. However, as yet, no hint of such a subclass has surfaced.

The new genome sequences have also increased the importance of consideration of bioinformatics in the analysis of RNA splicing, an area of research that has so far contributed only modestly to the subject. A reexamination of the information content of the consensus sequences of the U2-type introns of humans and the yeast *S. cerevisiae* suggests that these two systems, although fundamentally similar in regard to spliceosome function, differ significantly in regard to the critical sequences recognized for assembly of the spliceosome. For *S. cerevisiae*, recognition of the consensus sequences is probably sufficient for assembly of the spliceosome and excision of most introns. In human cells, recognition of sequences within exons and introns as well as the consensus sequences at the splice sites must be critical. It is probably also the case that interactions of factors involved in transcription, capping, and polyadenylation with those involved in splicing are important (see, e.g., McCracken et al. 1997). The integration of interactions over extended lengths of RNA sequences in assembly of the spliceosome may therefore be a defining feature of vertebrate splicing. A strong candidate for mediating these types of interactions is the arginine-serine (RS or SR) domain of proteins, a domain that is present in many proteins important for splicing in vertebrates but essentially absent from yeast (Tables 1–10). Proteins containing RS domains are critical for both early and late stages of spliceosome assembly in vertebrate cells (see, e.g., Manley and Tacke 1996; Valcárcel and Green 1996). The heterogeneous distribution of these proteins in the vertebrate nucleus suggests an intimate relationship between splicing and nuclear structure (see, e.g., Xing et al. 1993).

The absence of a polypyrimidine tract and the greater conservation of the splice site consensus sequences in U12-type introns suggest that the assembly of the U12-type spliceosome on the pre-mRNA is probably more dependent on snRNA-based recognition of the splice site sequences than is the case for the U2-type spliceosome. However, there have been so few biochemical studies of the U12-type system to date that this must be considered a conjecture. Future studies of protein factors specific to the U12-type and U2-type spliceosomes as well as factors common to both systems will provide insight into the evolution and functioning of the spliceosomes.

ACKNOWLEDGMENTS

We gratefully acknowledge the contributions of Ben Blencowe, Göran Baurén, Lee Lim, Patrick McCaw, and other members of the Sharp lab to the writing of this review. We specifically thank Margarita Siafaca, who prepared the manuscript. Dr. Thomas Tuschl is supported by a Merck fellowship through the Department of Biology, Massachusetts Institute of Technology. This work was supported by U.S. Public Health Service MERIT award R37-GM34277 from the National Institutes of Health to P.A.S. and partially by a Cancer Center Support (core) grant P30-CA14051 from the National Cancer Institute.

Table 1 SnRNP core proteins of the Sm family

Yeast protein	MW[a] (kD)	KO[b]	Human homolog[c]	Comments	Sequence motifs
Smb1/Yer029c	22.4	v	Sm B' (29 kD), Sm B (28 kD)[d]	U1, U2, U4/U6, U5; interacts with Smd3	Sm1 and 2, Pro-rich C terminus
Smd1/Spp92/Ygr074w	16.3	l	Sm D1 (16 kD)	U1, U2, U4/U6, U5	Sm 1 and 2
Smd2/Ylr275w	12.9	—	Sm D2 (16.5 kD)	U1, U2, U4/U6, U5	Sm 1 and 2
Smd3/Ylr147c	11.2	l	Sm D3 (18 kD)	U1, U2, U4/U6, U5	Sm 1 and 2
Sme1/Yor159c	10.4	l	Sm E (12 kD)	U1, U2, U4/U6, U5	Sm 1 and 2
Smx3/Ypr182w	9.7	—	Sm F (11 kD)	U1, U2, U4/U6, U5	Sm 1 and 2
Snp2/Smx2/Yfl018wa	8.5	l	Sm G (9 kD)	U1, U2, U4/U6, U5	Sm 1 and 2
Spb8/Yjl124c	20.3	—	unknown; sim. to ySmb1	U6, U4/U6	Sm 1 and 2
Snp3/Smx5/Ybl026w	11.2	—	unknown; sim. to ySmd1	U6, U4/U6	Sm 1 and 2
Smx4/Uss2/Ylr438.c-a	10.0	l	unknown; sim. to ySmd2	U6, U4/U6	Sm 1 and 2
Uss1/Yer112w	21.3	l	unknown; sim. to ySmd3	U6, U4/U6	Sm 1 and 2
Ylr146w	10.4	—	unknown; sim. to ySme1	U6, U4/U6	Sm 1 and 2
Ydr378c	13.8	—	unknown; sim. to ySmx3	U6, U4/U6	Sm 1 and 2
Ynl147w	12.1	—	unknown; sim. to ySnp2	U6, U4/U6	Sm 1 and 2

(Sm) Antibody epitope recognized by anti-Sm antibodies.
[a](MW) Molecular weights are given as predicted by the protein sequence.
[b](KO) Knockout phenotype: (v) viable; (l) lethal; (–) unknown.
[c]The apparent MWs of human homologs or splicing factors are given in parentheses.
[d]Human SmB' and SmB are alternatively spliced variants.

Three major databases for yeast proteins and function were screened for information about factors important for splicing. These databases keep references and links to mammalian homologs and can be easily accessed via the world wide web to retrieve bibliographic or sequence information.

1. Saccharomyces genome database (SGD) is a scientific database of the molecular biology and genetics of the yeast.
   URL: http://genome-www.stanford.edu/Saccharomyces/
2. Munich Information Centre for Protein Sequences (MIPS): A database for protein sequences, homology data, and yeast genome information.
   URL: http://www.mips.biochem.mpg.de/
3. The Yeast Protein Database (YPD™ by Proteome, Inc.) contains an up-to-date accumulation of knowledge on all the proteins of yeast, including curated literature, properties, and annotations for all yeast genes/proteins. YPD also lists precompiled yeast–human sequence alignments and covers functional genomics/proteomics such as gene expression profiling, large-scale knockout phenotyping, and large-scale 2-hybrid analysis.
   URL: http://www.proteome.com/YPDhome.html

Table 2 U1 SnRNP-specific or U1 snRNP-associated factors

Yeast factor	MW (kD)	KO	Human homolog	Comments	Sequence motifs
SNR19 (gene for U1 snRNA on chr. XIV, 230104-230671C)	568 nt	l	164 nt (GenBank K00788)	base-pairs with intron at the 5′ SS, displaced by U6 snRNA after spliceosome assembly	Sm binding site, $m_3^{2,2,7}$Gppp cap
Sm core proteins			B, B′, D1, D2, D3, E, F, G	see Table 1	Sm 1 and 2
Snp1/Yil061c	34.5	v	U1-70K (70 kD)	binds stem loop I, interacts with Brr2 and Nam7 (protein involved in decay of mRNA containing nonsense codons)	1 RRM, (RS)[a]
Mud1/Ybr119w	34.4	v	U1-A (34 kD)	binds stem-loop II, may interact with Mud2, Mud13	2 RRMs
Yhc1/U1c/Ylr298c	27.1	l	U1- C (22 kD)	affects level of commitment complexes, promotes base-pairing between 5′ SS and U1 snRNA	zinc finger
Prp39/Yml046w	74.8	l	unknown	may facilitate the interaction of U1 snRNP with the 5′ SS	TPRs
Prp40/Ykl012w	69.1	l	unknown	binds pre-mRNA near 5′ SS, interacts with Prp8 and Msl5	2 WWP
Snu56/Mud10/Ydr240c	56.5	l	unknown	affects the in vitro formation of commitment complex and spliceosomes and the in vivo splicing efficiency of certain introns	Ser-rich
Nam8/Mre2/Mud15/ Yhr086w	57.0	v	unknown	required for meiosis-specific splicing of MER2 pre-mRNA with noncanonical 5′ SS, hyper-stabilizes association of Prp42 to U1 snRNP	3 RRMs
Snu71/Ygr013w	71.3	l	unknown	weakly associated with U1 snRNP, shares 50% sequence similarity with Prp39	few RS-, RE-, and RD-dipeptides
Prp42/Snu65/ Mud16/Ydr235	65.1	l	unknown		TPRs

The yeast U1 snRNP (18S particle, 16 proteins) is larger and contains more proteins than its human counterpart (12S particle, 11 proteins) (Neubauer et al., 1997). Abbreviations: (Mud) mutant-U1-die (mutants that are synthetically lethal with a U1 snRNA mutation); (Prp) precursor RNA processing; (RD) arginine/aspartic acid repeat; (RE) arginine/glutamic acid repeat; (RRM) RNA recognition motif; (RS) arginine/serine repeat; (snRNA) small nuclear RNA; (snRNP) small nuclear ribonucleoprotein; (Snu) "snup"-associated protein; (TPRs) tetratrico peptide repeats; (WWP) amino acid motif present in single or multiple copies in various proteins including dystrophin.
[a](RS) indicates the presence of an arginine/serine-rich motif in the human homolog.

Table 3 U2 SnRNP-specific or U2 snRNP-associated factors

Yeast factor[a]	MW (kD)	KO	Human homolog	Comments	Sequence motifs
LSR1/SNR20 (gene for U2 snRNA on chr. II 680650–681820C)	1171 nt	l	188 nt (GenBank K02227)	base-pairs with branch site and with U6 after spliceosome assembly	Sm binding site, $m_3^{2,2,7}$/Gppp cap
Sm core proteins			B, B', D1, D2, D3, E, F, G U2-A' (31 kD)	see Table 1 required for binding of B'' to U2 snRNA	Sm 1 and 2 Leu-rich
Msl1/Yib9/Nam2/Yir009w	12.8	v	U2 B'' (28.5 kD), sim. to U1-A	binds to stem-loop IV	2 RRMs
Prp9/Ydl030w	63.0	l	SF3a60/SAP 61	interacts with the stem-loop IIa, associated with Prp11 and Prp21	zinc finger
Prp11/Rna11/Ydl043c	29.9	—	SF3a66/SAP 62	interacts with the stem-loop IIa and Mud2, associated with Prp9 and Prp21	zinc finger
Prp21/Spp91/Yjl203w	33.1	l	SF3a120/SAP 114	interacts with the stem-loop IIa, associated with Prp9 and Prp11	surp1 module (Drosophila SWAP family), Pro-rich
Hsh49/Yor319w	24.5	l	SF3b53/SAP 49	binds pre-mRNA, interacts with Cus1	2 RRMs, Pro-/Gly-rich
Cus1/Ymr240c	50.3	l	SF3b150/SAP 145	binds pre-mRNA near branch site, interacts with Hsh49	Pro-rich
			SF3b120/SAP 130	splicing factor	
			SF3b160/SAP 155	splicing factor	

[a]The biochemical fractionation of yeast U2 snRNP has not yet been described; yeast factors are listed based on homology with factors present in the human 17S form of U2-snRNP. (SAP) Spliceosome-associated protein.

Table 4 U5 SnRNP-specific or U5 snRNP-associated factors

Yeast factor[a]	MW (kD)	KO	Human homolog	Comments	Sequence motifs
SNR7 (gene for U5 snRNA on chr. VII 939451-939665C)	215 nt, 180 nt	1	102–121 nt (multiple gene products) (GenBank, e.g., U5A: X01691)	short and long transcripts from same gene in yeast, stem-loop I involved in positioning released 5′ exon at 3′ SS	Sm binding site, $m_3^{2,2,7}$Gppp cap
Sm core proteins			B, B′, D1, D2, D3, E, F, G	see Table 1	Sm 1 and 2
Prp8/Rna8/Yhr165c	279.5	1	hPrp8/U5-220 kD	required for association with U4/U6 snRNP and later steps of splicing; contacts pre-mRNA at 5′ and 3′ SS, interacts with Prp17, Msl5, Prp16, Prp18, Prp40, and Slu7	
Brr2/Snu246/Slt22/Yer172c	246.1	1	U5-200 kD	interacts with Snp1, Cus1; promotes conformational changes in RNA–RNA network essential for first step, dissociation of U4/U6 snRNP, U2/U6 is best ATPase substrate	2 RNA helicase motifs (DEIH and DxxH)
Gin10/Snu114/Ykl173w	114.0	1	U5-116 kD	GTP-binding factor that is related to ribosomal translocase EF-2	G domain
Prp6/Rna6/Ybr055c	104.2	—	unknown	required for tri-snRNP formation	TPRs, leucine repeat, zinc finger
Prp28/Ydr243c	66.6	1	U5-102 kD		
			U5-100 kD	required for first step; possibly involved in unwinding U4/U6	RNA helicase (DEAD), Leu zipper, (RS)[b]
			U5-52 kD		
			U5-40 kD		
			U5-15 kD		

(DEAD, DEAH, or DxxH) Helicase-like domains.
[a]The biochemical fractionation of yeast U5 snRNP has not yet been described; yeast factors are listed based on the homology with factors present in the human 20S form of U5-snRNP.
[b](RS) indicates the presence of an arginine/serine-rich motif in the human homolog.

Table 5. U4/U6 or U4/U6.U5 SnRNP or snRNP-associated factors

Yeast factor[a]	MW (kD)	KO	Human homolog	Comments	Sequence motifs
SNR14 (gene for U4 snRNA on chr. V 167426-167585C)	160 nt	1	145 nt (GenBank M15957)	base-pairs with U6 snRNA, interaction is resolved during spliceosome assembly	Sm binding site, $m_3^{2,2,7}$Gppp cap
SNR6 (gene for U6 snRNA on chr. XII 366236-366347W)	112 nt	1	107 nt (GenBank X07425)	base-pairs to U4 snRNA in U4/U6 or U4/U6.U5 snRNP, displaces U1 at the 5′ SS and base-pairs to U2 after U4 snRNP release	mppp cap
Sm core proteins			B, B′, D1, D2, D3, E, F, G, U6-specific homologs	see Table 1	Sm 1 and 2
Prp3/Rna3/Ydr473c	55.9	—	U4/U6-90 kD/SAP 90	U4/U6 component, associated with U4/U6.U5, important for spliceosome assembly	short homology to *E. coli* RNase III, including its double-stranded RNA-binding domain
Prp4/Rna4/Ypr178w	52.4	1	U4/U6-60 kD/SAP 60	U4/U6 component, associated with U4/U6.U5, required for spliceosome assembly	WD repeats
			U4/U6.U5-63 kD		
			U4/U6.U5-61 kD		
			U4/U6.U5-27 kD		
			U4/U6.U5-20 kD		
			U4/U6.U5-15.5 kD		RS
			U4/U6.U5-15.5 kD	peptidyl-prolyl *cis/trans*-isomerase	

(WD) Amino acid motif originally described in the beta-subunit of transducin and involved in protein–protein interactions.
[a]The biochemical fractionation of yeast di- or tri-snRNPs has not yet been described; yeast factors are listed based on homology with factors present in the human 25S form of U4/U6.U5 tri-snRNP.

Table 6 Splicing factors that bind during formation of the commitment complex

Yeast factor	MW (kD)	KO	Human homolog	Comments	Sequence motifs
U1 snRNP	18S		12S	binds to 5′ SS	
Mud2/Ykl074c	60.5	v	weak similarity to U2AF65 (65 kD)	binds polyY tract near 3′ SS; interacts with Msl5, Prp11, Prp39, Prp40	3 RRMs, (N-terminal RS in human homolog)
Msl5/Bbp1/Ylr116w	53.1	l	ortholog of splicing factor SF1/mBBP (75 kD)	binds branch site; interacts with Mud2, Prp40 and bridges the 5′ and 3′ SS	2 zinc fingers,[a] KH, Pro-rich
Mud13/Cbp20/Sae1/ Ypl178w	23.8	v	human CBP20 (20 kD)	binds to m^{7}G cap of pre-mRNAs enhancing association of U1 snRNP with 5′ SS	RRM
Gcr3/Cbp80/Sto1/ Ymr125w	99.5	v	human CBP80 (80 kD)	binds to m^{7}G cap of pre-mRNAs enhancing association of U1 snRNP with 5′ SS	
			U2AF35 (35 kD)	associated with U2AF65	RRM, RS[b]
			ASF/SF2 (33 kD)	interacts with 5′ SS and U1-70K and U1-C	2 RRMs, RS[b]
			SAPs 115, 92, 88, 72, 42	spliceosome-associated proteins, bind first bind during commitment complex formation	

Abbreviations: (KH) hnRNP H homology implicated in RNA binding; (polyY) polypyrimidine.
[a]Only one zinc finger motif is present in human mBPP.
[b]Two classes of RRM and RS-domain containing proteins are exclusively involved in mammalian splice site definition and coupling of splicing to other processes required for gene expression.

Table 7 Splicing factors required for ATP-dependent pre-spliceosome formation

Yeast factor	MW (kD)	KO	Human homolog	Comments	Sequence motifs
U2 snRNP			17 S U2 snRNP	binds to pre-mRNA branch site	
Sub2/Ydl084w	50.3	—	UAP56	suppressor of cs Brr1 mutants, important for U2 snRNP addition; UAP56 is associated with U2AF65	RNA helicase (DECD)
Prp5/Rna5/Ybr237w	96.4	—	unknown	interacts with Prp9, Prp11, Prp21, and with stem loop IIa of U2 snRNA, important for U2 addition	RNA helicase (DEAD)
Cus2/Ynl286w	32.3	v	unknown	suppressor of U2 snRNA mutations polyY binding protein, important for alternative splicing	2 RRMs
			PTB/hnRNP I (57–59 kD)[a]		4 RRMs
			p14	branch site interacting protein	RRM
			SAP 33	spliceosome-associated protein, binds first bind during pre-spliceosome formation	

(cs) Cold sensitive.
[a]Proteins of the hnRNP family spontaneously associate with pol II transcripts but are generally not considered to be constitutive splicing factors. HnRNP proteins, however, play an important role in alternative splicing.

Table 8 Splicing factors required for the formation of the catalytically active spliceosome

Yeast factor	MW (kD)	KO	Human homolog	Comments	Sequence motifs
U4/U6.U5 tri-snRNP			25S U4/U6.U5 tri-snRNP	formation of novel snRNA-pre-mRNA networks and the catalytic active sites	
Prp31/Ygr091w	56.3	l	unknown	recruitment of U4/U6.U5 to pre-spliceosome	
Prp19/Yll036c	56.6	l	unknown	interacts with Snt309, associates with spliceosome concomitant with or after dissociation of U4 snRNP	
Snt309/Ypr101w	20.7	v	unknown	interacts with Prp19, associates with spliceosome simultaneously with Prp19	
Prp2/Ynr011c	99.8	—	similar to RNA helicase HRH1	required for 1st step; interacts with Spp2 and transiently with the spliceosome before and during the first step	RNA helicase (DEAH)
Spp2/Yor148c	20.7	l	unknown	interacts with Prp2, associates with the spliceosome after spliceosome assembly, released from the spliceosome before ATP hydrolysis at the 1st step	
Prp16/Ykr086w	121.7	l	hPrp16 (140 kD)	required for second step; transiently associated with spliceosome and released from excised intron; interacts with Prp17, Slu7, Prp8, Prp18, and U5 snRNA	RNA helicase (DEAH)
Prp17/Cdc40/Slu4/Ydr364c	52.0	v	hPrp17 (66 kD)	required for second step; interacts with Prp16, Prp18, Slu7, and U5 snRNA, acts before Prp18 and Slu7	WD repeats

Slu7/Ydr088c	44.6	l	hSlu7	required for second step; interacts with 3' SS, Prp18, Smd1, Prp16, Prp17, and U5 snRNA; essential when BP to 3' SS distance is larger than 7 nt; remains bound to excised intron	zinc finger, PEST region
Prp18/Ygr006w	24.7	v	hPrp18 (42 kD)	required for second step; weakly associated with U5 or U4/U6.U5, interacts with Prp17, Prp8, Prp16, and Slu7	
Prp22/Yer013w	130.0	—	RNA helicase HRH1 (139 kD)	required for second step and release of mature mRNA from the spliceosome	RNA helicase (DEAH), S1
			PSF/SAP 102 (100 kD)	polyY-binding protein, possibly replaces U2AF65, required for 2nd step, SAP 68 is a proteolytic fragment of SAP 102, IBP (intron-binding protein) might be identical to SAP 102/68	2 RRMs, Pro-/Gln-rich
			69-kD protein, similarity to RNA-binding proteins TLS and EWS	associated with Sm core of all snRNPs, binds G- and U-rich sequences, recognizes the 3' splice site during the second step of pre-mRNA splicing	RRM, zinc finger, Arg-/Gly-rich
			SAPs 82, 55	spliceosome-associated proteins, required for tri-snRNP entry	
			SAPs 165, 95, 75, 70, 65, 58, 57, 52, 48, 45, 36, 35, 30	spliceosome-associated proteins, first detected in catalytically active spliceosome	

Abbreviations: (Slu) Synthetic lethal with U5 snRNA mutant; (S1) a ribosomal motif implicated in RNA binding.

Table 9. Factors important for snRNP formation, spliceosome disassembly, and recycling

Yeast factor	MW (kD)	KO	Human homolog	Comments	Sequence motifs
Prp38/Ygr075c	28.0	—	unknown	required for maximal U6 snRNA levels, affects U6 snRNA stability	
Rnt1/Ymr239c	54.1	l	unknown	ribonuclease (sim. to *E. coli* RNase III), participates in 3′ end processing of U5 snRNA	
Brr1/Ypr057w	39.9	v	SIP1 (found in a complex with the SMA disease gene product SMN)	associates with all snRNPs but is distinct in sequence from Sm core proteins, possibly interacts with Smd1; involved in 3′-end snRNA processing and affects snRNA stability	
Prp43/Ja1/Ygl120c	87.5	—	similar to RNA helicase HRH1 and homolog of mouse DEAH9	important for spliced intron release from the spliceosome	RNA helicase (DEAH)
Prp24/Ymr268c	50.9	l	unknown	reanneals U4 and U6 snRNA, important for snRNP recycling	3 RRMs
Dbr1/Prp26/Ykl149c	47.7	v	unknown	lariat RNA debranching enzyme, involved in spliced intron degradation	

Table 10 Factors possibly involved in splicing

Yeast factor	MW (kD)	KO	Mammalian homolog	Comments	Sequence motifs
Ja2/Ykl078w	82.7	—	similar to RNA helicase HRH1	possible RNA helicase based on sequence similarity to Prp2, Prp16, and Prp22	RNA helicase (DEAH)
Ded1/Spp81/Yor204w	65.6	1	Putative ATP-dependent RNA helicase (DEAD) of the mouse PL10 subfamily	important for translation, rather than splicing; Spp81 mutant alleles suppress the ts Prp8-1 mutant	RNA helicase (DEAD)
Spp41/Ydr464w	161.6	1	unknown	interacts with Rap1 transcription factor, possibly involved in negative transcriptional regulation of several PRP genes, suppresses cs mutants of Prp3, Prp4, and Prp11	
Ymr053c	55.5	1	similar to human breast tumor-associated autoantigen	interacts with Yjr022w, Smb1, Prp9, Prp11, Prp21	
Aar2/Ybl074c	41.7	v	unknown	required for splicing of two small introns in MATa1	Leu zipper
Npl3/Nop3/Mts1/ Ydr432w	45.4	1	similar to hnRNP proteins	weakly associated with all snRNPs, predominantly with U1	2 RRMs, RGG domain, few RS-, RE- and RD-dipeptides

(ts) Temperature-sensitive.

REFERENCES

Anderson K. and Moore M.J. 1997. Bimolecular exon ligation by the human spliceosome. *Science* **276:** 1712–1716.

Arenas J.E. and Abelson J.N. 1997. Prp43: An RNA helicase-like factor involved in spliceosome disassembly. *Proc. Natl. Acad. Sci.* **94:** 11798–11802.

Barabino S.M., Blencowe B.J., Ryder U., Sproat B.S., and Lamond A.I. 1990. Targeted snRNP depletion reveals an additional role for mammalian U1 snRNP in spliceosome assembly. *Cell* **63:** 293–302.

Baserga S.J. and Steitz J.A. 1993. The diverse world of small ribonucleoproteins. In *The RNA world* (ed. R. Gesteland and J. Atkins), pp. 359–381. Cold Spring Harbor Laboratory Press, Cold Spring Harbor, New York.

Bennet M. and Reed R. 1993. Correspondence between a mammalian spliceosome component and an essential yeast splicing factor. *Science* **262:** 105–108.

Brosi R., Hauri H.P., and Krämer A. 1993a. Separation of splicing factor SF3 into two components and purification of SF3a activity. *J. Biol. Chem.* **268:** 17640–17646.

Brosi R., Groning K., Behren S.E., Lührmann R., and Krämer A. 1993b. Interaction of mammalian splicing factor SF3a with U2 snRNP and relation of its 60-kD subunit to yeast PRP9. *Science* **262:** 102–105.

Brys A. and Schwer B. 1996. Splicing glue: A role for SR proteins in *trans* splicing? *Microb. Pathog.* **21:** 149–155.

Burge C.B. and Karlin S. 1997. Prediction of complete gene structures in human genomic DNA. *J. Mol. Biol.* **268:** 78–94.

———. 1998. Finding the genes in genomic DNA sequences. *Curr. Opin. Struct. Biol.* **8:** 346–354.

Burset M. and Guigo R. 1996. Evaluation of gene structure prediction programs. *Genomics* **34:** 353–367.

Chapon C. and Legrain P. 1992. A novel gene, spp91-1, suppresses the splicing defect and the pre-mRNA nuclear export in the prp9-1 mutant. *EMBO J.* **11:** 3279–3288.

Cheng S.C. and Abelson J. 1987. Spliceosome assembly in yeast. *Genes Dev.* **1:** 1014–1027.

Chiara M.D., Gozani O., Bennett M., Champion-Arnaud P., Palandjian L., and Reed R. 1996. Identification of proteins that interact with exon sequences, splice sites, and the branchpoint sequence during each stage of spliceosome assembly. *Mol. Cell. Biol.* **16:** 3317–3326.

Company M., Arenas J., and Abelson J. 1991. Requirement of the RNA helicase-like protein PRP22 for release of messenger RNA from spliceosomes. *Nature* **349:** 487–493.

Cover T.M. and Thomas J.A. 1991. *Elements of information theory.* Wiley, New York.

Crispino J., Blencowe B.J., and Sharp P.A. 1994. Complementation by SR proteins of pre-mRNA splicing reactions depleted of U1 snRNP. *Science* **265:** 1860–1869.

Crispino J.D., Mermoud J., Lamond A., and Sharp P.A. 1996. *Cis*-acting elements from the 5′ splice site promote U1-independent pre-mRNA splicing. *RNA* **2:** 664–673.

Dietrich R.C., Incorvaia R., and Padgett R.A. 1997. Terminal intron dinucleotide sequences do not distinguish between U2- and U12-dependent introns. *Mol. Cell* **1:** 151–160.

Fabrizio P., Laggerbauer B., Lauber J., Lane W.S., and Lührmann R. 1997. An evolutionarily conserved U5 snRNP-specific protein is a GTP-binding factor closely related to the ribosomal translocase EF-2. *EMBO J.* **16:** 4092–4106.

Fickett J.W. and Tung C.-S. 1992. Assessment of protein coding measures. *Nucleic Acids Res.* **20:** 6441–6450.

Gee S., Krauss S.W., Miller E., Aoyagi K., Arenas J., and Conboy J.G. 1997. Cloning of mDEAH9, a putative RNA helicase and mammalian homologue of *Saccharomyces cerevisiae* splicing factor Prp43. *Proc. Natl. Acad. Sci.* **94:** 11803–11807.

Hall S.L. and Padgett R.A. 1994. Conserved sequences in a class of rare eukaryotic introns with non-consensus splice sites. *J. Mol. Biol.* **239:** 357–365.

———. 1996. Requirement of U12 snRNA for the in vivo splicing of a minor class of eukaryotic nuclear pre-mRNA introns. *Science* **271:** 1716–1718.

Hertel K.J., Lynch K.W., and Maniatis T. 1997. Common themes in the function of transcription and splicing enhancers. *Curr. Opin. Cell Biol.* **9:** 350–357.

Hodges P.E., Jackson S.P., Brown J.D., and Beggs J.D. 1995. Extraordinary sequence conservation of the PRP8 splicing factor. *Yeast* **11:** 337–342.

Horowitz D.S. and Krainer A.R. 1997. A human protein required for the second step of pre-mRNA splicing is functionally related to a yeast splicing factor. *Genes Dev.* **11:** 139–151.

Horowitz D.S., Kobayashi R., and Krainer A.R. 1997. A new cyclophilin and the human homologues of yeast Prp3 and Prp4 form a complex associated with U4/U6 snRNPs. *RNA* **3:** 1374–1387.

Hwang D.Y. and Cohen J.B. 1996. U1 snRNA promotes the selection of nearby 5' splice sites by U6 snRNA in mammalian cells. *Genes Dev.* **10:** 338–350.

Jackson I.J. 1991. A reappraisal of non-consensus mRNA splice sites. *Nucleic Acids Res.* **19:** 3795–3798.

Jones M.H., Frank D.N., and Guthrie C. 1995. Characterization and functional ordering of Slu7p and Prp17p during the second step of pre-mRNA splicing in yeast. *Proc. Natl. Acad. Sci.* **92:** 9687–9691.

Kandels-Lewis S. and Seraphin B. 1993. Involvement of U6 snRNA in 5' splice site selection. *Science* **262:** 2035–2039.

Konarska M.M. and Sharp P.A. 1987. Interactions between small nuclear ribonucleoprotein particles in formation of spliceosomes. *Cell* **49:** 763–774.

Krämer A., Mulhauser F., Wersig C., Groning K., and Bilbe G. 1995. Mammalian splicing factor SF3a120 represents a new member of the SURP family of proteins and is homologous to the essential splicing factor PRP21p of *Saccharomyces cerevisiae*. *RNA* **1:** 260–272.

Krämer A. 1996. The structure and function of proteins involved in mammalian pre-mRNA splicing. *Annu. Rev. Biochem.* **65:** 367–409.

Legrain P. and Chapon C. 1993. Interaction between PRP1 and SPP91 yeast splicing factors and characterization of a PRP9-PRP11-SPP91 complex. *Science* **262:** 108–110.

Lesser C.F. and Guthrie C. 1993. Mutations in U6 snRNA that alter splice site specificity: Implications for the active site. *Science* **262:** 1982–1988.

Li Z. and Brow D.A. 1996. A spontaneous duplication in U6 spliceosomal RNA uncouples the early and late functions of the ACAGA element in vivo. *RNA* **2:** 879–894.

Long M., de Souza S.J., and Gilbert W. 1997. The yeast splice site revisited: New exon consensus from genomic analysis. *Cell* **91:** 739–740.

Lücke S., Klockner T., Palfi Z., Boshart M., and Bindereif A. 1997. *Trans* mRNA splicing in trypanosomes: Cloning and analysis of a PRP8-homologous gene from *Trypanosoma brucei* provides evidence for a U5-analogous RNP. *EMBO J.* **16:** 4433–4440.

MacMillan A.M., Query C.C., Allerson C.R., Chen S., Verdine G.L., and Sharp P.A. 1994. Dynamic association of proteins with the pre-mRNA branch region. *Genes Dev.* **8:** 3008–3020.

Madhani H.D. and Guthrie C. 1994. Dynamic RNA-RNA interactions in the spliceosome. *Annu. Rev. Genet.* **28:** 1–26.

Manley J.L. and Tacke R. 1996. SR proteins and splicing control. *Genes Dev.* **10:** 1589–1579.

McCracken S., Fong N., Rosonina E., Yankulov K., Brothers G., Siderovski D., Hessel A., Foster S., Shuman S., and Bentley D.L. 1997. 5'-Capping enzymes are targeted to pre-mRNA by binding to the phosphorylated carboxy-terminal domain of RNA polymerase II. *Genes Dev.* **11:** 3306–3318.

Mewes H.W., Albermann K., Bahr M., Frishman D., Gleissner A., Hani J., Heumann K., Kleine K., Maierl A., Oliver S.G., Pfeiffer F., and Zollner A. 1997. Overview of the yeast genome. *Nature* **387:** 7–65.

Montzka K.A. and Steitz J.A. 1988. Additional low-abundance human small nuclear ribonucleoproteins: U11, U12, etc. *Proc. Natl. Acad. Sci.* **85:** 8885–8889.

Moore M.J. and Sharp P.A. 1993. Evidence for two active sites in the spliceosome provided by stereochemistry of pre-mRNA splicing. *Nature* **365:** 364–368.

Moore M.J., Query C.C., and Sharp P.A. 1993. Splicing of precursors to messenger RNAs by the spliceosome. In *The RNA world* (ed. R. Gesteland and J. Atkins), pp. 303–357. Cold Spring Harbor Laboratory Press, Cold Spring Harbor, New York.

Neubauer G., Gottschalk A., Fabrizio P., Seraphin B., Lührmann R., and Mann M. 1997. Identification of the proteins of the yeast U1 small nuclear ribonucleoprotein complex by mass spectrometry. *Proc. Natl. Acad. Sci.* **94:** 385–390.

Nelson K.K. and Green M.R. 1989. Mamalian U2 snRNP has sequence-specific RNA-binding activity. *Genes Dev.* **3:** 1562–1571.

Newman A. 1997. RNA splicing: Out of the loop. *Curr. Biol.* **7:** 418–420.

Nilsen T.W. 1994. RNA-RNA interactions in the spliceosome: Unraveling the ties that bind. *Cell* **78:** 1–4.

———. 1998. RNA-RNA interactions in nuclear pre-mRNA splicing. In *RNA structure and function* (ed. R.W. Simons and M. Grunberg-Manago), pp. 279–307. Cold Spring Harbor Laboratory Press, Cold Spring Harbor, New York.

O'Day C.L., Dalbadie-McFarland G., and Abelson J. 1996. The *Saccharomyces cerevisiae* Prp5 protein has RNA-dependent ATPase activity with specificity for U2 small nuclear RNA. *J. Biol. Chem.* **271:** 33261–33267.

O'Keefe R.T. and Newman A.J. 1998. Functional analysis of the U5 snRNA loop 1 in the second catalytic step of yeast pre-mRNA splicing. *EMBO J.* **17:** 565–574.

O'Keefe R.T., Norman C., and Newman A.J. 1996. The invariant U5 snRNA loop 1 sequence is dispensable for the first catalytic step of pre-mRNA splicing in yeast. *Cell* **86:** 679–689.

Query C.C., McCaw P.S., and Sharp P.A. 1997. A minimal spliceosomal complex A recognizes the branch site and polypyrimidine tract. *Mol. Cell. Biol.* **17:** 2944–2953.

Reed R. 1996. Initial splice-site recognition and pairing during pre-mRNA splicing. *Curr. Opin. Genet. Dev.* **6:** 215–220.

Reese M.G., Eeckman F.H., Kulp D., and Haussler D. 1997. Improved splice site detection in Genie. *J. Comput. Biol.* **4:** 311–323.

Reyes J.L., Kois P., Konforti B.B., and Konarska M.M. 1996. The canonical GU dinucleotide at the 5' splice site is recognized by p220 of the U5 snRNP within the spliceosome. *RNA* **2:** 213–225.

Richard G.F., Fairhead C., and Dujon B. 1997. Complete transcriptional map of yeast chromosome XI in different life conditions. *J. Mol. Biol.* **268:** 303–321.

Roy J., Kim K., Maddock J.R., Anthony J.G., and Woolford Jr., J.L. 1995. The final stages of spliceosome maturation require Spp2p that can interact with the DEAH box protein Prp2p and promote step 1 of splicing. *RNA* **1:** 375–390.

Ruby S.W., Chang T.H., and Abelson J. 1993. Four yeast spliceosomal proteins (PRP5, PRP9, PRP11, and PRP21) interact to promote U2 snRNP binding to pre-mRNA. *Genes Dev.* **7:** 1909–1925.

Schwer B. and Gross C.H. 1998. Prp22, a DExH-box RNA helicase, plays two distinct roles in yeast pre-mRNA splicing. *EMBO J.* **17:** 2086–2094.

Schwer B. and Guthrie C. 1992. A conformational rearrangement in the spliceosome is dependent on PRP16 and ATP hydrolysis. *EMBO J.* **11:** 5033–5039.

Seraphin B. and Rosbash M. 1990. Exon mutations uncouple 5′ splice site selection from U1 snRNA pairing. *Cell* **63:** 619–629.

Sharp P.A. and Burge C.B. 1997. Classification of introns: U2–type or U12–type. *Cell* **91:** 875–879 (1997).

Smith C.W., Chu T.T., and Nadal-Ginard B. 1993. Scanning and competition between AGs are involved in 3′ splice site selection in mammalian introns. *Mol. Cell. Biol.* **13:** 4939–4952.

Smith C.W.J., Porro E.B., Patton J.G., and Nadal-Ginard B. 1989. Scanning from an independently specified branch point defines the 3′ splice site of mammalian introns. *Nature* **342:** 243–247.

Solovyev V.V., Salamov A.A., and Lawrence C.B. 1994. Predicting internal exons by oligonucleotide composition and discriminant analysis of spliceable open reading frames. *Nucleic Acids Res.* **22:** 5156–5163.

Sontheimer E.J., Sun S., and Piccirilli J.A. 1997. Metal ion catalysis during splicing of pre-messenger RNA. *Nature* **388:** 801–805.

Staley J.P. and Guthrie C. 1998. Mechanical devices of the spliceosome: Motors, clocks, springs, and things. *Cell* **92:** 315–326.

Stephens R.M. and Schneider T.D. 1992. Features of spliceosome evolution and function inferred from an analysis of the information at human splice sites. *J. Mol. Biol.* **228:** 1124–1136.

Sun J.S. and Manley J.L. 1995. A novel U2-U6 snRNA structure is necessary for mammalian mRNA splicing. *Genes Dev.* **9:** 843–854.

Tarn W.-Y. and Steitz J.A. 1994. SR proteins can compensate for the loss of U1 snRNP functions in vitro. *Genes Dev.* **8:** 2704–2717.

———. 1996a. A novel spliceosome containing U11, U12, and U5 snRNPs excises a minor class (AT-AC) intron in vitro. *Cell* **84:** 801–811.

———. 1996b. Highly diverged U4 and U6 small nuclear RNAs required for splicing rare AT-AC intron. *Science* **273:** 1824–1832.

Teigelkamp S., Newman A.J., and Beggs J.D. 1995a. Extensive interactions of PRP8 protein with the 5′ and 3′ splice sites during splicing suggest a role in stabilization of exon alignment by U5 snRNA. *EMBO J.* **14:** 2602–2612.

Teigelkamp S., Whittaker E., and Beggs J.D. 1995b. Interaction of the yeast splicing factor PRP8 with substrate RNA during both steps of splicing. *Nucleic Acids Res.* **23:** 320–326.

Teigelkamp S., McGarvey M., Plumpton M., and Beggs J.D. 1994. The splicing factor PRP2, a putative RNA helicase, interacts directly with pre-mRNA. *EMBO J.* **13:** 888–897.

Teigelkamp S., Achsel T., Mundt C., Göthel S.-F., Cronshagen U., Lane W.S., Marahiel M., and Lührmann R. 1998. The 20kD protein of human [U4/U6.U5] tri-snRNPs is a novel cyclophilin that forms a complex with the U4/U6–specific 60kD and 90kD proteins. *RNA* **4:** 127–141.

Umen J.G. and Guthrie C. 1995. A novel role for a U5 snRNP protein in 3′ splice site selection. *Genes Dev.* **9:** 855–868.

Valcárcel J. and Green M.R. 1996. The SR protein family: Pleiotropic functions in pre-mRNA splicing. *Trends Biochem. Sci.* **21:** 296–301.

Wang J. and Manley J.L. 1997. Regulation of pre-mRNA splicing in metazoa. *Curr. Opin. Genet. Dev.* **7:** 205–211.

Wells S.E. and Ares Jr., M. 1994. Interactions between highly conserved U2 small nuclear RNA structures and Prp5p, Prp9p, Prp11p, and Prp21p proteins are required to ensure integrity of the U2 small nuclear ribonucleoprotein in Saccharomyces cerevisiae. *Mol. Cell. Biol.* **14:** 6337–6349.

Wiest D.K., O'Day C.L., and Abelson J. 1996. In vitro studies of the Prp9.Prp11.Prp21 complex indicate a pathway for U2 small nuclear ribonucleoprotein activation. *J. Biol. Chem.* **271:** 33268–33276.

Will C.L. and Lührmann R. 1997. Protein functions in pre-mRNA splicing. *Curr. Opin. Cell Biol.* **9:** 320–328.

Wu Q. and Krainer A.R. 1996. U1-mediated exon definition interactions between AT-AC and GT-AG introns. *Science* **274:** 1005–1008.

———. 1997. Splicing of a divergent subclass of AT-AC introns requires the major spliceosomal snRNAs. *RNA* **3:** 586–601.

Wyatt J.R., Sontheimer E.J., and Steitz J.A. 1992. Site-specific cross-linking of mammalian U5 snRNP to the 5′ splice site before the first step of pre-mRNA splicing. *Genes Dev.* **6:** 2542–2553.

Xing Y., Johnson C.V., Dobner P.R., and Lawrence J.B. 1993. Higher level organization of individual gene transcription and RNA splicing. *Science* **259:** 1326–1330.

Zhou Z. and Reed R. 1998. Human homologs of yeast prp16 and prp17 reveal conservation of the mechanism for catalytic step II of pre-mRNA splicing. *EMBO J.* **17:** 2095–2106.

WWW RESOURCES

http://ccr-081.mit.edu/introns/Human.html A set of human U2-type introns.
http://ccr-081.mit.edu/introns/U12.html A set of 42 U12-type introns.
http://ccr-081.mit.edu/introns/Yeast.html A set of *S. cerevisiae* U2-type introns.
http://genome-www.stanford.edu/Saccharomyces/ *Saccharomyces* genome database.
http://www.hgc.lbl.gov/inf/genesets.html A representative human gene data set.
http://www.mips.biochem.mpg.de/ Munich Information Centre for protein sequences.
http://www.proteome.com/YPDhome.html The yeast protein database.

21

tRNA Splicing: An RNA World Add-on or an Ancient Reaction?

Christopher R. Trotta and John Abelson
Department of Biology
California Institute of Technology
Pasadena, California 91125

Introns are encoded in the genes for tRNA in organisms from all three kingdoms of life. Their removal is an essential step in the maturation of tRNA precursors. In Bacteria, introns are self-splicing and are removed by a group 1 splicing mechanism (Kuhsel et al. 1990; Reinhold-Hurek and Shub 1992; Biniszkiewicz et al. 1994). In Eukaryotes and Archaea, intron removal is mediated enzymatically by proteins. Recent progress in understanding both eukaryotic and archaeal tRNA splicing has revealed that the two processes, previously thought to be unrelated, are in fact similar. Insight gained from the comparison has provided a clearer understanding of intron recognition, the catalysis of intron removal, and has given new insight into the evolution of the tRNA splicing process.

INTRON-CONTAINING tRNA

tRNA Introns in Eukaryotes

Intervening sequences (introns) were discovered 20 years ago in the yeast genes for the tyrosine-inserting non-sense suppressor tRNA (Goodman et al. 1977) and for tRNA^{Phe} (Valenzuela et al. 1978). With the completion of the yeast genome, it is known that of the 274 yeast tRNA genes 61, or 20%, contain introns. Table 1 lists the tRNAs that contain introns. PCR cloning of tRNAs from higher eukaryotes has revealed a similar distribution of intron-containing tRNA (Stange and Beier 1986; Green et al. 1990; Schneider et al. 1993). The introns in all of the genes are small (14–60 bases), and they are all located in the same position, one base to the 3' side of the anticodon (Fig. 1A). Structure probing revealed that the common "cloverleaf" tertiary structure seen in the crystal structure of tRNA^{Phe} is maintained in intron-containing tRNAs and that the intron, including the splice sites, is the most exposed region of the molecule

Table 1 Yeast tRNA precursors containing introns

tRNA	Intron length (nucleotides)	No. of genes
$tRNA^{Ser}_{CGA}$	19	1
$tRNA^{Ser}_{GCU}$	19	4
$tRNA^{Lys}_{UUU}$	23	7
$tRNA^{Pro}_{UGG}$	31,30,33	7,2,1
$tRNA^{Trp}_{CCA}$	34	6
$tRNA^{Phe}_{GAA}$	18,19	3,7
$tRNA^{Leu}_{CAA}$	32,33	8,2
$tRNA^{Ile}_{UAU}$	60	2
$tRNA^{Leu}_{UAG}$	19	3
$tRNA^{Tyr}_{GUA}$	14	8
		Total = 61
	Total tRNA genes = 274	

(Swerdlow and Guthrie 1984; Lee and Knapp 1985). This led to a proposed model for the tertiary structure of pre-tRNA shown in Figure 1B. More recent experiments by Tocchini-Valentini and coworkers (Baldi et al. 1992) have demonstrated that a conserved base pair (the A-I base pair) between a base of the 5' exon (position 32) immediately following the anticodon stem and a base in the single-stranded loop of the intron (position –3) is required for correct excision at the 3' splice site (Fig. 1A). Thus, the distinctive feature in eukaryotic intron-containing pre-tRNAs is the position of the intron and the presence of the A-I base pair.

tRNA Introns in Archaea

tRNA introns are found in every archaeal species that has been studied (Thompson et al. 1989; Kleman-Leyer et al. 1997; Lykke-Anderson and Garrett 1997; Klenk et al. 1998). Although they are often similar, archaeal introns are different from their eukaryotic counterparts. Comparative sequence analysis of a host of archaeal introns has revealed that the 5' and 3' splice sites are always located in a 3-nucleotide bulge separated by a 4-bp helix: the bulge-helix-bulge (BHB) motif (Fig. 2) (Thompson et al. 1989). Intron length is variable, from a short intron of 33 nucleotides found in the $tRNA^{Met}$ gene of *Methanococcus jannaschii*, to 105 nucleotides of the $tRNA^{Trp}$ gene of *Haloferax volcanii* (Daniels et al. 1985), but the intron is required to base-pair with the 5' exon to allow formation

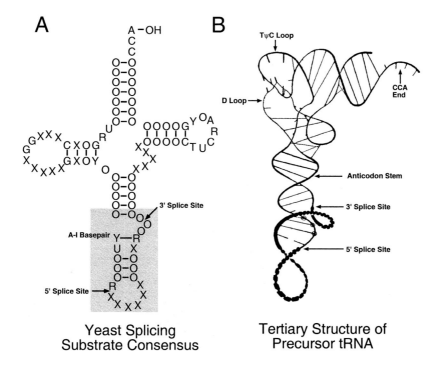

Figure 1 (*A*) Consensus of yeast pre-tRNA (Ogden et al. 1984). (O) Nonconserved base in a region of conserved length or secondary structure; (X) nonconserved base in a region of variable length or secondary structure. (G,C,A,U) Conserved base among the pre-tRNAs. (Y and R) Conserved pyrimidines and purines, respectively. The conserved anticodon-intron (A-I) base pair is depicted. (*B*) Model of the tertiary structure of pre-tRNAs based on the crystal structure of yeast tRNA^{Phe}. Intron is indicated in bold, dashed lines. (Reprinted, with permission, from Lee and Knapp 1985.)

of the BHB motif. In general, the introns reside one base 3' to the anticodon, but introns have been found elsewhere in the tRNA molecule. One such example is that of the tRNA^{Pro} gene of *Methanobacterium thermoautotrophicum*, where there are two introns in the anticodon stem that must be removed in an obligate order to produce the mature tRNA (C.R. Trotta, unpubl.). Another repository for introns in Archaea is the rRNA genes. In fact, one of the earliest discovered archaeal introns, the 622-nucleotide intron of the 23S rRNA of *Desulfurococcus mobilis*, was shown through extensive biochemical characterization to be removed by the archaeal tRNA splicing system with dependence on the BHB motif (Kjems and Garrett 1985; Kjems et al. 1989).

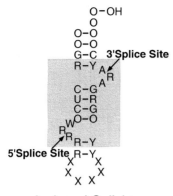

Archaeal Splicing Substrate Consensus

Figure 2 Consensus archaeal substrate. Positions depicted as in Fig. 1A. The bulge-helix-bulge (BHB) motif is shaded in gray.

THE ENZYMATIC GYMNASTICS OF tRNA SPLICING

The Eukaryotic Pathway

The accumulation of pre-tRNAs in the *rna1-1* mutant (Hopper et al. 1978) provided a source of pre-tRNA for assay of tRNA splicing in vitro (Knapp et al. 1978). The system was employed to work out the pathway of tRNA splicing (Fig. 3). The first step of the reaction is recognition and cleavage of the pre-tRNA substrate at the 5′ and 3′ splice sites by a tRNA splicing endonuclease. The cleavage by the endonuclease at the splice sites results in 2′,3′ cyclic phosphate and 5′-hydroxyl termini (Peebles et al. 1983). The tRNA 5′ and 3′ exons are then the substrate for a series of reactions catalyzed by the tRNA splicing ligase (for a detailed review of the ligation mechanism, see Arn and Abelson 1998). Ligase is a multifunctional enzyme possessing phosphodiesterase, polynucleotide kinase, and RNA ligase activities. In the ligase reaction, phosphodiesterase opens the 2′,3′ cyclic phosphate to give a 2′-phosphate. Next the 5′-hydroxyl is phosphorylated by transfer of the γ-phosphate from an NTP cofactor. Ligase is then adenylylated and the AMP moiety is transferred to the 5′-phosphate on the 3′-exon forming an activated phosphoanhydride. Finally, ligase joins the two exons by catalyzing an attack of the 3′-hydroxyl on the activated donor phosphoanhydride to form a 3′,5′ phosphodiester, 2′-monophosphoester bond with the release of AMP. This results in a pre-tRNA molecule with a 2′-phosphate at the splice junction. Removal of the 2′-phosphate is accomplished by the activity of a 2′-phosphotransferase

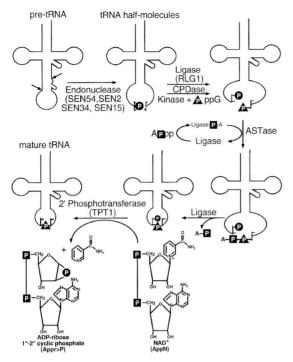

Figure 3 The tRNA splicing pathway of yeast. Gene names are given in parentheses for the proteins of the pathway. (CPDase) Cyclic phosphodiesterase; (ASTase) adenylyl synthetase. See text for details. (Reprinted, with permission, from Abelson et al. 1998.)

that catalyzes the transfer of the phosphate to NAD, releasing nicotinamide and yielding ADP-ribose $1''$-$2''$ cyclic phosphate (Appr>P) and mature tRNA (Fig. 3).

The Archaeal Pathway

Much less is known about the enzymes involved in the processing of archaeal pre-tRNA due to the difficulty in working with extracts from these organisms. As in the eukaryotic pathway, removal of introns requires the function of both an endonuclease and a ligase. The archaeal endonuclease recognizes substrates containing the consensus BHB motif and cleaves at the $5'$ and $3'$ splice sites to yield $2',3'$ cyclic phosphate and $5'$-hydroxyl termini (Thompson et al. 1989). Thus, the cleavage mechanism is identical to the eukaryotic mechanism, providing the earliest clue to the relatedness of the two enzymes (see below). Upon cleavage by the endonuclease, the tRNA half-molecules must be ligated to produce a mature tRNA, but the ligase activity has yet to be defined. It is clear that the yeast mechanism does not pertain since the γ-phosphate of ATP is not incorporated into spliced tRNA as seen in the yeast mechanism (Kjems and Garrett 1985; Gomes and Gupta 1997).

CHARACTERIZATION OF THE tRNA SPLICING ENZYMES

The Eukaryotic tRNA Endonuclease

The first step of tRNA splicing is the recognition and cleavage of the splice sites by the tRNA splicing endonuclease. Utilizing the in vitro system for tRNA splicing, the endonuclease was shown to be nuclear-membrane-associated (Peebles et al. 1983). The detergent, Triton X-100, at 0.9% is required to release the activity from the membrane fraction. In a particularly difficult purification, the endonuclease was purified one million fold from a yeast nuclear membrane preparation to homogeneity and appeared to be a αβγ trimer of 54-kD, 44-kD, and 34-kD subunits (Rauhut et al. 1990). Later, a fourth subunit of 15-kD was discovered (Trotta et al. 1997). Thus, the enzyme is a αβγδ heterotetramer. Cloning of the genes encoding each of these protein subunits was accomplished by both genetics and protein chemistry. Ho et al. (1990) screened a bank of temperature-sensitive mutants and isolated a yeast mutant that accumulated 2/3 tRNA precursor molecules that were not cleaved at the 5' splice site. The gene, SEN2 (*s*plicing *en*donuclease) (Winey and Culbertson 1988), was cloned by complementation, and antibodies to the product were shown to recognize the 44-kD subunit of the purified splicing endonuclease. Development of a simpler affinity purification utilizing a tagged SEN2 gene allowed preparation of sufficient material to identify the other three gene products, SEN54, SEN34, and SEN15 (Trotta et al. 1997).

Two of the subunits, SEN2 and SEN34, contain a homologous domain of approximately 130 amino acids. SEN2 contains the only transmembrane sequence and presumably mediates membrane association of the enzyme. All four subunits contain a nuclear localization sequence. Two-hybrid interactions were detected between Sen54-Sen2 and Sen34-Sen15. Since the enzyme is a stable tetramer, interaction must occur between these two presumed heterodimers although they are not detected in the two-hybrid assay (Trotta et al. 1997).

The Eukaryotic tRNA Ligase

The purification of the soluble tRNA ligase from yeast was simpler and quickly led to the isolation of a 95-kD polypeptide and the cloning of its essential gene, RLG1 (Phizicky et al. 1986, 1992). This single polypeptide carries out all three of the steps in the ligase mechanism (Fig. 3). Interestingly, partial proteolysis of the protein resulted in three fragments that were shown to contain distinct activities supporting a domain-like structure of the enzyme (Xu et al. 1990). The amino-terminal fragment is adenylylated at lys-114, and sequence comparisons have shown this lysine residue to be equivalent to the active lysine of the T4 RNA ligase (Xu et al. 1990; Apostol et al. 1991). The carboxy-terminal fragment contains the cyclic phosphodiesterase activity. The central domain then would be the likely locus of the kinase activity; however, an enzyme deleted in this domain paradoxically retains ligase activity. The enzyme has two nucleotide-binding sites, for GTP and for ATP (Belford et al. 1993). GTP is employed in the kinase step and ATP is used for the formation of the activated adenylyl-RNA intermediate. The in vivo function of such a complex NTP requirement is unclear, but it has been suggested that such a requirement could couple the splicing reaction to transcription and/or translation (Belford et al. 1993).

The complexity of the domain structure of ligase raises interesting questions as to the origin of this protein. The yeast tRNA ligase is mechanistically and structurally related to phage T4 RNA ligase and polynucleotide kinase (Apostol et al. 1991). The phage enzymes together contain the three activities present in the tRNA ligase, but in addition, T4 polynucleotide kinase contains a phosphatase that removes the $2'$-phosphate (Walker et al. 1975). Thus, T4 RNA ligase and T4 polynucleotide kinase can ligate pre-tRNA half-molecules produced by endonuclease, yielding a product that does not contain a $2'$-phosphate.

The $2'$-Phosphotransferase

Early investigation into the activity responsible for $2'$-phosphate removal at the splice junction demonstrated the requirement for two separate components present in yeast extracts (McCraith and Phizicky 1990). The first component was purified and determined to be the cellular cofactor NAD^+ (McCraith and Phizicky 1991). As described earlier, NAD^+ is the receptor in a reaction that involves the transfer of the $2'$-phosphate at the splice junction to the $2''$ position of the ribose of NAD^+, yielding the high-energy compound ADP-ribose $1''$-$2''$ cyclic phosphate (Culver et al. 1993). The second component was purified and shown to be a 26.2-kD polypeptide encoded by the TPT1 gene (Culver et al. 1997). Expressed in *Escherichia coli*, the single polypeptide could catalyze the transfer of the $2'$-phosphate to NAD^+. The gene was shown to be essential in yeast, and it was suggested that its essential role could be either removal of the 2'-phosphate from all intron-containing tRNA molecules or generation of the novel molecule Appr>p. To further understand the role of each of these potential functions, a conditional lethal phosphotransferase mutant was generated (Spinelli et al. 1997). Yeast expressing this mutant accumulated tRNA molecules with a 2'-phosphate at the splice junction. Interestingly, these tRNA molecules were undermodified at positions near the splice junction residue, whereas other modifications appeared to be normal. These results suggest that the removal of the 2'-phosphate is essential for correct modification of the residues near the splice junction and that tRNA containing the 2'-phosphate is not a substrate for the modification enzyme and likely inactive in carrying out its function in translation. The results also suggest a temporal order for the maturation of pre-tRNA with splicing and 2'-phosphate removal occurring before modification of certain positions in the tRNA molecule. The TPT1 gene was also found to have homologs in other eukaryotes and, surprisingly, in *E. coli* and some Archaea (Culver et al. 1997; Trotta et al. 1997). The function of this protein in Bacteria and Archaea is unclear, since a source of 2'-phosphates is not apparent (see the discussion on the archaeal ligase mechanism below). *E. coli* does not possess intron-containing tRNA or a eukaryotic-like RNA ligase. It will therefore be interesting to elucidate its role in bacteria, hopefully providing a clue as to the origin of this unique processing event.

The Archaeal tRNA Endonuclease

The endonuclease of members of the Archaea proved to be just as difficult to purify as those of Eukaryotes. Daniels and coworkers chose the halophilic archaeon *Haloferax volcanii* to carry out the purification. The enzyme is present in low abundance and is extremely difficult to stabilize during the purification protocol. When finally purified, the endonuclease of *H. volcanii* proved to be composed of a single 37-kD protein encoded by the EndA gene (Kleman-Leyer et al. 1997). Unlike the heterotetrameric yeast enzyme, the active enzyme is composed of a single subunit shown by gel filtration to behave as a homodimer in solution. The protein was shown to contain a 130-amino-acid domain homologous to the two yeast subunits SEN2 and SEN34. This was the observation that unified the two lines of research and led to the progress described below.

The Archaeal tRNA Ligase

The identity of the archaeal tRNA ligase is at present unknown; however, with the complete genome sequence of four members of the Archaea, it is clear that there is no homolog of the yeast tRNA splicing ligase (a remarkable statement to be able to make, now common in the genomic sequence era). Thus, it appears that ligation of tRNA half-molecules occurs via a different reaction mechanism from that of the eukaryotic ligation reaction.

All sequenced archaeal genomes do, however, contain an open reading frame encoding a protein homologous to the 2'-5' ligase of *E. coli* characterized by E. Arn (Arn and Abelson 1996). This protein functions in vitro as an RNA ligase joining tRNA half-molecules to form a 2'-5' phosphodiester bond at the splice junction. The exact mechanism of this reaction is unknown, but it does not require the addition of exogenous ATP. Thus, it is tempting to speculate that the homolog found in Archaea may function to ligate the tRNA half-molecules of the splicing reaction. 2'-5' linkages have not been detected in tRNA, but it is possible that the archaeal ligase has an altered specificity and catalyzes the formation of a 3'-5' linkage. Clearly, it will be interesting to further investigate the archaeal homologs of the 2'-5' ligase.

tRNA Splicing in Higher Eukaryotes

Like the yeast tRNA introns, the position of the introns in higher eukaryotes is conserved and the cleavage reaction operates in a manner identical to the yeast tRNA splicing endonuclease as exemplified by the *Xenopus* endonuclease. Ligation of the half-molecules to yield mature tRNA has been shown to proceed through a mechanism similar to the yeast ligase (Konarska et al. 1981; Schwartz et al. 1983; Zillmann et al. 1991). Using HeLa cell extracts, Zilmann et al. (1991) have demonstrated the incorporation of exogenous γ-phosphate from ATP upon ligation of yeast tRNA^{Phe}, a hallmark of the yeast tRNA ligase reaction mechanism. Furthermore, they demonstrated the presence of a 2'-phosphate at the splice junction. A 2'-phosphotransferase activity was detected in both HeLa extracts (Zillmann et al. 1992) and is implicated in the dephosphorylation of ligated tRNA in microinjected *Xenopus* oocytes (Culver et al. 1993). However, early studies implicated a different ligase in tRNA splicing in HeLa extracts and *Xenopus* oocytes. It was demonstrated that the product of the tRNA splicing reactions catalyzed by these extracts did not contain a 2'-phosphate at the splice junction. This novel ligase uses the cyclic phosphate at the end of the 5' half of tRNA to generate the 3',5'-phosphodiester bond in the mature tRNA, as there is no incorporation of exogenous phosphate at the splice junction (Nishikura and De Robertis 1981; Filipowicz et al. 1983; Laski et al. 1983). Thus, there appear to be redundant pathways for ligation of tRNA molecules in metazoans. The further characterization of the alternate eukaryotic ligase activity could conceivably shed light on the nature of the archaeal ligase mechanism.

RECOGNITON OF PRE-tRNA BY THE tRNA SPLICING ENDONUCLEASE

Eukaryotic Endonuclease: The Ruler Mechanism

Early speculation concerning the ability of the tRNA endonuclease to recognize and cleave tRNA substrates relied on comparison of the primary and tertiary sequences of the precursor molecules. As previously mentioned, there is no sequence conservation around the splice sites and the precursors all fold into the same tertiary structure. The introns are of variable sequence and can be altered by both addition and deletion of nucleotides as well as changes in the sequence with little effect on removal by endonuclease (Johnson et al. 1980; Strobel and Abelson 1986; Reyes and Abelson 1988). The exception to this generalization is the A-I base pair necessary for accurate cleavage at the 3′ splice site. Mutations that abolish this base pair are not substrates for the splicing endonuclease (Willis et al. 1984; Baldi et al. 1992). This led to the hypothesis that recognition involved specific interactions with the conserved tertiary structure present in the mature domain of the pre-tRNA. Experimental support for this hypothesis involved the engineering of mutant pre-tRNA, which altered distinct features of the primary, secondary, or tertiary structures of the molecule (for review, see Culbertson and Winey 1989). This suggested a ruler model for recognition and cleavage by the endonuclease in which the enzyme recognizes the mature domain and measures the conserved distance to the splice sites (Fig. 4) (Greer et al. 1987; Reyes and Abelson 1988). To test this model, tRNA molecules were constructed in which the conserved distance was altered by insertion and deletion of base pairs in the anticodon stem. Addition of a single base pair increased the length of the intron by 2 nucleotides, one at either end (Reyes and Abelson 1988) as predicted by the model.

Archaeal Endonuclease: The Bulge-helix-bulge Motif

In contrast to the yeast results, it was demonstrated that for the archaeal endonuclease, the pre-tRNA substrate could be drastically altered and cleavage would still occur. Synthetic tRNA molecules where almost the

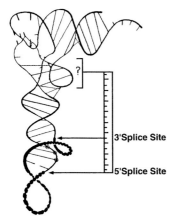

Figure 4 Ruler model for interaction of the yeast endonuclease with pre-tRNA (see text). (Reprinted, with permission, from Reyes and Abelson 1988 [copyright Cell Press].)

entire tRNA mature domain was removed were accurately cleaved by the endonuclease (Thompson and Daniels 1988). As in the case of the eukaryotic endonuclease, the intron plays a mostly passive role, as large deletions can be tolerated by the enzyme. There is, however, a strict requirement for sequence and structure present at the exon–intron boundaries (Thompson and Daniels 1988). This bulge-helix-bulge (BHB) motif is the only requirement for intron recognition and removal by the archaeal tRNA splicing endonuclease.

GENES FOR THE ENDONUCLEASE: A BRIDGE BETWEEN KINGDOMS

Once the genes for the eukaryotic and archaeal endonuclease were in hand, a comparison between the yeast subunits and the single *H. volcanii* endonuclease revealed that these proteins were related. A domain of 130 amino acids is found in the carboxyl terminus of the SEN2 and SEN34 genes of the yeast endonuclease and in the *H. volcanii* endonuclease. Furthermore, homologs of this domain were discovered in the Archaea *M. jannaschii* and *P. aerophilum*, where the domain represented the entirety of the protein (Fig. 5). This homology led to the suggestion that the two yeast subunits each contain an active site for the cleavage reaction. It had been demonstrated that a mutant in the SEN2 gene, *sen2-3*, was defective in 5′ splice-site cleavage (Ho et al. 1990). This mutation, Gly-292 to glutamate, lies within the conserved domain, suggesting that SEN2 contains

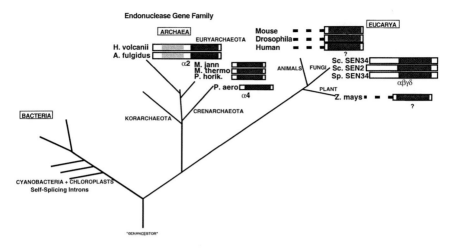

Figure 5 Graphic representation of archaeal and eukaryotic endonuclease proteins. The sizes of the various proteins (*open rectangle*) and location of the conserved sequence regions (*black box*) are indicated. Light gray blocks represent a degenerative domain repeat present in several archaeal endonucleases (see text). Proteins are placed on the Woesnian evolutionary tree. Endonuclease configuration is noted (see text for details): (α4) tetramer; (α2) dimer; (αβγδ) heterotetramer; (?) configuration unknown. Protein designations: (H. volcanii) *H. volcanii* EndA (AF001578); (A. fulgidus) *Archaeoglobus fulgidus* (AE001041); (M. jann) *M. jannaschii* (MJ1424); (M. thermo) *Methanobacterium thermoautotrophicum* (AF001577); (P. horik) *Pyrococcus horikoshii* (AB009474); (Sc Sen2) *Saccharomyces cerevisiae* Sen2 (P16658); (Sc Sen34) *S. cerevisiae* Sen34 (P39707); (Sp Sen34) *Schizosaccharomyces pombe* Sen34; (Z. mays) *Zea mays* open reading frame contained in the intron of HMG (X72692); Mouse, Drosophila, and Human represent homologs of the endonuclease domain that have been detected in a partially sequenced genomic database submission (C.R. Trotta, unpubl.).

the active site for 5' splice-site cleavage and, by extension, that SEN34 carries the active site for 3' splice-site cleavage. A mutant in SEN34 changing a conserved histidine at position 242 to alanine resulted in a marked decrease in cleavage at the 3' splice site, whereas cleavage at the 5' splice site was normal, strongly supporting the two-active-site model (Trotta et al. 1997).

In the dimeric *H. volcanii* endonuclease, identical subunits cleave the symmetrically disposed splice sites in BHB substrate. Thus, the arrangement of active sites in eukaryotic and archaeal endonucleases must be similar (as depicted in Fig. 6).

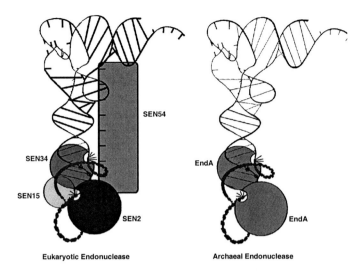

Figure 6 Comparison of the models proposed for the eukaryotic (yeast) and archaeal (*H.volcanii*) endonucleases (see text for details). (Reprinted, with permission, from Li et al. 1998 [copyright American Association for the Advancement of Science].)

tRNA SPLICING AT THE ATOMIC LEVEL

Given the relatedness of the tRNA splicing endonuclease from the two kingdoms, it seemed expedient to choose the archaeal system for structural work. The endonuclease of the archaeon *M. jannaschii* was chosen, because this enzyme consists of only the homologous domain implicated by the endonuclease sequence alignment. It is small and proved easy to express and purify.

Extensive biochemical characterization of this enzyme by Lykke-Andersen and Garrett (1997) showed it to be a homotetramer in solution, an observation confirmed by the crystal structure (Fig. 7B) (Li et al. 1998). Each monomer consists of two domains: the amino-terminal domain (residues 9–84), which is composed of three α helices and a mixed antiparallel/parallel β-pleated sheet of four strands, and the carboxy-terminal domain (residues 85–179), which contains two α helices flanking a five-stranded mixed β-sheet (Fig. 7A). The manner in which four of these monomers are brought together provides insight into the architecture and evolution of other members of the endonuclease family of both Archaea and eukaryotes.

Two sets of interactions are crucial to formation of the tetramer. The first interaction involves the formation of an isologous dimer between two monomers. Formation of this dimer is mediated by the interaction of β9 from one monomer and β9′ from another monomer. This tail-to-tail interaction is mediated by main-chain hydrogen bonds between monomers and leads to a two-stranded β-sheet spanning the subunit boundary. Loop L8 from both monomers interacts to form more hydrogen bonds, and together these interactions enclose a hydrophobic core at the

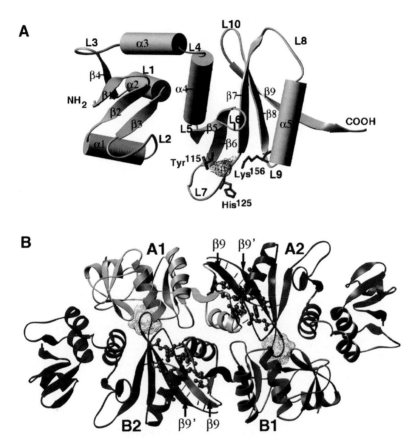

Figure 7 Crystal structure of the *M. jannaschii* tRNA splicing endonuclease. (*A*) Ribbon representation of the endonuclease monomer. The proposed catalytic triad residues are within 7 Å of one another and are shown in red ball-and-stick. Also shown is electron density detected near the putative catalytic triad, where the scissile phosphate is proposed to be located for cleavage. (*B*) Endonuclease tetramer. Each subunit is represented by a distinct label and color. The main-chain hydrogen bonds formed between β9 and β9' and between loops L8 and L8' for the formation of isologous dimers are shown as thin lines. Side chains of the hydrophobic residues enclosed at the dimer interface are shown as blue ball-and-stick models. The heterologous interaction that forms the tetramer is mediated by interaction between subunits A1 and B2 (or B1 and A2) through the acidic loops L10 and L8 and is highlighted by dotted surfaces. (Reprinted, with permission, from Li et al. 1998 [copyright American Association for the Advancement of Science].)

subunit interface. This leads to an extremely stable dimeric unit that Li et al. (1998) have suggested is conserved in endonucleases from other Archaea and eukaryotes (see below).

The second interaction involves the heterologous interaction between the two dimers that compose the tetramer. The main interaction is an insertion of acidic residues in loop L10 into a polar groove formed between the amino- and carboxy-terminal domains of the endonuclease monomer (Fig. 7B). This causes the two dimers to be translated relative to each other by about 20 Å and brings two subunits, A1 and B1, much closer together than the other two subunits, A2 and B2. This interaction is essential to the configuration of the two symmetrically disposed active sites and is also proposed to be conserved in the endonucleases from other organisms (see below).

Active Site of the tRNA Splicing Endonuclease

As discussed earlier, the product of the endonuclease cleavage reaction is a 2′,3′ cyclic phosphate and a 5-hydroxyl, a product identical to that of other ribonucleases such as RNase A. In a well-characterized acid–base-catalyzed reaction, RNase A utilizes a histidine, His-12, to abstract a proton from the 2′-hydroxyl of the ribose, leading to an in-line attack on the adjacent phosphodiester bond and the formation of a pentacovalent reaction intermediate stabilized by Lys-41. The general acid, His-119, then protonates the 5′-leaving group leading to the 2′,3′ cyclic phosphate and 5′-hydroxyl product. In a second step, a proton is abstracted from a water molecule, OH– attacks, and the 2′,3′ cyclic phosphate is hydrolyzed to the 3′-phosphate (Walsh 1979; Thompson and Raines 1994). Inspection of the sequence alignment between the conserved members of the endonuclease family showed a single histidine residue that is absolutely conserved. Recall that the histidine-to-alanine mutation in Sen34 causes a marked reduction in the catalytic efficiency of 3′ splice-site cleavage. Similar mutants have been created in this conserved histidine of *M. jannaschii* and *H. volcanii* with a similar reduction in the catalytic efficiency of the cleavage reaction (Lykke-Andersen and Garrett 1997; C.J. Daniels, pers. comm.). Thus, it would appear that the tRNA splicing endonucleases catalyze the cleavage of RNA by a general acid–base catalysis.

With the crystal structure for the *M. jannaschii* endonuclease, this histidine was localized at the atomic level. Figure 7A shows the environment surrounding His-125, which is found in a cluster with the absolutely conserved residues Tyr-115 (from loop L7) and Lys-156 (from α5), forming a pocket into which the scissile phosphate is proposed to fit. The three residues can be spatially superimposed with the catalytic triad of Rnase A

(Li et al. 1998), leading to the prediction that His-125 of the *M. jannaschii* enzyme is equivalent to the general base, Tyr-115 is equivalent to the general acid, and Lys-156 stabilizes the transition state. Preliminary experiments indicate a role for both His-125 and Lys-156, but the role of Tyr-115 has yet to be established (C.R. Trotta, unpubl.).

Because the *M. jannaschii* enzyme is a homotetramer, each monomer is expected to contain a separate active site for cleavage. However, it is expected that only two of these active sites function in cleavage of the symmetric substrate. Recently, Diener and Moore (1998) have solved the NMR solution structure of a true archaeal substrate molecule. This molecule and a similar modeled substrate based on the TAR RNA structure (Li et al. 1998) dock in plausible fashion with the enzyme (Fig. 8). The scissile phosphates are shown to fit nicely into the A1 and B1 active sites of the tetramer. The A2 and B2 active sites are too far apart to allow for interaction with the substrate.

ON THE EVOLUTION OF THE ENDONUCLEASES

The Archaeal Endonucleases

The archaeal endonuclease of *H. volcanii* appears to be configured in a different manner from the *M. jannaschii* endonuclease. The protein behaves as a homodimer, not a homotetramer, but must be arrayed in a similar manner due to the identical substrate specificity. Based on the observation that the *H. volcanii* endonuclease is actually an in-frame duplication of the endonuclease domain, Li et al. (1998) proposed a model to describe how this enzyme is configured (Fig. 9B). The *H. volcanii* enzyme can be thought of as a pseudo-dimer with the amino-terminal endonuclease domain repeat comprising one pseudo-monomer and the carboxy-terminal repeat comprising the other. The repeats are connected by a stretch of polypeptide and serve to form the pseudo-dimer by the $\beta 9$-$\beta 9'$ hydrophobic interaction that occurs between monomers in the *M. jannaschii* enzyme. Loop L10 present in the amino-terminal repeat allows dimerization of the pseudo-dimers to form the active enzyme. Careful examination of the sequence of the amino-terminal repeat shows the absence of the amino acids for the catalytic triad, thus, the dimeric enzyme contains only two active sites present in the carboxy-terminal repeat.

The occurrence of an in-frame gene duplication event explains the configuration of *H. volcanii* enzyme (Lykke-Andersen and Garrett 1997). This event has been localized to a set of closely related organisms of the Euryarchaeota branch of the Archaea (Fig. 5), which includes the genus *Haloferax*, *Archaeoglobus*, and *Methanosarcina*.

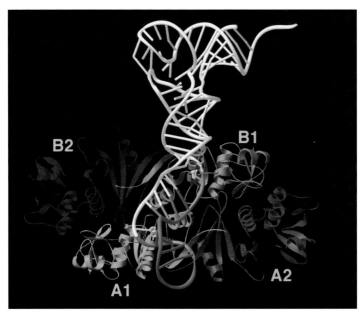

Figure 8 Endonuclease docked with a hypothetical pre-tRNA substrate. The hypothetical tRNA substrate was constructed from joining the known crystal structure of yeast tRNA^{Phe} with the archaeal substrate solved by Diener and Moore (1998). This represents the pre-tRNA$^{Arch-Euk}$ demonstrated to be a substrate for both the eukaryotic and archaeal endonucleases (Fabbri et al. 1998) (see text). The scissile phosphates are shown to dock directly in the active sites present on subunits A1 and B1 of the endonuclease.

The Yeast Endonuclease

Like the *M. jannaschii* endonuclease, the yeast tRNA splicing endonuclease is a tetramer. The two active-site-containing subunits, Sen2 and Sen34, must be configured in a similar manner to the A1 and B1 subunits of *M. jannaschii*. This arrangement is proposed to be facilitated by interactions with Sen54 and Sen15. Sen54 and Sen15 both contain a stretch of amino acids at their carboxyl terminus that is homologous to the *M. jannaschii* endonuclease carboxyl terminus (Lykke-Andersen and Garrett 1997). This homology includes the crucial regions involved in the two sets of interactions seen in formation of the tetramer: the β9-β9′ hydrophobic tail interaction and the loop L10 insertion sequence. Thus, it is proposed (Fig. 9C) that the Sen54-Sen2 and Sen34-Sen15 interactions detected in the two-hybrid assay are mediated by the β9-β9′ interaction. The loop L10 interactions are proposed as in the *M. jannaschii* structure to mediate dimer-dimer interactions.

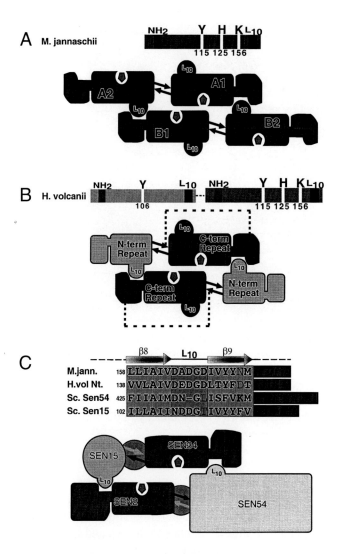

Figure 9 Model of the tRNA splicing endonucleases of *M. jannaschii*, *H. volcanii*, and *S. cerevisiae*. (*A*) The *M. jannaschii* endonuclease is graphically depicted. Several important features are shown in the primary sequence and by a model: loop L10 for tetramerization, the carboxy-terminal β9 strands (*arrows*) for dimerization, and the conserved catalytic residue His-125 (*pentagon*). (*B*) The *H. volcanii* endonuclease consists of a tandem repeat of the endonuclease domain. The amino-terminal repeat has degenerated to possess only the loop L10 and carboxy-terminal interaction domains present in the *M. jannaschii*. The carboxy-terminal repeat contains the groove for loop L10 insertion, as well as the catalytic triad including the conserved histidine (*pentagon*). The dashed line represents the polypeptide chain connecting the amino-terminal repeat and the carboxy-terminal

Interestingly, it has recently been shown that the eukaryotic endonuclease can recognize and cleave an archaeal bulge-helix-bulge containing tRNA substrate (Fabbri et al. 1998). This tRNA substrate was constructed by creating a hybrid tRNA containing an archaeal intron fused to yeast pre-tRNA^{Phe}. The eukaryotic enzyme can recognize and correctly cleave the 5′ and 3′ splice sites in this substrate. In doing so, the enzyme dispenses with the ruler mechanism. There is no change in the size or location of the intron released from the pre-tRNA upon addition or deletion of base pairs in the anticodon stem as was seen for endonucleolytic cleavage of eukaryotic substrates. This result proves that the disposition of the active sites is identical in the archaeal and eukaryotic endonucleases and that this architecture has been conserved since the divergence from a common ancestor. It must be through specialization of the subunits of the yeast enzyme that the ruler mechanism has evolved. The SEN2 gene has acquired a transmembrane sequence, which likely anchors the endonuclease in the inner nuclear membrane, perhaps near nuclear pore structures. This has been shown to be a primary localization of the tRNA splicing ligase (Clark and Abelson 1987), and the two enzymes likely act in concert at this site (Greer 1986).

ORIGIN OF tRNA INTRONS

The relatedness of the eukaryotic and archaeal enzymes almost certainly proves that the endonuclease gene was present in their last common ancestor. The simplest hypothesis is that its function then was to splice tRNA precursors, but it could also be that it had a different function and has been independently recruited in both lines to a tRNA splicing function.

Certainly recruitment is a theme that is a feature of the enzymes in this system. It has recently been shown that tRNA ligase functions in the ligation of an unusually spliced yeast messenger RNA (Cox and Walter 1996; Sidrauski et al. 1996). The mRNA encodes a transcription factor, HAC1, which up-regulates transcription of genes involved in the maintenance of

---

repeat. (*C*) Proposed structural model of the yeast endonuclease. (*Upper*) The sequences of the carboxyl terminus of the (M. jann.) *M. jannaschii*, (H vol. Nt.) *H. volcanii* amino-terminal repeat, and Sc. Sen54 and Sc. Sen15 are shown to be homologous. (*Lower*) These important interaction elements are modeled in the yeast endonuclease. The two interaction elements are modeled as loop L10 conserved in Sen54 and Sen15 and the carboxy-terminal interaction (*circled arrows*) (see text for details). (Adapted with permission, from Li et al. 1998 [copyright American Association for the Advancement of Science].)

unfolded protein in the endoplasmic reticulum (ER). The presence of unfolded proteins appears to be sensed by a receptor tyrosine kinase-like protein, IRE1, which spans the membrane of the ER. The Ire1 protein contains a nuclease domain capable of cleaving the Hac1 mRNA and releasing an intron (Sidrauski and Walter 1997). Ligation of the two Hac1 exons is dependent on the function of tRNA ligase both in vivo and in vitro (Sidrauski et al. 1996; Sidrauski and Walter 1997).

Two hypotheses have been postulated for the origin of tRNA introns. The first (Cavalier-Smith 1991) posits that existing proteins were recruited to splice introns resulting from the partial deletion of preexisting group I or group II self-splicing introns in tRNA or rRNA genes. Support for such a hypothesis is derived from the presence of group I and group II introns in the tRNA genes of a handful of bacterial species. In particular, a group I intron found in the tRNA^{Leu} gene of cyanobacteria and bacteria interrupts the anticodon loop at the same position as a protein-dependent intron found in some Archaea (Wich et al. 1987). A second hypothesis proposes the expansion of loops of the tRNA molecule. In both instances, the preexistence of compatible splicing machinery in the cell would ensure against lethality of the insertion (Belfort and Weiner 1997). It is unlikely that agreement on a time for the origin of tRNA introns is possible; however, the new addition to the ongoing dialogue is that at least the splicing endonuclease is ancient.

CONCLUDING REMARKS

Whatever the origin of tRNA introns, it is clear that they are here to stay. Both Archaea and eukaryotes have failed to displace all introns from tRNA genes. This would suggest that the introns serve some selective advantage for the organism and that maintenance of the splicing system is necessary for survival. It is clear from studies in yeast that one function of introns is to aid in the modification of tRNAs (for review, see Grosjean et al. 1997). Thus, the modification enzymes have evolved to depend on the presence of the intron for correct substrate recognition and modification.

By studying the enzymology of the removal process, we have begun to understand just how introns have evolved. It will be interesting to fully delineate the splicing pathway of the archaeal system through discovery of the ligase that must function to ligate the products of the endonuclease reaction. Perhaps we will again be afforded a glimpse into the RNA World that existed before the invention of protein enzymes.

ACKNOWLEDGMENTS

We thank Dr. Hong Li for advice concerning the crystal structure, in particular, Figure 8. This work is supported by a generous grant from the American Cancer Society.

REFERENCES:

Abelson J., Trotta C.R., and Li H. 1998. tRNA splicing. *J. Biol. Chem.* **273**: 12685–12688.

Apostol B.L., Westaway S.K., Abelson J., and Greer C.L. 1991. Deletion analysis of a multifunctional yeast tRNA ligase polypeptide: Identification of essential and dispensable functional domains. *J. Biol. Chem.* **266**: 7445–7455.

Arn E.A. and Abelson J.N. 1996. The 2′-5′ ligase from *Escherichia coli*: Purification, cloning and genomic disruption. *J. Biol. Chem.* **271**: 31145–31153.

———. 1998. RNA ligases: Function, mechanism and sequence conservation. In *RNA structure and function* (ed. R.W. Simons and M. Grunberg-Manago), pp. 695–726. Cold Spring Harbor Laboratory Press, Cold Spring Harbor, New York.

Baldi M.I., Mattoccia E., Bufardeci E., Fabbri S., and Tocchini-Valentini G.P. 1992. Participation of the intron in the reaction catalyzed by the *Xenopus* tRNA splicing endonuclease. *Science* **255**: 1404–1408.

Belford H.G., Westaway S.K., Abelson J., and Greer C.L. 1993. Multiple nucleotide cofactor use by yeast ligase in tRNA splicing: Evidence for independent ATP- and GTP-binding sites. *J. Biol. Chem.* **268**: 2444–2450.

Belfort M. and Weiner A. 1997. Another bridge between kingdoms: tRNA splicing in Archaea and eukaryotes. *Cell* **89**: 1003–1006.

Biniszkiewicz D., Cesnaviciene E., and Shub D.A. 1994. Self-splicing group-I intron in cyanobacterial initiator methionine tRNA: Evidence for lateral transfer of introns in bacteria. *EMBO J* **13**: 4629–4635.

Cavalier-Smith T. 1991. Intron phylogeny: A new hypothesis. *Trends Genet.* **7**: 145–148.

Clark M.W. and Abelson J. 1987. The subnuclear localization of tRNA ligase in yeast. *J. Cell Biol.* **105**: 1515–1526.

Cox J.S. and Walter P. 1996. A novel mechanism for regulating activity of a transcription factor that controls the unfolded protein response. *Cell* **86**: 391–404.

Culbertson M.R. and Winey M. 1989. Split tRNA genes and their products: A paradigm for the study of cell-function and evolution. *Yeast* **5**: 405–427.

Culver G.M., McCraith S.M., Consaul S.A., Stanford D.R., and Phizicky E.M. 1997. A 2′-phosphotransferase implicated in tRNA splicing is essential in *Saccharomyces cerevisiae*. *J. Biol. Chem.* **272**: 13203–13210.

Culver G.M., McCraith S.M., Zillmann M., Kierzek R., Michaud N., Lareau R.D., Turner D.H., and Phizicky E.M. 1993. An NAD derivative produced during tRNA splicing: ADP-ribose 1″-2″ cyclic phosphate. *Science* **261**: 206–208.

Daniels C.J., Gupta R., and Doolittle W.F. 1985. Transcription and excision of a large intron in the tRNA^{Trp} gene of an Archaebacterium, *Halobacterium volcanii*. *J. Biol. Chem.* **260**: 3132–3134.

Diener J. and Moore P. 1998. Solution structure of a substrate for the archaeal pre-tRNA splicing endonucleases: The bulge-helix-bulge motif. *Mol. Cell* **1**: 883–894.

Fabbri S., Fruscoloni P., Bufardeci E., Di Nicola Negri E., Baldi M.I., Gandini Attardi D., Mattoccia E., and Tocchini-Valentini G.P. 1998. Conservation of substrate recognition mechanisms by tRNA splicing endonucleases. *Science* **280:** 284–286.

Filipowicz W., Konarska M., Gross H., and Shatkin A.. 1983. RNA 3'-terminal phosphate cyclase activity and RNA ligation in HeLa cell extract. *Nucleic Acids Res.* **11:** 1405–1418.

Gomes I. and Gupta R. 1997. RNA splicing ligase activity in the archaeon *Haloferax volcanii. Biochem. Biophys. Rev.* **237:** 588–594.

Goodman H.M., Olson M.V., and Hall B.D. 1977. Nucleotide sequences of a mutant eukaryotic gene: The yeast tyrosine inserting suppressor, SUP4-0. *Proc. Natl. Acad. Sci.* **74:** 5453–5457.

Green C.J., Sohel I., and Vold B.S. 1990. The discovery of new intron-containing human tRNA genes using the polymerase chain reaction. *J. Biol. Chem.* **265:** 12139–12142.

Greer C.L. 1986. Assembly of a tRNA splicing complex: Evidence for concerted excision and joining steps in splicing *in vitro. Mol. Cell. Biol.* **6:** 635–644.

Greer C.L., Soll D., and Willis I. 1987. Substrate recognition and identification of splice sites by the tRNA-splicing endonuclease and ligase from *Saccharomyces cerevisiae. Mol. Cell. Biol.* **7:** 76–84.

Grosjean H., Szweykowska-Kulinska Z., Motorin Y., Fasiolo F., and Simos G. 1997. Intron-dependent enzymatic formation of modified nucleosides in eukaryotic tRNAs: A review. *Biochimie* **79:** 293–302.

Ho C.K., Rauhut R., Vijayraghavan U., and Abelson J. 1990. Accumulation of pre-tRNA splicing 2/3 intermediates in a *Saccharomyces cerevisiae* mutant. *EMBO J* **9:** 1245–1252.

Hopper A.K., Banks F., and Evangelidis V. 1978. A yeast mutant which accumulates precursor tRNAs. *Cell* **14:** 211–219.

Johnson J., Ogden R., Johnson P., Abelson J., Dembeck P., and Itakura K. 1980. Transcription and processing of a yeast tRNA gene containing a modified intervening sequence. *Proc. Natl. Acad. Sci.* **77:** 2564–2568.

Kjems J. and Garrett R.A. 1985. An intron in the 23S rRNA gene of the Archaebacterium *Desulfurococcus mobilis. Nature* **318:** 675–677.

Kjems J., Jensen J., Olesen T., and Garrett R.A. 1989. Comparison of tRNA and rRNA intron splicing in the extreme thermophile and Archaebacterium *Desulfurococcus mobilis. Can. J. Microbiol.* **35:** 210–214.

Kleman-Leyer K., Armbruster D.A., and Daniels C.J. 1997. Properties of *H. volcanii* tRNA intron endonuclease reveal a relationship between the archaeal and eucaryal tRNA intron processing systems. *Cell* **89:** 839–848.

Klenk H.P., Clayton R.A., Tomb J.F., White O., Nelson K.E., Ketchumm K.A., Dodson R.J., Gwinn M., Hickey E.K., Peterson J.D. 1998. The complete genome sequence of the hyperthermophilic, sulphate-reducing archaeon *Archaeoglobus fulgidus. Nature* **390:** 364–370.

Knapp G., Beckmann J.S., Johnson P.F., Fuhrman S.A., and Abelson J.N. 1978. Transcription and processing of intervening sequences in yeast tRNA genes. *Cell* **14:** 221–236.

Konarska M., Filipowicz W., Domdey H., and Gross H. 1981. Formation of a 2'-phosphomonoester, 3',5'-phosphodiester linkage by a novel RNA ligase in wheat germ. *Nature* **293:** 112–116.

Kuhsel M.G., Strickland R., and Palmer J.D. 1990. An ancient group I intron shared by eubacteria and chloroplasts. *Science* **250:** 1570–1573.

Laski F., Fire A., RajBhandary U., and Sharp P. 1983. Characterization of tRNA precursor splicing in mammalian extracts. *J. Biol. Chem.* **258:** 11974–11980.

Lee M.C. and Knapp G. 1985. tRNA splicing in *Saccharomyces cerevisiae*: Secondary and tertiary structures of the substrates. *J. Biol. Chem.* **260:** 3108–3115.

Li H., Trotta C.R., and Abelson J.N. 1998. Crystal structure and evolution of a tRNA splicing enzyme. *Science* **280:** 279–284.

Lykke-Andersen J. and Garrett R.A. 1997. RNA-protein interactions of an archaeal homotetrameric splicing endoribonuclease with an exceptional evolutionary history. *EMBO J.* **16:** 6290–6300.

McCraith S.M. and Phizicky E.M. 1990. A highly specific phosphatase from *Saccharomyces cerevisiae* implicated in tRNA splicing. *Mol. Cell. Biol.* **10:** 1049–1055.

———. 1991. An enzyme from *Saccharomyces cerevisiae* uses NAD^+ to transfer the splice junction 2′-phosphate from ligated tRNA to an acceptor molecule. *J. Biol. Chem.* **266:** 11986–11992.

Nishikura K. and De Robertis E. 1981. RNA processing in microinjected *Xenopus* oocytes. Sequential addition of base modification in the spliced transfer RNA. *J. Mol. Biol.* **145:** 405–420.

Ogden R.C., Lee M.C., and Knapp G. 1984. Transfer RNA splicing in *Saccharomyces cerevisiae*: Defining the substrates. *Nucleic Acids Res.* **12:** 9367–9382.

Peebles C.L., Gegenheimer P., and Abelson J. 1983. Precise excision of intervening sequences from precursor tRNAs by a membrane-associated yeast endonuclease. *Cell* **32:** 525–536.

Phizicky E.M., Schwartz R.C., and Abelson J. 1986. *Saccharomyces cerevisiae* tRNA ligase: Purification of the protein and isolation of the structural gene. *J. Biol. Chem.* **261:** 2978–2986.

Phizicky E.M., Consaul S.A., Nehrke K.W., and Abelson J. 1992. Yeast tRNA ligase mutants are nonviable and accumulate tRNA splicing intermediates. *J. Biol. Chem.* **267:** 4577–4582.

Rauhut R., Green P.R., and Abelson J. 1990. Yeast tRNA-splicing endonuclease is a heterotrimeric enzyme. *J Biol. Chem.* **265:** 18180–18184.

Reinhold-Hurek B. and Shub D.A. 1992. Self-splicing introns in tRNA genes of widely divergent bacteria. *Nature* **357:** 173–176.

Reyes V.M. and Abelson J. 1988. Substrate recognition and splice site determination in yeast tRNA splicing. *Cell* **55:** 719–730.

Schneider A., Perry-McNally K., and Agabian N. 1993. Splicing and 3′-processing of the tyrosine tRNA of *Trypanosoma brucei*. *J. Biol. Chem.* **268:** 21868–21874.

Schwartz R., Greer C., Gegenheimer P., and Abelson J. 1983. Enzymatic mechanism of an RNA ligase from wheat germ. *J. Biol. Chem.* **258:** 8374–8383.

Sidrauski C. and Walter P. 1997. The transmembrane kinase Ire1p is a site-specific endonuclease that initiates mRNA splicing in the unfolded protein response. *Cell* **90:** 1031–1039.

Sidrauski C., Cox J.S., and Walter P. 1996. tRNA ligase is required for regulated mRNA splicing in the unfolded protein response. *Cell* **86:** 405–413.

Spinelli S., Consaul S., and Phizicky E. 1997. A conditional lethal yeast phophotransferase (tpt1) mutant accumulates tRNAs with a 2′-phosphate and an undermodified base at the splice junction. *RNA* **3:** 1388–1400.

Stange N. and Beier H. 1986. A gene for the major cytoplasmic tRNA^{Tyr} from *Nicotiana rustica* contains a 13 nucleotides long intron. *Nucleic Acids Res.* **14:** 8691.

Strobel M. and Abelson J. 1986. Effect of intron mutations on processing and function of *Saccharomyces cerevisiae* SUP53 tRNA in vitro and in vivo. *Mol. Cell. Biol.* **6:** 2663–2673.

Swerdlow H. and Guthrie C. 1984. Structure of intron-containing tRNA precursors: analysis of solution conformation using chemical and enzymatic probes. *J. Biol. Chem.* **259:** 5197–5207.

Thompson J.E. and Raines R.T. 1994. Value of general acid-base catalysis to ribonucleaseA. *J. Am. Chem. Soc.* **116:** 5467–5468.

Thompson L.D. and Daniels C.J. 1988. A tRNA^{trp} intron endonuclease from *Halobacterium volcanii*: Unique substrate recognition properties. *J. Biol. Chem.* **263:** 17951–17959.

Thompson L.D., Brandon L.D., Nieuwlandt D.T., and Daniels C.J. 1989. Transfer RNA intron processing in the halophilic archaebacteria. *Can. J. Microbiol.* **35:** 36–42.

Trotta C.R., Miao F., Arn E.A., Stevens S.W., Ho C.K., Rauhut R., and Abelson J.N. 1997. The yeast tRNA splicing endonuclease: A tetrameric enzyme with two active site subunits homologous to the archaeal tRNA endonucleases. *Cell* **89:** 849–858.

Valenzuela P., Venegas A., Weinberg F., Bishop R., and Rutter W.J.. 1978. Structure of yeast phenylalanine-tRNA genes: An intervening DNA segment within the coding region for the tRNA. *Proc. Natl. Acad. Sci.* **75:** 190–194.

Walker G., Uhlenbeck O.C., Bedows E., and Gumport R.I. 1975. T4-induced RNA ligase joins single-stranded oligoribonucleotides. *Proc. Natl. Acad. Sci.* **72:** 122–126.

Walsh C. 1979. *Enzymatic reaction mechanisms*. W.H. Freeman, San Francisco, California.

Wich G., Leinfelder W., and Bock A. 1987. Genes for stable RNA in the extreme thermophile *Thermoproteus-tenax* introns and transcription signals. *EMBO J.* **6:** 523–528.

Willis I., Hottinger H., Pearson D., Chisholm V., Leupold U., and Soll D. 1984. Mutations affecting excision of the intron from a eukaryotic dimeric tRNA precursor. *EMBO J.* **3:** 1573–1580.

Winey M. and Culbertson M.R. 1988. Mutations affecting the tRNA splicing endonuclease activity of *Saccharomyces cerevisiae*. *Genetics* **118:** 49–63.

Xu Q., Teplow D., Lee T.D., and Abelson J. 1990. Domain structure in yeast tRNA ligase. *Biochemistry* **29:** 6132–6138.

Zillmann M., Gorovsky M.A., and Phizicky E.M. 1991. Conserved mechanism of tRNA splicing in Eukaryotes. *Mol. Cell. Biol.* **11:** 5410–5416.

———. 1992. Hela cells contain a 2′-phosphate-specific phosphotransferase similar to a yeast enzyme implicated in tRNA splicing. *J. Biol. Chem.* **267:** 10289–10294.

22
RNA Editing—An Evolutionary Perspective

Larry Simpson
Howard Hughes Medical Institute and
Departments of Molecular, Cell and Developmental Biology and
Medical Microbiology, Immunology and Molecular Genetics,
University of California, Los Angeles, California 90095

RNA editing is a term used to describe a variety of phenomena that involve the modification of nucleotide sequences of RNA transcripts in different organisms. In general, there are two basic types of editing, involving either insertions and deletions or substitutions of nucleotides. The genetic effect of the former, especially within coding regions of mRNAs, is of course more dramatic in that translation frameshifts can be created or corrected, but the effect of nucleotide substitutions can cause changes in the encoded amino acids and subsequent phenotypic consequences. The definition of RNA editing is actually somewhat historical in that nucleotide modifications of tRNAs and rRNAs described prior to the 1986 (Benne et al. 1986) discovery of the uridine (U) insertion/deletion modifications in trypanosome mitochondria are not usually included in this rubric. Editing is used to describe the diverse phenomena represented by trypanosome U-insertions and deletions; cytidine(C)-insertions in *Physarum* mitochondria; C-to-U substitutions in plant mitochondria and chloroplasts and mammalian apoB mRNA; nucleotide substitutions in *Acanthamoeba*, marsupial, and rat tRNAs; adenosine(A)-to-inosine(I) substitutions in mammalian glutamate receptor mRNA; and guanosine insertions in negative-strand RNA viruses. In this chapter, I review the trypanosome mitochondrial U-insertion/deletion editing in an evolutionary perspective.

The transcripts of 13 of the 18 structural genes of the maxicircle mitochondrial DNA in trypanosomatid protozoa are modified after transcription by the insertion and occasional deletion of uridine residues, mainly within coding regions (Fig. 1) (Benne et al. 1986; Simpson and Shaw 1989; Simpson and Emeson 1996). These sequence modifications overcome frameshifts (internal-editing), create multiple amino-terminal

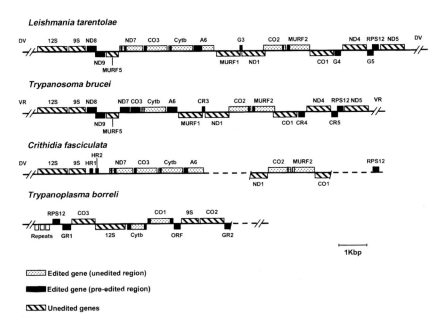

Figure 1 Comparative organization of maxicircle genomes in four kinetoplastid species. The maxicircles are shown linearized, with the genes above the line 5' to 3', left to right, and the genes below the line 5' to 3', right to left. Unedited genes and preedited and unedited regions of edited genes are indicted as shown. (DV) Divergent region of *L. tarentolae* and *C. fasciculata*. (VR) Variable region in *T. brucei*. The unsequenced portions of the *C. fasciculata* genome are indicated by the dashed line. (Reprinted, with permission, from Alfonzo et al. 1997 [copyright Oxford University Press].)

amino acid codons of the protein, sometimes including AuG methionine translation initiation codons (5'-editing), and in some cases, create entire translatable gene sequences (pan-editing) (Fig. 2) (Simpson and Shaw 1989). The sequence information is contained in short transcripts termed guide RNAs (gRNAs), which are encoded both in the maxicircle genome (Blume et al. 1990) and in the thousands of catenated minicircle DNA molecules (Pollard et al. 1990; Sturm and Simpson 1990a) that comprise the single network of mitochondrial DNA, which is termed kinetoplast DNA (Simpson 1972). Each trypanosomatid cell has a single tubular mitochondrion (Paulin 1983; Simpson and Kretzer 1997) with a single giant network of kinetoplast DNA situated in the matrix adjacent to the basal body of the flagellum. The kDNA network has a highly ordered structure in vivo, with 5–20,000 catenated minicircles, 0.4–2.5 kb in size depending on the species, aligned in a compact disk-like configuration. Approx-

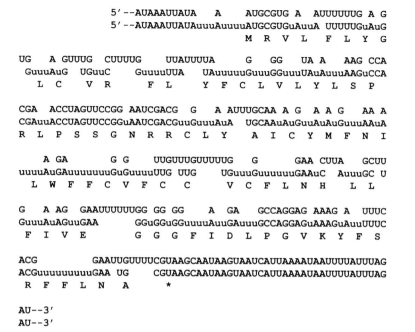

Figure 2 Pan-editing of the ribosomal protein S12 mRNA in *L. tarentolae*. The preedited genomic mRNA sequence is above and the edited sequence below. Us inserted by editing are shown in lower case. The translated amino acid sequence of the edited mRNA is also shown. There are three editing domains in this cryptogene.

imately 20–50 maxicircle molecules, 20–36 kb in size, are also catenated within the network. The genetic role of the minicircle DNA is the encoding of gRNAs involved in RNA editing (Blum et al. 1990). The site specificity of U-insertions and deletions is determined by base-pairing with the gRNAs, provided both canonical and G-U base pairs are allowed. The gRNAs form an anchor duplex of variable length just downstream from the region to be edited (the PER or pre-edited region), and can form complete duplexes with cognate regions of the mature edited RNA (Fig. 3). gRNAs also have a non-encoded 3′ oligo(U) tail of variable length, the role of which may be to stabilize the initial interaction of the gRNA and mRNA by hybridization to the G+A-rich PER (Blum and Simpson 1990; Seiwert et al. 1996).

The gRNAs do not strictly represent antisense mRNA sequences due to the G·U wobble base-pairing, but the sequence information for the insertion and deletion of Us does involve base-pairing. The discovery of

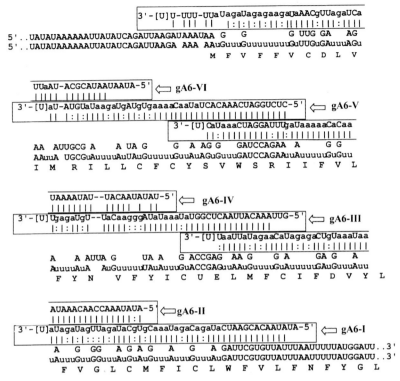

Figure 3 5' pan-editing of the ATPase 6 (MURF4) mRNA in *L. tarentolae*. There is a single editing domain with six overlapping gRNAs. See Fig. 2 for details. The gRNA/edited mRNA canonical base pairs are indicated by vertical lines and the G-U base pairs are indicated by colons.

gRNAs provided the solution to an apparent paradigm-breaking phenomenon but opened the door to an evolutionary enigma. This division of mitochondrial genetic information into two genomes, one containing both complete genes and incomplete purine-rich skeletons of genes, and the other containing segments of genes, including the pyrimidine information, is unprecedented in nature, and the evolutionary origin and biological distribution of such a genetic system are of great interest.

MECHANISM

A knowledge of the detailed mechanism of U-insertion/deletion editing will clearly provide some constraints for evolutionary speculation. Two basic models have been proposed for the mechanism. The enzyme cascade model invokes a series of protein enzyme-mediated reactions at each editing site involving cleavage, 3′-terminal U-addition, and ligation (Blum et al. 1990). The double transesterification model invokes a transfer of U-residues from the 3′ oligo(U) tail of the gRNA to the editing site by two successive transesterifications such as occur in RNA splicing, and involved gRNA/mRNA chimeric molecules as intermediates in which the 3′ end of the gRNA is covalently linked to the mRNA at an editing site (Blum et al. 1991). Variations of each model have been proposed; for example, one in which the Us are transferred from the 3′ oligo(U) tail of the gRNA by cleavage-ligation (Sollner-Webb 1992), or another in which the Us are derived directly from UTP by two transesterifications (Cech 1991). Evidence has accumulated recently indicating that the cleavage-ligation model in which Us are derived directly from UTP is essentially correct (Alfonzo et al. 1997; Stuart et al. 1997):

1. The predicted mitochondrial enzymatic activities have been shown to exist. A gRNA-dependent endoribonuclease that cleaves pre-edited mRNAs just upstream of the gRNA/mRNA anchor duplex (Piller et al. 1997), a 3′-terminal uridylyl transferase (TUTase) (Bakalara et al. 1989), a 3′-5′ exonuclease (Cruz-Reyes and Sollner-Webb 1996), and an RNA ligase activity (Bakalara et al. 1989; Sabatini and Hajduk 1995) were identified in mitochondrial lysates from *Trypanosoma brucei* and *Leishmania tarentolae*.
2. An editing-like activity was detected that was independent of the addition of exogenous gRNA and dependent on the secondary structure of the mRNA substrate (Frech et al. 1995; Connell et al. 1997).
3. The observation of an inversion of the stereochemical configuration of an inserted U-phosphorothioate (Frech and Simpson 1996) in this reaction eliminated the possibility of the involvement of gRNA as a reservoir for Us, which would require a net retention of the stereoconfiguration. These data were also inconsistent with the transesterification model, but entirely consistent with a cleavage-ligation model in which UTP was the source of the inserted Us.
4. A lack of involvement of endogenous gRNA in this reaction was established by showing that mutation of the added mRNA anchor sequence had no effect on the U-insertion activity (Connell et al. 1997).
5. The predicted intermediates were observed in a gRNA-dependent in vitro editing system using mitochondrial extracts from *T. brucei*: 5′ cleavage fragments with added or deleted Us at the 3′ end, and 3′ cleavage fragments without added Us at the 5′ end (Seiwert and Stuart 1994; Kable et al. 1996; Seiwert et al. 1996). The site of cleavage was exactly as predicted by the enzyme cascade model, 3′ of the first

non-base-paired nucleotide upstream of the anchor duplex (Cruz-Reyes and Sollner-Webb 1996). These cleavage fragments appeared in the reaction prior to gRNA/mRNA chimeras, suggesting that chimeras represent nonproductive by-products and not intermediates (Seiwert et al. 1996). In line with this, stabilization of the interaction of the 3′ end of the gRNA with the pre-edited region of the mRNA inhibited chimera formulation but had no effect on in vitro editing.

6. A gRNA-dependent in vitro editing activity was detected in *L. tarentolae* that was unaffected by blockage of the 3′ end of the gRNA by periodation (Byrne et al. 1996). This is strong evidence against a transfer of Us from the 3′ end of the gRNA to the editing site. However, in the *T. brucei* gRNA-dependent U-deletion system, blockage of the 3′ end of the gRNA led to inhibition of editing (Seiwert et al. 1996).

7. Both gRNA-independent (Peris et al. 1994) and gRNA-dependent (Corell et al. 1996) in vitro editing activities cosedimented with an RNA ligase-containing 20S complex (Peris et al. 1997). Consistent with an involvement of an RNA ligase in the in vitro editing reactions, α-β bond hydrolysis of ATP was required (Byrne et al. 1996; Cruz-Reyes and Sollner-Webb 1996).

In the original enzyme cascade model (Blum et al. 1990), the Us were added one at a time to the 3′ end of the 5′ cleavage fragment, and the number added was determined by base-pairing with guiding A or G nucleotides in the gRNA. However, evidence from both the *T. brucei* and the *L. tarentolae* in vitro systems has indicated that an initial untemplated addition of multiple Us to the 5′ cleavage fragment occurs, followed by a 3′ to 5′ exonucleolytic trimming of Us not base-paired to the gRNA and then by a ligation of the two cleavage fragments (Fig. 4) (Byrne et al. 1996; Kable et al. 1996; Seiwert et al. 1996). In the *L. tarentolae* system, it was shown that a minor proportion of the final edited products contain fewer or more than the number of templated Us at a specific site, and it was suggested that this misediting is a result of ligation prior to complete trimming or after excessive trimming (Byrne et al. 1996). This suggestion that there is an untemplated addition of Us followed by trimming is the major conceptual modification of the original enzyme cascade model. In the original model, trimming by a U-specific 3′ to 5′ exonuclease was also proposed, but only to remove the non-base-paired Us at a deletion site.

Despite the general acceptance of the enzyme cascade model, many uncertainties remain. The in vitro systems are very inefficient and do not show any signs of processivity past a single editing site. None of the

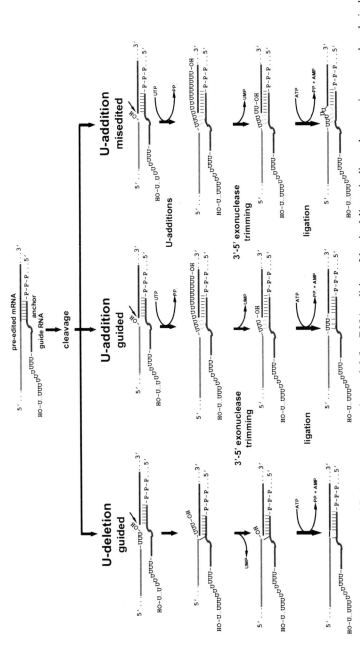

Figure 4 Diagram of the modified enzyme cascade model for RNA editing. Vertical lines indicate base-pairs; arrowheads indicate sites of cleavage. The 3' oligo(U) tail of the gRNA is shown as a single-stranded overhang, but it is possible that the tail can interact with the purine-rich preedited sequence (Blum and Simpson 1990; Seiwert et al. 1996) and the gRNA may have secondary structure (Schmid et al. 1995). In the U-deletion model, three unpaired Us to be deleted are shown as an example. It is possible that the U-addition activity adds Us to the 3' end of the cleavage fragment at the deletion site, which are then trimmed back, but this scenario is not indicated. In the U-addition model, 13 Us are shown added to the 5' fragment, but the evidence indicates that the number of added Us is actually heterogeneous (60). In the "guided" diagram, the exonuclease nucleotide trimming is complete, yielding the correct −3 or +3 guided products. If trimming is incomplete or excessive prior to ligation, gRNA-dependent misedited products are produced, as shown in the "misedited" diagram. (Reprinted, with permission, from Alfonzo et al. 1997 [copyright Oxford University Press].)

proposed enzymatic activities—endonuclease, TUTase, exonuclease, and RNA ligase—have yet been purified to homogeneity and characterized. Only the RNA ligase has been correlated with two adenylatable proteins, but neither has been purified. The only protein possibly involved in editing that has been isolated and that has had the gene cloned is a mitochondrial DEAD-box protein, mHel61p, in *T. brucei* (Missel et al. 1997), which was shown to be required for optimal steady-state abundance of edited transcripts in vivo by gene disruption in procyclic trypanosomes. The role of this protein, however, is uncertain since the knockouts showed no loss of a mitochondrial RNA helicase activity, which was thought to be the enzymatic role of this protein (Missel et al. 1995), and there was no effect on the in vitro U-deletion activity of mitochondrial extract.

Several mitochondrial proteins have been identified that exhibit binding to gRNA. The mitochondrial glutamate dehydrogenase in *L. tarentolae* was shown to interact with the oligo(U) tail of gRNA at the dinucleotide pocket, but the biological significance of this interaction is uncertain. The mitochondrial protein, gBP21, has been isolated from *T. brucei* and shows a specific binding to gRNA (Koller et al. 1997).

Finally, evidence has been recently obtained by Cruz-Reyes et al. (1998) for differences in nucleotide requirements of the endonucleolytic cleavages at sites of U-insertion and U-deletion, opening up the possibility that U-insertion and U-deletion may represent separate enzymatic pathways.

It is clear that editing reaction involves protein enzyme-mediated cleavage-ligation reactions, and there is no evidence for any involvement of a ribozyme. Previous speculation that this type of RNA editing was possibly as ancient as RNA-mediated RNA splicing due to a presumed similarity of the *trans*-esterification chemistry of the reaction (Blum et al. 1991; Cech 1991) can probably now be discarded.

POLARITY OF EDITING AND THE EXISTENCE OF MISEDITING

Editing was initially shown by Northern analysis to proceed 3' to 5' within a multiple gRNA-mediated domain (Abraham et al. 1988). This finding was confirmed and extended by sequence analysis of a large number of cDNAs of partially edited transcripts of the CYb 5'-edited cryptogene in *L. tarentolae*, the editing of which is mediated by two overlapping gRNAs (Sturm and Simpson 1990b). Almost all of the cDNAs could be arranged in a precise 3' to 5' polarity, with the downstream sites being correctly and completely edited prior to editing of the adjacent upstream sites. The observed 3' to 5' polarity of editing within a single gRNA-mediated editing block is a direct prediction of the enzyme cascade model in which the initial cleavage is limited to the 3' side of the first unpaired nucleotide upstream of the duplex formed by the gRNA and the edited mRNA. On the other hand, the overall 3' to 5' polarity of editing within an editing domain is due to the creation of anchor sequences for upstream gRNAs by downstream editing. This was shown for the A6 and RPS12 pan-edited mRNAs in *L. tarentolae*, in which six overlapping gRNAs mediate the former (see Fig. 3) and eight gRNAs that overlap within each of the three domains mediate the latter (Maslov and Simpson 1992). In *T. brucei* and *T. cruzi*, the situation is more complex in that there are multiple "redundant" gRNAs of different sequences that overlap extensively but encode the same editing information, but the explanation for the 3' to 5' polarity appears to be identical (Riley et al. 1994; Avila and Simpson 1995).

Sequence analysis of partially edited COIII cDNAs from *L. tarentolae*, however, showed a more complex situation. In the case of COIII editing, only 177 of the 304 cDNAs could be arranged in this manner, with the remainder showing a variably sized junction region of incorrectly edited sequences situated between correctly edited and unedited sequences. The presence of extensive "misedited" junction regions was also observed in CYb transcripts from *T. brucei* (Decker and Sollner-Webb 1990), and in each editing domain of the RPS12 (Maslov et al. 1992) and A6 cryptogenes from *L. tarentolae*. An apparent re-editing of misedited junction regions in a 3' to 5' polarity was also observed, indicating that misedited regions within the editing domain could be repaired by correct editing with the correct gRNA (Sturm et al. 1992).

Several models were proposed to explain the mechanism of misediting, both within a domain and within a single gRNA-mediated block. In one model, misediting was a result of editing occurring randomly within an editing domain, with the role of the gRNA being to hybridize with fortuitously fully edited sequence and prevent further editing from occurring, thereby giving rise to an apparent 3' to 5' polarity (Decker and Sollner-Webb 1990). However, the discovery that many misedited sequences could form duplexes with gRNAs encoding editing information for other blocks or even for other genes led to the hypothesis that one type of misediting was due to misguiding by noncognate gRNAs (Sturm et al. 1992). Chimeric molecules (Blum and Simpson 1992) were also observed consisting of misedited mRNAs attached to heterologous gRNAs, which had the potential of guiding the misedited sequences. This hypothesis was broadened to the concept that this type of misediting represents 3' to 5' editing occurring in an incorrect "guiding frame" (Sturm et al. 1992). An illustrative example of misediting was shown for the ND3 pan-edited transcript in *L. tarentolae*, in which some partially edited RNAs were correctly edited in block I and misedited upstream (Maslov et al. 1994a; Thiemann et al. 1994). In this case, the correct editing of block I in the UC strain could be attributed to the retention of the maxicircle-encoded gRNA for this block, and the upstream misediting could be attributed to the absence of any minicircle-encoded gRNAs for ND3 and the fortuitous hybridization of gRNAs for G4-IV and A6-IV, the sequences of which matched the misedited sequence of several cDNAs.

An additional mechanism of misediting was provided by the evidence discussed above for an untemplated addition of multiple Us to the 3' end of the 5' cleavage fragment followed by a 3' to 5' exonuclease trimming of non-base-paired Us as part of the normal editing process. The generation of gRNA-dependent misedited sequences at site 1 of the ND7 mRNA was observed to occur in vitro, in addition to a predominant gRNA-guided sequence (Byrne et al. 1996). The mechanism was presumed to be a religation of the two fragments prior to complete trimming of the oligo(U) 3' overhang. This misediting was dependent on the presence of gRNA, probably due to the requirement of base-paired gRNA for the initial cleavage event. The suppression of misediting (and guided editing) by a cognate gRNA, which lacked any guiding nucleotides for site 1, could also be due to a lack of the proper secondary structure required for the initial cleavage.

The existence of partially misedited mRNAs appears to be a consequence of the polarity and the complexity of the editing process. Whatever is the mechanism of misediting, it does appear to make editing a rather inefficient and energy-consuming process. However, misedited mRNAs are probably not deleterious to the cell, since most partially edited mRNAs probably do not load on polysomes and are therefore inactive in translation. This is due to the fact that the 5' end of the mRNA, which includes the methionine translation initiation codon and a short sequence just upstream (a ribosome-binding site?), is usually the last sequence to be created by editing. The 3' to 5' polarity of editing and the presence of partially edited mRNAs appears to have dramatic evolutionary consequences, as discussed below.

gRNA COMPLEXITY AND THE LOSS OF EDITING IN AN OLD LABORATORY STRAIN OF *L. tarentolae*

Several genes that are pan-edited in *T. brucei* (Souza et al. 1992, 1993; Corell et al. 1994) do not appear to be edited in an old laboratory strain of *L. tarentolae* (UC strain), which has been in culture for over 50 years, but are edited in a recently isolated strain (Thiemann et al. 1994). These genes were originally designated as "G-rich regions" due to an enrichment for G residues in the sense strand, and were conserved in location and polarity but not in sequence between *T. brucei* and *L. tarentolae* (Simpson and Shaw 1989). G6 proved to be a pan-edited cryptogene encoding ribosomal protein S12 (RPS12), the mRNA for which was correctly edited in both species (Maslov et al. 1992; Read et al. 1992). Partially or fully edited transcripts could not be detected for G1–G5 in the UC strain of *L. tarentolae*. In addition, determination of the complete gRNA genomic complexity for the UC strain of *L. tarentolae* led to the conclusion that appropriate gRNAs were lacking to mediate the editing of G1–G5 (Maslov and Simpson 1992). A total of 10 maxicircle-encoded gRNAs and 17 minicircle-encoded gRNAs were detected in the UC strain. Use of a freshly isolated strain of *L. tarentolae* (LEM125) led to the detection of 32 additional gRNAs that could partially mediate editing of the G1–G5 transcripts (Thiemann et al. 1994). It was estimated that an approximately equal number of gRNAs remain to be discovered in the LEM125 strain to account for all the known editing events.

It was speculated that specific minicircle sequence classes encoding gRNAs for the editing of these transcripts were lost during the long culture history of the UC strain, perhaps due to a lack of requirement for the protein products in culture (Maslov et al. 1994b; Simpson and Maslov 1994a,b). One possible mechanism for the selective loss of entire minicircle sequence classes could result from the mode of mitochondrial DNA replication and segregation in these cells. Circles are randomly decatenated from the network, replicated once per cell cycle, and recatenated to the network at two antipodal sites termed replisomes (Pérez-Morga and Englund 1993). In all species but *T. brucei*, the network appears to rotate within the mitochondrion, resulting in a peripheral distribution of replicated minicircles, whereas in *T. brucei*, the reattachment of the replicated minicircles results in a dumbbell-shaped network, which then divides into two daughter networks (Ferguson et al. 1994; Robinson and Gull 1994; Matthews et al. 1995). There is some evidence that minicircles are randomly distributed between daughter cells. Measurements of the relative copy numbers of different minicircle sequence classes showed a large heterogeneity and a plasticity. Copy numbers varied from as low as 10 copies per network to as high as 2,000 copies per network (Maslov and Simpson 1992). It was speculated that stochastic fluctuation of the copy number of specific sequence classes perhaps to missegregation at division of the network could result in the complete loss of a low-abundance class, resulting in the loss of gRNAs for a specific editing block (Simpson and Maslov 1994b).

It is of some interest that two minicircle sequence classes containing genes for nonfunctional gRNAs are present in the UC strain (Thiemann et al. 1994). One encodes a gRNA for editing block II of the G4 transcript and the other a gRNA for editing block IX of the G5 (ND3) transcript.

COMPARATIVE ANALYSIS OF EXTENT OF EDITING AND ORGANIZATION AND COMPLEXITY OF gRNA GENOMES IN THE KINETOPLASTIC PROTOZOA

The kinetoplastic protozoa consist of at least two major taxonomic groups, the trypanosomatids and the bodonids-cryptobiids. Within the trypanosomatids, there are approximately ten genera, some of which are monogenetic parasites of invertebrates and others of which are digenetic parasites of invertebrates and vertebrates (or plants). Analysis of alignments of nuclear small rRNA sequences by three different groups, using a *Euglena* small rRNA sequence as an outgroup, led to the conclusions that the trypanosomatids are paraphyletic and that the African trypanosomes represented the deepest lineage in the family. However, analysis of alignments of conserved protein sequences and also rRNA sequences with additional species included led to the conclusion that the trypanosomatids are monophyletic. As in the case of many other deeply rooted eukaryotic lineages with widely different rates of evolution, this represents an unresolved problem and requires additional analysis. It is clear, however, that *Bodo caudatus* and *Trypanoplasma borreli*, the two representatives of the bodonid-cryptobiid lineage, which have been studied at the molecular level, do represent an early diverged sister group to the trypanosomatids, as was previously proposed in classic protozoal taxonomy.

A comparative analysis of the extent of editing of the A6, COIII, and ND7 cryptogenes in several species of trypanosomatids (Maslov et al. 1994b) and in two strains of *T. borreli* (Lukes et al. 1994; Maslov and Simpson 1994) was performed. The results could be interpreted using a parsimony analysis as indicating that pan-editing is a primitive trait, and that several times during the evolution of these cells, pan-edited cryptogenes were substituted by genes edited at the 5' ends of the editing domains. In one case, a fully edited COIII gene was apparently substituted for a pan-edited ancestral gene. In each case, the loss of editing appeared to occur in units corresponding to gRNA-mediated blocks of editing (Fig. 5). We have speculated, on the basis of this distribution of editing patterns in the various species and of the previously discussed culture-induced loss of minicircle-encoded gRNA, that the evolution of RNA editing in kinetoplastids involves a replacement of the original pan-edited cryptogenes

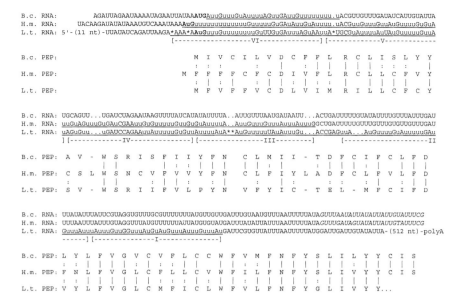

Figure 5 Alignment of edited mRNA sequences from *Blastocrithidia culicis*, *Herpetomonas muscarum*, and *L. tarentolae*. Alignments were made according to the amino acid sequence alignments. The *L. tarentolae* editing blocks are indicated below the RNA sequences. Us inserted by editing are shown in lowercase and Us deleted by asterisks. The editing domains in each mRNA are underlined. (Reprinted, with permission, from Maslov et al. 1994b [copyright Macmillan].)

by cRNA copies of partially edited RNAs (Maslov et al. 1994b; Simpson and Maslov 1994a,b; see also Landweber 1992). This hypothesis would predict the presence of reverse transcriptase activity in the mitochondrion of the kinetoplastids; this is made more plausible by the finding of reverse transcriptase activity in modern trypanosomatids (Gabriel and Boeke 1991; González et al. 1997). The loss of entire minicircle sequence classes encoding specific gRNAs may have been the selective pressure for these retroposition gene replacement events. This loss could have occurred by missegregation of low-copy-number minicircles at division of the single network. A dramatic change in minicircle sequence heterogeneity with drug pressure, which has been termed "transkinetoplastidy" (Lee et al. 1992; Lee et al. 1993, 1994; Chiang et al. 1996), has also been observed in culture, but the mechanism for this selective amplification of minicircle sequence classes is not known. In any case, cells that have undergone a gene replacement of a pan-edited gene with a partially edited gene would survive the loss of 3′-acting gRNAs.

In the trypanosomatids, the majority of the gRNAs are encoded in the catenated minicircles. Minicircles are organized into one or more conserved regions that contain the origins of replication for both strands, and an equal number of variable regions that contain the gRNA genes (Fig. 6). The number of gRNA genes varies in different species from one per minicircle in *L. tarentolae* (Sturm and Simpson 1990a) to three per minicircle in *T. brucei* (Pollard et al. 1990). In *T. cruzi*, there is a single gRNA gene in each of the four variable regions (Avila and Simpson 1995), and in *C. fasciculata*, there is a single gRNA gene in one of the two variable regions (Yasuhira and Simpson 1995). The genomic complexity of gRNAs varies between species. The African and South American trypanosomes, *T. brucei* and *T. cruzi*, apparently have a large gRNA complexity of over 600–900 gRNAs encoded in 200–300 minicircle sequence classes. Many of these gRNAs are redundant (Riley et al. 1994; Avila and Simpson

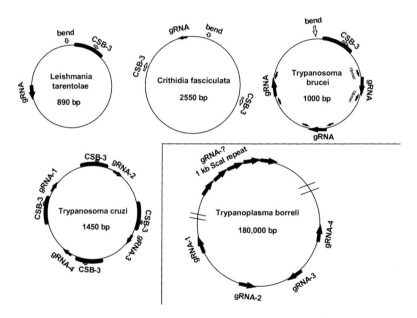

Figure 6 Comparison of gRNA gene organization in minicircle DNA from several trypanosomatid species and in the 180-kb DNA from *T. borreli*. The size of the minicircle for each species is indicated, as well as the location and polarity of the conserved region containing the CSB-3 sequence (Ray 1989), the bend (if present), and the gRNA genes. It is not established whether the identified gRNA genes in the *T. borreli* DNA are in the same molecule as the ScaI gRNA genes as shown, nor is it known whether the polarity of the gRNA genes is as diagramed. (Reprinted, with permission, from Simpson 1997 [copyright Elsevier Science].)

1995), in that, due to the allowed G-U base pairs, different gRNAs can encode the same editing information but have different sequences. In no case, however, has a complete set of overlapping gRNAs been identified in either species. in *L. tarentolae* LEM125, the total number of different gRNAs is approximately 60–80, of which 60 have already been identified. Only a few redundant gRNAs have been detected (Thiemann et al. 1994), and the entire sets of overlapping gRNAs have been described for two pan-edited genes. The large redundancy of gRNAs in *T. brucei* may serve the function of preventing loss of entire minicircle sequence classes by missegregation.

In *T. borreli* there is also a double genome in which the genes and cryptogenes are present in a 40- to 80-kb circular genome and the gRNAs are present as tandem repeats in 200-kb circular molecules (Maslov and Simpson 1994; Yasuhira and Simpson 1996). Further examples are required to apply parsimony analysis, but it is tempting to speculate that this represents the evolutionary precursor of the minicircular type of gRNA gene organization in the trypanosomatids. Each gRNA gene unit could have acquired an autonomous replicating sequence and become a self-replicating circular plasmid, and catenation of the circles into a network could have represented a mechanism to greatly limit the loss of individual molecules at cell division. Other roles for catenation have also been proposed, such as to enable the proper segregation of the maxicircle molecule. In both roles, the unique cytological feature of the single mitochondrion may have provided the selective pressure for the evolution of this genetic system.

Recently, two additional representatives of the Bodonina have been examined—*Cryptobia helicis* (Lukes et al. 1998) and *Bodo saltans* (Blom et al. 1998). In *C. helicis*, the mitochondrial DNA is organized into nodules spread throughout the mitochondrial lumen, a morphological organization that has been termed "pan-kinetoplastic" (Lukes et al. 1998). The DNA consists of approximately 8,400 4.2-kb noncatenated minicircles and a smaller number of 43-kb noncatenated maxicircles. The maxicircles contain structural genes, and the minicircles contain the bent helix and sequence blocks conserved in minicircles from trypanosomatids. It is likely, although not yet established, that editing of maxicircle transcripts occurs and that the gRNAs are encoded in the minicircle molecules. Lukes et al. (1998) have speculated that the minicircles in these cells might be descendants of a plasmid that invaded the mitochondrion of an ancient free-living bodonid and that these plasmids functioned as vehicles for gRNA genes.

B. saltans is a free-living bodonid, which from rRNA phylogenetic reconstructions is more closely related to the trypanosomatids than to *T.*

borreli and *C. helicis* (Blom et al. 1998). It contains a kDNA nucleoid body similar in morphology to that of the trypanosomatids and a 70-kD maxicircle component that has a gene organization similar to, but not identical with, that of the trypanosomatid maxicircle genome. Several edited genes were identified. *B. saltans* appears to lack a large DNA network composed of catenated circular molecules; instead the 1.4-kb minicircle molecules are only present in small catenanes. Clearly, more studies of the mitochondrial gene organization on bodonids are required in order to produce a meaningful hypothesis on the origin of this genetic system.

POSSIBLE ABSENCE OF EDITING IN *Euglena* MITOCHONDRIA

Phylogenetic reconstructions using rRNA sequences and morphological characters place the euglenoid flagellates as a sister group to the kinetoplastids. The presence of *trans*-splicing of cytosolic mRNAs, which occurs in both kinetoplastids and euglenoids, is consistent with this relationship. Since the above described analysis indicated that pan-editing was present in an ancestor of the entire kinetoplastid lineage, it was of interest to search for the presence of editing in the mitochondrion of *Euglena*. This proved to be a difficult task, as the isolated mitochondrial DNA of *Euglena* was highly fragmented. To date, only the COI gene has been cloned and sequenced from the mitochondrial genome of *Euglena*, and it was found to be unedited (Tessier et al. 1997; Yasuhira and Simpson 1997). In addition, no evidence for the existence of small gRNA-like molecules could be obtained (Yasuhira and Simpson 1997).

There is evidence that the chloroplast of euglenoid protozoa was derived from a secondary symbiosis of eukaryotic algae. This raises the possibility that the mitochondrial genome of the euglenoids is not monophyletic with the mitochondrion of the kinetoplastids and was obtained by lateral transfer from a eukaryotic algal endosymbiont. However, phylogenetic analysis of the mitochondrial-encoded COI gene and also the nuclear-encoded hsp60 gene, which is thought to have been derived from the original protomitochondrial genome, shows a phylogenetic affinity of mitochondria from euglenoids with those from kinetoplastids, and an affinity of both with the α-proteobacteria (Yasuhira and Simpson 1997). These results are in agreement with previous phylogenetic reconstructions indicating that mitochondria of all organisms are probably monophyletic.

The COI gene is unedited in all trypanosomatids examined but is pan-edited in *T. borreli*. The negative evidence for the absence of editing in the mitochondrion of *E. gracilis* suggests that editing may not occur in this species, but this must remain an open question pending analysis of additional mitochondrial genes. An absence of the U-insertion/deletion type of RNA editing in the Euglenoid mitochondrion may suggest that editing is an ancient but derived trait that appeared in the mitochondrion of an ancestor of the kinetoplastid protozoa. On the other hand, editing could have been present in a primitive euglenoid, but suffered a secondary loss. It is possibly relevant that editing has not been reported to occur in the α-proteobacteria, which are thought to represent the closest modern relatives to the original protomitochondrial endosymbiont (Gray 1989). One possible scenario for the appearance of editing as a derived trait is that of Covello and Gray (1993). In this hypothesis, there is a duplication of a mitochondrial gene, followed by the transposition of the gene copy back into the mitochondrial genome in an antisense direction behind a heterologous promoter. The antisense transcript would represent the primordial guide RNA, which, by utilization of an existing mismatch-repair mechanism based on base-pairing, could repair deletions or additions at the U locus in the transcript of the original gene. I would speculate further that once G-U base pairs between the "gRNA" and the "edited" mRNA were allowed to occur, this would have fixed the gRNA genome.

Cavalier-Smith (1997) has speculated on the origin of RNA editing in kinetoplastid mitochondria as resulting from a coevolution of glycosomes, which are derived peroxysomes containing bound glycolytic enzymes. In this hypothesis, the glycosomes were an adaptation of ancient free-living bodonids to facultative anaerobiosis, and editing of mitochondrial transcripts originated as a mechanism to correct errors in mitochondrial RNAs that accumulated during periods of anaerobiosis, during which time harmful mutations in mitochondrial DNA were not selected against.

PERSPECTIVES

The uridine insertion/deletion RNA editing that occurs in the mitochondrion of kinetoplastic protozoa is initially perceived as a striking and wasteful biological phenomenon and is often designated as bizarre. However, it is inherently no more bizarre than *cis*-splicing of introns in higher eukaryotes, and appears to have been maintained in this ancient family of protists for at least 500 million years of evolutionary history. Mitochondrial genomes are inherently incomplete, since most of the genes for mitochondrial proteins are in the nucleus. However, the presence of two separate genomes in the mitochondria of the kinetoplastids, one encoding genes and G+A-rich cryptogenes and the other the missing uridines in the form of the gRNA genes, is novel. The evolutionary origin of this genetic dichotomy is unclear, especially since a few gRNA genes are found in the maxicircle genome and one is actually situated *in cis* with the cognate preedited cryptogene. To date, two basic types of gRNA genomic organizations have been identified in the kinetoplastids, one of which comprises catenated minicircles with 1–4 gRNA genes and the other 180-kb circles with multiple gRNA genes. A more extensive comparative analysis of the variety of mitochondrial genomic organizations in the bodonid/cryptobiid lineages may prove illuminating in regard to the evolutionary origin of the gRNA genome.

Since loss of editing in the mitochondrial genome, probably by retroposition of partially or fully edited mRNAs, was found to occur in the course of evolution of the trypanosomatids, and loss of minicircles and the encoded gRNAs occurs with ease in the laboratory, it is clear that editing is a labile genetic trait and there must be a strong selective pressure to maintain it. The selective pressure may be the utilization of editing as a gene regulation mechanism for the mitochondrial biogenesis that occurs during the life cycles of these cells. However, editing also occurs in free-living kinetoplastids, and the selective advantage for these cells is not yet clear.

A cytidine-insertion type of editing occurs in the mitochondria of another lower eukaryotic cell, *P. polycephalum*, and in other members of the *Myxomycota* phylum (Mahendran et al. 1991, 1994; Miller et al. 1993). However, this editing is more complex in that it occurs with rRNAs and tRNAs in addition to mRNAs, and also involves the insertion of dinucleotides and the substitution of C to U nucleotides (Gott et al. 1993). In addition, this editing is closely coupled to transcription (Visomirski-Robic and Gott 1995, 1997a,b) and the mechanism for the specificity of this type of editing is unknown.

The C to U substitution editing that occurs in mitochondria and chloroplasts of plants involves a cytidine deamination reaction (Rajasekhar and Mulligan 1993), but the site-specificity mechanism is not yet understood. There is evidence from analysis of transgenic chloroplasts that there are *cis*-acting sequences flanking the sites and the involvement of a *trans*-acting factor, the nature of which is unknown (Bock et al. 1994, 1997; Bock and Maliga 1995; Bock and Koop 1997). A similar but more limited type of C to U substitution editing occurs in mammals in the ApoB mRNA. In this case, a downstream *cis*-acting "mooring sequence" has been identified as well as upstream enhancer elements (Backus and Smith 1992, 1994; Smith et al. 1997). A cytidine deaminase has been cloned and several auxiliary factors have been identified (Teng et al. 1993; Davidson 1994; Hadjiagapiou et al. 1994; MacGinnitie et al. 1995).

The A to I substitution editing that occurs in the mRNAs for several membrane receptors in higher organisms is due to the action of a widespread family of double-strand RNA-specific adenosine deaminase enzymes. The precise specificity mechanism is not understood in detail, but it is clear in the case of the GluR-B mRNA that duplex RNA structures produced by foldback of complementary upstream sequences are required for the observed specific editing events (Higuichi et al. 1993).

It is likely that RNA editing has evolved independently multiple times in different eukaryotic lineages as a mechanism to solve the problem of modifying the sequence of RNA molecules during or after transcription. In many cases, editing is regulated and has significant physiological consequences for the organism, as would be expected from the long-term retention of these genetic mechanisms in cells as diverse as trypanosomes, plants, and mammals. RNA editing represents a mode of gene regulation that was not conceived in the original genetic dogma scenario, but which, with hindsight, represents a highly functional way to modulate genetic information.

REFERENCES

Abraham J., Feagin J., and Stuart K. 1988. Characterization of cytochrome c oxidase III transcripts that are edited only in the 3' region. *Cell* **55:** 267–272.

Alfonzo J.D., Thiemann O., and Simpson L. 1997. The mechanism of U insertion/deletion RNA editing in kinetoplastid mitochondria. *Nucleic Acids Res.* **25:** 3751–3759.

Avila H. and Simpson L. 1995. Organization and complexity of minicircle-encoded guide RNAs from *Trypanosoma cruzi*. *RNA* **1:** 939–947.

Backus J.W. and Smith H.C. 1992. Three distinct RNA sequence elements are required for efficient apolipoprotein B (apoB) RNA editing *in vitro*. *Nucleic Acids Res.* **20:** 6007–6014.

———. 1994. Specific 3' sequences flanking a minimal apolipoprotein B (apoB) mRNA editing 'cassette' are critical for efficient editing in vitro. *Biochim. Biophys. Acta.* **1217:** 65–73.

Bakalara N., Simpson A.M., and Simpson L. 1989. The Leishmania kinetoplast-mitochondrion contains terminal uridylyltransferase and RNA ligase activities. *J. Biol. Chem.* **264:** 18679–18686.

Benne R., Van den Burg J., Brakenhoff J., Sloof P., Van Bloom J., and Tromp M. 1986. Major transcript of the frameshifted coxII gene from trypanosome mitochondria contains four nucleotides that are not encoded in the DNA. *Cell* **46:** 819–826.

Blom D., de Haan A., van den Berg M., Sloof P., Jirku M., Lukes J., and Benne R. (1998). RNA editing in the free-living bodonid *Bodo saltans*. *Nucleic Acids Res.* **26:** 1205–1213.

Blum B. and Simpson L. 1990. Guide RNAs in kinetoplastid mitochondria have a nonencoded 3' oligo-(U) tail involved in recognition of the pre-edited region. *Cell* **62:** 391–397.

———. 1992. Formation of gRNA/mRNA chimeric molecules in vitro, the initial step of RNA editing, is dependent on an anchor sequence. *Proc. Natl. Acad. Sci.* **89:** 11944–11948.

Blum B., Bakalara N., and Simpson L. 1990. A model for RNA editing in kinetoplastid mitochondria: "Guide" RNA molecules transcribed from maxicircle DNA provide the edited information. *Cell* **60:** 189–198.

Blum B., Sturm N.R., Simpson A.M., and Simpson L. 1991. Chimeric gRNA-mRNA molecules with oligo(U) tails covalently linked at sites of RNA editing suggest that U addition occurs by transesterification. *Cell* **65:** 543–550.

Bock R. and Koop H.U. 1997. Extraplastidic site-specific factors mediate RNA editing in chloroplasts. *EMBO J.* **16:** 3282–3288.

Bock R. and Maliga P. 1995. In vivo testing of a tobacco plastid DNA segment for guide RNA function in *psbL* editing. *Mol. Gen. Genet.* **247:** 439–443.

Bock R., Hermann M., and Fuchs M. 1997. Identification of critical nucleotide positions for plastid RNA editing site recognition. *RNA* **3:** 1194–1200.

Bock R., Kössel H., and Maliga P. 1994. Introduction of a heterologous editing site into the tobacco plastid genome: The lack of RNA editing leads to a mutant phenotype. *EMBO J.* **13:** 4623–4628.

Byrne E.M., Connell G.J., and Simpson L. 1996. Guide RNA-directed uridine insertion RNA editing in vitro. *EMBO J.* **15:** 6758–6765.

Cavalier-Smith T. 1997. Cell and genome coevolution: Facultative anaerobiosis, glycosomes and kinetoplastan RNA editing. *Trends Genet.* **13:** 6–9.

Cech T.R. 1991. RNA editing: World's smallest introns. *Cell* **64:** 667–669.

Chiang S.C., Yu J.C., and Lee S.T. 1996. Transkinetoplastidy in arsenite-resistant *Leishmania major*. *Mol. Biochem. Parasitol.*. **82:** 121–124.

Connell G.J., Byrne E.M., and Simpson L. 1997. Guide RNA-independent and guide RNA-dependent uridine insertion into cytochrome b mRNA in a mitochondrial extract from *Leishmania tarentolae*. *J. Biol. Chem.* **272:** 4212–4218.

Corell R.A., Myler P., and Stuart K. 1994. *Trypanosoma brucei* mitochondrial CR4 gene encodes an extensively edited mRNA with completely edited sequence only in bloodstream forms. *Mol. Biochem. Parasitol.* **64:** 65–74.

Corell R.A., Read L.K., Riley G.R., Nellissery J.K., Allen T.E., Kable M.L., Wachal M.D., Seiwert S.D., Myler P.J., and Stuart K.D. 1996. Complexes from *Trypanosoma brucei* that exhibit deletion editing and other editing-associated properties. *Mol. Cell. Biol.* **16:** 1410–1418.

Covello P.S. and Gray M.W. 1993. On the evolution of RNA editing. *Trends Genet.* **9:** 265–268.

Cruz-Reyes J. and Sollner-Webb B. 1996. Trypanosome U-deletional RNA editing involves guide RNA-directed endonuclease cleavage, terminal U exonuclease, and RNA ligase activities. *Proc. Natl. Acad. Sci.* **93:** 8901–8906.

Cruz-Reyes J., Rushe L., Piller K., and Sollner-Webb B. 1998. *T. brucei* RNA editing: Adenosine nucleotides inversely affect U-deletion and U-insertion reactions at mRNA cleavage. *Mol. Cell* **1:** 1–20.

Davidson N.O. 1994. RNA editing of the apolipoprotein B gene: A mechanism to regulate the atherogenic potential of intestinal lipoproteins. *Trends Cardiovasc. Med.* **4:** 231–235.

Decker C.J. and Sollner-Webb B. 1990. RNA editing involves indiscriminate U changes throughout precisely defined editing domains. *Cell* **61:** 1001–1011.

Ferguson M.L., Torri A.F., Pérez-Morga D., Ward D.C., and Englund P.T. 1994. Kinetoplast DNA replication: Mechanistic differences between *Trypanosoma brucei* and *Crithidia fasciculata*. *J. Cell Biol.* **126:** 631–639.

Frech G.C. and Simpson L. 1996. Uridine insertion into preedited mRNA by a mitochondrial extract from *Leishmania tarentolae*: Stereochemical evidence for the enzyme cascade model. *Mol. Cell. Biol.* **16:** 4584–4589.

Frech G.C., Bakalara N., Simpson L., and Simpson A.M. 1995. In vitro RNA editing-like activity in a mitochondrial extract from *Leishmania tarentolae*. *EMBO J.* **14:** 178–187.

Gabriel A. and Boeke J.D. 1991. Reverse transcriptase encoded by a retrotransposon from the trypanosomatid *Crithidia fasciculata*. *Proc. Nat. Acad. Sci.* **88:** 9794–9798.

González C.I., Thomas M.C., Martín F., Alcami J., Alonzo C., and López M.C. 1997. Reverse transcriptase-like activity in *Trypanosoma cruzi*. *Acta Trop.* **63:** 117–126.

Gott J.M., Visomirski L.M., and Hunter J.L. 1993. Substitutional and insertional RNA editing of the cytochrome *c* oxidase subunit 1 mRNA of *Physarum polycephalum*. *J. Biol Chem.* **268:** 25483–25486.

Gray M.N. 1989. The evolutionary origins of organelles. *Trends Genet.* **5:** 294–299.

Hadjiagapiou C., Giannoni F., Funahashi T., Skarosi S.F., and Davidson N.O. 1994. Molecular cloning of a human small intestinal apolipoprotein B mRNA editing protein. *Nucleic Acids Res.* **22:** 1874–1879.

Higuichi M., Single F.N., Kohler M., Sommer B., Sprengel R., and Seeburg P.H. 1993. RNA editing of AMPA receptor subunit GluR-B: A base-paired intron-exon structure determines position and efficiency. *Cell* **75:** 1361–1370.

Kable M.L., Seiwert S.D., Heidmann S., and Stuart K. 1996. RNA editing: A mechanism for gRNA-specified uridylate insertion into precursor mRNA. *Science* **273:** 1189–1195.

Koller J., Muller U.F., Schmid B., Missel A., Kruft V., Stuart K., and Goringer H.U. 1997. *Trypanosoma brucei* gBP21. An arginine-rich mitochondrial protein that binds to guide RNA with high affinity. *J. Biol. Chem.* **272:** 3749–3757.

Landweber L.F. 1992. The evolution of RNA editing in kinetoplastid protozoa. *BioSystems* **28:** 41–45.

Lee S.-T., Tarn C., and Chang K.-P. 1993. Characterization of the switch of kinetoplast DNA minicircle dominance during development and reversion of drug resistance in *Leishmania*. *Mol. Biochem. Parasitol.* **58:** 187–204.

Lee, S.-T., Liu H.Y., Lee S.P., and Tarn C. 1994. Selection for arsenite resistance causes reversible changes in minicircle composition and kinetoplast organization in *Leishmania mexicana*. *Mol. Cell. Biol.* **14:** 587–596.

Lee S.-Y., Lee S.-T., and Chang K.-P. 1992. Transkinetoplastidy—A novel phenomenon involving bulk alterations of mitochondrion-kinetoplast DNA of a trypanosomatid protozoan. *J. Protozool.* **39:** 190–196.

Lukes J., Jirku M., Avliyakulov N., and Benada O. 1998. Pankinetoplast DNA structure in a primitive bodonid flagellate, *Cryptobia helicis*. *EMBO J.* **17:** 838–846.

Lukes J., Arts G.J., van den Burg J., de Haan A., Opperdoes F., Sloof P., and Benne R. 1994. Novel pattern of editing regions in mitochondrial transcripts of the cryptobiid *Trypanoplasma borreli*. *EMBO J.* **13:** 5086–5098.

MacGinnitie A.J., Anant S., and Davidson N.O. 1995. Mutagenesis of apobec-1, the catalytic subunit of the mammalian apolipoprotein B mRNA editing enzyme, reveals distinct domains that mediate cytosine nucleoside deaminase, RNA binding, and RNA editing activity. *J. Biol. Chem.* **270:** 14768–14775.

Mahendran R., Spottswood M.R., and Miller D.L. 1991. RNA editing by cytidine insertion in mitochondria of *Physarum polycephalum*. *Nature* **349:** 434–438.

Mahendran R., Spottswood M.S., Ghate A., Ling M.-I., Jeng K., and Miller D.L. 1994. Editing of the mitochondrial small subunit rRNA in *Physarum polycephalum*. *EMBO J.* **13:** 232–240.
Maslov D.A. and Simpson L. 1992. The polarity of editing within a multiple gRNA-mediated domain is due to formation of anchors for upstream gRNAs by downstream editing. *Cell* **70:** 459–467.
———. 1994. RNA editing and mitochondrial genomic organization in the cryptobiid kinetoplastid protozoan, *Trypanoplasma borreli*. *Mol. Cell. Biol.* **14:** 8174–8182.
Maslov D.A., Thiemann O., and Simpson L. 1994a. Editing and misediting of transcripts of the kinetoplast maxicircle G5 (ND3) cryptogene in an old laboratory strain of *Leishmania tarentolae*. *Mol. Biochem. Parasitol.* **68:** 155–159.
Maslov D.A., Avila H.A., Lake J.A., and Simpson L. 1994b. Evolution of RNA editing in kinetoplastid protozoa. *Nature* **365:** 345–348.
Maslov D.A., Sturm N.R., Niner B.M., Gruszynski E.S., Peris M., and Simpson L. 1992. An intergenic G-rich region in *Leishmania tarentolae* kinetoplast maxicircle DNA is a pan-edited cryptogene encoding ribosomal protein S12. *Mol. Cell. Biol.* **12:** 56–67.
Matthews K.R., Sherwin T., and Gull K. 1995. Mitochondrial genome repositioning during the differentiation of the African trypanosome between life cycle forms is microtubule mediated. *J. Cell Sci.* **108:** 2231–2239.
Miller D., Mahendran R., Spottswood M., Costandy H., Wang S., Ling M.L., and Yang N. 1993. Insertional editing in mitochondria of *Physarum. Semin. Cell Biol.* **4:** 261–266.
Missel A., Norskau G., Shu H.H., and Goringer H.U. 1995. A putative RNA helicase of the DEAD box family from *Trypanosoma brucei*. *Mol. Biochem. Parasitol.* **75:** 123–126.
Missel A., Souza A.E., Norskau G., and Goringer H.U. 1997. Disruption of a gene encoding a novel mitochondrial DEAD-box protein in *Trypanosoma brucei* affects edited mRNAs. *Mol. Cell. Biol.* **17:** 4895–4903.
Paulin J. 1983. Conformation of a single mitochondrion in the trypomastigote stage of *Trypanosoma cruzi*. *J. Parasitol.* **69:** 242–244.
Pérez-Morga D.L. and Englund P.T. 1993. The attachment of minicircles to kinetoplast DNA networks during replication. *Cell* **74:** 703–711.
Pérez-Morga D. and Englund P.T 1993. The structure of replicating kinetoplast DNA networks. *J. Cell Biol.* **123:** 1069–1079.
Peris M., Simpson A.M., Grunstein J., Liliental J.E., Frech G.C., and Simpson L. 1997. Native gel analysis of ribonucleoprotein complexes from a *Leishmania tarentolae* mitochondrial extract. *Mol. Biochem. Parasitol.* **85:** 9–24.
Peris M., Frech G.C., Simpson A.M., Brigaud F., Byrne E., Bakker A., and Simpson L. 1994. Characterization of two classes of ribonucleoprotein complexes possibly involved in RNA editing from *Leishmania tarentola* mitochondria. *EMBO J.* **13:** 1664–1672.
Piller K.J., Rusche L.N., Cruz-Reyes J., and Sollner-Webb B. 1997. Resolution of the RNA editing gRNA-directed endonuclease from two other endonucleases of *Trypanosoma brucei* mitochondria. *RNA* **3:** 279–290.
Pollard V.W., Rohrer S.P., Michelotti E.F., Hancock K., and Hajduk S.L. 1990. Organization of minicircle genes for guide RNAs in *Trypanosoma brucei*. *Cell* **63:** 783–790.
Rajasekhar V.K. and Mulligan R.M. 1993. RNA editing in plant mitochondria: α-Phosphate is retained during C-to-U conversion in mRNAs. *Plant Cell* **5:** 1843–1852.
Ray D. 1989. Conserved sequence blocks in kinetoplast DNA minicircles from diverse species of trypanosomes. *Mol. Cell. Biol.* **9:** 1365–1367.

Read L.K., Myler P.J., and Stuart K. 1992. Extensive editing of both processed and pre-processed maxicircle CR6 transcripts in *Trypanosoma brucei. J. Biol. Chem.* **267:** 1123–1128.

Riley G.R., Corell R.A., and Stuart K. 1994. Multiple guide RNAs for identical editing of *Trypanosoma brucei* apocytochrome *b* mRNA have an unusual minicircle location and are developmentally regulated. *J. Biol. Chem.* **269:** 6101–6108.

Robinson D.R. and Gull K. 1994. The configuration of DNA replication sites within the *Trypanosoma brucei* kinetoplast. *J. Cell Biol.* **126:** 641–648.

Sabatini R. and Hajduk S.L. 1995. RNA ligase and its involvement in guide RNA/mRNA chimera formation. *J. Biol. Chem.* **270:** 7233–7240.

Schmid B., Riley G.R., Stuart K., and Goringer H.U. 1995. The secondary structure of guide RNA molecules from *Trypanosoma brucei. Nucleic Acids Res.* **23:** 3093–3102.

Seiwert S.D. and Stuart K. 1994. RNA editing: Transfer of genetic information from gRNA to precursor mRNA in vitro. *Science* **266:** 114–117.

Seiwert S.D., Heidmann S., and Stuart K. 1996. Direct visualization of uridylate deletion in vitro suggests a mechanism for kinetoplastid RNA editing. *Cell* **84:** 831–841.

Simpson L. 1972. The kinetoplast of the hemoflagellates. *Int. Rev. Cytol.* **32:** 139–207.

———. 1997. The genomic organization of guide RNA genes in kinetoplastid protozoa: Several conundrums and their solutions. *Mol. Biochem. Parasitol.* **86:** 133–141.

Simpson L. and Emeson R.B. 1996. The mitochondrion in dividing *Leishmania tarentolae* cells is symmetric and becomes a single asymmetric tubule in non-dividing cells due to division of the kinetoplast portion. *Mol. Biochem. Parasitol.* **87:** 71–78.

Simpson L. and Maslov D.A. 1994a. Ancient origin of RNA editing in kinetoplastid protozoa. *Curr. Opin. Genet. Dev.* **4:** 887–894.

———. 1994b. RNA editing and the evolution of parasites. *Science* **264:** 1870–1871.

Simpson L. and Shaw J. 1989. RNA editing and the mitochondrial cryptogenes of kinetoplastid protozoa. *Cell* **57:** 355–366.

Smith H.C., Gott J.M., and Hanson M.R. 1997. A guide to RNA editing. *RNA* **3:** 1105–1123.

Sollner-Webb B. 1992. RNA editing: Guides to experiments. *Nature* **356:** 743–744.

Souza A.E., Myler P.J., and Stuart K. 1992. Maxicircle CR1 transcripts of *Trypanosoma brucei* are edited, developmentally regulated, and encode a putative iron-sulfur protein homologous to an NADH dehydrogenase subunit. *Mol. Cell. Biol.* **12:** 2100–2107.

Souza A.E., Shu H.-H., Read L.K., Myler P.J., and Stuart K.D. 1993. Extensive editing of CR2 maxicircle transcripts of *Trypanosoma brucei* predicts a protein with homology to a subunit of NADH dehydrogenase. *Mol. Cell. Biol.* **13:** 6832–6840.

Stuart K., Allen T.E., Heidmann S., and Seiwert S.D. 1997. RNA editing in kinetoplastid protozoa. *Microbiol. Rev.* **61:** 105–120.

Sturm N.R. and Simpson L. 1990a. Kinetoplast DNA minicircles encode guide RNAs for editing of cytochrome oxidase subunit III mRNA. *Cell* **61:** 879–884.

———. 1990b. Partially edited mRNAs for cytochrome b and subunit III of cytochrome oxidase from *Leishmania tarentolae* mitochondria: RNA editing intermediates. *Cell* **61:** 871–878.

Sturm N.R., Maslov D.A., Blum B., and Simpson L. 1992. Generation of unexpected editing patterns in *Leishmania tarentolae* mitochondrial mRNAs: Misediting produced by misguiding. *Cell* **70:** 469–476.

Teng B., Burant C.F., and Davidson N.O. 1993. Molecular cloning of an apolipoprotein B messenger RNA editing protein. *Science* **260:** 1816–1819.

Tessier L.H., van der Speck H., Gualberto J.M., and Grienenberger J.M. 1997. The *cox1* gene from *Euglena gracilis*: A protist mitochondrial gene without introns and genetic code modifications. *Curr. Genet.* **31:** 208–213.

Thiemann O.H., Maslov D.A., and Simpson L. 1994. Disruption of RNA editing in *Leishmania tarentolae* by the loss of minicircle-encoded guide RNA genes. *EMBO J.* **13:** 5689–5700.

Visomirski-Robic L.M. and Gott J.M. 1995. Accurate and efficient insertional RNA editing in isolated *Physarum* mitochondria. *RNA* **1:** 681–691.

———. 1997a. Insertional editing in isolated *Physarum* mitochondria is linked to RNA synthesis. *RNA* **3:** 821–837.

———. 1997b. Insertional editing of nascent mitochondrial RNAs in *Physarum*. *Proc. Natl. Acad. Sci.* **94:** 4324–4329.

Yasuhira S. and Simpson L. 1995. Minicircle-encoded guide RNAs from *Crithidia fasciculata*. *RNA* **1:** 634–643.

———. 1996. Guide RNAs and guide RNA genes in the cryptobiid kinetoplastid protozoan, *Trypanoplasma borreli*. *RNA* **2:** 1153–1160.

———. 1997. Phylogenetic affinity of mitochondria of *Euglena gracilis* and kinetoplastids using cytochrome oxidase I and hsp60. *J. Mol. Evol.* **44:** 341–347.

23
Telomerase

Elizabeth H. Blackburn
Department of Microbiology and Immunology
University of California
San Francisco, California 94143-0414

According to the view of evolution in which the genetic material was first in the form of RNA, a transition to the present-day DNA genomes occurred. In the current world, although most of the DNA genome of a eukaryote is replicated by copying of the preexisting parental DNA strands by DNA-dependent DNA polymerase, portions have also resulted from copying RNA into DNA. One of these genomic components is the telomeric DNA, which requires for its continued maintenance the action of a ribonucleoprotein (RNP) enzyme, telomerase. The telomeric DNA sequence is specified by a template sequence within the telomerase RNA moiety, which is copied by reverse transcriptase action of the telomerase RNP. In this chapter, I discuss the properties of telomerase and suggest how they may shed light on two evolutionary transitions: from RNA-based to DNA-based genomes and from RNA-based enzymes to protein enzymes.

Telomeres were originally defined functionally as the natural ends of eukaryotic chromosomes, without which a chromosome is unstable (for review, see Blackburn 1984, 1991a,b, 1994; Zakian 1995). Telomeric DNA in the form of very simple, tandemly repeated sequences is a conserved feature throughout the eukaryotes, with some interesting exceptions in certain species (Table 1). A terminal stretch of this simple-sequence DNA, ranging in total length from under fifty base pairs in some ciliated protozoans, through a few hundred base pairs in yeasts and several other lower eukaryotes, to thousands of base pairs in mammalian cells, appears to supply sufficient *cis*-acting information to maintain a stable telomere. As shown in Table 1, these sequences usually result in the telomeric DNA having a strand composition bias: At each chromosome end the strand containing the characteristic clusters of G residues (G-strand) runs 5' to 3' toward the chromosomal terminus.

It is the G-strand of telomeric DNA that is synthesized in an unusual fashion by the ribonucleoprotein enzyme telomerase: As described below,

Table 1 Telomeric repeat sequences in eukaryotes

Telomeric repeat (5' to 3' toward chromosome end)	Representative species	Reference[a]
AG_{1-8}	*Dictyostelium discoideum*	
TTAGGGGG	*Cryptococcus*	Edman 1992
$TG_{2-3}(TG)_{1-3}$	*Saccharomyces cerevisiae*	
TTGGGG	*Tetrahymena thermophila*	
TT(T/G)GGG	*Paramecium aurelia*	
TAGGG	*Giardia lamblia*	Adam et al. 1991
TTAGGG	*Homo sapiens, Neurospora crassa, Trypanosoma brucei, Aspergillus nidulans*	Bhattacharyya and Blackburn 1997b
TTTAGGG	*Arabidopsis thaliana*	
TTTTAGGG	*Chlamydomonas reinhardtii*	
$TTAC(A)AG_{2-7}$	*Schizosaccharomyces pombe*	
TT(T/C)AGGG	*Plasmodium berghei*	
CTTAGG	*Ascaris lumbricoides*	Muller et al. 1991
ACGGATGTCTAAC TTCTTGGTGT	*Candida albicans*	
GGATTTGATTAGG TATGTGGTGTAC	*Kluyveromyces lactis*	McEachern and Blackburn 1994
HeT-A (several kb complex sequence)	*Drosophila melanogaster*	Pardue et al. 1996

[a]For references see Blackburn 1990a,b, except where shown otherwise.

a short RNA sequence within the telomerase RNA is used as the template for synthesis of this strand (for review, see Blackburn 1991b). Telomerase was thus recognized as a reverse transcriptase (RT) (Shippen-Lentz and Blackburn 1990) when it became apparent that its mechanism of action involves the copying of an RNA template into DNA.

THE POLYMERIZATION ACTIVITY OF TELOMERASE

The dynamic aspects of in vivo telomere behavior shared by diverse eukaryotes (for review, see Blackburn 1990a, 1991a,b), and the direct addition of yeast telomeric DNA to a *Tetrahymena* telomere introduced into yeast cells, led to the proposal that a new kind of polymerization activity exists (Shampay et al. 1984). This activity, subsequently called telomerase, was first identified in vitro in cell-free extracts of the ciliated protozoan *Tetrahymena thermophila* (Greider and Blackburn 1985). A synthetic G-rich telomeric DNA oligonucleotide representing the overhanging G-strand of the telomere was shown to act as a primer for synthesis of TTGGGG repeats, the telomeric sequence of *Tetrahymena* (Greider and Blackburn 1985, 1987, 1989). Telomerase elongates this primer from its 3'-OH end by direct polymerization, in the 5' to 3' direction, of deoxynucleoside triphosphate precursors into tandem repeats of the TTGGGG sequence. Telomeric DNA purified from natural telomeres have also been shown to be elongated by telomerase in vitro (Henderson and Blackburn 1989). Telomere repeat addition by telomerase onto preexisting telomeres in vivo has been directly confirmed in certain instances by sequencing telomeres from cells expressing various mutant telomerase RNAs that cause the synthesis of distinctive mutated telomeric DNA repeats (see below) (Yu et al. 1990; Yu and Blackburn 1991; McEachern and Blackburn 1995; Prescott and Blackburn 1997a). Although under various circumstances telomerase elongates oligonucleotides with a variety of sequences, significantly, it normally recognizes the sequence of the 3' end of the primer. Hence, for example with *Tetrahymena* and human telomerases, the correct next nucleotides are added to complete some cyclical permutation of, respectively, a perfect TTGGGG and TTAGGG repeat that includes the 3' end of the primer (Fig. 1) (Greider and Blackburn 1987, 1989; Blackburn et al. 1989; Morin 1989). This observation was important in formulating the mechanism of telomerase, as described below.

After the discovery of telomerase in *Tetrahymena*, comparable telomerase polymerization activities were subsequently found in vitro in widely divergent eukaryotes: the ciliated protozoans *Oxytricha* and *Euplotes* (Zahler and Prescott 1988; Shippen-Lentz and Blackburn 1989, 1990), human (Morin 1989); mouse (Prowse et al. 1993), the frog *Xenopus* (Mantell and Greider 1994), yeasts (Cohn and Blackburn 1995; Boswell-Fulton and Blackburn 1998), plants (Fitzgerald et al. 1996) and, recently, parasitic kinetoplastid protozoans (Cano et al. 1998). Each of these telomerase activities synthesizes one strand (the typically G-rich strand) of its species-specific telomeric sequence, and has primer recognition properties analogous to those of the *Tetrahymena* activity. In this chapter, I often use the *Tetrahymena* telomerase to exemplify various aspects of telomerases; as will be noted for specific cases, equivalent or additional findings have been made with telomerases of various other eukaryotes.

A. Tetrahymena

5'   ....GGGGTTGGGG ttggggttgggg.... 3'

       ....GGGGTTGGG gttggggttgggg....

B. Human

5'   ....AGGGTTAGGG ttagggttaggg... 3'

       ....AGGGTTAG ggttagggtt...

Figure 1 Examples of typical in vitro elongation reactions by telomerase. Shown are reactions carried out in vitro by the telomerases from *Tetrahymena* (*A*) and human (*B*) cells. The 3'-end region of the primer supplied to the reaction is shown in capital letters, with the nucleotide sequence at the extreme 3' end of the primer and the first nucleotides added to make a complete repeat unit underlined. Added nucleotides are shown as lowercase letters.

Telomerase synthesizes only one strand of telomeres. A scheme for the complete replication of telomeres incorporating the elongating action of telomerase has been proposed previously (Shampay et al. 1984). In this and related models, it was proposed that synthesis of the complementary C-rich strand is carried out by discontinuous synthesis by primase-polymerase, as is typical of lagging-strand synthesis during semi-conservative DNA replication, using as a template the extended G-rich strand made by telomerase. Telomerase in vitro requires a single-stranded primer for its elongating action (Lee et al. 1993; Lingner and Cech 1996; Wang and Blackburn 1997), raising the question of how, and when in the chromosomal DNA replication cycle, this substrate is generated in vivo (Lingner et al. 1995; Wellinger et al. 1996). In *Tetrahymena* and various lower eukaryotic species where direct analysis has been possible, the G-strand protrudes 12–16 nucleotides beyond the complementary C-rich strand (Henderson and Blackburn 1989 and references therein). Analyses of human and yeast telomeric termini suggest longer G-strand overhangs (Wellinger et al. 1993; Makarov et al. 1997). In all these systems, there remains a question of whether, during DNA preparation, loss of a portion of the complementary C-rich strand, which has been shown to have discontinuities in *Tetrahymena* and yeast (Blackburn and Gall 1978; Szostak and Blackburn 1982), generated these long overhangs in vitro (Blackburn et al. 1983).

It is possible that in vivo, telomerase acts on the DNA after the helicase accompanying the replication fork has unpaired the terminal region of the chromosomal DNA duplex. However, the relative timing of the actions of telomerase and the replication fork at the telomere are unknown.

IDENTIFICATION OF THE RNA AND PROTEIN MOIETIES OF TELOMERASE

The in vitro enzymatic activity of telomerase is destroyed by treatment with either micrococcal nuclease or ribonuclease A (RNase A), or with protease. These results showed that activity requires both RNA and protein components (Greider and Blackburn 1987). The essential protein component for telomerase catalytic activity is a reverse transcriptase of a newly identified class (Nakamura et al. 1997) called TERT (*te*lomerase *r*everse *t*ranscriptase) (Nakamura and Cech 1998). TERTs share certain conserved amino acid motifs with other reverse transcriptase proteins. However, interestingly, they also lack certain commonly conserved amino acids in these motifs, a notable example being the essentially universal phenylalanine or tyrosine located two positions away from the two catalytic aspartate residues in motif C of nontelomerase reverse transcriptases (Nakamura et al. 1997). The TERT component of telomerase was initially identified through a combination of a yeast genetic screen for defects in telomere maintenance, and biochemical purification and sequencing of a ciliate telomerase protein subunit (Lingner et al. 1997). The TERT homologs of fission yeast and mammals were then identified by gene sequence homology (Nakamura et al. 1997). Confirmation that this protein was required for telomerase came from the demonstration that the expected catalytic aspartate residues, based on the known reverse transcriptase motif similarities, were required for telomerase activity in vitro and in vivo (Counter et al. 1997; Harrington et al. 1997b; Lingner et al. 1997).

The 159-nucleotide RNA moiety TER (*te*lomerase *R*NA) of the telomerase of *T. thermophila* was the first telomerase RNA found. It was identified on the basis of its co-fractionation with telomerase activity, RNase sensitivity of telomerase activity, and the sensitivity of telomerase activity to RNase H cleavage of the RNA directed by DNA oligonucleotides complementary to a region of the telomerase RNA centered on the templating domain (Greider and Blackburn 1987, 1989). Template function of this and other species' telomerase RNA was confirmed as described below. Telomerase RNA sequences have diverged very rapidly among eukaryotes. Accordingly, identification of telomerase RNA has required a diversity of approaches. The 192-nucleotide telomerase RNA of the distantly related ciliate *Euplotes crassus* (whose telomeric sequence is T_4G_4 repeats) was first identified by its sensitivity to cleavage by RNase

H directed by a DNA oligonucleotide containing T_4G_4 repeats (Shippen-Lentz and Blackburn 1990). Additional telomerase RNAs from other ciliates (ranging in size from 147 to ~200 nucleotides) were identified by cross-hybridization of their single-copy genes to the *T. thermophila* gene, or by PCR-based cloning techniques (Romero and Blackburn 1991; Lingner et al. 1994; McCormick-Graham and Romero 1995; McCormick-Graham et al. 1997). The approximately 1.3-kb telomerase RNA of *Saccharomyces cerevisiae* was identified on the basis of a screen for altered telomere function (Singer and Gottschling 1994), and those of the budding yeasts *Kluyveromyces lactis* and *Candida albicans* (also >1 kb in length) through hybridization of telomeric repeat sequence probes with their long, telomere-complementary internal template region (McEachern and Blackburn 1995; E. Orr et al., in prep.). The identification of the human and mouse telomerase RNAs (~450 and ~400 bases, respectively) was initially on the basis of extensive copurification with telomerase activity for the human RNA and the finding of a homologous RNA in mouse, and identifications were then confirmed by functional tests (Blasco et al. 1995; Feng et al. 1995). Various terminologies have been used for each species' telomerase RNA and its gene: TER1 was initially used for *Tetrahymena* RNA (Yu et al. 1990), TER1, TLC1, hTR, and mTR for *K. lactis*, *S. cerevisiae*, human, and mouse, respectively (Singer and Gottschling 1994; Blasco et al. 1995; Feng et al. 1995; McEachern and Blackburn 1995). The original name TER is used here for the telomerase RNA, consistent with the recent TERT designation for the protein reverse transcriptase component of telomerase.

The *Tetrahymena* telomerase RNA, like several other types of small eukaryotic RNAs, was deduced to be an RNA polymerase III transcript from its lack of a 5′ cap, the presence in its gene of a run of T residues encompassing the position of the 3′ end of the RNA (Greider and Blackburn 1989), and its sensitivity to α-amanitin in run-off transcription experiments with isolated nuclei (Yu et al. 1990). Ciliate telomerase RNA genes share a pair of highly conserved upstream sequences and lack a conserved box A consensus sequence, suggesting that, like U6 RNA, they are transcribed by RNA polymerase III regulated by upstream *cis*-acting elements (Romero and Blackburn 1991). In contrast, telomerase RNA in yeasts and mammals is transcribed by RNA polymerase II and a fraction of the transcript is found polyadenylated, although a 3′-end-processed, nonpolyadenylated form appears to be the form in the telomerase enzyme RNP (McEachern and Blackburn 1995; Chapon et al. 1997). In mammals, a TATA box and CAAT box are conserved in the telomerase RNA promoter (Hinkley et al. 1998). Transcription of a particular RNA by RNA polymerase III in some species and by RNA polymerase II in others is unusual, but not unprecedented; U3 RNA is transcribed by different polymerases in animals and plants.

The active enzymatic telomerase RNP complex in the ciliate *Euplotes* and in human cells has been biochemically demonstrated to contain the TERT protein (Lingner and Cech 1996; Harrington et al. 1997b). In addition, proteins that associate specifically with the telomerase RNA have been identified. A p80 protein subunit binds *Tetrahymena* TER specifically (Collins et al. 1995), and mouse and human p80 homologs exist (Harrington ct al. 1997a). The mouse homolog also binds telomerase RNA as determined by a three-hybrid scheme (Harrington et al. 1997a; Nakayama et al. 1997), and is located in the same RNP particle as TERT protein and TER/mTR RNA (Harrington et al. 1997b). In *Tetrahymena*, TER1 RNA is found in at least two distinct RNP complexes resolved by nondenaturing gel electrophoresis (Bhattacharyya and Blackburn 1997a; Gilley and Blackburn 1999). One complex contains the p80 and p95 proteins; the other has relatively very weak or no reactivity with p80 or p95 antibodies and may contain the TERT protein.

THE TEMPLATE FUNCTION OF TELOMERASE RNA

In Vitro and Phylogenetic Evidence

Polymerization of the Telomeric DNA Sequence

A central function for the RNA component of telomerase was suggested by the identification within the *Tetrahymena* telomerase RNA of a nine-nucleotide sequence, 5' CAACCCCAA 3', which is complementary to the telomeric TTGGGG repeats synthesized by this enzyme (Greider and Blackburn 1989). Evidence suggesting that this sequence acts as the template came initially from experiments that tested the effect on telomerase activity of RNase H treatment in the presence of DNA oligonucleotides complementary to this RNA region (Greider and Blackburn 1989), leading to the model for the elongation reaction catalyzed by *Tetrahymena* telomerase shown schematically in Figure 2 (Greider and Blackburn 1989). Analogous models are applicable to other telomerases of other species. Initially, the 3' end of the DNA primer base-pairs with the template domain of the telomerase RNA (Fig. 2a). The primer is elongated, copying specific telomerase RNA templating residues, making first-round products (Fig. 2b). At the end of the telomerase RNA template domain (the 43rd position), the primer translocates and repositions for second-round synthesis (Fig. 2c). Thus, according to the model, some of the nine

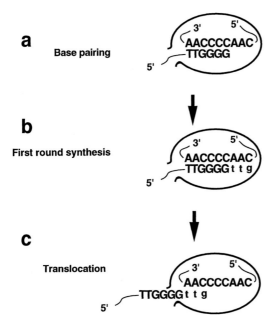

Figure 2 Synthesis of telomeric DNA by the ribonucleoprotein enzyme telomerase from *Tetrahymena*. (*a*) The 3' few nucleotides of the terminal chromosomal G-rich overhang (*thick line*; shown arbitrarily as ending in ..TTGGGG 3') basepair with the telomere-complementary sequence in the telomerase RNA. (*b*) The chromosomal end is extended by polymerization of dGTP and dTTP using the RNA as a template, resulting in the addition of telomeric nucleotides. (*c*) The extended DNA terminus unpairs from its RNA template, becoming available for another round of elongation by telomerase, and/or to primase-polymerase, which uses it in turn as the template for lagging-strand synthesis of the C-rich telomeric strand. (Adapted from Greider and Blackburn 1989.)

RNA residues within the template domain of *Tetrahymena* telomerase can be copied into the six-base telomeric DNA repeat of *Tetrahymena* (GGGGTT), whereas other bases allow alignment of a GGGGTT-repeat primer for synthesis of tandem repeat units.

This model rationalized the precise recognition of the 3'-end sequence of the substrate primer described above and has been amply supported by the identification and functional analysis of telomerase activities from several species (Zahler and Prescott 1988; Morin 1989; Shippen-Lentz and Blackburn 1990; Zahler et al. 1991; Mantell and Greider 1994; Cohn and Blackburn 1995). In vitro reconstitution of human telomerase with in vitro-transcribed hTR/hTER RNA and cotranslated hTERT protein has shown that these two components are the only ones likely to be required for basal telomerase catalytic activity (Weinrich et al. 1997). During in vitro reactions with the ciliate and human enzymes, multiple cycles of copying of the template occur, implying that the template RNA is not destroyed in the course of repeated cycles of template copying (Morin 1989; Lee and Blackburn 1993). Hence, telomerase appears not to have a ribonuclease H activity associated with it, in contrast to typical protein reverse transcriptases.

As indicated in Figure 2c, at the end of a cycle of templated synthesis up to 9 Watson-Crick base pairs of an RNA–DNA helix could potentially exist on the RNA template of the *Tetrahymena* telomerase. Similarly, with the *Euplotes* telomerase, results obtained in vitro using primers complementary to the region 3′ of the template and extending into the template suggested that a minimum of 11 bp might have to be dissociated after the initial round of elongation of such a primer, and at least 8 bp in each subsequent elongation round, as each GGGGTTTT repeat was added (Shippen-Lentz and Blackburn 1990). These observations raised the question of whether such a long helix is in fact made in vitro or in vivo. If so, unwinding it is expected to require energy. Telomerase activity in vitro has no requirement for deoxynucleoside triphosphates other than those incorporated into the DNA synthesized, or for ribonucleoside triphosphates such as ATP. Whether a helicase activity is either intrinsic to, or normally associated with, telomerase in vivo has not been proven. The situation could be different for telomerases depending on the repeat length synthesized. In vitro studies with partially purified *Tetrahymena* telomerase, which synthesizes 6-base repeats, are consistent with a much shorter (about 4-5 bp) RNA–template–DNA primer/product being formed (Lee et al. 1993; Wang et al. 1998). However, analyses of *K. lactis* telomerase, which synthesizes 25-base telomeric repeats using the 30-base templating domain of its RNA (McEachern and Blackburn 1995), suggest that under at least some in vitro conditions a much longer helix may have to be disrupted in order to complete synthesis of a full repeat (T. Boswell-Fulton and E.H. Blackburn 1998).

DNA Endonuclease Cleavage

A cleavage activity that normally cleaves the DNA primer residue copied from the 5′ residue of the template sequence was initially found in *Tetrahymena* telomerase (Collins and Greider 1993). This activity, which has also been found in telomerase from other species, is template-directed in the sense that it cleaves DNA substrates bound via the templating domain (Cohn and Blackburn 1995; Gilley et al. 1995; Lingner and Cech 1996; Melek et al. 1996; Prescott and Blackburn 1997). The activity was shown to be endonucleolytic in its action rather than exonucleolytic (Melek et al. 1996). Evidence that the activity is a part of the telomerase RNP itself and not a contaminating endonuclease comes from two primary observations: First, primers containing mismatches to the template RNA sequence are cleaved preferentially over matched primers, with the cleavage site reflecting the position of the mismatches relative to the complementary template (Melek et al. 1996; Prescott and Blackburn 1997a). Second, direct evidence that the cleavage activity is inherent to telomerase itself is that the properties of the endonuclease cleavage pattern reaction are altered in specific ways when telomerase contains a mutated form of telomerase RNA, as described below (Bhattacharyya and Blackburn, 1997a; H. Wang et al., unpubl.).

In Vivo Demonstration of the Template Function of Telomerase RNA

In several cases, alteration of the template RNA sequence of TER RNA has been shown to result in the synthesis of telomeric DNA with a correspondingly altered sequence. This was directly shown in vivo with *Tetrahymena* telomerase RNA (Yu et al. 1990; Yu and Blackburn 1991). The template sequence of the cloned *Tetrahymena thermophila* telomerase RNA gene was mutated by site-directed mutagenesis to produce different templates: one with an additional C residue, converting the CCCC sequence to CCCCC, which was predicted to specify the synthesis of G_5T_2 repeats, and another with the A at position 44 (see Fig. 3) substituted by G (predicting G_4TC repeats). Each mutated telomerase RNA gene was expressed in vivo by transformation of *Tetrahymena* with the altered gene. Expression of either of these two altered telomerase RNA genes resulted in the in vivo synthesis of telomeric DNA whose sequence corresponded to that of the altered template. Interestingly, the TER mutant predicted to cause G_5T_2 repeats also caused synthesis of G_6T_2, G_7T_2, and G_8T_2 repeats, suggesting slippage occurs along the expanded run of Cs. Such slippage was confirmed by in vitro studies with this mutated telomerase (Gilley and Blackburn 1996). These in vivo results proved that the 5'CAACCCCAA3' sequence in the *Tetrahymena* telomerase RNA gene is the template for telomere synthesis. In several other species as well, including the ciliate *Paramecium*, yeasts, mouse, and human, the identity of a putative telomerase RNA has also been confirmed by showing that mutating the template region causes synthesis of appropriately mutated telomeric DNA sequences, both in vitro and in vivo (Autexier and Grei-

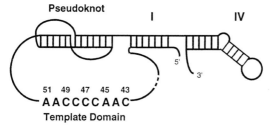

Figure 3 Model of the core conserved secondary structure of ciliate telomerase RNAs. The bases in the template domain sequence are shown as letters with numbering showing the distance from the 5' end of the RNA for *Tetrahymena thermophila*; the rest of the RNA is indicated as the thicker line. Thin lines represent base-pairing in secondary structural features whose existence is supported by co-variation among these RNAs; these include a long-range pairing close to the 5' end (helix I) enclosing the template domain and a pseudoknot, and the bulged stem-loop structure (helix IV). (Adapted from Blackburn 1997a,b.)

der 1994; Singer and Gottschling 1994; Blasco et al. 1995; Feng et al. 1995; Gilley et al. 1995; McEachern and Blackburn 1995; Gilley and Blackburn 1996; Marusic et al. 1997; McCormick-Graham et al. 1997; Prescott and Blackburn 1997b).

AN ESSENTIAL ROLE FOR TELOMERASE RNA, BESIDES TEMPLATING, IN ACTIVE-SITE FUNCTIONS OF TELOMERASE

The findings described above established that telomerase is a reverse transcriptase, which carries its own internal RNA template for DNA synthesis. The core telomerase activity appears to be supplied by TERT and TER (Weinrich et al. 1997). Telomerase RNAs have diverged very rapidly in evolution compared with, for example, ribosomal or some of the spliceosomal RNAs. Telomerase RNA sizes range from 147 in a ciliate to at least 1.3 kb in budding yeasts (Singer and Gottschling 1994; McCormick-Graham and Romero 1995; McEachern and Blackburn 1995). Yet when the secondary structures are compared within groups of related organisms, they are found to share a conserved structure despite their rapid primary sequence divergence (Romero and Blackburn 1991; Lingner et al. 1994; Singer and Gottschling 1994; McCormick-Graham and Romero 1995; McEachern and Blackburn 1995; McCormick-Graham et al. 1997). A striking example is evident for the >30 known ciliate RNAs, whose RNAs differ by more than 80% overall in primary sequence. All share a remarkably conserved common core structure, in which even the relative spacing between different secondary structure elements is largely preserved (Fig. 3). Such structural conservation suggests that telomerase RNA has other functions besides solely providing a short internal template. Some of these functions are likely to include features required for transcription, stability, and assembly of telomerase into a ribonucleoprotein. For example, mutating the pseudoknot or helix IV of *Tetrahymena* telomerase RNA interferes with assembly of this RNA into a normal RNP (Gilley and Blackburn 1999; T. Ware et al., unpubl.).

Requirement for Specific Telomerase RNA Bases for Polymerization

Strikingly, functional analyses of mutated telomerase RNAs in both yeast and *Tetrahymena* have provided evidence that the telomerase RNA is also required for aspects of the enzymatic activity itself. Mutagenesis of the *S. cerevisiae* telomerase RNA template domain demonstrated templating function for several of its residues, spanning a region over 11 of its 17 potentially copied bases (Kramer and Haber 1993; Prescott and Blackburn 1997a). However, a substitution of a specific trinucleotide in the template (the 476GUG mutation) unexpectedly destroys telomerase activity both in vivo and in vitro (Prescott and Blackburn 1997a,b). When it is the only telomerase RNA expressed in vivo, the 476GUG mutant RNA causes progressive telomere shortening, slow growth, and eventual cellular senescence. These are the phenotypes characteristic of yeast cells unable to replenish their telomeric DNA, and are seen by deleting much or all of the telomerase TER/TLC1 RNA or TERT subunits (Lundblad and Szostak 1989; Singer and Gottschling 1994; McEachern and Blackburn 1995, 1996; Lingner et al. 1997; Nakamura et al. 1997). In in vitro assays, no polymerization activity of the mutant 476GUG telomerase is detectable in telomerase preparations made from cells expressing only this mutant RNA. However, the 476GUG mutant RNA is as stable as wild type, accumulates to similar levels, and forms a telomerase RNP with fractionation and size properties indistinguishable from wild type (Prescott and Blackburn 1997b). A related set of findings has been made for the ~1.3-kb telomerase RNA of *K. lactis*: A small change in its sequence, this time located several hundred bases away from the templating domain, results in no detectable telomerase activity either in vitro or in vivo (Roy et al. 1998). Hence, in both these yeasts the mutant RNA is properly expressed and assembled into an RNP, but one which is lacking in any detectable enzymatic activity. Similar tests have also been done using partially purified telomerase preparations extracted from *Tetrahymena* cells that were first treated with nuclease to remove endogenous wild-type telomerase RNA, then reconstituted with an excess of exogenously added, in vitro-transcribed *Tetrahymena* telomerase RNA. Mutating the RNA in the loop of helix IV (see Fig. 3) caused greatly reduced polymerization activity (Autexier and Greider 1998). However, it was not shown that this RNA was capable of assembly into the RNP.

Various template mutations of the *Tetrahymena* telomerase RNA also cause highly aberrant telomerase activity in vivo and in vitro. Substituting the template cytosine (C) at position 48 of the *Tetrahymena* telomerase RNA (see Fig. 3) by U (48U mutation, predicting synthesis of GAGGTT repeats) causes telomere shortening and a severe cell senescence phenotype. Furthermore, rather than the mutated template directing the synthesis of the expected mutant telomeric repeat sequences in vivo as with some other template mutants, no GAGGTT repeat sequence was detectable in the telomeric DNA of 48U mutant cells (Yu et al. 1990). As well as constituting direct evidence that telomerase is essential for the long-term viability of *Tetrahymena*, and that its continued action is necessary to prevent cellular senescence, the unexpected failure of the 48U mutant RNA to sustain telomere maintenance in vivo suggested that telomerase RNA may have other functions.

In vitro assays reveal directly that the 48U and certain other single-base changes in *Tetrahymena* telomerase RNA cause marked impairment of active-site functions, including enzyme processivity and fidelity. These experiments were done using mutant telomerase RNA expressed and assembled in vivo into telomerase RNP, which was then partially purified and its action analyzed by in vitro enzyme reactions (Gilley et al. 1995; Gilley and Blackburn 1996). Hence, the effects of mutations could be shown to be exerted at the level of enzymatic function itself, rather than being effects on synthesis, stability, or assembly of the RNP. Notably, mutating either of the template positions 48 or 43 leads to premature product dissociation at position 45. Thus, these positions are involved in base-specific interactions that affect DNA–telomerase interactions; in the case of the 43 mutation, the effect of the mutation is exerted during polymerization even before the mutated template position is reached. The 48U mutant telomerase has up to 50% error rates: It both synthesizes and extends rC/dA mispairs, by incorporating dA with high efficiency where dG should be incorporated, opposite positions 47 and 46 on the template (Gilley et al. 1995). Thus, base-specific interactions by template residue 48 are also required for proper selection of the complementary base for incorporation at some positions along the template. Mutant telomerase with the template cytosine base at position 49 mutated to a G exhibits a high level of mispair 3′ extension, in which a rG/T mispair at the 3′ end of the primer is efficiently extended. The specific losses of fidelity caused by the 48U and

49G mutations could result from a distortion of the active site, or failure of the mutated RNA residue to contribute a critical functional group to the active site, which prevents a mispair from being distinguished from the correct local geometry of a Watson-Crick base pair, as has been described for DNA-dependent DNA polymerase (Doublie et al. 1998; Kiefer et al. 1998; Steitz 1998). Wild-type *Paramecium tetraurelia* telomerase RNA carries out a natural, unprecedented specific misincorporation: The single telomerase RNA species directs synthesis of GGGTTT, as well as the predicted GGGGTT, repeats in vivo from its 3'ACCCCAAC5' templating domain. The underlined C residue directs either addition of the predicted dG residue, or, at high frequencies, specific misincorporation of a T residue at this position in the repeat. Mutating the base next to the 3' end of the templating domain (G57A mutation) increases the level of T misincorporation (McCormick-Graham et al. 1997). In summary, multiple experiments provide compelling evidence that base-specific interactions involving the telomerase RNA play critical roles in essential active-site functions of telomerase.

The Role of Telomerase RNA in Defining the Boundaries of the Templating Region

Mutational analysis of telomerase RNA, particularly that of *Tetrahymena*, has allowed the templating nucleotides of the 9-nucleotide templating domain to be defined. Mutating the 49C of the *Tetrahymena* enzyme into 49G allows the mutated position 49 to be copied into a dC in vitro (Gilley and Blackburn 1996). Introducing a site-specifically mutated telomerase RNA gene with an A mutation at position 43 (Fig. 3) into cells causes GGGGTTT repeat sequences to be added to the telomeres. When this telomerase is extracted from cells overexpressing the 43A mutant RNA, the predicted 7-base GGGGTTT repeats are made in vitro instead of the usual 6-base GGGGTT repeats. Thus, positions 49 to 43 are copied by the mutated 43A enzyme (Gilley et al. 1995); this can be attributed to basepairing of the DNA product with the RNA template dictating the template sequence used in repeated rounds of synthesis. The analogous C to A mutation in the 5' position of the template sequence of *P. tetraurelia* telomerase RNA also causes the 7-base GGGGTTT repeat to be made in vivo (McCormick-Graham et al. 1997). Residues in the AACCCCAAC sequence that act as templating bases were also identified by in vitro reconstitution of the *Tetrahymena* enzyme. These experiments used partially purified protein fractions first treated with nuclease to remove endogenous wild-type telomerase RNA, then reconstituted with excess mutated telomerase RNA transcribed in vitro. Such in vitro reconstitution of a telomerase RNA with position 43 mutated to a U residue also shows that 43U is copied into dA (Autexier and Greider 1994).

A short sequence (GUCA) located two nucleotides to the 5' side of the templating domain is conserved in all known ciliate telomerase RNAs (Lingner et al. 1994; McCormick-Graham and Romero 1995). The template boundary of telomerase is altered when bases outside the template itself are mutated. Insertion or deletion of a base in the 2-nucleotide spacer separating the conserved GUCA and template sequence, and analysis of telomerase products with in vitro-reconstituted telomerase, suggest that a major determinant of the 5' boundary of the template is its spacing from the GUCA sequence: Copying stops 2 bases before this sequence is reached. Mutating the GUCA sequence to a different sequence causes copying beyond the 5' boundary of the template (Autexier and Greider 1995). Thus, the conserved GUCA sequence appears to act as a barrier pre-venting synthesis from approaching closer than 2 bases away. Mutating the 3-base sequence immediately abutting the 5' boundary of *S. cerevisiae* telomerase also causes copying in vitro to extend past the normal 5' boundary of the template (Prescott and Blackburn 1997a), although the sequence of the analogous region of telomerase RNA has diverged rapidly among budding yeasts (Y. Tzfati and E.H. Blackburn, unpubl.). A conserved long-range base-pairing of the sequence 5' to the template with another TER sequence defines the template boundary in *K. lactis* telomerase RNA (Y. Tzfati et al., in prep.).

Altering the ionic strength conditions of the elongation reaction of wild-type *Tetrahymena* telomerase causes strong elongation product dissociation after copying of position 44 instead of the usual position 43 (Lee et al. 1993). Hence, under some circumstances positions 44 to 49, instead of the usual 43 to 48, may be copied to generate 6-base repeats. When the C at position 48 is changed to a U (which also has effects on enzyme action; see above), the pattern of products made by the mutated enzyme under certain in vitro conditions suggests that this mutant telomerase synthesizes the predicted GAGGTT repeats by copying nucleotides 50 to 45 (Gilley et al. 1995). In an in vitro reconstitution system, circularly permuting the C and A residues within the template domain of the in vitro-transcribed exogenously added mutant RNAs leads to apparent copying of positions 51 to 45 (Autexier and Greider 1994). In addition, which region of the *Euplotes* telomerase template is copied depends on the concentration of dGTP in the reaction (Hammond and Cech 1997). In summary, there is some flexibility in which bases of the terminally repeated templating domain are copied, and in some cases changes in template usage are dictated by small changes in the telomerase RNA.

Influence of Telomerase RNA on the Endonuclease Activity of Telomerase

In addition to deleterious effects on templated polymerization, certain specific changes in telomerase RNA induce aberrant endonuclease activity. When the *Tetrahymena* RNA is substituted by telomerase RNA from the ciliate *Glaucoma*, the position of cutting by the endonuclease activity of this telomerase undergoes a shift relative to the template. *Glaucoma* and *Tetrahymena thermophila* RNA share an identical 23-base region centered on the template, but the rest of these ~160-nucleotide-long RNAs, which have virtually identical secondary structures, differ by 50% in primary sequence. The *Glaucoma* RNA-substituted telomerase RNP, extracted from *Tetrahymena* cells, exhibits endonuclease activity that now cuts a bound telomeric DNA primer at a position opposite the middle of the template, instead of at the normal 5′ end of the template (Bhattacharyya and Blackburn 1997a). Hence, RNA regions outside the template markedly influence the geometric properties of the endonuclease active site relative to the template-primer. In a converse experiment, the 9-base templating domain of *Tetrahymena* telomerase was substituted entirely by a 9-base alternating AU stretch, leaving the rest of the 159-nucleotide RNA completely wild type. Again, the endonucleolytic activity was aberrant, with cleavage of a bound DNA primer occurring at template positions near the center of the template (T. Ware and E.H. Blackburn, in prep.). These results show that RNA base identities strongly influence the endonuclease enzymatic action as well as the polymerization activity of telomerase.

Functional Interaction between Telomerase RNAs

As described above, the mutant *tlc1-476GUG* RNA of *S. cerevisiae* was incapable of supporting telomerase activity in vitro or in vivo. However, unexpectedly, co-expressing the mutant and wild-type telomerase RNAs caused restoration of activity of the *tlc1-476GUG* mutant telomerase RNA, both in vivo and in vitro (Prescott and Blackburn 1997a,b). This result suggested that the two telomerase RNAs can functionally interact. It was shown directly by biochemical experiments that telomerase in *S. cerevisiae* is active in an oligomeric (minimally dimeric) form containing at least two functional RNAs in a single telomerase RNP complex. Furthermore, the wild-type *TER1/TLC1* RNA had to be in the same oligomeric/dimeric RNP particle as the *tlc1-476GUG* RNA to restore the activity of *tlc1-476GUG* RNA telomerase (Prescott and Blackburn 1997b). Since the templating domain residues lie in the active site for polymerization, these results also imply that the active sites within the telomerase oligomer interact with each other, either directly or indirectly. In this regard, it is interesting that the patterns of interspersion of mutant and wild-type telomeric repeats in telomeres generated in vivo, in cells simultaneously expressing both a mutant RNA and the wild-type telomerase RNA, are consistent with telomerase functioning in vivo as a mixed-RNA dimer or oligomer (Prescott and Blackburn 1997b).

CELLULAR LOCATION AND DIFFERENT RNP FORMS OF TELOMERASE RNA

Following its in vitro polymerization reaction, *S. cerevisiae* telomerase stably binds its product after polymerization. Stabilization of this enzyme–product complex involves not only Watson-Crick base-pairing between the DNA product and the RNA template, but also other interactions between the product and RNA and/or protein on the telomerase RNP (Prescott and Blackburn 1997a). Previous concepts of telomerase function had implicitly assumed that the enzyme associates with its telomeric end substrate only for the duration of the polymerization reaction. However, surprisingly, in haploid yeast cells containing a mutant telomerase that was assembled into a stable RNP but was enzymatically inactive (the *tlc1-476GUG* mutant described above), cells began to undergo senescence when telomeres were still ~100 bp longer than the short but stable telomeres seen in other template mutant strains (Prescott and Blackburn 1997a). This observation suggested that reduced telomere length per se was not the cause of senescence onset, and rather that telomerase has some function in addition to enzymatic telomeric DNA addition, which in the absence of DNA product binding it was unable to fulfill. This function was proposed to be the binding of telomerase RNP to its newly elongated telomeric DNA product (Prescott and Blackburn 1997a). Furthermore, as mentioned above, the patterns of interspersion of repeats in telomeres of cells simultaneously expressing both mutant RNA and wild-type telomerase RNA are readily explained by proposing a dimeric telomerase in which the second template has a high probability of being the next one copied (Prescott and Blackburn 1997a,b). Such would be the case if telomerase remains stably bound to the telomere throughout all or part of the cell cycle. These properties of *S. cerevisiae* telomerase suggest that the telomerase RNP might itself be a structural component of a telomere "cap," helping to protect the newly synthesized 3′ overhanging DNA of the telomere from recombination and degradation activities. Telomerase might be displaced by a replication fork or helicase during DNA replication, or be replaced at some point in the cell cycle by putative end-binding proteins.

Telomerase has not been localized within the nucleus in *S. cerevisiae*, but in the somatic nucleus of the ciliate *Oxytricha nova*, which carries large numbers of short genomic DNAs, each ending with short telomeres, the TER RNA has been localized by in situ hybridization. In *Oxytricha* and other hypotrichous ciliates, DNA replication in the somatic nucleus takes place in a distinctive structure called the replication band. Some of the telomerase RNA was also localized to this band (Fang and Cech 1995). This finding provides the best indication that telomerase carries out its replication function in concert with general DNA replication in S-phase of the cell cycle. There was no evidence that high levels of telomerase colocalized with the nonreplicating telomeres throughout the macronucleus. Instead, the major telomerase RNA hybridization signal is found in large, round bodies in the nucleus which also contain other nuclear RNAs such as spliceosomal RNAs (Fang and Cech 1995). Such nuclear bodies have been proposed to be sites of RNP processing, storage, or assembly.

No other functions have been reported for a telomerase RNA-containing RNP besides those described above for the active telomerase enzyme. Yet telomerase RNA is found as two RNP forms in *Tetrahymena*, as seen by nondenaturing gel electrophoresis. One form contains the telomerase-associated proteins p80 and p95 as judged by reactivity to antibodies against each protein; the other RNP form has at most weak reactivity with antibodies to these proteins (Bhattacharyya and Blackburn 1997a; Gilley and Blackburn 1999). The location of telomerase RNA in two nuclear locations in *Oxytricha*, and its presence in two distinct RNPs, raise the possibility that the different complexes containing telomerase RNA are functionally distinct and differently located in the cell.

TELOMERASE ACTIVITY: MECHANISTIC CONSIDERATIONS FOR AN RNA–PROTEIN MACHINE

Having a built-in template poses certain challenges, which are probably reflected in the special properties of telomerase. The length of primer DNA–template RNA helix required for efficient polymerization by *Tetrahymena* telomerase varies according to which template position is being copied (Wang et al. 1998). In addition, a given nucleoside triphosphate analog inhibitor competes with the correct normal triphosphate substrate at very different efficiencies depending on which template position is being copied (Strahl and Blackburn 1994). A varying set of interactions and spatial relationships between telo-merase RNA and the active site, as polymerization occurs at each step along the template, is likely to underlie these highly position-specific properties of the telomerase active site.

If the three-dimensional arrangement of catalytic aspartates is relatively rigid in the TERT protein, as expected from the required geometry of an RT catalytic center involving metal ions correctly positioned and interacting with the aspartate side groups, then after each polymerization step the template moves through the catalytic center. In conventional RTs, in a single polymerization cycle the template-primer is in a fixed set of positions relative to all the RT active-site residues, which are the same for each successive polymerization cycle. In contrast, in telomerase, the template RNA is part of the complex and appears to be anchored at both ends of the template, as judged by inaccessibility of surrounding regions to RNase H and by physical RNA–protein interactions (Greider and Blackburn 1989; Zaug and Cech 1995). Hence, the template region has to be flexible in order to move through the active site. The RNA–protein interactions necessarily differ at each step of polymerization along the template, with a distinct new site being built up at each step. This may restrict the rigidity of the active-site amino acid contacts with template-primer and the incoming dNTP. Thus, telomerase RNA residues may have retained functions in stabilizing the active-site arrangements of the catalytic aspartates, such functions being assumed solely by amino acid residues in fully protein RTs.

IS TELOMERASE A MOLECULAR FOSSIL OF A TRANSITION FROM AN RNA REPLICASE TO A PURELY PROTEIN REVERSE TRANSCRIPTASE?

The conservation of the telomerase enzyme in the form of a ribonucleoprotein throughout a wide range of eukaryotes suggests that this enzyme was already present at the time of the separation of eukaryotes from other kingdoms. Telomerase RNA has a seemingly partly paradoxical set of properties. As described above, among eukaryotes these RNA sequences have diverged very rapidly in both primary sequence and overall size, and in their transcriptional modes. Yet significantly, specific residues in the RNA, both within and outside the template region, are crucially important for enzymatic function. The degree of functional involvement of telomerase RNA in telomerase's enzymatic action appears unlike the situation seen for conventional purely protein reverse transcriptases found in systems ranging from retroviruses to prokaryotes. These findings, together with the fact that the RNA component acts as a template for DNA synthesis, prompt the speculation that telomerase may be a relic from the time of, or even predating, the evolutionary transition from RNA to DNA genomes. Specifically, the striking ability of the group I self-splicing rRNA intron ribozyme of *Tetrahymena* to polymerize RNA using an internal RNA template (Cech 1990) has suggested the possibility that RNA replicases were once all-RNA catalysts. The evolution of the protein (TERT) component of telomerase has been discussed in other reviews (Eickbush 1997; Nakamura and Cech 1998). Here I propose a specific model for the evolution of telomerase that expands on the model proposed in the first edition of this book (Blackburn 1993). In this new model, telomerase is a molecular fossil, whose RNA moiety is a relic of an RNA replicase ribozyme that acquired protein and the ability to synthesize DNA.

The evolutionary model supposes an RNA World, as discussed elsewhere in this volume, in which the ancestor to telomerase was an example of an RNA catalytic replicase that copied itself, an RNA genome, into an RNA copy (Fig. 4a). As described for many RNA enzymes elsewhere in this volume, this RNA used bound divalent metal ions for catalysis. As an RNA-plus-protein world evolved, the ancestor of telomerase acquired a protein component, notably with the triad of catalytic aspartates that took over the binding of divalent metal ions for catalysis at the active site (Fig. 4b). However, the RNA was retained and contributed other func-

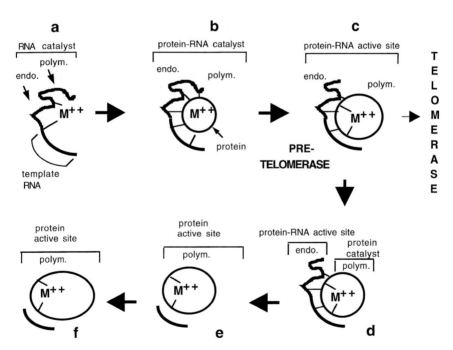

Figure 4 Hypothetical model for the evolutionary relationships between an ancestral RNA replicase, telomerase ancestor, telomerase, and reverse transcriptases. See text for explanation.

tional groups to building the active site necessary to place the catalytic aspartates in an active arrangement. At least a portion of the RNA also retained its template function. This RNA-plus-protein entity functioned as an RNA-dependent RNA polymerase copying RNA into RNA. A transition into a DNA-synthesizing activity took place. The transition from RNA to DNA synthesis may not have required a profound change in enzymatic properties. The structures of the catalytic DNA polymerase domains of the reverse transcriptase from the HIV retrovirus and of DNA polymerase I from *Escherichia coli* have been determined by X-ray crystallography, revealing a striking structural similarity between these two active sites (Steitz 1998). Modern telomerase can utilize RNA primers and ribonucleoside triphosphate substrates, albeit at much lower efficiencies than DNA primers and dNTPs (Collins and Greider 1995).

As this evolution took place, telomerase may have remained as a "molecular fossil" carrying still-functional remnants of an earlier RNA-based biology. The RNP ancestor of telomerase (pre-telomerase in Fig. 4c) also had an endonuclease activity associated with the polymerization active site. Such an endonuclease activity would be analogous to that in the current RNA polymerase, in which the endonucleolytic activity is also mediated by the polymerization active site (Rudd et al. 1994). The reverse transcriptase of the R2Bm ribosomal RNA gene intron of *Bombyx mori*, and group II intron-associated reverse transcriptase activities, each have an endonuclease action that cleaves a DNA target. Interestingly, this activity requires an RNA component as well as the protein reverse transcriptase (Zimmerly et al. 1995). As described above, *Tetrahymena* telomerase endonuclease activity that acts on template-bound DNA is strongly influenced by telomerase RNA residue identities.

It has been shown that the telomerase TERT family is more closely related by sequence alignments to the group II intron-encoded reverse transcriptases than to other reverse transcriptases (Nakamura et al. 1997). The model proposed here is that the mechanism of a common, telomerase-like ancestor was retained in the modern telomerase. This telomerase ancestor, like modern telomerases, had an RNA template and polymerization and endonucleolytic active sites built up of RNA functional groups as well as protein functional groups (Fig. 4c). In addition, the telomerase ancestor evolved into group II intron-encoded reverse transcriptases and the non-LTR reverse transcriptases, by a series of losses of RNA function (Fig. 4d–f). Loss of RNA involvement in the polymerization active site, but retention of RNA function in the endonuclease activity, led to the current group II intron-associated reverse transcriptase RNPs (Fig. 4d) (Zimmerly et al. 1995). Loss of the RNA residues involved in endonuclease function was accompanied by recruitment of protein residues into these active-site functions, leading to present-day non-LTR-class reverse transcriptases, which lack an endonuclease (Fig. 4e) (Eickbush 1997). Further evolution of these protein reverse transcriptases led to the retroposon and viral protein reverse transcriptases (Fig. 4f).

The existence of long telomeric repeats in *K. lactis*, *C. albicans*, and certain other budding yeasts demonstrates that their telomerases can copy a significantly longer region of its telomerase RNA than the telomerases of those eukaryotes with the more typical short telomeric repeats. Do these long telomeric repeats represent a more primitive state of telomerase? If an ancestor of telomerase RNA was an RNA replicase ribozyme that could in principle be copied completely, the long repeats may represent an intermediate stage in the evolution of telomerase toward enzymes that copy only very short regions of their RNA. By this model, it can be imagined that during evolution progressively shorter regions of the RNA were copied, so that most organisms have now evolved to the point where only a minimal region of the telomerase RNA is used as the template for reverse transcription.

A class of unusual telomeric sequences occurs in the insect *Drosophila melanogaster*. No conventional short-repeat type of telomeric sequence has been identified in this species. Instead, broken chromosomes can become healed by the addition of apparently retroelement-derived sequences called HeT-A sequences. The de novo addition of HeT-A sequences to broken chromosome ends has been shown to occur in two distinct situations in *D. melanogaster*: a broken end induced by movement of a transposon (Biessmann et al. 1990), and the new ends created by opening of a ring X chromosome (Traverse and Pardue 1988). These independent events suggest that HeT-A sequences, as well as another class of retroelements, TART sequences (Levis et al. 1993), may serve a telomeric role in this species. In *D. melanogaster*, the HeT-A elements that are added to the ends of broken chromosomes have all of the hallmarks of having been synthesized by reverse transcription of (often) defective retroelements (Pardue et al. 1996). In this way, the elongation of chromosome ends in *D. melanogaster* is analogous to the action of telomerase. Perhaps this situation represents an exploitation of a reverse transcriptase reaction to carry out a role usually accomplished by telomerase: the addition of DNA to the chromosome ends to balance the loss of terminal DNA sequences through incomplete replication. It is striking that the addition of typical short telomeric repeats and HeT-A sequences both involve reverse transcription. Thus, the essential difference between *D. melanogaster* and more typical eukaryotes may be only that in *D. melanogaster* the added DNA repeats are synthesized by copying a substantial part of the retroelement RNA, so that adding DNA segments to broken or other chromosome ends occurs in large increments, and therefore addition needs to occur only infrequently. On the other hand, in most eukaryotic species the typical telomerase copies only a very short region of RNA, so that lengthening of the chromosome occurs in small increments and thus must occur in most or all cell divisions. It can be imagined that *D. melanogaster* represents a late-derived situation: In the presence of a sufficiently efficient HeT-A mechanism, the requirement for telomerase was bypassed in *D. melanogaster* and telomerase was lost.

ACKNOWLEDGMENTS

I thank Mike McEachern, Renata Gallagher, Simon Chan, David Gilley, and John Prescott for helpful discussions of the manuscript, and the National Institutes of Health for research support.

REFERENCES

Adam R.D., Nash T.E., and Wellems T.E. 1991. Telomeric location of Giardia rDNA genes. *Mol. Cell. Biol.* **11:** 3326–3330.

Autexier C. and Greider C.W. 1994. Functional reconstitution of wild-type and mutant *Tetrahymena* telomerase. *Genes Dev.* **8:** 563–575.

———. 1995. Boundary elements of the *Tetrahymena* telomerase RNA template and alignment domains. *Genes Dev.* **9:** 2227–2239.

———. 1998. Mutational analysis of the *Tetrahymena* telomerase RNA: Identification of residues affecting telomerase activity in vitro. *Nucleic Acids Res.* **26:** 787–795.

Bhattacharyya A. and Blackburn E.H. 1997a. A functional telomerase RNA swap in vivo reveals the importance of nontemplate RNA domains. *Proc. Natl. Acad. Sci.* **94:** 2823–2827.

———. 1997b. *Aspergillus nidulans* maintains short telomeres throughout development. *Nucleic Acids Res.* **25:** 1426–1431.

Biessmann H., Mason J.M., Ferry K., d'Hulst M., Valgeirsdottir K., Traverse K.L., and Pardue M.L. 1990. Addition of telomere-associated HeT DNA sequences "heals" broken chromosome ends in *Drosophila. Cell* **61:** 663–673.

Blackburn E.H. 1984. The molecular structure of centromeres and telomeres. *Annu. Rev. Biochem.* **53:** 163–194.

———. 1990a. Telomeres and their synthesis. *Science* **249:** 489–490.

———. 1990b. Telomeres: Structure and synthesis. *J. Biol. Chem.* **265:** 5919–5921.

———. 1991a. Structure and function of telomeres. *Nature* **350:** 569–573.

———. 1991b. Telomeres. *Trends Biochem. Sci.* **16:** 378–381.

———. 1993. Telomerase RNA. In *The RNA world* (ed. R. Gesteland and J. Atkins), Cold Spring Harbor Laboratory Press, Cold Spring Harbor, New York.

———. 1994. Telomeres: No end in sight. *Cell* **77:** 621–623.

———. 1997. Telomerase RNA structure and function. In *RNA Structure and Function* (ed. R.W. Simons and M. Grunberg-Manago), Cold Spring Harbor Laboratory Press, Cold Spring Harbor, New York.

Blackburn E.H. and Gall J.G. 1978. A tandemly repeated sequence at the termini of the extrachromosomal ribosomal RNA genes in *Tetrahymena. J. Mol. Biol.* **120:** 33–53.

Blackburn E.H., Greider C.W., Henderson E., Lee M.S., Shampay J., and Shippen-Lentz D. 1989. Recognition and elongation of telomeres by telomerase. *Genome* **31:** 553–560.

Blackburn E.H., Budarf M.L., Challoner P.B., Cherry J.M., Howard E.A., Katzen A.L., Pan W.-C., and Ryan T. 1983. DNA termini in ciliate macronuclei. *Cold Spring Harbor Symp. Quant. Biol.* **47:** 1195–1207.

Blasco M.A., Funk W., Villeponteau B., and Greider C.W. 1995. Functional characterization and developmental regulation of mouse telomerase RNA. *Science* **269:** 1267–1270.

Boswell-Fulton T. and Blackburn E.H. 1998. Identification of *Kluyveromyces lactis* telomerase: Discontinuous synthesis along the thirty nucleotide-long templating domain. *Mol. Cell. Biol.* (in press).

Cano M.I., Dungan J., Agabian N., and Blackburn E.H. 1998. Telomerase in kinetoplastida parasitic protozoa. *Proc. Natl. Acad. Sci.* (in press).

Cech T.R. 1990. Self-splicing of group I introns. *Annu. Rev. Biochem.* **59:** 543–568.

Chapon C., Cech T.R., and Zaug A.J. 1997. Polyadenylation of telomerase RNA in budding yeast. *RNA* **3:** 1337–1351.

Cohn M. and Blackburn E.H. 1995. Telomerase in yeast. *Science* **269:** 396–400.

Collins K. and Greider C.W. 1993. *Tetrahymena* telomerase catalyzes nucleolytic cleavage and nonprocessive elongation. *Genes Dev.* **7:** 1364–1376.

———. 1995. Utilization of ribonucleotides and RNA primers by *Tetrahymena* telomerase. *EMBO J.* **14:** 5422–5432.

Collins K., Kobayashi R., and Greider C.W. 1995. Purification of *Tetrahymena* telomerase and cloning of genes encoding the two protein components of the enzyme. *Cell* **81:** 677–686.

Counter C.M., Meyerson M., Eaton E.N., and Weinberg R.A. 1997. The catalytic subunit of yeast telomerase. *Proc. Natl. Acad. Sci.* **94:** 9202–9207.

Doublie S., Tabor S., Long A.M., Richardson C.C., and Ellenberger T. 1998. Crystal structure of a bacteriophage T7 DNA replication complex at 2.2 Å resolution. *Nature* **391:** 251–258.

Edman J.C. 1992. Isolation of telomerelike sequences from *Cryptococcus neoformans* and their use in high-efficiency transformation. *Mol. Cell. Biol.* **12:** 2777–2783.

Eickbush T. 1997. Telomerase and retroposons: Which came first? *Science* **277:** 911–912.

Fang G. and Cech T.R. 1995. Telomerase RNA localized in the replication band and spherical subnuclear organelles in hypotrichous ciliates. *J. Cell Biol.* **130:** 243–253.

Feng J., Funk W.D., Wang S.S., Weinrich S.L., Avilion A.A., Chiu C.P., Adams R.R., Chang E., Allsopp R.C., Yu J., Yu J., Le S., West M.D., Harley C.B., Andrews W.H., Greider C.W., and Villeponteau B. 1995. The RNA component of human telomerase. *Science* **269:** 1236–1241.

Fitzgerald M.S., McKnight T.D., and Shippen D.E. 1996. Characterization and developmental patterns of telomerase expression in plants. *Proc. Natl. Acad. Sci.* **93:** 14422–14427.

Gilley D. and Blackburn E.H. 1996. Specific RNA residue interactions required for enzymatic functions of *Tetrahymena* telomerase. *Mol. Cell. Biol.* **16:** 66–75.

———. 1999. The telomerase RNA pseudoknot is critical for the stable assembly of the catalytically active ribonucleoprotein. *Proc. Natl. Acad. Sci.* **96:** 6621–6625.

Gilley D., Lee M.S., and Blackburn E.H. 1995. Altering specific telomerase RNA template residues affects active site function. *Genes Dev.* **9:** 2214–2226.

Greider C.W. and Blackburn E.H. 1985. Identification of a specific telomere terminal transferase activity in *Tetrahymena* extracts. *Cell* **43:** 405–413.
———. 1987. The telomere terminal transferase of *Tetrahymena* is a ribonucleoprotein enzyme with two kinds of primer specificity. *Cell* **51:** 887–898.
———. 1989. A telomeric sequence in the RNA of *Tetrahymena* telomerase required for telomere repeat synthesis. *Nature* **337:** 331–337.
Hammond P.W. and Cech T.R. 1997. dGTP-dependent processivity and possible template switching of Euplotes telomerase. *Nucleic Acids Res.* **25:** 3698–3704.
Harrington L., McPhail T., Mar V., Zhou W., Oulton R., Bass M.B., Arruda I., and Robinson M.O. 1997a. A mammalian telomerase-associated protein. *Science* **275:** 973–977.
Harrington L., Zhou W., McPhail T., Oulton R., Yeung D.S., Mar V., Bass M.B., and Robinson M.O. 1997b. Human telomerase contains evolutionarily conserved catalytic and structural subunits. *Genes Dev.* **11:** 3109–3115.
Henderson E.R. and Blackburn E.H. 1989. An overhanging 3′ terminus is a conserved feature of telomeres. *Mol. Cell Biol.* **9:** 345–348.
Hinkley C.S., Blasco M.A., Funk W.D., Villeponteau B., Greider C.W., and Herr W. 1998. The mouse telomerase RNA 5′-end lies just upstream of the telomerase template sequence. *Nucleic Acids Res.* **26:** 532–536.
Kiefer J.R., Mao C., Braman J.C., and Beese L.S. 1998. Visualizing DNA replication in a catalytically active *Bacillus* DNA polymerase crystal. *Nature* **391:** 304–307.
Kramer K.M. and Haber J.E. 1993. New telomeres in yeast are initiated with a highly selected subset of TG_{1-3} repeats. *Genes Dev.* **7:** 2345–2356.
Lee M.S. and Blackburn E.H. 1993. Sequence-specific DNA primer effects on telomerase polymerization activity. *Mol. Cell. Biol.* **13:** 6586–6599.
Lee M.S., Gallagher R.C., Bradley J., and Blackburn E.H. 1993. In vivo and in vitro studies of telomeres and telomerase. *Cold Spring Harbor Symp. Quant. Biol.* **58:** 707–718.
Levis R.W., Ganesan R., Houtchens K., Tolar L.A., and Sheen F.M. 1993. Transposons in place of telomeric repeats at a *Drosophila* telomere. *Cell* **75:** 1083–1093.
Lingner J. and Cech T.R. 1996. Purification of telomerase from *Euplotes aediculatus*: Requirement of a primer 3′ overhang. *Proc. Natl. Acad. Sci.* **93:** 10712–10717.
Lingner J., Cooper J.P., and Cech T.R. 1995. Telomerase and DNA end replication: No longer a lagging strand problem? *Science* **269:** 1533–1534.
Lingner J., Hendrick L.L., and Cech T.R. 1994. Telomerase RNAs of different ciliates have a common secondary structure and a permuted template. *Genes Dev.* **8:** 1984–1998.
Lingner J., Hughes T.R., Shevchenko A., Mann M., Lundblad V., and Cech T.R. 1997. Reverse transcriptase motifs in the catalytic subunit of telomerase. *Science* **276:** 561–567.
Lundblad V. and Szostak J.W. 1989. A mutant with a defect in telomere elongation leads to senescence in yeast. *Cell* **57:** 633–643.
Makarov V.L., Hirose Y., and Langmore J.P. 1997. Long G tails at both ends of human chromosomes suggest a C strand degradation mechanism for telomere shortening. *Cell* **88:** 657–666.
Mantell L.L. and Greider C.W. 1994. Telomerase activity in germline and embryonic cells of *Xenopus*. *EMBO J.* **13:** 3211–3217.
Marusic L., Anton M., Tidy A., Wang P., Villeponteau B., and Bacchetti S. 1997. Reprogramming of telomerase by expression of mutant telomerase RNA template in human cells leads to altered telomeres that correlate with reduced cell viability. *Mol. Cell. Biol.* **17:** 6394–6401.

McCormick-Graham M. and Romero D.P. 1995. Ciliate telomerase RNA structural features. *Nucleic Acids Res.* **23:** 1091–1097.

McCormick-Graham M., Haynes W.J., and Romero D.P. 1997. Variable telomeric repeat synthesis in *Paramecium tetraurelia* is consistent with misincorporation by telomerase. *EMBO J.* **16:** 3233–3242.

McEachern M.J. and Blackburn E.H. 1994. A conserved sequence motif within the exceptionally diverse telomeric sequences of budding yeasts. *Proc. Natl. Acad. Sci.* **91:** 3453–3457.

———. 1995. Runaway telomere elongation caused by telomerase RNA gene mutations. *Nature* **376:** 403–409.

———. 1996. Cap-prevented recombination between terminal telomeric repeat arrays (telomere CPR) maintains telomeres in *Kluyveromyces lactis* lacking telomerase. *Genes. Dev.* **10:** 1822–1834.

Melek M., Greene E.C., and Shippen D.E. 1996. Processing of nontelomeric 3′ ends by telomerase: Default template alignment and endonucleolytic cleavage. *Mol. Cell. Biol.* **16:** 3437–3445.

Morin G.B. 1989. The human telomere terminal transferase enzyme is a ribonucleoprotein that synthesizes TTAGGG repeats. *Cell* **59:** 521–529.

Muller F., Wicky C., Spicher A., and Tobler H. 1991. New telomere formation after developmentally regulated chromosomal breakage during the process of chromatin diminution in *Ascaris lumbricoides. Cell* **67:** 815–822.

Nakamura T.M. and Cech T.R. 1998. Reversing time: Origin of telomerase. *Cell* **92:** 587–590.

Nakamura T.M., Morin G.B., Chapman K.B., Weinrich S.L., Andrews W.H., Lingner J., Harley C.B., and Cech T.R. 1997. Telomerase catalytic subunit homologs from fission yeast and human (see comments). *Science* **277:** 955–959.

Nakayama J., Saito M., Nakamura H., Matsuura A., and Ishikawa F. 1997. TLP1: A gene encoding a protein component of mammalian telomerase is a novel member of WD repeats family. *Cell* **88:** 875–884.

Pardue M.L., Danilevskaya O.N., Lowenhaupt K., Slot F., and Traverse K.L. 1996. *Drosophila* telomeres: New views on chromosome evolution. *Trends Genet.* **12:** 48–52.

Prescott J. and Blackburn E.H. 1997a. Functionally interacting telomerase RNAs in the yeast telomerase complex. *Genes Dev.* **11:** 2790–2800.

———. 1997b. Telomerase RNA mutations in *Saccharomyces cerevisiae* alter telomerase action and reveal nonprocessivity in vivo and in vitro. *Genes Dev.* **11:** 528–540.

Prowse K.R., Avilion A.A., and Greider C.W. 1993. Identification of a nonprocessive telomerase activity from mouse cells. *Proc. Natl. Acad. Sci.* **90:** 1493–1497.

Romero D.P. and Blackburn E.H. 1991. A conserved secondary structure for telomerase RNA. *Cell* **67:** 343–353.

Roy J., Fulton T.B., and Blackburn E.H. 1998. Specific telomerase RNA residues distant from the template are essential for telomerase function. *Genes Dev.* **12:** 3286–3300.

Rudd M., Izban M.G., and Luse D.S. 1994. The active site of RNA polymerase II participates in transcript cleavage within arrested ternary complexes. *Proc. Natl. Acad. Sci.* **91:** 8057–8061.

Shampay J., Szostak J.W., and Blackburn E.H. 1984. DNA sequences of telomeres maintained in yeast. *Nature* **310:** 154–157.

Shippen-Lentz D. and Blackburn E.H. 1989. Telomere terminal transferase activity from

Euplotes crassus adds large numbers of TTTTGGGG repeats onto telomeric primers. *Mol. Cell. Biol.* **9:** 2761–2764.
———. 1990. Functional evidence for an RNA template in telomerase. *Science* **247:** 546–552.
Singer M.S. and Gottschling D.E. 1994. TLC1: Template RNA component of *Saccharomyces cerevisiae* telomerase. *Science* **266:** 404–409.
Steitz T.A. 1998. A mechanism for all polymerases. *Nature* **391:** 231–232.
Strahl C. and Blackburn E.H. 1994. The effects of nucleoside analogs on telomerase and telomeres in *Tetrahymena*. *Nucleic Acids Res.* **22:** 893–900.
Szostak J.W. and Blackburn E.H. 1982. Cloning yeast telomeres on linear plasmid vectors. *Cell* **29:** 245–255.
Traverse K.L. and Pardue M.L. 1988. A spontaneously opened ring chromosome of *Drosophila melanogaster* has acquired He-T DNA sequences at both new telomeres. *Proc. Natl. Acad. Sci.* **85:** 8116–8120.
Wang H. and Blackburn E.H. 1997. De novo telomere addition by *Tetrahymena* telomerase in vitro. *EMBO J.* **16:** 866–879.
Wang H., Gilley D., and Blackburn E.H. 1998. A novel specificity for the primer-template pairing requirement in *Tetrahymena* telomerase. *EMBO J.* **17:** 1152–1160.
Weinrich S.L., Pruzan R., Ma L., Ouellette M., Tesmer V.M., Holt S.E., Bodnar A.G., Lichtsteiner S., Kim N.W., Trager J.B., Taylor R.D., Carlos R., Andrews W.H., Wright W.E., Shay J.W., Harley C.B., and Morin G.B. 1997. Reconstitution of human telomerase with the template RNA component hTR and the catalytic protein subunit hTRT. *Nat. Genet.* **17:** 498–502.
Wellinger R.J., Wolf A.J., and Zakian V.A. 1993. *Saccharomyces* telomeres acquire single-strand TG_{1-3} tails late in S phase. *Cell* **72:** 51–60.
Wellinger R.J., Ethier K., Labrecque P., and Zakian V.A. 1996. Evidence for a new step in telomere maintenance. *Cell* **85:** 423–433.
Yu G.L. and Blackburn E.H. 1991. Developmentally programmed healing of chromosomes by telomerase in *Tetrahymena*. *Cell* **67:** 823–832.
Yu G.L., Bradley J.D., Attardi L.D., and Blackburn E.H. 1990. In vivo alteration of telomere sequences and senescence caused by mutated *Tetrahymena* telomerase RNAs. *Nature* **344:** 126–132.
Zahler A.M. and Prescott D.M. 1988. Telomere terminal transferase activity in the hypotrichous ciliate *Oxytricha nova* and a model for replication of the ends of linear DNA molecules. *Nucleic Acids Res.* **16:** 6953–6972.
Zahler A.M., Williamson J.R., Cech T.R., and Prescott D.M. 1991. Inhibition of telomerase by G-quartet DNA structures. *Nature* **350:** 718–720.
Zakian V.A. 1995. Telomeres: Beginning to understand the end. *Science* **270:** 1601–1607.
Zaug A.J. and Cech T.R. 1995. Analysis of the structure of *Tetrahymena* nuclear RNAs in vivo: Telomerase RNA, the self-splicing rRNA intron, and U2 snRNA. *RNA* **1:** 363–374.
Zimmerly S., Guo H., Perlman P., and Lambowitz A.M. 1995. Group II intron mobility occurs by target DNA-primed reverse transcription. *Cell* **82:** 1–20.

24
Dynamics of the Genetic Code

John F. Atkins
Howard Hughes Medical Institute and Department of Human Genetics
University of Utah
Salt Lake City, Utah 84112-5330;
Department of Genetics
Trinity College, Dublin, Ireland

August Böck
Lehrstuhl für Mikrobiologie der Universität München
80638 München, Germany

Senya Matsufuji
Department of Biochemistry II
Jikei University School of Medicine
Tokyo 105-8461, Japan

Raymond F. Gesteland
Howard Hughes Medical Institute and Department of Human Genetics
University of Utah
Salt Lake City, Utah 84112-5330

Shape is crucial for catalysis. In the hypothetical RNA World, the replicative RNAs, which constituted the hereditary information, also functioned as the "shapes" for catalysis. There was no need for decoding. When decoding originated, presumably discrimination between alternate coding possibilities was initially weak, but once one mode became predominant, there would have been selection to lock it in with increasing efficiency. Nontriplet translocation and nonstandard meaning of code words presumably generally approached a minimum compatible with speed and energy use optimization. In the present day, nonstandard decoding alternatives generally just contribute to a low level of translational errors, of which frameshifting errors (Atkins et al. 1972; Kurland 1992) are a grave type. However, some unknown proportion of genes in probably all organisms has special sites where efficient decoding alternatives are programmed into the mRNA. The group of mechanisms involved in redirection of decoding is called "recoding" (Gesteland et al. 1992). Either the evolution of the ability to perform recoding was coincident with evolution of the ability of the decoding apparatus to perform standard decoding, or it is a later sophistication. Of course, the answer is unknown, but the conservation of the required mRNA signals presented below indicates that recoding has been part of the decoding repertoire for at least several hundred million years and must therefore be favored by evolution. This is clearly distinct from the error rate that is a trade-off between energy expenditure, speed, and accuracy. In the latter case, evolution has optimized the balance as a whole. With recoding, it is the specific, nonstandard decoding events themselves that have led to selective advantage. In addition to the evolutionary questions posed by recoding, the mRNA signals involved reveal heretofore unsuspected roles of RNA sequences and structures in modern organisms.

Signals for recoding are carried in the mRNA sequence and include specification of a site in the coding sequence and stimulatory elements ranging from simple sequences on either side of the site to far removed and complex stem-loop structures. Little is known about the involvement of other cellular components, although in one case (insertion of selenocysteine in prokaryotes), a special protein factor is known to be essential.

Three classes of recoding are recognized. The meaning of specific code words can be redefined. The reading frame can be altered by ribosomes that switch from one overlapping reading frame to another, and blocks of nucleotides can be bypassed with, or without, a change in reading frame (for review, see Gesteland and Atkins 1996).

Recoding is in competition with standard decoding. For instance, the programmed ribosomal frameshifting that occurs two-thirds of the way through the *Escherichia coli dnaX* coding sequence is 50% efficient. Half the ribosomes frameshift and soon terminate at a stop codon in the new frame, whereas the other half traverse the shift site with standard decoding and synthesize a product with an extra domain. The two products are present in a 1:1 ratio in DNA polymerase III (Blinkova and Walker 1990; Flower and McHenry 1990; Tsuchihashi and Kornberg 1990; Kelman and O'Donnell 1995). The signals in other cases of recoding give different set efficiencies, sometimes for viral packaging purposes (Wickner 1989; Atkins et al. 1990; Levin et al. 1993), or the process can be regulated. The dynamic nature of recoding contrasts with the "hard-wired" derivatives of the universal code found in specialized niches such as mitochondria, where the meaning of a code word is altered wherever it occurs. In recoding, the rule change is dependent on mRNA context.

REDEFINITION

Redefinition of stop codons by recoding involves a variety of stimulatory signals. The first case of redefinition studied, the low-efficiency, but essential, readthrough of the UGA at the end of the phage Qβ coat protein gene (Weiner and Weber 1973; Hofstetter et al. 1974), is probably influenced only by the identity of a few flanking nucleotides. Tate and colleagues have shown that the efficiency of release factor 2 function at UGA is influenced by the identity of the flanking 3' nucleotide and, to a lesser extent, the following two nucleotides (Poole et al. 1998). The identity of corresponding Qβ nucleotides is expected to give somewhat inefficient termination. The identity of the last couple of amino acids encoded prior to the stop codon also influences termination efficiency (Mottagui-Tabar and Isaksson 1997). At the end of the Qβ coat protein gene, there appears to be a trade-off between amino acids that give efficient termination and 3' nucleotides that give moderately efficient readthrough, so that modest levels of readthrough result.

In tobacco mosaic virus, the recoding signal is the identity of six nucleotides 3' of the redefined UAG "stop" codon (Skuzeski et al. 1991; Zerfass and Beier 1992; Stahl et al. 1995). The counterpart for synthesis of the Gag-Pol precursor of murine leukemia virus is a nearby 3' pseudoknot (Fig. 1A) (ten Dam et al. 1990; Wills et al. 1991, 1994; Felsenstein and Goff 1992; Feng et al. 1992). Although not yet studied, a pseudoknot may also be important for the putative *gag-pro* analog readthrough of the *Dictyostelium* retrotransposon, Skipper (Leng et al. 1998). For UAG stop codon redefinition in decoding barley yellow dwarf virus, both close and distant signals are important (Brown et al. 1996). Given the widespread occurrence of redefinition in plant viruses, understanding the mechanism involved is especially interesting and worthwhile. For *Drosophila* trachea branching controller, *hdc*, high-level readthrough of UAA is required, but the location of the stimulatory signals is unknown (Steneberg et al. 1998). For *Drosophila* kelch, the recoding signal is also unknown, but the efficiency of redefinition is regulated (Robinson and Cooley 1997). For these cases of redefinition, the amino acid inserted is either known, or highly likely to be one of the standard 20 amino acids having their own unique codon(s). However, the 21st (excluding formyl methionine) directly encoded amino acid, selenocysteine, is encoded only by UGA, a stop codon in the standard code.

Selenocysteine

Selenocysteine was recognized as a constituent of special proteins in 1976 (Cone et al. 1976). Experiments in two different biological systems in 1986 showed that this nonstandard amino acid is inserted cotranslationally, directed by an in-frame UGA codon in the mRNA (Chambers et al. 1986; Zinoni et al. 1986). Thus, in a single mRNA, UGA can have two

MuLV - Redefinition

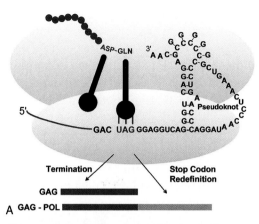

Procaryotic Selenocysteine Insertion

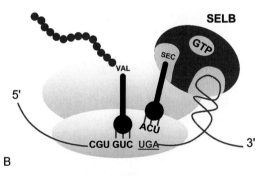

Figure 1 (*A*) Redefinition of the murine leukemia virus *gag* terminator required for synthesis of the Gag-Pol precursor. The size of loop 1 and stem 2 shown have been determined in recent work by N.M. Wills et al. (unpubl.). (*B*) Model of the ribosomal complex decoding UGA with selenocysteine. The model shows the quaternary complex between SelB, GTP, selenocysteyl-tRNA, and the SECIS element of the mRNA. Contact between SelB in the quaternary complex and the ribosome triggers GTP hydrolysis resulting in the release of selenocysteyl-tRNA (Hüttenhofer and Böck 1998b).

contrasting meanings, stop or selenocysteine insertion. This is in contrast to the use of UGA as a codon for tryptophan in *Mycoplasma* and mitochondria, where the new meaning is hard-wired and the stop function is lost. UGA thus is a "chameleon" among codons, which raises intriguing questions about whether the ancestral UGA encoded stop or an amino acid and how the evolution of the change of meaning occurred.

Distribution of Selenoproteins

Selenocysteine-containing proteins occur in all three lines of descent, but not in all organisms. Our own recent screening revealed that among twenty gram-positive and gram-negative bacteria, only five synthesized selenoproteins (S. Schorling et al., unpubl.). This finding of a nonubiquitous distribution is supported by the recent results of whole genome sequence analysis, which reveals a similar frequency. We were also unable to demonstrate the occurrence of selenoproteins in several plant cell cultures (Neuhierl and Böck 1996). Intriguingly, *Mycoplasma* species in which UGA codes only for tryptophan do not contain selenoproteins (Himmelreich et al. 1997). The numbers of selenoproteins synthesized are different among organisms. They can range from just two in the case of *Haemophilus influenzae* (Wilting et al. 1998) to three for *E. coli* (Baron and Böck 1995) to seven for *Methanococcus jannaschi* (Bult et al. 1996; Wilting et al. 1997) to an estimated more than thirty for mammals (Behne et al. 1996).

S and Se in Metabolism

Sulfur and selenium occur in the biosphere at a ratio between 10^3–10^5 to 1 and, with the exception of one major branch point leading to the specific biosynthesis of selenocysteine, they share the same metabolic paths. Free selenocysteine formed via the cysteine biosynthetic enzymes (Müller et al. 1997) can be aminoacylated onto tRNAs by cysteyl-tRNA synthetase and incorporated into any cysteine position of proteins (Müller et al. 1994). Low-molecular-weight selenocysteine is also the precursor for selenomethionine (Sliwkowski and Stadtman 1986). For our discussion, one has to keep in mind, therefore, that any cysteine in a protein is "contaminated" by selenocysteine at a ratio determined by the relative abundance of the two elements and the biochemical S/Se discrimination capacity of the respective organism. The major branch point mentioned above separates the fate of the two elements by a high-affinity metabolic route targeted to the efficient synthesis of selenocysteine under low trace element concentrations.

Selenocysteine Biosynthesis

The biosynthesis of selenocysteine in bacteria differs from that leading to cysteine as it takes place in a tRNA-bound state (Leinfelder et al. 1990; Forchhammer and Böck 1991). A specific tRNA (tRNA^{Sec}) is charged with L-serine by seryl-tRNA synthetase, and the seryl moiety is converted into the selenocysteyl residue by selenocysteine synthase with selenomonophosphate (SeP) as selenium donor. SeP itself is the reaction product of selenophosphate synthetase (Leinfelder et al. 1990; Ehrenreich et al. 1992; Veres et al. 1992). Genes coding for tRNA^{Sec} have been identified in many organisms within Bacteria, Archaea, and Eukarya, and their products share a number of characteristics differentiating them from ordinary elongator tRNAs. In addition to the UCA anticodon, complementary to UGA, they display sequence and architectural deviations from the consensus of classic elongator tRNAs (Baron et al. 1993; Sturchler et al. 1993). They have a 6-bp D-stem with a 4-bp loop, an extended aminoacyl-acceptor-T-stem axis of 13 bp and a large extra arm which makes these tRNAs the largest ones known. These sequence deviations have a role in maintaining novel tertiary interactions (Baron et al. 1993; Sturchler et al. 1993). tRNA^{Sec}, therefore, may constitute a different evolutionary line of elongator tRNAs. The structural differences are the basis of the additional functions which tRNA^{Sec} has compared to elongator tRNAs, namely, serving as an adapter for the biosynthesis of selenocysteine by selenocysteine synthase, binding in a selenocysteyl-specific manner to a specialized elongation factor (see below), and precluding binding to elongation factor Tu (Baron and Böck 1991).

Biosynthesis of selenocysteine resembles that of glutamine and asparagine, which also take place starting from a precursor in the tRNA-bound state (Ibba et al. 1997). In this context, it is intriguing that archaeal genomes sequenced thus far do not contain recognizable genes for essential cysteine biosynthetic enzymes or for a cysteyl-tRNA synthetase. Thus, either the sequences of these enzymes in Archaea are highly diverged, or cysteine biosynthesis could occur from a specialized seryl-tRNA by analogy with that of selenocysteyl-tRNA (Bult et al. 1996; Ibba et al. 1997; Smith et al. 1997). If so, this would add another example of aminoacyl transformations and fill an important gap in our knowledge of the connections between evolution of the genetic code and amino acid biosynthesis (Wong 1975; Di Giulio 1997).

Elements Involved in Decoding UGA as Selenocysteine

Discrimination between UGA as the selenocysteine-specific codon and UGA as a stop codon is by an mRNA secondary/tertiary structure (the SECIS element) which, in bacteria, is located at the immediate 3' side of UGA, i.e., within the reading frame. Swapping of the SECIS within bacteria is restricted by specific interaction of SECIS with SelB (see below) and the ribosome (Tormay and Böck 1997; Wilting et al. 1998). In Eukarya (Berry et al. 1991, 1993; Hill et al. 1993; Kollmus et al. 1996; Walczak et al. 1998) and Archaea (Wilting et al. 1997), the SECIS motif is positioned outside the reading frame in the 3'untranslated region and acts at

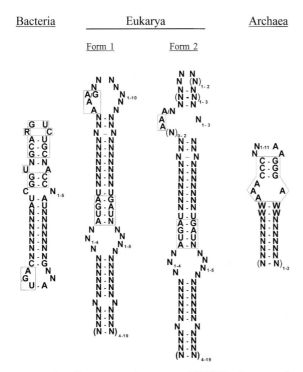

Figure 2 Consensus features of SECIS elements from Bacteria, Archaea, and Eukarya as adopted from Hüttenhofer et al. (1996), Low and Berry (1996), Wilting et al. (1997), Walczak et al. (1998), and M. Berry (pers. comm.). Consensus nucleotides are boxed and semi-conservative ones are printed in bold.

a distance of up to 2,700 bases. The SECIS motifs in bacteria are variable in sequence, based on the constraint of the reading frame, whereas those from mammalia and Archaea possess highly conserved motifs (Fig. 2). Mammalian SECIS elements are predicted to fall into two classes, which differ in the secondary structure around the conserved adenosines (Low and Berry 1996). Selenoprotein P has one of each class in its 3′ UTR, but no functional difference between the two is known (M. Berry, pers. comm.). The mammalian SECIS element can be transplanted and functions to cause selenocysteine insertion at UGA in a heterologous gene (Shen et al. 1993). Although the analysis is at an early stage, there appears to be remarkable conservation between the SECIS element(s) in lower metazoans and mammals. *Caenorhabditis elegans* thioredoxin reductase is almost certainly a selenoprotein very similar to its mammalian counterpart. The 3′ UTR of its mRNA shows a putative SECIS element that has only a single, nonsignificant, substitution from the highly conserved vertebrate-like SECIS (M. Berry, pers. comm.). This SECIS, however, is very different from those identified in Archaea.

SelB, first discovered in *E. coli* (Forchhammer et al. 1989), is a specialized translation factor that interacts with guanine nucleotides, selenocysteyl-tRNA, and the SECIS element in mRNA forming a quaternary complex (Fig. 1B) (Heider et al. 1992; Hüttenhofer et al. 1996). Within this complex, SelB attains a structure suitable for interaction with the ribosome; as a consequence, GTP is hydrolyzed and selenocysteyl-tRNA is released in the proximity of the ribosomal A site (Hüttenhofer and Böck 1998a,b). Despite considerable effort, a homolog to SelB has not yet been identified in Eukarya and Archaea.

The domains of translation factor SelB involved in selenocysteyl-tRNA binding and in mRNA binding can be separated. A 17-kD carboxy-terminal domain of SelB is responsible for binding to the SECIS element (Kromayer et al. 1996) and retains this property when separated from the rest of the protein. The amino-terminal part—separated by a linker domain—has considerable sequence similarity with EF-Tu, and can bind selenocysteyl-tRNA in vivo and in vitro. SelB thus is an elongation factor homologous to EF-Tu that is tethered to the mRNA by its carboxy-terminal extension.

An intriguing consequence of the mechanism used for the "localized" decoding of the UGA codon is that the substrate—selenocysteyl-tRNA—is bound to the translation factor together with the mRNA. It will be very interesting to see the spatial relationship of codon and anticodon within this complex. Is there an interaction between the two nucleic acids before the ribosome even arrives at the UGA?

Evolution of Selenocysteine Insertion

Selective Advantage of Selenocysteine

Selenoproteins that have identified functions are enzymes with selenocysteine in their active site. Natural variants containing a cysteine in this position have been identified for many of these enzymes, showing that selenocysteine per se, in most of the selenoproteins, does not possess an essential role. Mutational change of the selenocysteine to a cysteine also gives variants that are active (Axley et al. 1991; Berry et al. 1992) but have decreased overall catalytic efficiency by a factor of 300 to 400, mostly due to a reduction of the reaction velocity (Axley et al. 1991). It is clear from these studies that although selenocysteine confers a considerable catalytic advantage, it can be replaced by a cysteine in most enzymes.

Did UGA Previously Code for Cysteine?

The discovery of UGA encoding selenocysteine raises the question whether UGA was originally a "sense codon" specifying selenocysteine incorporation or whether this is a "new" development selected to expand the genetic code (Leinfelder et al. 1988). Osawa (1995) and Jukes (1990) in their "codon capture" hypothesis present a detailed pro and con discussion of the two alternatives, and they point out that the evolution of UGA from a selenocysteine to a stop codon may be very difficult to achieve. They assume that the UGN family box originally encoded both cysteine and selenocysteine, pairing with the anticodon UCA. After duplication of the tRNA gene, one of the siblings mutated to GCA, pairing with UGY codons (the present cysteine codons). The other one (UCA) was "captured" by the newly evolving amino acid tryptophan, and then changed to CCA that only pairs with UGG (the present tryptophan codon). The only remaining function of UGA then was coding for selenocysteine. In this scheme, UGA as a stop codon then appeared by mutation of UAA (Jukes 1990).

Several of these arguments are in accord with recent biochemical facts. First, it was shown that cysteine and selenocysteine are equally well accepted by cysteinyl-tRNA synthetase and incorporated into protein (Stadtman et al. 1989; Müller et al. 1994, 1997). The previous failure to detect unspecific selenocysteine incorporation might have been due to its chemical instability or to the fact that cystathionine-β-synthase has a higher affinity for selenocysteine than cysteinyl-tRNA-synthetase, relative to the substrate cysteine. The UGY codons thus can be considered to indiscriminately code for cysteine plus its selenium analog. Second, when the UGA codon in the fdhF mRNA was changed to a UGY codon, selenocysteine was still incorporated, although to a reduced extent, since selenocysteyl-tRNA bound to SelB was competing with cysteyl-tRNA.EF-Tu (Baron et al. 1989). This indicates that the UGY codon can pair with UCA of the selenocysteyl-tRNA and also leads to the conclusion that the switch from UGN to UGY and UGA would not have been detrimental, since it interchanged chemically very similar amino acids.

The assumption that the UGN codon family, and therefore UGA, originally coded for cysteine plus selenocysteine also supports speculations on why selenocysteine is incorporated at only a few specific sites and was not maintained at "neutral" positions. It is assumed that the switch to specific selenocysteine insertion with the exclusion of cysteine insertion was a continuous, step by step, optimization process, which allowed the development of all components of the insertion machinery, SECIS, tRNA^{Sec}, SelB, and the biosynthetic path.

A Possible Scenario

As discussed previously, selenocysteine might have been incorporated indiscriminately with cysteine, encoded by the UGN codon family and a cysteine-specific tRNA (UCA). Indiscriminate incorporation at certain positions may have conferred to that gene product a higher reactivity and thereby a selective advantage. After duplication of the anticodon and separation of the UGY (cysteine) and UGR families, UGA was maintained for the readout of cysteine and selenocysteine and UGG was "captured" by the new amino acid tryptophan (Jukes 1990). The selective advantage forced the development of the selenium biosynthesis and insertion machinery for these special positions. UGA at other positions may have been counterselected by the high reactivity of selenol residues leading to trapped folding intermediates or by oxidative inactivation due to the appearance of oxygen in the atmosphere (Leinfelder et al. 1988).

The existence of the SECIS element and of the specialized translation factor might not have been crucial at this stage, since there was no need for discrimination against chain termination. On the other hand, later development of the SECIS motif designated the special UGA as sense, specific for selenocysteine, and forced its maintenance. "Unprotected" UGA could disappear or gain a new function, e.g., termination. It is noteworthy that the SECIS elements of Bacteria, Archaea, and Eukarya bear no structural similarity, which supports the possibility of convergent evolution. It is also an open question whether the SECIS element within the coding sequence is original and the 3′ SECIS derived from it. An argument in favor of this view is that the influence of an mRNA structure on the recoding process may be easier mechanistically if the codon is in the vicinity. However, SECIS elements within coding regions are under sequence constraint and may not be suitable to direct the insertion of more than one selenocysteine residue, whereas 3′ SECIS elements are not under such constraints and have the capacity for multiple insertions (Low and Berry 1996; Wilting et al. 1997).

The human selP gene has 10 UGA codons, and recoding is promoted by two tandem SECIS elements in the 3′ UTR (Hill et al. 1993). It is not at all clear how 3′ SECIS elements can promote multiple insertions, especially in view of the apparent inefficiency of even single insertions (see below). It is difficult to imagine that a mechanism analogous to the prokaryotic example could suffice; the 3′ element would need to cycle aminoacyl tRNA from the element to each UGA as ribosomes progressed down the message.

An alternative model is that the 3′ element interacts with the 5′ end of the mRNA analogous to the well-known communication of 3′ and 5′ ends of eukaryotic mRNAs. Through this interaction, perhaps the initiating ribosome is modified so that for its transit of the mRNA, it reads each UGA codon as selenocysteine. In this case, the 3' element would provide a ribosome switch rather than a tRNA delivery system (Gesteland and Atkins 1996; Kollmus et al. 1996). Even if triggering the switch for recoding is inefficient, perhaps once the ribosomes are programmed for selenocysteine incorporation, the efficiency at each subsequent site would be high, and the overall efficiency of multiple and single selenocysteine incorporations may not be very different (Kollmus et al. 1996). However, recent experiments caution against a simple version of this model and can more readily be explained by the information being delivered to ribosomes at individual UGA codons. In deiodinase mRNA, there is a single selenocysteine-encoding internal UGA codon and a single SECIS element in the 3' UTR. Increasing the number of SECIS elements had no effect on the efficiency of selenocysteine incorporation, whereas it did when the number of UGA codons was artificially increased (S.C. Low et al., pers. comm.).

The mechanism of mammalian selenocysteine incorporation is unresolved, particularly as to how the distinction is made between termination and redefinition. Two additional results may be relevant. A UGA that is less than 55–110 nucleotides from a SECIS element functions as a terminator (Martin et al. 1996; Gu et al. 1997), and at least the second UGA in SelP mRNA may sometimes function efficiently as a terminator, since foreshortened forms of SelP protein are found (Himeno et al. 1996).

Efficiency

A recent study in *E. coli* shows that the efficiency of selenocysteine insertion is low. The normal decoding of UGA by SelB-GTP-selenocysteyl-tRNA was only 2% efficient compared to decoding of UCA by EF-Tu.GTP.selenocysteyl-tRNA (with a suitably mutant tRNA; S. Suppmann et al., unpubl.). The efficiency seems limited by nonsaturating amounts of charged tRNA^{Sec} and by the kinetics of the formation and resolution of SelB quaternary complex itself (S. Suppmann et al., unpubl.). This low efficiency in *E. coli* is similar to the 1–3% levels (Berry et al. 1992; Kollmus et al. 1996) measured for selenocysteine insertion in mammalia (these efficiencies are based on transient transfection experiments, but see Martin et al. 1996).

Evolution of Selenocysteine Biosynthesis

The mode of synthesis of selenocysteine is in accord with an hypothesis for the coevolution of the genetic code and amino acid biosynthesis

(Wong 1975, 1988). UGA belongs to the serine/cysteine codon family, and both cysteine and selenocysteine are synthesized from a serine precursor. Whereas cysteine is synthesized in the low-molecular state and charged to tRNA by a specific enzyme in Bacteria and Eukarya, biosynthesis of selenocysteine takes place in the tRNA-bound state. This is similar to the biosynthesis of glutaminyl-tRNA or asparaginyl-tRNA from the glutamyl or aspartyl precursors (Ibba et al. 1997) and may reflect coevolution as postulated by Wong (1975, 1988). On the other hand, additional forces may also have been involved in necessitating tRNA-bound biosynthesis: (1) Free selenocysteine is highly toxic; (2) the development of aminoacyl-tRNA synthetases with specific recognition of selenocysteine and cysteine may be difficult to reach in view of the known lack of discrimination by cysteyl-tRNA synthetase (Müller et al. 1994). With respect to our model presented above, selenocysteine biosynthesis had to switch from a co-synthesis via the cysteine biosynthetic path to the formation in the tRNA-bound state. This had to occur early in the sequence of events described, possibly after the split of the primordial UGN codon family into UGY and UGR.

Phylogeny of Sel Gene Products

The fact that selenoproteins occur in all three lines of descent and that selenocysteine is encoded in all cases by UGA supports, but by no means proves, an early evolutionary origin. A considerable number of sel genes have been cloned and sequenced in the past years allowing (with all reservations) some conclusions about relationships.

Selenocysteine Synthase. Selenocysteine synthase is a pyridoxal-phosphate-dependent enzyme. Alignment of the known sequences shows that the enzyme belongs to the α/γ-superfamily of PLP-dependent enzymes and that it has diverged very early from the γ-family. It is intriguing that the closest relatives of selenocysteine synthase are enzymes from sulfur metabolism, namely cystathionine-γ-lyase, O-acetylhomoserine sulfhydrylase, cystathionine-γ-synthase, and cystathionine-β-lyase (Tormay et al. 1998). Thus, selenocysteine synthase may have diverged early, possibly from some enzyme of sulfur metabolism.

Translation Factor SelB. A dendrogram of the known SelB sequences (Hilgenfeld et al. 1996) revealed that the part of the SelB protein that is homologous to EF-Tu displays a greater similarity in different organisms than it does to the EF-Tu sequence from the same organism. This also holds for the relationship with IF-2. One can conclude that SelB belongs to an individual class of translation factors that separated very early from other factors involved in protein synthesis (Hilgenfeld et al. 1996). Unfortunately, sequences of SelB homologs from archaeal or eukaryal species are not yet available for comparison.

tRNA^{Sec}. The predicted tRNA^{Sec} secondary and tertiary structures are much more conserved than the primary structure. Alignment of the sequences shows that the sequence relationships are parallel to those deduced from the 16S rRNA structures of the same organisms (Tormay et al. 1994; Baron and Böck 1995), although the small size of the molecule does not allow statistically significant conclusions

Generality of Redefinition

Sense codons can be redefined to function as start codons. GUG, UUG, and AUU specify valine, leucine, and isoleucine, respectively, when at internal positions of a coding region, but when they function as an initiator they specify methionine (or formyl methionine in *E. coli*). In *E. coli* and its phages, this redefinition requires an appropriately positioned, preceding Shine-Dalgarno sequence. The process is again dynamic; for instance, in the transposon IS911, one particular AUU acts sometimes as an initiator and sometimes as an internal sense codon (Polard et al. 1991), and in the RNA phage fr, a particular UUG behaves similarly (Adhin and van Duin 1990).

The fact that the meaning of specific codons can be redefined by mRNA context raises the important possibility of specific alteration of the meaning of one internal sense codon to another. Conventional protein chemistry could easily miss such events if their efficiency was below 10%.

An intriguing question is whether a redefinition strategy is used for the insertion of additional amino acids beyond the encoded 21, or perhaps could be experimentally exploited for the targeted insertion of normally nonencoded amino acids.

These examples of redefinition of codon meaning all use triplet translocation, the standard mechanism of mRNA readout. In contrast, the next type of recoding to be considered involves altering linear readout andx thus changing the reading frame.

REDIRECTION OF LINEAR READOUT

Frameshifting: Once-only Codon Anticodon Pairing Versus Dissociation and Re-pairing

The issue of entering and maintaining the desired reading frame must have been a significant one for the early translation apparatus. Triplet RNA:RNA interactions are inherently unstable even when the stabilizing topology of an anticodon loop is involved (for review, see Grosjean and Chantrenne 1980). This instability is important. The potential for cognate tRNA to dissociate at initial pairing at the A-site allows near-cognate tRNAs to dissociate and to be preferentially discarded (noncognates are less of an issue) (for review, see Yarus and Smith 1995). Dissociation at the P-site is essential for some types of programmed frameshifting and perhaps one way of dealing with translation errors (Menninger 1977; for review, see Heurgué-Hamard et al. 1996). The weak triplet RNA:RNA interaction is stabilized by events at the ribosome in an active way (for review, see Yarus and Smith 1995), which themselves favor discrimination. tRNA design is integral to this process. However, the instability of triplet RNA:RNA interactions, which is advantageous now, must have posed a problem for early decoding in the absence of the stabilizing role of a sophisticated ribosome and associated factors. If more codon–anticodon bases were paired in early decoding, there is a problem in comprehending how decoding could have evolved to triplet codon–anticodon pairing without wiping out the fruits of previously selected codons. One proposed scheme for early decoding (Crick et al. 1976) was that at any one time, five codon–anticodon bases were paired, but because of a ratcheting of the tRNA (Woese 1970), only triplet "decoding" was involved. An alternative, which would also not involve whole-scale scrambling of previous information, is that six codon–anticodon bases were initially involved in pairing. If this were so, a transition to triplet pairing would just result in interspersed amino acids. Another alternative is that decoding was triplet from the start, but that stacking interactions with protoribosomal RNA stabilized the pairing (Noller et al. 1986). Whatever the explanation for early decoding, it is highly likely that modern protein synthesis involves tRNA interactions with ribosomal components that stabilize codon:anticodon pairing, and presumably these have a major role, direct or indirect (Lodmell and Dahlberg 1997), in mediating framing. Even though pairing is stabilized, if a tRNA anticodon dissociates from pairing with its cognate codon within the ribosome and quickly repairs with the same codon, this would have been undetected in the experiments performed to date. If this happens, one might imagine that the function of some ribosomal component would be to minimize this dissociation. On the basis of what has been found in genetic studies with the large ribosomal protein L9 (Herbst et al. 1994; Adamski et al. 1996; C. Johnston, unpubl.), it is a candidate for having such a function in *E. coli*.

The discrimination at the initial selection of tRNAs at the ribosomal A site is impressive. However, this selectivity can get overwhelmed, with serious consequences for frame maintenance, if the balance of competing tRNAs is upset, especially with a small minority of tRNAs (Atkins et al. 1979; Gallant and Foley 1980; Gallant and Lindsley 1993). An imbalance can lead to acceptance of a noncognate, or near-cognate, tRNA for pairing of its anticodon with the codon. This can be important for frameshifting, even if on occasion the effect of framing is not manifest until the tRNA enters the P site (for review, see Farabaugh 1996). However, in many cases efficient programmed frameshifting follows after selection of the cognate tRNA. The tRNAs that mediate frameshifting in response to an imbalance, as well as in response to other signals, are not special in terms of their anticodon loop size; they have the same size as virtually all tRNAs. In general, this means that their anticodon size is also the standard three bases. Whether the shift tRNAs, in some cases, are special in terms of their base modifications or other features (Hatfield et al. 1989; Atkins and Gesteland 1995; Brierley et al. 1997) is currently being investigated. Following initial studies with model systems (Weiss et al. 1987), it has been found that most cases of programmed frameshifting involve tRNA dissociation and triplet re-pairing in a new frame. Because of this, weak initial pairing contributes to dissociation and so to frameshifting (Tsuchihashi and Brown 1992; Curran 1993).

A key ingredient for efficient programmed frameshifting is having an overlapping codon available for re-pairing by the P-site tRNA. For +1 frameshifting (quadruplet translocation) this is achieved by having the first base of the next zero-frame codon temporarily unoccupied. This means having an empty A site. As initially found in model systems (for review, see Gallant and Lindsley 1993), and later with yeast Ty programmed frameshifting (for review, see Farabaugh 1996), having the zero-frame A-site codon as a rare codon stimulates +1 frameshifting. The tRNAs for rare codons are themselves sparse. When one of these tRNAs is specified by the codon immediately 3' of a shift codon, the level of aminoacylation of the tRNA becomes critical for the level of frameshifting. Hence, shortage of the amino acid used to charge such a sparse tRNA can be revealed via programmed +1 frameshifting (Kawakami et al. 1993). Amino acid starvation must have been a problem in ancient times, and it is common today for many bacteria. It will be interesting to see if the expected attendant frameshifting has regulatory significance. Various movable elements, including the yeast Ty elements, become more mobile when "hard times" are encountered. It may be of selective advantage for their hosts and consequently themselves if they transpose out of important genes whose inactivation causes hard times, or if by inserting elsewhere they provide a promoter that activates expression of desirable, but heretofore silent, genes. Whatever the reason, synthesis of the transposase in Ty elements requires programmed +1 frameshifting that is responsive to the level of a particular aminoacylated, sparse tRNA. In addition, the yeast gene est3 (ever shorter telomeres 3), whose product is required for telomerase, has similar programmed frameshifting (Morris and Lundblad 1997). In contrast, the telomere-specific retrotransposon used for telomere maintenance in *Drosophila* apparently uses –1 frameshifting, and it is in the middle of its *gag* counterpart in distinction to the location of frameshifting in retrovirus decoding (Danilevskaya et al. 1994, 1998).

Another way to reduce competition in the A site is for the first base of the zero-frame codon 3' adjacent to the shift codon to be part of an efficient stop codon (Poole et al. 1998). This is illustrated by the programmed frameshifting required for *E. coli* release factor 2 expression, where codons 25 and 26 are CUU UGA (Craigen et al. 1985). tRNA^{Leu} pairs initially with the zero-frame CUU and some of the time dissociates and re-pairs with the overlapping UUU (underlined) to cause a shift to the +1 frame that encodes the rest of the release factor (Fig. 3A) (Weiss et al. 1987). Since termination at UGA is specifically mediated by release factor 2, low amounts of release factor 2 permit a greater chance of re-pairing in the +1 frame. Subsequent triplet reading of the +1 frame leads to synthesis of functional release factor 2. The converse is also true, giving an autoregulatory circuit (Craigen and Caskey 1986).

However, stops may not only stimulate shifting frame by diminishing competition from an incoming tRNA. Model systems have shown that they stimulate –1 frameshifting where a tRNA re-pairs with a triplet that overlaps the previous upstream codon (Weiss et al. 1987; 1990b; Horsfield et al. 1995). This is the basis for the programmed –1 frameshifting in decoding potato virus M that involves a single-shift tRNA (Gramstat et al. 1994), but the mechanism is unclear.

A very different way to make the last nucleotide of the previous zero-frame codon available for tRNA to re-pair with the overlapping –1 codon is for its corresponding tRNA (in the P site) to also shift –1; i.e., for tandem A-site and P-site –1 shifting. This was first discovered for retroviral, programmed –1 frameshifting (Jacks et al. 1988), and is a common type of programmed –1 frameshifting. Since both tRNAs re-pair, the characteristic shift sequence for this type of frameshifting is of the general form X-XXY-YYZ. In infectious bronchitis virus it is U-UUA-AAC (Brierley et al. 1992) and in *dnaX* it is A AAA AAG (see Fig. 3B). There have been several suggestions as to the details of re-pairing with respect to ribosomal A and P sites (Jacks et al. 1988; Weiss et al. 1989; Yelverton et al. 1994; Atkins and Gesteland 1995). However, not all programmed –1 frameshifts are tandem shifts.

Not all programmed frameshifting involves dissociation and re-pairing in an overlapping frame. In yeast Ty3 and mammalian antizyme +1 frameshifting, the evidence points to "once-only" pairing so that the first base of the next zero-frame codon is somehow unavailable for pairing with an incoming tRNA (Fig. 4A) (Farabaugh et al. 1993; Matsufuji et al. 1995). This means that finding potential frameshift sequences by looking for overlapping cognate codons will miss some examples.

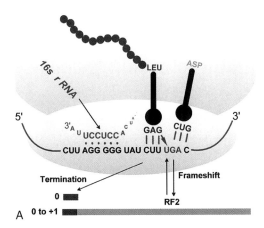

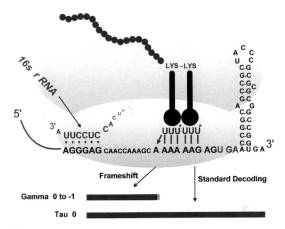

Figure 3 (*A*) The obligatory regulated +1 frameshifting required for synthesis of *E. coli* polypeptide chain release factor 2. (*B*) The tandem codon –1 frameshift required for synthesis of the γ subunit of *E. coli* DNA polymerase III.

FRAMESHIFTING: STIMULATORY SIGNALS

mRNA stimulatory signals are critical for efficient programmed frameshifting of either the "dissociation—re-pairing" or the "once-only pairing" types. In many cases, the mRNA signal(s) is 3′ of the shift site. A relatively simple stem-loop 3′ of the shift site is responsible for stimu-

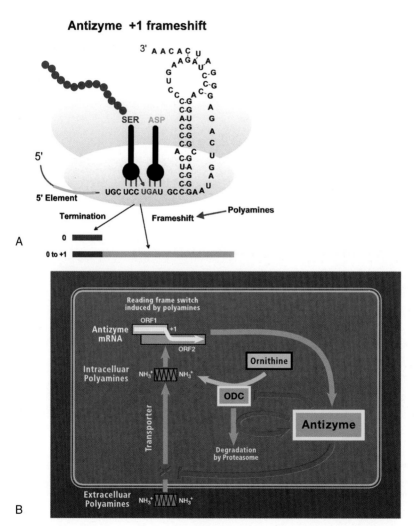

Figure 4 (*A*) The "once-only" pairing +1 frameshifting required for synthesis of human antizyme 1. (*B*) A key component of the autoregulatory circuit for polyamine homeostasis is the modulation by polyamines of the frameshifting required for antizyme synthesis.

lating the programmed frameshifting in decoding HIV *gag-pol* (Parkin et al. 1992; Bidou et al. 1997) and *E. coli dnaX* (Fig. 3B) (Larsen et al. 1997), whereas a complicated stem-loop is utilized in the bacterial transposable element, IS911 (Polard et al. 1991). In the case of *dnaX*, the frameshift efficiency is directly proportional to the predicted stability of the stem-loop structure.

The stimulatory mRNA structure is often a pseudoknot located within 8 bases 3' of the shift site (Brierley et al. 1989; ten Dam et al. 1990). Examples are found in Coronaviruses (Brierley et al. 1991), retroviruses, the double-stranded RNA virus L-A of yeast (Dinman and Wickner 1992; Tu et al. 1992), and mammalian antizyme (Matsufuji et al. 1995, Fig. 4). With mouse mammary tumor virus *gag-pro* frameshifting, a wedge base at the junction of the two pseudoknot stems that keeps them from stacking coaxially was shown to be important for stimulation (Chen et al. 1996). However, there is controversy as to the generality of this conformation (Du et al. 1997; Sung and Kang 1998). A number of these structures are likely to interact directly with the oncoming ribosome to influence frameshifting, but it is possible that some extraribosomal factors are involved in other cases.

Tantalizingly, distant sequences in the 3' UTR are important for programmed frameshifting in decoding barley yellow dwarf luteovirus (Miller et al. 1997), and phage T7 gene 10 (Condron et al. 1991). Phage T7 RNA polymerase transcribes faster than *E. coli* ribosomes translate and faster than *E. coli* RNA polymerase transcribes, so that distant 3' sequences will be transcribed before ribosomes reach the shift site. The possible importance of coupling of replication and translation in some RNA animal viruses also needs investigation (Lewis and Matsui 1996).

In the few cases tested (Tu et al. 1992; Somogyi et al. 1993), pseudoknots cause pausing of ribosomes that may be necessary but not sufficient for recoding. Insertion of a stem-loop with at least equal predicted stability to a pseudoknot does not lead to frameshifting, even though it still causes pausing, albeit less effectively (Somogyi et al. 1993).

Stimulatory signals 5' of shift sites are also found. In one case, a direct interaction with ribosomal RNA of the translocating ribosome has been characterized. Three bases 5' of the shift site in the *E. coli* release factor 2 coding sequence, there is a Shine-Dalgarno sequence which is important for the +1 frameshifting by ribosomes that initiated 25 codons upstream of the shift site (Fig. 3A) (Weiss et al. 1987; Curran and Yarus 1988). The anti-Shine-Dalgarno sequence near the 3' end of 16S rRNA in translocating ribosomes pairs with its mRNA complement and stimulates frameshifting (Weiss et al. 1988). Similar mRNA:16S rRNA pairing is important for the programmed –1 frameshifting in *E. coli dnaX* decoding, but here the Shine-Dalgarno sequence is 10 bases 5′ of the shift site (Fig. 3B) (Larsen et al. 1994). Spacing of the Shine-Dalgarno sequence influences directionality of the shift at the slippery sequence. Perhaps tension in the short region of 16S rRNA between the anti-Shine-Dalgarno sequence and the part of the 16S rRNA at the decoding site upsets the framing mechanism. Since Shine-Dalgarno interactions between mRNA and rRNA were discovered for initiation before they were found to be utilized by translocating ribosomes, one tends unconsciously to think that they first arose for initiation, but of course we don't know which came first.

Although Shine-Dalgarno interactions are not used for initiation by eukaryotic ribosomes, it is much too soon to write off the possibility that, at least for translocating ribosomes and programmed frameshifting, some type of mRNA–rRNA interaction may be involved. One place to start looking is at the 5′ signal for mammalian antizyme programmed +1 frameshifting (Fig. 4A) (Matsufuji et al. 1995; Ivanov et al. 1998a,c; S. Matsufuji, in prep.). It is also possible that interactions of ribosomal RNA in translating ribosomes with mRNA sequences are not confined just to mRNA sequences 5′ of the shift site. One candidate for such an interaction is the sequence 3′ of the Ty3 shift site (Farabaugh et al. 1993).

Ancient Programmed Frameshifting

Most of the known or suspected cases of programmed frameshifting and codon redefinition, other than selenocysteine, are in viruses or transposable elements. Frameshifting is rampant in the expression of plant virus genes and probably also for bacterial insertion sequences of the IS3 family (Chandler and Fayet 1993; Ohtsubo and Sekine 1996), where approximately 60 cases are suspected (O. Fayet, pers. comm.). It is also found in the expression of quite a number of animal viruses, especially retroviruses, and also their retrotransposon counterparts. Inferring the evolutionary relationships of the recoding involved in these cases is at an early stage. Our comments on this topic, other than selenocysteine discussed above, will be restricted to the programmed frameshifting used in the expression of two nonmobile chromosomal genes. The first example is the autoregulatory frameshifting involved in decoding the bacterial gene for release factor 2 which, as described above, mediates termination at UGA (Fig. 3A).

The early evidence that the release factor 2 programmed frameshifting signals were highly similar among divergent bacteria came from a sequence comparison of the shift signals from *Bacillus subtilis* and *E. coli*. The 12 nucleotides known to be important for the autoregulatory frameshifting are identical (Pel et al. 1992). A recent analysis of the sequences from 20 bacteria, several of them even more distant than *B. subtilis* is from *E. coli*, has led to the inference that this frameshift mechanism was present in the common ancient ancestor of a large group of divergent bacteria but was subsequently lost in three independent lineages (Persson and Atkins 1998).

The second case is the +1 frameshifting in decoding antizyme genes. The protein antizyme governs the intracellular level of polyamines by negatively impacting the intracellular synthesis, and extracellular uptake, of polyamines. It binds to, and inactivates, ornithine decarboxylase, which catalyzes the first step of the synthesis of polyamines and also inhibits the polyamine transporter (Fig. 4B). As discovered by Matsufuji and colleagues (for review, see Gesteland et al. 1992), the programmed frameshifting required for the synthesis of antizyme is in turn regulated by polyamines, thus completing an autoregulatory circuit. Following on from the original identification of a gene in rats (Miyazaki et al. 1992), a gene for antizyme has been detected in other mammals (Tewari et al. 1994; Kankare et al. 1997; Nilsson et al. 1997), in fowl (Drozdowski et al. 1998), in zebra fish (T. Saito et al.; I.P. Ivanov et al., both unpublished), in *Xenopus* (Ichiba et al. 1995), in *Drosophila melanogaster* (Ivanov et al. 1998c), in *Schizosaccharomyces pombe* yeast, and in *C. elegans* (I.P. Ivanov, unpubl.). When a cassette containing the mammalian antizyme shift site and recoding signal is introduced into the budding yeast, *Saccharomyces cerevisiae*, high levels of frameshifting to the +1 frame occur at the shift site. However, the product has an extra amino acid as the ribosomes shift –2 instead of +1 and the utilization of the recoding signals is very different from what it is in mammals (Matsufuji et al. 1996). In contrast, the same mammalian shift cassette directs mammalian-like +1 shifting in the fission yeast, *S. pombe* (Ivanov et al. 1998b).

Recently, a second antizyme gene has been identified in mammals, and its product, antizyme 2, is distinct from the previously known mammalian antizyme 1 (Ivanov et al. 1998a). Two antizymes are now also known in zebra fish (termed Short and Long to avoid implying correspondence with the respective mammalian antizymes 1 and 2) (T. Saito et al., unpubl.). Despite substantial divergence of overall nucleotide sequence, the UGA stop codon of ORF1—the first nucleotide of which is part of the shift site (Rom and Kahana 1994; Matsufuji et al. 1995)—and 16 out of 18 nucleotides immediately 5' of it are identical from *Drosophila* antizyme mRNA to mammalian antizymes 1 and 2 mRNAs (Ivanov et al. 1998a,c). This sequence includes much of the 5' element discussed above, which acts in an unknown manner to stimulate frameshifting (S. Matsufuji, unpubl.). The sequences of the stems of the stimulatory pseudoknot, 3' of the shift site, are highly conserved between mammalian antizymes 1 and 2, but the loop sequences have diverged (Ivanov et al. 1998a). A flanking 3' pseudoknot is not apparent in *Drosophila* by sequence inspection, but there is some sequence conservation with its mammalian counterparts in this region. It seems safe to discount convergent evolution in the case of antizyme and deduce that the shift signals have been used for efficient regulated frameshifting for hundreds of millions of years. As suggested by A.E. Dahlberg (pers. comm.), perhaps polyamines played a crucial role with primordial ribosomal RNA and, subsequently, ribosomal proteins displaced some of these roles. This raises the question of whether the sensing of polyamine levels by modern ribosomes is an evolutionary remnant.

SUBVERSION OF CONTIGUITY

Bypassing

As described above, codon–anticodon dissociation can lead to the anticodon re-pairing to an overlapping triplet resulting in frameshifting. However, the re-pairing can be elsewhere on the mRNA leading to bypassing of mRNA sequences. This was initially discovered with low efficiency (ca. 1%) to nearby sequences in special model systems (Weiss et al. 1987; O'Connor et al. 1989). However, with phage T4 gene 60 decoding, bypassing of 50 bases occurs with an efficiency of 50% from a so-called "take-off" codon to a "landing site" (Fig. 5A) (Huang et al. 1988; Weiss et al. 1990a; Maldonado and Herr 1998). The mechanism of this bypass involves 70S ribosome complexes, with peptidyl tRNA scanning the gap region to find the landing site (F. Adamski et al., unpubl.). Part of the nascent peptide, still within the ribosome, is important for this bypassing (Weiss et al. 1990a). The nascent peptide is cross-linkable to 50S subunit components (Choi and Brimacombe 1998) and appears flexible, perhaps partly folded, in an exit tunnel in that subunit. However, it is also cross-linkable to the 30S subunit, close to the decoding site. At least some of its role in bypassing may be mediated by direct contacts with the decoding area of the 30S subunit or with the tRNA–mRNA complex (Choi et al. 1998). In addition to the nascent peptide, a short stem-loop within the coding gap is important for bypassing. However, without these two special features exhibited by gene 60, efficient bypassing can occur over shorter distances if the codon following the take-off site is a rare codon and its cognate aminoacylated tRNA is limiting (J. Gallant and D. Lindsley, pers. comm.). The above-described translational bypassing is quite distinct from the shunting of 40S ribosomal subunits to another site within the 5 untranslated regions of cauliflower mosaic virus and adenovirus mRNAs. Here the intervening sequence is not traversed; rather, specific structures appear to pass the ribosomal subunit from one site to the other (Fütterer et al. 1993; Yueh and Schneider 1996; Hemmings-Mieszczak et al. 1997).

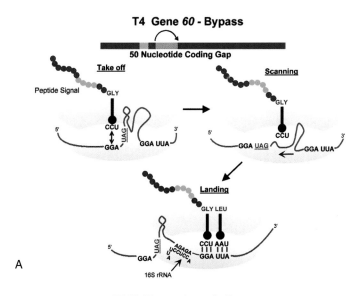

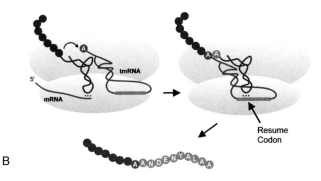

Figure 5 (*A*) The recoding signal for the 50% efficient translational bypass of 50 nucleotides in decoding phage T4 gene *60*. (*B*) Rescue of ribosomes stalled at the end of bacterial mRNAs lacking a terminator and degradation of the aberrant protein product, utilizes tmRNA which functions both as a tRNA and mRNA.

Clearly, if bypassing without special signals were rampant, decoding would be chaotic. Nonetheless, the current translational mechanism is capable of carrying out noncontiguous decoding of a message.

AN INTRAMOLECULAR RIBOSOME?

The primary (modern) function of mRNA is to be a linear tape, feeding through the ribosome readout machine. This linear property is, of course, a direct reflection of the information style of its DNA origins. However, as we have seen, mRNA sequences have additional roles in mRNA decoding. Do these observations provide an insight into primitive decoding early in evolution?

It is difficult to imagine how the early decoding apparatus could assemble amino acid chains according to a nucleic acid code. A major difficulty must have been maintaining high enough local concentrations of the reactants to drive reactions. However, the number of diffusible reactants might have been minimized by combined functions within a multifunctional mRNA molecule. Any such scheme would require evolution of complex RNA molecules, which seems counterintuitive. The trade-off between dealing with many reacting molecules and constructing a complex molecule may have been tilted toward the latter by evolution of RNA ligation activity.

The modern translation apparatus assembles the amino acid chain by sequential passage of aminoacyl tRNAs through A and P sites on the ribosome, as dictated by codon sequence in the mRNA. The success of in vitro protein synthesis experiments makes us think of the decoding apparatus operating in a soluble soup with a diffusible flow of substrate tRNAs into the ribosomal A site, where incorrect molecules are rejected until the correct one is identified. However, there is ample evidence that substrates and factors are not just free-floating, but rather are harbored in a ribosome megacomplex (Stapulionis et al. 1997; Kruse et al. 1998). Within this complex, aminoacyl-tRNAs are tested at the A site, but the volume available for free diffusion must be greatly reduced, aiding the reaction rate. The discharged tRNAs are likely recycled within the complex by resident acylating enzyme. The implication is that small molecules—ATP and amino acids—flow into the complex and the newly synthesized polypeptide chain emerges.

Could the megacomplex of current ribosomes be suggestive of an earlier strategy to deal with the substrate concentration problem in early evolution of a translation system? The ultimate megacomplex would have many functions and substrates in one molecule to maximize the number of unimolecular reactions. The multiple capabilities of RNA are intriguing for thinking about a primitive, unimolecular decoding complex in the absence of proteins.

First, tmRNA combines mRNA and tRNA functions in one molecule (Fig. 5B) (Tu et al. 1995; Keiler et al. 1996; Himeno et al. 1997), hence its name "tmRNA" (Jentsch 1996; Atkins and Gesteland 1996). This 363-nucleotide RNA has 3′ and 5′ ends that come together to form a partial tRNA-like structure that can be charged with alanine (Komine et al. 1994; Ushida et al. 1994; Felden et al. 1998). It can access a ribosomal A site that has no tRNA and no mRNA codon, such as at the end of an mRNA with no stop codon. The alanine is donated to the growing peptide chain in the P site just as if tmRNA was an ordinary tRNA. Then, remarkably, the ribosome reads out 10 codons from an internal part of tmRNA, adding an 11-amino-acid tag to the growing chain, targeting it for degradation (Keiler et al. 1996). tmRNA has a very complex shape with at least 4 pseudoknots (Williams and Bartel 1996; Felden et al. 1997) that must provide the conditions necessary for the tRNA ends to be in the P site and the internal coding sequence to be in the mRNA track for the A site. Although this interesting mechanism employs modern-day ribosomes, it does encourage thoughts about one molecule having multiple functions involved in decoding.

Second, peptide-bond catalysis by RNA is clearly possible (Zhang and Cech 1997), and it seems likely that the peptidyl transfer function of modern ribosomes is affected by ribosomal RNA (Nitta et al. 1998; see Chapter 8). We could imagine that at early times the catalytic center for peptide-bond formation and the informational sequence (mRNA) could be in one RNA molecule; each mRNA would need to have its own, resident catalytic center.

Third, we know that folded RNA structures within an mRNA can participate in bringing an appropriate substrate aminoacyl tRNA into the ribosome for recognition of its codon, as exemplified by selenocysteine insertion, in *E. coli*. In this case, a downstream stem-loop structure (within the coding sequence) tethers the tRNA^{Sec} (via a special EF-Tu protein) in order to deliver the tRNA to the waiting UGA codon in the ribosomal A site. Could we imagine that the primitive mRNA catalytic center molecule suggested above might also have an amino acid delivery system?

There are intriguing suggestions that the amino acid acceptor branch of tRNA (acceptor arm plus TψC arm) originated independently from the anticodon branch (anticodon arm plus DHU arm) (see Chapters 3 and 8) and a reason for the origin of the acceptor branch independent of its role in protein synthesis has been proposed (see Chapter 3). Perhaps the anticodon branch of tRNAs originated in folded structures in internal regions of primordial mRNAs that folded back and paired with "codons" in the same mRNA. One proto-anticodon branch would need to be capable of forming for each of a limited number of amino acids. How could pairing of the proto-anticodon trigger delivery of an amino acid on a proto-acceptor branch that is not contiguous with the proto-anticodon branch? Two alternatives can be considered. One is that a stereochemical, folded RNA pocket containing an amino acid is delivered by a structurally contiguous proto-anticodon branch due to an association between the two. If so, the pocket could hold the now-positioned amino acid at the catalytic site until the next amino acid is delivered. The amino acid-specific fold could then bind another amino acid, ready for delivery when its codon was required again. Thus, a series of "fingers" with bound amino acids could play back on coding sequences within one molecule to decode a part of the RNA sequence. In this scenario, activation of the amino acid takes place at the catalytic center. Alternatively, if a 3' end is involved in primordial aminoacylation, it could act repetitively to deliver amino acids to the catalytic site. In this scenario, the amino acid could not be held at the catalytic site by the delivery system and might be held by the catalytic site until the 3' end delivered the subsequent amino acid. Perhaps pairing of a particular anticodon branch with its codon influences, by way of tertiary interactions, the identity of the amino acid aminoacylated to the 3' end. (For a discussion of self-aminoacylation, see Chapter 7.) In either case, as each amino acid reaches the catalytic center, a peptide bond needs to form with the growing chain. If amino acids are delivered by folded internal pockets, then subsequent evolution of 3' end aminoacylation is a big step. However, synthesis of proteins by the internal delivery system would provide a different milieu for the subsequent but parallel development of the 3' end delivery system.

By this imaginary scheme a single RNA molecule is mRNA, peptide bond catalyst, and "tRNA" that acts as an amino acid collection and delivery system. Many interactions would be intramolecular; the diffusion-limited reactions would be the amino acids finding their binding pockets. Another possibility, maybe a step further in evolution, might be a two-component system, with the mRNA molecule separate from a "primitive ribosome" that had the peptide bond catalyst, the tRNAs, and the delivery system in one molecule. This "ribosome" could then act on a variety of mRNAs, much like the modern ribosome megacomplex. These scenarios are admittedly farfetched and do not deal with a number of crucial issues. However, they may be illustrative of ways to think about multifunctional RNA "mega" molecules.

PERSPECTIVE

With a more complete understanding of decoding and recoding, it may be possible to consider engineering organisms with an expanded repertoire of coding capacities to include nontraditional amino acids. The challenges are clearly formidable, but there is already an impressive start by specific manipulation of an aminoacyl tRNA synthetase (Liu et al. 1997).

Early decoding likely yielded many products from a single coding sequence, because of randomness in the mechanism. Presumably, the differing specificities of these products gave some molecules with activities that provided a survival advantage. As sophistication of the decoding system evolved, the repertoire of products from a single coding sequence must have become more limited, eventually reaching the current coding rules. Although these rules, in general, result in one protein product per mRNA, recoding examples tell us that there has been coevolution of specific mechanisms to produce more than one product. The big unknown is how many coding sequences have themselves evolved to take advantage of the diversity of expression offered by recoding. Might there be many messages where 5% of the ribosomes bypass the terminator, reading codons in what is normally considered to be the 3'UTR? How commonly does decoding of mRNAs involve frameshifting akin to mammalian antizyme? Could it be that, in some mRNAs, ribosomes bypass codons by scanning from one codon to the next cognate one, as seen in T4 gene 60? Each of these acts would result in a different protein with perhaps a new or additional function. Although current methods for protein analysis are very powerful, rarely would protein variants at the level of 5% be discovered, except by a fortuitous observation or by serious digging. This also holds true for posttranslational modifications. Again, we know a great deal about some proteins, but we have little appreciation for the overall picture. A serious attack on the "proteome" is needed to begin to understand the full diversity of the products of genes.

ACKNOWLEDGMENTS

We thank Marla Berry for characteristically helpful comments and sharing unpublished data. J.F.A. is supported by National Institutes of Health grant RO1-GM48152 and Japan Health Science Foundation grant K-1005; A.B. by Deutsche Forschungsgemeinschaft and Fonds der Chemischen Industrie; S.M. by Grants-in-Aid from the Ministry of Education, Science and Culture in Japan, and the Human Frontier Sciences Program; and R.F.G. is an investigator of the Howard Hughes Medical Institute.

REFERENCES

Adamski F.M., Atkins J.F., and Gesteland R.F. 1996. Ribosomal protein L9 interactions with 23 S rRNA: The use of a translational bypass assay to study the effect of amino acid substitutions. *J. Mol. Biol.* **261:** 357–371.

Adhin, M.R. and van Duin, J. 1990. Scanning model for translational reinitiation in eubacteria. *J. Mol. Biol.* **213:** 811–818.

Atkins J.F. and Gesteland R.F. 1995. Discontinuous triplet decoding with or without repairing by peptidyl tRNA. In *tRNA: Structure, biosynthesis and function* (ed. D. Söll and U.L. RajBhandary) pp. 471–490. ASM Press, Washington, D.C.

———. 1996. A case for *trans* translation. *Nature* **379:** 769–770.

Atkins J.F., Elseviers D., and Gorini L. 1972. Low activity of β-galactosidase in frameshift mutants of *Escherichia coli*. *Proc. Natl. Acad. Sci.* **69:** 1192–1195.

Atkins J.F., Weiss R.B., and Gesteland R.F. 1990. Ribosome gymnastics—Degree of difficulty 9.5, style 10.0. *Cell* **62:** 413–423.

Atkins J.F., Gesteland R.F., Reid B.R., and Anderson C.W. 1979. Normal tRNAs promote ribosomal frameshifting. *Cell* **18:** 1119–1131.

Axley M.J, Böck A., and Stadtman T.C. 1991. Catalytic properties of an *Escherichia coli* formate dehydrogenase mutant in which sulfur replaces selenium. *Proc. Natl. Acad. Sci.* **88:** 8450–8454.

Baron C. and Böck A. 1991. The length of the aminoacyl acceptor stem of the selenocysteine-specific tRNA^{Sec} of *Escherichia coli* is the determinant for binding to elongation factor SELB or Tu. *J. Biol. Chem.* **266:** 20375–20379.

———. 1995. The selenocysteine-inserting tRNA species: Structure and function. In *tRNA: Structure, biosynthesis and function* (ed. D. Söll and U.L. RajBhandary), pp. 529–544. ASM Press, Washington, D.C.

Baron C., Heider J., and Böck A. 1989. Mutagenesis of selC, the gene for the selenocysteine-inserting tRNA species in *E. coli*: Effects on *in vivo* function. *Nucleic Acids Res.* **18:** 6761–6766.

Baron C., Westhof E., Böck A., and Giege R. 1993. Solution structure of selenocysteine-inserting tRNA^{Sec} from *Escherichia coli*. Comparison with canonical tRNA^{Sec}. *J. Mol. Biol.* **231:** 274–292.

Behne D., Kyriakopoeulos A., Weiss-Nowak C., Kalcklösch M., Westphal C., and Gessner H. 1996. Newly found selenium-containing proteins in the tissues of the rat. *Biol. Trace Elem. Res.* **55:** 99–110.

Berry M.J., Banu L., Harney J.W., and Larsen P.R. 1993. Functional characterization of the eukaryotic SECIS elements which direct selenocysteine insertion at UGA codons. *EMBO J.* **12:** 3315–3322.

Berry M.J., Harney J.W., Ohama T., and Hatfield D.L. 1994. Selenocysteine insertion or termination: Factors affecting UGA codon fate and complementary anticodon: codon mutations. *Nucleic Acids Res.* **22:** 3753–3759.

Berry M.J., Mai A.L., Kieffer J.D., Harney J.W., and Larsen P.R. 1992. Substitution of cysteine for selenocysteine in type I iodothyronine deiodinase reduces the catalytic efficiency of the protein but enhances its translation. *Endocrinology* **131:** 1848–1852.

Berry M.J., Banu L., Chen Y., Mandel S.J., Kieffer J.D., Harney J.W., and Larsen P.R. 1991. Recognition of UGA as a selenocysteine codon in type I deiodinase requires sequences in the 3' untranslated region. *Nature* **353:** 273–276.

Bidou L., Stahl G., Grima B., Liu H., Cassan M., and Rousset J.-P. 1997. *In vivo* HIV-1 frameshifting efficiency is directly related to the stability of the stem-loop stimulatory signal. *RNA* **3:** 1153–1158.

Blinkowa A.L. and Walker J.R. 1990. Programmed ribosomal frameshifting generates the *Escherichia coli* DNA polymerase III γ subunit from within the τ subunit reading frame. *Nucleic Acids Res.* **18:** 1725–1729.

Brierley I., Digard P., and Inglis S.C. 1989. Characterization of an efficient Coronavirus frameshifting signal: Requirement for an RNA pseudoknot. *Cell* **57:** 537–547.

Brierley I., Jenner A.J., and Inglis S.C. 1992. Mutational analysis of the "slippery-sequence" component of a Coronavirus ribosomal frameshifting signal. *J. Mol. Biol.* **227:** 463–479.

Brierley I., Meredith M.R., Bloys A.J., and Hagervall T.G. 1997. Expression of a Coronavirus ribosomal frameshift signal in *Escherichia coli*: Influence of tRNA anticodon modification on frameshifting. *J. Mol. Biol.* **271:** 1–14.

Brierley I., Rolley N.J., Jenner A.J., and Inglis S.J. 1991. Mutational analysis of the RNA pseudoknot component of a Coronavirus frameshifting signal. *J. Mol. Biol.* **220:** 889–902.

Brown C.M., Dinesh-Kumar S.P., and Miller A. 1996. Local and distant sequences are required for efficient read-through of the barley yellow virus-PAV coat protein gene stop codon. *J. Virol.* **70:** 5884–5892.

Bult C.J., White O., Olsen G.J., Zhou L., Fleischmann R.D., Sutton G.G., Blake J.A., FitzGerald L.M., Clayton R.A., Gocayne J.D et al. 1996. Complete genome sequence of the methanogenic archaeon, *Methanococcus jannaschii*. *Science* **273:** 1058–1072.

Chambers I., Frampton J., Goldfarb P, Affara N., McBain W., and Harrison P.R. 1986. The structure of the mouse glutathione peroxidase gene: The selenocysteine in the active site is encoded by the "termination" codon TGA. *EMBO J.* **5:** 1221–1227.

Chandler M. and Fayet O. 1993. Translational frameshifting in the control of transposition in bacteria. *Mol. Microbiol.* **7:** 497–503.

Chen X., Kang H., Shen L.X., Chamorro M., Varmus H.E., and Tinoco I, Jr. 1996. A characteristic bent conformation of RNA pseudoknots promotes –1 frameshifting during translation of retroviral RNA. *J. Mol. Biol.* **260:** 479–483.

Choi K.M. and Brimacombe R. 1998. The path of the growing peptide chain through the 23S rRNA in the 50S ribosomal subunit; a comparative cross-linking study with three different peptide families. *Nucleic Acids Res.* **26:** 887–895.

Choi K.M., Atkins J.M., Gesteland R.F., and Brimacombe R. 1998. Flexibility of the nascent polypeptide chain within the ribosome. Contacts from the peptide N-terminus to a specific region of the 30S subunit. *Eur. J. Biochem.* **255:** 409–413.

Condron B.G., Gesteland R.F., and Atkins J.F. 1991. An analysis of sequences stimulating frameshifting in the decoding of gene 10 of bacteriophage T7. *Nucleic Acids Res.* **19:** 5607–5612.

Cone J.E., del Rio M., Davis J.N., and Stadtman T.C. 1976. Chemical characterization of the selenoprotein component of clostridial glycine reductase: Identification of selenocysteine as the organoselenium moiety. *Proc. Natl. Acad. Sci.* **73:** 2659–2663.

Craigen W.J. and Caskey C.T. 1986. Expression of peptide chain release factor 2 requires high-efficiency frameshift. *Nature* **322:** 273–275.

Craigen W.J., Cook R.G., Tate W.P., and Caskey C.T. 1985. Bacterial peptide chain release factors: Conserved primary structure and possible frameshift regulation of release factor 2. *Proc. Natl. Acad. Sci.* **82:** 3616–3620.

Crick F.H.C., Brenner S., Klug A., and Pieczenik G. 1976. A speculation on the origin of protein synthesis. *Origins Life* **7:** 389–397.

Curran J.F. 1993. Analysis of effects of tRNA: Message stability on frameshift frequency at the *Escherichia coli* RF2 programmed frameshift site. *Nucleic Acids Res.* **21:** 1837–1843.

Curran J.F. and Yarus M. 1988. Use of tRNA suppressors to probe regulation of *Escherichia coli* release factor 2. *J. Mol. Biol.* **203:** 75–83.

Danilvskaya O., Slot F., Pavlova M., and Pardue M.-L. 1994. Structure of the *Drosophila* HeT-A transposon: A retrotransposon-like element forming telomeres. *Chromosoma* **103:** 215–224.

Danilevskaya O.N., Tan C., Wong J., Alibhai M., and Pardue M.-L. 1998. Unusual features of the *Drosophila melanogaster* telomere transposable element HeT-A are conserved in *Drosophila yakuba* telomere elements. *Proc. Natl. Acad. Sci.* **95:** 3770–3775.

Di Guilio M. 1997. The origin of the genetic code. *Trends Biochem. Sci.* **22:** 49.

Dinman J.D. and Wickner R.B. 1992. Ribosomal frameshifting efficiency and *gag/gag-pol* ratio are critical for yeast M_1 double-stranded RNA virus propagation. *J. Virol.* **66:** 3669–3676.

Drozdowski B., Gong T.W.-L., and Lomax M.I. 1998. The chicken cDNA for ornithine decarboxylase antizyme. *Biochim. Biophys. Acta* **1396:** 21–26.

Du Z., Holland J.A., Hansen M.R., Giedroc D.P., and Hoffman D.W. 1997. Base-pairings within the RNA pseudoknot associated with the simian retrovirus-1 *gag-pro* frameshift site. *J. Mol. Biol.* **270:** 464–470.

Ehrenreich A., Forchhammer K., Tormay P., Vepreck B., and Böck A. 1992. Selenoprotein synthesis in *E. coli*. Purification and characterization of the enzyme catalyzing selenium activation. *Eur. J. Biochem.* **206:** 767–773.

Farabaugh P.J. 1996. Programmed translational frameshifting. *Microbiol. Rev.* **60:** 103–134.

Farabaugh P.J., Zhao H., and Vimaladithan A. 1993. A novel programmed frameshift expresses the *POL3* gene of retrotransposon Ty3 of yeast: Frameshifting without tRNA slippage. *Cell* **74:** 93–103.

Felden B., Himeno H., Muto A., McCutcheon J.P., Atkins J.F., and Gesteland R.F. 1997. Probing the structure of the *Escherichia coli* 10Sa (tmRNA). *RNA* **3:** 89–104.

Felden B., Hanawa K., Atkins J.F., Himeno H., Muto A., Gesteland R.F., McCloskey J.A., and Crain P.F. 1998. Presence and location of modified nucleotides in *E. coli* tmRNA: Structural mimicry with tRNA acceptor branches. *EMBO J.* **17:** 3188–3196.

Felsenstein K.M. and Goff S.P. 1992. Mutational analysis of the *gag-pol* junction of Moloney murine leukemia virus: Requirements for expression of the *gag-pol* fusion protein. *J. Virol.* **66:** 6601–6608.

Feng Y., Yuan H., Rein A., and Levin J.G. 1992. Bipartite signal for read-through suppression in murine leukemia virus mRNA: An eight nucleotide purine-rich sequence immediately downstream of the *gag* termination codon followed by an RNA pseudoknot. *J. Virol.* **66:** 5127–5132.

Flower A.M. and McHenry C.S. 1990. The γ subunit of DNA polymerase III holoenzyme of *Escherichia coli* is produced by ribosomal frameshifting. *Proc. Natl. Acad. Sci.* **87:** 3713–3717.

Forchhammer K. and Böck A. 1991. Selenocysteine synthase from *Escherichia coli*. Analysis of the reaction sequence. *J. Biol. Chem.* **266:** 6324–6328.

Forchhammer K., Leinfelder W., and Böck A. 1989. Identification of a novel translation factor necessary for the incorporation of selenocysteine into proteins. *Nature* **342**: 453–456.

Fütterer J., Kiss-László Z., and Hohn T. 1993. Nonlinear ribosome migration on cauliflower mosaic virus 35S RNA. *Cell* **73**: 789–802.

Gallant J. and Foley D. 1980. On the causes and prevention of mistranslation. In *Ribosomes: Structure, function and genetics* (ed. G. Chambliss et al.), pp. 615–638. University Park Press, Baltimore.

Gallant J. and Lindsley D. 1993. Ribosome frameshifting at hungry codons: Sequence rules, directional specificity and possible relationship to mobile element behaviour. *Biochem. Soc. Trans.* **21**: 817–821.

Gesteland R.F. and Atkins J.F. 1996. Recoding: Dynamic reprogramming of translation. *Annu. Rev. Biochem.* **65**: 741–768.

Gesteland R.F., Weiss R.B., and Atkins J.F. 1992. Recoding: Reprogrammed genetic decoding. *Science* **257**: 1640–1641.

Gramstat A. Prüfer D., and Rohde W. 1994. The nucleic acid-binding zinc finger protein of potato virus M is translated by internal initiation as well as by ribosomal frameshifting involving a shifty stop codon and a novel mechanism of P-site slippage. *Nucleic Acids Res.* **22**: 3911–3917.

Grosjean H. and Chantrenne H. 1980. On codon-anticodon interactions. *Mol. Biol. Biochem. Biophys.* **32**: 347–367.

Gu Q.-P., Beilstein M.A., Vendeland S.C., Lugade A., Ream W., and Whanger P.D. 1997. Conserved features of selenocysteine insertion sequence (SECIS) elements in selenoprotein W cDNAs from five species. *Gene* **193**: 187–196.

Hatfield D., Feng Y.X., Lee, B.J., Rein A., Levin J.G., and Oroszlan S. 1989. Chromatographic analysis of the aminoacyl tRNAs which are required for translation of codons at and around the ribosomal frameshift sites of HIV, HTLV-1, and BLV. *Virology* **173**: 736–742.

Heider J., Baron C., and Böck A. 1992. Coding from a distance: Dissection of the mRNA determinants required for the incorporation of selenocysteine into proteins. *EMBO J.* **11**: 3759–3766.

Hemmings-Mieszczak M., Steger G., and Hohn T. 1997. Alternative structures of the cauliflower mosaic virus 35S RNA leader: Implications for viral expression and replication. *J. Mol. Biol.* **267**: 1075–1088.

Herbst K.L., Nichols L.M., Gesteland R.F., and Weiss R.B. 1994. A mutation in ribosomal protein L9 affects ribosomal hopping during translation of gene 60 from bacteriophage T4. *Proc. Natl. Acad. Sci.* **91**: 12525–12529.

Heurgué-Hamard V., Mora L., Guarneros G., and Buckingham R.H. 1996. The growth defect in *Escherichia coli* deficient in peptidyl-tRNA hydrolyase is due to starvation for Lys-tRNA^{Lys}. *EMBO J.* **15**: 2826–2833.

Hilgenfeld R., Wilting R., and Böck A. 1996. Structural model for the selenocysteine-specific elongation factor SelB. *Biochimie* **78**: 971–978.

Hill K.E., Lloyd R.S., and Burk R.F. 1993. Conserved nucleotide sequences in the open reading frame and 3′ untranslated region of selenoprotein P mRNA. *Proc. Natl. Acad. Sci.* **90**: 537–541.

Himeno S., Chittum H.S., and Burk R.F. 1996. Isoforms of selenoprotein P in rat plasma. *J. Biol. Chem.* **271**: 15769–15775.

Himeno H., Sato M., Tadaki T., Fukushima M., Ushida C., and Muto A. 1997. *In vitro trans* translation mediated by alanine-charged 10Sa RNA. *J. Mol. Biol.* **268:** 803–808.

Himmelreich R., Plagens H., Hilbert H., Reiner B., and Herrmann R. 1997. Comparative analysis of the genomes of the bacteria *Mycoplasma pneumoniae* and *Mycoplasma genitalium*. *Nucleic Acids Res.* **25:** 701–712.

Hofstetter H., Monstein H.-J., and Weissman C. 1974. The readthrough protein A_1 is essential for the formation of viable Qβ particles. *Biochim. Biophys. Acta* **374:** 238–251.

Horsfield J.A., Wilson D.N., Mannering S.A., Adamski F.M., and Tate W.P. 1995. Prokaryotic ribosomes recode the HIV-1 *gag-pol*–1 frameshift sequence by an E/P site post-translocation simultaneous slippage mechanism. *Nucleic Acids Res.* **23:** 1487–1494.

Huang W.M., Ao S.-Z., Casjens S., Orlandi R., Zeikus R., Weiss R., Winge D., and Fang M. 1988. A persistent untranslated sequence within bacteriophage T4 DNA topoisomerase gene 60. *Science* **239:** 1005–1012.

Hüttenhofer A. and Böck A. 1998a. RNA structures involved in selenoprotein synthesis. In *RNA structure and function* (ed. R. Simons and M. Grunberg-Manago), pp. 603–639. Cold Spring Harbor Laboratory Press, Cold Spring Harbor, New York.

———. 1998b. Selenocysteine-inserting RNA elements mediate GTP hydrolysis by elongation factor SelB. *Biochemistry* **37:** 885–890.

Hüttenhofer A., Heider J., and Böck A. 1996. Interaction of the *Escherichia coli* fdhF mRNA hairpin promoting selenocysteine incorporation with the ribosome. *Nucleic Acids Res.* **20:** 3903–3910.

Ibba M., Curnow A.W., and Söll D. 1997. Aminoacyl-tRNA synthesis: Divergent routes to a common goal. *Trends Biochem. Sci.* **22:** 39–42.

Ichiba T., Matsufuji S., Miyazaki Y., and Hayashi S. 1995. Nucleotide sequence of ornithine decarboxylase antizyme cDNA from *Xenopus laevis*. *Biochim. Biophys. Acta* **1262:** 83–86.

Ivanov I.P., Gesteland R.F., and Atkins J.F. 1998a. A second mammalian antizyme: Conservation of programmed ribosomal frameshifting. *Genomics* (in press).

Ivanov I.P., Gesteland R.F., Matsufuji S., and Atkins J.F. 1998b. Programmed frameshifting in the synthesis of mammalian antizyme is +1 in mammals, predominantly +1 in fission yeast but –2 in budding yeast. *RNA* (in press).

Ivanov I.P., Simin K., Letsou, A., Atkins J.F., and Gesteland R.F. 1998c. The *Drosophila* gene for antizyme requires ribosomal frameshifting for expression and contains an intronic gene for snRNP Sm D3 on the opposite strand. *Mol. Cell. Biol.* **18:** 1553–1561.

Jacks T., Madhani T.H.D., Masiarz F.R., and Varmus H.E. 1988. Signals for ribosomal frameshifting in the Rous sarcoma virus *gag-pol* region. *Cell* **55:** 447–458.

Jentsch S. 1996. When proteins receive deadly messages at birth. *Science* **271:** 955–956.

Jukes T.H. 1990. Genetic code 1990. Outlook. *Experientia* **46:** 1149–1157.

Kankare K., Uusi-Oukari M., and Jänne O.A. 1997. Structure, organization and expression of the mouse ornithine decarboxylase antizyme gene. *Biochem. J.* **324:** 807–813.

Kawakami K., Pande S., Faiola B., Moore D.P., Boeke J.D., Farabaugh P.J., Strathern J.N., Nakamura Y., and Garfinkel D.J. 1993. A rare tRNA-Arg(CCU) that regulates Ty1 element ribosomal frameshifting is essential for Ty1 retrotransposition in *Saccharomyces cerevisiae*. *Genetics* **135:** 309–320.

Keiler K.C., Waller P.R.H., and Sauer R.T. 1996. Role of a peptide tagging system in degradation of proteins synthesized from damaged messenger RNA. *Science* **271:** 990–993.

Kelman Z. and O'Donnell M. 1995. DNA polymerase III: Structure and function of a chromosomal replicating machine. *Annu. Rev. Biochem.* **64:** 171–200.

Kollmus H., Flohe L., and McCarthy J.E.G. 1996. Analysis of eukaryotic mRNA structure directing cotranslational incorporation of selenocysteine. *Nucleic Acids Res.* **24:** 1195–1201.

Komine Y., Kitabatake M., Yokogawa T., Nishikawa K., and Inokuchi H. 1994. A tRNA-like structure is present in 10Sa RNA, a small stable RNA from *Escherichia coli*. *Proc. Natl. Acad. Sci.* **91:** 9223–9227.

Kromayer M., Wilting R., Tormay P., and Böck A. 1996. Domain structure of the prokaryotic selenocysteine-specific elongation factor SELB. *J. Mol. Biol.* **262:** 413–420.

Kruse C., Grünweller A., Willkomm D.K., Pfeiffer T., Hartmann R.K., and Müller P.K. 1998. tRNA is entrapped in similar, but distinct, nuclear and cytoplasmic ribonucleoprotein complexes, both of which contain vigilin and elongation factor 1α. *Biochem. J.* **329:** 615–621.

Kurland C.G. 1992. Translational accuracy and the fitness of bacteria. *Annu. Rev. Genet.* **26:** 29–50.

Larsen B., Gesteland R.F., and Atkins J.F. 1997. Structural probing and mutagenic analysis of the stem-loop required for *E. coli dnaX* ribosomal frameshifting: Programmed efficiency of 50%. *J. Mol. Biol.* **271:** 47–60.

Larsen B., Wills N.M., Gesteland R.F., and Atkins J.F. 1994. rRNA-mRNA base pairing stimulates a programmed –1 ribosomal frameshift. *J. Bacteriol.* **176:** 6842–6851.

Leinfelder W., Zehelein E., Mandrand-Berthelot M.A., and Böck A. 1988. Gene for a novel tRNA species that accepts L-serine and cotranslationally inserts selenocysteine. *Nature* **331:** 723–725.

Leinfelder W., Forchhammer K., Veprek B., Zehelein E., and Böck A. 1990. In vitro synthesis of selenocysteyl-tRNA$_{UCA}$ from seryl-tRNA$_{UCA}$: Involvement and characterization of the selD gene product. *Proc. Natl. Acad. Sci.* **87:** 543–547.

Leng P., Klatte D.H., Schumann G., Boeke J.D., and Steck T.L. 1998. Skipper, an LTR retrotransposon of *Dictyostelium*. *Nucleic Acids Res.* **26:** 2008–2015.

Levin M.E., Hendrix R.W., and Casjens S.R. 1993. A programmed translational frameshift is required for the synthesis of a bacteriophage λ tail assembly protein. *J. Mol. Biol.* **234:** 124–139.

Lewis T.L. and Matsui S.M. 1996. Astrovirus ribosomal frameshifting in an infection-transfection transient expression system. *J. Virol.* **70:** 2869–2875.

Liu D.R., Magliery T.J., Pastrnak M., and Schultz P.G. 1997. Engineering a tRNA and aminoacyl-tRNA synthetase for the site-specific incorporation of unnatural amino acids into proteins *in vivo*. *Proc. Natl. Acad. Sci.* **94:** 10092–10097.

Lodmell J.S. and Dahlberg A.E. 1997. A conformational switch in *Escherichia coli* 16S ribosomal RNA during decoding of messenger RNA. *Science* **277:** 1262–1267.

Low S.C. and Berry M.J. 1996. Knowing when not to stop: Selenocysteine incorporation in eukaryotes. *Trends Biochem. Sci.* **21:** 203–208.

Maldonado R. and Herr A.J. 1998. The efficiency of T4 gene 60 translational bypassing. *J. Bacteriol.* **180:** 1822–1830.

Martin G.W., Harney J.W., and Berry M.J. 1996. Selenocysteine incorporation in eukaryotes: Insights into mechanism and efficiency from sequence, structure, and spacing proximity studies of the type I deiodinase SECIS element. *RNA* **2:** 171–182.

Matsufuji S., Matsufuji T., Wills N.M., Gesteland R.F., and Atkins J.F. 1996. Reading two bases twice: Mammalian antizyme frameshifting in yeast. *EMBO J.* **15:** 1360–1370.

Matsufuji S., Matsufuji T., Miyazaki Y., Atkins J.F., Gesteland R.F., and Hayashi S. 1995. Autoregulatory frameshifting in decoding mammalian ornithine decarboxylase antizyme. *Cell* **80:** 51–60.

Menninger J.R. 1977. Ribosome editing and the error catastrophe hypothesis of cellular aging. *Mech. Ageing Dev.* **6:** 131–142.

Miller W.A., Brown C.M., and Wang S. 1997. New punctuation for the genetic code: Luteovirus gene expression. *Semin. Virol.* **8:** 3–13.

Miyazaki Y., Matsufuji S., and Hayashi S. 1992. Cloning and characterization of a rat gene encoding ornithine decarboxylase antizyme. *Gene* **113:** 191–197.

Morris D.K. and Lundblad V. 1997. Programmed translational frameshifting in a gene required for yeast telomere replication. *Curr. Biol.* **7:** 969–976.

Mottagui-Tabar S., and Isaksson L.A. 1997. Only the last amino acids in the nascent peptide influence translation termination in *Escherichia coli* genes. *FEBS Lett.* **414:** 165–170.

Müller S., Heider J., and Böck A. 1997. The path of unspecific incorporation of selenium in *Escherichia coli*. *Arch. Microbiol.* **168:** 421–427.

Müller S., Senn H., Gsell B., Vetter W., Baron C., and Böck A. 1994. The formation of diselenide bridges in proteins by incorporation of selenocysteine residues: Biosynthesis and characterization of (Se2)-thioredoxin. *Biochemistry* **33:** 3404–3412.

Neuhierl B. and Böck A. 1996. On the mechanism of selenium tolerance in selenium-accumulating plants. Purification and properties of a specific selenocysteine methytransferase from cultured cells of *Astragalus bisulcatus*. *Eur. J. Biochem.* **239:** 235–238.

Nilsson J., Koskiniemi S., Persson K., Grahn B., and Holm I. 1997. Polyamines regulate both transcription and translation of the gene encoding ornithine decarboxylase antizyme in mouse. *Eur. J. Biochem.* **250:** 223–231.

Nitta I., Ueda T., and Watanabe K. 1998. Possible involvement of *Escherichia coli* 23S ribosomal RNA in peptide bond formation. *RNA* **4:** 257–267.

Noller H.F., Asire M., Barta A., Douthwaite S., Goldstein T., Gutell R.R., Moazed D., Normanly J., Prince J.B., Stern S., Triman K., Turner S., van Stolk B., Wheaton V., Weiser B., and Woese C.R. 1986. Studies on the structure and function of ribosomal RNA. In *Structure, function and genetics of ribosomes* (ed. B. Hardesty and G. Kramer), pp. 143–163. Springer-Verlag, New York.

O'Connor M., Gesteland R.F., and Atkins J.F. 1989. tRNA hopping: Enhancement by an expanded anticodon. *EMBO J.* **8:** 4315–4323.

Ohtsubo E. and Sekine Y. 1996. Bacterial insertion sequences. *Curr. Top. Microbiol. Immunol.* **204:** 1–26.

Osawa S. 1995. *Evolution of the genetic code*. Oxford University Press, New York.

Parkin N.T., Chamorro M., and Varmus H.E. 1992. Human immunodeficiency virus type 1 *gag-pol* frameshifting is dependent on downstream mRNA secondary structure: Demonstration by expression *in vivo*. *J. Virol.* **66:** 5147–5151.

Pel H.J., Rep M., and Grivell L.A. 1992. Sequence comparison of new prokaryotic and mitochondrial members of the polypeptide chain release factor family predicts a five-domain model for release factor structure. *Nucleic Acids Res.* **20:** 4423–4428.

Persson B.C. and Atkins J.F. 1998. Does the disparate occurrence of the autoregulatory programmed frameshifting in decoding the release factor 2 gene reflect an ancient origin with loss in independent lineages? *J. Bacteriol.* **180:** 3462–3466.

Polard P., Prère M.F., Chandler M., and Fayet O. 1991. Programmed translational frameshifting and initiation at an AUU codon in gene expression of bacterial insertion sequence IS911. *J. Mol. Biol.* **222:** 465–477.

Poole E.S., Major L.L., Mannering S.A., and Tate W.P. 1998. Translation termination in *Escherichia coli*: Three bases following the stop codon crosslink to release factor 2 and affect the decoding efficiency of UGA-containing signals. *Nucleic Acids Res.* **26:** 954–960.

Robinson D.N. and Cooley L. 1997. Examination of the function of two kelch proteins generated by stop codon suppression. *Development* **124:** 1405–1417.

Rom E. and Kahana C. 1994. Polyamines regulate the expression of ornithine decarboxylase antizyme *in vitro* by inducing ribosomal frame-shifting (correction, p. 9195). *Proc. Natl. Acad. Sci.* **91:** 3959–3963.

Shen Q., Chu F.F., and Newburger P.E. 1993. Sequences in the 3'-untranslated region of the human cellular glutathione peroxidase gene are necessary and sufficient for selenocysteine incorporation at the UGA codon. *J. Biol. Chem.* **268:** 11463–11469.

Skuzeski J.M., Nichols L.M., Gesteland R.F., and Atkins J.F. 1991. The signal for a leaky UAG stop codon in several plant viruses includes the two downstream codons. *J. Mol. Biol.* **218:** 365–373.

Sliwkowski M.X. and Stadtman T.C. 1985. Incorporation and distribution of selenium into thiolase from *Clostridium kluyveri*. *J. Biol. Chem.* **260:** 3140–3144.

Smith D.R., Doucette-Stamm L.A., Deloughery C., Lee H., Dubois J., Aldredge T., Bashirzadeh R., Blakely D., Cook R., Gilbert K., Harrison D., Hoang L., Keagle P., Lumm W., Pothier B., Qiu D., Spadafora R., Vicaire R., Wang Y., Wierzbowski J., Gibson R., Jiwani N., Caruso A., Bush D., Reeve J.N. et al. 1997. Complete genome sequence of *Methanobacterium thermoautotrophicum* ΔH: Functional analysis and comparative genomics. *J. Bacteriol.* **179:** 7135–7155.

Somogyi P., Jenner A.J., Brierley I., and Inglis S.C. 1993. Ribosomal pausing during translation of an RNA pseudoknot. *Mol. Cell. Biol.* **13:** 6931–6940.

Stadtman T.C., Davis J.N., Zehelin E., and Böck A. 1989. Biochemical and genetic analysis of *Salmonella typhimurium* and *Escherichia coli* mutants defective in specific incorporation of selenium into formate dehydrogenase and tRNAs. *Biofactors* **2:** 35–44.

Stahl G., Bidou L., Rousset J.-P., and Cassan M. 1995. Versatile vectors to study recoding: Conservation of rules between yeast and mammalian cells. *Nucleic Acids Res.* **23:** 1557–1560.

Stapulionis R., Kolli S., and Deutscher M.P. 1997. Efficient mammalian protein synthesis requires an intact F-actin system. *J. Biol. Chem.* **272:** 24980–24986.

Steneberg P., Englund C., Kronhamn J., Weaver T.A., and Samakovlis C. 1998. Translational readthrough in the *hdc* mRNA generates a novel branching inhibitor in the *Drosophila* trachea. *Genes Dev.* **12:** 956–967.

Sturchler C., Westhof E., Carbon P., and Krol A. 1993. Unique secondary and tertiary structural features of the eukaryotic selenocysteine-tRNA^{Sec}. *Nucleic Acids Res.* **21:** 1073–1079.

Sung D. and Kang H. 1998. Mutational analysis of the RNA pseudoknot involved in efficient ribosomal frameshifting in simian retrovirus-1. *Nucleic Acids Res.* **26:** 1369–1372.

ten Dam E.B., Pleij C.W.A., and Bosch L. 1990. RNA pseudoknots; translational frameshifting and readthrough on viral RNAs. *Virus Genes* **4:** 121–136.

Tewari D.S., Qian Y., Thornton R.D., Pieringer J., Taub R., Mochan E., and Tewari M. 1994. Molecular cloning and sequencing of a human cDNA encoding ornithine decarboxylase antizyme. *Biochim. Biophys. Acta* **1209:** 293–295.

Tormay P. and Böck A. 1997. Barriers to heterologous expression of a selenoprotein gene in bacteria. *J. Bacteriol.* **179:** 576–582.

Tormay P., Wilting R., Heider J., and Böck A. 1994. Genes coding for the selenocysteine-inserting tRNA species from *Desulfomicrobium baculatum* and *Clostridium thermoaceticum*: Structural and evolutionary implications. *J. Bacteriol.* **176:** 1268–1274.

Tormay P., Wilting R., Lottspeich F., Mehta P. Christen P., and Böck A. 1998. Bacterial selenocysteine synthase: Enzyme properties. *Eur. J. Biochem.* **254:** 655–661.

Tsuchihashi Z. and Brown, P.O. 1992. Sequence requirements for efficient translational frameshifting in the *Escherichia coli dnaX* gene and the role of an unstable interaction between tRNA^{Lys} and an AAG lysine codon. *Genes Dev.* **6:** 511–519.

Tsuchihashi Z. and Kornberg A. 1990. Translational frameshifting generates the γ subunit of DNA polymerase III holoenzyme. *Proc. Natl. Acad. Sci.* **87:** 2516–2520.

Tu C., Tzeng T.-H., and Bruenn J.A. 1992. Ribosomal movement impeded at a pseudoknot required for frameshifting. *Proc. Natl. Acad. Sci.* **89:** 8636–8640.

Tu G.F., Reid G.E., Zhang J.G., Moritz R.L., and Simpson R.J. 1995. C-terminal extension of truncated recombinant proteins in *Escherichia coli* with 10Sa RNA decapeptide. *J. Biol. Chem.* **270:** 9322–9326.

Ushida C., Himeno H., Watanabe T., and Muto A. 1994. tRNA-like structures in 10Sa RNAs of *Mycoplasma capricolum* and *Bacillus subtilis*. *Nucleic Acids Res.* **22:** 3392–3396.

Veres Z., Tsai L., Scholz T.D., Polotino M., Balaban R.S., and Stadtman T.C. 1992. Synthesis of 5-methylaminomethyl-2–selenouridine in tRNAs: 31P NMR studies show the labile selenium donor synthesized by the selD gene product contains selenium bonded to phosphorus. *Proc. Natl. Acad. Sci.* **89:** 2975–2979.

Walczak R., Carbon P., and Krol A. 1998. An essential non-Watson-Crick base pair motif in 3'UTR to mediate selenoprotein translation. *RNA* **4:** 74–84.

Weiner A.M. and Weber K. 1973. A single UGA functions as a natural termination signal in the coliphage Qβ coat protein cistron. *J. Mol. Biol.* **80:** 837–855.

Weiss R.B., Huang W.M., and Dunn D.M. 1990a. A nascent peptide is required for ribosomal bypass of the coding gap in bacteriophage T4 gene 60. *Cell* **62:** 117–126.

Weiss R.B., Dunn D.M., Atkins J.F., and Gesteland R.F. 1987. Slippery run, shifty stops, backward steps, and forward hops: –2, –1, +1, +2, +5 and +6 ribosomal frameshifting. *Cold Spring Harbor Symp. Quant. Biol.* **52:** 687–693.

———. 1990b. Ribosomal frameshifting from –2 to +50 nucleotides. *Prog. Nucleic Acid Res. Mol. Biol.* **39:** 159–183.

Weiss R.B., Dunn D.M., Dahlberg A.E., Atkins J.F., and Gesteland R.F. 1988. Reading frame switch caused by base-pair formation between the 3' end of 16S rRNA and the mRNA during elongation of protein synthesis in *Escherichia coli*. *EMBO J.* **7:** 1503–1507.

Weiss R.B., Dunn D.M., Shuh M., Atkins J.F., and Gesteland R.F. 1989. *E. coli* ribosomes re-phase on retroviral frameshift signals at rates ranging from 2 to 50 percent. *New Biol.* **1:** 159–169.

Wickner R.B. 1989. Yeast virology. *FASEB J.* **3:** 2257–2265.

Williams K.P. and Bartel D.P. 1996. Phylogenetic analysis of tmRNA secondary structures. *RNA* **2:** 1306–1310.

Wills N.M., Gesteland R.F., and Atkins J.F. 1991. Evidence that a downstream pseudoknot is required for translational read-through of the Moloney murine leukemia virus *gag* stop codon. *Proc. Natl. Acad. Sci.* **88:** 6991–6995.

———. 1994. Pseudoknot-dependent read-through of retroviral gag termination codons: Importance of sequences in the spacer and loop 2. *EMBO J.* **13:** 4137–4144.

Wilting R., Vamvakidou K., and Böck A. 1998. Functional expression in *Escherichia coli* of the *Haemophilus influenzae* gene coding for selenocysteine-containing selenophosphate synthetase. *Arch. Microbiol.* **169:** 71–75.

Wilting R., Schorling S., Persson B.C., and Böck A. 1997. Selenoprotein synthesis in archaea: Identification of an mRNA element of *Methanococcus jannaschii* probably directing selenocysteine insertion. *J. Mol. Biol.* **266:** 637–641.

Woese C.R. 1970. Molecular mechanics of translocation: A reciprocating ratchet mechanism. *Nature* **226:** 817–820.

Wong J.T.-F. 1975. A co-evolution theory of the genetic code. *Proc. Natl. Acad. Sci.* **72:** 1909–1912.

———. 1988. Evolution of the genetic code. *Microbiol. Sci.* **5:** 174–181.

Yarus M. and Smith D. 1995. tRNA on the ribosome. In *tRNA: Structure, biosynthesis and function* (ed. D. Söll and U.L. RajBhandary), pp. 443–469. ASM Press, Washington, D.C.

Yelverton E., Lindsley D., Yamauchi P., and Gallant J. 1994. The function of a ribosomal frameshifting signal from human immunodeficiency virus-1 in *Escherichia coli*. *Mol. Microbiol.* **11:** 303–313.

Yueh A. and Schneider R.J. 1996. Selective translation initiation by ribosome jumping in adenovirus-infected and heat-shocked cells. *Genes Dev.* **10:** 1557–1567.

Zerfass K. and Beier H. 1992. Pseudouridine in the anticodon GψA of plant cytoplasmic tRNA^{Tyr} is required for UAG and UAA suppression in the TMV-specific context. *Nucleic Acids Res.* **20:** 5911–5918.

Zhang B. and Cech T.R. 1997. Peptide bond formation by in vitro selected ribozymes. *Nature* **390:** 96–100.

Zinoni F., Birkmann A., Stadtman T.C., and Böck A. 1986. Nucleotide sequence and expression of the selenocysteine-containing polypeptide of formate-dehydrogenase (formate-hydrogen-lyase-linked) from *Escherichia coli*. *Proc. Natl. Acad. Sci.* **83:** 4560–4564.

APPENDIX 1: Structures of Base Pairs Involving at Least Two Hydrogen Bonds

Provided by Mark E. Burkard and Douglas H. Turner
Department of Chemistry, University of Rochester
Rochester, New York 14627-0216

Ignacio Tinoco, Jr.
Department of Chemistry, University of California, Berkeley
Structural Biology Division, Lawrence Berkeley National Laboratory
Berkeley, California 94720-1460

The structures of 29 possible base pairs that involve at least two hydrogen bonds are given in Figures 1–5 (for further descriptions, see Saenger, in *Principles of nucleic acid structure*, p. 120. Springer-Verlag [1984]). A base pair that is not a Watson-Crick pair or a G·U wobble pair is called a base-base mismatch, or an internal loop of two nucleotides. All the base pairs can be divided into two classes: normal and flipped. The normal class is defined by the arrangement of the Watson-Crick base pairs. The hydrogen bonding occurs for nucleotides with antiparallel strands and *anti* orientation of the bases relative to the ribose rings. The 11 base pairs that can be made with this same arrangement of nucleotides are called normal; they are shown in Figures 1 and 2. The remaining 18 base pairs require that one of the bases be flipped (inverted) by either reversing the direction of the strand or by switching the base from *anti* to *syn*. (Figs. 3–5). Normal base-base mismatches are found more often than flipped mismatches.

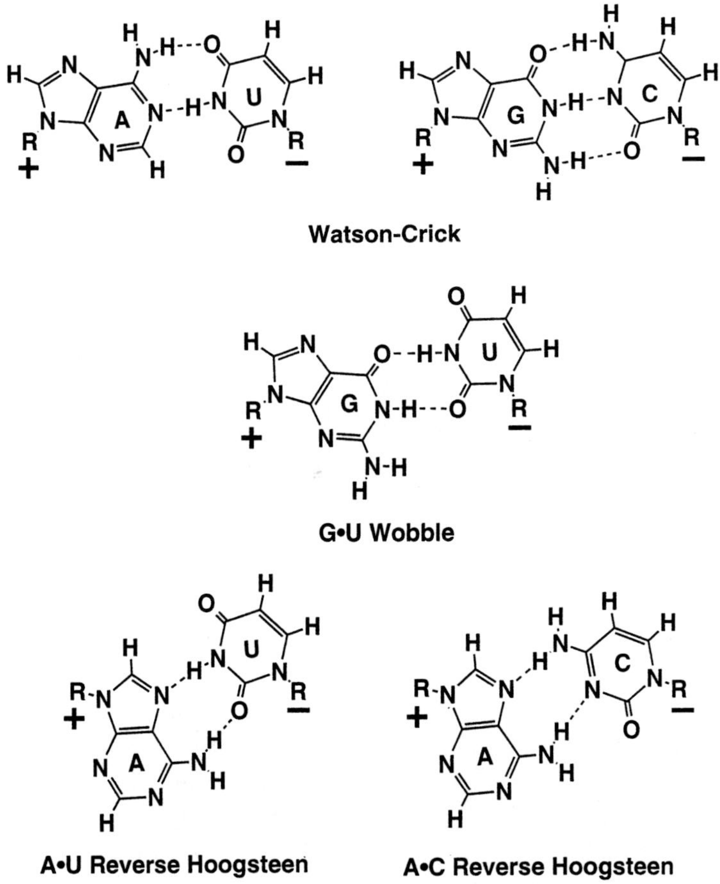

Figure 1 Five possible **normal** purine-pyrimidine base pairs. The Watson-Crick A·U, Watson-Crick G·C, and G·U wobble pairs fit into a double helix with very little distortion; A·U and A·C reverse Hoogsteen are mismatches. The plus and minus signs represent the direction of the strands (antiparallel) for *anti* nucleotides. The same orientation of each base can be obtained by reversing the direction of the strand and rotating the base around the glycosidic bond to *syn*.

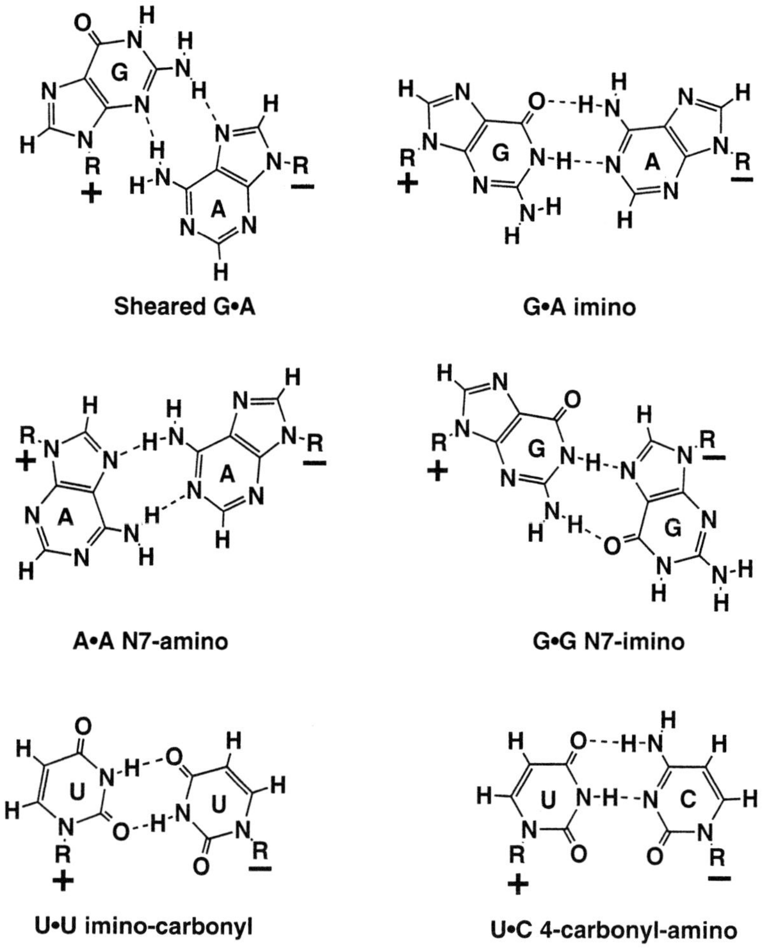

Figure 2 Six possible **normal** purine-purine and pyrimidine-pyrimidine base pairs. The plus and minus signs represent the direction of the strands for *anti* nucleotides. The significance of the pluses and minuses is clear from the fact that no rotation of the bases in the plane of the figure can superimpose, for example, a +A base on a –A base. To superimpose two bases, either one strand must be reversed or the base must be changed from *anti* to *syn*. Sheared and imino G·A mismatches are found often in RNA structures.

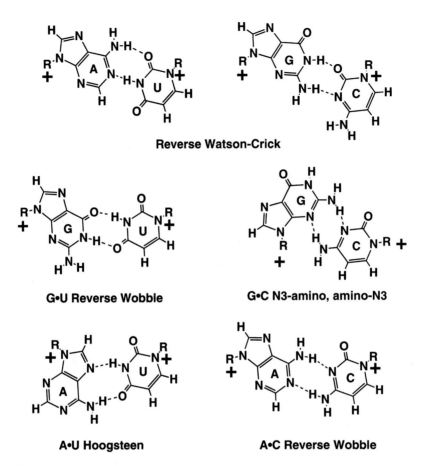

Figure 3 Six possible **flipped** purine-pyrimidine mismatches. Note that all the nucleotides are labeled **+**. Each base pair could as well have been rotated 180° around an axis in the plane of the figure and labeled **−**. These base pairs can be formed from parallel strands with *anti* bases, or from antiparallel strands with one base changed to *syn*. *Syn* bases are higher energy conformations than *anti* bases.

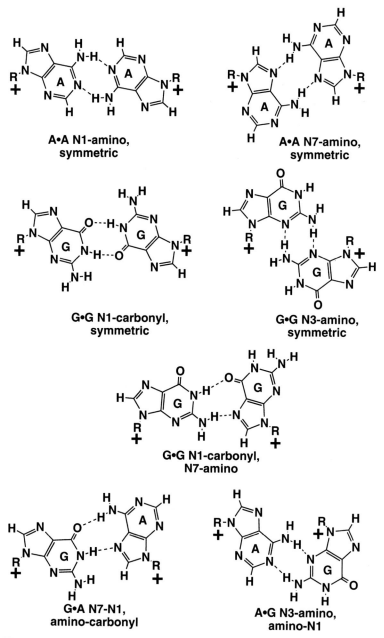

Figure 4 Seven possible **flipped** purine-purine mismatches. Note that all the nucleotides are labeled +. Each base pair could as well have been rotated 180° around an axis in the plane of the figure and labeled –. These base pairs can be formed from parallel strands with *anti* bases, or from antiparallel strands with one of the purine bases changed to *syn*. *Syn* purines are higher energy than *anti* purines, but they have been found in several RNA molecules.

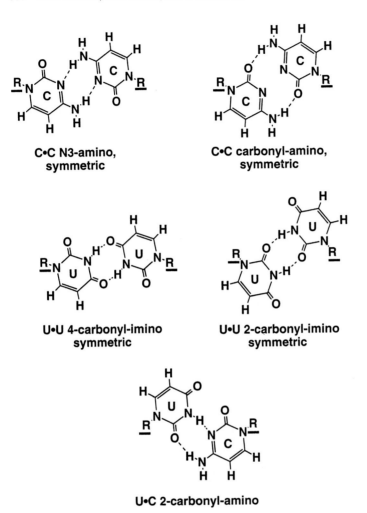

Figure 5 Five possible **flipped** pyrimidine-pyrimidine mismatches. Note that all the nucleotides are labeled –. Each base pair could as well have been rotated 180° around an axis in the plane of the figure and labeled +. These base pairs can be formed from parallel strands with *anti* bases, or from antiparallel strands with one of the pyrimidine bases changed to *syn*. *Syn* pyrimidines are high-energy conformations that have been very rarely identified, but they do occur.

APPENDIX 2: Schematic Diagrams of Secondary and Tertiary Structure Elements

Provided by Mark E. Burkard and Douglas H. Turner
Department of Chemistry, University of Rochester
Rochester, New York 14627-0216

Ignacio Tinoco, Jr.
Department of Chemistry, University of California, Berkeley
Structural Biology Division, Lawrence Berkeley National Laboratory
Berkeley, California 94720-1460

Figure 1 Elements of RNA secondary structure. A secondary structure can be divided into single-strand regions, helices, bulges, hairpin loops, internal loops, and junctions. The distinction between a single-strand region and a bulge, loop, or junction is that in a single strand the ends are not constrained. In contrast, the ends of bulges, loops, or junctions must be in a tightly limited volume. Single-strand regions next to helices are dangling ends; the dangling nucleotide may be on a 5′ end, or a 3′ end. A dangling mismatch is produced by apposing 5′ and 3′ dangling nucleotides. The figure is a slightly modified version of Figure 1, with permission from Nowakowski and Tinoco, *Seminars in Virology* **8:** 153–165 (1997).

Appendix 2 **683**

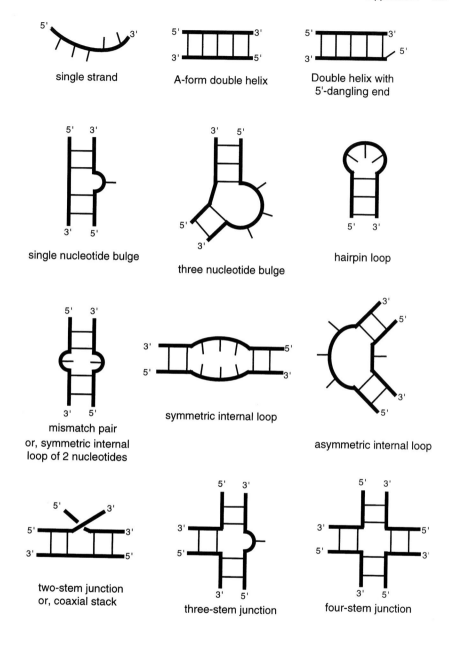

Figure 2 Some elements of RNA tertiary structure. Secondary structure elements can interact to form tertiary structure elements. (*a*) A pseudoknot is formed when a single strand folds to base-pair with a hairpin loop. In the pseudoknot shown (from mouse mammary tumor virus [Shen and Tinoco, *J. Mol. Biol.* **247:** 963–978 [1995]) both loop 1 and loop 1.5 consist of a single adenylate nucleotide. (*b*) Two hairpin loops can base-pair to form kissing hairpins; the kissing hairpin can be intramolecular or intermolecular. The sequence shown is the TAR loop from HIV-1 paired to a complementary loop (Chang and Tinoco, *J. Mol. Biol.* **269:** 52–66 [1997]). (*c*) A hairpin loop can interact with an internal loop as shown for the GAAA tetraloop docked into its receptor from the P4-P6 domain of the *Tetrahymena thermophyla* group I intron (Cate et al., *Science* **273:** 1678–1685 [1996]). The interaction involves stacking of the loop As on the A-platform of the receptor, plus base triple formation. The figure is a slightly modified version of Figure 2, with permission from Nowakowski and Tinoco, *Seminars in Virology* **8:** 153–165 (1997).

a)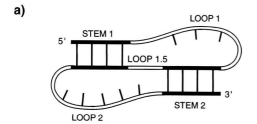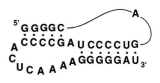

Pseudoknot

b)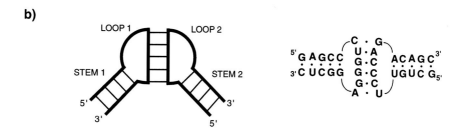

Kissing hairpins

c)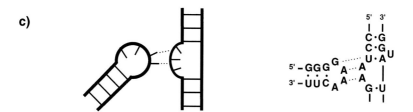

Hairpin loop - bulge contact

APPENDIX 3: Reactions Catalyzed by RNA and DNA Enzymes

Gerald F. Joyce
Departments of Chemistry and Molecular Biology
The Skaggs Institute for Chemical Biology
La Jolla, California 92037

Table 1 Reactions catalyzed by RNA and DNA enzymes

Reaction[a]	Enzyme[b]	Catalytic activity[c]			Reference
		k_{cat} (min^{-1})	K_m (μM)	k_{cat}/k_{uncat}	
Phosphoester transfer	R-nat	0.1	1×10^{-3}	10^{11}	Herschlag and Cech (1990)
	R-lab	0.3	0.02	10^{13}	Tsang and Joyce (1996)
Phosphoester cleavage	R-nat	1	0.05	10^6	Fedor and Uhlenbeck (1992)
	R-lab	0.1	0.03	10^5	Vaish et al. (1998)
	D-lab	3	8×10^{-4}	10^6	Santoro and Joyce (1997)
Polynucleotide ligation	R-nat	4	3	10^6	Hegg and Fedor (1995)
	R-lab	100	9	10^9	Ekland et al. (1995)
	D-lab	0.04	100	10^4	Cuenoud and Szostak (1995)
Polynucleotide phosphorylation	R-lab	0.3	40	$>10^5$	Lorsch and Szostak (1994)
Mononucleotide polymerization	R-lab	0.3	5×10^3	$>10^7$	Ekland and Bartel (1996)
Polynucleotide aminoacylation	R-lab	1	9×10^3	10^6	Illangasekare and Yarus (1997)
Aminoacyl ester hydrolysis	R-nat	0.02	0.5	10	Piccirilli et al. (1992)
Aminoacyl transfer	R-lab	0.2	0.05	10^3	Lohse and Szostak (1996)

Reaction	Type	K_m	k_{cat}/K_m	k_{cat}	Reference
Amide bond cleavage	R-lab			10^2	Dai et al. (1995)
Amide bond formation	R-lab[d]	0.04	2	10^5	Wiegand et al. (1997)
Peptide bond formation	R-lab	0.05	200	10^6	Zhang and Cech (1997)
N-alkylation	R-lab	0.6	1×10^3	10^7	Wilson and Szostak (1995)
S-alkylation	R-lab			10^3	Wecker et al. (1996)
Oxidative DNA cleavage	D-lab			$>10^6$	Carmi et al. (1996)
Biphenyl rotation	R-lab	3×10^{-5}	500	10^2	Prudent et al. (1994)
Porphyrin metallation	R-lab	0.9	10	10^3	Conn et al. (1996)
Porphyrin metallation	D-lab	0.2	3×10^3	10^3	Li and Sen (1996)
Diels-Alder cycloaddition	R-lab[d]	>0.1	>500	10^3	Tarasow et al. (1997)

[a] One example is listed for each class of reaction and each type of enzyme. In some cases additional examples have been reported.
[b] (R-nat) RNA enzyme derived from a naturally occurring catalytic RNA; (R-lab) RNA enzyme obtained by in vitro evolution; (D-lab) DNA enzyme obtained by in vitro evolution.
[c] Values for k_{cat} and K_m are listed to one significant digit even if more precise data were reported. Not all k_{cat} values reflect the chemical step of the reaction.
[d] Contains 5-substituted uridine analogs that are essential for catalysis.

REFERENCES

Carmi N., Shultz L.A., and Breaker R.R. 1996. *In vitro* selection of self-cleaving DNAs. *Chem. Biol.* **3:** 1039–1046.
Conn M.M., Prudent J.R., and Schultz P.G. 1996. Porphyrin metalation catalyzed by a small RNA molecule. *J. Am. Chem. Soc.* **118:** 7012–7013.
Cuenoud B. and Szostak J.W. 1995. A DNA metalloenzyme with DNA ligase activity. *Nature* **375:** 611–614.
Dai X., De Mesmaeker A., and Joyce G.F. 1995. Cleavage of an amide bond by a ribozyme. *Science* **267:** 237–240; for published erratum see *Science* **272:** 18–19, 1996.
Ekland E.H. and Bartel D.P. 1996. RNA-catalysed RNA polymerization using nucleoside triphosphates. *Nature* **382:** 373–376.
Ekland E.H., Szostak J.W., and Bartel D.P. 1995. Structurally complex and highly active RNA ligases derived from random RNA sequences. *Science* **269:** 364–370.
Fedor M.J. and Uhlenbeck O.C. 1992. Kinetics of intermolecular cleavage by hammerhead ribozymes. *Biochemistry* **31:** 12042–12054.
Hegg L.A. and Fedor M.J. 1995. Kinetics and thermodynamics of intermolecular catalysis by hairpin ribozymes. *Biochemistry* **34:** 15813–15828.
Herschlag D. and Cech T.R. 1990. Catalysis of RNA cleavage by the *Tetrahymena thermophila* ribozyme. 1. Kinetic description of the reaction of an RNA substrate complementary to the active site. *Biochemistry* **29:** 10159–10171.
Illangasekare M. and Yarus M. 1997. Small-molecule–substrate interactions with a self-aminoacylating ribozyme. *J. Mol. Biol.* **268:** 631–639.
Li Y. and Sen D. 1996. A catalytic DNA for porphyrin metallation. *Nat. Struct. Biol.* **3:** 743–747.
Lohse P.A. and Szostak J.W. 1996. Ribozyme-catalysed amino-acid transfer reactions. *Nature* **381:** 442–444.
Lorsch J. and Szostak J.W. 1994. *In vitro* evolution of new ribozymes with polynucleotide kinase activity. *Nature* **371:** 31–36.
Piccirilli J.A., McConnell T.S., Zaug A.J., Noller H.F., and Cech T.R. 1992. Aminoacyl esterase activity of the *Tetrahymena* ribozyme. *Science* **256:** 1420–1424.
Prudent J.R., Uno T., and Schultz P.G. 1994. Expanding the scope of RNA catalysis. *Science* **264:** 1924–1927.
Santoro S.W. and Joyce G.F. 1997. A general purpose RNA-cleaving DNA enzyme. *Proc. Natl. Acad. Sci.* **94:** 4262–4266.
Tarasow T.M., Tarasow S.L., and Eaton B.E. 1997. RNA-catalysed carbon–carbon bond formation. *Nature* **389:** 54–57.
Tsang J. and Joyce G.F. 1996. Specialization of the DNA-cleaving activity of a group I ribozyme through *in vitro* evolution. *J. Mol. Biol.* **262:** 31–42.
Vaish N.K., Heaton P.A., Fedorova O., and Eckstein F. 1998. *In vitro* selection of a purine nucleotide-specific hammerhead-like ribozyme. *Proc. Natl. Acad. Sci.* **95:** 2158–2162.
Wecker M., Smith D., and Gold L. 1996. *In vitro* selection of a novel catalytic RNA: Characterization of a sulfur alkylation reaction and interaction with a small peptide. *RNA* **2:** 982–994.
Wiegand T.W., Janssen R.C., and Eaton B.E. 1997. Selection of RNA amide synthases. *Chem. Biol.* **4:** 675–683.
Wilson C. and Szostak J.W. 1995. *In vitro* evolution of a self-alkylating ribozyme. *Nature* **374:** 777–782.
Zhang B. and Cech T.R. 1997. Peptide bond formation by *in vitro* selected ribozymes. *Nature* **390:** 96–100.

APPENDIX 4: Visualization of Elongation Factor Tu on the 70S *E. coli* Ribosome

Holger Stark and Marin van Heel
Imperial College of Science, Medicine and Technology
Department of Biochemistry
London SW7 2AY, United Kingdom

Marina Rodnina and Wolfgang Wintermeyer
Institut für Molekularbiologie, Universität Witten/Herdecke
D-58448 Witten, Germany

Richard Brimacombe
Max-Planck-Institut für Molekulare Genetik
D-14195 Berlin, Germany

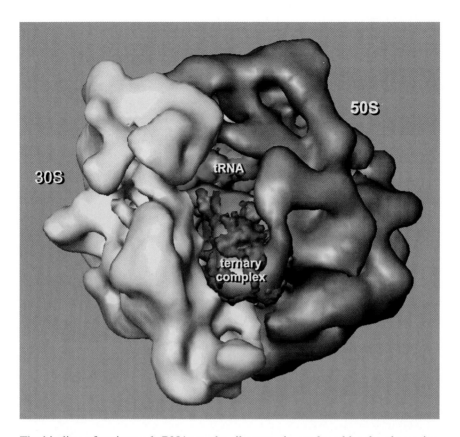

The binding of aminoacyl-tRNAs to the ribosome is catalyzed by the elongation factor Tu (EF-Tu). The release of the ternary complex (EF-Tu/tRNA/GTP) from the ribosome usually occurs after GTP hydrolysis and can be prevented by the antibiotic kirromycin. Those ribosomes were embedded in amorphous ice by rapidly freezing the sample in its natural environment on a holey carbon grid and imaged in an electron microscope either at liquid helium or liquid nitrogen temperature. Three-dimensional structures of ribosomes in different functional states can then be calculated (Agrawal et al. 1996; Stark et al. 1997b) using advanced image processing procedures (van Heel et al. 1996). The structure of the kirromycin-stalled ribosome was determined to a resolution of 18 Å (Stark et al. 1997a). The kirromycin-stalled ribosome is shown here with the 50S subunit on the right and the 30S subunit on the left-hand side. The crystal structures of the ternary complex in red and the P-site tRNA in green were fitted to the cryo-EM density map. Domain 1 of the EF-Tu is bound to the L7/L12 stalk of the 50S subunit, whereas domain 2 is oriented toward the S12 region on the 30S subunit. The ternary complex is thus spanning the gap between the two subunits with the acceptor region of the tRNA reaching into the decoding center on the 30S subunit.

REFERENCES

Agrawal R.K., Penczek P., Grassucci R.A., Li Y., Leith A., Nierhaus K.H., and Frank J. 1996. Direct visualization of A-, P-, and E-site transfer RNAs in the *Escherichia coli* ribosome. *Science* **271:** 1000–1002.

Stark H., Rodina M.V., Rinke-Appel J., Brimacombe R., Wintermeyer W., and van Heel M. 1997a. Visualization of elongation factor Tu on the *Escherichia coli* ribosome. *Nature* **389:** 403–406.

Stark H., Orlova E.V., Rinke-Appel J., Jünke N., Mueller F., Rodnina M., Wintermeyer W., Brimacombe R., and van Heel M. 1997b. Arrangement of tRNAs in pre- and post-translational ribosomes revealed by electron cryomicroscopy. *Cell* **88:** 19–28.

van Heel M., Harauz G., Orlova E.V., and Schmidt R. 1996. A new generation of the IMAGIC image processing system. *J. Struct. Biol.* **116:** 17–24.

APPENDIX 5: The Large Ribosomal Subunit from *H. marismortui* at 9 Å Resolution

Nenad Ban
Department of Molecular Biophysics and Biochemistry
Yale University
New Haven, Connecticut 06520-8114

Peter B. Moore
Department of Chemistry
Department of Molecular Biophysics and Biochemistry
Yale University
New Haven, Connecticut 06520-8107

Thomas A. Steitz
Department of Molecular Biophysics and Biochemistry
Department of Chemistry
Howard Hughes Medical Institute
Yale University
New Haven, Connecticut 06520-8114

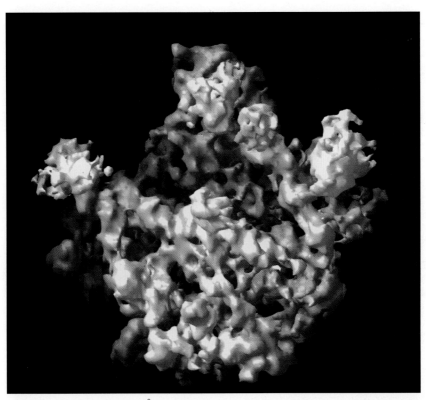

A surface rendering of a 9 Å resolution X-ray map of the large ribosomal subunit shown in the crown view from the side that interacts with the 30S subunit. The projection on the left is ribosomal protein L1, and the RNA arm that supports it. The central protuberance, which includes 5S rRNA is in the top center of the image, and the projection on the right contains ribosomal proteins L7/L12. (Ban et al., *Cell 93:* 1105 [1998].)

Index

Adenine, prebiotic synthesis, 67–68
Aldehydes, formation in Löb system, 4–5, 7, 12, 21
Alkaline phosphatase, reaction mechanism homology with *Tetrahymena* ribozyme, 334–336
AMBER, solvent modeling, 238, 385
Amino acids, polymers as prebiotic genetic material, 71
Aminoacylation. *See* Aminoacyl-tRNA synthetase; Self-aminoacylating RNA
Aminoacyl-tRNA synthetase. *See also* CYT-18; Nam2p; Self-aminoacylating RNA
  classification into two groups, 89, 183–184
  evolution from ribozymes, 89, 183–184, 193–195
  ligand recognition, roles
    major groove, 431–432
    minor groove, 438–439
    modified bases, 443, 445–446
    sequence-dependent conformations, 434–436, 446–447
    single-stranded regions, 440–443, 447
    water, 437
  reactions
    activation, 185
    aminoacyl transfer, 185–186
    posttransfer editing, 186
    pretransfer editing, 185
Aminoglycoside. *See* A-site ribosomal RNA—paromycin complex; Tobramycin aptamer
Ammonia, formation from nitrite ion, 3–4
Antienzyme, frameshifting in decoding, 657–658
Apatite, formation, 35
Aptamer
  evolutionary pressure, 409
  ligand recognition mechanisms, 420–422
  nuclear magnetic resonance, structure determination of ligand complexes
    A-site ribosomal RNA—paromycin complex, 410–412
    ATP aptamer, 404, 406
    box B–N-protein complex, 417–418
    flavin mononucleotide aptamer, 404–405
    Rev response element—Rev complex, 417, 419
    TAR complexes
      arginine complex, 408–409
      BIV Tat complex, 414–415, 419
      HIV Tat complex, 413–414, 419
    theophylline aptamer, 404, 408
    tobramycin aptamer, 412–413
  screening for ligand binding, 405

Aptamer (*continued*)
  structural principles, 419–420
Archaea
  phylogeny studies of ribosomal RNA, 4, 129–133
  transfer RNA introns, 562–563
A-site ribosomal RNA–paromycin complex, structure, 410–412
AT-AC spliceosome. *See* Spliceosome
Atmosphere
  carbon dioxide cycle and greenhouse effects in early atmosphere, 13–15
  cold plasma regions and organic compound synthesis, 9, 11–12
  components of primordial atmosphere, 6
  lightning and organic compound synthesis, 12
  prebiotic soup, 4–9
  sprites, 11
  time scale of evolution, 2–3
ATP binding, RNA engineering studies, 171–172, 404, 406

Bacteria, phylogeny studies of ribosomal RNA, 133–134
Base pairing
  noncanonical pairs, 117, 119, 430
  protein recognition of RNA sequences, 427
  RNA shaping, 242–243, 420
  structures involving at least two hydrogen bonds, 675–680
  variations in RNA World, 167–168
Base stacking
  RNA shaping, 239–242, 254, 419–420
  thermodynamics, 239–242
Biopoesis. *See* Hydrothermal biopoesis
Bonded interactions, RNA shaping, 234–235
Box B–N-protein complex, structure, 417–418
Bulge loop, RNA structure, 252–253
Bypassing
  elements, 658–659
  take-off codon and landing site, 658
  types, 658–660

C5 protein, ribonuclease P cofactor, 357–358, 367, 369, 373

Carbon, isotopic record of life, 36–37
Carbon dioxide
  partial pressures over time, 15–16
  weathering of rock, 13–14
Catalytic RNA. *See* Ribozyme
Cbp2p, group I intron splicing role, 461–462
CCA-adding enzyme, as first telomerase, 86–87
Chemofossils, 26
Coaxial stack, RNA structure, 247–248
Cofactors
  nucleotides as evidence for RNA World, 164, 223
  proteins as cofactors, 226–227, 322
  ribozyme catalysis, 167, 223, 226
Comparative analysis
  molecular biology applications, 113–115
  phylogeny studies of ribosomal RNA
    Archaea, 129–133
    Bacteria, 133–134
    Eucarya, 134–135
    history of study, 128–129
    Universal Ancestor characterization, 135–137
  RNA structure determinations
    history, 114–115, 389–390
    interpretative value with physical studies, 124, 126, 128
    noncanonical pairs, 117, 119
    overview of comparative process, 115–117, 390
    physical chemistry interplay, 121
    pseudoknot, 119
    ribosomal RNA, 117, 119–121
    ribozymes, 122, 124, 126
    tetraloop, 119–121
  U-insertion/deletion editing in trypanosome mitochondria
    extent of editing, 596–597
    guide RNA genome complexity and organization, 598–600
COSMIC—LOPER biopolymers
  backbone modifications, effects in RNA, 175
  definition, 173
  implications for Darwinian evolution, 173–174, 176
  polyelectrolyte structure requirement, 175–176
Coulomb's law, 235–237

Counterion condensation, RNA shaping, 243–244
Covalent catalysis, ribozymes, 334–336
Cross-linking
    major spliceosome studies, 493–495
    photoaffinity cross-linking in ribozyme active site identification, 126
    RNA structural probing, 392
Cyanide, formation in primordial atmosphere, 6–7
Cyanohydrin, RNA precursor, 19–21
CYT-18, group I intron splicing role, 457–458, 460

Darwinian evolution
    COSMIC–LOPER biopolymers, 173–174, 176
    RNA replicase ribozyme, 62, 144, 159
    RNA/RNA interactions, implications, 254–256
    variation and selection, 221
Darwinian pond, 7–9
Decoding
    definition, 637–638
    UGA as selenocysteine, 642–644
Diels-Alder reaction, catalysis by RNA, 171, 325
DNA
    comparison to RNA in origin of life theories, 221–222
    incorporation into RNA–protein world, 228–230
DNA ligase, rate constant for catalysis, 326
DNAzyme, general acid–base catalysis, 336
Domains, RNA structure, 384–385
Double helix, RNA
    comparison with DNA, 428, 430
    major groove in protein recognition, 430–434
    minor groove in protein recognition, 430, 438–440
    structure, 245–246, 382, 428
Duplex stability
    evolutionary implications, 255–257
    RNA replicase and strand dissociation, 156–158

EBV. *See* Epstein–Barr virus

Editing. *See* RNA editing
Eigen limit problem, 225
Electrostatic interactions
    ligand binding, 421
    RNA shaping, 235–237, 420
Elongation factors
    elongation factor Tu visualization in 70S ribosome, 690
    23S rRNA interactions, 206–209
Emergence of life, water requirement, 30–32
Endonuclease, tRNA splicing
    Archaea
        bulge–helix–bulge motif, 570–571
        evolution, 576
        genes, 571–572
        molecular mechanism of *M. jannaschii* enzyme, 573–576
        reaction overview, 565
        structure, 568
    eukaryotes
        evolution of yeast enzyme, 577, 579
        genes, 571–572
        reaction overview, 564
        ruler mechanism, 570
        structure, 566
Entropy
    RNA shaping, 238–239
    translational versus conformational, 238
Epstein–Barr virus (EBV), small nuclear ribonucleoproteins, 512
Equilibrium constant, RNA strand conformations, 223–224
Eucarya
    phylogeny studies of ribosomal RNA, 134–135
    transfer RNA introns, 561–562
Evolution. *See* Darwinian evolution
Exon shuffling, protein incorporation into RNA World, 228

Flavin mononucleotide aptamer, 404–405
Folding, RNA. *See* Secondary structure, RNA; *specific RNAs*
Formaldehyde
    comet composition, 21
    RNA precursor, 17, 19–21, 67
Formose reaction, 21
Frameshifting
    ancient programmed frameshifting, 656–658

Frameshifting (*continued*)
  once-only codon anticodon pairing versus dissociation and repairing, 649–654
  recoding, 637–638
  stimulatory signals, 654–656
Free-energy change, definition and RNA shaping, 234, 239, 241

General acid–base catalysis
  DNAzyme, 336
  tRNA splicing endonucleases, 575
Genetic takeover, 70
Genomic tag hypothesis. *See also Neurospora* retroplasmid; Ribonuclease P; *Tetrahymena* telomerase
  aminoacylation similarity with RNA polymerization, 87–88
  explanatory power, 82
  minihelix structure of tags, 83, 91
  molecular fossils, 81–82, 92–94
  NCCA sequence structure, 84–85
  overview, 80–81, 83–85, 371
  replication initiation, 83–84
  sequence of replication apparatus evolution, 105
  telomere functions of tags, 84, 86–87
  transitional genomes in evolution, 92–98
  viruses as models of ancient genomic replication, 103–105
Geologic record, incompleteness, 31–34
Glycine, formation in Löb system, 4–5, 13
Glycoaldehyde
  formation in Löb system, 4–5, 7, 12, 21
  phosphate
    absorption into double-layer hydroxide minerals, 23, 67
    phosphorylation reactions, 19, 21–22
    sugar precursor, 67
  RNA precursor, 17
Glycolonitrile, RNA precursor, 20
Group I intron. *See also SunY* ribozyme; *Tetrahymena* group I ribozyme
  binding and orienting reactive groups, 330–331

DNA endonuclease
  encoding, 451–452
  functions, 454–457
  LAGLIDADG motif, 454–457
  ectopic transposition, 452, 454
  evolution of protein-assisted splicing reactions, 474–476
  homing, 452
  host-encoded splicing factors
    Cbp2p, 461–462
    CYT-18, 457–458, 460
    Mss116p, 463
    Nam2p, 460–461
    Pet54p, 463
    S12, 462
    StpA, 462
  2′-hydroxyls in splice site recognition, 340
  nucleotide and amino acid binding, 87, 145
  reverse splicing, 451, 454
  self-splicing reaction, 323
  solvent-inaccessible core, 337–342
  substrate specificity, 327
  turnover of excised RNAs, 472
Group II intron. *See also Tetrahymena* group II ribozyme
  binding and orienting reactive groups, 331
  domains, 463
  evolution of protein-assisted splicing reactions, 474–476
  host-encoded splicing factors
    Mrs2p, 468–469
    Mss116p, 468
  reverse splicing, 451
  reverse transcriptase, encoding and functions, 451, 464, 466–468
  ribonucleoprotein particles in intron mobility, 469, 471–472
  self-splicing reaction, 323
  solvent-inaccessible core, stabilizing interactions, 339–341
  turnover of excised RNAs, 472
Group III intron, structure, 463–464
Guide RNA, RNA editing
  complexity
    comparative analysis, 598–600
    loss of editing in *L.tarentolae* UC strain, 594–596
  mechanism, 586–590, 592
Hairpin loop, RNA structure, 251–252, 382–383, 395–396

Hairpin ribozyme
  kinetics, 265–266, 268–271
  metal requirement, 271–274, 294, 333
  minimal core sequence, 266–268
  minimal kinetic scheme, 268–269, 271
  pH dependence, 272–273
  reaction catalyzed, 265
  structure, 266–267
Hammerhead ribozyme
  kinetics, 265–266, 268–271
  metal requirement
    binding sites, characterization, 296–297, 301–302, 307
    catalysis, 271–274, 294, 333
  minimal core sequence, 266–268
  minimal kinetic scheme, 268–269, 271
  noncatalytic conformation of crystal structures and conformational switching, 277–282, 343
  pH dependence, 272–273
  rate constant for catalysis, 326
  reaction catalyzed, 265
  structure, 266–267, 274–277
  substrate specificity, 329
Hepatitis delta virus ribozyme
  kinetics, 265–266, 268–271
  metal requirement, 271–274, 294
  minimal core sequence, 266–268
  minimal kinetic scheme, 268–269, 271
  pH dependence, 272–273
  reaction catalyzed, 265
  structure, 266–267
Herpes simplex virus, latency associated transcripts, 513
Herpesvirus saimiri, small nuclear ribonucleoproteins, 512–513
HeT-A retrotransposon. See Heterochromatin-associated retrotransposon
Heterochromatin-associated (HeT-A) retrotransposon, relationship with telomerases, 97–98
Hexacyanoferroate, RNA precursor, 19–21
Hexose-2,4,6-triphosphates, formation by abiotic processes, 23–24, 67
Histone, phylogenetic analysis, 132–133
Hot springs, biopoesis, 2–4
Human immunodeficiency virus reverse transcriptase, transfer RNA as primer, 101

Hydrogen bonding. See Base pairing
Hydrosphere
  Darwinian pond, 7–9
  organic reactions, 7
  pH, 14
  time scale of evolution, 2–3, 15
Hydrothermal biopoesis, 2–4, 7–8

Internal loop, RNA structure, 248–251, 382–383
Interstellar medium. See Space
Intron—exon structure, RNA World, 225–228
Intron splicing. See Group I intron; Group II intron; Spliceosome; Transfer RNA splicing
In vitro selection
  random sequence generation of polymerase-like ribozymes, 150, 153, 155
  ribozyme activities generated, 325
  theophylline-binding RNA, 327

Junction, RNA structure, 253

Kaposi sarcoma-associated herpesvirus (KSHV), small nuclear ribonucleoproteins, 513
Kinetoplast DNA, 586–587
KSHV. See Kaposi sarcoma-associated herpesvirus

Late heavy bombardment, 24–27
Ligase, tRNA splicing
  Archaea, 568–569
  eukaryotes, 564, 566–567, 569–570
London attractions, RNA shaping, 237, 240
Long terminal direct repeats (LTRs)
  origin, 102
  retroviruses, 101–102
LTRs. See Long terminal direct repeats
Luminescence spectroscopy, metal binding site characterization in RNA, 307
Lunar impact record, late heavy bombardment, 25–26

M1 RNA, ribonuclease P cleavage, 355, 358, 360, 362, 364, 368–373

Major groove, protein recognition role, 430–434
Mars
  geologic record, 31–32
  late heavy bombardment, 26
  meteorite fossils, 176–177
  organic molecules, 17
Maturase, group I intron encoding and functions, 451–452, 454–457
MC-SYM, model-building of RNA structure, 392–393
MeiRNA, polyadenylation, 515
Melting temperature (T_m)
  determination of stem-loop diagrams, 388–389
  ion effects, 244
Metal ion
  abundance in oceans and biological fluids, 287–288
  catalytic functions in RNA, 332
  charge neutralization, 291
  coordination in RNA, 244–245
  hardness versus softness, 288, 291
  magnesium binding and uptake studies on various RNAs, 291–292
  physical properties, table, 289–290
  ribozyme catalysis, examples and mechanisms, 167, 223, 271–274, 313–314, 331–334
  RNA-binding sites
    comparison with protein metal-binding sites, 312–313
    competitive inhibition experiments, 311
    crystallographically characterized sites, 295, 299–306
    hydration, 299, 312–313
    identification, overview of techniques, 294–295
    ion-specific cleavage studies, 308
    luminescence spectroscopy characterization, 307
    nuclear magnetic resonance characterization, 306–307
    nuclearity of sites, 295
    specificity, 291, 293
    sulfur modification of oxygen ligands, rescue experiments, 308–311
  RNA shaping, 254
  self-aminoacylating RNA, metal requirements, 187–188

solvent-inaccessible core, stabilizing interactions in ribozymes, 341–342
MFOLD, RNA structure prediction, 388
Minor groove, protein recognition role, 430, 438–440
Mitochondria. *See* RNA editing
Molecular fossils
  genomic tag hypothesis, 81–82, 92–94
  telomerase, 627–631
  viruses as models of ancient genomic replication, 103–105
Montmorillonite, nonenzymatic polymerization of nucleotides, 52
Moon
  lunar impact record, 25–26
  origins, 27, 29
  platinum group elements, 29
Morphofossils, 34
Mrs2p, group II intron splicing role, 468–469
Mss116p
  group I intron splicing role, 463
  group II intron splicing role, 468

Nam2p, group I intron splicing role, 460–461
Neurospora retroplasmid
  genomic tag, 92–93, 104
  replication, 93–94
  reverse transcriptase
    comparison with telomerase, 96–97
    primers, 99
Neurospora VS ribozyme
  kinetics, 265–266, 268–271
  metal requirement, 271–274, 294
  minimal core sequence, 266–268
  minimal kinetic scheme, 268–269, 271
  pH dependence, 272–273
  reaction catalyzed, 265
  structure, 266–267
NMR. *See* Nuclear magnetic resonance
Nonbonded interactions, RNA shaping, 235–238
NTT RNA, polyadenylation, 515
Nuclear magnetic resonance (NMR)
  metal-binding site characterization in RNA, 306–307
  structure determination of RNA–ligand complexes
    A-site ribosomal RNA–paromycin complex, 410–412

ATP aptamer, 404, 406
box B–N-protein complex, 417–418
flavin mononucleotide aptamer, 404–405
Rev response element–Rev complex, 417, 419
TAR complexes
  arginine, 408–409
  BIV Tat complex, 414–415, 419
  HIV Tat complex, 413–414, 419
theophylline aptamer, 404, 408
tobramycin aptamer, 412–413
Nucleoside-5′-phosphorimidazolides, nonenzymatic polymerization, 51–52
Nucleotide triple
  alignment of helices in junctions, 253
  solvent-inaccessible core, stabilizing interactions in ribozymes, 339

Omega-n, RNA polyadenylation, 515
Origin of life. *See* Emergence of life; RNA World
Outer space. *See* Space
Oxiranecarbonitrile, RNA precursor, 17, 19

P4-P6 domain. *See Tetrahymena* group I ribozyme
Pentose-2,4-diphosphates, formation by abiotic processes, 23–24, 67
Peptide nucleic acid (PNA)
  helix chimeras, 69–70
  properties, 69–70
Peptide, superiority to RNA as catalysts, 165–166
Pet54p, group I intron splicing role, 463
Phosphates
  biological importance, 34–35
  condensation, 8, 22–23
  phosphatic biomineralization, 35
  prebiotic chemistry role, 16–17
  sugar phosphate formation by abiotic processes, 23–24
Phosphorothioate rescue experiments, metal-binding site characterization in RNA, 308–311, 342
2′-Phosphotransferase, tRNA splicing, 564–565, 567–568

Phylogenetic analysis. *See* Comparative analysis
PNA. *See* Peptide nucleic acid
Polymerization, RNA
  activation of nucleotides, 50–51
  nonenzymatic polymerization, 51–53, 63
  polymerase subunit specificity, 99–101
  transfer RNA aminoacylation similarity with RNA polymerization, 87–88
Prebiotic chemistry
  early studies, 4–6
  functionalized nucleobases, 172–173
Proteinoid, formation in Oparin system, 5
Pseudoknot
  comparative analysis, 119
  formation, 254
  frameshifting, 654–655
  polymerase readthrough, 639

Qβ replicase
  initiation of replication, 84, 93
  subunits, 100

Recoding. *See also* Frameshifting
  classes, 638
  definition, 637
  generality of redefinition, 649
  messenger RNA signals, 638
  redefinition of stop codons, 638–639
Release factor, 2, frameshifting in decoding, 656–657
Replication, RNA. *See also* RNA replicase ribozyme
  preceding of protein synthesis, 49–50
  template-directed synthesis, nonenzymatic schemes, 53–55, 62–63
Rev response element–Rev complex, 417, 419
Reverse transcriptase. *See* Group II intron; Human immunodeficiency virus reverse transcriptase; *Neurospora* retroplasmid; Telomerase
Ribonuclease MRP, evolution, 372

Ribonuclease P
  C5 protein, enhancement of catalysis, 357–358, 367, 369, 373
  CCA-adding enzyme, 86–87
  CCA sequence, 354, 358, 361
  cleavage of modern genomic tags, 82, 85, 91
  coevolution with substrate structure, 86, 368–373
  functions, 85, 322–323, 351–354
  metal-binding requirement, 294, 359
  minimal catalytic structure, 122, 124
  minimal RNA structure, 369
  photoaffinity cross-linking in active-site identification, 126
  rate-limiting step, 358
  sequence and cleavage site recognition, 362–363, 367–368
  solvent-inaccessible core, 339–340
  structure modeling of active site, 368
  subcellular transport, 353–354
  substrates
    M1 RNA, 355, 358, 360, 362, 364, 368–373
    specificity, 354, 356–362
    substitution studies in model compounds, 360–361
    types, 352–353, 368–369
  subunits, 351–354
Ribonucleotides
  enantiomers in prebiotic synthesis, 68
  formation by abiotic processes, 23–24, 67
  pyranosyl analog properties, 69–70
Ribose zipper, 340
Ribosomal RNA (rRNA)
  base modifications and possible roles in RNA World, 167–168, 173
  comparative analysis
    phylogeny studies
      Archaea, 129–133
      Bacteria, 133–134
      Eucarya, 134–135
      history of study, 128–129
      Universal Ancestor characterization, 135–137
    structure, 117, 119–121
  processing overview, 504–505
  5S rRNA
    metal-binding site characterization, 295, 298, 304–305
    minor groove in protein recognition, 440
  16S rRNA
    conformational switch, 200–201
    site-directed mutagenesis studies of function, 200
    structure, 197–198
    transfer RNA interactions, 201–204
  23S rRNA
    elongation factor interactions, 206–209
    peptide bond formation by engineered molecules, 201, 212
    transfer RNA interactions, 204, 206
  small nucleolar ribonucleoproteins in ribosome biogenesis
    intron-encoded RNAs and functions, 510–511
    pre-rRNA cleavage role, 508–510
    pre-rRNA targeting for modification, 505–508
  translation functions, 199–201
Ribosome. *See also* Bypassing; Frameshifting
  complexity of structure, 197, 351
  elongation factor Tu visualization in 70S ribosome, 690
  evolution from RNA
    feasibility of RNA ribosome, 660–663
    protodomains, 209–213
    proto-ribosome, 197, 212–214
  large subunit from *H. marismortui*, three-dimensional structure, 694
  proteins as cofactors, 226–227
  small nucleolar ribonucleoproteins in biogenesis
    intron-encoded RNAs and functions, 510–511
    pre-rRNA cleavage role, 508–510
    pre-rRNA targeting for modification, 505–508
Ribozyme. *See also specific ribozymes*
  catalysis versus information storage, contradictory requirements, 173–176
  conformational flexibility, 343
  covalent catalysis, 334–336
  general acid–base catalysis, 336
  intron–exon structure, 225
  limitations of RNA as catalysts, 165

proteins as cofactors, 226–227, 322
rate constants for catalysis, 326–327
reactions catalyzed, overview,
    322–325, 686–687
RNA World, reactions catalyzed and
    functions, 223–224,
    321–322
size limitations, 224–225
solvent-inaccessible core, 337–342
substrate destabilization, 336–337
transfer RNA synthetase evolution
    from ribozymes, 89
RNA editing
  A to I substitution for membrane
      receptor mRNA, 602–603
  C to U substitution and cytidine
      deaminase reaction, 602
  types, 585
  U-insertion/deletion editing in
      trypanosome mitochondria
    absence in Euglenoid mitochondria,
        600–601
    comparative analysis
      extent of editing, 596–597
      guide RNA genome complexity
          and organization, 598–600
    comparative organization of
        maxicircle genomes,
        585–586
    functions, 601–603
    guide RNA
      complexity and loss of editing in
          L.tarentolae UC strain,
          594–596
      mechanism of editing, 586–590,
          592
    kinetoplast DNA network, 586–587
    mechanistic models, 588–590, 592
    misediting mechanisms, 592–594
    polarity of editing, 592–594
RNA polymerase. *See also* Reverse
    transcriptase; RNA replicase
    ribozyme; Telomerase; T7
    RNA polymerase
  phylogenetic analysis, 132–133
RNA replicase ribozyme
  Darwinian evolution, 62, 144, 159
  error threshold and survival, 56–60,
      144
  mechanisms of catalysis, 64–66
  modeling of intrinsic properties
      fidelity, 145, 148–149, 154–155,
      159

oligonucleotide assembly by *sunY*
    ribozyme, 146–148
primer extension by *Tetrahymena*
    ribozyme
  guanosine-binding site
      alteration, 145
  polymerase engineering,
      144–145, 149–150
  primer–template binding, 148–150
random sequence generation of
    polymerase-like ribozymes
  in vitro selection and evolution,
      150, 153, 155
  leaving groups in
      polymerization, 150–151
  primer extension, 153
  self-ligation reaction, 153–154
  strand dissociation, 155–158
  template binding, 155
multimer substrates, 61
plausibility in RNA World, 62, 143,
    222–224
quasispecies in evolution, 61–62
requirements, overview, 55–56, 144
sequence specificity, 63–64
size, 60–61, 166
RNA World. *See also* Replication, RNA
  definition, 1, 49, 143, 164
  evidence for, 1–2, 39–40, 164
  genetic takeover, 70
  intron–exon structure, 225–228
  limitations in modeling, 2
  protein introduction, 226–228
  ribozymes
    overview of functions in RNA
        World, 321–322
    reactions catalyzed in RNA World,
        223–224
    small ribozyme functions, 282–283
  time of existence, 2
roX1, polyadenylation, 514
rRNA. *See* Ribosomal RNA

S12, group I intron splicing role, 462
10Sa RNA, tmRNA functions, 90
SAPSSARN, RNA structure prediction,
    389
SECIS. *See* Selenocysteine
Secondary structure, RNA. *See also*
    specific RNAs
  assembly into tertiary suprahelical
      structures, 253–254, 384,
      419, 683–684

Secondary structure (*continued*)
  biochemical probing, 390–392
  bulge loop, 252–253
  coaxial stack, 247–248
  comparative analysis of structures
    history, 114–115, 389–390
    interpretative value with physical studies, 124, 126, 128
    noncanonical pairs, 117, 119
    overview of comparative process, 115–117, 390
    physical chemistry interplay, 121
    pseudoknot, 119
    ribosomal RNA, 117, 119–121
    ribozymes, 122, 124, 126
    tetraloop, 119–121
  diagrams, 682
  domains, 384–385
  double helix
    comparison with DNA, 428, 430
    structure, 245–246, 382
  explanatory power of models, 394
  folding events
    early events, 395–396
    late events, 396–397
    pathways, 396–397
  hairpin loop, 251–252, 382–383, 395–396
  internal loop, 248–251, 382–383
  junction, 253
  model-building, 392–393
  nomenclature comparison to protein structure, 383–385
  prediction from sequences
    ab initio computation, 385–388
    difficulty, 381
    limitations of molecular dynamics programs, 385–387
    pairwise additivity of programs, 386
    semi-empirical approaches, 388–389
  recognition by proteins, 427–428, 430
  reproducibility of models, 393–394
  sequence-dependent alternative conformations in protein recognition, 434–436
  terminal loop, 383
SelB. *See* Selenocysteine
Selenocysteine
  biosynthesis
    evolution, 647–648
    overview, 641–642
  phylogeny of selenocysteine synthase, 648
  codon, 639–640
  efficiency of insertion, 647
  evolution of insertion
    models, 646–647
    selective advantage, 644
    UGA codon usage, 645
  SECIS–SelB elements in UGA decoding, 642–644, 646–647
  selenoprotein distribution, 641
  sulfur and selenium metabolism, 641
Self-aminoacylating RNA
  active-site structure, 189–191
  amino acid specificity, 191–193
  comparison to aminoacyl-tRNA synthetases, 193–195
  hydrophobic derivatization, 186–187
  kinetics, 188–189, 191–193
  mechanism, 187–189, 194–195
  metal requirements, 187–188
Self-replication. *See* RNA replicase ribozyme
Shine–Dalgarno sequence, frameshifting, 655–656
Short-range labeling reagents, RNA structural probing, 392
Small nuclear ribonucleoproteins (snRNPs). *See also* Spliceosome
  AT-AC spliceosome ribonucleoproteins
    evolution, 499–500
    functions, 498–499
    introns, 498–499
    Sm-binding sites, 498
    types, 497
  autoantibody targeting of protein components, 487, 496
  factors in formation, 554
  major spliceosome ribonucleoproteins
    assembly of spliceosome, 491, 493
    base-pairing homology with group II and group II introns, 495–496
    cross-linking studies, 493–495
    modified bases, 495
    secondary structure, 491
    transesterification step in splicing, 493, 495
  polyadenylation of RNAs, 514–515
  pre-messenger RNA
    processing, 489–491

U7 measurement of histone RNAs, 502–504
7SK, 512
Sm proteins, 526, 545
small nucleolar ribonucleoproteins in ribosome biogenesis
  intron-encoded RNAs and functions, 510–511
  pre-rRNA cleavage role, 508–510
  pre-rRNA targeting for modification, 505–508
  subcellular localization, 487
  trans-splicing, requirements and functions, 500–502
  types, 488–489
U1-associated factors, 528, 546
U2-associated factors, 528, 547
U4-associated factors, 549
U5-associated factors, 529, 548
U6-associated factors, 549
viral ribonucleoproteins and host protein binding, 512–513
snRNPs. *See* Small nuclear ribonucleoproteins
Solvent-inaccessible core, ribozymes stabilizing interactions
  2′-hydroxyl group, 340–341
  long-range base pairs and triples, 339–340
  metal ions, 341–342
  Tetrahymena group I ribozyme and P4-P6 domain, 337–342
Space
  late heavy bombardment, 24–27
  limitations in organic molecule synthesis, 9
  meteorites, organic compounds, 9
Spliceosome. *See also* Small nuclear ribonucleoproteins
  AT-AC spliceosome ribonucleoproteins
    evolution, 499–500
    functions, 498–499
    introns, 498–499
    Sm-binding sites, 498
    types, 497
  catalytic core, 539
  evolution of types, 537–538
  factors in disassembly and recycling, 554
  overview of assembly and splicing, 526, 528–530
  RS domain proteins in assembly, 543–544
  sequence specificity of splicing, 525, 539–543
  Sm proteins, 526, 545
  splicing factors
    ATP-dependent pre-spliceosome formation, 551
    catalytically active complex formation, 552–553
    commitment complex formation, 550
  splicing reaction, 555
  transesterification reactions in splicing, 493, 495, 525
  U2-type major spliceosome
    assembly, 491, 493
    base-pairing homology with group II and group II introns, 495–496
    cross-linking studies, 493–495
    introns, 530, 532–534
    modified bases, 495
    RNA secondary structure, 491, 535–537
    splice signals, 539–541
  U12-type minor spliceosome
    introns, 530, 532–534
    RNA secondary structure, 535–537
    species distribution, 525
    splice signals, 541–542
Stem-loop. *See* Hairpin loop
StpA, group I intron splicing role, 462
Structure, RNA. *See* Secondary structure, RNA; *specific RNAs*
Sugar phosphate backbone
  base attachment under abiotic conditions, 17
  formation of sugar phosphates by abiotic processes, 23–24, 67
Sun, evolution with Earth, 14–15
SunY ribozyme, oligonucleotide assembly, 146–148

T7 DNA polymerase, primer–template binding, 149
T7 RNA polymerase, rate constant for catalysis, 326
TAR complexes, structures
  arginine complex, 408–409
  BIV Tat complex, 414–415, 419
  HIV Tat complex, 413–414, 419
  major groove in protein recognition, 432–433

TART. *See* Telomere-associated retrotransposon
Telomerase. *See also Tetrahymena* telomerase
  CCA-adding enzyme as first telomerase, 86–87
  DNA endonuclease cleavage activity, 617–618
  evolution, 627–631
  heterochromatin-associated retrotransposon relationship with telomerases, 97–98
  localization in cells, 626
  Neurospora retroplasmid reverse transcriptase comparison with telomerase, 96–97
  polymerization, 611–613, 626–627
  primer, 611–612, 615–617
  product binding in yeast, 625
  reverse transcriptase
    evolution, 628–630
    function, 609–610
    mechanism, 626–627
    structure, 613, 615
  RNA
    boundaries of templating region, 622–623
    effect on endonuclease activity, 624
    functional interaction between telomerase RNAs, 624–625
    polymerization role, essential bases, 620–622
    sequence conservation, 619–620
    structure, 613–615
    template function, 615–619
  telomere-associated retrotransposon relationship with telomerases, 97–98
Telomere. *See also* telomerase genomic tags, 84, 94–95
  Drosophila sequences, 630–631
  G quartets, 95–96
  repeat sequences in eukaryotes, 609–610
Telomere-associated retrotransposon (TART), relationship with telomerases, 97–98
Terminal loop, RNA structure, 383
Tetrahymena group I ribozyme
  alkaline phosphatase, reaction mechanism homology, 334–336
  conformational flexibility, 343

engineering of reaction specificity, 324–325
  metal binding
    folding function, 332–333
    site characterization, 297, 303–304, 307, 333
    specificity, 294, 304
  primer extension
    guanosine-binding site alteration, 145
    polymerase engineering, 144–145, 149–150
  rate constant for catalysis, 326–327
  solvent-inaccessible core and P4-P6 domain, 337–342
  substrate
    destabilization, 337
    specificity, 327–328, 330
Tetrahymena group II ribozyme, metal binding specificity, 294
Tetrahymena telomerase
  discovery, 610–611
  DNA endonuclease cleavage activity, 617–618, 624
  forms, 626
  genomic tag, 94–95
  polymerization, 611, 615–617
  protein components, 95
  RNA
    boundaries of templating region, 622–623
    effect on endonuclease activity, 624
    polymerization role, essential bases, 620–622
    structure, 613–615
    template function, 615–619
Tetraloop, comparative analysis, 119–121
Theophylline aptamer
  in vitro selection and binding specificity, 327
  structure, 404, 408
T_m. *See* Melting temperature
tmRNA
  modeling of RNA ribosome, 661–662
  10Sa RNA, tmRNA functions, 90
Tobramycin aptamer, structure, 412–413
Transfer RNA (tRNA)
  base modifications and possible roles in RNA World, 167–168, 173
  genetic punctuation in RNA processing and translation, 89–90

genomic tag hypothesis. *See also*
 Neurospora retroplasmid;
 Ribonuclease P;
 Tetrahymena telomerase
 aminoacylation similarity with RNA
  polymerization, 87–88
 explanatory power, 82
 minihelix structure of tags, 83, 91
 molecular fossils, 81–82, 92–94
 NCCA sequence structure, 84–85
 overview, 80–81, 83–85
 replication initiation, 83–84
 sequence of replication apparatus
  evolution, 105
 telomere functions of tags, 84, 86–87
 transitional genomes in evolution,
  92–98
 viruses as models of ancient
  genomic replication,
  103–105
metal-binding site characterization,
 296, 299–301
priming of retroviral reverse
 transcription, 99
ribosomal RNA interactions
 16S rRNA, 201–204
 23S rRNA, 204, 206
RNA replication roles, 80–82
structural domains and evolution of
 ends, 79, 91–92, 199
Transfer RNA splicing
 Archaea introns
  endonuclease
   bulge–helix–bulge motif,
    570–571
   evolution, 576
   genes, 571–572
   molecular mechanism of *M.
    jannaschii* enzyme, 573–576
   reaction overview, 565
   structure, 568
  features, 562–563
  ligase, 568–569
  origin, 579–580
  pathway for splicing, 565–566
 eukaryote introns
  endonuclease

evolution of yeast enzyme, 577,
 579
genes, 571–572
reaction overview, 564
ruler mechanism, 570
structure, 566
features, 561–562
ligases, 564, 566–567, 569–570
origin, 579–580
pathway for splicing, 564–565
2′-phosphotransferase, 564–565,
 567–568
 group I mechanism in Bacteria, 561
Transfer RNA synthetase. *See*
 Aminoacyl-tRNA synthetase
tRNA. *See* Transfer RNA
Trypanosome mitochondria. *See* RNA
 editing
Two-biopolymer system, 163

U2-type spliceosome. *See* Spliceosome
U12-type spliceosome. *See* Spliceosome
UGA codon
 decoding as selenocysteine, 642–644
 SelB phylogeny, 648–649
 transfer RNA phylogeny, 649
 usage, 639–640
U-insertion/deletion editing. *See* RNA
 editing
Universal Ancestor, characterization,
 135–137

van der Waals interactions, RNA
 shaping, 237–238

Water. *See also* Hydrosphere
 requirement for life, 30–32
 role in protein recognition of RNA,
  436–438
Whitlockite, formation, 22, 35

Xist, RNA polyadenylation, 514
X-ray crystallography, 295, 299–306,
 690, 694